Pediatric Cancers

Pediatric Cancers

Editors

Saurabh Agarwal
Jianhua Yang

MDPI

Basel • Beijing • Wuhan • Barcelona • Belgrade • Novi Sad • Cluj • Manchester

Editors

Saurabh Agarwal
Department of
Pharmaceutical Sciences
St. John's University
New York
United States

Jianhua Yang
School of Medicine and
Health Science
The George Washington
University
Houston
United States

Editorial Office
MDPI
St. Alban-Anlage 66
4052 Basel, Switzerland

This is a reprint of articles from the Special Issue published online in the open access journal *Cancers* (ISSN 2072-6694) (available at: https://www.mdpi.com/journal/cancers/special_issues/Pediatric_Tumor).

For citation purposes, cite each article independently as indicated on the article page online and as indicated below:

Lastname, A.A.; Lastname, B.B. Article Title. *Journal Name* **Year**, *Volume Number*, Page Range.

ISBN 978-3-0365-9024-0 (Hbk)
ISBN 978-3-0365-9025-7 (PDF)
doi.org/10.3390/books978-3-0365-9025-7

Contents

About the Editors

Saurabh Agarwal

Dr. Saurabh Agarwal is an Assistant Professor in the Department of Pharmaceutical Sciences, College of Pharmacy and Health Sciences, at St. John's University, New York. Dr. Agarwal's research is focused on deciphering the molecular and epigenetic regulations in pediatric cancer neuroblastoma. His lab discovered a neuroblastoma cancer stem cell subpopulation and is now developing effective targeted therapeutic approaches using pre-clinical models. Dr. Agarwal has published numerous high-impact as well as highly cited research publications and received numerous awards, recognitions, and philanthropy grants for his research. He serves as an Editorial Board Member and Topical Editor for journals such as *PLOS One, Scientific Reports, Cancers, Pharmaceutics, Cells,* and *Current Oncology*. Dr. Agarwal is an active member of the AACR, ANRA, AAPS, ASPET, and ASBMB, and regularly participates in social fundraising events to promote pediatric cancer research.

Jianhua Yang

Jianhua Yang, Ph.D., is an Associate Professor of pediatrics at the George Washington University School of Medicine and Health Sciences and a faculty member of the Center for Cancer and Immunology Research at the Children's National Research Institute, Children's National Hospital, Washington DC. Dr. Yang's current research focuses on elucidating the molecular mechanism of neuroblastoma development and developing novel therapeutics against multiple molecular targets identified in his laboratory.

Preface

This reprint book is a collection of articles published in a Special Issue on pediatric cancers in Cancers. This reprint consists of 25 articles, comprising an editorial, 16 original research articles, 6 reviews, and 2 systematic reviews covering molecular insights into different pediatric cancers. Overall, this Special Issue reprint on pediatric cancers brings together international leaders in the field of cancers to discuss and present the current knowledge, understanding, management, subtypes, and molecular features of different pediatric cancers. This reprint also includes research articles on experimental therapeutics for different pediatric cancers and the development of therapies of the future to cure pediatric cancers. This preface is an invitation to learn, share, and stand together in the fight against pediatric cancers. Let these articles and research be a testament to our commitment and unwavering hope for a future where no child will suffer from cancer.

Saurabh Agarwal and Jianhua Yang
Editors

cancers

Editorial

Pediatric Cancers: Insights and Novel Therapeutic Approaches

Saurabh Agarwal

Department of Pharmaceutical Sciences, College of Pharmacy and Health Sciences, St. John's University, New York, NY 11439, USA; agarwals@stjohns.edu ; Tel.: +1-718-990-1623

Pediatric cancers cast a dark shadow over the lives of countless children and their families and represent a leading cause of mortality among children worldwide. Despite significant advancements in medical research and treatment modalities, the battle against pediatric cancers remains a challenging and urgent priority. Every day, numerous children are diagnosed with cancers such as leukemia, brain tumors, neuroblastoma, sarcomas, and others, thrusting them into a grucling battle against an invisible enemy. Treatments of pediatric cancers present unique challenges compared to adult cancers, due to the harsh side effects of chemotherapy and radiation, and delayed diagnosis. It is important to understand the unique challenges and insights of different pediatric cancers and to develop novel therapeutic approaches. This editorial covers the unique special edition on Pediatric Cancers published in *Cancers*, which consists of 24 articles presented by international leaders in the pediatric cancers field. This Special Issue comprises 16 original research articles, 6 reviews, and 2 systematic reviews covering insights on different pediatric cancers.

Conventional chemotherapies play a vital role in pediatric cancer treatment, especially in low- and middle-income countries. Several chemo drugs have been approved by the FDA for pediatric cancers; however, multi-drug resistance (MDR) and transporters mediating MDR pose serious obstacles to drug efficacy and require dose escalation, which leads to side effects in pediatric patients [1]. The review by Bo et al. summarizes the mechanisms involving MDR and different drug transporters found in pediatric tumors. Understanding the mechanisms of MDR transporters will enhance the efficacy of pediatric chemotherapies to improve overall survival and reduce treatment toxicity [1]. A systematic review of the effect of chemotherapy in the management of a very rare pediatric neoplasm Melanotic Neuroectodermal Tumor of Infancy (MNTI) highlights the effects of chemotherapy treatments in supporting surgical resections in inoperable, metastatic, and recurrent cases of MNTI [2]. Another retrospective study showed substantial infectious morbidity in pediatric sarcomas patients during neoadjuvant chemotherapy treatment [3]. This study emphasizes developing better risk stratification protocols for preventing and managing febrile neutropenia and infections to maintain quality of life and better chemotherapy treatment outcomes [3]. Understanding the interrelationships between childhood cancers and their treatment with chronic stress in patients throughout their lifespan is very important for effectively managing survivorship. White et al. review the physical, neurological, and psychological effects that lead to chronic stress in childhood cancer survivors and advocate for effective stress management for overall better outcomes [4].

The most common childhood cancer is acute leukemia, which accounts for almost 28% of all pediatric cancers. The most common types in children are acute lymphocytic leukemia (ALL) and acute myeloid leukemia (AML) [5]. An acute leukemia data study of 690 patients showed that the incidence of hyperleukocytosis was 16.6% in ALL and 20.3% in AML patients and was associated with higher morbidity rates and worse survival outcomes. This study suggests modifying the treatment regimen and improving the early-stage monitoring [5]. Myeloid sarcomas (MS), commonly referred to as chloromas, are extramedullary tumors of AML with varying incidence and influence on outcomes [6]. The review article by Zorn et al. summarizes the current understanding of pediatric MS and its

Citation: Agarwal, S. Pediatric Cancers: Insights and Novel Therapeutic Approaches. *Cancers* **2023**, *15*, 3537. https://doi.org/10.3390/cancers15143537

Received: 28 June 2023
Accepted: 4 July 2023
Published: 8 July 2023

biological drivers to spark the development of effective therapeutic strategies for MS and AML [6].

The second most common pediatric cancers are brain and spinal cord tumors, accounting for about 26% of childhood cancers. There are many types of brain and spinal cord tumors, and the treatment and pathogenesis for each are different. The review article by Thorbinson and Kilday summarizes the molecular landscape, prognosis, current therapies, and novel therapeutic approaches for the three most common pediatric brain tumors medulloblastoma, high-grade gliomas, and ependymoma [7]. Further, Chen et al. review the immune microenvironment and immunotherapy clinical trials for the diffuse intrinsic pontine glioma (DIPG) [8]. This glial tumor, DIPG, accounts for 10–15% of pediatric brain tumors, with no effective treatment in the clinic. Therefore, developing novel immunotherapies for transforming DIPG tumors from cold to hot holds a high potential for effective DIPG treatment [8]. Ependymoma is the third most prevalent pediatric CNS tumor with considerable molecular and clinical diversity. The review by Zaytseva et al. underscores the importance of comprehensive molecular profiling to identify (epi)genetic variants for advanced risk stratification of patients [9]. This profiling will support better management and will be pivotal for the development of novel therapeutic strategies for ependymoma [9].

The third most common pediatric cancer is neuroblastoma (NB) which accounts for almost 15% of all pediatric cancer deaths with an overall 5-year survival rate of less than 50% [10]. Therefore, developing novel therapeutic approaches targeting the molecular mechanisms that drive NB progression is very important. In this Special Issue, we have published six research articles on developing effective therapeutic strategies for NB using in vitro and in vivo tumor models. A dual HDAC and PI3K inhibitor CUDC-907 [10] and an ERK inhibitor Ulixertinib [11] were found to be effective treatment approaches for high-risk NB. Further, SOX4 is shown to mediate NB cell differentiation, and SOX4 knockdown partially reversed retinoic acid-induced differentiation in NB cells [12]. The LPA-LPAR1 axis is shown to have migration-inhibitory effects on NB cells, and knockdown of LPRA1 promotes NB cell migration and metastasis [13]. The featured paper published in this Special Issue researches the role of Serine-threonine kinase receptor-associated protein (STRAP) in NB [14]. Bownes et al. found that the genetic knockout of STRAP inhibits NB cell proliferation in vitro and tumor growth in vivo, and overall concluded that STRAP plays a role in NB stemness and tumorigenesis [14]. A retrospective clinical study in 217 high-risk NB patients found that oral metronomic maintenance treatment can improve overall survival in patients not treated with autologous stem cell transplantation or anti-GD2 antibody therapy [15].

Nephroblastoma or Wilms tumor is the most common kidney tumor in childhood and accounts for almost 5% of all pediatric cancers [16]. A retrospective clinical study in 2927 Wilms tumor patients advocates for considering cancer predisposition syndrome with causative genetic or epigenetic variants for effective genetic counseling and management [16]. Pediatric sarcomas represent a diverse group of rare bone and soft tissue malignancies comprising almost 13% of all pediatric cancers [17]. The most common bone sarcomas are osteosarcomas and Ewing's sarcoma, while rhabdomyosarcoma is the most common soft-tissue pediatric sarcoma [17]. A clinical study of 620 survivors of pediatric bone sarcomas in Nordic countries reveals that these patients are at increased risk of developing somatic diseases and long-term adverse health effects [18]. This study emphasizes intervention strategies for optimal patient counseling and follow-up care for the survivors of osteosarcoma and Ewing's sarcoma [18]. Kim et al. developed a radio-genomics predictive model that incorporates both imaging features and gene expression to accurately predict metastasis and chemotherapy responses to improve pediatric osteosarcoma patient outcomes [19]. Chen et al. identified RNA-binding protein level patterns as a prognostic model to identify Ewing's sarcoma patients with a higher mortality risk [20]. An interesting research study screening natural product extracts identified fungal metabolite altertoxin II (ATXII) inhibiting Ewing's sarcoma growth both in vitro and in vivo [21]. ATXII was found to induce DNA double-strand breaks and cell cycle S phase accumulation without directly binding to the DNA [21]. In rhabdomyosarcoma, increased EZH2 protein levels are associated with poor prognosis and increased metastatic potential [22]. Direct and indirect targeting of EZH2 showed differential efficacy due to divergent epigenetic and cellular signaling regulations in different rhabdomyosarcoma cell subtypes [22]. A research study by Lavoie et al. using cell surface proteomics, transcriptomics, and tissue specimens uncovers B7-H3 as a major immunoregulatory molecule expressed by rhabdomyosarcomas and not by normal human tissues [23]. Furthermore, B7-H3 knockout in rhabdomyosarcoma tumor cells increases T-cell-mediated cytotoxicity, indicating B7-H3 as a potential target for developing next-generation immunotherapies for rhabdomyosarcoma [23]. Fuchs et al. used meta-analysis data and suggest developing a uniform classification for pediatric rhabdoid liver tumors and small cell undifferentiated liver tumors, due to the evidence of overlapping histopathology and significantly better survival [24]. Desmoplastic small round cell tumor (DSRCT) is a rare and aggressive soft tissue sarcoma, characterized by a chromosomal translocation resulting in the *EWSR1-WT1* gene fusion [25]. Blejis et al. develop a novel primary patient-derived DSRCT in vitro model recapitulating the original tumor to study disease progression and drug sensitivity [25].

Overall, this Special Issue on pediatric cancers brings together the current understanding of different pediatric cancers, their management, sub-types, and molecular features. Additionally, this Special Issue also includes research articles on experimental therapeutics for different pediatric cancers. In conclusion, this Special Issue book will provide the readership with a comprehensive knowledge of pediatric cancers and hope for developing more effective and less-toxic therapeutic strategies for children battling with cancers.

Conflicts of Interest: The author declare no conflict of interest.

References

1. Bo, L.; Wang, Y.; Li, Y.; Wurpel, J.N.D.; Huang, Z.; Chen, Z.-S. The Battlefield of Chemotherapy in Pediatric Cancers. *Cancers* **2023**, *15*, 1963. [CrossRef] [PubMed]
2. Styczewska, M.; Krawczyk, M.A.; Brecht, I.B.; Haug, K.; Izycka-Świeszewska, E.; Godziński, J.; Raciborska, A.; Ussowicz, M.; Kukwa, W.; Cwalina, N.; et al. The Role of Chemotherapy in Management of Inoperable, Metastatic and/or Recurrent Melanotic Neuroectodermal Tumor of Infancy—Own Experience and Systematic Review. *Cancers* **2021**, *13*, 3872. [CrossRef] [PubMed]
3. Willmer, D.; Zöllner, S.K.; Schaumburg, F.; Jürgens, H.; Lehrnbecher, T.; Groll, A.H. Infectious Morbidity in Pediatric Patients Receiving Neoadjuvant Chemotherapy for Sarcoma. *Cancers* **2021**, *13*, 1990. [CrossRef] [PubMed]
4. White, G.E.; Caterini, J.E.; McCann, V.; Rendall, K.; Nathan, P.C.; Rhind, S.G.; Jones, H.; Wells, G.D. The Psychoneuroimmunology of Stress Regulation in Pediatric Cancer Patients. *Cancers* **2021**, *13*, 4684. [CrossRef]
5. Kittivisuit, S.; Jongthitinon, N.; Sripornsawan, P.; Songthawee, N.; Chavananon, S.; Limratchapong, C.; McNeil, E.B.; Chotsampancharoen, T. Hyperleukocytosis in Childhood Acute Leukemia: Early Complications and Survival Outcomes. *Cancers* **2023**, *15*, 3072. [CrossRef]
6. Zorn, K.E.; Cunningham, A.M.; Meyer, A.E.; Carlson, K.S.; Rao, S. Pediatric Myeloid Sarcoma, More than Just a Chloroma: A Review of Clinical Presentations, Significance, and Biology. *Cancers* **2023**, *15*, 1443. [CrossRef]
7. Thorbinson, C.; Kilday, J.-P. Childhood Malignant Brain Tumors: Balancing the Bench and Bedside. *Cancers* **2021**, *13*, 6099. [CrossRef]
8. Chen, Y.; Zhao, C.; Li, S.; Wang, J.; Zhang, H. Immune Microenvironment and Immunotherapies for Diffuse Intrinsic Pontine Glioma. *Cancers* **2023**, *15*, 602. [CrossRef]
9. Zaytseva, M.; Papusha, L.; Novichkova, G.; Druy, A. Molecular Stratification of Childhood Ependymomas as a Basis for Personalized Diagnostics and Treatment. *Cancers* **2021**, *13*, 4954. [CrossRef]
10. Chilamakuri, R.; Agarwal, S. Dual Targeting of PI3K and HDAC by CUDC-907 Inhibits Pediatric Neuroblastoma Growth. *Cancers* **2022**, *14*, 1067. [CrossRef]
11. Yu, Y.; Zhao, Y.; Choi, J.; Shi, Z.; Guo, L.; Elizarraras, J.; Gu, A.; Cheng, F.; Pei, Y.; Lu, D.; et al. ERK Inhibitor Ulixertinib Inhibits High-Risk Neuroblastoma Growth In Vitro and In Vivo. *Cancers* **2022**, *14*, 5534. [CrossRef]
12. Zhang, D.; Gong, B.; Zhao, Q.; Li, Z.; Tan, X.; Hua, Z. SOX4 Mediates ATRA-Induced Differentiation in Neuroblastoma Cells. *Cancers* **2022**, *14*, 5642. [CrossRef] [PubMed]
13. Liu, X.; Pei, M.; Yu, Y.; Wang, X.; Gui, J. Reduction of LPAR1 Expression in Neuroblastoma Promotes Tumor Cell Migration. *Cancers* **2022**, *14*, 3346. [CrossRef] [PubMed]

14. Bowness, L.V.; Williams, A.P.; Marayati, R.; Quinn, C.H.; Hutchins, S.C.; Stewart, J.E.; Vu, T.; Easlick, J.L.; Mroczek-Musulman, E.; Crossman, D.K.; et al. Serine-Threonine Kinase Receptor-Associated Protein (STRAP) Knockout Decreases the Malignant Phenotype in Neuroblastoma Cell Lines. *Cancers* **2021**, *13*, 3201. [CrossRef] [PubMed]
15. Sun, X.; Zhen, Z.; Guo, Y.; Gao, Y.; Wang, J.; Zhang, Y.; Zhu, J.; Lu, S.; Sun, F.; Huang, J.; et al. Oral Metronomic Maintenance Therapy Can Improve Survival in High-Risk Neuroblastoma Patients Not Treated with ASCT or Anti-GD2 Antibodies. *Cancers* **2021**, *13*, 3494. [CrossRef]
16. Welter, N.; Wagner, A.; Furtwängler, R.; Melchior, P.; Kager, L.; Vokuhl, C.; Schenk, J.-P.; Meier, C.M.; Siemer, S.; Gessler, M.; et al. Correction: Welter et al. Characteristics of Nephroblastoma/Nephroblastomatosis in Children with a Clinically Reported Underlying Malformation or Cancer Predisposition Syndrome. *Cancers* **2021**, *13*, 5016. *Cancers* **2021**, *13*, 5743. [CrossRef]
17. Anderson, J.L.; Denny, C.T.; Tap, W.D.; Federman, N. Pediatric sarcomas: Translating molecular pathogenesis of disease to novel therapeutic possibilities. *Pediatr Res.* **2012**, *72*, 112–121. [CrossRef]
18. Pedersen, C.; Rechnitzer, C.; Andersen, E.A.W.; Kenborg, L.; Norsker, F.N.; Bautz, A.; Baad-Hansen, T.; Tryggvadottir, L.; Madanat-Harjuoja, L.-M.; Holmqvist, A.S.; et al. Somatic Disease in Survivors of Childhood Malignant Bone Tumors in the Nordic Countries. *Cancers* **2021**, *13*, 4505. [CrossRef]
19. Kim, B.-C.; Kim, J.; Kim, K.; Byun, B.H.; Lim, I.; Kong, C.-B.; Song, W.S.; Koh, J.-S.; Woo, S.-K. Preliminary Radiogenomic Evidence for the Prediction of Metastasis and Chemotherapy Response in Pediatric Patients with Osteosarcoma Using 18F-FDG PET/CT.; EZRIN.; and KI67. *Cancers* **2021**, *13*, 2671. [CrossRef]
20. Chen, Y.; Su, H.; Su, Y.; Zhang, Y.; Lin, Y.; Haglund, F. Identification of an RNA-Binding-Protein-Based Prognostic Model for Ewing Sarcoma. *Cancers* **2021**, *13*, 3736. [CrossRef]
21. Robles, A.J.; Dai, W.; Haldar, S.; Ma, H.; Anderson, V.M.; Overacker, R.D.; Risinger, A.L.; Loesgen, S.; Houghton, P.J.; Cichewicz, R.H.; et al. Altertoxin II, a Highly Effective and Specific Compound against Ewing Sarcoma. *Cancers* **2021**, *13*, 6176. [CrossRef] [PubMed]
22. Schmidt, A.; Behrendt, L.; Eybe, J.; Warmann, S.W.; Schleicher, S.; Fuchs, J.; Schmid, E. The Effect of Direct and Indirect EZH2 Inhibition in Rhabdomyosarcoma Cell Lines. *Cancers* **2022**, *14*, 41. [CrossRef]
23. Lavoie, R.R.; Gargollo, P.C.; Ahmed, M.E.; Kim, Y.; Baer, E.; Phelps, D.A.; Charlesworth, C.M.; Madden, B.J.; Wang, L.; Houghton, P.J.; et al. Surfaceome Profiling of Rhabdomyosarcoma Reveals B7-H3 as a Mediator of Immune Evasion. *Cancers* **2021**, *13*, 4528. [CrossRef] [PubMed]
24. Fuchs, J.; Murtha-Lemekhova, A.; Kessler, M.; Ruping, F.; Günther, P.; Fichtner, A.; Sturm, D.; Hoffmann, K. A Systematic Review and Meta-Analysis of Malignant Rhabdoid and Small Cell Undifferentiated Liver Tumors: A Rational for a Uniform Classification. *Cancers* **2022**, *14*, 272. [CrossRef] [PubMed]
25. Bleijs, M.; Pleijte, C.; Engels, S.; Ringnalda, F.; Meyer-Wentrup, F.; van de Wetering, M.; Clevers, H. EWSR1-WT1 Target Genes and Therapeutic Options Identified in a Novel DSRCT In Vitro Model. *Cancers* **2021**, *13*, 6072. [CrossRef] [PubMed]

cancers MDPI

Article

Dual Targeting of PI3K and HDAC by CUDC-907 Inhibits Pediatric Neuroblastoma Growth

Rameswari Chilamakuri and Saurabh Agarwal *

Department of Pharmaceutical Sciences, College of Pharmacy and Health Sciences, St. John's University, New York, NY 11439, USA; rameswari.chilamakuri19@my.stjohns.edu
* Correspondence: agarwals@stjohns.edu; Tel.: +1-718-990-1623

Simple Summary: High-risk neuroblastoma (NB) is an aggressive cancer of very young children and accounts for almost 15% of all pediatric cancer deaths. Current therapies include high-dose chemotherapy and radiation, which have long-term toxic side effects. Despite these intensive therapies, the overall 5-year survival rate of NB is less than 50%. Therefore, developing novel therapeutic approaches targeting the molecular mechanisms that drive NB progression is very important. In the present study, we repurpose CUDC-907, a dual inhibitor of PI3K and histone deacetylases. These regulators are known to regulate MYCN expression, a key prognostic marker of NB. CUDC-907 potently inhibits NB growth and 3D spheroid tumor growth by inhibiting PI3K, HDAC, and MYCN. Overall, our pre-clinical data demonstrate that repurposing CUDC-907 as a single drug is a novel and effective therapeutic approach for NB.

Abstract: The dysregulation of PI3K, HDACs, and MYCN are well known for promoting multiple cancer types, including neuroblastoma (NB). Targeting the upstream regulators of MYCN, including HDACs and PI3K, was shown to suppress cancer growth. In the present study, we analyze different NB patient datasets to reveal that high *PI3K* and *HDAC* expression is correlated with overall poor NB patient survival. High *PI3K* level is also found to be associated with high *MYCN* level and NB stage progression. We repurpose a dual inhibitor CUDC-907 as a single agent to directly target both PI3K and HDAC in NB. We use in vitro methodologies to determine the efficacy and selectivity of CUDC-907 using six NB and three control fibroblast cell lines. Our results show that CUDC-907 significantly inhibits NB proliferation and colony growth, induces apoptosis, blocks cell cycle progression, inhibits MYCN, and enhances H3K9Ac levels by inhibiting the PI3K/AKT signaling pathway and HDAC function. Furthermore, CUDC-907 significantly inhibits NB tumor growth in a 3D spheroid tumor model that recapitulates the in vivo tumor growth. Overall, our findings highlight that the dual inhibition of PI3K and HDAC by CUDC-907 is an effective therapeutic strategy for NB and other MYC-dependent cancers.

Keywords: drug repurposing; PI3K pathway; epigenetics; Fimepinostat; neuroblastoma; pediatric cancer

Citation: Chilamakuri, R.; Agarwal, S. Dual Targeting of PI3K and HDAC by CUDC-907 Inhibits Pediatric Neuroblastoma Growth. *Cancers* **2022**, *14*, 1067. https://doi.org/10.3390/cancers14041067

Academic Editor: Miguel Segura Ginard

Received: 10 January 2022
Accepted: 17 February 2022
Published: 20 February 2022

1. Introduction

High-risk neuroblastoma (NB) is the most common extracranial solid tumor that accounts for almost 10% of all childhood-related cancers [1]. Despite major advancements in intensive multi-modal therapies, the overall 5-year survival rate of NB is less than 50% [2]. Current NB therapy follows an induction chemotherapy regimen that is often associated with disease comorbidities and increased risk of secondary malignancies [3]. Therefore, developing novel single-agent therapeutic strategies targeting multiple oncogenic pathways such as PI3K/AKT, histone deacetylases, and MYCN, is important for NB and other cancers.

The pathologic activation of *MYCN* plays a central role in NB, with *MYCN* amplification identified in approximately 25% of primary NB tumors [4]. *MYCN* amplification drives rapid metastasis, relapse, and drug resistance, and patients with relapsed *MYCN* amplified tumors have less than 5% overall survival rate [5]. Direct MYCN targeting strategies were shown to have limited effect, thus indirectly targeting the oncogenic activation of *MYCN* transcription is an important therapeutic approach for MYCN-driven cancers [4]. Among others, histone deacetylases (HDAC) and phosphoinositide-3 kinase (PI3K) are upstream regulators of *MYCN* and are also known to promote NB pathogenesis [6]. PI3K/AKT pathway is one of the most important intracellular signaling pathways regulating cell growth, proliferation, angiogenesis, motility, survival, and metabolism [7]. The PI3K signaling pathway also regulates multiple cellular proteins, such as mTORC1, S6 kinase, GSK3β, FOXO, MDM2, MYCN, P27, and BAD [8–10]. The dysregulation of PI3K has been reported in multiple cancer types, including prostate cancer, breast cancer, ovarian cancer, and NB [11,12].

Epigenetic alterations, such as histone acetylation and methylation, play an essential role in tumor initiation and progression [6]. HDACs are key transcription cofactors that regulate histone and non-histone protein substrates through epigenetic or non-epigenetic modifications, thus affecting multiple signaling networks, including MYC, p53, and STAT3 [13]. HDACs also play an important role in cell proliferation, differentiation, cellular homeostasis, and stemness maintenance [13]. Studies showed that the H3K9 acetylation (H3K9Ac) of active promoters positively correlates with gene expression, and HDAC inhibitors induce these acetylation marks by inhibiting HDAC enzymes [14]. HDAC inhibitors alone or in combination with other anti-cancer compounds showed promising preclinical results in the treatment of different cancers [15]. Preclinical studies have shown promising anti-cancer activity for combining PI3K inhibitors with HDAC inhibitors in various cancers [16].

In the present study, we use a dual PI3K and HDAC inhibitor, CUDC-907 or Fimepinostat, and show its potency in inhibiting NB growth. CUDC-907 is a first-in-class, oral small molecule dual HDAC (class I and II) and PI3K (class Iα, β, and δ) inhibitor that can simultaneously target multiple oncogenic signaling pathways [17,18]. CUDC-907 is effective in multiple cancer types, including acute myeloid leukemia, relapsed or refractory diffuse large B-cell lymphoma, prostate cancer, thyroid cancer, and multiple myeloma [19–21]. FDA designated CUDC-907 for fast track development for patients with relapsed diffuse large B-cell lymphoma [17]. Our preclinical results in NB demonstrate that CUDC-907 potently inhibits NB growth by inhibiting the PI3K signaling pathway and HDACs. We also found that CUDC-907 inhibits MYCN and enhances the H3K9 histone acetylation. Overall, our study highlights the efficacy of CUDC-907 as a single drug in targeting multiple pathways and as a novel therapeutic approach for NB. As CUDC-907 is currently under extensive clinical trials for multiple cancers, our study will further pave the way for the effective clinical translation of CUDC-907 for NB patients.

2. Materials and Methods

2.1. Cell Culture and Reagents

Human NB cell lines NGP, LAN5, CHLA-255-MYCN (*MYCN*-amplified), SH-SY5Y, SK-N-AS, and CHLA-255 (*MYCN* non-amplified), and normal fibroblast control cell lines WI-38, NIH-3T3, and COS-7 were routinely cultured and maintained as described previously [22,23]. All cell lines used in this study were routinely tested for mycoplasma and validated via short-tandem repeat analysis for genotyping. In this study, the experiments were performed using different numbers of cell lines as required for the assay. Primary antibodies anti-PI3K (4292S), anti-p-PI3K (4228S), anti-PDK1 (3062S), anti-p-PDK1 (Ser241; 3438S), anti-AKT (9272S), anti-p-AKT (Thr308; 9275S), anti- p70S6 kinase (S6K; 9202S), anti-p-p70-S6 kinase (pS6K; Thr389; 9205S), anti-MYCN (9405S), anti-H3 (4499S), anti-H3K9Ac (9649S), anti-Cyclophilin B (43603S), and anti-rabbit IgG HRP-linked secondary antibody (7074S) were purchased from Cell Signaling Technology. CUDC-907 was purchased from MedChem Express, Monmouth Junction, NJ, USA.

2.2. Clinical Patient Dataset

NB patient datasets were analyzed using the publicly available patient databases from the R2: Genomic Analysis and Visualization Platform. Available online: https://hgserver1.amc.nl/cgi-bin/r2/main.cgi (accessed on 1 February 2022). This unique platform provides a multi-parametric analysis of NB patient outcomes with correlation to gene expression and microarray profiles of their primary tumors. In the present study, we analyzed a total of 1235 primary NB patient data by analyzing Versteeg, Kocak, and SEQC datasets. Dataset analyses were performed by including patients' confounding factors.

2.3. Cell Viability and Clonogenic Assay

Cell viability assays were performed as described previously [23]. Briefly, cells were treated with different drug concentrations for 72 h, followed by incubation with CellTiter 96 AQueous One Solution from the Cell Proliferation Assay kit (G3582; Promega, Madison, WI, USA) as per the manufacturer's instructions. Cell viability was measured using a spectrophotometer (SpectraMax iD3, Molecular Device, San Jose, CA, USA) at 490 nm. The data were analyzed using GraphPad Prism 9 software, and IC_{50} values were calculated for individual cell lines. Clonogenic cell colony formation assays were performed as described previously [23,24].

2.4. Apoptosis and Cell Cycle Assay

Apoptosis assays were performed using the Muse Annexin V and Dead Cell Kit (MCH100105; Luminex Corp., Austin, TX, USA), and cell cycle analysis was performed using Muse Cell Cycle Kit reagent (MCH100106; Luminex Corp), according to the manufacturer's instructions and as described previously [23]. NB cell lines were treated for 16 h with different concentrations of CUDC-907. Apoptosis and cell cycle samples were analyzed using the Guava Muse cell analyzer (Luminex Corp) to determine the percentage of early apoptosis and cell cycle changes, respectively.

2.5. 3D Spheroid Tumor Assay

Three-dimensional spheroidal tumor assay was performed using three-dimensional spheroidal 96-well microplates (4515; Corning Inc., Glendale, AZ, USA), according to the manufacturer's instructions and as described previously [23]. Briefly, NB spheroids of about 250 µm were formed, randomized, and treated with regular drug replenishment. Spheroid images were captured, and size was measured using a Leica DMi1 microscope using the LASX software suite from Leica Microsystems. This software suite provides tools to analyze the captured images and measure size. Each treatment cohort includes at least 6 spheroid tumors. A Viability/Cytotoxicity Assay Kit for Animal Live and Dead Cells (Biotium Inc., Fremont, CA, USA) was used to measure spheroid cell viability as per the manufacturer's instructions. Calcein AM stains live cells, and ethidium homodimer III stains dead cells and yields green and red fluorescence, respectively. Fluorescent spheroid images were captured using the EVOS FL imaging system (Thermo Fisher Scientific, Waltham, MA, USA), and fluorescent quantification was performed using a SpectraMax iD3 microplate reader (Molecular Device) at 517 nm to detect Calcein AM and at 625 nm to detect EthD-III dye. Furthermore, the CellTiter-Glo 3D Cell Viability Assay (G968; Promega) dye was used according to the manufacturer's instructions to determine the viability of spheroid cells.

2.6. RNA Extraction and Quantitative Real-Time RT-PCR

Gene expression analysis was performed using the RT-qPCR method, as described previously [23]. NB cells were treated with different drug concentrations for 6 h. Total RNA was extracted by using RNeasy plus mini kit (74134; Qiagen, Hilden, Germany), followed by cDNA synthesis using a high-capacity cDNA reverse transcription kit (4368814; Thermo Fisher Scientific), as per manufacturer's instructions. Further, the cDNA was used in RT-qPCR reactions for individual genes using SYBR Green dye (4385610; Thermo Fisher Scientific) and QuantStudio 3 Real-Time PCR System (Thermo Fisher Scientific). The expression of genes was normalized to the expression of GAPDH as a housekeeping gene. Primers used in this study are listed in Supplementary Table S1.

2.7. Immunoblotting Assays

Immunoblotting assays were performed by treating NB cells with different concentrations of CUDC-907 for 6 h, followed by cell lysis using RIPA buffer (89900; Thermo Fisher Scientific) supplemented with protease inhibitor cocktail (Complete mini EDTA free, Roche, Basel, Switzerland) and phosphatase inhibitor cocktail (PhosSTOP, Roche). Protein samples were quantified with Bradford assay (5000205; Bio-Rad, Hercules, CA, USA) and equal amounts of total protein were separated on 4–12% SDS-PAGE gels, followed by transfer to PVDF membranes using a Bio-Rad Trans-Blot Turbo TM system, then blocking with 5% BSA solution, and probing with corresponding primary antibodies (1:1000 dilution) overnight at 4 °C [25]. This was followed by washing and incubating the membranes with either anti-mouse or anti-rabbit IgG HRP-conjugated secondary antibody (1:10,000 dilution) for 2 h. Protein bands on membranes were developed using Clarity ECL Western substrate (Bio-Rad) and visualized using the ChemiDoc XRS Plus system (Bio-Rad). Densitometric analysis of the protein bands was performed using the ImageJ software.

2.8. Statistical Analysis

In the present study, three technical replicates and appropriate controls were performed for all the assays. Results are presented as the mean ± standard error mean (SEM) of the replicates. Two-tailed Student's t-test and ANOVA statistical significance tests were used among different groups in experiments. The p values were calculated for fold differences among different comparative groups and $p < 0.05$ was considered statistically significant. Patient survival analyses were performed using the Kaplan–Meier method and two-sided log-rank tests.

3. Results

3.1. PI3K Expression Strongly Correlates with Poor NB Prognosis

To investigate the role of PI3K in NB, we analyzed different NB patient datasets (total 1235 primary NB patients) using the R2 dataset and determined the correlation of the PI3K gene (*PIK3C2A*) with overall NB patient outcome. Kaplan–Meier survival analysis revealed that expression of the *PIK3C2A* gene inversely correlates with the overall survival of NB patients (Figure 1A–C). In contrast, the low expression of *PIK3C2A* gene showed significantly better prognosis and overall survival in all the analyzed patient datasets (Kocak N = 649, $p = 1.3 \times 10^{-9}$; SEQC N = 498, $p = 1.2 \times 10^{-6}$; Versteeg N = 88, $p = 1.5 \times 10^{-4}$; Figure 1A–C). Additionally, aggressive higher stage NB tumors showed higher *PIK3C2A* expression levels (Figure 1D–F), suggesting a role of PI3K in NB progression. Further, we observed a strong correlation between *PI3K* and *MYCN* genes in NB (data not shown). Additionally, *HDAC2* gene expression also showed an inverse correlation with overall and event-free survival of NB patients (Supplementary Figure S1). These findings suggest the oncogenic role of PI3K and HDAC in NB and highlight an effective therapeutic targeting strategy of using a dual inhibitor CUDC-907 for NB.

3.2. CUDC-907 Inhibits NB Proliferation

Based on the patient dataset analysis, we used CUDC-907, a dual PI3K and HDAC inhibitor in NB. Cytotoxicity assays for CUDC-907 in different NB cell lines and control fibroblast cell lines were performed (Figure 2). We used a total of six NB cell lines, including three *MYCN*-amplified and three *MYCN* non-amplified cell lines. Additionally, to determine the selectivity of CUDC-907 for cancer cells, we used three control fibroblast cell lines (Figure 2A). The results show that CUDC-907 significantly inhibits the cell proliferation of both *MYCN*-amplified (Figure 2B) and non-amplified cell lines (Figure 2C) in a dose-dependent manner and in contract to control fibroblast cell lines, with IC_{50} values ranging from 0.91 μM ($p < 0.01$) for LAN-5 to 1.94 ($p < 0.05$) μM for NGP (Figure 2).

The results of the cytotoxicity assays on control cell lines clearly show the selectivity and potency of CUDC-907 in inhibiting NB proliferation (Figure 2A). To further validate the anti-proliferative effect of CUDC-907, we performed clonogenic assays using four different NB cell lines, including two *MYCN*-amplified (LAN-5, NGP) and two non-amplified (SH-SY5Y, SK-N-AS) cell lines. Our data showed that CUDC-907 significantly inhibits overall NB colony formation capacity in contrast to control treatment in a dose-dependent manner (Figure 3). These results further demonstrate the inhibitory effect of CUDC-907 on NB growth and proliferation.

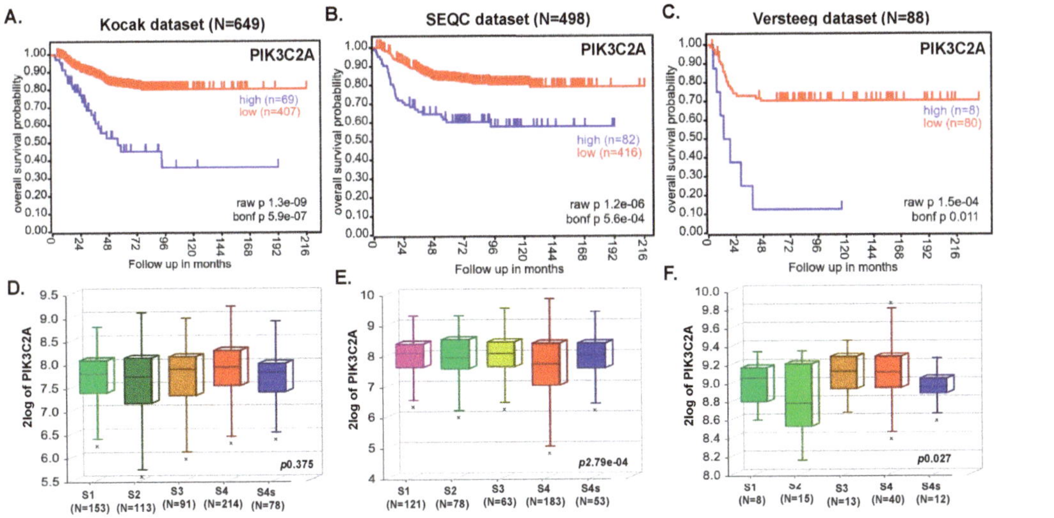

Figure 1. Expression of *PIK3C2A* gene inversely correlates with overall NB patients' survival. Kaplan–Meier survival analysis of *PIK3C2A* gene expression shows an overall inverse probability of NB patients' survival. (**A**) Kocak (N = 649). (**B**) SEQC (N = 498). (**C**) Versteeg (N = 88). (**D–F**) International Neuroblastoma Staging System (INSS)-based NB stage analysis showing the correlation of *PIK3C2A* expression with NB progression: (**D**) Kocak dataset, (**E**) SEQC dataset, (**F**) Versteeg dataset. x represents statistical significance within the group.

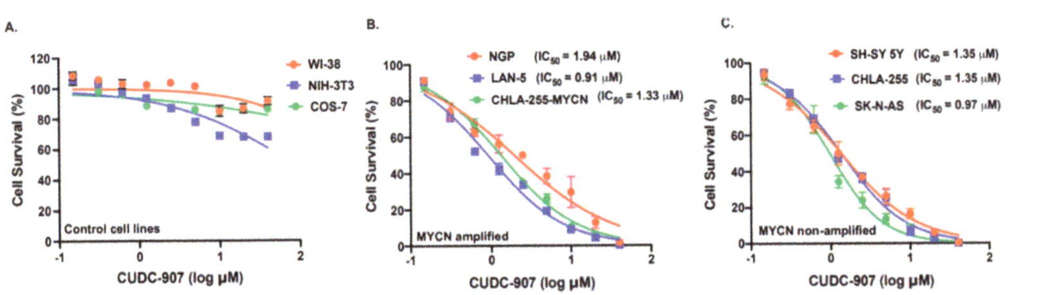

Figure 2. CUDC-907 inhibits NB proliferation. In vitro cytotoxicity assays in response to CUDC-907 using fibroblast cell lines and human NB cell lines. (**A**) WI-38, NIH-3T3, and COS-7 fibroblast cell lines. (**B**) NGP, LAN-5, and CHLA-255-MYCN as *MYCN*-amplified NB cell lines. (**C**) SH-SY-5Y, SK-N-AS, and CHLA-255 as *MYCN* non-amplified NB cell lines. The nonlinear variable slope regression method was used to determine IC_{50} values for individual NB cell lines.

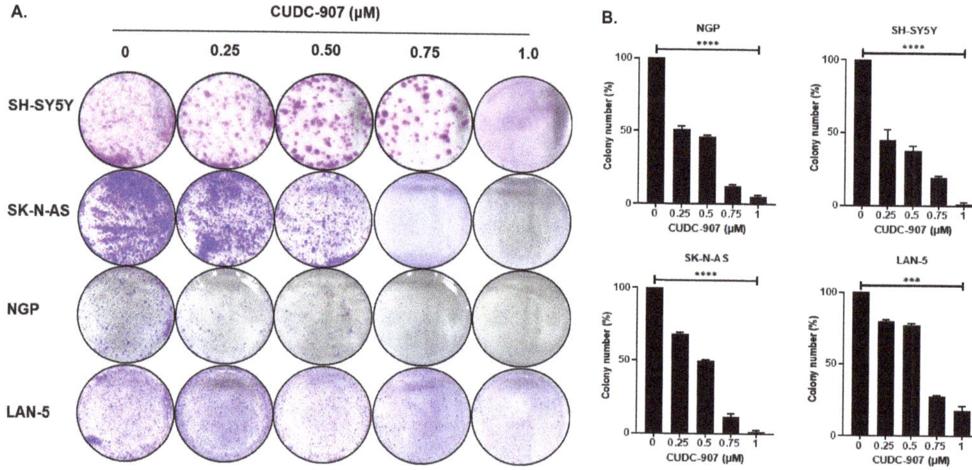

Figure 3. CUDC-907 inhibits NB colony formation and growth. Colony formation assays in response to CUDC-907 in four NB cell lines. (**A**) Representative images of colony formation assays in *MYCN*-amplified cell lines NGP and LAN-5, and *MYCN* non-amplified cell lines SH-SY-5Y and SK-N-AS in response to CUDC-907 treatment. Cells were treated with CUDC-907 for 48 h. (**B**) Survival index graphs showing the quantitation of the relative inhibition of colony formation in the respective cell lines in response to CUDC-907 treatment. Colony numbers are normalized to control. *** $p < 0.001$, **** $p < 0.0001$.

3.3. CUDC-907 Induces Apoptosis and Blocks Cell Cycle Progression in NB

Next, we performed apoptosis and cell cycle assays using different NB cell lines in response to CUDC-907. The results show that CUDC-907 significantly and in a dose-dependent manner induces apoptosis in both NGP and SH-SY5Y NB cell lines (Figure 4A,B). Specifically, CUDC-907 treatment increases the percentage of early apoptotic cells by 1.45- and 1.74-fold in SH-SY5Y cells, and 1.46- and 1.86-fold in NGP cells, in response to 0.5 µM and 1 µM, respectively, and in contrast to control treatment (Figure 4A,B).

Furthermore, we observed that CUDC-907 treatment significantly inhibits NB cell cycle progression by inhibiting the S phase (Figure 4C). In response to CUDC-907 treatment, the percentage of cells in the S phase decreased by 0.88- and 0.62-fold ($p < 0.05$), while the percentage of cells in the G2/M phase increased by 1.6- and 1.7-fold ($p < 0.05$), for 0.5 µM and 1 µM, respectively, and in contrast to control treatment (Figure 4C). These data further validate the efficacy of CUDC-907 in inducing cytotoxicity and inhibiting NB cell growth by arresting the cell cycle and by inducing apoptosis.

3.4. CUDC-907 Inhibits NB Spheroid Tumor Growth

To further determine the effect of CUDC-907, we developed a 3D spheroid tumor model of NB by using the SH-SY5Y cell line. These 3D spheroids strongly mimic the in vivo physiological growth patterns of solid tumor NB by generating anchorage-independent spheroid tumor mass. We have developed spheroid tumors of similar sizes, randomized them, and treated them with increasing doses of CUDC-907 (Figure 5). The size and growth of individual spheroid tumors were measured and imaged regularly. The results show significant and dose-dependent inhibition of NB tumor growth in response to CUDC-907 and in contrast to control treatment (Figure 5A,B). Furthermore, we observed a significant reduction of live cells in spheroid tumors in response to CUDC-907 treatments, emphasizing the effect of CUDC-907 in inhibiting tumor mass by blocking NB cell growth (Figure 5C,E,F). These results were further validated by a live-cell ATP release assay, which demonstrated that CUDC-907 significantly and in a dose-dependent manner inhibits the number of live cells in NB spheroid tumors to overall inhibiting tumor growth (Figure 5D). These qualitative and quantitative assays in 3D spheroid tumors further confirmed the efficacy of CUDC-907 as a single drug in inhibiting NB growth.

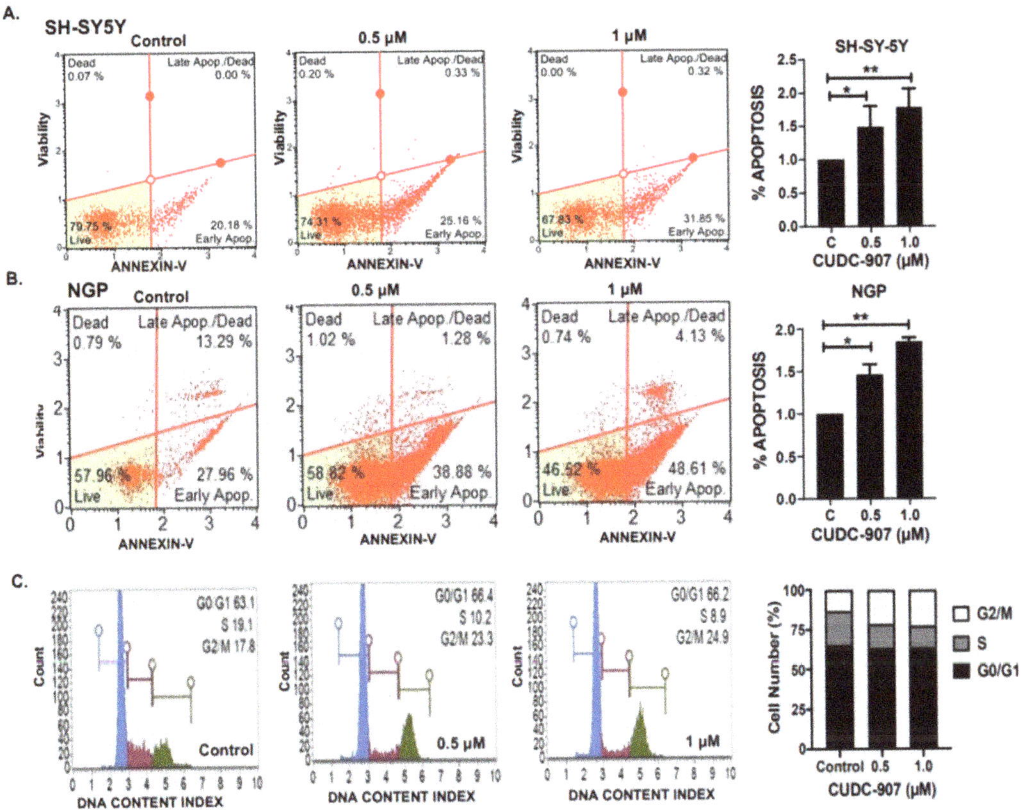

Figure 4. CUDC-907 induces apoptosis and blocks cell cycle progression in NB cells. (**A**,**B**) Representative images showing the percentage of apoptosis in NB cell lines in response to CUDC-907 treatment for 16 h. (**A**) SH-SY5Y and (**B**) NGP. Early apoptosis cells from three independent experiments are plotted and normalized to control treatments. C= control, * $p < 0.05$, ** $p < 0.01$. (**C**) Representative images of cell cycle assay in SH-SY5Y cell line in response to CUDC-907 treatment for 16 h. Percentage of cells in G0/G1, S, and G2/M cell cycle phases are represented, showing the effect of CUDC-907 on NB cell cycle phases.

3.5. *CUDC-907 Inhibits PI3K/AKT Pathway*

To further determine the molecular mechanisms by which CUDC-907 inhibits NB growth, we performed gene and protein expression profiling. Gene expression analysis showed that CUDC-907 significantly inhibits mRNA expression of key PI3K/AKT pathway genes, including *PIK3C2A*, *AKT1*, *TBK1*, and *MTOR* (Figure 6A). For the *PIK3C2A* gene, we observed about three-fold inhibition of mRNA expression in response to 1 μM of CUDC-907, and in contrast to the control treatment (Figure 6A). Additionally, we found that CUDC-907 leads to significant inhibition of the phosphorylation and activation of the PI3K/AKT pathway proteins, including p-PI3K (Tyr458/Tyr199), p-PDK1 (S241), p-AKT (T308), and p-S6K (T389). The dose-dependent inhibition of phosphorylation was observed in response to CUDC-907 treatment and in contrast to the control treatment (Figure 6B). As expected, the total protein levels of PI3K, PDK1, AKT, and S6K were found to be slightly reduced in response to CUDC-907 treatments (Figure 6B). Furthermore, we observed a dose-dependent inhibition of MYCN expression at both mRNA and protein levels in response to the CUDC-907 treatment (Figure 6A,B). These data highlight the effects of CUDC-907 in inhibiting the PI3K/AKT pathway and MYCN at both mRNA and protein levels.

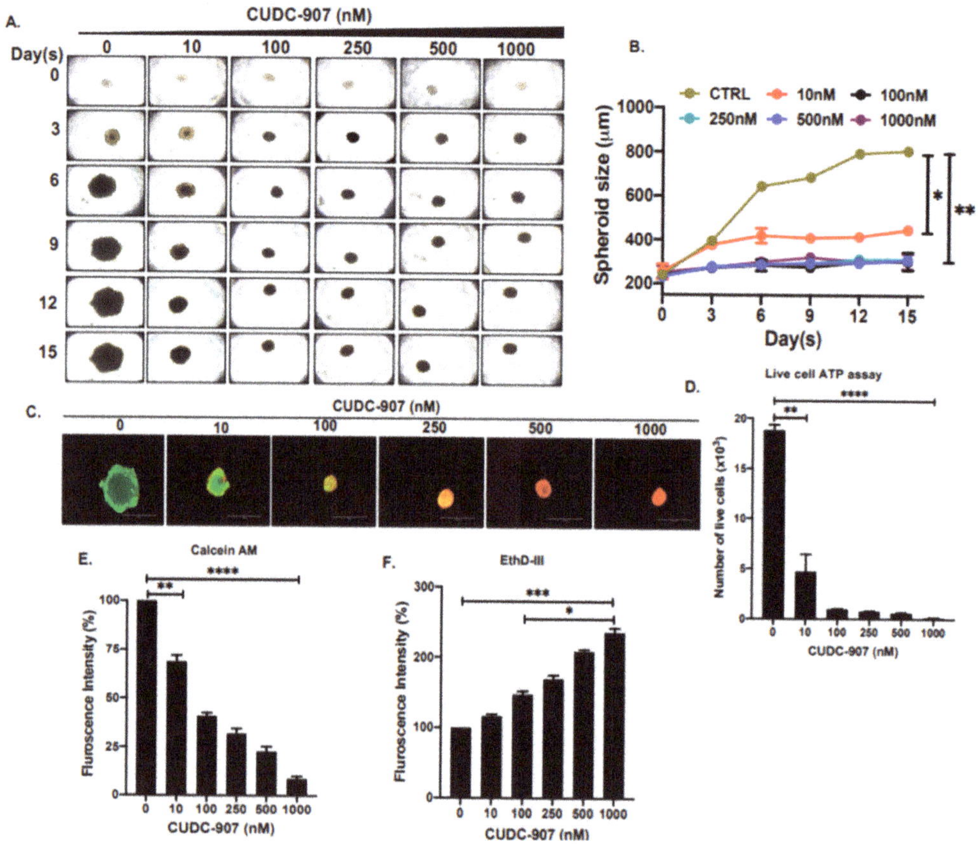

Figure 5. CUDC-907 inhibits NB spheroid tumor growth. (**A**) 3D spheroid tumors developed using SH-SY5Y cell line at different days of growth in response to increasing concentrations of CUDC-907 treatment. (**B**) Spheroid tumor growth measurements as displayed in (**A**), show a significant decrease in spheroidal mass in response to CUDC-907 treatments. (**C**) Representative images of terminal spheroids stained with Calcein AM (green; live cells) and EthD-III (red; dead cells) fluorescence dyes. (**D**) Quantitative assessment of the number of live cells in 3D spheroid tumors using live-cell ATP release assay. (**E**) Quantitative representation of the percentage of cells stained with Calcein AM shows a significant decrease in the number of live cells. (**F**) Quantitative representation of the percentage of cells stained with EthD-III shows a significant increase in the number of dead cells. * $p < 0.05$, ** $p < 0.01$. *** $p < 0.001$, **** $p < 0.0001$.

3.6. CUDC-907 Inhibits HDAC and Induces Histone Acetylation at H3K9

To further understand the effect of CUDC-907 at the epigenetic level, we analyzed the gene expression of histone deacetylases *HDAC1* and *HDAC2* in response to CUDC-907. Our results show an inhibition of *HDAC1* mRNA by 0.2- and 0.4-fold and *HDAC2* mRNA by 0.6- and 0.7-fold in response to 0.5 and 1.0 µM, respectively, and in contrast to control treatment (Figure 7A). Additionally, to determine the effect of HDAC inhibition, we further analyzed the overall H3K9Ac levels in NB cells. The results show that CUDC-907 in a dose-dependent manner significantly increases H3K9Ac levels in contrast to the control (Figure 7B). These data highlight the epigenetic effects of CUDC-907 by inhibiting HDACs and therefore enhancing H3K9Ac levels in NB.

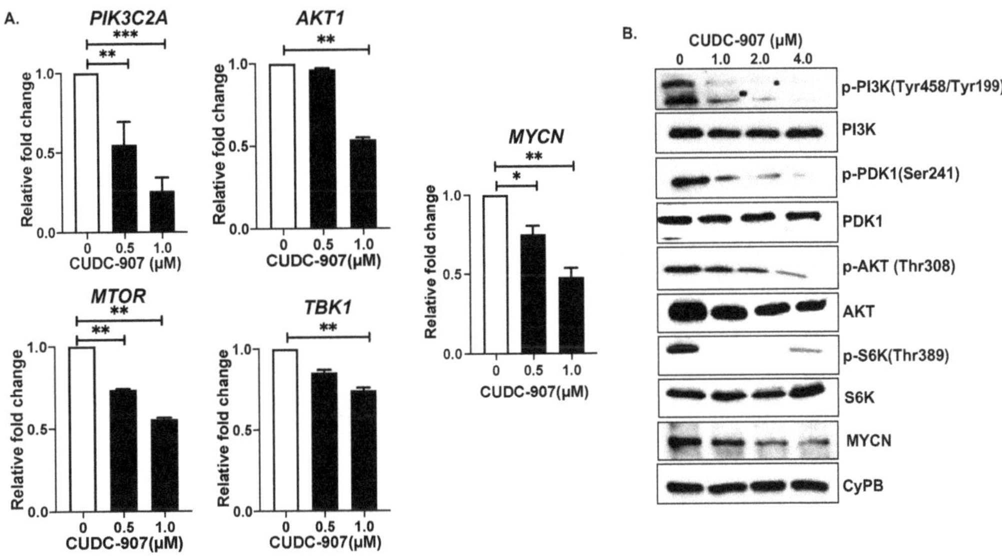

Figure 6. CUDC-907 inhibits PI3K/AKT pathway: (**A**) Gene expression analysis of *PIK3C2A*, *AKT1*, *MTOR*, *TBK1*, and *MYCN* in response to CUDC-907 treatments in SH-SY5Y cells. * $p < 0.05$, ** $p < 0.01$, and *** $p < 0.001$. (**B**) Western blot analysis of different PI3K pathway proteins in response to increasing concentrations of CUDC-907 treatments. CyPB was used as a loading control. Full blots and densitometric data are available in Supplementary Figure S2.

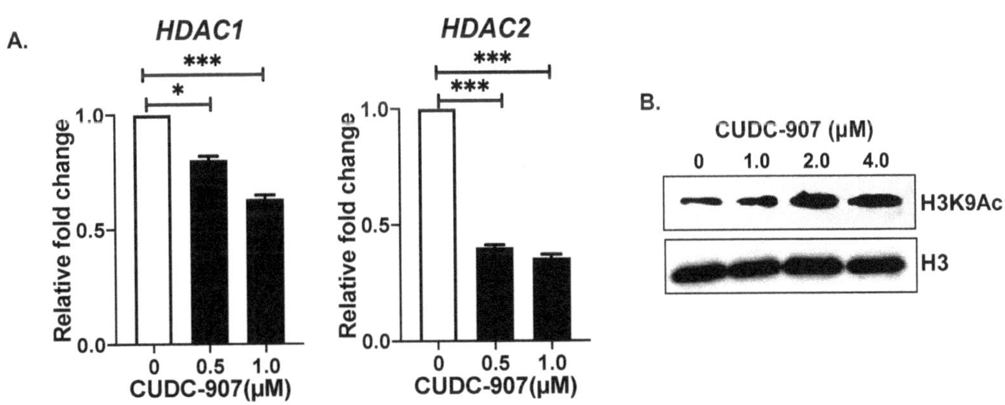

Figure 7. CUDC-907 inhibits HDAC: (**A**) Gene expression analysis of *HDAC1* and *HDAC2* in response to CUDC-907 treatments in SH-SY5Y cells. * $p < 0.05$, *** $p < 0.001$. (**B**) Western blot analysis of H3K9Ac and total H3 in response to increasing concentrations of CUDC-907 treatments. H3 was used as a loading control. Full blots and densitometric data are available in Supplementary Figure S2.

4. Discussion

The direct targeting of PI3K and HDAC enzymes has been shown as an effective therapeutic approach in various cancers, including NB [7,26,27]. To date, five PI3K inhibitors (Copanlisib, Idelalisib, Umbralisib, Duvelisib, and Alpelisib) and four HDAC inhibitors (Romidepsin, Vorinostat, Belinostat, and Panobinostat) received FDA approval for treating different types of leukemia and lymphoma, multiple myeloma, and breast cancer [28,29]. This highlights the clinical importance of both PI3K and HDAC inhibitors for the development of targeted therapeutic approaches for hard-to-treat cancers, such as NB [28]. A combination of PI3K and HDAC inhibitors has been shown to synergistically inhibit tumor cell growth and induce apoptosis in diffuse large B-cell lymphoma and glioblastoma multiforme [30]. Drug combinational approaches require individual administration of two or more drugs for their targets and are expected to have synergistic or additive effects [31]. Another combination therapeutic approach is to develop single-agent dual inhibitor drugs that can target two different pathways simultaneously. This approach has several advantages, such as low toxicity, greater therapeutic efficacy, better pharmacokinetic and pharmacodynamic properties, and fewer solubility-related and drug–drug interaction related issues [18].

In the present study, we used a dual inhibitor CUDC-907 that targets both PI3K and HDAC [19,32]. Pre-clinical studies in colorectal cancer, B cell lymphoma, thyroid cancer, refractory lymphoma, and multiple myeloma have shown that CUDC-907 significantly inhibits both the PI3K pathway and different HDAC enzymes, to inhibit cancer proliferation and growth [33,34]. Similarly, our results in NB also showed that CUDC-907, in a dose-dependent manner, inhibits NB cell proliferation and 3D spheroid tumor growth. CUDC-907 has been reported to inhibit pancreatic adenocarcinoma [35], glioblastoma [36], prostate cancer [21], and acute myeloid leukemia [20] by inhibiting HDACs, DNA damage response, cell cycle proteins, and by inducing apoptosis. In NB, we demonstrated that CUDC-907 induces apoptosis and inhibits the cell cycle S phase. Similar results of the CUDC-907-mediated blockage of the S and G2/M phases were shown for pancreatic and thyroid cancers via the downregulation of the cell cycle regulators cyclin B1, AURKA, and PLK1 [33,35]. In glioblastoma, CUDC-907 induces G1 cell cycle arrest through *CDKN1A* promoter hyperacetylation-driven transcriptional activation and downregulation of CDK1 [36], while in lung fibroblast cells, CUDC-907 blocks G1 and S phase [37]. In breast cancer, CUDC-907 enhances TRAIL-induced apoptosis through the upregulation of cell survival proteins, including XIAP, Bcl-xL, and Bcl-2 [38]. The oncogenic activation of PI3K/AKT signaling pathway regulates multiple parallel signaling pathways, which are known to be involved in metabolism, proliferation, motility, and autophagy in different cancers [39,40]. Our results also showed that CUDC-907-mediated PI3K and HDAC inhibition leads to inhibiting NB growth.

In the past decade, most preclinical studies were based on the 2D cell culture models despite its limitations, including difficulty in simulating three-dimensional physiological conditions [41,42]. In vitro 3D tumor models filled this gap by enumerating the growth patterns of in vivo solid tumors. These 3D spheroidal tumor models became a research standard by replacing time-consuming and expensive animal studies to determine preliminary anti-cancer drug effects [42,43]. Three-dimensional cell culture models have higher invasiveness and resistance to cytotoxic drugs in comparison to two-dimensional monolayer studies [44]. Three-dimensional spheroidal tumor studies were shown as a substitute for in vivo animal models in different cancers, including NB [45]. In our study, we observed a dose-dependent inhibition of NB spheroidal tumor growth by CUDC-907. Similar results were observed in in vitro hepatocarcinoma, pancreatic cancer, and thyroid cancer spheroidal models, and were found comparable with in vivo studies in these cancers [17,33,35]. Our results demonstrated the effect of CUDC-907 in inhibiting the PI3K/AKT, HDACs, and enhancing the H3K9Ac levels in NB. Numerous reports in different cancers have shown similar results of the CUDC-907-mediated inhibition of PI3K, HDACs, and related proteins [20,21,32,35,37,38]. *MYCN* is one of the most important prognostic factors for the NB progression [46]. Our patient dataset analysis revealed a strong correlation of *PI3K* and *MYCN* levels in primary NB patients. *PI3K* expression inversely correlates with the poor overall survival of NB patients, and *PI3K* expression is found to be significantly high in *MYCN* amplified tumors. Similarly, *HDAC2* expression has been shown to correlate with MYCN levels, and multiple reports show the development of alternative strategies to target MYCN via the HDAC inhibition [6]. These analyses indicate that the dual targeting of PI3K and HDAC by CUDC-907 may inhibit MYCN levels. Our results showed that CUDC-907 significantly inhibits MYCN at both gene and protein levels. Similar results have been shown for the CUDC-907-mediated inhibition of cancer cell growth by inhibiting MYC levels in pancreatic adenocarcinoma, acute myeloid leukemia, and prostate cancer [20,21,35].

Overall, our results showed the utility of drug repurposing by demonstrating the efficacy of CUDC-907 in inhibiting NB growth. CUDC-907 is currently in phase I and phase II clinical trials for different tumor types, including pediatric cancers (NCT02307240, NCT02674750, NCT02909777, and NCT03002623). In phase I clinical studies, CUDC-907 has been considered as a moderately safe compound with minor gastrointestinal and hematologic side effects [47,48]. Therefore, our current study will provide further preclinical evidence for repurposing CUDC-907 for NB, to develop a novel and effective therapeutic approach for NB patients.

5. Conclusions

Our present study highlights the efficacy and potency of the dual inhibitor CUDC-907 in inhibiting NB growth. Our data demonstrated the inverse correlation between PI3K and HDAC levels and overall NB patient outcomes. To conclude, CUDC-907 as a single drug directly targets both PI3K and HDAC to inhibit the PI3K/AKT cell signaling pathway and to enhance H3K9Ac levels that lead to the downregulation of MYCN and inhibition of overall NB growth. As CUDC-907 is currently under advanced clinical trials for both pediatric and adult cancers, our pre-clinical results will further pave the way for a successful clinical translation of CUDC-907 for treating NB patients.

Supplementary Materials: The following supporting information can be downloaded at: https://www.mdpi.com/article/10.3390/cancers14041067/s1: Figure S1: *HDAC2* expression correlates with poor overall and event-free survival of NB patients; Figure S2: Full blots and densitometric analysis of Western blots; Table S1: RT-qPCR primers used in this study.

Author Contributions: Conception and design: S.A.; Development of methodology: S.A. and R.C.; Acquisition of data: R.C. and S.A.; Analysis and interpretation of data: R.C. and S.A.; Writing and review of the manuscript: S.A. and R.C.; Administrative, technical, or material support: S.A. All authors have read and agreed to the published version of the manuscript.

Funding: This research was funded by the St. Baldrick's Foundation Career Development Award and funding from the St. John's University to Agarwal.

Institutional Review Board Statement: Not applicable.

Informed Consent Statement: Not applicable.

Data Availability Statement: Publicly available datasets were analyzed in this study. This data can be found at R2 Genomic Analysis and Visualization Platform. Available online: https://hgserver1.amc.nl/cgi-bin/r2/main.cgi (accessed on 1 February 2022).

Acknowledgments: This work was supported by the Oliver Wells Hero Fund and David's Warriors Hero Fund by the St. Baldrick's Foundation, and funding from the St. John's University to Agarwal. R.C. is supported by the St. John's University teaching assistantship program. The graphical abstract of this manuscript was prepared using BioRender.com.

Conflicts of Interest: Authors declare no conflict of interest.

References

1. Louis, C.U.; Shohet, J.M. Neuroblastoma: Molecular pathogenesis and therapy. *Annu. Rev. Med.* **2015**, *66*, 49–63. [CrossRef] [PubMed]
2. Smith, V.; Foster, J. High-Risk Neuroblastoma Treatment Review. *Children* **2018**, *5*, 114. [CrossRef] [PubMed]
3. Zafar, A.; Wang, W.; Liu, G.; Wang, X.; Xian, W.; McKeon, F.; Foster, J.; Zhou, J.; Zhang, R. Molecular targeting therapies for neuroblastoma: Progress and challenges. *Med. Res. Rev.* **2021**, *41*, 961–1021. [CrossRef] [PubMed]
4. Puissant, A.; Frumm, S.M.; Alexe, G.; Bassil, C.F.; Qi, J.; Chanthery, Y.H.; Nekritz, E.A.; Zeid, R.; Gustafson, W.C.; Greninger, P.; et al. Targeting MYCN in neuroblastoma by BET bromodomain inhibition. *Cancer Discov.* **2013**, *3*, 308–323. [CrossRef]
5. Canete, A.; Gerrard, M.; Rubie, H.; Castel, V.; Di Cataldo, A.; Munzer, C.; Ladenstein, R.; Brichard, B.; Bermudez, J.D.; Couturier, J.; et al. Poor survival for infants with MYCN-amplified metastatic neuroblastoma despite intensified treatment: The International Society of Paediatric Oncology European Neuroblastoma Experience. *J. Clin. Oncol.* **2009**, *27*, 1014–1019. [CrossRef] [PubMed]
6. Phimmachanh, M.; Han, J.Z.R.; O'Donnell, Y.E.I.; Latham, S.L.; Croucher, D.R. Histone Deacetylases and Histone Deacetylase Inhibitors in Neuroblastoma. *Front. Cell Dev. Biol.* **2020**, *8*, 578770. [CrossRef]
7. Yang, J.; Nie, J.; Ma, X.; Wei, Y.; Peng, Y.; Wei, X. Targeting PI3K in cancer: Mechanisms and advances in clinical trials. *Mol. Cancer* **2019**, *18*, 26. [CrossRef]
8. Papa, A.; Pandolfi, P.P. The PTEN(-)PI3K Axis in Cancer. *Biomolecules* **2019**, *9*, 153. [CrossRef]
9. Nalairndran, G.; Hassan Abdul Razack, A.; Mai, C.W.; Fei-Lei Chung, F.; Chan, K.K.; Hii, L.W.; Lim, W.M.; Chung, I.; Leong, C.O. Phosphoinositide-dependent Kinase-1 (PDPK1) regulates serum/glucocorticoid-regulated Kinase 3 (SGK3) for prostate cancer cell survival. *J. Cell. Mol. Med.* **2020**, *24*, 12188–12198. [CrossRef]
10. Balasuriya, N.; Davey, N.E.; Johnson, J.L.; Liu, H.; Biggar, K.K.; Cantley, L.C.; Li, S.S.; O'Donoghue, P. Phosphorylation-dependent substrate selectivity of protein kinase B (AKT1). *J. Biol. Chem.* **2020**, *295*, 8120–8134. [CrossRef]
11. Shorning, B.Y.; Dass, M.S.; Smalley, M.J.; Pearson, H.B. The PI3K-AKT-mTOR Pathway and Prostate Cancer: At the Crossroads of AR, MAPK, and WNT Signaling. *Int. J. Mol. Sci.* **2020**, *21*, 4507. [CrossRef] [PubMed]
12. Ghoneum, A.; Said, N. PI3K-AKT-mTOR and NFkappaB Pathways in Ovarian Cancer: Implications for Targeted Therapeutics. *Cancers* **2019**, *11*, 949. [CrossRef] [PubMed]
13. Wang, P.; Wang, Z.; Liu, J. Role of HDACs in normal and malignant hematopoiesis. *Mol. Cancer* **2020**, *19*, 5. [CrossRef] [PubMed]
14. Li, Y.; Seto, E. HDACs and HDAC Inhibitors in Cancer Development and Therapy. *Cold Spring Harb. Perspect. Med.* **2016**, *6*, a026831. [CrossRef]
15. Lemoine, M.; Younes, A. Histone deacetylase inhibitors in the treatment of lymphoma. *Discov. Med.* **2010**, *10*, 462–470.
16. Rahmani, M.; Yu, C.; Reese, E.; Ahmed, W.; Hirsch, K.; Dent, P.; Grant, S. Inhibition of PI-3 kinase sensitizes human leukemic cells to histone deacetylase inhibitor-mediated apoptosis through p44/42 MAP kinase inactivation and abrogation of p21(CIP1/WAF1) induction rather than AKT inhibition. *Oncogene* **2003**, *22*, 6231–6242. [CrossRef]
17. Liao, W.; Yang, W.; Xu, J.; Yan, Z.; Pan, M.; Xu, X.; Zhou, S.; Zhu, Y.; Lan, J.; Zeng, M.; et al. Therapeutic Potential of CUDC-907 (Fimepinostat) for Hepatocarcinoma Treatment Revealed by Tumor Spheroids-Based Drug Screening. *Front. Pharmacol.* **2021**, *12*, 658197. [CrossRef]
18. Ranganna, K.; Selvam, C.; Shivachar, A.; Yousefipour, Z. Histone Deacetylase Inhibitors as Multitarget-Directed Epi-Drugs in Blocking PI3K Oncogenic Signaling: A Polypharmacology Approach. *Int. J. Mol. Sci.* **2020**, *21*, 8198. [CrossRef]
19. Qiao, X.; Ma, J.; Knight, T.; Su, Y.; Edwards, H.; Polin, L.; Li, J.; Kushner, J.; Dzinic, S.H.; White, K.; et al. The combination of CUDC-907 and gilteritinib shows promising in vitro and in vivo antileukemic activity against FLT3-ITD AML. *Blood Cancer J.* **2021**, *11*, 111. [CrossRef]
20. Li, X.; Su, Y.; Madlambayan, G.; Edwards, H.; Polin, L.; Kushner, J.; Dzinic, S.H.; White, K.; Ma, J.; Knight, T.; et al. Antileukemic activity and mechanism of action of the novel PI3K and histone deacetylase dual inhibitor CUDC-907 in acute myeloid leukemia. *Haematologica* **2019**, *104*, 2225–2240. [CrossRef]
21. Hu, C.; Xia, H.; Bai, S.; Zhao, J.; Edwards, H.; Li, X.; Yang, Y.; Lyu, J.; Wang, G.; Zhan, Y.; et al. CUDC-907, a novel dual PI3K and HDAC inhibitor, in prostate cancer: Antitumour activity and molecular mechanism of action. *J. Cell Mol. Med.* **2020**, *24*, 7239–7253. [CrossRef] [PubMed]
22. Agarwal, S.; Lakoma, A.; Chen, Z.; Hicks, J.; Metelitsa, L.S.; Kim, E.S.; Shohet, J.M. G-CSF Promotes Neuroblastoma Tumorigenicity and Metastasis via STAT3-Dependent Cancer Stem Cell Activation. *Cancer Res.* **2015**, *75*, 2566–2579. [CrossRef] [PubMed]
23. Chilamakuri, R.; Rouse, D.C.; Yu, Y.; Kabir, A.S.; Muth, A.; Yang, J.; Lipton, J.M.; Agarwal, S. BX-795 inhibits neuroblastoma growth and enhances sensitivity towards chemotherapy. *Transl. Oncol.* **2022**, *15*, 101272. [CrossRef] [PubMed]
24. Guan, S.; Lu, J.; Zhao, Y.; Yu, Y.; Li, H.; Chen, Z.; Shi, Z.; Liang, H.; Wang, M.; Guo, K.; et al. MELK is a novel therapeutic target in high-risk neuroblastoma. *Oncotarget* **2018**, *9*, 2591–2602. [CrossRef]
25. Agarwal, S.; Ghosh, R.; Chen, Z.; Lakoma, A.; Gunaratne, P.H.; Kim, E.S.; Shohet, J.M. Transmembrane adaptor protein PAG1 is a novel tumor suppressor in neuroblastoma. *Oncotarget* **2016**, *7*, 24018–24026. [CrossRef]
26. Stopsack, K.H.; Huang, Y.; Tyekucheva, S.; Gerke, T.A.; Bango, C.; Elfandy, H.; Bowden, M.; Penney, K.L.; Roberts, T.M.; Parmigiani, G.; et al. Multiplex Immunofluorescence in Formalin-Fixed Paraffin-Embedded Tumor Tissue to Identify Single-Cell-Level PI3K Pathway Activation. *Clin. Cancer Res.* **2020**, *26*, 5903–5913. [CrossRef]

27. Mohlin, S.; Hansson, K.; Radke, K.; Martinez, S.; Blanco-Apiricio, C.; Garcia-Ruiz, C.; Welinder, C.; Esfandyari, J.; O'Neill, M.; Pastor, J.; et al. Anti-tumor effects of PIM/PI3K/mTOR triple kinase inhibitor IBL-302 in neuroblastoma. *EMBO Mol. Med.* **2019**, *11*, e10058. [CrossRef]
28. Brown, J.R. Phosphatidylinositol 3 Kinase delta Inhibitors: Present and Future. *Cancer J.* **2019**, *25*, 394–400. [CrossRef]
29. Raedler, L.A. Farydak (Panobinostat): First HDAC Inhibitor Approved for Patients with Relapsed Multiple Myeloma. *Am. Health Drug Benefits* **2016**, *9*, 84–87.
30. Meng, W.; Wang, B.; Mao, W.; Wang, J.; Zhao, Y.; Li, Q.; Zhang, C.; Ma, J. Enhanced efficacy of histone deacetylase inhibitor panobinostat combined with dual PI3K/mTOR inhibitor BEZ235 against glioblastoma. *Nagoya J. Med. Sci.* **2019**, *81*, 93–102. [CrossRef]
31. De Lera, A.R.; Ganesan, A. Epigenetic polypharmacology: From combination therapy to multitargeted drugs. *Clin. Epigenetics* **2016**, *8*, 105. [CrossRef] [PubMed]
32. Ma, L.; Bian, X.; Lin, W. Correction to: The dual HDAC-PI3K inhibitor CUDC-907 displays single-agent activity and synergizes with PARP inhibitor olaparib in small cell lung cancer. *J. Exp. Clin. Cancer Res.* **2021**, *40*, 7. [CrossRef] [PubMed]
33. Kotian, S.; Zhang, L.; Boufraqech, M.; Gaskins, K.; Gara, S.K.; Quezado, M.; Nilubol, N.; Kebebew, E. Dual Inhibition of HDAC and Tyrosine Kinase Signaling Pathways with CUDC-907 Inhibits Thyroid Cancer Growth and Metastases. *Clin. Cancer Res.* **2017**, *23*, 5044–5054. [CrossRef]
34. Zhong, L.; Li, Y.; Xiong, L.; Wang, W.; Wu, M.; Yuan, T.; Yang, W.; Tian, C.; Miao, Z.; Wang, T.; et al. Small molecules in targeted cancer therapy: Advances, challenges, and future perspectives. *Signal. Transduct. Target. Ther.* **2021**, *6*, 201. [CrossRef] [PubMed]
35. Fu, X.H.; Zhang, X.; Yang, H.; Xu, X.W.; Hu, Z.L.; Yan, J.; Zheng, X.L.; Wei, R.R.; Zhang, Z.Q.; Tang, S.R.; et al. CUDC-907 displays potent antitumor activity against human pancreatic adenocarcinoma in vitro and in vivo through inhibition of HDAC6 to downregulate c-Myc expression. *Acta Pharmacol. Sin.* **2019**, *40*, 677–688. [CrossRef] [PubMed]
36. Pal, S.; Kozono, D.; Yang, X.; Fendler, W.; Fitts, W.; Ni, J.; Alberta, J.A.; Zhao, J.; Liu, K.X.; Bian, J.; et al. Dual HDAC and PI3K Inhibition Abrogates NFkappaB- and FOXM1-Mediated DNA Damage Response to Radiosensitize Pediatric High-Grade Gliomas. *Cancer Res.* **2018**, *78*, 4007–4021. [CrossRef] [PubMed]
37. Zhang, W.; Zhang, Y.; Tu, T.; Schmull, S.; Han, Y.; Wang, W.; Li, H. Dual inhibition of HDAC and tyrosine kinase signaling pathways with CUDC-907 attenuates TGFbeta1 induced lung and tumor fibrosis. *Cell Death Dis.* **2020**, *11*, 765. [CrossRef]
38. Li, Z.J.; Hou, Y.J.; Hao, G.P.; Pan, X.X.; Fei, H.R.; Wang, F.Z. CUDC-907 enhances TRAIL-induced apoptosis through upregulation of DR5 in breast cancer cells. *J. Cell Commun. Signal.* **2020**, *14*, 377–387. [CrossRef]
39. Westhoff, M.A.; Faham, N.; Marx, D.; Nonnenmacher, L.; Jennewein, C.; Enzenmuller, S.; Gonzalez, P.; Fulda, S.; Debatin, K.M. Sequential dosing in chemosensitization: Targeting the PI3K/Akt/mTOR pathway in neuroblastoma. *PLoS ONE* **2013**, *8*, e83128. [CrossRef]
40. Lin, A.; Piao, H.L.; Zhuang, L.; Sarbassov, D.D.; Ma, L.; Gan, B. FoxO transcription factors promote AKT Ser473 phosphorylation and renal tumor growth in response to pharmacologic inhibition of the PI3K-AKT pathway. *Cancer Res.* **2014**, *74*, 1682–1693. [CrossRef]
41. Jensen, C.; Teng, Y. Is It Time to Start Transitioning From 2D to 3D Cell Culture? *Front. Mol. Biosci.* **2020**, *7*, 33. [CrossRef] [PubMed]
42. Daunys, S.; Janoniene, A.; Januskeviciene, I.; Paskeviciute, M.; Petrikaite, V. 3D Tumor Spheroid Models for In Vitro Therapeutic Screening of Nanoparticles. *Adv. Exp. Med. Biol.* **2021**, *1295*, 243–270. [CrossRef] [PubMed]
43. Gilazieva, Z.; Ponomarev, A.; Rutland, C.; Rizvanov, A.; Solovyeva, V. Promising Applications of Tumor Spheroids and Organoids for Personalized Medicine. *Cancers* **2020**, *12*, 2727. [CrossRef] [PubMed]
44. Edmondson, R.; Broglie, J.J.; Adcock, A.F.; Yang, L. Three dimensional cell culture systems and their applications in drug discovery and cell-based biosensors. *Assay Drug Dev. Technol.* **2014**, *12*, 207–218. [CrossRef]
45. Kumar, A.; Fan, D.; Dipette, D.J.; Singh, U.S. Sparstolonin B, a novel plant derived compound, arrests cell cycle and induces apoptosis in N-myc amplified and N-myc nonamplified neuroblastoma cells. *PLoS ONE* **2014**, *9*, e96343. [CrossRef]
46. Schneiderman, J.; London, W.B.; Brodeur, G.M.; Castleberry, R.P.; Look, A.T.; Cohn, S.L. Clinical significance of MYCN amplification and ploidy in favorable-stage neuroblastoma: A report from the Children's Oncology Group. *J. Clin. Oncol.* **2008**, *26*, 913–918. [CrossRef]
47. Oki, Y.; Kelly, K.R.; Flinn, I.; Patel, M.R.; Gharavi, R.; Ma, A.; Parker, J.; Hafeez, A.; Tuck, D.; Younes, A. CUDC-907 in relapsed/refractory diffuse large B-cell lymphoma, including patients with MYC-alterations: Results from an expanded phase I trial. *Haematologica* **2017**, *102*, 1923–1930. [CrossRef]
48. Chen, I.C.; Sethy, B.; Liou, J.P. Recent Update of HDAC Inhibitors in Lymphoma. *Front. Cell Dev. Biol.* **2020**, *8*, 576391. [CrossRef]

cancers

MDPI

Article

ERK Inhibitor Ulixertinib Inhibits High-Risk Neuroblastoma Growth In Vitro and In Vivo

Yang Yu [1,†], Yanling Zhao [2,†], Jongmin Choi [3], Zhongcheng Shi [4], Linjie Guo [2], John Elizarraras [2], Andy Gu [2], Feng Cheng [1], Yanxin Pei [1,5], Dai Lu [6], Muller Fabbri [1,5], Saurabh Agarwal [7], Chunchao Zhang [2], Sung Yun Jung [4], Jennifer H. Foster [2,*] and Jianhua Yang [1,5,*]

1 Center for Cancer and Immunology Research, Children's National Research Institute, Children's National Hospital, Washington, DC 20010, USA
2 Texas Children's Hospital, Department of Pediatrics, Dan L. Duncan Cancer Center, Baylor College of Medicine, Houston, TX 77030, USA
3 Advanced Technology Cores/Office of Research, Baylor College of Medicine, Houston, TX 77030, USA
4 Department of Molecular and Cellular Biology, Baylor College of Medicine, Houston, TX 77003, USA
5 Department of Pediatrics, The George Washington University School of Medicine and Health Sciences, Washington, DC 20010, USA
6 Rangel College of Pharmacy, Texas A&M University, Kingsville, TX 78363, USA
7 Department of Pharmaceutical Sciences, College of Pharmacy and Health Sciences, St. John's University, Queens, NY 11439, USA
* Correspondence: jhfoster@bcm.edu (J.H.F.); jyang1@childrensnational.org (J.Y.); Tel.: +1-832-822-4556 (J.H.F.); +1-202-476-5772 (J.Y.)
† These authors contributed equally to this work.

Simple Summary: Neuroblastoma (NB) is the most common extracranial solid tumor in children, and the majority of patients with high-risk disease are unable to be cured. There is an urgent need to design novel therapeutics for NB patients. In NB, the RAS-MAPK pathway plays a crucial role in essential processes such as cell proliferation, survival, and chemoresistance. In the present study, we determined the therapeutic potential of the ERK inhibitor ulixertinib in NB using a panel of NB cell lines, patient-derived xenograft (PDX) cell lines, and NB xenograft mouse models. Ulixertinib significantly and potently inhibited NB cell proliferation and tumor growth, as well as prolonged survival in the treated mice. Additionally, ulixertinib synergistically sensitized NB cells to the conventional chemotherapeutic drug doxorubicin. This study provides proof-of-concept pre-clinical evidence for exploring ulixertinib as a novel therapeutic approach for NB.

Abstract: Neuroblastoma (NB) is a pediatric tumor of the peripheral nervous system. Approximately 80% of relapsed NB show RAS-MAPK pathway mutations that activate ERK, resulting in the promotion of cell proliferation and drug resistance. Ulixertinib, a first-in-class ERK-specific inhibitor, has shown promising antitumor activity in phase 1 clinical trials for advanced solid tumors. Here, we show that ulixertinib significantly and dose-dependently inhibits cell proliferation and colony formation in different NB cell lines, including PDX cells. Transcriptomic analysis revealed that ulixertinib extensively inhibits different oncogenic and neuronal developmental pathways, including EGFR, VEGF, WNT, MAPK, NGF, and NTRK1. The proteomic analysis further revealed that ulixertinib inhibits the cell cycle and promotes apoptosis in NB cells. Additionally, ulixertinib treatment significantly sensitized NB cells to the conventional chemotherapeutic agent doxorubicin. Furthermore, ulixertinib potently inhibited NB tumor growth and prolonged the overall survival of the treated mice in two different NB mice models. Our preclinical study demonstrates that ulixertinib, either as a single agent or in combination with current therapies, is a novel and practical therapeutic approach for NB.

Keywords: neuroblastoma; ulixertinib; ERK inhibition; combination therapy; c-Myc/N-Myc

Citation: Yu, Y.; Zhao, Y.; Choi, J.; Shi, Z.; Guo, L.; Elizarraras, J.; Gu, A.; Cheng, F.; Pei, Y.; Lu, D.; et al. ERK Inhibitor Ulixertinib Inhibits High-Risk Neuroblastoma Growth In Vitro and In Vivo. *Cancers* **2022**, *14*, 5534. https://doi.org/10.3390/cancers14225534

Academic Editor: David Wong

Received: 28 September 2022
Accepted: 5 November 2022
Published: 10 November 2022

1. Introduction

NB is the most common extracranial tumor in children, and accounts for 15% of childhood malignancy-related deaths [1]. Although low- and intermediate-risk NB is highly likely to be cured, high-risk NB usually recurs with incurable disease despite intensive treatment with multiple modalities [2]. High-risk NB can develop sustained drug resistance during chemotherapy, which contributes to relapsed disease [3]. Therefore, finding more effective targeted therapies that can reduce drug resistance has been the focus of NB cancer research for decades.

The MAPK pathway is the central signal cascade that promotes cell proliferation. This pathway can be activated by growth factors binding to the corresponding receptors, leading to the activation of the RAS, RAF, MEK, and ERK pathways [4]. ERK is a major effector kinase of the MAPK pathway that activates various substrates through phosphorylation and triggers multiple cellular responses, including cell proliferation, by upregulating cell cycle genes [5]. Overactivation of the MAPK pathway due to the mutations of many pathway-related components is observed in numerous cancers, including relapsed NB [6]. Therefore, targeting the ERK/MAPK pathway has been an attractive therapeutic strategy in treating NB.

In addition to promoting cell proliferation, aberrant activation of the Raf/MEK/ERK pathway has been demonstrated to induce drug resistance in cancer cells [7]. The hyperactivated ERK pathway can be suppressed by both BRAF and MEK inhibitors, which has shown impressive results thus far [8,9]. However, tumors in most patients develop resistance after about one year of treatment with these inhibitors, establishing ERK, the terminal master kinase, as an ideal target for small-molecule inhibitors [10–12]. It has been demonstrated that ERK activation and subsequent RSK activation suppress GSK-3β activity [13]. Active GSK-3β can promote MYC phosphorylation and degradation in an SCF FBW7-dependent manner, which may also apply to N-Myc destabilization [14]. ERK activation can also upregulate c-Myc and N-Myc in NB [15,16]. Inhibition of ERK activation may thus reduce c-Myc and N-Myc levels and destabilize both proteins. It has been known that increased levels of c-Myc or N-Myc result in the development of drug resistance in multiple cancers, such as NB [17], leukemia [18], endometrial cancer [19], hepatocellular carcinoma (HCC) [20], human small cell lung carcinoma [21], and pancreatic cancer [22]. Ulixertinib, a novel selective ERK inhibitor, has shown promising preclinical activities in treating several cancers in vitro and in clinical trials [23]. However, whether ulixertinib can effectively treat NB or reduce the chemoresistance of NB cells remains unclear.

The present study demonstrates that ulixertinib inhibits NB cell proliferation in vitro and NB tumor growth in vivo. We further delineated ulixertinib-induced transcriptomic and proteomic alterations in multiple signaling pathways in NB cells by using RNA-Seq and mass spectrometry analyses. Additionally, ulixertinib sensitizes NB cells to doxorubicin-induced apoptosis, suggesting ulixertinib as a novel and effective treatment for relapsed and chemo-resistant NB.

2. Materials and Methods

2.1. Reagents

Ulixertinib (BVD-523) (CT-VRT752) was purchased from ChemieTek (Indianapolis, IN, USA). Fetal bovine serum (FBS) (35-011-CV), RPMI-1640 medium (10-040-CV), penicillin/streptomycin (30-002-CI), and 0.25% trypsin (25-050-CI) were purchased from Corning Incorporated (Corning, NY, USA). Bovine serum albumin (BSA) (A7906), dimethyl sulfoxide (DMSO) (D8418), and 3-(4, 5-dimethylthiazol-yl)-2,5-diphenyltetrazolium bromide (MTT) (M5655), were purchased from Sigma-Aldrich (Saint Louis, MO, USA). Anti-c-Myc (9E10) (sc-40), anti-α-tubulin (10D8) (sc-53646), and anti-N-Myc (NCM II 100) (sc-56729) were purchased from Santa Cruz Biotechnology (Dallas, TX, USA). ERK1/2 (#9102L), phospho-RSK (#9341), RSK (#9355), Cleaved Caspase 3 (#9664), anti-rabbit (#7074S), and anti-mouse (#7076S) secondary antibodies were purchased from Cell Signaling Technology (Danvers, MA, USA).

2.2. NB Cell Lines and Patient-Derived Xenograft Cells

Nine human NB cell lines were used in this study, including five MYCN-amplified NB cell lines, LAN-1, NGP, SK-N-BE(2), CHLA136, IMR-32, and four MYCN-non-amplified NB cell lines, CHLA255, SH-SY5Y, SK-N-AS, and LAN-6. Human bone marrow stromal cell line HS-5 was purchased from ATCC. All cell lines were maintained in RPMI-1640 medium, supplemented with 20% (*v/v*) heat-inactivated fetal bovine serum (FBS), 100 units/mL penicillin, and 100 µg/mL streptomycin at 37 °C in 5% CO_2. All cell lines were authenticated via short tandem repeat (STR) analysis. Mycoplasma testing was performed by the LookOut® Mycoplasma PCR Detection Kit (MP0035, Sigma-Aldrich). Patient-derived xenograft (PDX) COG-N-519x, COG-N-564x, COG-N-618x, and COG-N-700x cells were obtained from the Children's Oncology Group (COG) Cell Culture and Xenograft Repository (www.COGcell.org) (Accessed on 4 March 2022) and maintained in Iscove's modified Dulbecco's medium supplemented with 20% fetal bovine serum, 4 mM L-glutamine, 1X ITS (5 µg/mL insulin, 5 µg/mL transferrin, 5 ng/mL selenous acid) (Biotechne, AR013, Minneapolis, MN, USA).

2.3. Cell Proliferation Assay

NB cell lines, NB PDX cells, and HS-5 cells were seeded in 96-well plates (3000 cells per well) in culture media. After incubating overnight, ulixertinib was added at indicated concentrations for five days. Cell proliferation was measured using Cell Counting kit-8 (CK04-20) purchased from Dojindo Molecular Technologies Inc. Each experiment was performed in three replicates.

2.4. Anchorage-Independent Colony Formation Assay

The soft agar assay was performed as described previously [24]. Briefly, 0.5% agar in cell culture media suspension was plated in 6-well plates and allowed to solidify, followed by adding an upper layer of agar (0.3%) containing 3×10^3 NB cells/well. Ulixertinib was added at the indicated concentrations to the soft agar surface the next day. After four weeks, the cells were stained with 500 µL of 0.5% (*w/v*) MTT solution. Images of the colonies were captured and counted using VersaDoc Imaging System (Bio-Rad). Each assay was performed in triplicate.

2.5. Immunoblotting

After administration of ulixertinib at the indicated concentration and time point, cells were harvested and lysed using RIPA buffer containing 50 mM Tris-HCl (pH 7.4), 1% NP-40 (IGEPAL CA-630) (#I8896, Sigma-Aldrich), 0.25% sodium deoxycholate, 150 mM NaCl, 1 mM EDTA, 0.1 mM sodium orthovanadate, 0.5 mM PMSF, 1 mM DTT, 10 µg/mL leupeptin, 10 µg/mL aprotinin, 1 mM benzamidine, and phosphatase inhibitor cocktail 2 and 3 (#P5726 and #P0044, Sigma). Total cell lysate (50 µg) was used for SDS-PAGE and transferred onto PVDF membranes (Millipore). After blocking with 5% non-fat milk, membranes were incubated with primary antibodies at 4°C overnight. HRP conjugated secondary antibodies and ECL Western blotting Kit (GE Health) were used for signal detection. All the whole western blot figures can be found in the supplementary materials.

2.6. Synergy Studies

About 3×10^3 cells were seeded in each well of 96-well plates for CHLA255, SK–N-BE(2), SK-N-AS, and NGP cells, then treated with ulixertinib, doxorubicin, or a combination of both drugs. The CI values and dose-reduction indices (DRIs) were calculated using the Chou–Talalay method for drug interactions, with CompuSyn software for the different fractions affected [25]. CI < 1, =1, and >1 indicate synergism, additive effect, and antagonism, respectively. DRI > 1 and <1 indicate a favorable and an unfavorable dose reduction, respectively.

2.7. NB Xenograft Mouse Model

The NB xenograft mouse model was established as described by other studies [26,27]. NSG mice were purchased from The Jackson Laboratory and maintained at the Baylor College of Medicine animal care facility. Eight- to twelve-week-old sex-matched mice were intravenously injected with 1×10^6 NB cells with firefly luciferase expression (CHLA136-Fluc and CHLA255-Fluc). Two weeks post-injection, mice were divided into two groups and intraperitoneally injected with ulixertinib (50 mg/kg) or vehicle for three weeks. Mice were assessed by a Xenogen IVIS 100 instrument every week. Xenogen images of mice were taken 15 min after injection of 1.5 mg D-luciferin (#122799, Perkin Elmer) intraperitoneally. For all experiments, the xenogen exposure time was set at 3 min. Mouse survival time was defined as the length of time from tumor cell injection until the end of the study or to euthanasia due to severe symptoms caused by tumor progression (for example, over 15% weight loss, weakness, seizures, inability to eat or drink, inability to stand, immobility, and/or paralysis). Percent survival was determined using a Kaplan–Meier analysis and a log-rank (Mantel-Cox) test. All animal procedures were approved by the Institutional Animal Care and Use Committee of Baylor College of Medicine.

2.8. RNA-Seq Analysis

NGP cells were treated with ulixertinib at 7.5 µmol/L for 24 h, the same volume of DMSO was used as the control. RNA was harvested using TRIzol™ Reagent (Thermo Fisher, Waltham, MA, USA). About 100 ng of total RNA was used for the construction of sequencing libraries. RNA libraries for RNA-seq were prepared using SMARTER mRNA-Seq Library Prep Kit following the manufacturer's protocols. RNA-Seq experiments were conducted by using Illumina paired-end sequencing technology. Raw reads were adaptor removed, and sequencing quality was assessed with the bioinformatics software FastQC (version 0.11.2) (Babraham Bioinformatics, Babraham Institute, Cambridge, UK). Sequence quality scores, sequence duplication, and adaptor content were evaluated to decide if further filtering should be applied before the genome mapping. Reads were mapped against the human reference genome (GRCh38.p13 assembly) using the aligner, HISAT2 (v2.1.0) (Lyda Hill Department of Bioinformatics, University of Texas Southwestern Medical Center, Dallas, TX, USA). The mapped reads were subsequently assembled into transcripts or genes using the assembler, StringTie (v1.3.5) (The Center for Computational Biology at Johns Hopkins University, Baltimore, MD, USA).

2.9. Mass Spectrometry Data Analysis

In order to determine proteomic changes in NB cells (NGP) with the ulixertinib treatment, three biological replicates of each condition were harvested after DMSO vehicle or ulixertinib treatment for 24 h. Harvested cells were washed with cold PBS and lysed with 10 sample volumes of 50 mM ammonium bicarbonate with 1 mM CaCl2. Cell suspensions were lysed and double-trypsinized as described previously [26]. Double-digested peptides were fractionated into 15 fractions and pooled into 5 fractions in the Stage-tip C18 column, as described previously [26]. Pooled fractions of peptide were enriched on a 2 cm trap column (100 µm i.d.) and separated by 5 cm analytical column (150 µm i.d.) containing Reprosil-Pur Basic C18 (1.9 µm, Dr. Maisch GmbH, Germany). A nanoLC-1200 (Thermo Scientific, Lenexa, KS, USA) delivered a 75 min discontinuous gradient of 4 to 24 % of acetonitrile/0.1% formic acid at a flow rate of 800 nL/min. A Orbitrap Fusion mass spectrometer (Thermo Scientific) was operated in data dependent acquisition mode in the following parameters, precursor was scanned by at 300–1400 m/z range, 120 k resolution at 400 m/z, AGC target of 5×105 (50 ms maximum injection time). Cycle time was top 3 s selected MS1 signal using Quadrupole filter in 2 m/z isolation window, 15 s exclusion time. The HCD fragmented MS2 ions were detected by ion trap with rapid scan, 5×10^3 AGC target, and 35 ms of maximum injection time. Proteome Discoverer 2.1 software (Thermo Fisher) with the Mascot 2.4 search engine (Matrix Science, Chicago, IL, USA) was used for spectra analysis using a target-decoy Human RefSeq proteome database with a 1%

false discovery rate (FDR) of percolator validation based on q-value. The acetylation of the N-terminus and oxidation of methionine were allowed for dynamic modification. A 20 ppm of precursor mass tolerance, 0.5 Da of fragment mass tolerance, and a maximum of two missed cleavages was allowed. A label-free quantification strategy (iBAQ) was adopted for identified proteins quantification calculated by the gpGrouper algorithm [27]. Protein abundances are median normalized across the experiments.

2.10. Statistical and Bioinformatics Analysis

Data were represented as the means ± standard error. All experiments were repeated at least three times. A two-tailed Student's *t*-test was used to determine the statistical significance of in vitro assay between drug treatment and the control groups. For survival analysis, the statistical significance was determined using a log-rank (Mantel-Cox) test. Differential gene expression (DGE) analysis of RNA-Seq was conducted with the R statistical package DESeq2 (v1.28.1) (Genome Biology Unit, European Molecular Biology Laboratory, Heidelberg, Germany). Genes were considered significantly altered if they had more or less than two-fold changes ($FC \geq 2$ or $FC \leq 0.5$) with an adjusted p-value less than 0.05. Only coding genes were retained on the list. For DGE analysis of proteins from MS, an unpaired *t*-test was adopted and a p-value < 0.05 was considered statistically significant. We performed gene set analysis on DGEs by using an over-representation test on the ConsensusPathDB [28] web-portal. Genes or proteins detected in our experiments were selected on the background list. KEGG and Reactome pathway databases were considered. Differentially expressed proteins were submitted to the STRING [29] database (https://string-db.org/) to evaluate global protein–protein interaction networks. The STRING version is 11.5 (Accessed on 12 August 2021). All parameters are default settings but confident scores. We only keep interaction scores higher than 0.70 (high confidence). The interaction score is a combined scored calculated by combining the probabilities from the different source and scaled between 0 and 1.

3. Results

3.1. Ulixertinib Significantly Inhibits NB Cell Proliferation

To test whether ulixertinib can inhibit cell proliferation and promote apoptosis in human NB cells, we first treated a panel of NB cell lines and PDX cell lines, including five *MYCN*-amplified (LAN-1, IMR-32, NGP, SK-N-BE(2), and CHLA136), four *MYCN* non-amplified (CHLA255, SH-SY5Y, SK-N-AS, and LAN-6), four NB PDX *MYCN*-amplified (COG-N-519x), and *MYCN* non-amplified (COG-N-564x, COG-N-618x, and COG-N-700x) cell lines. Our data demonstrate that ulixertinib significantly inhibited cell proliferation of all NB and PDX cell lines tested in a dose-dependent manner regardless of the *MYCN* status (Figure 1A–C). Treatment of a noncancerous bone marrow stromal fibroblast cell line HS-5 with ulixertinib showed inhibition of cell proliferation at a very high drug dose, in contrast to NB cell lines (Figure 1A). Different cell lines displayed a wide range of half-maximal inhibitory concentrations (IC_{50}), in which CHLA255 was most sensitive, and LAN-1 and COG-N-618x were found to be least sensitive to ulixertinib, respectively (Figure 1D–F). Furthermore, ulixertinib significantly and dose-dependently inhibited different NB cells' anchorage-independent colony formation (Figure 1G). Ulixertinib at the respective IC_{50} concentrations dramatically reduced colony sizes and numbers compared to controls. Across all cell lines, several colonies were observed in higher concentrations (5-fold IC_{50}) (Figure 1H–K). Our data show that ulixertinib can suppress NB cell proliferation and anchorage-independent colony formation capacity in vitro.

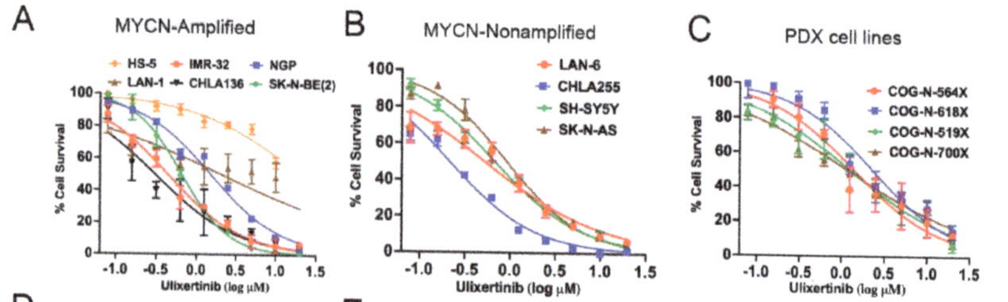

D

	Cell Line	IC_{50} (μM)
MYCN-Amplified	HS-5	20.70
	IMR-32	0.44
	NGP	1.52
	SK-N-BE(2)	0.65
	LAN-1	2.08
	CHLA136	0.30

E

	Cell Line	IC_{50} (μM)
MYCN-Non Amplified	LAN-6	0.52
	CHLA255	0.21
	SH-SY5Y	0.68
	SK-N-AS	0.92

F

	Cell Line	IC_{50} (μM)
PDX	COG-N-564X	1.39
	COG-N-618X	2.29
	COG-N-519X	1.25
	COG-N-700X	1.15

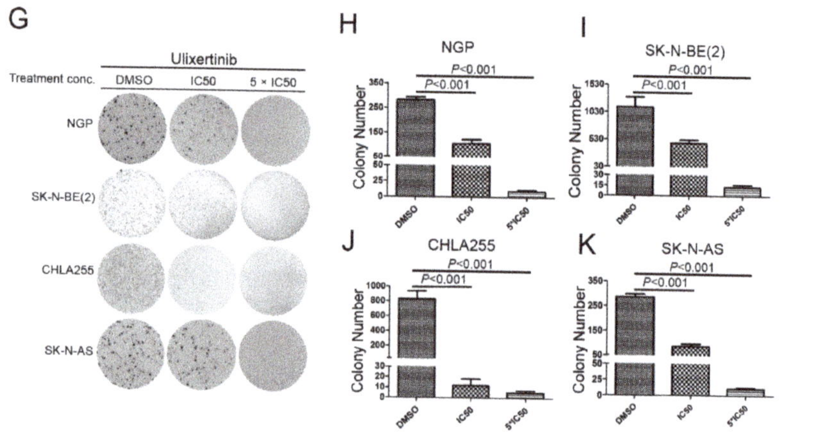

Figure 1. Ulixertinib inhibits NB cell proliferation and anchorage-independent growth in vitro. (**A**) Five MYCN amplified NB cell lines (IMR-32, NGP, SK-N-BE(2), LAN-1, and CHLA136) and a noncancerous bone marrow stromal fibroblast cell line, HS-5, were treated with the indicated concentrations of ulixertinib for 5 days. Cell proliferation was measured by CCK-8 assays. (**B**,**C**) Four MYNC- non-amplified NB cell lines (LAN-6, CHLA255, SH-SY5Y, and SK-N-AS) (**B**) and four PDX cell lines (COG-N-519x, COG-N-564x, COG-N-618x, and COG-N-700x) (**C**) were also treated with unlixertinib as in (A). Cell proliferation was measured by CCK-8 assays. (**D**–**F**) IC50 values were calculated based on the data collected in the CCK-8 assay from (**A**–**C**). (**G**) The effects of ulixertinib on anchorage-independent cell proliferation. Four NB cell lines (NGP, SK-N-BE(2), CHLA255, and SK-N-AS) were grown in soft agar with DMSO, 1 × IC50, and 5 × IC50 of ulixertinib for 4 weeks. Cells were stained with 0.5% (*w/v*) MTT solution to visualize colonies and photographed. (**H**–**K**) Colonies from (**G**) were counted in NGP (**H**), SK-N-BE(2) (**I**), CHLA255 (**J**), and SK-N-AS (**K**) cells. Colony numbers were represented as mean ± S.D. A two-tailed Student's *t*-test was used to determine the statistical significance.

3.2. Ulixertinib Inhibits RSK Phosphorylation and Downregulates c-Myc/N-Myc Protein Levels in NB Cells

Ulixertinib has been shown to inhibit RSK phosphorylation induced by ERK activation in other cancer cell types [23]. We hypothesized that ulixertinib has the same effects on ERK signaling in NB cells. Ulixertinib treatment in all tested NB cells significantly inhibited RSK1 phosphorylation (Figure 2). It suggests that ulixertinib also blocks ERK activation in NB cells. *MYCN* and *MYC* play a crucial role in NB tumor development. The previous study shows that both are the downstream targets of ERK [15,16]. We then explored whether ulixertinib-mediated ERK inhibition affects c-Myc or N-Myc protein levels. Our data show that ulixertinib treatment, in a time-dependent manner, significantly down-regulated N-Myc protein levels in *MYCN* amplified cell lines (NGP, SK-N-BE(2), CHLA136, and LAN-1) (Figure 2A–D), and c-Myc protein levels in *MYCN* non-amplified cell lines (LAN-6 and CHLA255) (Figure 2E,F). Since MYCN is an important prognostic factor in NB and is known to drive NB tumorigenesis and progression, our data on ulixertinib-mediated N-Myc inhibition highlight the potential of ulixertinib as a novel therapeutic approach for high-risk NB.

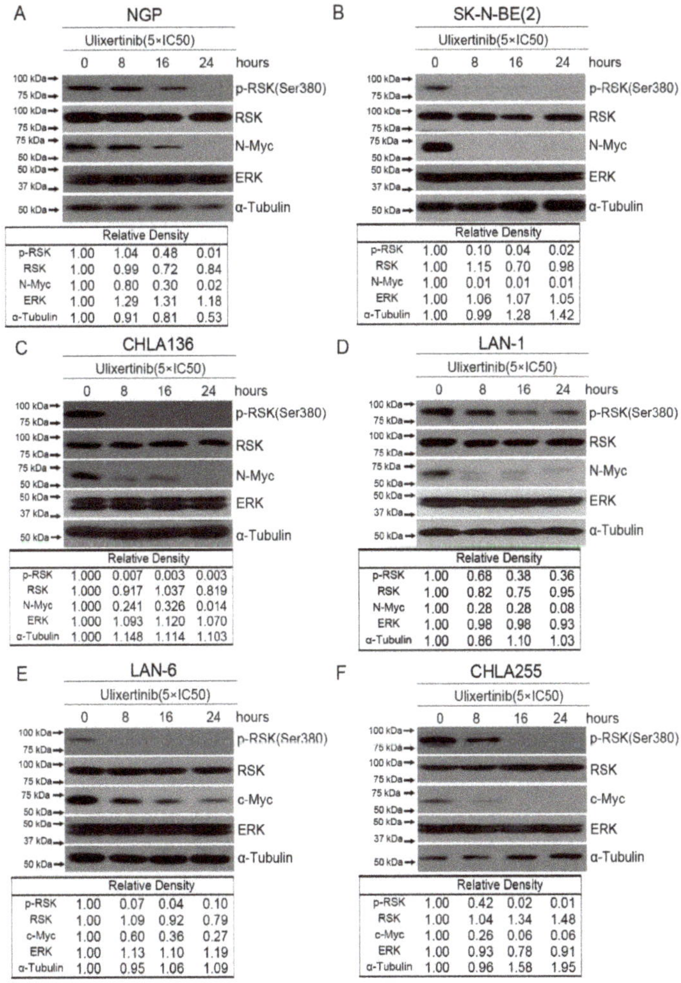

Figure 2. Ulixertinib suppresses ERK activation and reduces MYC/N-MYC levels in NB cells. Six NB cell lines, NGP (**A**), SK-N-BE(2) (**B**), CHLA136 (**C**), LAN-1 (**D**), LAN-6 (**E**), and CHLA255 (**F**) were treated with indicated doses of ulixertinib for variable periods (0 h, 8 h, 16 h, and 24 h). The proteins were extracted and subjected to SDS-PAGE for immunoblotting with the antibodies indicated. Gel densitometrical analysis was performed using ImageJ software (Version 1.52a, accessed on 23 April 2018) (Wayne Rasband, National Institutes of Health, USA).

3.3. *Ulixertinib Induces Extensive Transcriptomic Changes in NB*

To uncover how ulixertinib inhibits cell proliferation in NB, we performed a genome-wide transcriptomic analysis with RNA-Seq in ulixertinib-treated NB cells (Figure 3A). This analysis revealed 120 upregulated and 187 downregulated genes in response to ulixertinib treatment compared to the controls (Figure 3A and Table S1). Gene set enrichment analysis by the over-representation test has revealed inhibition of multiple pathways and their cross-talks (Figure 3B,C). As expected, MAPK signaling was significantly decreased as well as EGFR, VEGF, and WNT signaling pathways, demonstrated by multiple shared key components (Figure 3C, left panel). Overexpression and dominance of these survival pathways drive NB tumorigenesis and the malignant transformation [30,31]. NGF-stimulated transcription and NTRK1 signaling (Figure 3C, right panel) are the other two pathways significantly inhibited by ulixertinib, both of which are required for neuronal growth or nervous system development [32]. Among the upregulated genes, MBNL2 [33], SEPT4 [34], REPS2 [35], OTUD5 [36], and TXNIP [37] were shown to play a tumor-suppressive role in many tumors. To determine the correlation between these upregulated genes with NB patient prognosis, we analyzed NB patient datasets in the R2 database (https://hgserver1.amc.nl/cgi-bin/r2/main.cgi, access on 1 April 2021) and found that decreased expression of these genes predicted a significantly poor outcome (Supplementary Figure S1). These results suggest that these genes upregulated by ulixertinib treatment exert a tumor-suppressive function in NB cells.

3.4. *Ulixertinib Exerts Extensive Proteomic Changes in NB*

To further delineate the mechanisms of ulixertinib-mediated ERK pathway inhibition in NB and to determine the proteomic effects of ulixertinib in NB, we performed mass spectrometry analysis in NB cells (NGP) with and without ulixertinib treatment. Our proteomic analysis data identified a total of 157 differentially expressed proteins out of a total of 6105 measurable proteins (Table S2), with a threshold of $p < 0.05$ and a log fold change of expression with an absolute value of at least 1.0 (Figure 3D). Ulixertinib treatment led to the upregulation of 72 proteins and the downregulation of 85 proteins. The complete lists of gene names are shown in Table S2. In our analysis, ulixertinib is found to significantly inhibit cell cycles or cell cycle-related pathways and DNA replication/synthesis pathways, as revealed by gene set analysis using an over-representation test (Figure 3E). STRING analysis (https://string-db.org/, access on 12 August 2021) of ulixertinib-inhibited proteins revealed global protein–protein interaction networks and core interactomes (Figure 3F). The core components are ATAD2, AURKB, CENPF, CENPM, CDCA5, FAM64A, KIF20A, KIF22, KIFC1, KIF2C, KPNA2, MK167, NUSAP1, PRC1, RRM2, SPDL1, TOP2A, TPX2, UBE2T, and UHRF1. Among these proteins, MK167 is a cellular marker for proliferation [38]. KIF (kinesin superfamily) proteins (KIF20A, KIF22, KIFC1, and KIF2C) are microtubule-dependent molecular motors that play vital roles in cellular transport and cell division [39]. Moreover, high expression of some of these identified proteins inhibited by ulixertinib, including CDCA5, ATAD2, CDC25A, CDK2, ANAPC10, RRM2, LRG5, SKP2, CENPM, and UBE2T, are strongly associated with worse overall survival of NB patients (Supplementary Figure S2). Ulixertinib treatment enhanced the PRUNE2 level (Table S2), and PRUNE2 has been identified as a pro-apoptotic effector in NB [40]. Decreased expression of PRUNE2 predicted poor NB patients' outcomes (Supplementary Figure S1). Our proteomic data indicate that ulixertinib inhibits cell proliferation and cell cycle by regulating the expression levels of multiple factors involved in NB development.

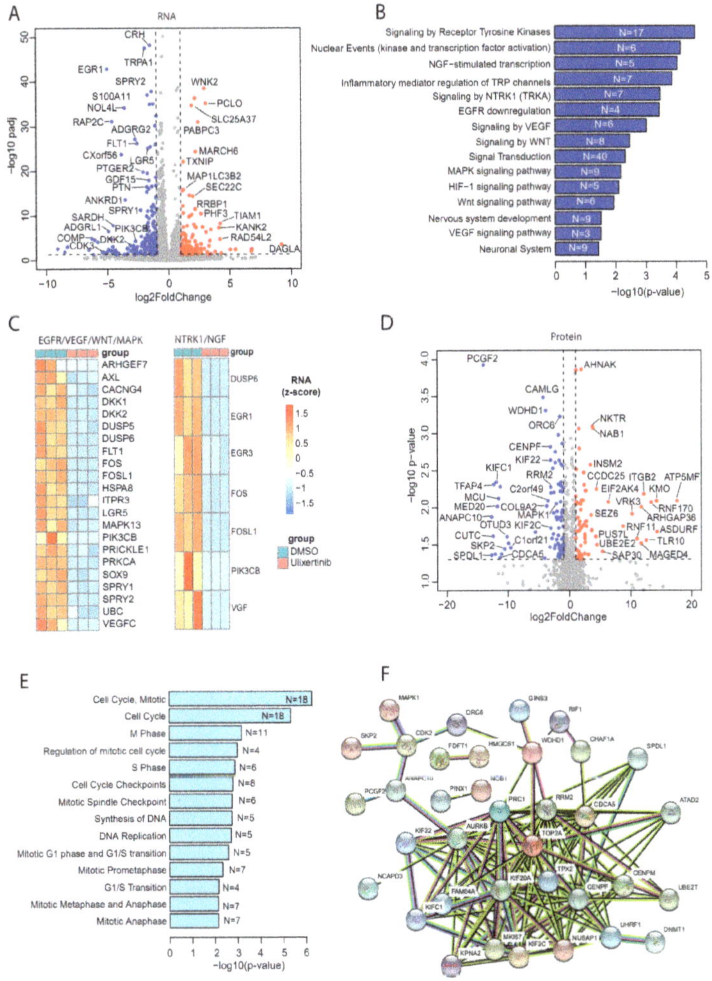

Figure 3. Transcriptomics and proteomics reveal ulixertinib-induced gene changes in NGP cells. (**A**) The differentially expressed genes revealed by DESeq2 analysis of RNA-Seq read counts. They are represented in the volcano plot in terms of their measured expression change (log2 of fold changes, *x*-axis) and the significance of the change (−log10 of adjusted *p* values, *y*-axis). The dotted lines represent the thresholds to define the differentially expressed genes. Red dots are up-regulated genes after the drug treatment (fold changes $\geq$ 2 and adjusted $p < 0.05$). Blue dots are down-regulated genes with the drug treatment (fold changes $\leq$ 0.5 and adjusted $p < 0.05$). (**B**) Top inhibited pathways revealed by down-regulated genes from RNA-Seq. N is number. (**C**) Key components inhibited and their associated pathways. (**D**) The differentially expressed proteins from MS were represented in the volcano plot in terms of their measured expression change (log2 of fold changes, *x*-axis) and the significance of the change (-log10 of *p* values, *y*-axis; unpaired *t*-test). The dotted lines represent the thresholds to define the differentially expressed proteins. Red dots are up-regulated proteins after the drug treatment (fold changes $\geq$ 2 and $p < 0.05$). Blue dots are down-regulated proteins with the drug treatment (fold changes $\leq$ 0.5 and $p < 0.05$). (**E**) Top inhibited pathways revealed by downregulated proteins from MS. N is number. (**F**) Protein–protein interaction networks analysis of down-regulated proteins from MS.

3.5. Ulixertinib Sensitizes NB to Chemotherapeutic Agent Doxorubicin

Doxorubicin, a chemotherapy agent standardly used in the treatment of high-risk NB, exhibited stronger cytotoxicity in combination with Raf inhibition in NB cells [41]. To determine whether blocking the ERK pathway using ulixertinib could enhance doxorubicin-induced apoptosis in NB cells, we treated four NB cell lines, including SK-N-AS, NGP, SK-N-BE(2), and CHLA255, with a combination of doxorubicin and ulixertinib and determined the combination index (CI) using the Chou–Talalay method. CI values < 1 signify synergism, =1, additive effects, and >1, antagonism [42]. A strong synergy, which was evaluated by different effective doses (ED: ED50, ED75, ED90, and ED95), was observed in the combination of doxorubicin and ulixertinib in all four NB cell lines (CI < 0.72 at ED50; CI < 0.19 at ED95) (Figure 4A–D). In addition, enhanced apoptosis in the cells with the treatment combination of ulixertinib and doxorubicin was observed using an immunoblotting assay to assess cleaved Caspase 3, an apoptosis marker, compared to single agent treatment (Figure 4E,F). To further confirm the results, apoptosis of different treatment groups was determined by flow cytometry detection of their sub-G0 DNA contents using propidium iodine (PI) staining. Results showed that ulixertinib and doxorubicin combination treatment increased the sub-G0 population compared to ulixertinib or doxorubicin treatment alone (Supplementary Figure S3). Our data indicate that ulixertinib significantly sensitizes NB cells to doxorubicin-induced apoptosis and could serve as a promising therapeutic agent for treating chemo-resistant NB when combined with conventional chemotherapeutic agents.

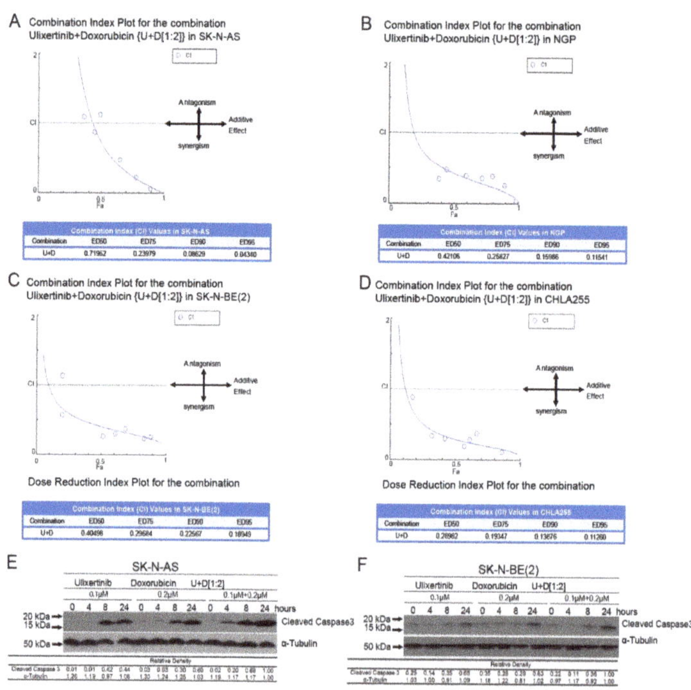

Figure 4. Ulixertinib enhances the anti-tumor activity of doxorubicin in human neuroblastoma cells. (**A–D**) CI plots for SK-N-AS, NGP, CHLA255, and SK-N-BE (2). Four NB cell lines, including SK-N-AS (**A**), NGP (**B**), SK-N-BE(2) (**C**), and CHLA255 (**D**) cells, were treated with ulixertinib, doxorubicin, and their combinations at a ratio of 1:2 respectively, for 72 h. The combination index (CI) values were calculated by the Chou–Talalay method for drug interactions using Compusyn software. Values of CI < 1, =1, and >1 indicate synergism, additive effects, and antagonism, respectively. The obtained CI

values for the combination of ulixertinib and doxorubicin at different effective doses (ED50, ED75, ED90, and ED95). (**E**,**F**) SK-N-AS (**E**) and SK-N-BE(2) (**F**) cells were treated with either doxorubicin alone, ulixertinib alone, or their combinations for 4 h, 8 h, and 24 h. Then the whole-cell lysates were subjected to SDS-PAGE and immunoblotted with the cleaved-caspase 3 antibody. α-tubulin was used as a loading control in all samples. Gel densitometrical analysis was performed using ImageJ software (https://imagej.nih.gov/ij/index.html, accessed on 25 October 2022).

3.6. *Ulixertinib Inhibits NB Tumor Growth in Xenograft Mouse Models*

To further evaluate the therapeutic potential of ulixertinib in treating NB, we tested its effect on tumor growth using two xenograft NB mouse models. We used CHLA136-Fluc (*MYCN* amplified) and CHLA255-Fluc (c-Myc overexpressed) to generate these xenograft mouse NB tumor models. Randomized mice were grouped in different cohorts and treated with ulixertinib or vehicle daily for three weeks (Figures 5A and 6A). Ulixertinib daily injection was well tolerated at a concentration of 50 mg/kg without apparent bodyweight loss and other adverse effects. We found that treatment with ulixertinib significantly inhibited the overall NB tumor growth compared to vehicle (Figure 5B,C and Figure 6B,C). We also found that ulixertinib-treated xenograft NB mice survived significantly longer when compared to control mice (Figures 5D and 6D). Overall, this data clearly demonstrate the potency and efficacy of ulixertinib in inhibiting NB tumor growth and prolonging overall survival.

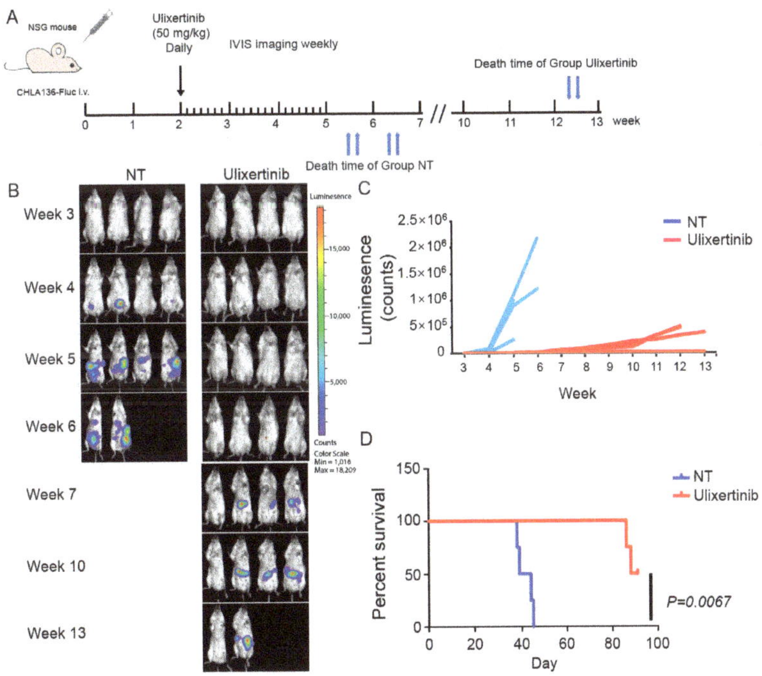

Figure 5. Ulixertinib inhibits tumor growth in the CHLA136-Fluc xenograft NB mouse model. (**A**) Treatment strategy of the xenograft NB mouse model. (**B**) Bioluminescent imaging photos of CHLA136-Fluc xenograft NB mice from the DMSO control group and the ulixertinib-treated group (50 mg/kg). (**C**) The values of bioluminescent imaging of xenograft NB mice were monitored during tumor growth. (**D**) The survival rate of CHLA136-Fluc xenograft NB mice treated with ulixertinib and control. Statistical analysis was performed by a Log-rank test. (χ^2 = 7.344; p = 0.0067).

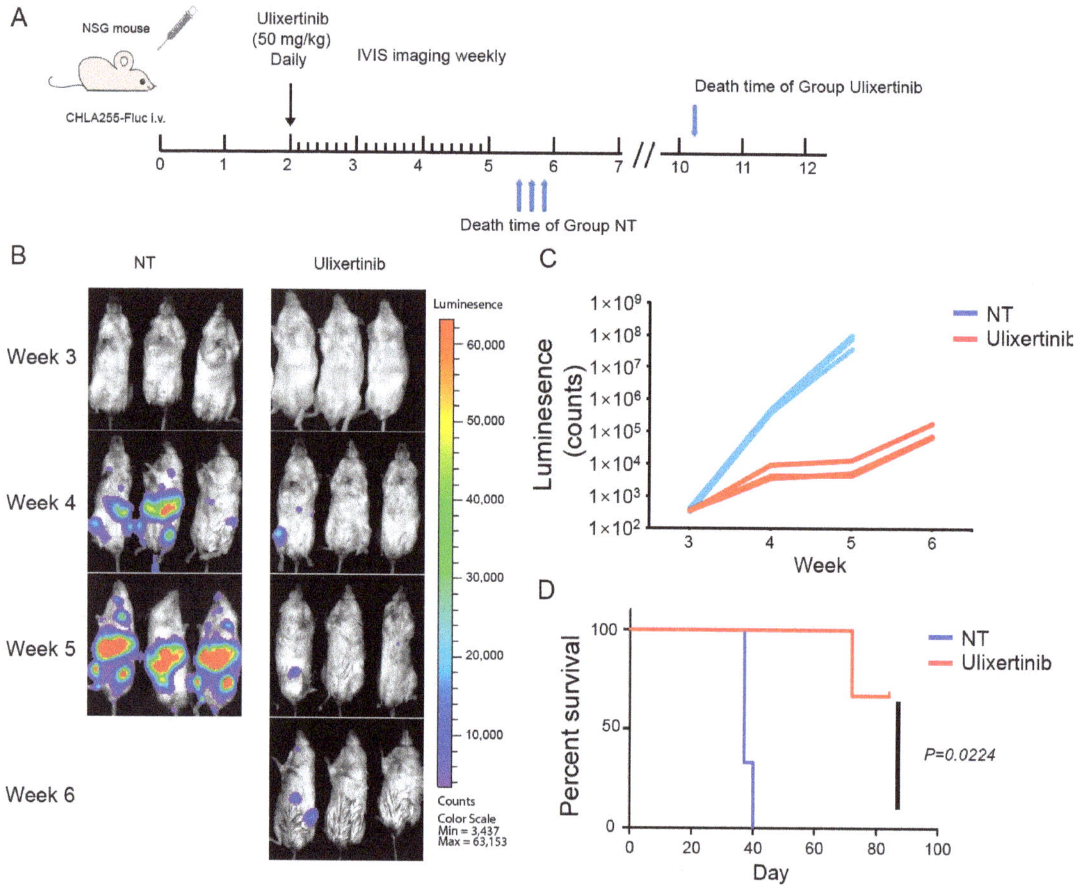

Figure 6. Ulixertinib inhibits tumor growth in the CHLA255-Fluc xenograft NB mouse model. (**A**) Treatment strategy of the xenograft NB mouse model. (**B**) Bioluminescent imaging photos of CHLA255 xenograft NB mice from the DMSO control group and the ulixertinib treated group (50 mg/kg). (**C**) The values of bioluminescent imaging of xenograft NB mice were monitored during tumor growth. (**D**) The survival rate of CHLA255 xenograft NB mice treated with ulixertinib and control. Statistical analysis was performed by a Log-rank test. ($\chi^2 = 5.213$; $p = 0.0224$).

4. Discussion

Aberrant RAS-MAPK-ERK activity can lead to uncontrolled cell proliferation, immortalization, and tumorigenesis. In NB, activation of the RAS-MAPK-ERK pathway has been associated with enhanced tumor cell proliferation and chemoresistance, possibly contributing to the recurrent disease that more than half of NB patients experience [6]. Targeting the RAS-MAPK-ERK pathway, especially ERK, provides a promising solution for improving survival in high-risk NB. In this study, we demonstrated that ulixertinib, a specific ERK inhibitor, effectively inhibits NB cell proliferation, promotes apoptosis, and sensitizes NB to doxorubicin, partially through downregulating c-Myc/N-Myc and other key signaling pathways. For the first time, we showed that in preclinical models ulixertinib inhibits NB tumor growth. We further delineated the underlying mechanisms of ulixertinib as a therapeutic agent for NB through the analysis of transcriptomics and proteomics. This could provide a solid foundation and clinical application of ulixertinib for use in the treatment of NB, especially for patients with refractory or recurrent disease.

Targeting ERK using small molecules has been demonstrated to effectively inhibit the Ras/Raf/MEK/ERK pathway in several models of cancers, including NB. Five ERK inhibitors have been tested in clinical trials, but only GDC-0994 and ulixertinib have shown antitumor activity [43]. Compared to GDC-0994, ulixertinib has exhibited anticancer activities for broader cancer types. More importantly, in clinical settings, cancer patients tolerated much higher doses of ulixertinib than GDC-0994 (1200 mg vs. 120–400 mg daily) [11,23,44,45]. Our study illustrates the efficacy of ulixertinib in treating NB in vitro and in vivo. Therefore, ulixertinib may be an ideal therapeutic agent to treat NB.

ERK inactivation by ulixertinib has been demonstrated in multiple cancers, including pancreatic cancer [46], melanoma, colon cancer [23], and glioma [47]. In these cancer cells, ulixertinib treatment significantly suppressed the phosphorylation of its target RSK [46]. Our study consistently revealed the same results in all tested NB cells. Various ERK inhibitors have been shown to have different effects on ERK phosphorylation. For instance, ERK phosphorylation was increased with ulixertinib; this was not observed with SCH772984 (an ERK inhibitor) in pancreatic cancer cells [46]. However, RSK phosphorylation is constantly suppressed by all ERK inhibitors tested, suggesting that RSK phosphorylation should be a reliable ERK activation marker regardless of cell types and inhibitors in NB.

In addition to blocking ERK activity, ulixertinib treatment also inhibited the Wnt/β-catenin pathway by downregulating LGR5 transcription (Figure 3A and Table S1). Tumor samples from high-risk NB patients with and without *MYCN* amplification show the expression of canonical Wnt pathway target genes at high levels, indicative of Wnt pathway deregulation in NB [31]. High expression of LGR5 is reported in human colorectal adenomas and cancers [48], hepatocellular carcinoma, and basal cell carcinoma [49,50]. LGR5 is highly expressed in NB, which modulates Wnt signaling and is associated with increased proliferation [51]. Importantly, LGR5 also regulates MEK/ERK and Akt pro-survival signaling, pathways that are frequently activated in primary NBs [51]. This suggests that LGR5 could be one of the key components in ulixertinib-induced inhibition of pro-survival signaling. It has been well documented that overexpression of MYC or MYCN promotes NB cell proliferation [17]. Ulixertinib-induced ERK inhibition can efficiently inhibit NB cell proliferation by modulating various downstream targets, including c-Myc/N-Myc, required for the transcription of critical cell cycle genes. Marco Ciró et al. showed that ATAD2 is highly expressed in multiple types of human cancers and cooperates with MYC to activate MYC target genes transcriptionally [52]. Depletion of ATAD2 could reduce the mRNA expression of Myc and the key genes in hepatocellular carcinoma [53]. Our study demonstrates that ulixertinib downregulated ATAD2, ensuring an effective blocking of MYC/MYCN transcriptional activation of target oncogenes in NB cells. The exact roles of ATAD2 in NB development need to be further explored, given high ATAD2 expression is closely associated with worse survival outcomes. AURKB, one of the core components of ulixertinib-inhibited global protein–protein interaction networks (Figure 3F and Table S2), promotes MYC protein stability by regulating its serine 67 phosphorylation [54]. Taken together, our findings indicate that ulixertinib might decrease MYCN/MYC levels by downregulating the expression of ATAD2 and AURKB.

Another possible mechanism of ulixertinib suppressing cell proliferation could be through RSK kinase inactivation-mediate $p27^{KiP1}$ nuclear translocation, which can further inhibit the activation of cyclin A/E-CDK2 complexes that are required for G1 progression [55]. In addition to reduced activation of cyclin A/E-CDK2 complexes, the mass spectrometry results revealed that ulixertinib treatment reduces CDK2 protein levels, further suppressing G1 progression. Moreover, ulixertinib also reduces the levels of other cell cycle proteins, such as CDCA5, ANAPC10, SKP2, and CENPM. Therefore, ulixertinib may inhibit cell cycle progression by downregulating multiple key regulators. In human cervical carcinoma cells, downregulation of CDK2 has been shown to induce tumor cell cycle arrest and cell apoptosis [56]. We therefore expect that inhibition of CDK2 might also be one mechanism by which ulixertinib promotes apoptosis in NB cells.

Transcriptomic and proteomics analyses also revealed some upregulated genes following ulixertinib treatment in NB cells, such as TXNIP, MBNL2, and SEPT4. TXNIP is considered to be a potential tumor suppressor gene in multiple tumors, including breast cancer [57], hepatocellular carcinoma [58], non-small-cell lung cancer [59], and renal cell carcinoma [60]. The expression of TXNIP is at a low level in these tumor types, and the overexpression of TXNIP inhibits the proliferation of cancer cells. Lee YH, et al. found that the expression of MBNL2 was lost in the late stage of hepatocellular carcinoma development [33]. MBNL2-positive correlates with better 5-year overall survival [33]. Overexpression of MBNL2 inhibits cell proliferation, migration in vitro, and tumor growth in vivo [33]. Genetic evidence shows that SEPT4 suppresses tumor development by antagonizing the function of inhibitor of apoptosis proteins (IAPs) [34]. Consistently, our findings indicate that patients with elevated expression of these three genes had significantly longer overall survival compared to patients bearing tumors with low levels (Supplementary Figure S1). The exact function of these genes in NB development requires further exploration.

Chemoresistance has long been a challenge in treating high-risk NB. Among ERK downstream targets, c-Myc and N-Myc have been demonstrated to promote chemoresistance of NB cells [61,62]. In our study, ulixertinib inhibits c-Myc/N-Myc levels in NB cells regardless of *TP53* status and *MYCN* status. This further enhanced doxorubicin sensitivity on all tested NB cells. Moreover, in addition to small molecules, RNA silencing of *MYCN* has been demonstrated to restore doxorubicin sensitivity in NB cells [63]. Enhancing the effectiveness of doxorubicin, a key component of high-risk NB induction chemotherapy, may help to improve end of induction disease response and thus potentially survival as well. Compared to the RNA silencing approach, treatments with small molecules are more efficient and controllable in in vivo studies and clinical applications. We also found that ulixertinib significantly reduced the levels of other promoters of chemoresistance, such as RRM2 and UBE2T. RRM2, a downstream target of ERK [64] and positive regulator of BCL2 [65], is a key determinant in suppressing apoptosis and promoting chemoresistance. In addition, high RRM2 expression has also been demonstrated to promote epithelial-mesenchymal transition (EMT) and angiogenesis in prostate cancer, resulting in poor patient outcomes [66]. So far, UBE2T has not been linked to being a downstream target of ERK signaling. Therefore, we speculate that ulixertinib could not only enhance the sensitivity of NB cells to chemotherapeutic drugs but may also suppress EMT and angiogenesis in NB. We chose to study doxorubicin given the existing pre-clinical data in combination with RAF inhibitors. One disadvantage of doxorubicin however is that it is not standardly used for patients with relapsed disease. Future studies can explore the impact ulixertinib has on chemotherapy regimens typically used for patients with relapsed disease. In the present study, treatment with ulixertinib induced the expression of PRUNE2 in NB cells (Table S2). PRUNE2 is associated with a favorable prognosis in NB, and is found in the cytoplasm of favorable NB cells but not in unfavorable ones with *MYCN* amplification [67]. PRUNE2, triggered by DNA damage, also exerts a pro-apoptotic role by interacting with BCL2 and suppressing AKT pathway activity in NB [40,68]. A recent study shows that EPO, NGF, and HGF signaling pathways are upregulated in NB patients with no or partial response to chemotherapy [69]. Chemical inhibitors potentiate strong ERK signaling activation by EPO and NGF, providing a protective effect to NB cells [69]. Our study suggests that ulixertinib also inhibits the NGF-induced signaling pathway (Figure 3C). However, we also found the tumor relapse after treatment with ulixertinib alone for 3 weeks. This suggests that ulixertinib has to be used in combination with current chemotherapy to treat NB. Valencia-Sama I. et al. also reported that ulixertinib was synergistic and showed reversed resistance to SHP2 inhibition in neuroblastoma in vitro and in vivo [70]. Considering that a phase 2 clinical trial using ulixertinib to treat solid tumors, including NB, is in progress (NCT03698994), ulixertinib could be closer to clinical use compared to other therapeutic strategies for treating NB. However, more NB clinical trials using the combination of ulixertinib and chemotherapeutic drugs are needed to verify the efficacy of this novel regime.

5. Conclusions

Our study evaluated the therapeutic potential of ulixertinib-based treatment in NB. Ulixertinib significantly inhibits cell proliferation in NB cell lines and PDX cells in vitro. Moreover, ulixertinib inhibits tumor growth and prolongs the overall survival in preclinical tumor models. Transcriptomics and proteomic analysis revealed tumor-suppressive properties of ulixertinib on the expression of genes involved in NB development. Importantly, we also demonstrated the efficacy of ulixertinib in sensitizing NB cells to standard chemotherapy. Ulixertinib has the potential either as a single agent or in combination with standard chemotherapy to improve survival for patients with high-risk NB.

Supplementary Materials: The following supporting information can be downloaded at: https://www.mdpi.com/article/10.3390/cancers14225534/s1. Figure S1: High expression of targets induced by ulixertinib treatment in NB cells correlates with a better NB patient survival. Figure S2. High expression of targets inhibited by ulixertinib treatment in NB cells correlates with a poor NB patient survival. Figure S3. Ulixertinib sensitizes human NB cells to doxorubicin-induced cell apoptosis. Table S1. Differential gene expression (DGE) analysis of RNA-Seq. Table S2. Differential gene expression (DGE) analysis of proteins from MS.

Author Contributions: Conceptualization, J.Y. and J.H.F.; methodology, Y.Z., J.C. and Y.Y.; software, C.Z. and S.Y.J.; validation, J.E. and A.G.; formal analysis, D.L. and S.A.; investigation, Y.Z., Y.Y. and F.C.; resources, L.G.; data curation, J.Y. and S.A.; writing—original draft preparation, Y.Y., Y.Z. and Z.S.; writing—review and editing, J.Y., C.Z., Y.P., M.F., J.H.F. and S.A.; visualization, C.Z.; supervision, J.Y. and J.H.F.; project administration, J.Y.; funding acquisition, J.Y. and J.H.F. All authors have read and agreed to the published version of the manuscript.

Funding: This research was funded by the Dan L Duncan Comprehensive Cancer Center (DLDCC) Pilot Award NIH grant P30, grant number CA125123 (to J.H.F.), the Elsa U. Pardee Foundation Award (to J.H.F.), and NIH-NINDS grant, grant number 5R01NS118008 (to J.Y.)

Institutional Review Board Statement: The animal study protocol was approved by the Institutional Review Board of Baylor College of Medicine (AN7589, Approval valid from 24 March 2021 through 23 March 2024).

Informed Consent Statement: Not applicable.

Data Availability Statement: The mass spectrometry data for global proteomics profiling have been deposited to MASSIVE repository (MassIVE MSV000087511). RNA-Seq raw files are deposited onto NCBI Sequence Read Archive (SRA) database and the GEO accession number is GSE213153.

Acknowledgments: We thank the technical support from the Cancer Prevention and Research Institute of Texas (CPRIT RP180734). We thank Leonid S. Metelitsa for providing the CHLA255-Fluc and CHLA136-Fluc cells and thank He Shi for language editing of the manuscript. We thank Christopher Lazarski for his assistance with our flow cytometry analysis.

Conflicts of Interest: The authors declare no conflict of interest.

Abbreviations

NB	neuroblastoma
ERK	extracellular signal–regulated kinases
MAPK	mitogen-activated protein kinase
RAF	rapidly accelerated fibrosarcoma
RSK	ribosomal S6 kinase
MEK	mitogen-activated protein kinase
BRAF	B-Raf proto-oncogene, serine/threonine kinase
GSK-3β	glycogen synthase kinase 3 beta
SCF	Skp1-Cullin 1-F-box proteins
FBW7	F-box and WD repeat domain containing 7
CDK2	cyclin dependent kinase 2
ANAPC10	anaphase promoting complex subunit 10

SKP2	S-phase kinase associated protein 2
CENPM	centromere protein M
UBE2T	ubiquitin conjugating enzyme E2T
EMT	Epithelial–mesenchymal transition
BCL2	B cell lymphoma 2
E2F1	E2F transcription factor 1
KIFs	kinesin superfamily proteins
LGR5	Leucine-rich repeat-containing G protein-coupled receptor 5
RRM2	ribonucleotide reductase regulatory subunit M2

References

1. Brodeur, G.M. Neuroblastoma: Biological insights into a clinical enigma. *Nat. Rev. Cancer* **2003**, *3*, 203–216. [CrossRef]
2. Chakrabarti, L.; Abou-Antoun, T.; Vukmanovic, S.; Sandler, A.D. Reversible adaptive plasticity: A mechanism for neuroblastoma cell heterogeneity and chemo-resistance. *Front. Oncol.* **2012**, *2*, 82. [CrossRef] [PubMed]
3. Elizabeth, R.; Tucker, E.P. Louis Chesler, Targeting mycn and alk in resistant and relapsing neuroblastoma. *Cancer Drug Resist.* **2019**, *2*, 803–812.
4. Wellbrock, C.; Karasarides, M.; Marais, R. The raf proteins take centre stage. *Nat. Rev. Mol. Cell Biol.* **2004**, *5*, 875–885. [CrossRef] [PubMed]
5. Kohno, M.; Pouyssegur, J. Targeting the erk signaling pathway in cancer therapy. *Ann. Med.* **2006**, *38*, 200–211. [CrossRef]
6. Eleveld, T.F.; Oldridge, D.A.; Bernard, V.; Koster, J.; Colmet Daage, L.; Diskin, S.J.; Schild, L.; Bentahar, N.B.; Bellini, A.; Chicard, M.; et al. Relapsed neuroblastomas show frequent ras-mapk pathway mutations. *Nat. Genet.* **2015**, *47*, 864–871. [CrossRef]
7. Abrams, S.L.; Steelman, L.S.; Shelton, J.G.; Wong, E.W.; Chappell, W.H.; Basecke, J.; Stivala, F.; Donia, M.; Nicoletti, F.; Libra, M.; et al. The raf/mek/erk pathway can govern drug resistance, apoptosis and sensitivity to targeted therapy. *Cell Cycle* **2010**, *9*, 1781–1791. [CrossRef]
8. Smorodinsky-Atias, K.; Soudah, N.; Engelberg, D. Mutations that confer drug-resistance, oncogenicity and intrinsic activity on the erk map kinases-current state of the art. *Cells* **2020**, *9*, 129. [CrossRef]
9. Hu-Lieskovan, S.; Mok, S.; Homet Moreno, B.; Tsoi, J.; Robert, L.; Goedert, L.; Pinheiro, E.M.; Koya, R.C.; Graeber, T.G.; Comin-Anduix, B.; et al. Improved antitumor activity of immunotherapy with braf and mek inhibitors in braf(v600e) melanoma. *Sci. Transl. Med.* **2015**, *7*, 279ra241. [CrossRef]
10. Flemington, V.; Davies, E.J.; Robinson, D.; Sandin, L.C.; Delpuech, O.; Zhang, P.; Hanson, L.; Farrington, P.; Bell, S.; Falenta, K.; et al. Azd0364 is a potent and selective erk1/2 inhibitor that enhances antitumor activity in kras-mutant tumor models when combined with the mek inhibitor, selumetinib. *Mol. Cancer Ther.* **2021**, *20*, 238–249. [CrossRef]
11. Varga, A.; Soria, J.C.; Hollebecque, A.; LoRusso, P.; Bendell, J.; Huang, S.A.; Wagle, M.C.; Okrah, K.; Liu, L.; Murray, E.; et al. A first-in-human phase i study to evaluate the erk1/2 inhibitor gdc-0994 in patients with advanced solid tumors. *Clin. Cancer Res.* **2020**, *26*, 1229–1236. [CrossRef] [PubMed]
12. Yuen, J.S.; Sim, M.Y.; Sim, H.G.; Chong, T.W.; Lau, W.K.; Cheng, C.W.; Ong, R.W.; Huynh, H. Combination of the erk inhibitor azd6244 and low-dose sorafenib in a xenograft model of human renal cell carcinoma. *Int. J. Oncol.* **2012**, *41*, 712–720. [CrossRef] [PubMed]
13. Ding, Q.; Xia, W.; Liu, J.C.; Yang, J.Y.; Lee, D.F.; Xia, J.; Bartholomeusz, G.; Li, Y.; Pan, Y.; Li, Z.; et al. Erk associates with and primes gsk-3beta for its inactivation resulting in upregulation of beta-catenin. *Mol. Cell* **2005**, *19*, 159–170. [CrossRef] [PubMed]
14. Knoepfler, P.S.; Kenney, A.M. Neural precursor cycling at sonic speed: N-myc pedals, gsk-3 brakes. *Cell Cycle* **2006**, *5*, 47–52. [CrossRef] [PubMed]
15. Wang, Z.; Ge, L.; Wang, M.; Carr, B.I. Phosphorylation regulates myc expression via prolonged activation of the mitogen-activated protein kinase pathway. *J. Cell Physiol.* **2006**, *208*, 133–140. [CrossRef] [PubMed]
16. Ren, Y.; Chan, H.M.; Li, Z.; Lin, C.; Nicholls, J.; Chen, C.F.; Lee, P.Y.; Lui, V.; Bacher, M.; Tam, P.K. Upregulation of macrophage migration inhibitory factor contributes to induced n-myc expression by the activation of erk signaling pathway and increased expression of interleukin-8 and vegf in neuroblastoma. *Oncogene* **2004**, *23*, 4146–4154. [CrossRef]
17. Gogolin, S.; Dreidax, D.; Becker, G.; Ehemann, V.; Schwab, M.; Westermann, F. Mycn/myc-mediated drug resistance mechanisms in neuroblastoma. *Int. J. Clin. Pharmacol. Ther.* **2010**, *48*, 489–491. [CrossRef]
18. Li, L.; Osdal, T.; Ho, Y.; Chun, S.; McDonald, T.; Agarwal, P.; Lin, A.; Chu, S.; Qi, J.; Li, L.; et al. Sirt1 activation by a c-myc oncogenic network promotes the maintenance and drug resistance of human flt3-itd acute myeloid leukemia stem cells. *Cell Stem Cell* **2014**, *15*, 431–446. [CrossRef]
19. Liu, L.; Zhang, J.; Yang, X.; Fang, C.; Xu, H.; Xi, X. Sall4 as an epithelial-mesenchymal transition and drug resistance inducer through the regulation of c-myc in endometrial cancer. *PLoS ONE* **2015**, *10*, e0138515. [CrossRef]
20. Liu, R.; Li, Y.; Tian, L.; Shi, H.; Wang, J.; Liang, Y.; Sun, B.; Wang, S.; Zhou, M.; Wu, L.; et al. Gankyrin drives metabolic reprogramming to promote tumorigenesis, metastasis and drug resistance through activating beta-catenin/c-myc signaling in human hepatocellular carcinoma. *Cancer Lett.* **2019**, *443*, 34–46. [CrossRef]
21. Van Waardenburg, R.C.; Prins, J.; Meijer, C.; Uges, D.R.; De Vries, E.G.; Mulder, N.H. Effects of c-myc oncogene modulation on drug resistance in human small cell lung carcinoma cell lines. *Anticancer Res.* **1996**, *16*, 1963–1970. [PubMed]

22. Parasido, E.M.; Avetian, G.S.; Brody, J.; Winter, J.; Londin, E.; Pishvaian, M.; Glasgow, E.; Byers, S.; Narla, G.; Albanese, C. Targeting c-myc and mapk pathway to overcome pancreatic cancer drug resistance. *Cancer Res.* **2019**, *79* (Suppl. S13), 1283. [CrossRef]
23. Germann, U.A.; Furey, B.F.; Markland, W.; Hoover, R.R.; Aronov, A.M.; Roix, J.J.; Hale, M.; Boucher, D.M.; Sorrell, D.A.; Martinez-Botella, G.; et al. Targeting the mapk signaling pathway in cancer: Promising preclinical activity with the novel selective erk1/2 inhibitor bvd-523 (ulixertinib). *Mol. Cancer Ther.* **2017**, *16*, 2351–2363. [CrossRef] [PubMed]
24. Borowicz, S.; Van Scoyk, M.; Avasarala, S.; Karuppusamy Rathinam, M.K.; Tauler, J.; Bikkavilli, R.K.; Winn, R.A. The soft agar colony formation assay. *J. Vis. Exp.* **2014**, *92*, e51998. [CrossRef] [PubMed]
25. Chou, T.C.; Talaly, P. A simple generalized equation for the analysis of multiple inhibitions of michaelis-menten kinetic systems. *J. Biol. Chem.* **1977**, *252*, 6438–6442. [CrossRef]
26. Xu, X.; Huang, W.; Heczey, A.; Liu, D.; Guo, L.; Wood, M.; Jin, J.; Courtney, A.N.; Liu, B.; Di Pierro, E.J.; et al. Nkt cells coexpressing a gd2-specific chimeric antigen receptor and il15 show enhanced in vivo persistence and antitumor activity against neuroblastoma. *Clin. Cancer Res.* **2019**, *25*, 7126–7138. [CrossRef]
27. Moghimi, B.; Muthugounder, S.; Jambon, S.; Tibbetts, R.; Hung, L.; Bassiri, H.; Hogarty, M.D.; Barrett, D.M.; Shimada, H.; Asgharzadeh, S. Preclinical assessment of the efficacy and specificity of gd2-b7h3 synnotch car-t in metastatic neuroblastoma. *Nat. Commun.* **2021**, *12*, 511. [CrossRef]
28. Kamburov, A.; Wierling, C.; Lehrach, H.; Herwig, R. Consensuspathdb—A database for integrating human functional interaction networks. *Nucleic Acids Res.* **2009**, *37*, D623–D628. [CrossRef]
29. Szklarczyk, D.; Gable, A.L.; Lyon, D.; Junge, A.; Wyder, S.; Huerta-Cepas, J.; Simonovic, M.; Doncheva, N.T.; Morris, J.H.; Bork, P.; et al. String v11: Protein-protein association networks with increased coverage, supporting functional discovery in genome-wide experimental datasets. *Nucleic Acids Res.* **2019**, *47*, D607–D613. [CrossRef]
30. Megison, M.L.; Gillory, L.A.; Beierle, E.A. Cell survival signaling in neuroblastoma. *Anticancer Agents Med. Chem.* **2013**, *13*, 563–575. [CrossRef]
31. Liu, X.; Mazanek, P.; Dam, V.; Wang, Q.; Zhao, H.; Guo, R.; Jagannathan, J.; Cnaan, A.; Maris, J.M.; Hogarty, M.D. Deregulated wnt/beta-catenin program in high-risk neuroblastomas without mycn amplification. *Oncogene* **2008**, *27*, 1478–1488. [CrossRef] [PubMed]
32. Funke, L.; Bracht, T.; Oeck, S.; Schork, K.; Stepath, M.; Dreesmann, S.; Eisenacher, M.; Sitek, B.; Schramm, A. Ntrk1/trka signaling in neuroblastoma cells induces nuclear reorganization and intra-nuclear aggregation of lamin a/c. *Cancers* **2021**, *13*, 5293. [CrossRef]
33. Lee, Y.H.; Jhuang, Y.L.; Chen, Y.L.; Jeng, Y.M.; Yuan, R.H. Paradoxical overexpression of mbnl2 in hepatocellular carcinoma inhibits tumor growth and invasion. *Oncotarget* **2016**, *7*, 65589–65601. [CrossRef] [PubMed]
34. Garcia-Fernandez, M.; Kissel, H.; Brown, S.; Gorenc, T.; Schile, A.J.; Rafii, S.; Larisch, S.; Steller, H. Sept4/arts is required for stem cell apoptosis and tumor suppression. *Genes Dev.* **2010**, *24*, 2282–2293. [CrossRef] [PubMed]
35. Oosterhoff, J.K.; Penninkhof, F.; Brinkmann, A.O.; Anton Grootegoed, J.; Blok, L.J. Reps2/pob1 is downregulated during human prostate cancer progression and inhibits growth factor signalling in prostate cancer cells. *Oncogene* **2003**, *22*, 2920–2925. [CrossRef] [PubMed]
36. Li, F.; Sun, Q.; Liu, K.; Zhang, L.; Lin, N.; You, K.; Liu, M.; Kon, N.; Tian, F.; Mao, Z.; et al. Otud5 cooperates with trim25 in transcriptional regulation and tumor progression via deubiquitination activity. *Nat. Commun.* **2020**, *11*, 4184. [CrossRef]
37. Pan, M.; Zhang, F.; Qu, K.; Liu, C.; Zhang, J. Txnip: A double-edged sword in disease and therapeutic outlook. *Oxid. Med. Cell. Longev.* **2022**, *2022*, 7805115. [CrossRef]
38. Scholzen, T.; Gerdes, J. The ki-67 protein: From the known and the unknown. *J. Cell Physiol.* **2000**, *182*, 311–322. [CrossRef]
39. Miki, H.; Setou, M.; Kaneshiro, K.; Hirokawa, N. All kinesin superfamily protein, kif, genes in mouse and human. *Proc. Natl. Acad. Sci USA* **2001**, *98*, 7004–7011. [CrossRef]
40. Islam, M.S.; Takano, R.; Yokochi, T.; Akter, J.; Nakamura, Y.; Nakagawara, A.; Tatsumi, Y. Programmed expression of pro-apoptotic bmcc1 during apoptosis, triggered by DNA damage in neuroblastoma cells. *BMC Cancer* **2019**, *19*, 542. [CrossRef]
41. Li, H.; Yu, Y.; Zhao, Y.; Wu, D.; Yu, X.; Lu, J.; Chen, Z.; Zhang, H.; Hu, Y.; Zhai, Y.; et al. Small molecule inhibitor agerafenib effectively suppresses neuroblastoma tumor growth in mouse models via inhibiting erk mapk signaling. *Cancer Lett.* **2019**, *457*, 129–141. [CrossRef] [PubMed]
42. Li, H.; Krstin, S.; Wang, S.; Wink, M. Capsaicin and piperine can overcome multidrug resistance in cancer cells to doxorubicin. *Molecules* **2018**, *23*, 557. [CrossRef] [PubMed]
43. Chin, H.M.; Lai, D.K.; Falchook, G.S. Extracellular signal-regulated kinase (erk) inhibitors in oncology clinical trials. *J. Immunother. Precis. Oncol.* **2019**, *2*, 10–16. [CrossRef]
44. Mendzelevski, B.; Ferber, G.; Janku, F.; Li, B.T.; Sullivan, R.J.; Welsch, D.; Chi, W.; Jackson, J.; Weng, O.; Sager, P.T. Effect of ulixertinib, a novel erk1/2 inhibitor, on the qt/qtc interval in patients with advanced solid tumor malignancies. *Cancer Chemother. Pharmacol.* **2018**, *81*, 1129–1141. [CrossRef] [PubMed]
45. Weekes, C.; Lockhart, A.; LoRusso, P.; Murray, E.; Park, E.; Tagen, M.; Singh, J.; Sarkar, I.; Mueller, L.; Dokainish, H.; et al. A phase ib study to evaluate the mek inhibitor cobimetinib in combination with the erk1/2 inhibitor gdc-0994 in patients with advanced solid tumors. *Oncologist* **2020**, *25*, 833-e1438. [CrossRef]

46. Jiang, H.; Xu, M.; Li, L.; Grierson, P.; Dodhiawala, P.; Highkin, M.; Zhang, D.; Li, Q.; Wang-Gillam, A.; Lim, K.H. Concurrent her or pi3k inhibition potentiates the antitumor effect of the erk inhibitor ulixertinib in preclinical pancreatic cancer models. *Mol. Cancer Ther.* **2018**, *17*, 2144–2155. [CrossRef]
47. Sigaud, R.; Rosch, L.; Gatzweiler, C.; Benzel, J.; von Soosten, L.; Peterziel, H.; Selt, F.; Najafi, S.; Ayhan, S.; Gerloff, X.F.; et al. The first-in-class erk inhibitor ulixertinib shows promising activity in mapk-driven pediatric low-grade glioma models. *Neuro Oncol.* **2022**, noac183. [CrossRef]
48. Fan, X.S.; Wu, H.Y.; Yu, H.P.; Zhou, Q.; Zhang, Y.F.; Huang, Q. Expression of lgr5 in human colorectal carcinogenesis and its potential correlation with beta-catenin. *Int. J. Colorectal Dis.* **2010**, *25*, 583–590. [CrossRef]
49. Tanese, K.; Fukuma, M.; Yamada, T.; Mori, T.; Yoshikawa, T.; Watanabe, W.; Ishiko, A.; Amagai, M.; Nishikawa, T.; Sakamoto, M. G-protein-coupled receptor gpr49 is up-regulated in basal cell carcinoma and promotes cell proliferation and tumor formation. *Am. J. Pathol.* **2008**, *173*, 835–843. [CrossRef]
50. Yamamoto, Y.; Sakamoto, M.; Fujii, G.; Tsuiji, H.; Kenetaka, K.; Asaka, M.; Hirohashi, S. Overexpression of orphan g-protein-coupled receptor, gpr49, in human hepatocellular carcinomas with beta-catenin mutations. *Hepatology* **2003**, *37*, 528–533. [CrossRef]
51. Vieira, G.C.; Chockalingam, S.; Melegh, Z.; Greenhough, A.; Malik, S.; Szemes, M.; Park, J.H.; Kaidi, A.; Zhou, L.; Catchpoole, D.; et al. Lgr5 regulates pro-survival mek/erk and proliferative wnt/beta-catenin signalling in neuroblastoma. *Oncotarget* **2015**, *6*, 40053–40067. [CrossRef] [PubMed]
52. Ciro, M.; Prosperini, E.; Quarto, M.; Grazini, U.; Walfridsson, J.; McBlane, F.; Nucifero, P.; Pacchiana, G.; Capra, M.; Christensen, J.; et al. Atad2 is a novel cofactor for myc, overexpressed and amplified in aggressive tumors. *Cancer Res.* **2009**, *69*, 8491–8498. [CrossRef]
53. Wu, G.; Lu, X.; Wang, Y.; He, H.; Meng, X.; Xia, S.; Zhen, K.; Liu, Y. Epigenetic high regulation of atad2 regulates the hh pathway in human hepatocellular carcinoma. *Int. J. Oncol.* **2014**, *45*, 351–361. [CrossRef] [PubMed]
54. Jiang, J.; Wang, J.; Yue, M.; Cai, X.; Wang, T.; Wu, C.; Su, H.; Wang, Y.; Han, M.; Zhang, Y.; et al. Direct phosphorylation and stabilization of myc by aurora b kinase promote t-cell leukemogenesis. *Cancer Cell* **2020**, *37*, 200–215 e205. [CrossRef] [PubMed]
55. Fujita, N.; Sato, S.; Tsuruo, T. Phosphorylation of p27kip1 at threonine 198 by p90 ribosomal protein s6 kinases promotes its binding to 14-3-3 and cytoplasmic localization. *J. Biol. Chem.* **2003**, *278*, 49254–49260. [CrossRef]
56. Xiao, H.; Gong, W.; Cao, J.; Li, X.; Tao, D.; Hu, J.; Gong, J. Influence of cdk1 and cdk2 sirna interference on tumor cell cycle and cell apoptosis. *Chin-Ger. J. Clin. Oncol.* **2009**, *8*, 371–374. [CrossRef]
57. Iqbal, M.A.; Chattopadhyay, S.; Siddiqui, F.A.; Ur Rehman, A.; Siddiqui, S.; Prakasam, G.; Khan, A.; Sultana, S.; Bamezai, R.N. Silibinin induces metabolic crisis in triple-negative breast cancer cells by modulating egfr-myc-txnip axis: Potential therapeutic implications. *FEBS J.* **2021**, *288*, 471–485. [CrossRef]
58. Hamilton, J.P.; Potter, J.J.; Koganti, L.; Meltzer, S.J.; Mezey, E. Effects of vitamin d3 stimulation of thioredoxin-interacting protein in hepatocellular carcinoma. *Hepatol. Res.* **2014**, *44*, 1357–1366. [CrossRef]
59. Hong, S.Y.; Yu, F.X.; Luo, Y.; Hagen, T. Oncogenic activation of the pi3k/akt pathway promotes cellular glucose uptake by downregulating the expression of thioredoxin-interacting protein. *Cell Signal.* **2016**, *28*, 377–383. [CrossRef]
60. Jiao, D.; Huan, Y.; Zheng, J.; Wei, M.; Zheng, G.; Han, D.; Wu, J.; Xi, W.; Wei, F.; Yang, A.G.; et al. Uhrf1 promotes renal cell carcinoma progression through epigenetic regulation of txnip. *Oncogene* **2019**, *38*, 5686–5699. [CrossRef]
61. Pan, X.N.; Chen, J.J.; Wang, L.X.; Xiao, R.Z.; Liu, L.L.; Fang, Z.G.; Liu, Q.; Long, Z.J.; Lin, D.J. Inhibition of c-myc overcomes cytotoxic drug resistance in acute myeloid leukemia cells by promoting differentiation. *PLoS ONE* **2014**, *9*, e105381. [CrossRef] [PubMed]
62. Nakagawara, A.; Kadomatsu, K.; Sato, S.; Kohno, K.; Takano, H.; Akazawa, K.; Nose, Y.; Kuwano, M. Inverse correlation between expression of multidrug resistance gene and n-myc oncogene in human neuroblastomas. *Cancer Res.* **1990**, *50*, 3043–3047. [PubMed]
63. Li, Y.; Zhuo, B.; Yin, Y.; Han, T.; Li, S.; Li, Z.; Wang, J. Anti-cancer effect of oncolytic adenovirus-armed shrna targeting mycn gene on doxorubicin-resistant neuroblastoma cells. *Biochem. Biophys. Res. Commun.* **2017**, *491*, 134–139. [CrossRef] [PubMed]
64. Zou, Y.; Li, W.; Zhou, J.; Zhang, J.; Huang, Y.; Wang, Z. Erk inhibitor enhances everolimus efficacy through the attenuation of dntp pools in renal cell carcinoma. *Mol. Ther. Nucleic Acids* **2019**, *14*, 550–561. [CrossRef]
65. Rahman, M.A.; Amin, A.R.; Wang, D.; Koenig, L.; Nannapaneni, S.; Chen, Z.; Wang, Z.; Sica, G.; Deng, X.; Chen, Z.G.; et al. Rrm2 regulates bcl-2 in head and neck and lung cancers: A potential target for cancer therapy. *Clin. Cancer Res.* **2013**, *19*, 3416–3428. [CrossRef]
66. Mazzu, Y.Z.; Armenia, J.; Chakraborty, G.; Yoshikawa, Y.; Coggins, S.A.; Nandakumar, S.; Gerke, T.A.; Pomerantz, M.M.; Qiu, X.; Zhao, H.; et al. A novel mechanism driving poor-prognosis prostate cancer: Overexpression of the DNA repair gene, ribonucleotide reductase small subunit m2 (rrm2). *Clin. Cancer Res.* **2019**, *25*, 4480–4492. [CrossRef]
67. Machida, T.; Fujita, T.; Ooo, M.L.; Ohira, M.; Isogai, E.; Mihara, M.; Hirato, J.; Tomotsune, D.; Hirata, T.; Fujimori, M.; et al. Increased expression of proapoptotic bmcc1, a novel gene with the bnip2 and cdc42gap homology (bch) domain, is associated with favorable prognosis in human neuroblastomas. *Oncogene* **2006**, *25*, 1931–1942. [CrossRef]
68. Tatsumi, Y.; Takano, R.; Islam, M.S.; Yokochi, T.; Itami, M.; Nakamura, Y.; Nakagawara, A. Bmcc1, which is an interacting partner of bcl2, attenuates akt activity, accompanied by apoptosis. *Cell Death Dis.* **2015**, *6*, e1607. [CrossRef]

69. Lebedev, T.; Vagapova, E.; Spirin, P.; Rubtsov, P.; Astashkova, O.; Mikheeva, A.; Sorokin, M.; Vladimirova, U.; Suntsova, M.; Konovalov, D.; et al. Growth factor signaling predicts therapy resistance mechanisms and defines neuroblastoma subtypes. *Oncogene* **2021**, *40*, 6258–6272. [CrossRef]
70. Valencia-Sama, I.; Ladumor, Y.; Kee, L.; Adderley, T.; Christopher, G.; Robinson, C.M.; Kano, Y.; Ohh, M.; Irwin, M.S. Nras status determines sensitivity to shp2 inhibitor combination therapies targeting the ras-mapk pathway in neuroblastoma. *Cancer Res.* **2020**, *80*, 3413–3423. [CrossRef]

cancers

MDPI

Article

Infectious Morbidity in Pediatric Patients Receiving Neoadjuvant Chemotherapy for Sarcoma

Denise Willmer [1], Stefan K. Zöllner [1,2], Frieder Schaumburg [3], Heribert Jürgens [1], Thomas Lehrnbecher [4] and Andreas H. Groll [1,*]

1 Infectious Disease Research Program, Center for Bone Marrow Transplantation, Department of Pediatric Hematology/Oncology, University Children's Hospital Münster, 48149 Münster, Germany; d_will06@uni-muenster.de (D.W.); stefan.zoellner@uk-essen.de (S.K.Z.); jurgh@ukmuenster.de (H.J.)
2 Department of Pediatric Oncology & Hematology, Pediatrics III, University Hospital of Essen, 45147 Essen, Germany
3 Institute of Medical Microbiology, University Hospital Münster, 48149 Münster, Germany; frieder.schaumburg@ukmuenster.de
4 Pediatric Hematology and Oncology, Hospital for Children and Adolescents, Johann Wolfgang Goethe-University, 60590 Frankfurt am Main, Germany; thomas.lehrnbecher@kgu.de
* Correspondence: grollan@ukmuenster.de; Tel.: +49-251-834-7742; Fax: +49-251-834-7828

Simple Summary: Infections are an important cause of morbidity and mortality in childhood cancer treatment. The aim of our retrospective study was to assess the infectious burden in pediatric sarcoma patients during neoadjuvant chemotherapy administered according to the EWING 2008, CWS SoTiSaR and EURAMOS clinical trial or registry. Our analyses indicate a substantial infectious morbidity in this group of patients, with 58.8% experiencing at least one episode of febrile neutropenia (FN) and 20.6% at least one microbiologically documented infection (MDI). We also identified parameters that impact on the occurrence of FN and MDIs, including treatment protocol, patient age, and mucositis. These findings may contribute to a better risk stratification for prevention and management of FN and infections as well as for maintaining quality of life, cost control, and optimum outcomes of anticancer treatment.

Citation: Willmer, D.; Zöllner, S.K.; Schaumburg, F.; Jürgens, H.; Lehrnbecher, T.; Groll, A.H. Infectious Morbidity in Pediatric Patients Receiving Neoadjuvant Chemotherapy for Sarcoma. *Cancers* **2021**, *13*, 1990. https://doi.org/10.3390/cancers13091990

Academic Editors: Saurabh Agarwal and Yang Jianhua

Received: 8 March 2021
Accepted: 19 April 2021
Published: 21 April 2021

Abstract: The purpose of this retrospective, single-center cohort study was to assess the infectious burden in pediatric sarcoma patients during neoadjuvant chemotherapy. The review included all patients with a new diagnosis of Ewing sarcoma, osteosarcoma or soft tissue sarcoma between September 2009 and December 2018 who were enrolled in the EWING 2008, CWS SoTiSaR and EURAMOS clinical trial or registry. Primary endpoints were the occurrence of febrile neutropenia (FN) and microbiologically documented infection (MDI). Parameters with a potential impact on FN and MDI were also analyzed. A total of 170 sarcoma patients (median age: 13 years, range: 0–21; 96 m/74 f) received 948 chemotherapy courses (median: 6; range: 2–8). Of these patients, 58.8% had $\geq$1 FN episode and 20.6% $\geq$ 1 MDI. FN occurred in 272/948 courses (28.7%) with fever of unknown origin (FUO) in 231 courses and 45 MDI and 19 clinically documented infections (CDI) occurring in a total of 57 courses. Patients enrolled in EWING 2008 had significantly more FN ($p < 0.001$), infections ($p = 0.02$) and MDI ($p = 0.035$). No infection-related deaths were observed. Younger age, tumor type and localization, and higher median and maximum mucositis grades were significantly associated with higher numbers of FN ($p < 0.001$), and younger age ($p = 0.024$) and higher median mucositis grade ($p = 0.017$) with MDI. The study shows substantial infectious morbidity in sarcoma patients during neoadjuvant chemotherapy treatment and opportunities to improve prevention and management.

Keywords: infections; cancer; children; bacteremia; pneumonia; sarcoma; solid tumor; outcome

1. Introduction

Febrile neutropenia and documented infections are important causes of morbidity in pediatric patients receiving chemotherapy for cancer [1,2], and optimal infectious disease supportive care in this population relies largely on a precise knowledge of their epidemiology and risk factors. Most information about febrile neutropenia and documented infections stems from studies conducted in patients undergoing treatment for hematological malignancies or allogeneic hematopoietic cell transplantation (HSCT) [3–5]. Although prior studies have suggested no substantial differences in the course or outcome of febrile neutropenia and infections between leukemia and solid tumor groups [6], very few studies have focused on febrile neutropenia and infections in pediatric patients with solid tumors, in particular sarcomas [5,7–11]. Given current approaches to antibiotic stewardship and risk-based strategies for management of febrile neutropenia, differentiating the risk for febrile neutropenia and infection for separate entities of solid tumors has become more important.

Sarcoma treatment involves the use of intensive neoadjuvant chemotherapy regimens prior to local therapy that result in transient myelosuppression, which is often complicated by fever and infection [12–16]. Although the safety assessment of induction chemotherapy in the treatment of Ewing sarcoma in the EURO-E.W.I.N.G. 99 clinical trial showed a high incidence of infections [12], little is known about the incidence, risk factors, and outcome of febrile neutropenia and infections during the neoadjuvant chemotherapy treatment phase of pediatric sarcomas. In order to contribute to a better understanding of the burden of infections in this patient population, we analyzed prevalence, risk factors, patterns and outcome of febrile neutropenia and documented infections in pediatric patients with Ewing sarcoma, soft tissue sarcoma, and osteosarcoma who received neoadjuvant chemotherapy according to internationally conducted treatment protocols.

2. Patients and Methods

2.1. Study Design

The study was a retrospective single-center cohort study of children and adolescents enrolled at the Department of Pediatric Hematology and Oncology, University Children's Hospital of Muenster, in the EWING 2008 clinical trial and registry, EURAMOS 1 clinical trial and EURAMOS-COSS registry, and CWS SoTiSaR registry treatment protocols between September 2009 and December 2018. Patients received neoadjuvant chemotherapy for Ewing sarcoma, osteosarcoma, or soft tissue sarcoma, respectively. The last follow-up was on 31 January 2020.

The Department serves an area of five million people in the Northwest of Germany and provides specialized care for the entire spectrum of pediatric hematological and oncological disorders. Each year, 140 to 160 patients with a new diagnosis of cancer, 20 to 40 patients with recurrent cancer and approximately 35 patients scheduled to undergo allogeneic HSCT are admitted, accounting for approximately 1200 hospital admissions and close to 15,000 outpatient or day clinic contacts [17].

Patients were identified through the registries of the Department's clinical research office. Inclusion criteria were: (1) medical care at the Department of Pediatric Hematology and Oncology, Muenster, Germany, (2) a diagnosis of either Ewing sarcoma, osteosarcoma, or soft tissue sarcoma, and (3) receipt of neoadjuvant chemotherapy according to either EWING 2008, CWS SoTiSaR or EURAMOS protocols. Patients not treated according to these protocols, diagnosed in Muenster but treated elsewhere or with an incomplete medical record were excluded. Patient demographics and clinical data were retrieved from the medical information system according to the study protocol by a pseudonymized standardized electronic case report form.

The primary endpoints were episodes of febrile neutropenia (FN) and microbiologically documented infections (MDI). Clinically documented infections (CDI), episodes of fever of unknown origin (FUO), unscheduled hospitalizations, treatment delay, admission to intensive care unit (ICU), and survival through January 2020 were further endpoints.

Parameters with a potential impact on FN and MDI were also analyzed. Analyses were performed both on the basis of patients and on the basis of chemotherapy courses, as appropriate.

Written informed consent for data collection and analysis was obtained within the consent procedure for cancer treatment. The study was conducted in accordance with the Declaration of Helsinki, and the protocol was reviewed and approved by the joint Ethics Committee of the University of Muenster and the Chamber of Physicians Westfalen-Lippe (Az 2017-728-f-S; 5 January 2018).

2.2. Neoadjuvant Chemotherapy Regimens

Detailed information on the neoadjuvant chemotherapy regimens can be found in the respective protocols [18–20]. In brief, the EWING 2008 clinical trial and registry (EWING 2008) protocol stipulated a total of six neoadjuvant courses of vincristine, ifosfamide, doxorubicine, etoposide (VIDE) administered in 21-day intervals depending on hematological recovery. Granulocyte colony-stimulating factor (G-CSF) was recommended, especially when the previous course was complicated by FN, but not mandatory, and MESNA was administered for prevention of urothelial toxicity. G-CSF-supported stem cell mobilization and harvesting was planned after the third course, if indicated. Local therapy was recommended to be performed 21 days after course six, depending on hematological recovery and clinical status [18].

Patients with soft tissue sarcoma enrolled in the CWS SoTiSaR registry (CWS SoTiSaR) were treated according to CWS-Guidance, which recommends courses of different combinations and dosages of actinomycin D (A), adriamycin (Ad), vincristine (V), ifosfamide (I), carboplatinum (C) and etoposide (E) according to risk stratification. G-CSF could be considered, but use was at the discretion of the attending physician. Local therapy was recommended after four courses of neoadjuvant chemotherapy in the majority of risk groups, but more courses were possible [20].

The EURAMOS 1 clinical trial and EURAMOS-COSS registry (EURAMOS) for osteosarcoma patients stipulated a neoadjuvant regimen of MAP, consisting of two cycles each of adriamycin/cisplatinum (AP) in week one plus two separate doses of high-dose methotrexate (M) in weeks 4 and 5, respectively. MAP is sometimes considered as one course, but as chemotherapy feasibility was assessed before each administration of M or AP, we considered each M and each AP as a separate course. G-CSF was recommended when a previous AP cycle had been complicated by FN or prolonged hospitalization (>7 days). Local therapy was planned after the completion of 2 MAP cycles. If needed for logistical reasons, up to two more doses of M were allowed prior to local therapy [19].

2.3. Standard of Care

Outpatients presenting with FN or signs of infection were generally hospitalized for diagnostics, treatment, and clinical observation. Patients with FN were evaluated by vital signs, physical examination, interim history, full blood count, c-reactive protein (CRP), liver and renal function, electrolytes, blood and urine cultures prior to the start of antimicrobial therapy. Further diagnostic, e.g., imaging or additional microbiological or biochemical diagnostics, was guided by clinical and laboratory findings.

Initial empirical antibacterial therapy for FN consisted of ceftazidime plus gentamycin until 2016 and was then replaced by piperacillin/tazobactam. Unstable patients were started with meropenem plus vancomycin and were subsequently de-escalated when feasible, based on microbiology laboratory reports. This regimen was also used for escalation in patients with fever persisting for more than 48–72 h or a new fever after defervescence, with or without additional empirical antifungal therapy at the discretion of the attending physician. Patients older than two years of age in stable condition with an anticipated duration of neutropenia of <7 days could be treated with ceftriaxone with the option of outpatient treatment after definite defervescence. Empirical antibacterial therapy was continued until defervescence for at least 48 h and recovery of the neutrophil count (ANC) above

500/μL. Suspected or proven infections were treated according to current management recommendations [21,22].

As standard of care, patients received a surgically implanted central venous catheter prior to the initiation of neoadjuvant chemotherapy. All patients received trimethoprim/sulfamethoxazole 8 mg/msqu twice weekly as prophylaxis for prevention of *Pneumocystis jirovecii* pneumonia, and topical polyenes or azoles for prevention of oropharyngeal candidiasis.

2.4. Definitions

Neutropenia was defined as an ANC of <500/μL or assumed when the white blood cell count (WBC) was <1000/μL. Fever was defined as a single oral temperature ≥38.3 °C or temperatures ≥38.0 °C during an 1-h-period [23]. FN was defined as every distinct episode of fever that occurred during neutropenia. FUO was defined as the presence of fever without a clinically or microbiologically identified focus. Organ infection was defined according to the International Pediatric Sepsis Consensus Conference in 2005 [24]. The diagnosis of a blood stream infection (BSI) was based on the detection of the organism in ≥1 blood culture bottle. Invasive fungal disease (IFD) was defined according to the revised definitions of invasive fungal disease by the EORTC/MSGERC consensus group 2020 [25]. Mucositis was defined according to the Common Toxicity Criteria (CTC) for adverse events version 4.03 [26]. Unscheduled hospitalization was defined as any admission to hospital not planned by the treatment protocol. Treatment delay was defined as difference in days between planned and the actual start of the next chemotherapy course. Underweight was defined as a body weight <10th percentile and overweight as a body weight >90th percentile, according to the age- and gender-adjusted German KiGGS reference percentiles [27].

2.5. Statistical Analysis

Statistical analysis was performed using Microsoft Excel and IBM SPSS Statistics (version 26, IBM, Armonk, New York, NY, USA) software. Categorial variables were analyzed by chi-square test or Fisher's exact test, where applicable. Metric variables were analyzed by Mann–Whitney U and Kruskal–Wallis test as appropriate. Parameters with potential impact on FN and MDI were analyzed by bivariate analysis; due to the small sample size and strong correlations between the respective factors in explorative analysis, multivariate analyses were not performed. Similar to the study of Haupt et al. [7], we calculated infection incidence rates for MDI per 100 person-months at risk. Overall and event-free survival were analyzed by Kaplan–Meier curves. All p-values are two-tailed and considered statistically significant at <0.05.

3. Results

A total of 195 patients were identified in the study registries, of whom 170 were included in the final analysis. Of the 25 patients excluded, 10 patients were excluded due to the administration of medical treatment at their referring hospitals, nine due to missing indication for chemotherapy, four due to individual treatment concepts and two due to an incomplete medical record.

In a further nine patients, a number of the neoadjuvant chemotherapy courses (14 in total) were not analyzed due to intermittent treatment at hospitals close to home.

3.1. Patient Demographics

The demographic and clinical characteristics of the 170 patients are shown in Table 1. Of the 170 patients, 58 (34.1%) were treated in EWING 2008, 50 (29.4%) in the CWS SoTiSaR and 62 (36.5%) in the EURAMOS trial or registry. The median age at diagnosis was 13 years (range 0–21); 96 (56.5%) of the patients were male and 74 (43.5%) were female, and the majority had a body mass index (BMI) between the 10th and the 90th percentile at baseline. Most tumors (62.4%) were localized in the extremities, and approximately one third (37.3%)

of patients had metastases at diagnosis. None of the patients had abnormal granulocyte counts at presentation, and, while the case report form did not capture steroid use, only a very few patients may have received glucocorticosteroids to reduce tumor-associated edema. During neoadjuvant chemotherapy, a total of 948 chemotherapy courses were administered (median per patient: 6, range: 2–8) (Table 1).

Table 1. Demographics and clinical characteristics of 170 patients.

Characteristic	No. (%) or Median (Range)			
	All (*n* = 170)	**EWING 2008** (*n* = 58)	**CWS SoTiSaR** (*n* = 50)	**EURAMOS** (*n* = 62)
Age	13 (0–21)	14 (0–21)	9.5 (0–18)	13.5 (2–18)
0–4 years	18 (10.6)	3 (5.2)	13 (26.0)	2 (3.2)
5–9 years	27 (15.9)	9 (15.5)	12 (24.0)	6 (9.7)
10–14 years	72 (42.4)	24 (41.4)	15 (30.0)	33 (53.2)
15–19 years	51 (30.0)	20 (34.5)	10 (20.0)	21 (33.9)
≥20 years	2 (1.2)	2 (3.4)	0	0
Sex				
male	96 (56.5)	36 (62.1)	22 (44.0)	38 (61.3)
female	74 (43.5)	22 (37.9)	28 (56.0)	24 (38.7)
BMI at baseline	18.9 (10.0–38.1)	20.0 (10.0–34.3)	17.6 (11.4–27.0)	19.1 (12.9–38.1)
BMI < P10	26 (15.3)	8 (13.8)	9 (18.0)	9 (14.5)
BMI P10–P90	126 (74.1)	41 (70.7)	38 (76.0)	47 (75.8)
BMI > P90	18 (10.6)	9 (15.5)	3 (6.0)	6 (9.7)
Underlying condition				
Ewing sarcoma	53 (31.2)	53 (91.4)	0	0
osteosarcoma	62 (36.5)	0	0	62 (100)
rhabdomyosarcoma	31 (18.2)	0	31 (62.0)	0
NOS	5 (2.9)	2 (3.4)	3 (6.0)	0
Ewing-like-sarcoma	2 (1.2)	2 (3.4)	0	0
synovial sarcoma	8 (4.7)	0	8 (16.0)	0
PNET	1 (0.6)	1 (1.7)	0	0
infantile fibrosarcoma	1 (0.6)	0	1 (2.0)	0
liposarcoma	1 (0.6)	0	1 (2.0)	0
DSRCT	1 (0.6)	0	1 (2.0)	0
MPNST	2 (1.2)	0	2 (4.0)	0
pleuropulmonary blastoma	2 (1.2)	0	2 (4.0)	0
neuroblastoma [1]	1 (0.6)	0	1 (2.0)	0
Tumor localization				
extremities	106 (62.4)	30 (51.7)	17 (34.0)	59 (95.2)
trunk	46 (27.1)	24 (41.4)	20 (40.0)	2 (3.2)
other	18 (10.6)	4 (6.9)	13 (26.0)	1 (1.6)
Any metastases at diagnosis	63 (37.3)	29 (50.0)	15 (30.0)	19 (31.1)
distant metastases	56 (33.1)	26 (44.8)	11 (22.0)	19 (31.1)
lung	30 (53.6)	12 (46.2)	3 (27.3)	15 (78.9)
bone	10 (17.9)	6 (23.1)	2 (18.2)	2 (10.5)
lymph nodes	2 (3.6)	0	2 (18.2)	0
other	2 (3.6)	1 (3.8)	1 (9.1)	0
multiple (≥2 organs)	12 (21.4)	7 (26.9)	3 (27.3)	2 (10.5)
regional lymph node infiltration	23 (13.5)	14 (24.1)	8 (16.0)	1 (1.6)
skip lesions	9 (5.3)	6 (10.5)	0	3 (4.8)

Table 1. *Cont.*

Number of chemotherapy courses [2] per patient	948 6 (2–8)	341 6 (3–7)	239 5 (2–8)	368 6 (3–8)
G-CSF administration [3]	67 (39.4)	57 (98.3)	9 (18.0)	1 (1.6)

BMI, Body mass index (kg/m^2); P10, 10th perentile and P90, 90th percentile, adapted to age and sex; NOS, not other specified sarcoma; PNET, peripheral neuroectodermal tumour; DSRCT, desmoplastic small round cell tumour; MPNST, malignant peripheral nerve sheath tumour; G-CSF, granulocyte-colony stimulating factor. [1] In one patient treated in the CWS protocol diagnosis was changed to neuroblastoma after local therapy. [2] VIDE (vincristine, ifosphamide, doxorubicin, etoposide) in 342 (36.1%); M (high-dose methotrexate) in 247 (26.1%), AP (adriamycin, cisplatinum) in 120 (12.7%); I^{2}VA (ifosphamide, vincristine, actinomycin D) in 137 (14.5%) and various other courses of CWS guidance in 102 (10.6%) courses. [3] G-CSF was administered in a total of 291 (30.7%) of the courses (41.5% of non-M courses).

3.2. *Overall Infectious and Non-Infectious Patient Morbidity*

Table 2 provides an overview of FN episodes, MDIs, CDIs, ICU-admission, weight loss, treatment delay and survival throughout neoadjuvant chemotherapy until local therapy (i.e., surgery or start of radiotherapy). The majority of the patients (n = 100; 58.8%) experienced at least one FN episode, and 63 (37.1%) experienced two or more. Comparison across the three different treatment protocols revealed significant differences in both occurrence and frequency of FN episodes with the highest FN morbidity in patients enrolled in EWING 2008, followed by those enrolled in CWS SoTiSaR and those enrolled in EURAMOS ($p < 0.001$). A similar trend across treatment protocols was observed for MDIs that occurred in a total of 20.6% of patients. Of note, there was no apparent difference in MDIs and CDIs between metastatic and non-metastatic disease. Median loss of weight during neoadjuvant chemotherapy was 7.0% of the baseline weight and the median treatment delay accounted for seven days with wide variability but no significant differences among treatment protocols for both parameters. ICU admission occurred in five patients, and all patients survived through local therapy (Table 2).

Table 2. Febrile neutropenia, microbiologically and clinically documented infections, weight loss, treatment delay and survival throughout neoadjuvant chemotherapy until local therapy in 170 patients.

Characteristic	No. (%) or Median (Range)				chi ^{2} Test
	All (n = 170)	EWING 2008 (n = 58)	CWS SoTiSaR (n = 50)	EURAMOS (n = 62)	p-Value
Febrile neutropenia					
≥1 episode of FN	100 (58.8)	49 (84.5)	32 (64.0)	19 (30.6)	<0.001
≥2 episodes of FN	63 (37.1)	40 (69.0)	17 (34.0)	6 (9.7)	<0.001
Microbiologically documented infection					
≥1 episode of MDI	35 (20.6)	17 (29.3)	10 (20.0)	8 (12.9)	0.084
≥2 episodes of MDI	6 (3.5)	4 (6.9)	1 (2.0)	1 (1.6)	0.230
Clinically documented infection					
≥1 episode of CDI	18 (10.6)	8 (13.8)	5 (10.0)	5 (8,1)	0.587
≥2 episodes of CDI	1 (0.6)	1 (1.7)	0	0	0.379
ICU admission	5 (2.9)	2 (3.4)	3 (6.0)	0	0.168
Weight loss during chemotherapy in %	7.0 (0–31.2)	7.2 (0–25.1)	4.1 (0–17.9)	8.4 (0–31.2)	
Overall treatment delay in days	7.0 (0–61)	8.5 (0–61)	4.0 (0–23)	9.0 (0–48)	
Survival through local therapy	165 (100) [1,2,3]	57 (100) [1]	49 (100) [2]	59 (100) [3]	

FN, febrile neutropenia; MDI, microbiologically documented infection; CDI, clinically documented infection; ICU, intensive care unit. [1] 1 missing value, 1 censored due treatment discontinuation. [2] 1 missing value. [3] 1 censored due to treatment discontinuation, 1 censored due to treatment modification.

3.3. Febrile Neutropenia and Infections during Chemotherapy

The episodes of febrile neutropenia and infections in the 948 chemotherapy courses are presented in Table 3. Neutropenia was observed in 519 (54.9%) of the 948 chemotherapy courses. Febrile neutropenia occurred in 272 (28.7%) courses with FUO as the final diagnosis in 231 and MDIs or CDIs in 41 courses. In an additional 16 courses, infection was diagnosed in absence of fever, and in 12 of these, the patient was non-neutropenic. In total, 64 documented infections occurred in 57 of the 948 courses (6.0%), of which 45 were microbiologically and 19 clinically documented. Significant differences in frequency between the different treatment protocols were observed for neutropenia ($p < 0.001$), febrile neutropenia ($p < 0.001$), FUO ($p < 0.001$), documented infections ($p = 0.02$) and MDI ($p = 0.035$) with the highest rates in EWING 2008, followed by CWS SoTiSaR and EURAMOS (Table 3).

Table 3. Neutropenia, febrile neutropenia, FUO, documented infections, unscheduled hospitalisation, ICU admission and mucositis in 948 courses.

Characteristic	No. (%) of Courses or Median (Range)				chi [2] Test
	All	**EWING 2008**	**CWS SoTiSaR**	**EURAMOS**	***p*-Value**
	(*n* = 948)	**(*n* = 341)**	**(*n* = 239)**	**(*n* = 368)**	
Neutropenia	519 (54.9)	321 (94.1)	157 (65.7)	41 (11.2)	<0.001
Febrile neutropenia	272 (28.7)	179 (52.5)	67 (28.0)	26 (7.1)	<0.001
FUO	231 (24.4)	152 (44.6)	55 (23.0)	24 (6.5)	<0.001
Documented infections [1]	64 (6.0 [1])	32 (8.5 [1])	18 (6.3 [1])	14 (3.5 [1])	0.020
MDIs [1]	45 (4.4 [1])	23 (6.5 [1])	13 (4.6 [1])	9 (2.4 [1])	0.035
CDIs	19 (2.0)	9 (2.6)	5 (2.1)	5 (1.4)	0.475
Unscheduled hospitalization	354 (37.4)	192 (56.3)	79 (33.2)	83 (22.6)	<0.001
duration in days	7.0 (1–23)	7.0 (2–21)	5.0 (2–21)	7.0 (1–23)	
ICU admission	5 (0.5)	2 (0.6)	3 (1.3)	0	0.112
duration in days	9.0 (2–21)	5.0 (2–8)	11.0 (8–21)		
infectious cause	2	2	0		
Mucositis					<0.001
no mucositis	662 (69.8)	109 (32.0)	225 (94.1)	328 (89.1)	
CTC grade 1–2	149 (15.7)	138 (40.5)	3 (1.3)	8 (2.2)	
CTC grade 3–4	137 (14.5)	94 (27.6)	11 (4.6)	32 (8.7)	

FUO, fever of unknown origin; MDI, microbiologically documented infection; CDI, clinically documented infection; ICU, intensive care unit; CTC, common toxicity criteria. [1] 5 courses with 2 infectious episodes, 1 course with 3 infectious episodes; in 16 courses infections occurred in absence of FN (2 EWING, 3 CWS, 11 EURAMOS); of those in 12 in absence of neutropenia (0 EWING, 1 CWS, 11 EURAMOS), please see text.

Among the MDIs, there were 21 BSI, one case of disseminated invasive aspergillosis and 23 infections at various body sites. More than one pathogen was isolated in two BSI episodes. Seventeen of the 23 isolates were Gram-positive organisms with coagulase-negative staphylococci (CoNS) as predominant isolate, and six were Gram-negative rods including two cases of extended-spectrum ß-lactamase (ESBL) producing *Klebsiella* ssp (Table 4). The most common sites for the 23 MDI organ infections were the abdomen, followed by the urogenital system and bone and soft tissues for bacterial infections and the upper respiratory tract followed by the otorhinolaryngeal system for viral infections. Among the 19 CDI organ infections, the upper respiratory tract, the abdomen and bone and soft tissues were the predominantly affected sites (Table S1).

Table 4. Overview of 45 * MDI infections and 19 CDI infections in 948 chemotherapy courses.

Infection	MDI (No.)			CDI (No.)
	Bacterial Gram+/Gram−	Fungal	Viral	
Blood stream infection	17/6 [1]	-	-	-
Systemic infection	0/0	1 [6]	1 [7]	-
Organ infection	-	-	-	-
Central nervous system	1/0 [2]	-	-	-
Oropharynx	-	-	3 [8]	2 [10]
Upper respiratory tract	-	-	7 [9]	5 [11]
Lung	-	-	-	2 [12]
Abdomen and gastrointestinal tract	7/0 [3]	-	-	4 [13]
Urogenital tract	2/2 [4]	-	-	2 [14]
Bone and soft tissues	1/3 [5]	-	-	4 [15]

MDI, microbiologically documented infection; CDI, clinically documented infection. * 4 infections with multiple pathogens. [1] for detail, consult Supplementary Materials, Table S1. [2] *Staphylococcus* hominis. [3] *Clostridioides difficile* colitis (6), *enteropathogenic Staphylococcus aureus* enteritis (1). [4] *Enterococcus faecalis* (2), *Pseudomonas aeruginosa* (1), *Enterobacter cloacae* (1). [5] wound infection with *Enterococcus faecalis*, *Pseudomonas luteola* and *Bacteroides sp.* (1), axillary abscess with *Escherichia coli* (1). [6] invasive pulmonary aspergillosis with dissemination in liver and spleen (1). [7] *primary Epstein-Barr virus infection*. [8] *Herpes simplex virus stomatitis* (3). [9] *human Metapneumovirus* (1), *Influenza A virus* (1), *Parainfluenza 2 virus* (1), *Rhinovirus* (1), *Enterovirus* (1), *Respiratory syncycial virus* (2). [10] tonsillitis (1), dental abscess (1). [11] unspecific upper respiratory tract infections (5). [12] pneumonia (2; 1 with respiratory insufficiency). [13] gastroenteritis (1), colitis (2), migratory peritonitis (1). [14] urinary tract infections (2; 1 with transurethral indwelling catheter). [15] port catheter site infection (1), panaritium (3); without documented microbacterial cause.

The overall infection rate for MDIs was 7.2 per 100 person-months at risk; the rates for patients treated for Ewing sarcoma, soft tissue sarcoma and osteosarcoma were 9.5, 7.0, and 4.6, respectively. The overall rate for BSI/IFD was 3.5 per 100 person-month at risk and 6.6, 2.7 and 0.5 for the respective entities. Of note, no infection-related deaths occurred during neoadjuvant chemotherapy.

3.4. Hospitalization, ICU Admission and Mucositis

Unscheduled hospitalizations for treatment-related adverse events occurred in 37.4% (354/948) of all courses with a median duration of seven days and were significantly more often in EWING 2008 than in CWS SoTiSaR than in EURAMOS ($p < 0.001$). In five courses (0.5%), patients were admitted to the ICU for a median duration of nine days. Admission was for infectious causes in two cases only. As marker of cytotoxicity, CTC grade 1–2 mucositis was recorded in 149 (15.7%) and CTC grade 3–4 in 137 (14.5%) courses. There were significant differences between treatment protocols with higher rates of CTC grade 3–4 mucositis in EWING 2008 than in EURAMOS than in CWS SoTiSaR (Table 3).

3.5. Factors Associated with FN and MDI

Younger age was significantly associated with higher number of FN episodes (weak association, Tau $b = -0.208$; $p < 0.001$). Higher number of FN episodes was also significantly associated with longer duration of unscheduled hospitalization (strong association, Tau $b = 0.665$, $p < 0.001$). Significant associations were also found for treatment regimen ($p < 0.001$; EWING 2008 > CWS SoTiSaR > EURAMOS), tumor localization ($p < 0.001$; trunk > extremities), and maximal and median mucositis grade ($p < 0.001$; mucositis grade 3–4 > grade 1–2 > no mucositis) during treatment. No associations were found for sex, metastases, weight at diagnosis, overall treatment delay and weight loss during treatment. Due to its predominant use in patients enrolled in EWING 2008, the effects of G-CSF were not included in the bivariate analyses (Table S2).

Significant associations with the occurrence of MDI by bivariate analysis were found for younger age ($p = 0.024$) and maximal and median mucositis grade ($p = 0.003$ and $p = 0.017$, respectively) with CTC grade 3–4 > CTC grade 1–2 > no mucositis. Presence of MDI was also significantly associated with longer treatment delay ($p = 0.006$) and longer duration of unscheduled hospitalization ($p = 0.001$). No association was found for sex, treatment regimen, tumor localization, metastases, weight at diagnosis and weight loss during therapy (Table S3).

Due to methodological limitations (small sample size and strong correlations between the respective factors in explorative analysis), we decided to abstain from performing multivariate analyses to further substantiate the statistical associations obtained by bivariate analyses.

3.6. Overall and Event-Free Survival

After a median duration of follow-up of 4.23 years (range 0.3–10.5), overall cumulative survival at last follow-up one year after the inclusion of the last patient was 68.8% (63.4%, 71.3%, and 70.6% for patients receiving treatment on EWING 2008, CWS SoTiSaR, and EURAMOS, respectively).

The cumulative event-free survival at last follow-up was 59.2% (55.8%, 59.4%, and 62.1% for patients receiving treatment on EWING 2008, CWS SoTiSaR, and EURAMOS, respectively).

4. Discussion

Apart from a recent publication on infectious complications in children with malignant bone tumors [11], detailed information on prevalence, risk factors, and outcome of febrile neutropenia and infections in pediatric patients undergoing chemotherapy for sarcoma is scarce. The results presented here document substantial infectious morbidity during the neoadjuvant chemotherapy treatment phase. The majority of the 170 patients (58.8%) experienced at least one FN episode, and 20.6% had at least one MDI. Febrile neutropenia occurred in 272 (28.7%) of the 948 treatment courses, and in 57 (6.0%) courses, a total of 45 microbiologically and 19 clinically documented infections were recorded.

The overall prevalence of FN in our cohort (58.8%) corresponds to data reported by others for pediatric solid tumors [9,28,29]., Patients enrolled in EWING 2008 had a significantly higher prevalence of FN (84.5%) than those enrolled in CWS SoTiSaR and EURAMOS, respectively (Table 3). This rate is also higher than that previously reported in unselected patients with solid tumors [8]. While the relative distribution of FUO and MDI or CDI is overall comparable to that reported in the literature [6,30,31], there was a significantly higher proportion of documented infections in patients enrolled in EWING 2008 relative to patients enrolled in CWS SoTiSaR and EURAMOS.

We observed an overall incidence rate of 7.2 MDI per 100 person-months at risk and of 3.5 BSI/IFD per 100 person-months at risk. These rates are similar to those reported by Haupt et al. who noted an infection rate of 3.2 per 100 person-months at risk for BSI/IFD in their cohort of solid tumors and a rate of 3.4 per 100 person-months at risk for sarcoma patients [7]; these are also similar to data presented by Calton et al. who reported 3.96 BSI per 100 person-months at risk for solid tumors [10]. However, when analyzed by tumor type, we found numerically higher infection rates for patients with Ewing sarcoma (9.5 MDI and 6,6 BSI/IFD per 100 person-months at risk) and lower rates for osteosarcoma patients (4.6 MDI and 0.5 BSI/IFD per 100 person-months at risk), while rates for soft tissue sarcoma patients were similar to the overall infection rate (7.0 MDI and 2,7 BSI/IFD per 100 person-months at risk). A recent retrospective multicenter survey also showed a high risk for infectious complications in patients treated for malignant bone and soft tissue tumors, particular for patients with Ewing sarcoma, during neoadjuvant and adjuvant chemotherapy with higher cumulative incidence rates as in patients with acute lymphoblastic leukemia, non-Hodgkin's lymphoma and Hodgkin's disease, and comparable incidence rates to those in patients with acute myeloid leukemia [11]. A study conducted at our center in patients undergoing autologous hematopoietic cell rescue for solid tumors and lymphoma showed an occurrence of non-fungal MDI in 16.5% of the patients [32]. In the analysis presented here, MDI occurred in 20.6% of the patients overall and in 29.3%, 20.0% and 12.9% of patients enrolled in EWING 2008, CWS SoTiSaR and EURAMOS, respectively. Compared to patients undergoing dose intensive chemotherapy with autologous hematopoietic stem cell rescue, who are generally considered to be at high risk of infection, the data from the present study show a comparable, and, in the case of Ewing sarcoma patients, an even higher risk of MDI. However, our findings stand in contrast to the results of a meta-analysis of global individual participant data regarding the prediction of MDI for FN episodes. By univariate analysis, a diagnosis of Ewing sarcoma and osteosarcoma was associated with a decreased risk of MDI [33].

The differences in FN and MDI between the different sarcoma types and treatments and also, tumor localization, may be explained by the differential dose-intensity and mucosal toxicity of the respective regimens [6,30,34], with VIDE combination therapy for Ewing sarcoma being particularly intense [12,14]. Multiagent combination chemotherapy, as used in EWING 2008 and CWS SoTiSaR, may also contribute [35]. An analysis of the potential influence of different chemotherapy courses or individual chemotherapy agents and their cumulative dosages on FN and MDI rates might have provided further associations but was not possible due to methodological limitations. Considering that patients treated for Ewing and soft tissue sarcoma have higher infection rates than the overall population of pediatric solid tumors, and that these rates are comparable to those observed in patients with hematological malignancies, the outcome in our study was favorable, with no infection-related death during neoadjuvant chemotherapy, and only two patients requiring intensive care unit admission for infection-related supportive care.

The spectrum of bacterial species in the 21 episodes of blood stream infection was predominated by CoNS (9/21, Table S1), which is in line with other reports [36]. However, the detection of CoNS raises the question of whether it reflects a true infection or contamination. There is good reason to assume that the detection of CoNS corresponds to true infection as the majority were *S. epidermidis* mainly found in neutropenic patients, both factors that have been associated with an increased likelihood of true infection [37]. Similar to other studies in patients with solid tumors, bone tumors, and solid tumors with autologous hematopoietic stem cell rescue [7,11,32], we observed a low frequency of fungemia and other forms of invasive fungal diseases in sarcoma patients. The only IFD in our cohort was a case of proven pulmonary and disseminated invasive aspergillosis in a patient with Ewing sarcoma following a period of profound granulocytopenia exceeding a duration of twenty days.

Apart from younger age, bivariate analysis identified an association of mucositis with FN and MDI: Higher grades of mucositis correlated with more FN episodes and higher infection rates. These findings are in concordance with the results of other studies that found associations of mucositis with FN [8,38] and severe infections [39] in bivariate analysis. Indeed, mucositis is the result of a complex series of biological events involving inflammation, alteration of the local tissue response, apoptosis of the basal cell layer and loss of mucosal integrity that facilitates migration of mucosal microorganisms and makes the individual prone to invasive bacterial, fungal and viral infection [40–44]. Patients with FN and mucositis are clearly at higher risk of infection, and the presence of mucositis should be considered in approaches to stratify patient management in low and high-risk categories. In addition, strategies to avoid or ameliorate mucositis and the development of treatments that do not cause mucositis should be prioritized in order to reduce infection risks.

Higher numbers of FN and the presence of MDIs were significantly associated with longer duration of unscheduled hospitalization. Longer duration of hospitalization reflects the burden of therapy-induced morbidity and accounts for important use of health care resources. There also was a significant association of MDIs with longer treatment delay, which in turn may contribute to worse event-free survival in the affected patients as has been demonstrated by others for osteosarcoma patients [45,46].

5. Conclusions

In conclusion, the data presented here document substantial infectious morbidity in sarcoma patients during the neoadjuvant chemotherapy treatment phase. Prevention and appropriate management of FN and infections are essential to maintain quality of life and cost control, and to avoid treatment delays that may compromise the outcomes of anticancer treatment.

Supplementary Materials: The following are available online at https://www.mdpi.com/article/10.3390/cancers13091990/s1, Table S1: Overview of 21 blood stream infections in 19 patients. Table S2: Bivariate analysis of potential factors associated to the number of FN episodes during neoadjuvant chemotherapy in 170 patients. Table S3: Bivariate analysis of potential factors associated with development of ≥ MDI episode during neoadjuvant chemotherapy in 170 patients.

Author Contributions: D.W., T.L. and A.H.G. conceived the concept of the study and developed the study protocol; D.W., S.K.Z. and F.S. accrued the data; D.W. and A.H.G. wrote the manuscript draft. D.W., A.H.G., T.L., S.K.Z., F.S. and H.J. revised the manuscript and provided final approval. All authors have read and agreed to the published version of the manuscript.

Funding: This research received no external funding.

Institutional Review Board Statement: The study was conducted in accordance with the Declaration of Helsinki, and the protocol was reviewed and approved by the joint ethics committee of the University of Muenster and the Chamber of Physicians Westfalen-Lippe (Az 2017-728-f-S; 5 January 2018).

Informed Consent Statement: Written informed consent for data collection and analysis was obtained within the consent procedure for cancer treatment.

Data Availability Statement: The data presented in this study are available on request from the corresponding author.

Acknowledgments: The authors are indebted to Ulrike von Hehn for consulting and performing part of the statistical evaluation of the data. The authors thank Christine Jürgens for reviewing and editing English grammar and style as a native speaker.

Conflicts of Interest: The authors declare no conflict of interest.

References

1. Ammann, R.A.; Tissing, W.J.E.; Phillips, B. Rationalizing the approach to children with fever in neutropenia. *Curr. Opin. Infect. Dis.* **2012**, *25*, 258–265. [CrossRef]
2. Bodey, G.P.; Buckley, M.; Sathe, Y.S.; Freireich, E.J. Quantitative relationships between circulating leukocytes and infection in patients with acute leukemia. *Ann. Intern. Med.* **1966**, *64*, 328–340. [CrossRef]
3. Katsimpardi, K.; Papadakis, V.; Pangalis, A.; Parcharidou, A.; Panagiotou, J.P.; Soutis, M.; Papandreou, E.; Polychronopoulou, S.; Haidas, S. Infections in a pediatric patient cohort with acute lymphoblastic leukemia during the entire course of treatment. *Support. Care Cancer* **2006**, *14*, 277–284. [CrossRef]
4. Srinivasan, A.; McLaughlin, L.; Wang, C.; Srivastava, D.K.; Shook, D.R.; Leung, W.; Hayden, R.T. Early infections after autologous hematopoietic stem cell transplantation in children and adolescents: The St. Jude experience. *Transpl. Infect. Dis.* **2014**, *16*, 90–97. [CrossRef] [PubMed]
5. Choi, Y.B.; Yi, E.S.; Kang, J.-M.; Lee, J.W.; Yoo, K.H.; Kim, Y.-J.; Sung, K.W.; Koo, H.H. Infectious Complications during Tandem High-Dose Chemotherapy and Autologous Stem Cell Transplantation for Children with High-Risk or Recurrent Solid Tumors. *PLoS ONE* **2016**, *11*, e0162178. [CrossRef] [PubMed]
6. Koçak, U.; Rolston, K.V.I.; Mullen, C.A. Fever and neutropenia in children with solid tumors is similar in severity and outcome to that in children with leukemia. *Support. Care Cancer* **2002**, *10*, 58–64. [CrossRef] [PubMed]
7. Haupt, R.; Romanengo, M.; Fears, T.R.; Viscoli, C.; Castagnola, E. Incidence of septicaemias and invasive mycoses in children undergoing treatment for solid tumours: A 12-year experience at a single Italian institution. *Eur. J. Cancer* **2001**, *37*, 2413–2419. [CrossRef]
8. Castelán-Martínez, O.D.; Rodríguez-Islas, F.; Vargas-Neri, J.L.; Palomo-Colli, M.A.; López-Aguilar, E.; Clark, P.; Castañeda-Hernádez, G.; Rivas-Ruiz, R. Risk Factors for Febrile Neutropenia in Children With Solid Tumors Treated With Cisplatin-based Chemotherapy. *J. Pediatr. Hematol. Oncol.* **2016**, *38*, 191–196. [CrossRef]
9. Garrido, M.M.; Garrido, R.Q.; Cunha, T.N.; Ehrlich, S.; Martins, I.S. Comparison of epidemiological, clinical and microbiological characteristics of bloodstream infection in children with solid tumours and haematological malignancies. *Epidemiol. Infect.* **2019**, *147*, e298. [CrossRef]
10. Calton, E.A.; Doaré, K.L.; Appleby, G.; Chisholm, J.C.; Sharland, M.; Ladhani, S.N. Invasive bacterial and fungal infections in paediatric patients with cancer: Incidence, risk factors, aetiology and outcomes in a UK regional cohort 2009–2011. *Pediatr. Blood Cancer* **2014**, *61*, 1239–1245. [CrossRef]

11. Czyzewski, K.; Galazka, P.; Zalas-Wiecek, P.; Gryniewicz-Kwiatkowska, O.; Gietka, A.; Semczuk, K.; Chelmecka-Wiktorczyk, L.; Zak, I.; Salamonowicz, M.; Fraczkiewicz, J.; et al. Infectious complications in children with malignant bone tumors: A multicenter nationwide study. *Infect. Drug Resist.* **2019**, *12*, 1471–1480. [CrossRef] [PubMed]
12. Juergens, C.; Weston, C.; Lewis, I.; Whelan, J.; Paulussen, M.; Oberlin, O.; Michon, J.; Zoubek, A.; Juergens, H.; Craft, A. Safety assessment of intensive induction with vincristine, ifosfamide, doxorubicin, and etoposide (VIDE) in the treatment of Ewing tumors in the EURO-E.W.I.N.G. 99 clinical trial. *Pediatr. Blood Cancer* **2006**, *47*, 22–29. [CrossRef]
13. Whelan, J.S.; Bielack, S.S.; Marina, N.; Smeland, S.; Jovic, G.; Hook, J.M.; Krailo, M.; Anninga, J.; Butterfass-Bahloul, T.; Böhling, T.; et al. EURAMOS-1, an international randomised study for osteosarcoma: Results from pre-randomisation treatment. *Ann. Oncol.* **2015**, *26*, 407–414. [CrossRef]
14. Strauss, S.J.; McTiernan, A.; Driver, D.; Hall-Craggs, M.; Sandison, A.; Cassoni, A.M.; Kilby, A.; Michelagnoli, M.; Pringle, J.; Cobb, J.; et al. Single center experience of a new intensive induction therapy for ewing's family of tumors: Feasibility, toxicity, and stem cell mobilization properties. *J. Clin. Oncol.* **2003**, *21*, 2974–2981. [CrossRef] [PubMed]
15. Gupta, A.A.; Anderson, J.R.; Pappo, A.S.; Spunt, S.L.; Dasgupta, R.; Indelicato, D.J.; Hawkins, D.S. Patterns of chemotherapy-induced toxicities in younger children and adolescents with rhabdomyosarcoma: A report from the Children's Oncology Group Soft Tissue Sarcoma Committee. *Cancer* **2012**, *118*, 1130–1137. [CrossRef]
16. Paulussen, M.; Fröhlich, B.; Jürgens, H. Ewing tumour: Incidence, prognosis and treatment options. *Paediatr. Drugs* **2001**, *3*, 899–913. [CrossRef] [PubMed]
17. Universitätsklinikum Münster: Klinik für Kinder-und Jugendmedizin—Pädiatrische Hämatologie und Onkologie. Klinik in Zahlen. Available online: https://www.ukm.de/index.php?id=4372 (accessed on 18 December 2020).
18. EU Clinical Trials Register. EWING 2008: EduraCT Number 2008-003658-13. Available online: https://www.clinicaltrialsregister.eu/ctr-search/trial/2008-003658-13/DE (accessed on 30 December 2020).
19. EU Clinical Trials Register. EURAMOS 1: EduraCT Number 2004-000242-20. Available online: https://www.clinicaltrialsregister.eu/ctr-search/trial/2004-000242-20/GB (accessed on 30 December 2020).
20. Cooperative Weichteilsarkom Studiengruppe der GPOH. Registry for Soft Tissue Sarcoma and Other Soft Tissue Tumours in Children, Adolescents, and Young Adults: CWS-SoTiSaR. Available online: https://www.kinderkrebsinfo.de/health_professionals/clinical_trials/pohkinderkrebsinfotherapiestudien/cws_sotisar/index_eng.html (accessed on 19 January 2021).
21. *DGPI-Handbuch: Infektionen bei Kindern und Jugendlichen*; Berner, R.; Bialek, R.; Forster, J.; Härtel, C.; Heininger, U.; Huppertz, H.-I.; Liese, J.G.; Nadal, D.; Simon, A. (Eds.) 7., Vollständig Überarbeitete Auflage; Georg Thieme Verlag: Stuttgart, NY, USA, 2018; ISBN 3132407909.
22. Deutsche Gesellschaft für Pädiatrische Infektiologie (DGPI), Gesellschaft für Pädiatrische Onkologie und Hämatologie. AWMF S2K Leitlinie—Diagnostik und Therapie bei Kindern mit Onkologischer Grunderkrankung, Fieber und Granulozytopenie (mit Febriler Neutropenie) Außerhalb der Allogenen Stammzelltransplantation: AWMF-Registernummer 048/14. Available online: https://www.awmf.org/leitlinien/detail/ll/048-014.html (accessed on 28 December 2020).
23. Freifeld, A.G.; Bow, E.J.; Sepkowitz, K.A.; Boeckh, M.J.; Ito, J.I.; Mullen, C.A.; Raad, I.I.; Rolston, K.V.; Young, J.-A.H.; Wingard, J.R. Clinical practice guideline for the use of antimicrobial agents in neutropenic patients with cancer: 2010 update by the infectious diseases society of america. *Clin. Infect. Dis.* **2011**, *52*, e56–e93. [CrossRef]
24. Goldstein, B.; Giroir, B.; Randolph, A. International pediatric sepsis consensus conference: Definitions for sepsis and organ dysfunction in pediatrics. *Pediatr. Crit. Care Med.* **2005**, *6*, 2–8. [CrossRef]
25. Donnelly, J.P.; Chen, S.C.; Kauffman, C.A.; Steinbach, W.J.; Baddley, J.W.; Verweij, P.E.; Clancy, C.J.; Wingard, J.R.; Lockhart, S.R.; Groll, A.H.; et al. Revision and Update of the Consensus Definitions of Invasive Fungal Disease From the European Organization for Research and Treatment of Cancer and the Mycoses Study Group Education and Research Consortium. *Clin. Infect. Dis.* **2020**, *71*, 1367–1376. [CrossRef]
26. National Cancer Institute. Common Terminology Criteria for Adverse Events (CTCAE). Available online: https://ctep.cancer.gov/protocolDevelopment/electronic_applications/ctc.htm#ctc_40 (accessed on 4 April 2018).
27. Robert Koch-Institut—Referenzperzentile für Anthropometrische Maßzahlen und Blutdruck aus der Studie zur Gesundheit von Kindern und Jugendlichen in Deutschland (KiGGS). Available online: https://www.rki.de/DE/Content/Gesundheitsmonitoring/Gesundheitsberichterstattung/GBEDownloadsB/KiGGS_Referenzperzentile.html (accessed on 24 November 2020).
28. Ammann, R.A.; Aebi, C.; Hirt, A.; Ridolfi Lüthy, A. Fever in neutropenia in children and adolescents: Evolution over time of main characteristics in a single center, 1993–2001. *Support. Care Cancer* **2004**, *12*, 826–832. [CrossRef] [PubMed]
29. Alexander, S.W.; Wade, K.C.; Hibberd, P.L.; Parsons, S.K. Evaluation of Risk Prediction Criteria for Episodes of Febrile Neutropenia in Children with Cancer. *J. Pediatr. Hematol. Oncol.* **2002**, *24*, 38–42. [CrossRef]
30. Castagnola, E.; Fontana, V.; Caviglia, I.; Caruso, S.; Faraci, M.; Fioredda, F.; Garrè, M.L.; Moroni, C.; Conte, M.; Losurdo, G.; et al. A prospective study on the epidemiology of febrile episodes during chemotherapy-induced neutropenia in children with cancer or after hemopoietic stem cell transplantation. *Clin. Infect. Dis.* **2007**, *45*, 1296–1304. [CrossRef]
31. Agyeman, P.; Aebi, C.; Hirt, A.; Niggli, F.K.; Nadal, D.; Simon, A.; Ozsahin, H.; Kontny, U.; Kühne, T.; Beck Popovic, M.; et al. Predicting bacteremia in children with cancer and fever in chemotherapy-induced neutropenia: Results of the prospective multicenter SPOG 2003 FN study. *Pediatr. Infect. Dis. J.* **2011**, *30*, e114–e119. [CrossRef]

32. Linke, C.; Tragiannidis, A.; Ahlmann, M.; Fröhlich, B.; Wältermann, M.; Burkhardt, B.; Rossig, C.; Groll, A.H. Epidemiology and management burden of invasive fungal infections after autologous hematopoietic stem cell transplantation: 10-year experience at a European Pediatric Cancer Center. *Mycoses* **2019**, *62*, 954–960. [CrossRef]
33. Phillips, R.S.; Sung, L.; Ammann, R.A.; Riley, R.D.; Castagnola, E.; Haeusler, G.M.; Klaassen, R.; Tissing, W.J.E.; Lehrnbecher, T.; Chisholm, J.; et al. Predicting microbiologically defined infection in febrile neutropenic episodes in children: Global individual participant data multivariable meta-analysis. *Br. J. Cancer* **2016**, *114*, 623–630. [CrossRef]
34. Bagnasco, F.; Haupt, R.; Fontana, V.; Valsecchi, M.G.; Rebora, P.; Caviglia, I.; Caruso, S.; Castagnola, E. Risk of repeated febrile episodes during chemotherapy-induced granulocytopenia in children with cancer: A prospective single center study. *J. Chemother.* **2012**, *24*, 155–160. [CrossRef] [PubMed]
35. Ouyang, Z.; Peng, D.; Dhakal, D.P. Risk factors for hematological toxicity of chemotherapy for bone and soft tissue sarcoma. *Oncol. Lett.* **2013**, *5*, 1736–1740. [CrossRef] [PubMed]
36. Simon, A.; Furtwängler, R.; Graf, N.; Laws, H.J.; Voigt, S.; Piening, B.; Geffers, C.; Agyeman, P.; Ammann, R.A. Surveillance of bloodstream infections in pediatric cancer centers—what have we learned and how do we move on? *GMS Hyg. Infect. Control* **2016**, *11*, Doc11. [CrossRef] [PubMed]
37. García-Vázquez, E.; Fernández-Rufete, A.; Hernández-Torres, A.; Canteras, M.; Ruiz, J.; Gómez, J. When is coagulase-negative Staphylococcus bacteraemia clinically significant? *Scand. J. Infect. Dis.* **2013**, *45*, 664–671. [CrossRef]
38. Rondinelli, P.I.P.; Ribeiro, K.d.C.B.; de Camargo, B. A proposed score for predicting severe infection complications in children with chemotherapy-induced febrile neutropenia. *J. Pediatr. Hematol. Oncol.* **2006**, *28*, 665–670. [CrossRef]
39. Badiei, Z.; Khalesi, M.; Alami, M.H.; Kianifar, H.R.; Banihashem, A.; Farhangi, H.; Razavi, A.R. Risk factors associated with life-threatening infections in children with febrile neutropenia: A data mining approach. *J. Pediatr. Hematol. Oncol.* **2011**, *33*, e9–e12. [CrossRef]
40. Al-Dasooqi, N.; Sonis, S.T.; Bowen, J.M.; Bateman, E.; Blijlevens, N.; Gibson, R.J.; Logan, R.M.; Nair, R.G.; Stringer, A.M.; Yazbeck, R.; et al. Emerging evidence on the pathobiology of mucositis. *Support. Care Cancer* **2013**, *21*, 3233–3241. [CrossRef]
41. Sobue, T.; Bertolini, M.; Thompson, A.; Peterson, D.E.; Diaz, P.I.; Dongari-Bagtzoglou, A. Chemotherapy-induced oral mucositis and associated infections in a novel organotypic model. *Mol. Oral Microbiol.* **2018**, *33*, 212–223. [CrossRef]
42. Villa, A.; Sonis, S.T. Mucositis: Pathobiology and management. *Curr. Opin. Oncol.* **2015**, *27*, 159–164. [CrossRef] [PubMed]
43. Hong, J.; Park, H.-K.; Park, S.; Lee, A.; Lee, Y.-H.; Shin, D.-Y.; Koh, Y.; Choi, J.-Y.; Yoon, S.-S.; Choi, Y.; et al. Strong association between herpes simplex virus-1 and chemotherapy-induced oral mucositis in patients with hematologic malignancies. *Korean J. Intern. Med.* **2020**, *35*, 1188–1198. [CrossRef] [PubMed]
44. Katagiri, H.; Fukui, K.; Nakamura, K.; Tanaka, A. Systemic hematogenous dissemination of mouse oral candidiasis is induced by oral mucositis. *Odontology* **2018**, *106*, 389–397. [CrossRef] [PubMed]
45. Vasquez, L.; Silva, J.; Chavez, S.; Zapata, A.; Diaz, R.; Tarrillo, F.; Maza, I.; Sialer, L.; García, J. Prognostic impact of diagnostic and treatment delays in children with osteosarcoma. *Pediatr. Blood Cancer* **2020**, *67*, e28180. [CrossRef] [PubMed]
46. Abou Ali, B.; Salman, M.; Ghanem, K.M.; Boulos, F.; Haidar, R.; Saghieh, S.; Akel, S.; Muwakkit, S.A.; El-Solh, H.; Saab, R.; et al. Clinical Prognostic Factors and Outcome in Pediatric Osteosarcoma: Effect of Delay in Local Control and Degree of Necrosis in a Multidisciplinary Setting in Lebanon. *J. Glob. Oncol.* **2019**, *5*, 1–8. [CrossRef] [PubMed]

cancers

Article

Preliminary Radiogenomic Evidence for the Prediction of Metastasis and Chemotherapy Response in Pediatric Patients with Osteosarcoma Using ^{18}F-FDG PET/CT, *EZRIN*, and *KI67*

Byung-Chul Kim [1,†], Jingyu Kim [2,†], Kangsan Kim [3], Byung Hyun Byun [1], Ilhan Lim [1], Chang-Bae Kong [4], Won Seok Song [4], Jae-Soo Koh [5] and Sang-Keun Woo [2,3,*]

1 Department of Nuclear Medicine, Korea Institute of Radiological and Medical Sciences, Seoul 01812, Korea; xikian@kirams.re.kr (B.-C.K.); nmbbh@kirams.re.kr (B.H.B.); ilhan@kirams.re.kr (I.L.)
2 Radiological & Medico-Oncological Sciences, University of Science & Technology, Seoul 34113, Korea; jingyu8754@kirams.re.kr
3 Division of Applied RI, Korea Institute of Radiological and Medical Science, Seoul 01812, Korea; krmount@kirams.re.kr
4 Department of Orthopaedic Surgery, Seoul National University Hospital, Seoul 03080, Korea; cbkongmd@gmail.com (C.-B.K.); wssongmd@gmail.com (W.S.S.)
5 Department of Pathology, Korea Institute of Radiological and Medical Sciences, Seoul 01812, Korea; jskoh@kirams.re.kr
* Correspondence: skwoo@kirams.re.kr; Tel.: +82-2-970-1659
† These authors equally contributed to this work.

Simple Summary: Pediatric osteosarcoma is one of the most aggressive cancers, and predictions of metastasis and chemotherapy response have a significant impact on pediatric patient survival. Radiogenomics, as methods of analyzing gene expression or image texture features, have previously been used for the diagnosis of chemotherapy responses and metastasis and can reveal the current state of cancer. In this study, we aimed to generate a predictive model using gene expression and ^{18}F-FDG PET/CT image texture features in pediatric osteosarcoma in relation to metastasis and chemotherapy response. A predictive model using radiogenomics technology that incorporates both imaging features and gene expression can accurately predict metastasis and chemotherapy responses to improve patient outcomes.

Abstract: Chemotherapy response and metastasis prediction play important roles in the treatment of pediatric osteosarcoma, which is prone to metastasis and has a high mortality rate. This study aimed to estimate the prediction model using gene expression and image texture features. ^{18}F-fluorodeoxyglucose positron emission tomography/computed tomography (^{18}F-FDG PET/CT) images of 52 pediatric osteosarcoma patients were used to estimate the machine learning algorithm. An appropriate algorithm was selected by estimating the machine learning accuracy. ^{18}F-FDG PET/CT images of 21 patients were selected for prediction model development based on simultaneous *KI67* and *EZRIN* expression. The prediction model for chemotherapy response and metastasis was estimated using area under the curve (AUC) maximum image texture features (AUC_max) and gene expression. The machine learning algorithm with the highest test accuracy in chemotherapy response and metastasis was selected using the random forest algorithm. The chemotherapy response and metastasis test accuracy with image texture features was 0.83 and 0.76, respectively. The highest test accuracy and AUC of chemotherapy response with AUC_max, *KI67*, and *EZRIN* were estimated to be 0.85 and 0.89, respectively. The highest test accuracy and AUC of metastasis with AUC_max, *KI67*, and *EZRIN* were estimated to be 0.85 and 0.8, respectively. The metastasis prediction accuracy increased by 10% using radiogenomics data.

Keywords: *KI67*; *EZRIN*; ^{18}F-FDG PET/CT; random forest; radiogenomics; chemotherapy response; metastasis

Citation: Kim, B.-C.; Kim, J.; Kim, K.; Byun, B.H.; Lim, I.; Kong, C.-B.; Song, W.S.; Koh, J.-S.; Woo, S.-K. Preliminary Radiogenomic Evidence for the Prediction of Metastasis and Chemotherapy Response in Pediatric Patients with Osteosarcoma Using ^{18}F-FDG PET/CT, *EZRIN*, and *KI67*. *Cancers* **2021**, *13*, 2671. https://doi.org/10.3390/cancers13112671

Academic Editors: Saurabh Agarwal and Yang Jianhua

Received: 7 May 2021
Accepted: 26 May 2021
Published: 28 May 2021

1. Introduction

Pediatric osteosarcoma is well-known as one of the most aggressive cancers [1]. Predictions of metastasis and chemotherapy response have a significant impact on pediatric patient survival because metastasis progresses rapidly, and treatment is difficult after the progression of metastasis in pediatric osteosarcoma [2,3]. Chemotherapy responses and cancer metastasis have a profound relationship with gene expression, and the current state of cancer can be identified and predicted by analyzing changes in gene expression [4,5]. Methods of analyzing gene expression or image texture features have previously been used for the diagnosis of chemotherapy responses and metastasis [6–8]. *KI67* is a well-known cancer metastasis marker [9]. It is mainly used to indicate cancer metastasis because it is primarily involved in cell division, an important function of metastasis, and an increase in expression was observed when metastasis was actively progressing and the number of cancer cells increased. *KI67* overexpression has been identified in pediatric osteosarcoma [10]. It has also been used as a marker of chemotherapy response [11]. *EZRIN* is a protein constituting the ERM (*EZRIN*-radixin-moesin) family that exists on the cell surface. *EZRIN* plays many roles, including acting as a signaling tube between metastasis-related cell surface molecules and signaling components. Similar to *KI67*, *EZRIN* has been used as a marker for cancer metastasis and chemotherapy response [12,13]. *EZRIN* expression provides an early survival advantage for cancer cells and plays an important role in the invasion of other tissues in pediatric osteosarcoma [14].

Nuclear medicine images, such as positron emission tomography/computed tomography (PET/CT), have also been used to analyze the results of metastasis and chemotherapy responses [15]. The phenotype of cancerous tissues from images or text image features obtained through image analysis can be used to analyze the results of chemotherapy response or cancer metastasis. The combination of genetic expression analysis and nuclear imaging texture features has been used to analyze pre-chemotherapy or chemotherapy responses and metastasis [16,17]. This can be done by observing the phenotype and postponing the change in a given gene because genetic changes in cells lead to changes in the phenotype. A predictive model was estimated using image texture features obtained by analyzing PET/CT images with machine and deep learning [18,19]. In a recent study, the associations between image texture features from tumor ^{18}F-FDG PET/CT image texture features and genetic alterations in patients were identified as lung cancer [20]. The related factors were investigated by analyzing the image texture characteristics of the ^{18}F-FDG PET/CT image of the gene phenotype.

Radiogenomics technology has also been used to determine whether cancer metastasizes in liver cancer and to estimate a metastasis prediction model [21]. Radiogenomics studies can reveal the current state of cancer by analyzing genetic expression and image texture features. In addition, it is possible to estimate predictive models using machine or deep learning because numerical analysis results, such as gene expression levels and quantitative image texture features, can be derived. In one study, a prediction model was estimated using the machine learning algorithm with a combination analysis of CT images and genetic expression in breast cancer [22]. In another study, image texture features from CT images and gene expression in pancreatic ductal adenocarcinoma were estimated using a prediction model [23].

In this study, we aimed to estimate a predictive model using *KI67*, *EZRIN*, and ^{18}F-FDG PET/CT image texture features in pediatric osteosarcoma, which are expressed in relation to metastasis and chemotherapy response. Machine and deep learning techniques were used to construct various predictive models. The accuracy of each model was compared to evaluate a predictive model suitable for metastasis and chemotherapy responses in pediatric osteosarcoma.

2. Materials and Methods

2.1. Pediatric Osteosarcoma Patient Data

Data from a total of 52 pediatric osteosarcoma patients consisted of 31 male and 21 female children aged <14 years. All of the patients with osteosarcoma received neoadjuvant chemotherapy over four weeks, which involved a combination of methotrexate (a dose of 8–12 g/m^2), adriamycin (a dose of 60 mg/m^2), and cisplatin (a dose of 100 mg/m^2) at intervals of three weeks. The surgery was performed three weeks after the end of the second neoadjuvant chemotherapy. A total of 21 patients expressed both *EZRIN* and *KI67* (Figure S1). *KI67* expression > 15% was classified as *KI67*-positive and <15% was classified as *KI67*-negative. *EZRIN* expression was classified as *EZRIN*-positive or -negative with no *EZRIN* expression. Cancer tissues were collected from the femur, tibia, humerus, and pelvis. All cancer tissues were classified into 2A, 2B, IIA, and unknown according to the American Joint Committee on Cancer stage classification method. The pathologic subtypes of each cancer tissue were identified as osteoblastic (OB), chondroblastic (CB), or others (Table 1). necrosis of 90% or more tumor region indicated a good histological response, and less than 90% tumor region necrosis indicated a poor histological response [24]. A total of 25 patients showed a good histological response to chemotherapy, whereas the remaining 27 patients had no response. In addition, 37 patients had no metastasis, whereas 15 patients had metastasis. Overall, 18 patients had a good histological chemotherapy response and no metastasis, 19 patients had a poor histological chemotherapy response and no metastasis, seven patients had a good histological chemotherapy response and metastasis, and eight patients had a poor histological chemotherapy response and metastasis.

Table 1. Patient information.

Characteristic	Value
Sex, *n* (%)	21 (40.38%)
Female	31 (59.61%)
Male	
Age, *n* (%)	52 (100%)
Years $\leq$ 14	
Location of primary tumor, *n* (%)	33 (63.46%)
Femur	16 (30.76%)
Tibia	2 (3.84%)
Humerus	1 (1.92%)
Pelvis	
AJCC stage, *n* (%)	13 (25%)
2A	16 (30.76%)
2B	4 (7.69%)
IIA	19 (36.53%)
Unknown	
Pathologic subtype, *n* (%)	39 (75%)
OB (Osteoblastic)	10 (19.23%)
CB (Chondroblastic)	3 (5.76%)
Others	

2.2. ^{18}F-FDG PET/CT Image Texture Features

A total of 52 patient ^{18}F-FDG PET/CT images were used for analysis. The ^{18}F-FDG PET/CT images were acquired before chemotherapy to confirm the prediction of chemotherapy treatment response in pediatric osteosarcoma patients. Radiomic features were extracted by texture analysis of the acquired ^{18}F-FDG PET/CT images. LiFEx (version 4.0) was used for radiomics feature extraction of the ^{18}F-FDG PET/CT images. Overall, 47 image texture features were classified as first-order, second-order, and high-order. Figure 1 shows a flow diagram of prediction model generation using image texture features and gene expression.

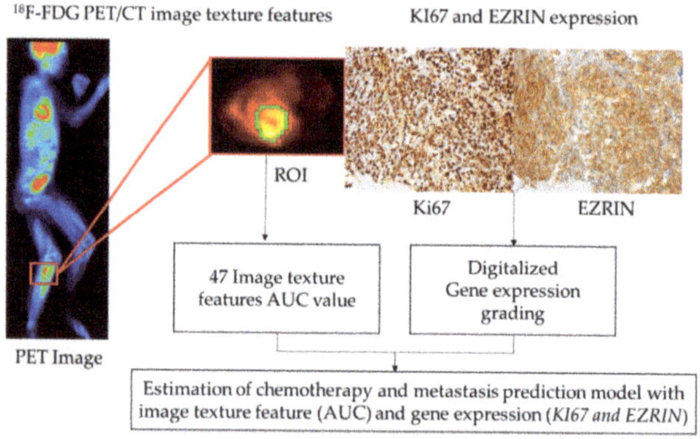

Figure 1. Diagram of the process for generation of the prediction model with image texture features and gene expression.

2.3. Feature Selection for the Prediction Model

Among the 47 imaging features, the area under the curve (AUC) values of 0.6 or higher were identified to improve the accuracy of chemotherapy treatment response and metastasis prediction in pediatric osteosarcoma patients. The AUC values of the imaging features were evaluated by analyzing the ^{18}F-FDG PET/CT images based on *EZRIN* and *KI67* expression levels. The image texture features for radiogenomics were selected by maximizing the AUC value (AUC_max). Medcalc (version 19.4.1) was used to determine the AUC value of each image feature obtained by extracting the features of the ^{18}F-FDG PET/CT images.

2.4. Prediction Model Development Using Machine and Deep Learning

Random forest and gradient boosting algorithms were used to predict the treatment response of pediatric osteosarcoma patients. To achieve this goal, the ratio of machine learning training data to test data was set to 7:3. However, owing to the lack of patient datasets, it is difficult to consider any input pre-processing involving the deletion of some data. Cross-validation was performed 10 times to increase the statistical reliability of the performance measurements. A convolutional neural network (CNN; Keras 2.3.1) was used to calculate the accuracy of the prediction model. The CNN consisted of an input layer, an output layer, two convolution layers, two pooling layers, and three fully connected layers. Maximum pooling was used to conserve each layer's properties. A fully connected layer was used to flatten the two-dimensional layer to a one-dimensional layer. A feature map was extracted from the output layer of the deep learning results. The feature map data were classified as 0 or 1 for the t-distributed stochastic neighbor embedding (t-SNE) plot.

2.5. Radiogenomics Data Analysis

Machine learning was performed to evaluate the predictive model for chemotherapy response and metastasis. For the chemotherapy response prediction model, *EZRIN*, *KI67*, image texture features (AUC > 0.6, 7 features) + *EZRIN* + *KI67*, and AUC_max + *EZRIN* + *KI67* were used as inputs. For the metastasis prediction model, *EZRIN*, *KI67*, image texture features (AUC > 0.6, 17 features) + *EZRIN* + *KI67*, and AUC_max + *EZRIN* + *KI67* were used as inputs.

3. Results

3.1. Image Texture Feature Extraction from ^{18}F-FDG PET/CT Images

A total of 47 imaging features (Table S1) were acquired by drawing the region of interest of the tumor site on each ^{18}F-FDG PET/CT image from responders/non-responders to chemotherapy and metastasis (Figure 2). Seven of the 47 imaging features had an AUC value of 0.6 or higher for evaluating the chemotherapy response (Table S2). 17 of the 47 imaging features had an AUC value of 0.6 or higher for evaluating the metastasis (Table S3). The image feature with the highest AUC was Neighborhood Gray-Level Different Matrix (NGLDM)_Contrast, for which the value was 0.652 (Table S2). After dimension reduction with t-SNE, the texture features of 47 images from the chemotherapy response and metastasis prediction models did not allow a clear separation of each image (Figure 3).

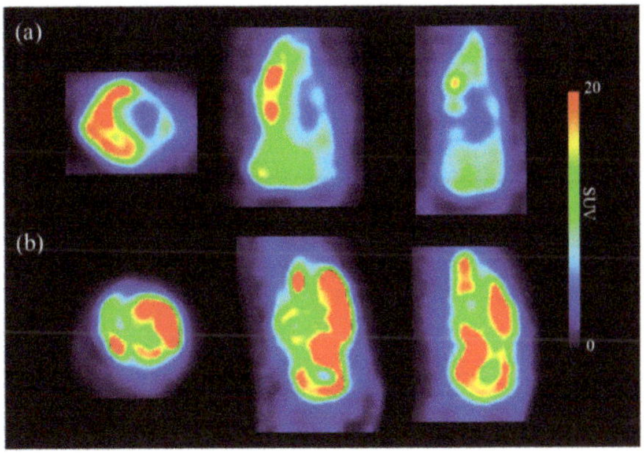

Figure 2. Transverse, coronal, and sagittal sections from osteosarcoma ^{18}F-FDG PET/CT images. (**a**) Images from a chemotherapy responder; (**b**) images from a chemotherapy non-responder.

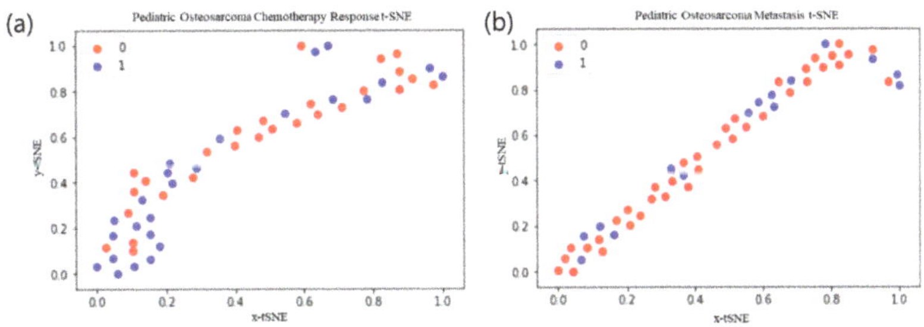

Figure 3. t-Distributed stochastic neighbor embedding (t-SNE) of texture features from 47 images. The 0/1 values represent non-responders and respondents, respectively. (**a**) Chemotherapy response t-SNE; (**b**) metastasis t-SNE.

3.2. Machine and Deep Learning Algorithms Using ^{18}F-FDG PET/CT Images

The sensitivity, specificity, AUC, train accuracy, and test accuracy of the prediction models for chemotherapy response and metastasis were calculated using the random forest algorithm, gradient boosting algorithm, and deep learning. The random forest algorithm prediction model test accuracy using total text features (47) and text features (AUC > 0.6) was 0.71 and 0.83 for chemotherapy response, respectively. In the gradient boosting prediction model, the test accuracy using total text features (47) and text features (AUC > 0.6) were 0.81 and 0.81, respectively (Table 2). In the deep learning prediction model, the test accuracy was 0.975 (Figure 4). The accuracy and loss function of chemotherapy response and metastasis were represented in Figure S2. The random forest algorithm prediction model test accuracy using total text features (47) and text features (AUC > 0.6) was 0.72 and 0.76 for metastasis, respectively. In the gradient boosting prediction model, the test accuracy using total text features (47) and text features (AUC > 0.6) was 0.61 and 0.76, respectively. In the deep learning prediction model, the test accuracy was 0.983 (Figure 4). Thus, the prediction models using the random forest algorithm and deep learning showed the highest accuracy for chemotherapy response and metastasis (Table 2).

Table 2. Chemotherapy response and metastasis prediction model with various machine learning algorithms and deep learning.

Chemotherapy Response	Random Forest		Gradient Boosting		Deep Learning
	Text Feature (47)	AUC > 0.6 (7)	Text Feature (47)	AUC > 0.6 (7)	Train (37): Test (15)
Sensitivity	0.76	0.79	0.85	0.84	0.956
Specificity	0.74	0.82	0.94	0.88	0.964
AUC	0.76	0.80	0.88	0.86	0.917
Train accuracy	0.71	0.83	0.77	0.83	0.978
Test accuracy	0.71	0.83	0.81	0.81	0.975
Metastasis	**Random Forest**		**Gradient Boosting**		**Deep Learning**
	Text Feature (47)	**AUC > 0.6 (17)**	**Text Feature (47)**	**AUC > 0.6 (17)**	**Train (37): Test (15)**
Sensitivity	0.77	0.80	0.76	0.85	0.958
Specificity	0.74	0.66	0.76	0.73	0.990
AUC	0.73	0.85	0.74	0.72	0.970
Train accuracy	0.72	0.76	0.77	0.67	0.986
Test accuracy	0.72	0.76	0.61	0.76	0.983

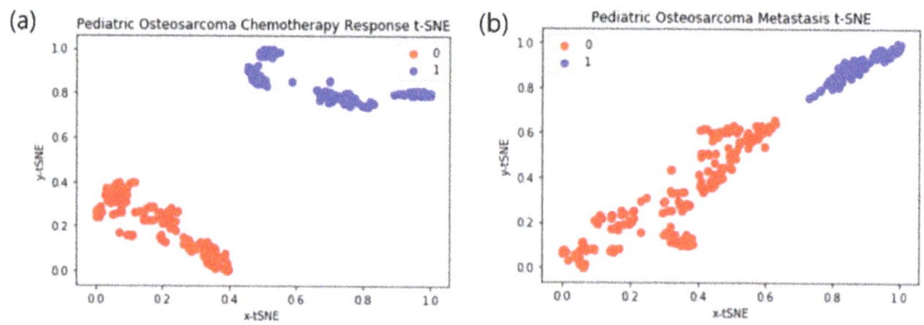

Figure 4. Deep learning t-distributed stochastic neighbor embedding (t-SNE) results. The 0/1 values represent non-responders and respondents, respectively. (**a**) Chemotherapy response t-SNE; (**b**) metastasis t-SNE.

3.3. Deep Learning Interpretation: t-SNE Plots

As shown in Figure 4, after dimension reduction with t-SNE, image texture features from the chemotherapy response and metastasis prediction models were separated into two classes. In the two cases, the classes were clearly separated. We obtained a relatively high precision rate for the chemotherapy response and metastasis prediction model class because the chemotherapy response and metastasis clusters were pure.

3.4. Radiogenomics Machine Learning Model

The random forest algorithm was confirmed to be used as a prediction model for the chemotherapy response and metastasis of pediatric osteosarcoma with a combination of gene expression data and image features. The sensitivity, specificity, AUC, train accuracy, and test accuracy of the prediction model were calculated.

The chemotherapy response prediction model AUCs using *EZRIN*, *KI67*, image texture features (7, AUC > 0.6) + *EZRIN* + *KI67*, and NGLDM_Contrast (AUC_max) + *EZRIN* + *KI67* were 0.58, 0.57, 0.77, and 0.89, respectively. The accuracy of the chemotherapy response prediction model using *EZRIN*, *KI67*, image texture features + *EZRIN* + *KI67*, and NGLDM_Contrast (AUC_max) + *EZRIN* + *KI67* was 0.53, 0.52, 0.73, and 0.85, respectively. The metastasis prediction model AUCs using *EZRIN*, *KI67*, image texture features (17, AUC > 0.6) + *EZRIN* + *KI67*, and Gray-Level Co-occurrence Matrix (GLCM)_Correlation (AUC_max) + *EZRIN* + *KI67* were 0.56, 0.57, 0.76, and 0.80, respectively. The metastasis prediction model test accuracy using *EZRIN*, *KI67*, image texture features + *EZRIN* + *KI67*, and GLCM_Correlation (AUC_max) + *EZRIN* + *KI67* was 0.54,

0.52, 0.74, and 0.85, respectively. The prediction model using AUC_max, *EZRIN*, and *KI67* with the random forest algorithm showed the highest accuracy (Table 3).

Table 3. Chemotherapy Response and metastasis prediction model with gene expressions, combination of gene expression image texture features and gene expression and combination of area under curve max (AUC_max) image texture and gene expressions.

Chemotherapy Response	***EZRIN***	***KI67***	**Image Texture Feature + *EZRIN* + *KI67***	**NGLDM_Contrast + *EZRIN*+ *KI67***
Sensitivity	0.59	0.57	0.84	0.87
Specificity	0.44	0.68	0.75	0.85
AUC	0.58	0.57	0.77	0.89
Train accuracy	0.53	0.52	0.73	0.85
Test accuracy	0.53	0.52	0.73	0.85
Metastasis	***EZRIN***	***KI67***	**Image Texture Feature+ *EZRIN* + *KI67***	**GLCM_Correlation + *EZRIN*+ *KI67***
Sensitivity	0.61	0.54	0.77	0.91
Specificity	0.42	0.65	0.55	0.6
AUC	0.56	0.57	0.76	0.8
Train accuracy	0.54	0.52	0.74	0.85
Test accuracy	0.54	0.52	0.74	0.85

3.5. Machine Learning Prediction Model with the Random Forest Algorithm

The receiver operating characteristic curves of the chemotherapy response and metastasis prediction models are shown in Figure 4. The AUCs for chemotherapy prediction using *KI67*, *EZRIN*, image texture features + *EZRIN* + *KI67*, and NGLDM_Contrast + *EZRIN* + *KI67* were 0.58, 0.57, 0.77, and 0.89, respectively. The AUCs for metastasis prediction using *KI67*, *EZRIN*, image texture features + *EZRIN* + *KI67*, and GLCM_Correlation + *EZRIN* + *KI67* were 0.56, 0.57, 0.76, and 0.8, respectively (Figure 5).

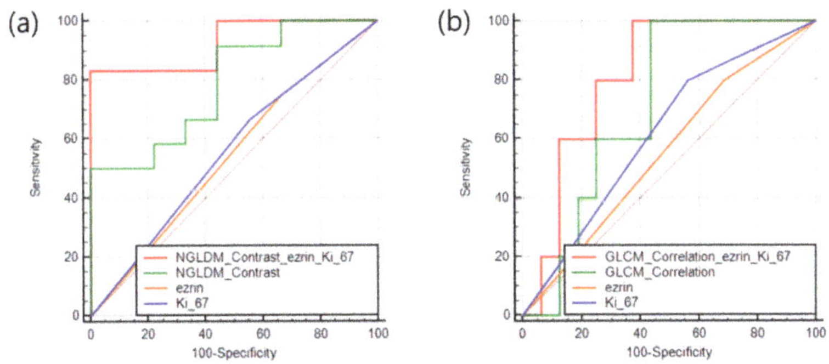

Figure 5. Receiver operating characteristic curves of the prediction features for patient outcomes. (**a**) Chemotherapy response. (**b**) Metastasis.

4. Discussion

In this study, we evaluated a predictive model that can predict the chemotherapy response and metastasis of pediatric osteosarcoma by analyzing gene expression and ^{18}F-FDG PET/CT image texture features. Several appropriate algorithms were selected from machine learning algorithms that have shown good predictive performance. Imaging features that are associated with metastasis and chemotherapy response were extracted. A predictive model showing high accuracy was estimated using the extracted image features, gene expression, and the previously selected algorithm.

KI67 and *EZRIN* are clinically used as biomarkers to determine metastasis or chemotherapy responses [25–28]. Rejniak et al. reported that *KI67* expression is associated with pediatric osteosarcoma metastasis and chemotherapy response [29], and Bacci et al. reported that *EZRIN* expression is associated with pediatric osteosarcoma metastasis and chemotherapy response [30]. In other studies, it is well-known that the expression levels of *KI67* and *EZRIN* are associated with metastasis and chemotherapy responses in pediatric osteosarcoma [31–36]. As described previously, these two genes were previously used as biomarkers for pediatric osteosarcoma, but the prediction model test accuracy was low in our study (test accuracy ~ 0.53). Low accuracy was estimated despite the use of well-known biomarkers for chemotherapy response and metastasis. The predictive model of chemotherapy response and metastasis suggests that accurate prediction is difficult using the expression of a single gene.

A total of 47 image features were extracted from the ^{18}F-FDG PET/CT images. Imaging features related to chemotherapy response and metastasis were classified by the AUC value. The image texture feature that was most closely related to chemotherapy response was NGLDM_Contrast, and related to metastasis was GLCM_Correlation. The prediction model with AUC_max showed low accuracy in chemotherapy response and metastasis. The chemotherapy results showed a high predictive accuracy than metastasis from the prediction models estimated using image texture features.

The predictive model using imaging texture features showed an accuracy of 83% for chemotherapy response and 76% for metastasis. For clinical applications, it is necessary to generate a predictive model with higher accuracy. We used a radiogenomics technique that evaluates both gene expression and imaging factors to improve the accuracy of the predictive model for both conditions. The predictive model using the radiogenomic technique showed high predictive ability in both chemotherapy response and metastasis (Figure 4). The accuracy was improved by about 10% or more when the AUC_max value was used in both conditions.

The predictive model using images to predict chemotherapy responses showed good results. In metastasis, the predictive model that used images and genetic information displayed improved performance. Deep learning has shown high predictive performance with image texture features, but it is difficult to apply genetic information with image texture features to improve accuracy. Machine learning predictive models can be applied to data with a variety of properties, such as gene expression and image texture features, as in this study. Additionally, factors other than gene expression and image factors can be used when conducting research on machine learning predictive models with digitalization of data.

The predictive model of chemotherapy using machine learning showed a high accuracy of 83% when estimated by ^{18}F-FDG PET/CT imaging alone and 85% when analyzed after adding gene expression. The chemotherapy response reaction process is related to the heterogeneity of cancer cells, and since this can be confirmed by imaging, it is possible to obtain higher accuracy with the analysis results of the image. It is difficult to obtain high accuracy through image analysis alone because of the challenge in determining the metastasis process through simple image analysis. The accuracy of the metastasis process was 76%, and when the genetic analysis results were added, it increased to 85%. A more accurate predictive model was estimated by adding the gene expression results to the image analysis of the metastasis process.

A limitation of this study is that owing to the rarity of pediatric osteosarcoma, a patient set for extra validation could not be obtained. The predictive model was generated using the radiogenomics technique, but the accuracy of the predictive model was not high even in other patient groups through additional verification. A larger population of pediatric osteosarcoma patients is needed to evaluate the accuracy of the predictive model. The data in this study may not be reliable because of the small sample size. However, it can be used as preliminary evidence to estimate the probability of the predictive model. Even though the number of patient groups is small, the analysis method using image texture features and gene expression data have been shown to be applicable to chemotherapy and metastasis prediction models. Additional data from other pediatric patients with osteosarcoma could improve the accuracy of the model for predicting chemotherapy response and metastasis.

5. Conclusions

Predictive models using the random forest algorithm showed the best accuracy for predicting metastasis and chemotherapy responses in our pediatric osteosarcoma dataset. The predictive model that combined *KI67*, *EZRIN*, and image texture features was estimated to have a higher accuracy than the predictive models using each factor separately. The accuracy of metastasis prediction increased by 10% using radiogenomics data. High accuracy was estimated using a radiogenomics technique that uses both gene expression and imaging texture features for metastasis prediction. Thus, a predictive model using radiogenomics technology that incorporates both imaging features and gene expression can accurately predict metastasis and chemotherapy responses to improve patient outcomes.

Supplementary Materials: The following are available online at https://www.mdpi.com/article/10.3390/cancers13112671/s1, Figure S1. Immunohistochemical staining of KI67 (a) and EZRIN (b). Figure S2. ^{18}F-FDG PET/CT image deep learning accuracy and loss value, (a) Chemotherapy response prediction, (b) Metastasis prediction. Table S1. The AUC values of image texture features for chemotherapy response from 52 pediatric osteosarcoma. Table S2. The AUC values of image texture features for metastasis from 52 pediatric osteosarcoma. Table S3. 47 image texture features from ^{18}F-FDG PET/CT.

Author Contributions: Conceptualization, B.-C.K. and J.K.; methodology, J.K.; software, J.K. and K.K.; validation, J.K. and S.-K.W.; formal analysis, J.K. and K.K.; investigation, J.K.; resources, B.H.B., C.-B.K., W.S.S., J.-S.K.; data curation, J.K.; writing—original draft preparation, B.-C.K.; writing—review and editing, B.-C.K., I.L. and S.-K.W.; visualization, B.-C.K.; supervision, S.-K.W.; project administration, S.-K.W.; funding acquisition, S.-K.W. All authors have read and agreed to the published version of the manuscript.

Funding: This study was supported by a National Research Foundation of Korea (NRF) grant funded by the Korean government (Ministry of Science and ICT; No. 2020M2D9A1094070).

Institutional Review Board Statement: This study was approved by the institutional review board of the Korea Institute of Radiological and Medical Sciences (IRB; e-IRB number: KIRAMS 2021-02-005).

Informed Consent Statement: Informed consent was obtained from all subjects involved in the study.

Data Availability Statement: Data sharing is not applicable to this article.

Conflicts of Interest: The authors declare no conflict of interest.

References

1. Brueffer, C.; Vallon-Christersson, J.; Grabau, D.; Ehinger, A.; Häkkinen, J.; Hegardt, C.; Malina, J.; Chen, Y.; Bendahl, P.-O.; Manjer, J. Clinical value of RNA sequencing-based classifiers for prediction of the five conventional breast cancer biomarkers: A report from the population-based multicenter sweden cancerome analysis network—Breast initiative. *JCO Precis. Oncol.* **2018**, *2*, 1–18. [CrossRef] [PubMed]
2. Petri, B.J.; Klinge, C.M. Regulation of breast cancer metastasis signaling by miRNAs. *Cancer Metastasis Rev.* **2020**, *39*, 837–886. [CrossRef]
3. Ryu, T.Y.; Kim, K.; Kim, S.-K.; Oh, J.-H.; Min, J.-K.; Jung, C.-R.; Son, M.-Y.; Kim, D.-S.; Cho, H.-S. SETDB1 regulates SMAD7 expression for breast cancer metastasis. *BMB Rep.* **2019**, *52*, 139. [CrossRef] [PubMed]
4. Kim, J.; Piao, H.-L.; Kim, B.-J.; Yao, F.; Han, Z.; Wang, Y.; Xiao, Z.; Siverly, A.N.; Lawhon, S.E.; Ton, B.N. Long noncoding RNA MALAT1 suppresses breast cancer metastasis. *Nat. Genet.* **2018**, *50*, 1705–1715. [CrossRef] [PubMed]
5. Kim, K.; Son, M.-Y.; Jung, C.-R.; Kim, D.-S.; Cho, H.-S. EHMT2 is a metastasis regulator in breast cancer. *Biochem. Biophys. Res. Commun.* **2018**, *496*, 758–762. [CrossRef]
6. Niu, Y.; Bao, L.; Chen, Y.; Wang, C.; Luo, M.; Zhang, B.; Zhou, M.; Wang, J.E.; Fang, Y.V.; Kumar, A. HIF2-induced long noncoding RNA RAB11B-AS1 promotes hypoxia-mediated angiogenesis and breast cancer metastasis. *Cancer Res.* **2020**, *80*, 964–975. [CrossRef] [PubMed]
7. McDaniel, J.M.; Varley, K.E.; Gertz, J.; Savic, D.S.; Roberts, B.S.; Bailey, S.K.; Shevde, L.A.; Ramaker, R.C.; Lasseigne, B.N.; Kirby, M.K. Genomic regulation of invasion by STAT3 in triple negative breast cancer. *Oncotarget* **2017**, *8*, 8226. [CrossRef] [PubMed]

8. Jia, Y.; Shi, L.; Yun, F.; Liu, X.; Chen, Y.; Wang, M.; Chen, C.; Ren, Y.; Bao, Y.; Wang, L. Transcriptome sequencing profiles reveal lncRNAs may involve in breast cancer (ER/PR positive type) by interaction with RAS associated genes. *Pathol. Res. Pract.* **2019**, *215*, 152405. [CrossRef]
9. Yan, N.; Xu, H.; Zhang, J.; Xu, L.; Zhang, Y.; Zhang, L.; Xu, Y.; Zhang, F. Circular RNA profile indicates circular RNA VRK1 is negatively related with breast cancer stem cells. *Oncotarget* **2017**, *8*, 95704. [CrossRef] [PubMed]
10. Cai, Y.; Mei, J.; Xiao, Z.; Xu, B.; Jiang, X.; Zhang, Y.; Zhu, Y. Identification of five hub genes as monitoring biomarkers for breast cancer metastasis in silico. *Hereditas* **2019**, *156*, 1–12. [CrossRef] [PubMed]
11. Seo, C.H.; Kim, J.-R.; Kim, M.-S.; Cho, K.-H. Hub genes with positive feedbacks function as master switches in developmental gene regulatory networks. *Bioinformatics* **2009**, *25*, 1898–1904. [CrossRef]
12. Langfelder, P.; Horvath, S. WGCNA: An R package for weighted correlation network analysis. *BMC Bioinform.* **2008**, *9*, 1–13. [CrossRef]
13. Beam, A.L.; Kohane, I.S. Big data and machine learning in health care. *JAMA* **2018**, *319*, 1317–1318. [CrossRef]
14. Shrestha, S.; Sengupta, P.P. Machine learning for nuclear cardiology: The way forward. *J. Nucl. Cardiol.* **2019**, *26*, 1755–1758. [CrossRef]
15. Huang, Y.Q.; Liang, C.H.; He, L.; Tian, J.; Liang, C.S.; Chen, X.; Ma, Z.L.; Liu, Z.Y. Development and validation of a radiomics nomogram for preoperative prediction of lymph node metastasis in colorectal cancer. *J. Clin. Oncol.* **2016**, *34*, 2157–2164. [CrossRef]
16. Dobin, A.; Davis, C.A.; Schlesinger, F.; Drenkow, J.; Zaleski, C.; Jha, S.; Batut, P.; Chaisson, M.; Gingeras, T.R. STAR: Ultrafast universal RNA-seq aligner. *Bioinformatics* **2013**, *29*, 15–21. [CrossRef] [PubMed]
17. Anders, S.; Pyl, P.T.; Huber, W. HTSeq—A Python framework to work with high-throughput sequencing data. *Bioinformatics* **2015**, *31*, 166–169. [CrossRef]
18. Wagner, G.P.; Kin, K.; Lynch, V.J. Measurement of mRNA abundance using RNA-seq data: RPKM measure is inconsistent among samples. *Theory Biosci.* **2012**, *131*, 281–285. [CrossRef] [PubMed]
19. Love, M.; Anders, S.; Huber, W. Differential analysis of count data–the DESeq2 package. *Genome Biol.* **2014**, *15*, 10.1186.
20. Zheng, G.; Ma, Y.; Zou, Y.; Yin, A.; Li, W.; Dong, D. HCMDB: The human cancer metastasis database. *Nucleic Acids Res.* **2018**, *46*, D950–D955. [CrossRef] [PubMed]
21. Tang, Y.; Horikoshi, M.; Li, W. ggfortify: Unified interface to visualize statistical results of popular R packages. *R J.* **2016**, *8*, 474. [CrossRef]
22. Tang, J.; Kong, D.; Cui, Q.; Wang, K.; Zhang, D.; Gong, Y.; Wu, G. Prognostic genes of breast cancer identified by gene co-expression network analysis. *Front. Oncol.* **2018**, *8*, 374. [CrossRef]
23. Rivera, C.; Zandonadi, F.S.; Sánchez-Romero, C.; Soares, C.D.; Granato, D.C.; González-Arriagada, W.A.; Leme, A.F.P. Agrin has a pathological role in the progression of oral cancer. *Br. J. Cancer* **2018**, *118*, 1628–1638. [CrossRef] [PubMed]
24. Coffin, C.M.; Lowichik, A.; Zhou, H. Treatment Effects in Pediatric Soft Tissue and Bone Tumors. *Anat. Pathol.* **2005**, *123*, 75–90. [CrossRef] [PubMed]
25. Lee, J.; Yesilkanal, A.E.; Wynne, J.P.; Frankenberger, C.; Liu, J.; Yan, J.; Elbaz, M.; Rabe, D.C.; Rustandy, F.D.; Tiwari, P. Effective breast cancer combination therapy targeting BACH1 and mitochondrial metabolism. *Nature* **2019**, *568*, 254–258. [CrossRef]
26. Way, G.P.; Allaway, R.J.; Bouley, S.J.; Fadul, C.E.; Sanchez, Y.; Greene, C.S. A machine learning classifier trained on cancer transcriptomes detects NF1 inactivation signal in glioblastoma. *BMC Genom.* **2017**, *18*, 127. [CrossRef] [PubMed]
27. Wu, S.-G.; Sun, J.-Y.; Tong, Q.; Li, F.-Y.; He, Z.-Y. Clinical features of brain metastases in breast cancer: An implication for hippocampal-sparing whole-brain radiation therapy. *Ther. Clin. Risk Manag.* **2016**, *12*, 1849. [CrossRef] [PubMed]
28. Nargis, H.; Nawaz, H.; Ditta, A.; Mahmood, T.; Majeed, M.; Rashid, N.; Muddassar, M.; Bhatti, H.; Saleem, M.; Jilani, K. Raman spectroscopy of blood plasma samples from breast cancer patients at different stages. *Spectrochim. Acta Part A Mol. Biomol. Spectrosc.* **2019**, *222*, 117210. [CrossRef]
29. Dhanya, K.; Menon, A.; Rai, L.S. In-vitro Models in Anticancer Screening. In *Phytochemistry: An In-Silico and In-Vitro Update*; Springer: Berlin/Heidelberg, Germany, 2019; pp. 251–265.
30. Lim, S.; Park, Y.; Hur, B.; Kim, M.; Han, W.; Kim, S. Protein interaction network (pin)-based breast cancer subsystem identification and activation measurement for prognostic modeling. *Methods* **2016**, *110*, 81–89. [CrossRef] [PubMed]
31. Kabiraj, S.; Akter, L.; Raihan, M.; Diba, N.J.; Podder, E.; Hassan, M.M. Prediction of Recurrence and Non-recurrence Events of Breast Cancer using Bagging Algorithm. In Proceedings of the 2020 11th International Conference on Computing, Communication and Networking Technologies (ICCCNT), Kharagpur, India, 1–3 July 2020; pp. 1–5.
32. Shao, B.; Conrad, T. Epithelial-mesenchymal transition regulatory network-based feature selection in lung cancer prognosis prediction. In Proceedings of the International Conference on Bioinformatics and Biomedical Engineering, Granada, Spain, 20–22 April 2016; pp. 135–146.
33. Zhang, Y.; Deng, Q.; Liang, W.; Zou, X. An efficient feature selection strategy based on multiple support vector machine technology with gene expression data. *BioMed Res. Int.* **2018**, *2018*, 7538204. [CrossRef]
34. Ganggayah, M.D.; Taib, N.A.; Har, Y.C.; Lio, P.; Dhillon, S.K. Predicting factors for survival of breast cancer patients using machine learning techniques. *BMC Med. Inform. Decis. Mak.* **2019**, *19*, 1–17. [CrossRef] [PubMed]
35. Weigelt, B.; Glas, A.M.; Wessels, L.F.; Witteveen, A.T.; Peterse, J.L.; van't Veer, L.J. Gene expression profiles of primary breast tumors maintained in distant metastases. *Proc. Natl. Acad. Sci. USA* **2003**, *100*, 15901–15905. [CrossRef] [PubMed]

36. Pei, G.; Chen, L.; Zhang, W. WGCNA application to proteomic and metabolomic data analysis. In *Methods in Enzymology*; Elsevier: Amsterdam, The Netherlands, 2017; Volume 585, pp. 135–158.

cancers MDPI

Article

Serine-Threonine Kinase Receptor-Associated Protein (STRAP) Knockout Decreases the Malignant Phenotype in Neuroblastoma Cell Lines

Laura V Bownes [1], Adele P Williams [1], Raoud Marayati [1], Colin H Quinn [1], Sara C Hutchins [1], Jerry E Stewart [1], Trung Vu [2], Juliet L Easlick [3], Elizabeth Mroczek-Musulman [4], David K Crossman [5], Joshua C Anderson [6], Christopher D Willey [6], Pran K Datta [2,*] and Elizabeth A Beierle [1,*]

1 Division of Pediatric Surgery, Department of Surgery, University of Alabama at Birmingham, Birmingham, AL 35233, USA; lbownes@uabmc.edu (L.V.B.); awil26@lsuhsc.edu (A.P.W.); rmarayati@uabmc.edu (R.M.); chquinn@uab.edu (C.H.Q.); schutchins@uabmc.edu (S.C.H.); jessy@uab.edu (J.E.S.)
2 Division of Hematology and Oncology, Department of Medicine, University of Alabama at Birmingham, Birmingham, AL 35233, USA; ttvu@uab.edu
3 Division of Transplantation, Department of Surgery, University of Alabama at Birmingham, Birmingham, AL 35233, USA; julieteaslick@uabmc.edu
4 Department of Pathology, Children's of Alabama, Birmingham, AL 35233, USA; Elizabeth.Mroczek-Musulman@childrensal.org
5 Department of Genetics, University of Alabama at Birmingham, Birmingham, AL 35233, USA; dkcrossm@uab.edu
6 Department of Radiation Oncology, University of Alabama at Birmingham, Birmingham, AL 35233, USA; janders7@uab.edu (J.C.A.); cwilley@uab.edu (C.D.W.)
* Correspondence: prandatta@uabmc.edu (P.K.D.); elizabeth.beierle@childrensal.org (E.A.B.)

Citation: Bownes, L.V.; Williams, A.P.; Marayati, R.; Quinn, C.H.; Hutchins, S.C.; Stewart, J.E.; Vu, T.; Easlick, J.L.; Mroczek-Musulman, E.; Crossman, D.K.; et al. Serine-Threonine Kinase Receptor-Associated Protein (STRAP) Knockout Decreases the Malignant Phenotype in Neuroblastoma Cell Lines. *Cancers* **2021**, *13*, 3201. https://doi.org/10.3390/cancers13133201

Academic Editors: Saurabh Agarwal and Yang Jianhua

Received: 14 June 2021
Accepted: 23 June 2021
Published: 26 June 2021

Simple Summary: Neuroblastoma is the most common extra-cranial tumor in children and despite medical advancements in cancer treatment, five-year survival for high-risk neuroblastoma remains less than 50%. Investigation of the mechanisms responsible for aggressive disease is necessary to identify novel therapeutic targets and improve survival. Serine-threonine kinase receptor associated protein (STRAP) is upregulated in several malignancies and plays an important role in tumor growth and metastasis. The role of STRAP in pediatric malignancies and specifically in neuroblastoma has not been explored. We sought to determine whether STRAP functions assisted to promote the malignant phenotype in neuroblastoma and could provide a potential target for future therapies.

Abstract: Background: Serine-threonine kinase receptor-associated protein (STRAP) plays an important role in neural development but also in tumor growth. Neuroblastoma, a tumor of neural crest origin, is the most common extracranial solid malignancy of childhood and it continues to carry a poor prognosis. The recent discovery of the role of STRAP in another pediatric solid tumor, osteosarcoma, and the known function of STRAP in neural development, led us to investigate the role of STRAP in neuroblastoma tumorigenesis. **Methods:** STRAP protein expression was abrogated in two human neuroblastoma cell lines, SK-N-AS and SK-N-BE(2), using transient knockdown with siRNA, stable knockdown with shRNA lentiviral transfection, and CRISPR-Cas9 genetic knockout. STRAP knockdown and knockout cells were examined for phenotypic alterations in vitro and tumor growth in vivo. **Results:** Cell proliferation, motility, and growth were significantly decreased in STRAP knockout compared to wild-type cells. Indicators of stemness, including mRNA abundance of common stem cell markers Oct4, Nanog, and Nestin, the percentage of cells expressing CD133 on their surface, and the ability to form tumorspheres were significantly decreased in the STRAP KO cells. In vivo, STRAP knockout cells formed tumors less readily than wild-type tumor cells. **Conclusion:** These novel findings demonstrated that STRAP plays a role in tumorigenesis and maintenance of neuroblastoma stemness.

Keywords: serine-threonine kinase receptor-associated protein; neuroblastoma; CRISPR-Cas9

1. Introduction

Despite advancements in pediatric cancer care, neuroblastoma, a childhood malignancy resulting from abnormal neural crest cell development, remains the cause of over 15% of pediatric cancer related deaths [1]. Current treatment regimens for high-risk disease include chemotherapy, surgical resection, autologous stem cell transplant, radiation, immunotherapy, and retinoic acid [2]. Despite these extensive therapies, outcomes remain dismal for patients with high-risk disease as less than 50% will attain a 5 year survival [3]. It is crucial to continue to investigate this disease in order to develop new therapies, especially for the cohort of patients with high-risk disease.

Serine-threonine kinase receptor-associated protein (STRAP) is a scaffolding protein that facilitates protein to protein interactions [4]. Investigators began to study the role of STRAP in cancer when they found that STRAP inhibited the tumor suppressor, TGF-β [5]. STRAP is overexpressed in several malignancies, including colorectal [6] and lung cancer [7] and the pediatric bone cancer, osteosarcoma [8]. In these malignancies, STRAP enhanced cancer cell proliferation and tumor growth.

STRAP expression supports tumorigenicity by promoting the Wnt/β-catenin pathway in colorectal cancer. In cells overexpressing STRAP, silencing the STRAP gene with short hairpin RNAs (shRNA) resulted in decreased invasion and metastasis [6]. In lung cancer, STRAP led to tumor progression by downregulating tumor suppressors, E-cadherin, and the CDK inhibitor, p21Cip1, through the modulation of the transcription factor Sp1 [9]. Pruksakorn et al. found STRAP was upregulated in osteosarcoma and inhibiting STRAP with small interfering RNA (siRNA) decreased migration and invasion; this suggests that STRAP contributed to osteosarcoma metastasis [8]. These discoveries lend support for investigating STRAP in neuroblastoma. We hypothesized that inhibition of STRAP protein expression in neuroblastoma would result in a decrease in the malignant phenotype in vitro and in vivo. In the current study, we demonstrate STRAP inhibition with siRNA, shRNA, and CRISPR-Cas9 knockout (KO) decreased tumor cell viability, proliferation, growth, stemness, and motility in vitro and tumor growth in vivo. To our knowledge, researchers have not performed extensive study of STRAP in neuroblastoma and our findings provide evidence for STRAP as a potential driver for neuroblastoma tumorigenesis.

2. Materials and Methods

2.1. Cells and Cell Culture

The human neuroblastoma cell lines SK-N-AS (AS, CRL-2137, and MYCN non-amplified), SK-N-BE (2) (BE, CRL-2271, and MYCN amplified), and SH-SY5Y (CRL-2266 and MYCN non-amplified) were obtained from the American Type Culture Collection (ATCC, Manassas, VA, USA). The isogenic SH-EP (SHEP) and WAC (2) (WAC), MYCN non-amplified and amplified cell lines, respectively, were a gift from M. Schwab (Deutsches Krebsforschungszentrum, Heidelberg, Germany) [10]. Cells were maintained under standard culture conditions at 37 °C and 5% CO_2. AS cells were maintained in Dulbecco's modified Eagle's medium (DMEM, 30-2601, ATCC) containing 10% fetal bovine serum (FBS, Hyclone, Suwanee, GA, USA), 4 mM L-glutamine (Thermo Fisher Scientific Inc., Waltham, MA, USA), 1 μM non-essential amino acids, and 1 μg/mL penicillin/streptomycin (Gibco, Carlsbad, CA, USA), which is referred to as media. BE cells were maintained in a 1:1 mixture of minimum Eagle medium and Ham F-12 medium (30–2004, ATCC) with 10% FBS (Hyclone), 1 μM non-essential amino acids, 2 mM l-glutamine (Thermo Fisher Scientific Inc.), and 1/μg/mL penicillin/streptomycin (Gibco). SH-EP and WAC (2) were maintained in Roswell Park Memorial Institute (RPMI) 1640 medium (30–2001, ATCC) with 10% FBS (Hyclone), 2/mM l-glutamine (Thermo Fisher Scientific Inc.), and 1 μg/mL penicillin/streptomycin (Gibco). All cell lines were verified within the last 12 months using short tandem repeat analysis (University of Alabama at Birmingham (UAB) Genomics Core, Birmingham, AL, USA) and tested for and deemed free of mycoplasma infection.

2.2. Small Interfering RNA (siRNA) Inhibition of STRAP

AS or BE cells (1 × 10^6) were plated in 6-well plates and transfected for 72 h with Lipofectamine RNAiMax (Thermo Fisher Scientific) or siRNA directed to either control (siNeg), siSTRAP1, siSTRAP2, or a combination of the two STRAP siRNAs at 20 µM concentration with Lipofectamine RNAiMax (Thermo Fisher Scientific). SiNeg (ON-TARGETplus Non-targeting siRNA #1) was obtained from Dharmacon (GE Dharmacon, Thermo Fisher Scientific, Lafayette, CO, USA). The siSTRAP1 (SASI_Hs01_00016957) and siSTRAP2 (SASI_Hs02_00343131) were obtained from Sigma Aldrich (St. Louis, MO, USA).

2.3. Short Hairpin RNA (shRNA) Inhibition of STRAP

The shScramble and shSTRAP cell lines were generously provided by Dr. Pran Datta's laboratory and were established as previously described [11]. The shScramble served as the control for shSTRAP cells. AS cells that underwent lentiviral transfection with shScramble or shSTRAP were cultured in AS media with the addition of puromycin (5 µM, P8833, Sigma-Aldrich, St. Louis, MO, USA) for selection.

2.4. CRISPR-Cas9 Knockout (KO) of STRAP

The CRISPR vector, pSpCas9(BB)-2A-GFP (pX458), was a gift from Dr. A. Joseph Tector and developed by Dr. Feng Zhang (Addgene plasmid #48138) [12]. We used Geneious software (Biomatters, Auckland, New Zealand) to design guide RNAs (gRNAs) to regions from the 5′ untranslated region through exon 3 of the STRAP gene. The selected gRNAs were evaluated with the MIT CRISPR Design Tool (http://crispr.mit.edu/) to assess for potential off-target sequences. The oligonucleotides (5′-CACCGTTGGGGTGCAACACTGAATA-3′ and 5′-CAACCCCACGTTGTGACTTATCAAAA-3′) were annealed at 37 °C for 30 min, 95 °C for 5 min, and ramped down to 25 °C at 5 °C per minute. Annealed oligonucleotides were cloned into the CRISPR plasmid by digesting 1 µg of plasmid pX458 with BbsI (New England Biolabs, Ipswich, MA, USA) in the presence of annealed oligonucleotides, T7 ligase, and ATP in a MyCycler™ thermal cycler (Bio-Rad, Hercules, CA, USA) for 37 °C for 5 min and 23 °C for 5 min for 6 cycles. Ligation reaction was transformed into Invitrogen MAX Efficiency™ DH5α™ competent *E. coli* cells (Invitrogen) following the manufacturer's protocol. Plasmids were isolated from one colony per treatment using the QIAprep® Miniprep (Qiagen, Germantown, MD, USA). DNA was isolated using QIAprep Spin Miniprep Kit (Qiagen) and sequenced by the UAB Genomics Core. AS or BE cells (2 × 10^6) were plated in 6-well plates and 24 h later, transfection was carried out using FuGENE® HD Transfection Reagent (Promega, Madison, WI, USA) per the manufacturer's protocol. The STRAP gRNA plasmid was incubated for 15 min at room temperature in OptiMEM™ media (Thermo Fisher Scientific) with FuGENE® HD Transfection Reagent (Promega) in 3:2 ratio of the transfection reagent to DNA. Plasmid DNA (10 µg) was added to the cells while swirling the flask. Forty-eight hours after transfection, cells from the plasmid transfection were sorted based on green fluorescent protein (GFP) expression using a FACSAria II cell sorter (BD Biosciences, San Jose, CA, USA) into 96-well plates with a single cell per well (Comprehensive Flow Cytometry Core, UAB, Birmingham, AL, USA). In order to screen for CRISPR-Cas9 mediated deletions of the STRAP gene, genomic DNA was isolated using the DNeasy Blood & Tissue Kit (Qiagen) from AS or BE wild-type (WT) cells and those clones that had grown to confluence and survived passage into larger flasks. Pwo SuperYield DNA Polymerase, dNTPack (Sigma Aldrich) with GC-rich solution were utilized per manufacturer's protocol to amplify the region of interest within the STRAP gene using the following primers, (forward: 5′-TTAGTGCCTTCAGTGGGTGG-3′, reverse: 5′-GGTGGGATCAAACATGCGTTC-3′), which were designed using Primer-Blast (https://www.ncbi.nlm.nih.gov/tools/primer-blast/ (accessed on 3 September 2018)). PCR products were assessed using gel electrophoresis on a 1% agarose gel. Individual bands were cut and DNA was purified using the QIAquick Gel Extraction Kit (Qiagen). Nucleotide sequences of these DNA fragments were analyzed by Sanger sequencing (UAB Genomics Core) and aligned to the human reference sequence using the basic local alignment search tool (BLAST, NCBI, https://pubmed.ncbi.nlm.nih.gov/22708584/ (accessed on 27 February 2019)). STRAP protein expression in the AS or BE WT cells and selected STRAP KO clones was assessed by Western blotting to confirm the absence of the STRAP protein.

2.5. Rescue of STRAP Expression

In order to validate that the phenotypic changes noted with STRAP KO cells were not secondary to off-target effects of the CRISPR-Cas9 system, we performed rescue experiments by transfecting AS STRAP KO cells with c-Flag pc-DNA empty vector (EV) or pc-DNA STRAP plasmid [13]. STRAP KO cells (3×10^3) were plated in 6-well plates and transfected with either FuGENE® HD, FuGENE® HD and EV plasmid, or FuGENE® HD and STRAP plasmid for 72 h. Stable rescue cells were selected and maintained in AS media with the addition of G418 (600 µg/mL, A1720, Sigma-Aldrich). Cells were used in proliferation studies with CellTiter 96® assay as described below to examine the phenotype of the KO cells following the re-introduction of STRAP.

2.6. Reagents and Antibodies

Trypan blue stain was obtained from Life Technologies Corporation (Grand Island, NY, USA). Primary antibodies used for Western blotting included the following: polyclonal rabbit anti-STRAP (18277-1-AP) from Proteintech (Rosemont, IL, USA), monoclonal rabbit anti-PGDFRβ (#3169) and monoclonal rabbit anti-vinculin (E1E9V, #13901) from Cell Signaling (Danvers, MA, USA), and monoclonal mouse anti-FLAG (F3165) and monoclonal mouse anti-β-actin (A1978) from Sigma Aldrich. Antibodies were used according to manufacturers' suggestions.

2.7. Immunoblotting

Cells or homogenized tumor specimens were lysed on ice in a buffer consisting of 50 mM Tris-HCl (pH 7.4), 150 mM NaCl, 1 mM EDTA, 1% Triton x-100, 1% sodium deoxycholate, 0.1% SDS, phosphatase inhibitor (P5726, Sigma Aldrich), protease inhibitor (P8340, Sigma Aldrich), and phenylmethylsulfonyl fluoride (PMSF, P7626, Sigma Aldrich) for 30 min. Lysates were centrifuged at $17{,}000\times g$ for 30 min at 4 °C. Protein concentrations were determined using a Micro BCA™ Protein Assay Kit (Thermo Fisher Scientific). Proteins were separated on SDS-PAGE gels by electrophoresis and transferred to Immobilon®-P polyvinylidene fluoride (PVDF) transfer membrane (EMD Millipore, Burlington, MA, USA). In order to confirm the expected size of target proteins, Precision Plus Protein Kaleidoscope Standards (161-0375, Bio-Rad) molecular weight markers were used. Antibodies were used per the manufacturers' recommended protocol. Luminata Classico or Luminata Crescendo (EMD Millipore) substrates were used to visualize immunoblots by enhanced chemiluminescence (ECL) of horseradish peroxidase (HPR)-conjugated secondary antibodies. β-actin or vinculin served as a control to ensure equal protein loading. We performed densitometry of Western blots using ImageJ software (Ver 1.49, http://imagej.nih.gov/ij (accessed on 7 July 2018).

2.8. Proliferation

Proliferation was examined using the CellTiter 96® Aqueous One Solution Cell Proliferation assay (Promega). Cells (5×10^3 cells) were plated onto 96-well plates. After 24 h, CellTiter 96® dye (10 µL) was added to each well and the absorbance was measured at 490 nm using a microplate reader (Epoch Microplate Spectrophotometer, BioTek Instruments, Winooski, VT, USA). For the siRNA experiments, we transfected the cells with siRNA for 72 h as described above, then we lifted and plated transfected cells onto 96-well plates for 24 h, added CellTiter96® dye (Promega), and read the plates. Proliferation experiments were completed with at least three biologic replicates and data reported as fold change $\pm$ standard error of the mean (SEM).

2.9. Growth Curve

Cells (5×10^4) were plated in a 12-well plate in adherent conditions in 12-well plates. In order to measure cell growth over time, cells were lifted and live cells were counted after being stained with trypan blue at 1, 2, 3, and 4 days for AS cells and at 5, 7, and 9 days for BE cells.

2.10. Cell Cycle

AS WT and STRAP KO cells were plated in 6-well plates and maintained in AS media with decreased FBS (4%). After 24 h, cells (5×10^5) were washed with PBS and fixed with 100% ethanol at 4 °C for at least 30 min. After a second wash with PBS, cells were stained with propidium iodide (Invitrogen), 0.1% TritonX (Active Motif, Carlsbad, CA, USA), and RNAse A (0.1 mg/mL, Qiagen) and cell cycle data were obtained using the FACSCalibur™ Flow Cytometer (BD Biosciences) and analyzed using the FlowJo software (FlowJo, LLC, Ashland, OR, USA).

2.11. Migration and Invasion

The effect of STRAP KO or shRNA inhibition on migration was assessed using modified Boyden chamber assays. Cells (4×10^4) were seeded onto 8 μM pore inserts (TransWell®, Corning, Corning, NY, USA) and were allowed to migrate for 24 h with laminin (10 μg/mL, 100 μL, Trevigen, Gaithersburg, MD, USA) used in the outer well as a chemoattractant. Inserts were fixed with 4% paraformaldehyde for 10 min, stained with 1% crystal violet for 15 min, and photographed. Photographs were analyzed using ImageJ software (Ver 1.49, http://imagej.nih.gov/ij (accessed on 17 March 2019) to quantitate migration.

Invasion was evaluated similarly, except for a layer of Matrigel™ (1 mg/mL, 50 μL, BD Biosciences) which was used to coat the top of the insert membrane. Cells (4×10^4) were seeded in the upper chamber and allowed to invade through the Matrigel™ layer for 24 h toward the laminin chemoattractant in the outer well. Inserts were then fixed, stained, and photographed as described for migration. Photographs were analyzed using ImageJ to quantitate invasion.

2.12. Anchorage-Independent Growth

A soft agar assay was utilized to assess for anchorage-independent growth. A base layer of 1% noble agar mixed with culture media was established in 60 mm culture dishes. AS WT or STRAP KO cells (1×10^4) were plated in the top layer in the agar and culture media mixture. After 6 weeks, colonies were stained with crystal violet, imaged, and quantified using ImageJ.

2.13. RNA Sequencing and Analysis

Total cellular RNA was extracted from AS WT and AS STRAP KO cells using the RNAeasy kit (Qiagen) according to the manufacturer's protocol. UAB Genomics Core performed sample quality control, library preparation, and RNA sequencing. The Agilent 2100 Bioanalyzer was used to assess the total RNA quality, which was followed by two rounds of poly A+ selection and conversion to cDNA. The NEBNext® Ultra™ Directional RNA Library Prep Kit for Illumina® library generation kit (New England Biolabs) was used per the manufacturer's instructions. qPCR in a Roche LightCycler 480 with the Kapa Biosystems kit (Kapa Biosystems, Woburn, MA, USA) was used for library quantitation. The Illumina NextSeq500 was used to perform the sequencing using the latest versions of the sequencing reagents and flow cells with single-end 75 bp reads. Raw and processed data were deposited in the Gene Expression Omnibus (GEO, Accession GSE169322) [14].

STAR (version 2.7.3a) was used to align the raw RNA-Seq fastq reads to the human reference genome (GRCh38 p13 Release 32) from Gencode using parameters the following parameters: outReadsUnmapped Fastx; outSAMtype BAM SortedByCoordinate; outSAMattributes All –outFilterIntronMotifs RemoveNoncanonicalUnannotated [15].

Following alignment, Cufflinks (version 2.2.1) was used to assemble transcripts, estimate their abundances, and test for differential expression and regulation using parameters—library-type fr-firststrand-G–L [16]. Cuffmerge, which is part of Cufflinks, merged the Cufflinks transcripts across multiple samples using the default parameters. Cuffdiff found significant changes in transcript expression, splicing, and promoter usage using default parameters. For generating pathway analysis of biological processes, a data set containing gene identifiers and corresponding expression values was uploaded into Reactome Pathway Database [17]. Differentially expressed genes that met the fold change cutoff of ± 2 were considered for the analysis. Each identifier is mapped to its corresponding molecule in the Reactome database and pathways identified at a false discovery rate (FDR) of 0.05. Entities ratio is defined as the proportion of Reactome pathway molecules represented in the dataset.

2.14. *Real-Time PCR (qPCR)*

iScript cDNA Synthesis kit (Bio-Rad) was used to synthesize cDNA with 1 µg of RNA used in a 20 µL reaction. The reverse transcription products were stored at −20 °C until further use. SsoAdvanced™ SYBR® Green Supermix (Bio-Rad) was utilized according to the manufacturer's protocol for quantitative real-time PCR (qPCR). Primers specific for Octamer-binding transcription factor 4 (Oct4), homeobox protein Nanog, and β-actin were utilized (Applied Biosystems, Foster City, CA, USA). Nestin primers (forward: 5′-TCCAGGAACGGAAAATCAAG-3′, reverse: 5′-GCCTCCTCATCCCCTACTTC-3′) were designed using Primer3 web version 4.1.0 [18] and examined for non-specific binding using BLAST (NCBI). qPCR was performed with 10 ng cDNA in 20 µL reaction volume. Amplification was performed using an Applied Biosystems 7900HT cycler (Applied Biosystems). Cycling conditions were 95 °C for 2 min, followed by 39-cycle amplification at 95 °C for 5 s and 60 °C for 30 s. β-actin was utilized as an internal control. Gene expression was calculated using the ΔΔCT method [19] and reported as mean fold change ± SEM.

2.15. *CD133 Cell Surface Expression*

AS WT or STRAP KO cells (1×10^6) were labeled with CD133/1 (AC133)-APC (Miltenyi Biotec, San Diego, CA, USA) according to the manufacturer's instructions. Unlabeled cells served as negative controls. The percent of cells positive for APC was determined via flow cytometry using the FACSCalibur™ Flow Cytometer (BD Biosciences) and analyzed using the FlowJo software (FlowJo, LLC FlowJo, Ashland, OR, USA).

2.16. *Extreme Limiting Dilution Analysis*

AS WT or STRAP KO cells were plated in conditioned media in 96-well plates with a decreasing number of cells in each row of 12 wells (1000 to 10 cells). After one week, each well was assessed for formation of tumorspheres. The number of wells containing spheres in each group was counted and analyzed using the extreme limiting dilution analysis (ELDA) software [20] and a plot of the log proportion of negative wells versus the number of cells plated was generated. The slope of the line is the estimated log-active sphere-forming fraction. Tables showing estimated and 95% confidence intervals for the 1/(stem cell frequency) for each group were also generated.

2.17. *Animal Statement*

Animal experiments were approved by the University of Alabama at Birmingham (UAB) Institutional Animal Care and Use Committee (IACUC-09363) and were conducted within institutional, national, and NIH guidelines and in compliance with the Animal Research: Reporting of In Vivo Experiments (ARRIVE) guidelines.

2.18. *In Vivo Tumor Growth*

AS WT or STRAP KO cells (1.8×10^6 cells in 25% Matrigel™, Corning, Inc.) were injected into the right flank of 6-week-old female athymic nude mice ($n = 9$ per group, Envigo, Prattville, AL, USA). Tumors were measured three times weekly, and tumor volumes calculated with the formula [(width^2 × length)/2] mm^3 with width being the smallest measurement. After 21 days post-injection or when animals met IACUC parameters for euthanasia, the animals were humanely euthanized in their home cages with CO_2 and cervical dislocation. Tumors were harvested and prepared for further study.

2.19. Immunohistochemistry

In order to evaluate proliferation in the tumors, immunohistochemistry staining for Ki67 [21] was performed. Formalin-fixed paraffin-embedded samples of AS WT and STRAP KO flank tumors were cut (5 μm sections), baked for 1 h at 70 °C, deparaffinized, rehydrated, and steamed. The sections were quenched with 3% hydrogen peroxide and blocked with PBS-blocking buffer (BSA, powdered milk, Triton X-100, PBS) for 30 min at 4 °C. Ki67 staining was completed by adding the primary antibody anti-Ki67 (rabbit polyclonal, 1:100, AB9260, Millipore Sigma) and incubated overnight at room temperature. After PBS washing, secondary antibody for rabbit (R.T.U. biotinylated universal antibody, Vector Laboratories, Burlingame, CA, USA) was added for 1 h at 22 °C. VECTASTAIN Elite ABC reagent (PK-7100, Vector Laboratories) and Metal Enhanced DAB Substrate (Thermo Fisher Scientific) was used to develop the staining reaction. Slides were counterstained with hematoxylin. For each run, a negative control (rabbit IgG, 1 μg/mL, EMD Millipore) was included. A board-certified pediatric pathologist (EMM), blinded to the treatment groups, evaluated the Ki67 staining which was quantified by counting the number of Ki67 positive cells per 500 cells in a representative section of each tumor [21]. The mean was calculated and results reported as mean ± SEM.

2.20. Statistical Analysis

ImageJ was utilized to perform densitometry of immunoblots with each protein band being normalized to the background (http://www.yorku.ca/yisheng/Internal/Protocols/ImageJ.pdf and the protein band in question then normalized to the internal control (vinculin or β-actin). Normalized bands were compared to that of WT cells. All experiments were performed with a minimum of three biologic replicates. Data were reported as mean ± SEM of separate experiments. Student's *t* test or analysis of variance (ANOVA) were used where appropriate. Statistical significance was defined as $p \leq 0.05$.

3. Results

3.1. STRAP Knockdown Decreased Proliferation, Growth, and Motility

Immunoblotting confirmed the presence of STRAP in five long-term passage neuroblastoma cell lines (Supplemental Figure S1). STRAP knockdown (KD) was accomplished with transient transfection of siRNA. Immunoblotting confirmed decreased STRAP protein expression in AS and BE cells following siSTRAP transfection (Figure 1A). Knockdown with siSTRAP2 was more marked than with siSTRAP1. STRAP expression was not affected by transfection with RNAiMax or siNeg (Figure 1A). STRAP band intensity was quantified with densitometry (Figure 1A). We wished to determine if STRAP KD affected neuroblastoma proliferation. We chose siSTRAP2 (SASI_Hs02-00343131) for these studies since it demonstrated the most knockdown in AS cells and in BE cells by densitometry (Figure 1A). STRAP KD via siRNA significantly decreased proliferation in the AS cells by 26 ± 1% ($p \leq 0.001$) and in the BE cells by 4 ± 1% ($p \leq 0.05$), when compared to the siNeg transfected cells (Figure 1B).

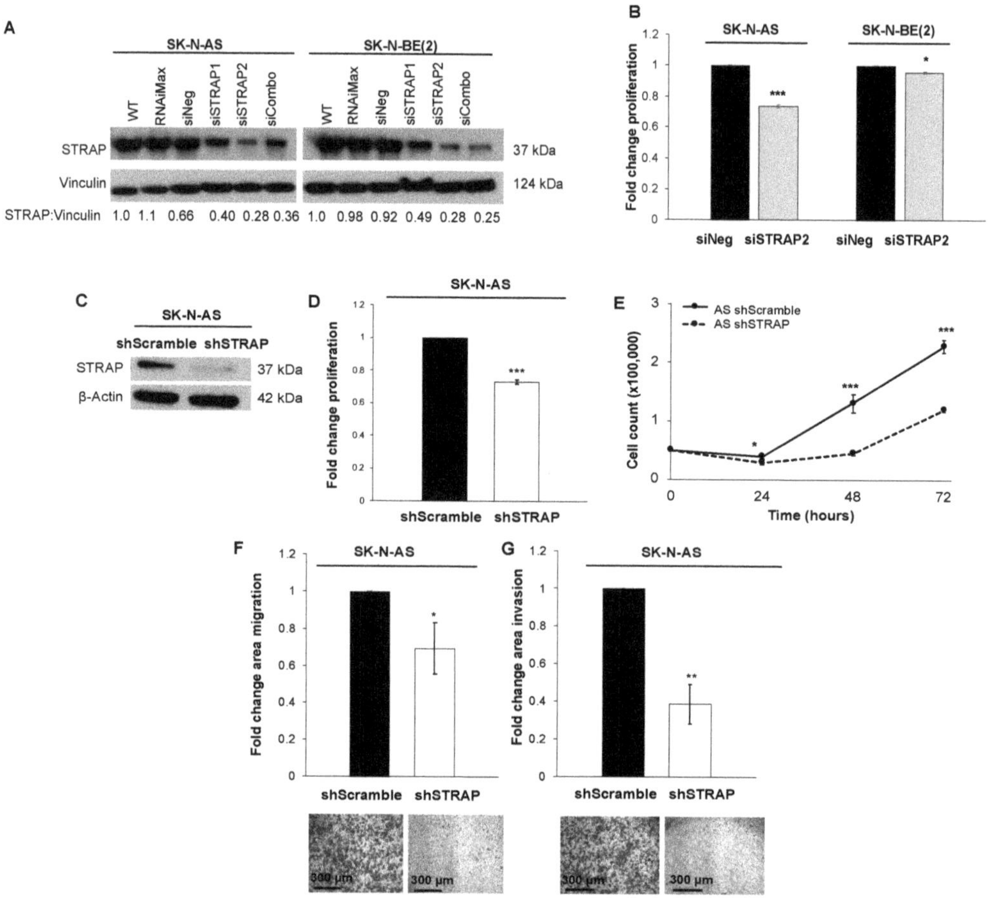

Figure 1. STRAP knockdown decreased neuroblastoma cell proliferation, growth, and motility. (**A**) SK-N-AS (AS) and SK-N-BE(2) (BE) cells were transfected with STRAP siRNA for 3 days. Immunoblotting of whole cell lysates revealed STRAP knockdown after transfection with siSTRAP1 or siSTRAP2 or both siRNAs (siCombo) compared to control siRNA (siNeg) or RNAiMax transfection reagent. Densitometry was used to quantify STRAP knockdown. Transfection with siSTRAP2 resulted in the most marked knockdown in STRAP expression in both the AS and BE cell lines. Therefore, siSTRAP2 was chosen for further studies. Vinculin was used as a loading control. (**B**) In both AS and BE cell lines, STRAP knockdown demonstrated a significant decrease in proliferation, measured by CellTiter 96® assay, compared to control siNeg cells. (**C**) Stable transfection of shScramble (control) and shSTRAP was established in AS cells. Immunoblotting of whole cell lysates confirmed the knockdown of STRAP in the shSTRAP cells. B-actin was used as a loading control. (**D**) Knockdown of STRAP with shRNA resulted in decreased proliferation in AS shSTRAP compared to AS shScramble cells. (**E**) The shScramble and shSTRAP cells were plated and counted at 24 h time points up to 72 h. The shSTRAP cells demonstrated significantly decreased growth over time compared to shScramble control cells. (**F**) To assess the effect of STRAP knockdown on motility, cells were seeded into modified Boyden chambers and allowed to migrate through an 8 μm micropore membrane for 24 h. There was a significant decrease in migration in shSTRAP cells compared to shScramble control cells. (**G**) Invasion was assessed similarly to migration with the addition of a layer of Matrigel™ to the top side of the insert. The shSTRAP cells demonstrated a significantly decreased ability to invade through the Matrigel™ layer compared to the shScramble control cells. Representative photomicrographs of stained migration and invasion inserts are shown in the graphs below and scale bars are shown in the bottom left corner of representative photographs. Experiments were repeated with at least three biologic replicates and data reported as mean fold change ± SEM. * $p \leq 0.05$, ** $p \leq 0.01$, and *** $p \leq 0.001$.

In order to further validate the siRNA findings, we next investigated STRAP knockdown (KD) using stable lentiviral transfection of AS cells with shScramble and shSTRAP plasmids. Due to the more rapid growth rate of the AS cell line versus BE, the focus of our phenotypic studies on STRAP were performed using AS cells to optimize experimentation and to produce the important conclusions discussed in this paper on the role of STRAP in neuroblastoma. Lentiviral transfection of AS cells resulted in successful STRAP KD (Figure 1C). The shSTRAP cells demonstrated a 27 $\pm$ 1% decrease in proliferation compared to the control shScramble cells ($p \leq 0.001$, Figure 1D). When examining cell growth over 72 h, we found a significant decrease in growth over time in the shSTRAP cells compared to the shScramble controls ($p \leq 0.05$, Figure 1E).

Other researchers have demonstrated the effects of STRAP on cancer cell motility [6,8,22]. In order to determine whether neuroblastoma cell motility was affected by STRAP KD, we utilized modified Boyden chamber assays to assess migration and invasion. The ability of the shSTRAP cells to migrate (65 $\pm$ 10% in shSTRAP cells compared to shScramble, $p \leq 0.05$, Figure 1F) and invade (38 $\pm$ 10% in shSTRAP cells compared to shScramble, $p \leq 0.01$, Figure 1G) was significantly decreased. These results indicate STRAP inhibition with shRNA affected motility in neuroblastoma.

3.2. STRAP KO Decreased Proliferation and Growth

Due to the potential for off-target effects seen with si/shRNA technology [23], we wished to determine the effect of decreased STRAP on the neuroblastoma cell phenotype following knockout of the STRAP gene. Using CRISPR-Cas9 gene editing technology, we established stable STRAP KO cells in two neuroblastoma cell lines, AS and BE. DNA gel showed the unedited WT DNA band to be at the expected 416 bp while the STRAP KO DNA band yielded lower molecular weight DNA fragments, confirming a cut in the DNA in both cell lines (Figure 2A,E). DNA fragments were excised and sent for Sanger sequencing. Using BLAST (https://pubmed.ncbi.nlm.nih.gov/22708584/, sequencing revealed that the unedited WT DNA aligned to the human reference gene, while the STRAP KO band possessed a gap which corresponded to the position of the STRAP gRNA that had been introduced. These findings confirmed the successful genetic knockout of STRAP in AS (Supplemental Figure S2A) and BE cells (Supplemental Figure S3A). In addition, immunoblotting confirmed the resulting absence of STRAP protein expression in the STRAP KO cells of both cell lines (Figure 2B,F).

Other investigators have shown the effects of STRAP on cancer cell growth [11,22]. Proliferation, as measured by CellTiter 96®, demonstrated a 23% decrease in the STRAP KO compared to AS WT cells ($p \leq 0.01$, Figure 2C). Examining cell growth over time, STRAP KO had significantly decreased growth compared to AS WT cells ($p \leq 0.05$, Figure 2D). In order to further validate the role of STRAP in these functions, we performed the same experiments in an additional neuroblastoma cell line, BE, with stable STRAP CRISPR-Cas9 KO. Similar to the findings in the AS cells, STRAP KO in BE cells resulted in decreased proliferation (52%, $p \leq 0.01$, Figure 2G) and growth ($p \leq 0.01$, Figure 2H) compared to BE WT cells.

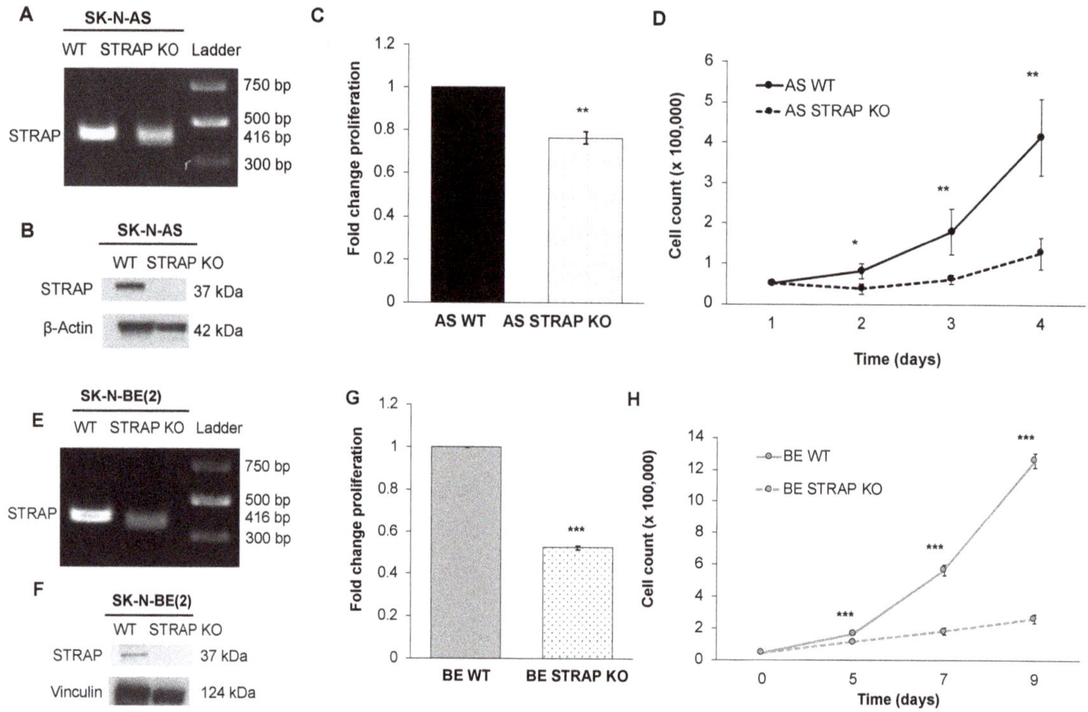

Figure 2. CRISPR-Cas9 gene knockout of STRAP decreased proliferation and growth in SK-N-AS and SK-N-BE (2) cells. (**A**) Gel electrophoresis was used to examine PCR products following amplification of genomic DNA isolated from clones transfected with STRAP gRNA and using STRAP primers. The STRAP band was found at the expected size of 416 bp in unedited SK-N-AS (AS) WT cells. The AS STRAP KO demonstrated two DNA fragment bands at a lower molecular weight than the WT, indicating a cut in the DNA following CRISPR-Cas9 gene editing. (**B**) Immunoblotting of whole cell lysates confirmed the absence of detectable STRAP protein in AS STRAP KO cells. (**C**) AS WT and STRAP KO cells were plated and proliferation was measured with CellTiter 96® assay at 24 h. AS STRAP KO demonstrated significantly decreased proliferation compared to AS WT. (**D**) AS WT and AS STRAP KO cells were plated and counted for 72 h. The AS STRAP KO cells had significantly decreased cell growth over time compared to AS WT cells. (**E**) Similar to AS cells following CRISPR-Cas9 gene editing, a DNA gel confirmed the presence of the STRAP band at the expected 416 bp in SK-N-BE (2) (BE) WT cells and two DNA fragments in the BE STRAP KO cells. (**F**) Immunoblotting of whole cell lysates confirmed the absence of detectable STRAP protein in BE STRAP KO cells. Similar to AS STRAP KO cells, BE STRAP KO cells had decreased (**G**) proliferation and (**H**) growth over time compared to BE WT cells. Data reported as mean fold change ± SEM. Experiments were repeated with at least three biologic replicates. * $p \leq 0.05$, ** $p \leq 0.01$, and *** $p \leq 0.001$.

3.3. STRAP KO Resulted in Failure to Progress through the Cell Cycle

In order to investigate a potential mechanism by which STRAP KO decreased cell proliferation, we examined the cell cycle. There was a significant decrease in the percentage of AS STRAP KO cells in the S phase (17.5% vs. 28.2%, STRAP KO vs. AS WT, $p \leq 0.01$, Figure 3A) and an increase in the percentage of AS STRAP KO cells in the G1 phase (62.7% vs. 53.8%, STRAP KO vs. AS WT, $p = 0.08$, Figure 3A). Data from three biologic replicates are presented in tabular form in Supplemental Table S1. The decrease in S phase provides evidence that STRAP KO cells do not progress through the cell cycle as well as their WT counterparts. These results support our findings of a decrease in proliferation in the STRAP KO cells (Figure 2C). Representative histograms of AS WT and AS STRAP KO cells from a single experiment are presented in Figure 3B.

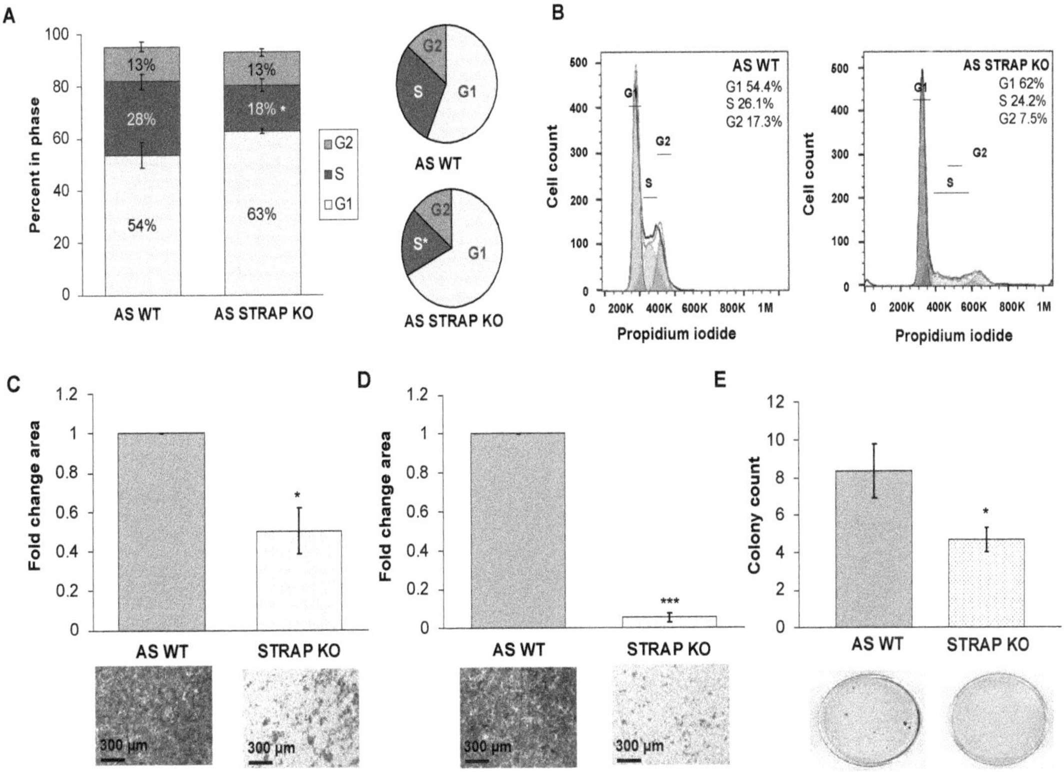

Figure 3. STRAP knockout diminished cell cycle progression and motility. (**A**) AS WT and AS STRAP KO cells were serum starved for 24 h then plated in routine culture media and stained with propidium iodide after 24 h. Flow cytometry was used to analyze progression through the cell cycle. AS STRAP KO had significantly decreased the percentage of cells in the S phase and an associated increased percentage of cells was observed in the G1 phase. Values represent mean percentage of cells in phase from three independent biologic experiments. (**B**) Representative histograms of AS WT and AS STRAP KO cells from a single experiment. (**C**) AS WT and AS STRAP KO cells were allowed to migrate for 24 h through a micropore membrane. Representative photomicrographs of migration inserts stained with crystal violet are shown beneath the graphs. Scale bars are shown in the bottom left corner of the representative photographs. AS STRAP KO cells had a significant decrease in migration compared to AS WT cells. Migration was reported as mean fold change area of migration ± SEM. (**D**) Similarly, AS STRAP KO cells demonstrated significantly decreased invasion after 24 h compared to AS WT. Representative photomicrographs of stained invasion inserts are shown below the graphs and scale bars are shown in the bottom left corner of representative photographs. Invasion was reported as mean fold change area of invasion ± SEM. (**E**) In order to assess for anchorage-independent growth, AS WT and AS STRAP KO cells were grown in a soft agar for 6 weeks and colonies were stained and quantified. Colony count was significantly lower in AS STRAP KO than in AS WT cells, which indicates a decrease in anchorage-independent growth in the AS STRAP KO cells. Data reported as mean colony count ± SEM. Experiments were repeated with at least three biologic replicates. * $p \leq 0.05$ and *** $p \leq 0.001$.

3.4. *STRAP KO Decreased Migration and Invasion*

STRAP inhibition with shRNA technology decreased neuroblastoma motility (Figure 1E,F). We wished to determine whether the absence of STRAP would similarly affect the motility of neuroblastoma cells. As in the shRNA experiments, we employed modified Boyden chamber assays for migration and invasion. AS STRAP KO cells had significantly decreased migration (50% of cells migrated compared to AS WT, $p \leq 0.05$, Figure 3C) and invasion

(5% of cells invaded compared to AS WT, $p \leq 0.001$, Figure 3D). Representative photographs of the migration and invasion inserts are displayed below the graphs (Figure 3C,D).

Anchorage-independent growth is another common metric to determine the metastatic potential of cancer cells [24]. In order to assess anchorage independent growth, AS WT and AS STRAP KO cells were cultured in anchorage-independent conditions using soft agar assays. AS STRAP KO cells formed fewer colonies than AS WT under these conditions ($p \leq 0.05$, Figure 3E). These results suggest STRAP affected neuroblastoma cell motility and potential for metastasis.

3.5. Transcriptomic Effects of the Loss of STRAP

In order to evaluate the effect of STRAP KO on neuroblastoma at the transcriptomic level, we utilized RNA-seq to investigate changes in gene expression in AS STRAP KO compared to AS WT cells. We utilized a volcano plot to represent the respective differences in gene expression in AS STRAP KO compared to AS WT cells (Figure 4A). Based on RNA-seq, there was a significant downregulation by at least two-fold of 367 genes in AS STRAP KO vs. AS WT, while 363 genes were significantly upregulated by at least two-fold ($p \leq 0.05$, Figure 4B). We utilized the Reactome Pathway Database to evaluate pathways associated with the genes downregulated by at least two-fold in the AS STRAP KO cells and found that genes downregulated following STRAP KO were associated with metabolic pathways, cell signaling transduction, gene transcription, and cell cycle progression (Figure 4C). We further investigated genes commonly associated with the malignant phenotype, such as apoptosis [25] and metastasis [26,27]. We found that STRAP KO downregulated genes were involved in stemness, metastasis, and multiple oncogenic signaling pathways such as TGF-β and WNT/β-catenin signaling, while genes associated with apoptosis and differentiation were upregulated in the AS STRAP KO cells (Figure 4D). We were also interested in genes associated with promoting tumor growth [28] and found several that were downregulated in the AS STRAP KO cells, notably platelet-derived growth factor receptor ß (PDGFRβ) (Figure 4D). PDGFRβ is a molecule of interest in neuroblastoma [29–31]. We examined PDGFRβ protein expression using immunoblotting and found a decrease in PDGFRβ expression in the AS STRAP KO compared to AS WT cells (Figure 4E), further validating our findings at the protein level.

3.6. STRAP KO Decreased Tumor Cell Stemness

STRAP increased cancer cell stemness in colorectal [11] and hepatocellular carcinoma [32]. After examining the transcriptomic data described, we found gene expression of the stemness markers NANOG and SOX 2 to be downregulated in the STRAP KO cells (Figure 4D). Therefore, we wished to determine if STRAP played a role in maintaining a stem cell-like phenotype in neuroblastoma. We performed qPCR to examine the mRNA abundance of the common neuroblastoma stem cell markers, Oct4, Nanog [33], and Nestin [34]. The AS STRAP KO cells had a decreased abundance of mRNA of these three stem cell markers compared to AS WT cells ($p \leq 0.001$, Figure 5A). We also investigated another feature associated with neuroblastoma stemness, which is the expression of the cell surface marker CD133 [35], using flow cytometry. CD133 cell surface expression decreased by 71% in AS STRAP KO compared to AS WT cells ($p \leq 0.01$, Figure 5B).

The ability of cells to form tumorspheres in low attachment serum-free conditions is a hallmark of stemness. An extreme limiting dilution analysis was employed to examine the effect of AS STRAP KO on tumorsphere formation. AS STRAP KO cells possessed significantly decreased ability to form tumorspheres compared to AS WT cells (Figure 5C). The stem cell frequency of AS STRAP KO cells was 1/796 (95% confidence interval of 1/1054–1/601) vs. 1/24 (1/30–1/20) in AS WT ($p \leq 0.001$, Supplemental Table S2). These results demonstrate STRAP KO decreases stemness in neuroblastoma.

3.7. Re-Introduction of STRAP Demonstrated Rescue of the Malignant Phenotype

In order to alleviate concerns of off-target effects resulting in the phenotype noted with the CRISPR-Cas9 generated STRAP KO cells, we performed STRAP rescue experiments. AS STRAP KO cells (3×10^3 cells) were plated and transfected with either FuGENE®, FuGENE® and EV pc-DNA plasmid, or FuGENE® and STRAP pc-DNA plasmid. Immunoblotting confirmed the absence and subsequent rescue of STRAP protein expression in the AS STRAP KO cells (Supplemental Figure S4A). In order to assess the phenotype of STRAP rescue cells, we evaluated proliferation. AS STRAP KO cells had significantly decreased proliferation ($p \leq 0.05$, Supplemental Figure S4B), which is consistent with data shown in the current study (Figure 3D). The re-introduction of STRAP (rescue) in the AS STRAP KO cells increased proliferation relative to the baseline level seen in AS WT cells, which demonstrates reconstitution of the malignant phenotype by re-introducing STRAP in the STRAP KO cells (Supplemental Figure S4B). These findings demonstrate STRAP conferred the phenotype seen in the AS STRAP KO cells.

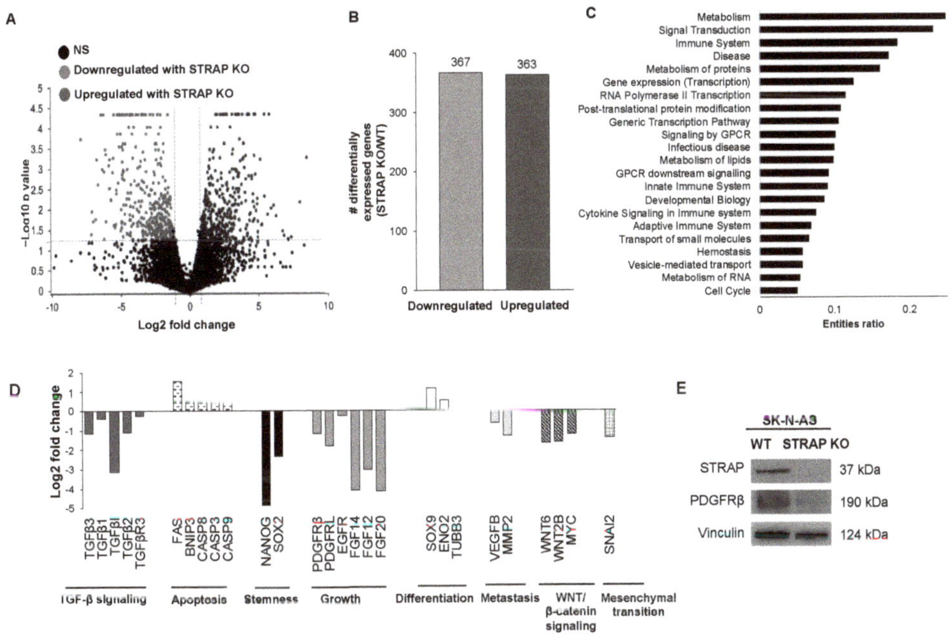

Figure 4. Knockout of STRAP resulted in changes in expression of oncogenic genes. (**A**) RNA sequencing was performed on AS WT and AS STRAP KO cells. A volcano plot of expressed genes is shown with the dotted horizontal line representing a p value of 0.05 and the vertical dotted lines representing a Log2 fold change of ± 1. Dark grey dots represent genes that were significantly upregulated by at least two-fold following STRAP KO and the light grey dots representing those that were significantly downregulated by at least two-fold. Non-significant (NS) genes (p value > 0.05) are shown in black. (**B**) Differentially expressed genes following STRAP KO of AS neuroblastoma cells are presented in a bar graph. There were 367 genes that were significantly downregulated while 363 genes were significantly upregulated in AS STRAP KO cells. (**C**) Reactome Pathway Database was used to evaluate downregulated genes in AS STRAP KO and pathways with the highest entities ratio presented. The ratio indicates the number of gene entities from the dataset that map to the Reactome pathway divided by the total number of entities in that pathway. Genes downregulated following STRAP KO were associated with metabolic pathways, cell signaling transduction, gene transcription, and cell cycle progression. (**D**) Fold change expression of selected genes in STRAP KO compared to AS WT cells is shown. STRAP KO of AS neuroblastoma cells downregulated genes involved in stemness, growth, metastasis, mesenchymal transition, and multiple oncogenic signaling pathways such as TGF-β and WNT/β-catenin signaling. On the contrary, genes linked to apoptosis and differentiation were upregulated following STRAP KO. (**E**) PDGFRβ protein expression was evaluated with immunoblotting and found to be decreased in AS STRAP KO cells compared to AS WT cells.

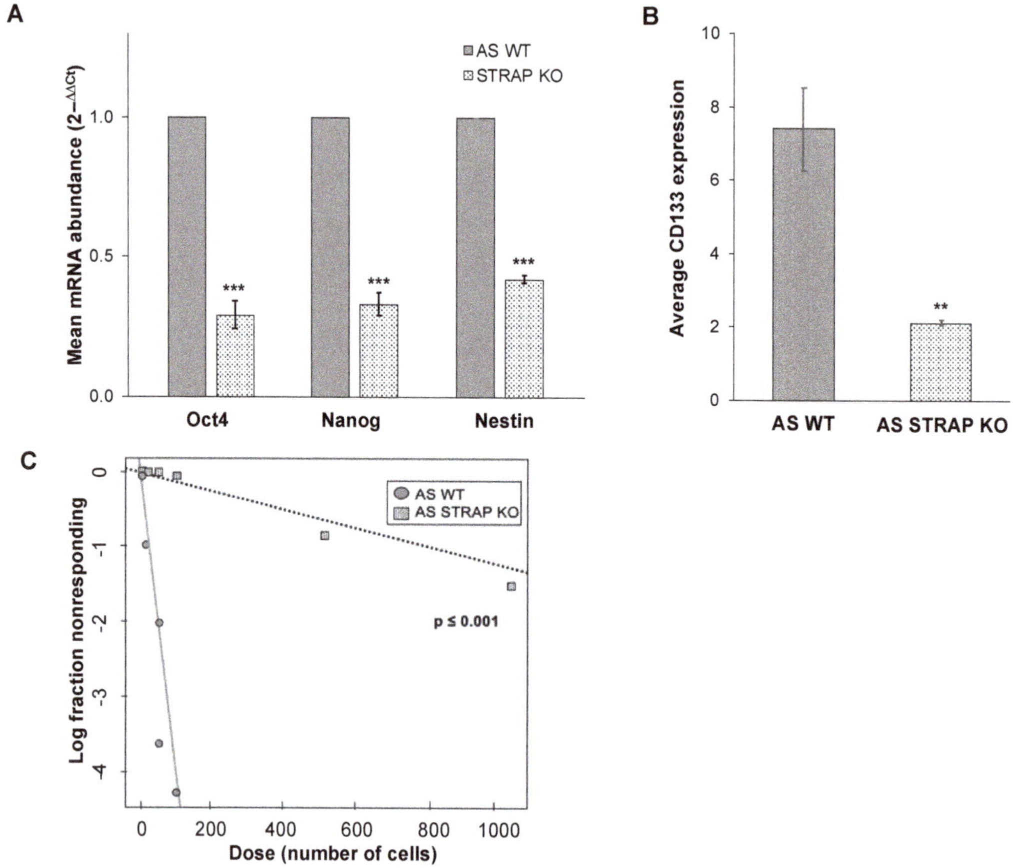

Figure 5. STRAP knockout decreased neuroblastoma cell stemness. (**A**). Quantitative real-time PCR was utilized to evaluate mRNA abundance of the common neuroblastoma stem cell markers Oct4, Nanog, and Nestin in AS WT and AS STRAP KO cells. AS STRAP KO cells had significantly decreased mRNA abundance of these stemness markers compared to AS WT. Gene expression was normalized to β-actin and calculated as fold change to AS WT using the ΔΔCt method. (**B**) AS WT and AS STRAP KO cells were stained with CD133 antibody and CD133 cell surface expression was determined using flow cytometry. STRAP KO cells had significantly decreased CD133 expression compared to WT cells. (**C**) In order to further evaluate stemness, an extreme limiting dilution analysis was utilized to assess tumorsphere formation. Cells were plated in conditioned media in non-adherent conditions at decreasing cell concentrations per well. Wells with tumorspheres present were counted and analyzed. AS STRAP KO cells had significantly decreased ability to form tumorspheres compared to AS WT cells. Data reported as mean ± SEM. Experiments were repeated with at least three biologic replicates. ** $p \leq 0.01$, *** $p \leq 0.001$.

3.8. STRAP KO Decreased Tumor Growth In Vivo

Based on the in vitro data, we proceeded to an in vivo model. AS WT or AS STRAP KO cells (1.8×10^6 cells in Matrigel™) were injected into the flank of athymic nude mice (n = 9 per group). Tumors were measured three times per week with calipers and tumor volumes were calculated as described. Animals with AS STRAP KO tumors had significantly decreased relative tumor growth (Figure 6A) and mean tumor volume (Supplemental Figure S5) compared to those with AS WT tumors. Immunoblotting confirmed the knockout of STRAP in the tumors from the animals injected with AS STRAP KO cells (Figure 6B). Similar to the in vitro studies AS STRAP KO tumors (Figure 6B) exhibited decreased protein expression of PDGFRβ, which is known to promote tumor growth in neuroblastoma [31]. Utilizing immunohistochemistry to detect Ki67, which is a marker of proliferation, we found that AS STRAP KO tumors had significantly less Ki67 staining than AS WT tumors and this signifies decreased proliferation (36 ± 24 vs. 159 ± 14 Ki67 positive cells out of 500 total cells, AS STRAP KO vs. AS WT, $p \leq 0.001$, Figure 6C,D). IgG negative staining control was completed with each IHC run and is shown in the following (Figure 6C, right lower corner inset).

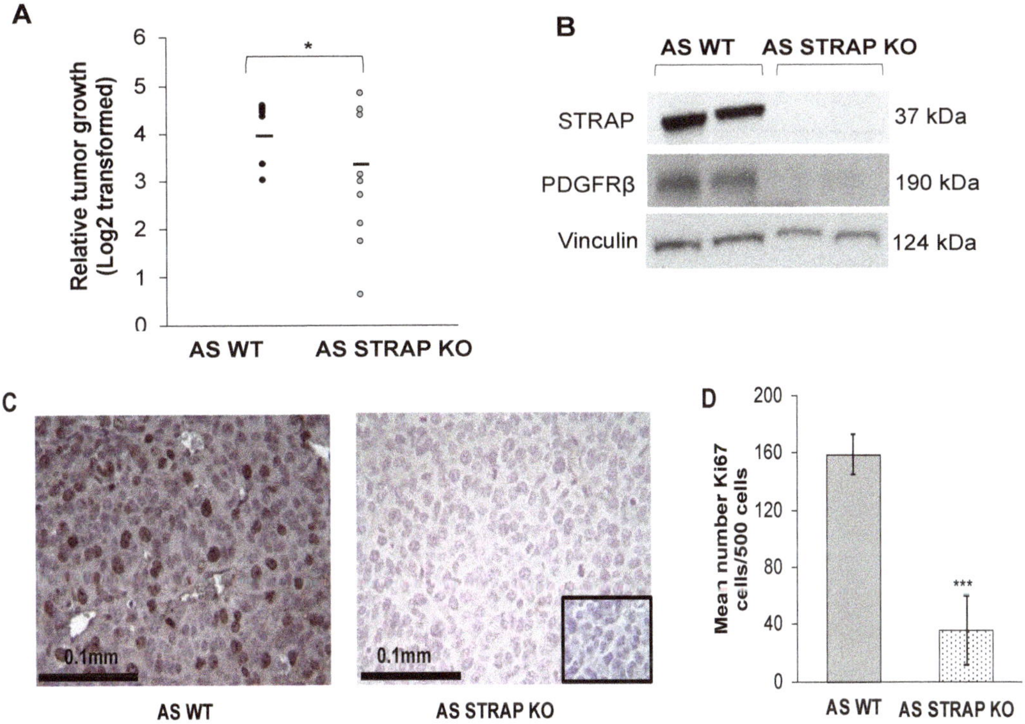

Figure 6. STRAP knockout decreased neuroblastoma tumor growth and proliferation in vivo. (**A**) AS WT or AS STRAP KO cells (1.8 × 10^6) were injected into the flanks of athymic nude mice ($n = 9$ per group). Tumor volumes were measured three times a week and tumors were harvested when IACUC parameters were met. STRAP KO tumors had a significantly decreased relative tumor growth compared to WT tumors. Results are reported as mean (Log2 transformed) ± SEM. (**B**) Using immunoblotting of tumor lysates, STRAP KO was confirmed in the STRAP KO tumors. Immunoblotting was also utilized to investigate PDGFRβ protein expression in the tumors and demonstrated decreased expression in the AS STRAP KO tumors compared to AS WT tumors. (**C**) Immunohistochemistry staining for Ki67, which is a marker for proliferation, was used on formalin-fixed paraffin-embedded AS WT and AS STRAP KO tumors. Representative pictures of Ki67 stained slides from each group are shown. A negative IgG control was included in each run (inset, right lower corner). (**D**) The number of Ki67 positive cells were counted per 500 cells. AS STRAP KO tumors had significantly decreased Ki67 positive cells compared to AS WT, indicating decreased proliferation. Results are reported as mean ± SEM. * $p \leq 0.05$ and *** $p \leq 0.001$.

4. Discussion

High-risk neuroblastoma continues to carry a poor prognosis and continued investigations are crucial to improve our understanding of the disease. STRAP is overexpressed in several malignancies including another pediatric cancer, osteosarcoma [7,8,32,36]. Recently, Jin et al. found that STRAP functions in neuronal development and STRAP knockout in Xenopus resulted in neural tube defects [37]. These findings were relevant to the present study, since neuroblastoma is an embryologic tumor derived from neural crest cells [2].

These previous studies supported the investigation of the potential for oncogenic function of STRAP in neuroblastoma. In the current study, we demonstrated that STRAP knockdown decreased neuroblastoma proliferation, growth, and motility. By establishing two stable neuroblastoma cell lines with the genetic knockout of STRAP and examining their malignant phenotype as well as the effect on the transcriptome, we provide evidence to suggest a role for STRAP in promoting neuroblastoma tumorigenicity in vitro and in vivo as well as maintaining the stem cell-like phenotype.

RNA interference is a method for transiently inhibiting gene expression and/or translation by targeting the corresponding RNA. Small interfering RNA (siRNA) and short hairpin RNA (shRNA) are two applications of RNA interference which have been utilized to achieve specific knockdown of a specific protein [38]. STRAP siRNA knockdown decreased migration and invasion of osteosarcoma cells [8]. However, Wu and colleagues found that knockdown of STRAP utilizing siRNA promoted hepatocellular carcinoma tumorgenicity in vitro and in vivo [36], suggesting that STRAP's role in cancer may not be tumor-specific. We initially investigated the knockdown of STRAP by utilizing siRNA in two neuroblastoma cell lines, which are AS (MYCN non-amplified) and BE (MYCN amplified), and found that STRAP knockdown resulted in decreased proliferation. These findings led us to further explore the role of STRAP in neuroblastoma.

Although siRNA and shRNA may achieve similar knockdown, the mechanism of action is different. SiRNA is a transient transfection, whereas shRNA carrying cells may be selected with an antibiotic for stable transfection. It is hypothesized that shRNA may have higher potency and fewer off-target effects than siRNA [38], enabling us to investigate the knockdown of STRAP utilizing shRNA in addition to siRNA. Datta et al. demonstrated decreased tumorigenicity in colorectal cancer cells following transfection with STRAP shRNA [6]. Similarly, we found that shRNA-mediated stable knockdown of STRAP resulted in decreased neuroblastoma cell proliferation, growth, and motility. The same investigators also demonstrated an increase in the sensitivity to 5-FU and oxaliplatin in colorectal cancer following shRNA STRAP knockdown [11]. Studying the effects of STRAP knockdown in combination with neuroblastoma chemotherapeutics will be an exciting avenue of future studies.

RNA interference has several limitations including a decrease in expression of the protein of interest, but not a total absence of that protein [39]. Due to these limitations, we proceeded to investigate the genetic knockout of STRAP. Using CRISPR-Cas9 gene editing technology, we established a stable cell line of STRAP KO cells in the neuroblastoma cell line SK-N-AS. Wang et al. also utilized CRISPR-Cas9 gene editing technology to establish a stable line of STRAP KO cells in hepatocellular carcinoma and demonstrated that STRAP KO led to decreased Wnt/β-catenin signaling, which was associated with decreased colony formation and stemness markers in vitro [32]. In order to further validate our findings of STRAP's role in neuroblastoma, we established an additional CRISPR-Cas9 stable knockout of STRAP in SK-N-BE (2) human neuroblastoma cells. We found that STRAP KO resulted in decreased growth, proliferation, motility, and anchorage-independent growth.

Using the findings from genetic sequencing of AS STRAP KO and AS WT cells, we found that numerous pathways were affected by the loss of STRAP. Specifically, pathways with the highest entities ratio that were associated with downregulated genes following the STRAP KO included those involved in metabolism, signal transduction, and cell cycles. We also found genes associated with stemness, growth, metastasis, and the known oncogenic pathways previously shown to be associated with STRAP by other investigators, such as TGF-β [22] and WNT/β-catenin [6], to be downregulated while apoptotic and differentiation genes were upregulated in the STRAP KO cells. Many of these pathways are important in neuroblastoma. Prior investigators showed that by stimulating neuroblastoma with TGF-β, there was an increase in cell migration and invasion [40]. Similarly, Tran et al. demonstrated the restoration of natural killer cells' cytotoxicity following treatment with a TGF-β inhibitor in neuroblastoma [41]. SiRNA knockdown of the WNT/β-catenin signaling pathway has been shown to decrease neuroblastoma viability, growth [42], and motility [43]. Therefore, the inhibition of these oncogenic signaling pathways observed with STRAP KO provides support for the inhibition of STRAP in neuroblastoma as a therapeutic strategy.

Platelet-derived growth factor receptors (PDGFRs) have been associated with growth, angiogenesis, cell viability, and proliferation in numerous malignancies, including glioma, prostate, breast, and pancreatic cancers [44]. By inhibiting PDGFRβ signaling, other investigators decreased cell viability, increased apoptosis, and induced cell cycle arrest in mesothelioma [45]. In colorectal cancer, lower PDGFRβ expression was associated with better prognosis and PDGFRβ knockdown using siRNA resulted in decreased colorectal cancer proliferation, growth, and invasion [46]. PDGFRβ was one of the downregulated genes observed with AS STRAP KO and we confirmed decreased protein expression in the AS STRAP KO cells. These studies provide evidence that STRAP may exert its phenotypic effects on cell proliferation and growth in neuroblastoma through its effect on expression of PDGFRβ. PDGFRβ has become a receptor of interest in neuroblastoma. Targeting PDGFRs with imantinib, which is a tyrosine kinase inhibitor, decreased neuroblastoma survival and proliferation as well as enhanced apoptosis when combined with doxorubicin [47]. Similarly, other investigators inhibited neuroblastoma growth in vivo and decreased tumor angiogenesis using SU11657, a tyrosine kinase inhibitor that targets PDGFRs [30]. In the current study, when STRAP was knocked out, we observed a decrease in PDGFRβ which suggests the presence of a potential mechanism by which STRAP may promote the malignant phenotype.

Stem cell-like cancer cells (SCLCCs) are a subpopulation of cancer cells that have been shown to be important for neuroblastoma progression, therapeutic resistance, and disease recurrence [34]. Therefore, therapies that target SCLCCs could greatly affect the course of the disease. STRAP has been implicated in maintaining colorectal cancer cell stemness. Jin and colleagues found that STRAP inhibition with shRNA decreased CD133 positive colorectal cancer cells and tumorsphere forming ability [11]. STRAP promoted the stem cell-like characteristics of colorectal cancer cells by activating the NOTCH pathway and the silencing of STRAP in these cells resulted in decreased stem cell phenotype [11]. In addition, STRAP knockout in hepatocellular carcinoma decreased mRNA abundance of stemness markers and liver progenitor genes, including AXIN2, LGR5, CD133, and CD44 [32]. In the current study, genetic evaluation demonstrated a downregulation of genes associated with neuroblastoma stemness following the knockout of STRAP. Further phenotypic evaluation of AS STRAP KO cells demonstrated a significant decrease in mRNA abundance of stemness markers, a decrease in CD133 cell surface expression, and decreased ability to form tumorspheres; all of these indicate that STRAP plays a role in promoting neuroblastoma cancer cell stemness.

Other investigators have examined STRAP knockdown and its effects on in vivo tumor growth. In colorectal cancer, STRAP knockdown with shRNA led to decreased tumor growth [11] and metastasis [6] in vivo. In addition, the knockdown of STRAP with shRNA in hepatocellular carcinoma resulted in decreased tumorigenicity and metastasis [36]. To our knowledge, the current study is the first to investigate the effect of STRAP knockdown and CRISPR KO on neuroblastoma and the first investigation of STRAP CRIPSPR KO on neuroblastoma tumor growth in vivo. Similar to results described in other in vivo studies, we demonstrated that the genetic knock out of STRAP resulted in decreased neuroblastoma tumor growth with significantly decreased proliferation in the KO tumors.

One limitation of CRISPR-Cas9 genetic knockout is the possibility for off-target activity or non-specific gene editing [12,48,49]. Validation of the target gene's role in the observed phenotype may be achieved by reintroducing the knocked-out gene in a rescue experiment. Other investigators have utilized this method. Chen and colleagues showed that CRISPR-Cas9 knockout of EZH2 in neuroblastoma resulted in decreased cell viability. They subsequently re-introduced EZH2, resulting in the rescue of the phenotype observed following EZH2 deletion. These findings confirmed EZH2′s importance in neuroblastoma survival [50]. In the present study, STRAP re-expression in the AS STRAP KO cells restored the phenotype to that observed in AS WT cells.

In the current study, we demonstrate that the knockdown and knockout of STRAP resulted in a decrease in the malignant phenotype in neuroblastoma. Our pre-clinical findings support the potential use of STRAP as a therapeutic target; however, there are currently no STRAP inhibitors available. Therefore, collaborations and development of inhibitors targeting STRAP are areas of interest and future studies. Interestingly, knocking down STRAP using siRNA had a more drastic effect on proliferation in AS cells compared to BE. One explanation for these findings is the difference in MYCN amplification in the two cell lines, which is related with a more aggressive phenotype in neuroblastoma [51]. An association between STRAP and MYCN has not yet been explored, providing us an additional avenue for future studies. STRAP is overexpressed in other cancers and has been associated with a more aggressive phenotype [6,8]; this provides an exciting avenue for future work to investigate STRAP expression levels in patient neuroblastoma specimens as a potential predictor for high-risk disease.

5. Conclusions

The data presented in the current study provide evidence that the inhibition of STRAP led to decreased neuroblastoma viability, proliferation, and motility in vitro and decreased tumor growth in vivo. Furthermore, we demonstrated STRAP's role in promoting cancer cell stemness as STRAP KO resulted in decreased mRNA abundance of stemness markers, CD133 expression, and tumorsphere formation. These findings provide an exciting avenue to continue the investigation of STRAP's role in neuroblastoma oncogenesis and as a potential therapeutic target for this disease.

Supplementary Materials: The following are available online at https://www.mdpi.com/article/10.3390/cancers13133201/s1, Figure S1: STRAP protein is present in neuroblastoma cell lines. Figure S2: Sanger sequencing confirmed successful knockout of STRAP in SK-N-AS cells. Figure S3: Sanger sequencing confirmed successful knockout of STRAP in SK-N-BE(2) cells. Figure S4: Re-introduction of STRAP DNA rescued the malignant phenotype in STRAP knockout cells. Figure S5: STRAP knockout decreased neuroblastoma tumor growth in vivo. Table S1: STRAP knockout diminished progression through the cell cycle. Table S2: STRAP knockout decreased neuroblastoma tumorsphere formation.

Author Contributions: Conceptualization, L.V.B., A.P.W., R.M., C.H.Q., S.C.H., P.K.D. and E.A.B.; methodology, L.V.B., A.P.W. and T.V.; software, D.K.C. and J.C.A.; validation, L.V.B. and A.P.W.; formal analysis, L.V.B., A.P.W., R.M., C.H.Q., T.V., E.M.-M., D.K.C. and J.C.A.; investigation, L.V.B., A.P.W., R.M., C.H.Q., J.E.S. and T.V.; resources, T.V., J.L.E., C.D.W. and P.K.D.; data curation, L.V.B. and A.P.W.; writing—original draft preparation, L.V.B.; writing—review and editing, L.V.B., A.P.W., R.M., C.H.Q., S.C.H., T.V., D.K.C., J.C.A. and E.A.B.; visualization, L.V.B., A.P.W. and E.A.B.; supervision, P.K.D. and E.A.B.; project administration, J.E.S.; funding acquisition, E.A.B. All authors have read and agreed to the published version of the manuscript.

Funding: This work was partially funded by institutional grants from the National Institutes of Health including T32 CA229102 (LV Bownes, R Marayati), T32 CA183926 (AP Williams), T32 CA091078 (LL Stafman), 5T32GM008361: Medical Scientist Training Program (CH Quinn), P30 AR048311 and P30 AI27667 to the Flow Cytometry Core at the University of Alabama at Birmingham, and P30 CA013148 to the O'Neal Comprehensive Cancer Center for support of the UAB Genomics Core. Further funding was provided by the Veterans Affairs Merit Review Award (1I01BX005143), UAB U54 Pilot Project (CA 118948) (PK Datta), Sid Strong Foundation, Elaine Roberts Foundation, Lombardi Cancer Research Fund/Starr Children's Fund, and Open Hearts Overflowing Hands (EA Beierle). The funding sources had no role in study design, analysis, or interpretation.

Institutional Review Board Statement: Animal experiments were approved by the University of Alabama at Birmingham (UAB) Institutional Animal Care and Use Committee (IACUC-09363) and were conducted within institutional, national, and NIH guidelines and in compliance with the Animal Research: Reporting of In Vivo Experiments (ARRIVE) guidelines.

Informed Consent Statement: Not applicable.

Data Availability Statement: Raw and processed data were deposited in the Gene Expression Omnibus (GEO, Accession GSE169322).

Acknowledgments: The authors wish to thank Pran Datta's laboratory for their collaboration, for providing plasmids, and shRNA cell lines. We would like to thank Anita Hjemeland's laboratory for their assistance with the qPCR. Furthermore, we would like to thank the UAB Genomics Core for assistance with genetic sequencing as well as Sagar Hanumanthu and UAB Comprehensive Flow Cytometry Core for his assistance with flow cytometry.

Conflicts of Interest: The authors declare no conflict of interest. The funders had no role in the design of the study; in the collection, analyses, or interpretation of data; in the writing of the manuscript or in the decision to publish the results.

References

1. Maris, J.M. The biologic basis for neuroblastoma heterogeneity and risk stratification. *Curr. Opin. Pediatr.* **2005**, *17*, 7–13. [CrossRef] [PubMed]
2. Salazar, B.M.; Balczewski, E.A.; Ung, C.Y.; Zhu, S. Neuroblastoma, a Paradigm for Big Data Science in Pediatric Oncology. *Int. J. Mol. Sci.* **2016**, *18*, 37. [CrossRef] [PubMed]
3. Smith, V.; Foster, J. High-Risk Neuroblastoma Treatment Review. *Children* **2018**, *5*, 114. [CrossRef]
4. Manoharan, R.; Seong, H.A.; Ha, H. Dual Roles of Serine-Threonine Kinase Receptor-Associated Protein (STRAP) in Re-dox-Sensitive Signaling Pathways Related to Cancer Development. *Oxid. Med. Cell. Longev.* **2018**, *2018*, 5241524. [CrossRef] [PubMed]
5. Datta, P.K.; Chytil, A.; Gorska, A.E.; Moses, H.L. Identification of STRAP, a novel WD domain protein in transforming growth factor-beta signaling. *J. Biol. Chem.* **1998**, *273*, 34671–34674. [CrossRef]
6. Yuan, G.; Zhang, B.; Yang, S.; Jin, L.; Datta, A.; Bae, S.; Chen, X.; Datta, P.K. Novel role of STRAP in progression and metastasis of colorectal cancer through Wnt/β-catenin signaling. *Oncotarget* **2016**, *7*, 16023–16037. [CrossRef]
7. Halder, S.K.; Anumanthan, G.; Maddula, R.; Mann, J.; Chytil, A.; Gonzalez, A.L.; Washington, M.K.; Moses, H.L.; Beauchamp, R.D.; Datta, P.K. Oncogenic Function of a Novel WD-Domain Protein, STRAP, in Human Carcinogenesis. *Cancer Res.* **2006**, *66*, 6156–6166. [CrossRef]
8. Pruksakorn, D.; Klangjorhor, J.; Lirdprapamongkol, K.; Teeyakasem, P.; Sungngam, P.; Chaiyawat, P.; Phanphaisarn, A.; Settakorn, J.; Srisomsap, C. Oncogenic roles of serine–threonine kinase receptor-associated protein (STRAP) in osteosarcoma. *Cancer Chemother. Pharmacol.* **2018**, *82*, 1039–1047. [CrossRef] [PubMed]
9. Jin, L.; Datta, P.K. Oncogenic STRAP functions as a novel negative regulator of E-cadherin and p21(Cip1) by modulating the tran-scription factor Sp1. *Cell Cycle* **2014**, *13*, 3909–3920. [CrossRef] [PubMed]
10. Schweigerer, L.; Breit, S.; Wenzel, A.; Tsunamoto, K.; Ludwig, R.; Schwab, M. Augmented MYCN Expression Advances the Malignant Phenotype of Human Neuroblastoma Cells: Evidence for Induction of Autocrine Growth Factor Activity. *Cancer Res.* **1990**, *50*, 4411–4416.
11. Jin, L.; Vu, T.; Yuan, G.; Datta, P.K. STRAP Promotes Stemness of Human Colorectal Cancer via Epigenetic Regulation of the NOTCH Pathway. *Cancer Res.* **2017**, *77*, 5464–5478. [CrossRef]
12. Ran, F.A.; Hsu, P.D.; Wright, J.; Agarwala, V.; Scott, D.A.; Zhang, F. Genome engineering using the CRISPR-Cas9 system. *Nat. Protoc.* **2013**, *8*, 2281–2308. [CrossRef]
13. Datta, K.P.; Moses, H.L. STRAP and Smad7 synergize in the inhibition of transforming growth factor beta signaling. *Mol. Cell. Biol.* **2000**, *20*, 3157–3167. [CrossRef]
14. Edgar, R.; Domrachev, M.; Lash, A.E. Gene Expression Omnibus: NCBI gene expression and hybridization array data repository. *Nucleic Acids Res.* **2002**, *30*, 207–210. [CrossRef]
15. Dobin, A.; Davis, C.A.; Schlesinger, F.; Drenkow, J.; Zaleski, C.; Jha, S.; Batut, P.; Chaisson, M.; Gingeras, T.R. STAR: Ultrafast universal RNA-seq aligner. *Bioinformatics* **2013**, *29*, 15–21. [CrossRef] [PubMed]
16. Trapnell, C.; Hendrickson, D.G.; Sauvageau, M.; Goff, L.; Rinn, J.L.; Pachter, L. Differential analysis of gene regulation at transcript resolution with RNA-seq. *Nat. Biotechnol.* **2013**, *31*, 46–53. [CrossRef]
17. Fabregat, A.; Sidiropoulos, K.; Viteri, G.; Forner-Martinez, O.; Marin-Garcia, P.; Arnau, V.; D'Eustachio, P.; Stein, L.; Hermjakob, H. Reactome pathway analysis: A high-performance in-memory approach. *BMC Bioinform.* **2017**, *18*, 142. [CrossRef] [PubMed]
18. Rozen, S.; Skaletsky, H. Primer3 on the WWW for General Users and for Biologist Programmers. *Methods Mol. Biol.* **2000**, *132*, 365–386. [CrossRef]
19. Winer, J.; Jung, C.K.S.; Shackel, I.; Williams, P.M. Development and validation of real-time quantitative reverse transcriptase-polymerase chain reaction for monitoring gene expression in cardiac myocytes in vitro. *Anal. Biochem.* **1999**, *270*, 41–49. [CrossRef]
20. Hu, Y.; Smyth, G.K. ELDA: Extreme limiting dilution analysis for comparing depleted and enriched populations in stem cell and other assays. *J. Immunol. Methods* **2009**, *347*, 70–78. [CrossRef]
21. Graham, D.; Magee, H.; Kierce, B.; Ball, R.; Dervan, P.; O'Meara, A. Evaluation of Ki-67 Reactivity in Neuroblastoma Using Paraffin Embedded Tissue. *Pathol. Res. Pract.* **1995**, *191*, 87–91. [CrossRef]

22. Reiner, J.E.; Datta, P.K. TGF-beta-dependent and -independent roles of STRAP in cancer. *Front. Biosci. Landmark Ed.* **2011**, *16*, 105–115. [CrossRef]
23. Putzbach, W.; Gao, Q.Q.; Patel, M.; Van Dongen, S.; Haluck-Kangas, A.; Sarshad, A.A.; Bartom, E.; Kim, K.-Y.A.; Scholtens, D.M.; Hafner, M.; et al. Many si/shRNAs can kill cancer cells by targeting multiple survival genes through an off-target mechanism. *eLife* **2017**, *6*, 6. [CrossRef]
24. Mori, S.; Chang, J.T.; Andrechek, E.; Matsumura, N.; Baba, T.; Yao, G.; Kim, J.W.; Gatza, M.L.; Murphy, S.; Nevins, J.R. Anchorage-independent cell growth signature identifies tumors with metastatic potential. *Oncogene* **2009**, *28*, 2796–2805. [CrossRef]
25. Mohammad, R.M.; Muqbil, I.; Lowe, L.; Yedjou, C.; Hsu, H.-Y.; Lin, L.-T.; Siegelin, M.D.; Fimognari, C.; Kumar, N.B.; Dou, Q.P.; et al. Broad targeting of resistance to apoptosis in cancer. *Semin. Cancer Biol.* **2015**, *35*, S78–S103. [CrossRef]
26. Apte, R.S.; Chen, D.; Ferrara, N. VEGF in Signaling and Disease: Beyond Discovery and Development. *Cell* **2019**, *176*, 1248–1264. [CrossRef]
27. Deryugina, E.I.; Quigley, J.P. Tumor angiogenesis: MMP-mediated induction of intravasation- and metastasis-sustaining neovasculature. *Matrix Biol.* **2015**, *44–46*, 94–112. [CrossRef]
28. Borriello, L.; Seeger, R.C.; Asgharzadeh, S.; DeClerck, Y.A. More than the genes, the tumor microenvironment in neuroblastoma. *Cancer Lett.* **2016**, *380*, 304–314. [CrossRef] [PubMed]
29. Matsui, T.; Sano, K.; Tsukamoto, T.; Ito, M.; Takaishi, T.; Nakata, H.; Nakamura, H.; Chihara, K. Human neuroblastoma cells express alpha and beta platelet-derived growth factor receptors coupling with neurotrophic and chemotactic signaling. *J. Clin. Investig.* **1993**, *92*, 1153–1160. [CrossRef] [PubMed]
30. Bäckman, U.; Christofferson, R. The Selective Class III/V Receptor Tyrosine Kinase Inhibitor SU11657 Inhibits Tumor Growth and Angiogenesis in Experimental Neuroblastomas Grown in Mice. *Pediatr. Res.* **2005**, *57*, 690–695. [CrossRef] [PubMed]
31. Wang, M.; Liu, Y.; Zou, J.; Yang, R.; Xuan, F.; Wang, Y.; Gao, N.; Cui, H. Transcriptional co-activator TAZ sustains proliferation and tumorigenicity of neuroblastoma by targeting CTGF and PDGF-β. *Oncotarget* **2015**, *6*, 9517–9530. [CrossRef]
32. Wang, W.; Li, S.; Liu, P.; Sideras, K.; Van De Werken, H.J.; Van Der Heide, M.; Cao, W.; Lavrijsen, M.; Peppelenbosch, M.P.; Bruno, M.; et al. Oncogenic STRAP Supports Hepatocellular Carcinoma Growth by Enhancing Wnt/β-Catenin Signaling. *Mol. Cancer Res.* **2019**, *17*, 521–531. [CrossRef]
33. Monajemzadeh, M.; Soleimani, V.; Vasei, M.; Koochakzadeh, L.; Karbakhsh, M. Expression and prognostic significance of Oct4 and Nanog in neuroblastoma. *APMIS* **2013**, *122*, 734–741. [CrossRef]
34. Bahmad, H.F.; Chamaa, F.; Assi, S.; Chalhoub, R.; Abou-Antoun, T.; Abou-Kheir, W. Cancer Stem Cells in Neuroblastoma: Expanding the Therapeutic Frontier. *Front. Mol. Neurosci.* **2019**, *12*, 131. [CrossRef]
35. Kamijo, T. Role of stemness-related molecules in neuroblastoma. *Pediatr. Res.* **2012**, *71*, 511–515. [CrossRef]
36. Wu, Q.; Yu, S.; Chen, J.; Li, Y.; Gao, Y. Downregulation of STRAP promotes tumor growth and metastasis in hepatocellular carcinoma via reducing PTEN level. *IUBMB Life* **2018**, *70*, 120–128. [CrossRef]
37. Jin, L.; Chen, Y.; Crossman, D.K.; Datta, A.; Vu, T.; Mobley, J.A.; Basu, M.K.; Scarduzio, M.; Wang, H.; Chang, C.; et al. STRAP regulates alternative splicing fidelity during lineage commitment of mouse embryonic stem cells. *Nat. Commun.* **2020**, *11*, 5941. [CrossRef]
38. Rao, D.D.; Vorhies, J.S.; Senzer, N.; Nemunaitis, J. siRNA vs. shRNA: Similarities and differences. *Adv. Drug Deliv. Rev.* **2009**, *61*, 746–759. [CrossRef]
39. Chen, S.-H.; Zhaori, G. Potential clinical applications of siRNA technique: Benefits and limitations. *Eur. J. Clin. Investig.* **2010**, *41*, 221–232. [CrossRef]
40. Lynch, J.; Fay, J.; Meehan, M.; Bryan, K.; Watters, K.M.; Murphy, D.M.; Stallings, R.L. MiRNA-335 suppresses neuroblastoma cell invasiveness by direct targeting of multiple genes from the non-canonical TGF-β signalling pathway. *Carcinogenesis* **2012**, *33*, 976–985. [CrossRef]
41. Tran, H.C.; Wan, Z.; Sheard, M.A.; Sun, J.; Jackson, J.R.; Malvar, J.; Xu, Y.; Wang, L.; Sposto, R.; Kim, E.S.; et al. TGFβR1 Blockade with Galunisertib (LY2157299) Enhances Anti-Neuroblastoma Activity of the Anti-GD2 Antibody Dinutuximab (ch14.18) with Natural Killer Cells. *Clin. Cancer Res.* **2017**, *23*, 804–813. [CrossRef] [PubMed]
42. Zhang, L.; Li, K.; Lv, Z.; Xiao, X.; Zheng, J. The effect on cell growth by Wnt1 RNAi in human neuroblastoma SH-SY5Y cell line. *Pediatr. Surg. Int.* **2009**, *25*, 1065–1071. [CrossRef] [PubMed]
43. Zins, K.; Schäfer, R.; Paulus, P.; Dobler, S.; Fakhari, N.; Sioud, M.; Aharinejad, S.; Abraham, D. Frizzled2 signaling regulates growth of high-risk neuroblastomas by interfering with β-catenin-dependent and β-catenin-independent signaling pathways. *Oncotarget* **2016**, *7*, 46187–46202. [CrossRef] [PubMed]
44. Heldin, C.-H. Targeting the PDGF signaling pathway in tumor treatment. *Cell Commun. Signal.* **2013**, *11*, 97. [CrossRef] [PubMed]
45. Melaiu, O.; Catalano, C.; De Santi, C.; Cipollini, M.; Figlioli, G.; Pellè, L.; Barone, E.; Evangelista, M.; Guazzelli, A.; Boldrini, L.; et al. Inhibition of the platelet-derived growth factor receptor beta (PDGFRB) using gene silencing, crenolanib besylate, or imatinib mesylate hampers the malignant phenotype of mesothelioma cell lines. *Genes Cancer* **2017**, *8*, 438–452. [CrossRef] [PubMed]
46. Fujino, S.; Miyoshi, N.; Ohue, M.; Takahashi, Y.; Yasui, M.; Hata, T.; Matsuda, C.; Mizushima, T.; Doki, Y.; Mori, M. Platelet-derived growth factor receptor-β gene expression relates to recurrence in colorectal cancer. *Oncol. Rep.* **2018**, *39*, 2178–2184. [CrossRef]

47. Palmberg, E.; Johnsen, J.I.; Paulsson, J.; Gleissman, H.; Wickström, M.; Edgren, M.; Östman, A.; Kogner, P.; Lindskog, M. Metronomic scheduling of imatinib abrogates clonogenicity of neuroblastoma cells and enhances their susceptibility to selected chemotherapeutic drugs in vitro and in vivo. *Int. J. Cancer* **2009**, *124*, 1227–1234. [CrossRef]
48. Zhang, X.-H.; Tee, L.Y.; Wang, X.-G.; Huang, Q.-S.; Yang, S.-H. Off-target effects in CRISPR/Cas9-mediated genome engineering. *Mol. Ther. Nucleic Acids* **2015**, *4*, e264. [CrossRef]
49. Peretz, L.; Besser, E.; Hajbi, R.; Casden, N.; Ziv, D.; Kronenberg, N.; Gigi, L.B.; Sweetat, S.; Khawaled, S.; Aqeilan, R.; et al. Combined shRNA over CRISPR/cas9 as a methodology to detect off-target effects and a potential compensatory mecha-nism. *Sci. Rep.* **2018**, *8*, 93. [CrossRef]
50. Chen, L.; Alexe, G.; Dharia, N.V.; Ross, L.; Iniguez, A.B.; Conway, A.S.; Wang, E.J.; Veschi, V.; Lam, N.; Qi, J.; et al. CRISPR-Cas9 screen reveals a MYCN-amplified neuroblastoma dependency on EZH2. *J. Clin. Investig.* **2017**, *128*, 446–462. [CrossRef]
51. Lee, J.W.; Son, M.H.; Cho, H.W.; Ma, Y.E.; Yoo, K.H.; Sung, K.W.; Koo, H.H. Clinical significance of MYCN amplification in patients with high-risk neuroblastoma. *Pediatr. Blood Cancer* **2018**, *65*, e27257. [CrossRef] [PubMed]

cancers

Article

Oral Metronomic Maintenance Therapy Can Improve Survival in High-Risk Neuroblastoma Patients Not Treated with ASCT or Anti-GD2 Antibodies

Xiaofei Sun [1,2,*,†], Zijun Zhen [1,2,†], Ying Guo [1,3], Yuanhong Gao [1,4], Juan Wang [1,2], Yu Zhang [1,5], Jia Zhu [1,2], Suying Lu [1,2], Feifei Sun [1,2], Junting Huang [1,2], Ruiqing Cai [1,2], Yizhuo Zhang [1,2], Juncheng Liu [6], Zizheng Xiao [1,7], Sihui Zeng [1,8] and Zhuowei Liu [1,9,*]

1 State Key Laboratory of Oncology in South China, Guanghzou 510060, China; zhenzj@sysucc.org.cn (Z.Z.); guoying@sysucc.org.cn (Y.G.); gaoyh@sysucc.org.cn (Y.G.); wangjuan@sysucc.org.cn (J.W.); zhangy@sysucc.org.cn (Y.Z.); zhuj@sysucc.org.cn (J.Z.); lusy@sysucc.org.cn (S.L.); sunff@sysucc.org.cn (F.S.); huangjt@sysucc.org.cn (J.H.); cairq@sysucc.org.cn (R.C.); zhangyz@sysucc.org.cn (Y.Z.); xiaozzh@sysucc.org.cn (Z.X.); zengsh@sysucc.org.cn (S.Z.)
2 Department of Pediatric Oncology, Sun Yat-sen University Cancer Center, Guanghzou 510060, China
3 Department of Clinical Research, Sun Yat-sen University Cancer Center, Guanghzou 510060, China
4 Department of Radiotherapy, Sun Yat-sen University Cancer Center, Guanghzou 510060, China
5 Department of Pathology, Sun Yat-sen University Cancer Center, Guanghzou 510060, China
6 Department of Pediatric Surgery, First Affiliated Hospital of Sun Yat-sen University, Guangzhou 510060, China; liujch@mail.sysu.edu.cn
7 Department of Nuclear Medicine, Sun Yat-sen University Cancer Center, Guanghzou 510060, China
8 Department of Radiology, Sun Yat-sen University Cancer Center, Guanghzou 510060, China
9 Department of Urology Surgery, Sun Yat-sen University Cancer Center, Guanghzou 510060, China
* Correspondence: sunxf@sysucc.org.cn (X.S.); liuzhw@sysucc.org.cn (Z.L.)
† Both are Co-first authors.

Citation: Sun, X.; Zhen, Z.; Guo, Y.; Gao, Y.; Wang, J.; Zhang, Y.; Zhu, J.; Lu, S.; Sun, F.; Huang, J.; et al. Oral Metronomic Maintenance Therapy Can Improve Survival in High-Risk Neuroblastoma Patients Not Treated with ASCT or Anti-GD2 Antibodies. *Cancers* **2021**, *13*, 3494. https://doi.org/10.3390/cancers13143494

Academic Editors: Saurabh Agarwal and Yang Jianhua

Received: 4 June 2021
Accepted: 7 July 2021
Published: 13 July 2021

Simple Summary: Low-dose metronomic chemotherapy has anti-angiogenic activity and inhibits tumor growth. Therefore, we investigated the benefits of low-dose metronomic maintenance therapy (MT) in high-risk neuroblastoma (NB) patients who are unable to undergo autologous stem cell transplantation (ASCT) or anti-GD2 antibody therapy. A total of 217 high-risk NB patients were enrolled. One hundred and eighty-five (85%) had a complete/very good partial remission/partial remission (CR/VGPR/PR) to treatment, of them, 167 patients with stage 4, that did or did not receive oral metronomic MT, 3 years of event-free survival (EFS) were 42.5% versus 29.4%, and overall survival (OS) was 71.1% versus 59.4%, respectively. Totally, 117 high-risk patients with oral metronomic MT had an EFS rate of 42.7%. The results were similar to those of ASCT from other studies. The toxicities of metronomic MT were lower. Our study showed that oral metronomic MT is an optimal option for high-risk NB patients without ASCT or anti-GD2 antibody therapy.

Abstract: Despite aggressive treatment, the prognosis of high-risk NB patients is still poor. This retrospective study investigated the benefits of metronomic maintenance treatment (MT) in high-risk NB patients without ASCT or GD2 antibody therapy. Patients aged $\leq$ 21 years with newly diagnosed high-risk NB were included. Patients with complete/very good partial remission (CR/VGPR/PR) to conventional treatment received, or not, oral metronomic MT for 1 year. Two hundred and seventeen high-risk NB patients were enrolled. One hundred and eighty-five (85%) had a CR/VGPR/PR to conventional treatment, of the patients with stage 4, 106 receiving and 61 not receiving oral metronomic MT, and the 3-year event-free survival (EFS) rate was 42.5 $\pm$ 5.1% and 29.6 $\pm$ 6%, respectively ($p = 0.017$), and overall survival (OS) rate was 71.1 $\pm$ 4.7% and 59.4 $\pm$ 6.4%, respectively ($p = 0.022$). A total of 117 high-risk patients with oral metronomic MT had EFS rate of 42.7 $\pm$ 4.8%. The toxicity of MT was mild. For high-risk NB patients without ASCT or anti-GD2 antibody therapy, stage 4, MYCN amplication and patients with stage 4 not receiving oral metronomic MT after CR/VGPR/PR were independent adverse prognostic factors. Oral metronomic MT can improve survival in high-risk NB patients in CR/VGPR/PR without ASCT or anti-GD2 antibodies therapy.

Keywords: neuroblastoma; metronomic chemotherapy; maintenance therapy; high-risk

1. Introduction

Neuroblastoma (NB) is the most common non-central nervous system pediatric malignant solid tumor. Children with low- or intermediate-risk NB have an excellent prognosis. However, high-risk NB patients have a poor prognosis, and their 3-year event-free survival (EFS) rate is about 40% among patients who receive induction chemotherapy, surgery, single high-dose chemotherapy with autologous stem cell transplantation (ASCT), and radiotherapy plus cis-retinoic acid maintenance therapy (MT) [1–6]. The survival rate of high-risk NB has increased to 65–75% when ASCT is combined with anti-GD2 antibody in the modern immunotherapy era [7–10]. Therefore, anti-GD2 antibodies immunotherapy after ASCT has been considered as the standard of care for high-risk NB disease.

ASCT is one of the consolidation therapies options for high-risk NB patients. Three large randomized controlled studies have shown an improvement in the 3-year EFS after ASCT (31% to 47%) compared to that after 4 courses of conventional chemotherapy or oral chemotherapy or no further treatment (22% to 31%) [1–3]. Single ASCT may increase EFS by about 10% in high-risk NB patients. High-risk NB patients can benefit from ASCT in terms of 3 to 5 years of EFS, while there is no OS benefit [1–4]. But long-term outcomes of the GPOH NB97 trial showed 10 years of OS were in favor of the ASCT groups [5], yet the other randomized studies showed that the difference in 10 years of OS was still no statistically significant [6]. However, few families in China have access to or can afford expensive anti-GD2 antibody therapy, and about 70% of high-risk NB patients who have achieved complete/very good partial remission (CR/VGPR) could not undergo ASCT for various reasons, including cost, poor physical tolerance, insufficient supportive care, or family reluctance. The high recurrence rate after ASCT is one of the main reasons that many families do not choose ASCT. For these patients, the survival rate is only 20–30%, thus the need to develop other treatments to delay recurrence and prolong survival. The continuous administration of oral low-dose metronomic MT is one of the options worthy of exploring.

A number of studies have shown that low-dose metronomic chemotherapy has anti-angiogenic activity and inhibit tumor growth, while stimulating the immune system with minimal toxicity, and has a potential to develop a tumor maintenance therapy [11–14]. The continuous administration of low-dose metronomic MT has been successfully used for decades in pediatric patients with acute lymphoblastic leukemia (ALL) [15]. The HD CWS-96 trial compared children with metastatic soft tissue sarcoma undergoing oral MT and high-dose chemotherapy and found that an oral MT was feasible and tolerable, with surprisingly a better outcome than high-dose salvage chemotherapy (5-year OS 52% versus 27%) [16]. Recent results from a multicenter, open-label, randomized, phase 3 trial from the European Pediatric Soft Tissue Sarcoma Study Group for patients with high-risk rhabdomyosarcoma showed that vinorelbine and continuous low-dose cyclophosphamide as MT improved the survival of patients with high-risk rhabdomyosarcoma. This approach will be the new standard of care for patients with high-risk rhabdomyosarcoma in the future [17]. This study has put low-dose metronomic MT in high-risk solid tumors back into the spotlight. However, there are few reports about low-dose metronomic MT in high-risk NB patients.

Since 2013, we have administered oral low-dose metronomic MT to high-risk NB patients with CR/VGPR/PR who are not treated with ASCT and/or anti-GD2 antibodies. The purpose of study was to determine whether those patients can benefit from oral metronomic MT. The prognostic factors among high-risk NB in this unique cohort Chinese children were also analyzed.

2. Materials and Methods

2.1. Patients

Children younger than 21 years with previously untreated high-risk NB who were treated at the Sun Yat-sen University Cancer Center (SYSUCC) between January 2013 and December 2018 were eligible. The high-risk NB patients who did not receive ASCT and/or anti-GD2 immunotherapy were enrolled to study. High-risk NB was defined as follows: (1) INSS stage 4 disease and age $\geq$ 18 months; (2) INSS stage 2, 3, and 4S disease and *MYCN* amplification; and (3) INSS stage 3 disease and age > 18 months with unfavorable pathology without *MYCN* amplification [18]. Informed consent was obtained from patients and their parents, and complete treatment and follow-up data were available. This study was approved by the Ethics Committee of SYSUCC.

The extent of tumor spread was evaluated by CT and/or MRI, bilateral bone marrow examination, ^{99}Tc bone scan, or ^{18}F-fluorodeoxy-D-glucose positron emission tomography/CT (MIBG scan was not available in China). Clinical stage was determined based on the International Neuroblastoma Staging System [19]. The Pathology Classification system was used for pathologic classification of tumors [18]. MYCN amplification was examined using fluorescence in situ hybridization.

2.2. Treatment Protocol

Patients received eight cycles of induction chemotherapy, surgery, local radiotherapy, followed by MT when they could not receive ASCT and/or anti-GD2 antibodies. For induction chemotherapy, the CAV (cyclophosphamide, pirarubicin, vincristine) and VIP (etoposide, ifosfamide, cisplatin) regimens were administered alternatively at 3-week intervals. Surgery was usually performed after four to six cycles of chemotherapy. If the tumor was deemed unresectable, patients were treated with another two cycles of CAV/VIP or changed to second-line chemotherapy consisting of vincristine, irinotecan, and temozolomide (VIT). Local radiotherapy 25–30 Gy was administered to all patients after surgery. Patients who achieved CR/VGPR/PR after comprehensive treatment were treated with low dose oral metronomic anti-angiogenic drugs (cyclophosphamide, vinorelbine, etoposide and/or topotecan, and celecoxib) as MT for 1 year (Table 1). The starting time of oral metronomic MT was that peripheral white blood cells reached 3×10^9/L after end of comprehensive therapy. Extent of disease evaluation was performed every 3 months during MT.

2.3. Evaluation of Response and Toxicity

Treatment response was evaluated every two cycles according to the international response standard for NB [19]. CR was defined as the absence of tumors with normal catecholamine levels. VGPR was defined as a 90% ~ 99% reduction in primary tumors, elimination of all measurable metastatic disease, normal catecholamine levels, with or without residual ^{99}Tc bone changes. PR was defined as a reduction of > 50% in primary and metastatic tumors. Stable disease (SD) was defined as the absence of new lesions or an increase of < 25% in existing lesions. Progressive disease (PD) was defined as any new lesion or any measurable lesions increasing by > 25%. Toxicity was assessed according to the Common Terminology Standard for Adverse Events version 4.03.

2.4. Statistical Analysis

Statistical analyses were performed using the SPSS, version 22 (IBM Corp, New York, NY, USA) and Stata software, version 15.1 (Stata Corp LLC TX, USA). Survival end points were described by their rate at specific time points with a 95% CI. Both EFS and OS were analyzed using the Kaplan-Meier method and compared using the log-rank test. Hazard ratios were calculated using univariable and multivariable Cox proportional hazards regression analyses. All statistical tests were two-sided, and a difference was considered significant when the p-value was < 0.05. Event-free survival was calculated as the time from enrollment to the first occurrence of relapse, progression, death from any cause, secondary cancer, or the time of the last contact if no event has occurred. Overall survival was calculated as the time from diagnosis to death or the last examination when patient remained alive. The last update on 31 October 2020 was used for this analysis.

Table 1. Chemotherapy regimen for newly diagnosed high-risk NB.

Chemotherapy Regimens	Drugs Dosage and Administration
Induction Chemotherapy:	
CAV (Cycle 1, 3, 5, and 7)	Cyclophosphamide (CTX) 1 g/m^2, iv drip for 0.5 h, d1–2 Mesna 330 mg/m^2 0, 4, and 8 h after CTX, iv, d1–2 Vincristine 1.5 mg/m^2, iv, d1 Pirarubicin 50 mg/m^2, iv, d1
VIP (Cycle 2, 4, 6, and 8)	Cisplatin 25 mg/m^2, iv drip for 3 h, d1–4 Etoposide 100 mg/m^2, iv drip for 3 h, d1–4 Ifosfamide (IFO) 1.5 g/m^2, iv drip for 3 h, d1–4 Mesna 300 mg/m^2, iv 0, 4, 8 h after IFO, d1–4
Second-line chemotherapy:	
VIT	Vincristine 1.5 mg/m^2, iv, d1 Irinotecan 50 mg/m^2, iv drip for 1.5 h, d1–5 Temozolomide 100 mg/m^2, po, d1–5
Maintenance therapy:	
Oral metronomic anti-angiogenic agents	At months 1, 3, 5, 7, 9, and 11: Cyclophosphamide 25–50 mg/m^2, po, d1–30 Vinorelbine 40 mg/m^2, po, qw × 3 Etoposide 25 mg/m^2, po, d1–21 Celecoxib 200 mg/m^2, po, bid d1–30 At months 2, 4, 6, 8, 10, and 12: Cyclophosphamide 25-50 mg/m^2, po, d1–30 Vinorelbine 40 mg/m^2, po, qw × 3 Topotecan [a] 1.4 mg/m^2, po, d1–5 Celecoxib 200 mg/m^2, po, bid d1–30

[a]: Since December 2017, topotecan has been discontinued due to no oral topotecan supply; NB: neuroblastoma; CAV: cyclophosphamide, pirarubicin, and vincristine; iv: intravenous injection; VIP: etoposide, ifosfamide, and cisplatin; po: oral administration; VIT: vincristine, irinotecan, and temozolomide.

3. Results

3.1. Patient Characteristics

A total of 217 patients newly diagnosed with high-risk NB were included in this study. The median age was 3.7 years (range: 0.5–21 years). Among them, one hundred and ninety-eight patients (91.2%) had stage 4 disease, and 172 (79.3%) had bone marrow/bone metastases. MYCN amplification was assessed in 160 tumors (73.7%), of which 46 (28.8%) were positive. The clinical characteristics of the patients are shown in Table 2.

3.2. Treatment Outcome

Among 217 patients with high-risk NB, 171 (78.8%) were treated with chemotherapy, surgery, and radiotherapy, 29 (13.4%) were treated with chemotherapy plus surgery, and 17 (7.8%) were treated with chemotherapy alone. After induction therapy, the extent of resection of the primary tumor was $\geq$ 90% in 164 (75.5%) patients. Fifty-nine patients achieved CR, 126 achieved VGPR/PR, 14 had SD, and 18 had PD to comprehensive therapies. The overall response rate was 85.2%. Among the 185 patients with CR/VGPR/PR, a total of 117 received oral metronomic MT, and 68 patients did not receive oral MT; of them, 28 cases received isotretinoin, and the remaining 40 cases received Chinese herbal medicine or other therapy, or no therapy. In 167 patients with stage 4, 106 received oral metronomic MT, and 61 did not receive oral metronomic MT (Figure 1).

Table 2. Clinical features of 217 high-risk NB patients without ASCT or anti-GD2 antibody therapy.

Feature	Entire Cohort (*n* = 217)	Cohort 1 (*n* = 167)	Cohort 2 (*n* = 117)
Sex			
Male	145 (66.8%)	122 (65.9%)	76 (65%)
Female	72 (33.2%)	63 (34.1%)	41 (35%)
Age			
Age ≥ 18 months at diagnosis	200 (92.2%)	171 (92.4%)	107 (91.5%)
Age < 18 months at diagnosis	17 (7.8%)	14 (7.6%)	10 (8.5%)
Stage			
INSS Stage 4	198 (91.2%)	167 (90.3%)	106 (90.6%)
INSS Stage 3	18 (8.3%)		10 (8.5%)
INSS Stage 4S	1 (0.5%)		1 (0.9%)
Primary tumor site			
Adrenal gland primary	121 (55.8%)	106 (57.3%)	72 (61.5%)
Retroperitoneal primary	71 (32.7%)	59 (31.9%)	36 (30.8%)
Mediastinum primary	17 (7.8%)	12 (6.5%)	5 (4.3%)
Other primary sites	8 (3.7%)	8 (4.3%)	4 (3.4%)
Metastatic sites			
Bone marrow/bone metastasis	172 (79.3%)	146 (78.9%)	94 (80.3%)
Other metastasis sites	45 (20.7%)	39 (21.1%)	23 (19.7%)
MYCN status			
MYCN amplified	46 (28.8%)	39 (28.3%)	24 (27.6%)
MYCN not amplified	114 (71.3%)	99 (53.5%)	63 (72.4%)
MYCN unknown	57 (26.3%)	47 (25.4%)	30 (25.6%)

NB: neuroblastoma. ASCT: autologous stem cell transplantation; INSS: International Neuroblastoma Staging System; Cohort 1: Stage 4 Patients with CR/VGPR/PR received or did not receive oral metronomic MT; Cohort 2: Patients with CR/VGPR/PR received oral metronomic MT.

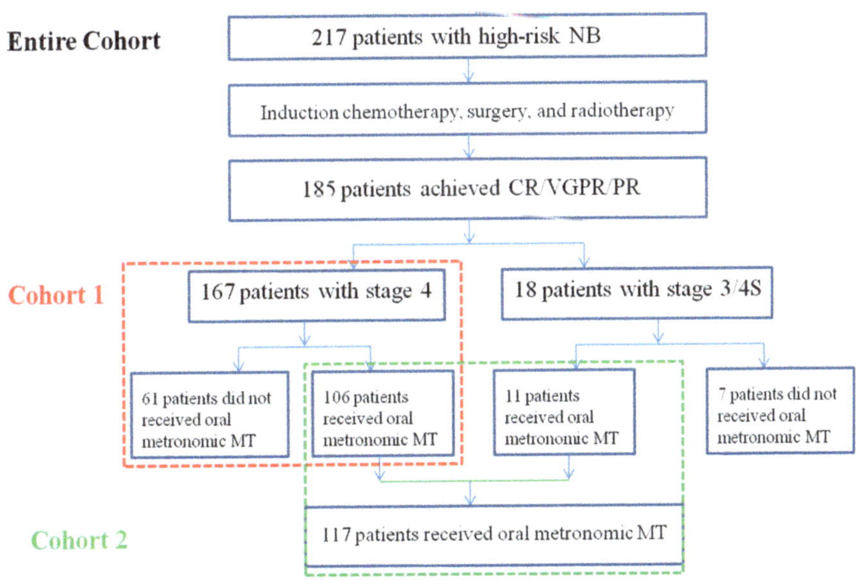

Figure 1. Therapy flow chart and cohorts of the patients with high-risk NB. NB: neuroblastoma; CR: complete response; VGPR: very good partial response; PR: partial response. MT: maintenance therapy.

A total of 126 patients experienced tumor recurrence or progression. Of these, 40 abandoned further therapy, and 86 patients received various salvage treatments, including chemotherapy, surgery, and radiotherapy. Anti-angiogenesis drugs, such as apatinib or arotinib, were administered alone or in combination with the salvage chemotherapy regimens.

3.3. *Survival*

The median follow-up was 41.3 months (range: 6.8–88.0 months). A total of 126 patients experienced tumor recurrence or progression, and 104 patients died. In the entire cohort, the 3-year EFS and OS rates were 36.3 ± 3.4% and 63.1 ± 3.4%, respectively. The EFS rates of patients with stage 4 disease was significantly lower than that of patients with stage 3/4S disease (33.8 ± 3.5% versus 62.2 ± 11.4%, $p = 0.011$). The EFS rates of patients with and without MYCN amplification were 28.6 ± 7.0% and 44.3 ± 4.8%, respectively ($p = 0.038$), and the corresponding OS rates were 48.6 ± 8.0% and 74 ± 4.3%, respectively ($p = 0.008$) (Table 3 and Figure 2).

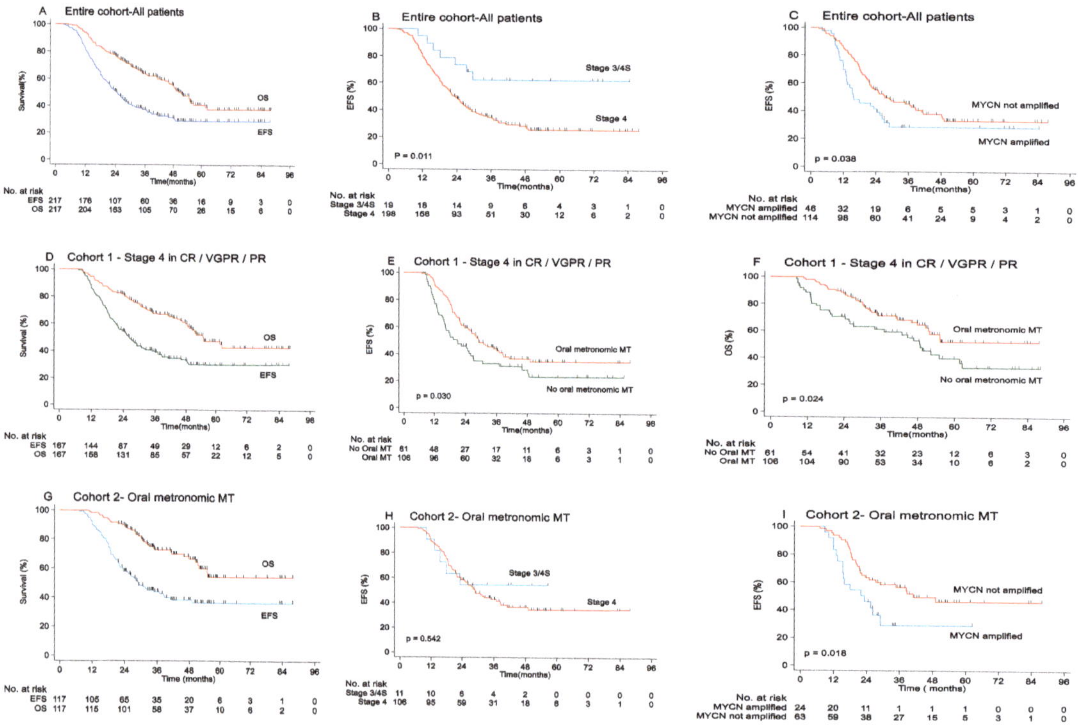

Figure 2. Survival curves of high-risk NB not treated with ASCT or anti-GD2 antibodies. (**A**) EFS and OS survival of all patients in the entire Cohort; (**B**) EFS of patients with different stage in the entire Cohort; (**C**) EFS of patients with MYCN amplified or no amplified in the entire Cohort; (**D**) EFS and OS survival in Cohort 1; (**E**) EFS of patients with or without metronomic MT in Cohort 1; (**F**) OS of patients with or without metronomic MT in Cohort 1; (**G**) EFS and OS survival in Cohort 2; (**H**) EFS of patients with different stage in Cohort 2; (**I**) EFS of patients with MYCN amplified or no amplified in Cohort 2.

Table 3. Treatment outcome of high-risk NB patients not treated with ASCT or anti-GD2 antibody.

	N	3-Year EFS (%)	*p*-Value	3-Year OS (%)	*p*-Value
Entire Cohort	217	36.3 ± 3.4		63.1 ± 3.4	
Age					
Age ≥ 18 months	200	35.5 ± 3.5	0.63	63.6 ± 3.5	0.575
Age < 18 months	17	45.3 ± 12.4		57.4 ± 12.3	
Sex					
Male	145	36.4 ± 4.1%	0.752	64.1 ± 4.1	0.965
Female	72	36.0 ± 6.0%		61.1 ± 6.1	
Stage					
INSS Stage 4	198	33.8 ± 3.5	0.011	60.3 ± 3.6	0.008
INSS Stage 3/4S	19	62.2 ± 11.4		83.3 ± 11.2	
MYCN status					
MYCN amplified	46	28.6 ± 7.0	0.038	48.6 ± 8.0	0.008
MYCN no amplified	114	44.3 ± 4.8		74.0 ± 4.3	
Metastatic sites					
BM/bone metastasis	172	32.6 ± 3.7	0.102	60.4 ± 3.9	0.097
Other metastasis sites	45	47.0 ± 7.7		70.5 ± 7.3	
Cohort 1 [a]	167	37.8 ± 3.9		66.9 ± 3.8	
Age					
Age ≥ 18 months	154	36.0 ± 0.4	0.212	67.5 ± 3.9	0.910
Age < 18 months	13	59.8 ± 14		59.2 ± 14.1	
Sex					
Male	110	37.2 ± 4.8	0.953	68.2 ± 4.6	0.652
Female	57	38.8 ± 6.9		64.2 ± 4.7	
MYCN status					
MYCN amplified	27	12.3 ± 9.9	0.019	37.8 ± 10.9	0.000
MYCN not amplified	93	46.1 ± 5.3		76.7 ± 4.6	
Oral MT					
Oral metronomic MT	106	42.5 ± 5.1	0.017	71.1 ± 6.7	0.022
No oral metronomic MT	61	29.6 ± 6		59.4 ± 6.4	
Cohort 2 [b]	117	42.7 ± 4.8		72.1 ± 4.5	
Age					
Age ≥ 18 months	107	41.4 ± 5.0	0.261	72.2 ± 4.6	0.563
Age < 18 months	10	68.6 ± 15.1		77.8 ± 13.9	
Sex					
Male	76	46.8 ± 5.9	0.513	70.5 ± 5.5	0.600
Female	41	36.2 ± 8.3		76.6 ± 7.4	
MYCN status					
MYCN amplified	24	16.8 ± 13	0.026	56.3 ± 13.4	0.042
MYCN not amplified	63	53.9 ± 6.5		78.1 ± 5.7	
Stage					
Stage 4	106	42.5 ± 5.1	0.556	71.1 ± 4.7	0.167
Stage 3/4S	11	54.5 ± 15		88.9 ± 10.5	

Abbreviations: NB: neuroblastoma; ASCT: autologous stem cell transplantation; BM: bone marrow; EFS: event free survival; OS: overall survival; [a]: Cohort 1 included patients in stage 4 with CR/VGPR/PR received or did not receive oral metronomic MT; [b]: Cohort 2 included patients with CR/VGPR/PR received oral metronomic MT.

The 3-years EFS rates of the 167 high-risk patients with stage 4 who received or did not receive oral metronomic MT after achieving CR/VGPR/PR (Cohort 1) were 42.5 ± 5.1% and 29.6 ± 6%, respectively (p = 0.017), and OS was 71.1 ± 4.7% and 59.4 ± 6.4%, respectively (p = 0.022). The patients with MYCN amplification had worse 3-year EFS and OS than patients without MYCN amplification (12.3% versus 46.1%, p = 0.019, and 37.8% versus 76.7%, p = 0.000) (Table 3 and Figure 2).

The 3-year EFS and OS rates of the 117 high-risk patients who received oral low dose metronomic MT after achieving CR/VGPR/PR (Cohort 2) were 42.7% ± 4.8% and 72.1% ± 4.5%, respectively. Further, among these patients, EFS and OS rates of those with MYCN amplification were worse than those without MYCN amplification (16.8% versus 53.9%, p = 0.026, and 56.3% versus 78.1%, p = 0.042) (Table 3 and Figure 2).

In univariate analysis, stage 4 disease, MYCN amplification and stage 4 without oral metronomic MT were significant adverse prognostic factors for EFS and OS. Multivariate analysis showed that the stage 4 disease (hazard ratio (HR) 0.246, p = 0.002), MYCN amplification (HR 0.462, p = 0.001), and stage 4 without oral metronomic MT (HR 0.559, p = 0.014) were predictive of an adverse EFS prognosis. For OS, the stage 4 disease (HR 0.147, p = 0.003), MYCN amplification (HR 0.351, p = 0.000), and stage 4 without oral metronomic MT (HR 0.366, p = 0.001) were significant adverse prognostic factors. Other factors, such as age, sex, and metastatic sites, had no significant impact on EFS or OS (Table 4).

Table 4. Univariate and multivariate analyses of prognostic factors of high-risk NB patients [a].

Category	No	3-Year Event-Free Survival			3-Year Overall Survival		
		Univariate	Multivariate [b]		Univariate	Multivariate [b]	
		p-Value	HR (95% CI)	*p*-Value	*p*-Value	HR (95% CI)	*p*-Value
Age							
≥18 months	200	0.631	0.680 (0.335, 1.379)	0.285	0.576	1.075 (0.481, 2.402)	0.859
<18 months	17						
Sex							
Male	145	0.752	1. 218 (0.798, 1.858)	0.362	0.965	1.125 (0.677, 1.869)	0.649
Female	72						
Stage							
Stage 4	198	0.014	0.246 (0.102, 0.594)	0.002	0.014	0.147 (0.041, 0.525)	0.003
Stage 3/4S	19						
MYCN amplified							
Yes	46	0.040	0.462 (0.292, 0.732)	0.001	0.009	0.351 (0.209, 0.590)	0.000
No	114						
Metastasis sites							
BM/Bone	172	0.104	1.151 (0.672, 1.972)	0.608	0.100	1.026 (0.550, 1.914)	0.937
Other	45						
Stage 4 in CR/VGPR/PR							
Oral MT [c]	106	0.030	0.559 (0.351, 0.891)	0.014	0.024	0.366 (0.197, 0.680)	0.001
No oral MT	61						

Abbreviations: BM: bone marrow; [a]: High-risk NB patients not treated with ASCT or anti-GD2 antibody; [b]: Multivariate analysis was adjusted for age, sex, stage, metastasis sites, MYCN amplified and stage 4 with oral metronomic MT in CR/VGPR/PR; [c]: Oral metronomic agents maintenance therapy.

Of the 126 patients with tumor recurrence or progression, 40 abandoned further treatment, all of whom died. Of the 86 patients who received salvage treatments, 47 died of tumor progression, and 38 were still alive at the end of the study. The 2-year OS after recurrence was 52.1 ± 6.0%.

3.4. The Toxicity of Oral Metronomic MT

The overall tolerance of oral metronomic MT was acceptable. MT was administered on an outpatient basis, and there was no treatment-related death. The most common toxicities were grade 1–2 hematological toxicity, accounting for about 80%. Grade 3 hematotoxicity accounted for 9%, no grade 4 toxicity. Grade 1–2 non-hematologic toxicity included transaminase elevation, nausea, gastritis, and creatinine elevation (Table 5). Oral metronomic MT was discontinued after 1–3 months in 5 patients, for various reasons.

Table 5. Toxicity of oral metronomic drugs maintenance therapy.

	Toxicity Grading [*n* (%)]				
	0	1	2	3	4
Hemoglobin	27 (23.1)	43 (36.8)	45 (38.5)	2 (1.7)	0 (0)
White blood cell	13 (11.1)	46 (39.3)	49 (41.9)	9 (7.7)	0 (0)
Platelets	116 (99.1)	1 (0.9)	0 (0)	0 (0)	0 (0)
Transaminase	104 (88.9)	10 (8.5)	2 (1.7)	1 (0.9)	0 (0)
Nausea	114 (97.4)	1 (0.9)	2 (1.7)	0 (0)	0 (0)
Creatinine	113 (96.6)	4 (3.4)	0 (0)	0 (0)	0 (0)
Gastritis	114 (97.4)	1 (0.9)	2 (1.7)	0 (0)	0 (0)

4. Discussion

This study was to probe a new approach to improve the outcomes of patient with high-risk NB inaccessible to ASCT or GD2 antibody therapy. Among the 217 high-risk NB patients, MYCN amplification was positive in 28.8% of in the detected patients. Patients with stage 4 accounted for 91.2%. The overall response rate (CR/VGPR/PR) was 85.2% to comprehensive treatment. None of the patients received the treatment of ASCT or anti-GD2 antibodies. The 3-year EFS and OS rates for the entire cohort were 36.3% and 63.1%. Surprisingly, patients with stage 4 receiving oral metronomic MT after achieving CR/VGPR/PR had better 3-year EFS and OS rates than those without oral metronomic MT (42.5% vs. 29.6%, and 71.1% vs. 59.4%); The patients with oral metronomic MT including stage 3/4S patients had 3-year EFS and OS rates of 42.7% and 72.1%. MYCN amplification and stage 4 disease were significant adverse prognostic factors.

It is widely known that the EFS rate of high-risk NB patients treated with ASCT is slightly higher than that of conventional chemotherapy or no further therapy [1–4]. A study from the Beijing Children Hospital in China reported that the 3-year EFS rates for high-risk NB patients who underwent ASCT (196 patients), or not (296 patients), were 43.7% and 36.7%, respectively ($p = 0.010$); the corresponding 3-year OS rates were 57.6% and 53.5%, respectively ($p = 0.153$) [20]. However, there are few studies on the impact of oral metronomic MT on the survival of high-risk NB patients without ASCT treatment. A randomized study from Germany comparing ASCT versus oral MT as consolidation therapy in patients with high-risk NB showed that ASCT had increased 3-year EFS compared with those allocated MT (47% vs. 31%) but did not have significantly increased 3-year overall survival. Moreover, the patients in the oral MT group only received oral cyclophosphamide at the dose of 150 mg/m2 per day on days 1 to 8 monthly for three consecutive months [2]. However, this therapy does not really meet the criteria of metronomic oral MT.

Another German retrospective study was conducted to compare the long-term survival outcomes of anti-GD2-antibody ch14.18 group with oral MT group (without ASCT) or no consolidation therapy group (without ASCT and oral MT). The EFS was better in the anti-GD2 antibody group than those in the oral MT group and in non-consolidation group (5-year EFS 50.5%, 34.1%, and 25.9%, respectively). Multivariable Cox regression analysis revealed anti-GD2 antibody ch14.18 consolidation improved outcome compared to no consolidation. However, no difference exists between the MT and anti-GD2-treated patients. Further, although anti-GD2 treatment may have prevented late relapses in this study, oral MT also appeared to be effective at preventing relapses [21]. A retrospective non-randomized study analysis from Memorial Sloan Kettering Cancer Center questioned the importance of ASCT in the modern era. Among the patients with high-risk NB treated with anti-GD2 antibody 3F8/GM-CSF in 1st CR/VGPR, 60 had ASCT, and 110 had none prior to immunotherapy. Five-year EFS rate for ASCT versus non-ASCT patients was 65% versus 51% (p = 0.128), while OS rate was 76% versus 75% (p = 0.975), respectively. In multivariate analysis, ASCT was not prognostic. Despite the limitations of nonrandomized retrospective single institutional study, this sizable cohort did raise a hypothesis that ASCT may not be necessary for all patients, especially if there is an effective anti-GD2 immunotherapy for consolidation after induction therapy [22].

In our study, all high-risk NB patients did not undergo ASCT or anti-GD2 antibodies. Among them (cohort 2), 60% of patients with CR/VGPR/PR received oral metronomic MT, which led to 3-year EFS and OS rates of 42.7% and 72.1%. Patients in stage 4 receiving oral metronomic MT had a better survival rate than those who did not receive oral metronomic MT (cohort 1). These results were similar to those of ASCT from other studies but lower than those of anti-GD2 antibody immunotherapy [1–10]. The toxicities of oral metronomic MT were well tolerated, and cost was lower. The most common toxicities were grade 1–2 hematological toxicity, accounting for about 80%. Grade 3 hematological toxicities accounted for about 9%, no grade 4 toxicity. The starting time of oral metronomic MT was that peripheral white blood cells reached 3×10^9/L after end of comprehensive therapy. Because all patients did not receive ASCT, bone marrow function recovered faster after the end of comprehensive treatment, and oral MT could be started quickly. Our study suggests that oral metronomic MT is an optimal option for high-risk NB patients who have achieved CR/VGPR/PR after comprehensive therapies without undergoing ASCT or anti-GD2 antibody therapy.

These metronomic MT drugs used in our study included low-dose cyclophosphamide, vinorelbine, topotecan/etoposide, and celecoxib as selective cyclooxygenase-2 inhibitors, which are potential anti-angiogenic agents [12]. Metronomic chemotherapy is defined as the continuous use of low dose chemotherapeutic agents. It was observed in vivo that tumors resistant to dose-intense therapy responded to the same agents at low continuous dosing [10]. Metronomic chemotherapy has been used in many refractory/relapsed solid tumors. Many studies have shown that low-dose metronomic chemotherapy combined with anti-angiogenic drugs can markedly enhance the anti-angiogenic efficacy [11–14,16,17]. Zapletalova reported the results of a multicenter study on metronomic chemotherapy with the Combined Oral Metronomic Biodifferentiating Antiangiogenic Treatment (COMBAT) regimen in advanced pediatric malignancies. The COMBAT regimen includes low-dose daily temozolomide, etoposide, celecoxib, vitamin D, fenofibrate, and retinoic acid. They found that COMBAT was a feasible and effective treatment option for patients with relapsed/refractory malignancies, and it was well tolerated with low toxicity. They proposed introducing metronomic chemotherapy at earlier phases of antitumor treatment to allow further exploitation of the potential of low-dose metronomic chemotherapy, for instance, as MT [11]. Kieran et al. demonstrated the feasibility of combining low-dose metronomic chemotherapy (alternating cyclophosphamide and etoposide) and anti-angiogenic agents (celecoxib and thalidomide) in children with recurrent or progressive solid tumors. Their results showed that regimen was well tolerated and had some activity in some tumors [23]. These studies suggest that low-dose metronomic chemotherapy combined with anti-angiogenic drugs as MT may be beneficial to patients with high-risk solid tumors. Recent result from a randomized clinical trial has demonstrated low-dose metronomic MT can improve outcomes in patients with pediatric high-risk rhabdomyosarcoma (RMS) [17]. Our study also showed low-dose oral metronomic MT can benefit high-risk NB patients without ASCT or anti-GD2 antibodies therapy.

Based on the Children's Oncology Group (COG) neuroblastoma high-risk group assignment criteria, high-risk NB is defined as stage 4 disease with age $\geq$ 18 months; stage 2, 3, and 4S disease with MYCN amplification; and stage 3 disease with age >18 months with unfavorable pathology without MYCN amplification. High-risk NB patients receive the same treatment regimen and have a poor prognosis. As far as we know, MYCN amplification strongly predicts a poor prognosis in NB. A study from South Korea reported that there was no survival difference between patients with and without MYCN amplified in high-risk NB [24]. But, inferior EFS and OS in high-risk NB patients with MYCN amplification have been reported from the COG and other studies [25–28]. One study from Japan showed MYCN amplification was the most favorable prognostic factor for EFS of high-risk NB [29]. Our results showed that the patients with MYCN amplified had worse the EFS and OS rates than those without MYCN amplified in high-risk NB. For stage 4 NB patients with MYCN amplification, new treatments need to be explored.

Another prognostic factor is staging. Most high-risk NB patients are in the stage 4, but the stage 3 patients with MYCN amplification or unfavorable pathology are also defined as high-risk patients. In our study, the patients with stage 4 accounted for 91.2%, while stage 3 patients with MYCN amplification or unfavorable pathology accounted for only 8.3%, and the EFS and OS rates in the stage 4 patients were worse than those in the stage 3 patients with MYCN amplification or unfavorable pathology (33.8% vs. 62.2%; 60.3% vs. 83.3%). In addition, stage 4 patients with CR/VGPR/PR who did not receive oral metronomic MT had worse survival than those with oral metronomic MT. Univariate and multivariate analysis also confirmed these results. Thus, our study demonstrated that stage 4 disease and MYCN amplification were independent adverse prognostic factors for high-risk NB patients. These results are similar to those reported in other studies [25–28,30].

Recurrence in patients initially defined as high-risk NB have a very poor prognosis, with an overall survival of only 11.0 months, and the reported 5-year overall survival rate for children after the initial relapse of NB is 6–20% [31–33]. The OS rate is closely related to whether the patients receive salvage treatment after recurrence. In this study, 126 patients experienced tumor recurrence or progression, and 40 of them abandoned further treatment and died. The remaining 86 patients received active salvage chemotherapy, surgery, radiotherapy, or other treatment, and most of these patients had effective tumor control after salvage therapy. Some patients relapsed repeatedly and were treated repeatedly. The 2-year overall survival rate after recurrence was 52.1%, and some patients survived for more than 5 years after recurrence. This may explain the reason of the high 3-year overall survival rate as 60–70% in our study. These relapsed patients who had not previously received ASCT tolerated salvage chemotherapy well. Our salvage chemotherapy regimen included some drugs that were not used as initial treatment and were effective in patients with recurrent NB. Therefore, our results suggest that the survival of high-risk NB patients with recurrence can be prolonged by salvage treatment.

5. Conclusions

Our study suggests that oral metronomic MT is an optimal option for high-risk NB patients without ASCT or anti-GD2 antibodies therapy. Stage 4, *MYCN* amplification, and stage 4 patients not receiving oral metronomic MT after CR/VGPR/PR were independent adverse prognostic factors for high-risk NB patients without ASCT or anti-GD2 antibodies therapy. Additionally, high-risk NB patients with recurrence can obtain a survival benefit from salvage therapy.

Author Contributions: X.S. Conceptualization; methodology; formal analysis; investigation; writing; writing—review and editing; supervision; and project administration. Z.Z. Conceptualization; methodology; formal analysis; investigation; writing—review and editing; visualization and project administration. Y.G. (Ying Guo) Software; validation; formal analysis; data curation; editing and visualization. Y.G. (Yuanhong Gao) Conceptualization; methodology; investigation; supervision; and project administration. J.W. Investigation; and data curation. Y.Z. (Yu Zhang) Investigation. J.Z. Investigation. S.L. Investigation. F.S. Investigation. J.H. Investigation. R.C. Investigation. Y.Z. (Yizhuo Zhang) Investigation. J.L. Investigation. Z.X. Investigation. S.Z. Investigation. Z.L. Conceptualization; methodology; investigation; supervision; and project administration. All authors have read and agreed to the published version of the manuscript.

Funding: This research received no external funding.

Institutional Review Board Statement: The study was conducted according to the guidelines of the Declaration of Helsinki, and approved by the Ethics Committee of Sun Yat-sen University Cancer Center (protocol code B2020-094-01).

Informed Consent Statement: Informed consent was obtained from all subjects involved in the study.

Data Availability Statement: The data that support the findings of this study are openly available in Research Data Deposit at https://www.researchdata.org.cn, reference number RDDA2021002014 (accessed on 3 June 2021).

Conflicts of Interest: The authors declare no conflict of interest.

References

1. Matthay, K.K.; Villablanca, J.G.; Seeger, R.C.; Stram, D.O.; Harris, R.E.; Ramsay, N.K.; Swift, P.; Shimada, H.; Black, C.T.; Brodeur, G.M.; et al. Treatment of High-Risk Neuroblastoma With Intensive Chemotherapy, Radiotherapy, Autologous Bone Marrow Transplantation, and 13-Cis-Retinoic Acid. *N. Engl. J. Med.* **1999**, *341*, 1165–1173. [CrossRef]
2. Berthold, F.; Boos, J.; Burdach, S.; Erttmann, R.; Henze, G.; Hermann, J.; Klingebiel, T.; Kremens, B.; Schilling, F.H.; Schrappe, M.; et al. Myeloablative megatherapy with autologous stem-Cell rescue versus oral maintenance chemotherapy as consolidation treatment in patients with high-risk neuroblastoma: A randomized controlled trial. *Lancet Oncol.* **2005**, *6*, 649–658. [CrossRef]
3. Pritchard, J.; Cotterill, S.J.; Germond, S.M.; Imeson, J.; De Kraker, J.; Jones, D.R. High dose melphalan in the treatment of advanced neuroblastoma: Results of a randomized trial (ENSG-1) by the European Neuroblastoma Study Group. *Pediatr. Blood Cancer* **2005**, *44*, 348–357. [CrossRef] [PubMed]
4. Matthay, K.K.; Reynolds, C.P.; Seeger, R.C.; Shimada, H.; Adkins, E.S.; Haas-Kogan, D.; Gerbing, R.B.; London, W.B.; Villablanca, J.G. Long-term results for children with high-risk neuroblastoma treated on a randomized trial of myeloablative therapy followed by 13-cis-retinoic acid: A children's oncology group study. *J. Clin. Oncol.* **2009**, *27*, 1007–1013. [CrossRef]
5. Berthold, F.; Ernst, A.; Hero, B.; Klingebiel, T.; Kremens, B.; Schilling, F.H.; Simon, T. Long-term outcomes of the GPOH NB97 trial for children with high-risk neuroblastoma comparing high-dose chemotherapy with autologous stem cell transplantation and oral chemotherapy as consolidation. *Br. J. Cancer* **2018**, *119*, 282–290. [CrossRef]
6. Yalçin, B.; Kremer, L.C.; van Dalen, E.C. High-dose chemotherapy and autologous haematopoietic stem cell rescue for children with high-risk neuroblastoma. *Cochrane Database Syst. Rev.* **2015**, *10*, CD006301. [CrossRef] [PubMed]
7. Yu, A.L.; Gilman, A.L.; Ozkaynak, M.F.; London, W.B.; Kreissman, S.G.; Chen, H.X.; Smith, M.; Anderson, B.; Villablanca, J.G.; Matthay, K.K.; et al. Anti-GD2 antibody with GM-CSF, interleukin-2, and isotretinoin for neuroblastoma. *N. Engl. J. Med.* **2010**, *363*, 1324–1334. [CrossRef]
8. Cheung, N.K.; Cheung, I.Y.; Kushner, B.H.; Ostrovnaya, I.; Chamberlain, E.; Kramer, K.; Modak, S. Murine anti-GD2 monoclonal antibody 3F8 combined with granulocyte-macrophage colony-stimulating factor and 13-cis-retinoic acid in high-risk patients with stage 4 neuroblastoma in first remission. *J. Clin. Oncol.* **2012**, *30*, 3264–3270. [CrossRef]
9. Ladenstein, R.; Pötschger, U.; Valteau-Couanet, D.; Luksch, R.; Castel, V.; Yaniv, I.; Laureys, G.; Brock, P.; Michon, J.M.; Owens, C.; et al. Interleukin 2 with anti-GD2 antibody ch14.18/CHO (dinutuximab beta) in patients with high-risk neuroblastoma (HR-NBL1/SIOPEN): A multicentre, randomised, phase 3 trial. *Lancet Oncol.* **2018**, *19*, 1617–1629. [CrossRef]
10. Yu, A.L.; Gilman, A.L.; Ozkaynak, M.F.; Naranjo, A.; Diccianni, M.B.; Gan, J.; Hank, J.A.; Batova, A.; London, W.B.; Tenney, S.C.; et al. Long-Term Follow-up of a Phase III Study of ch14.18 (Dinutuximab) + Cytokine Immunotherapy in Children with High-Risk Neuroblastoma: COG Study ANBL0032. *Clin. Cancer Res.* **2021**, *27*, 2179–2189. [CrossRef]
11. Zapletalova, D.; André, N.; Deak, L.; Kyr, M.; Bajciova, V.; Mudry, P.; Dubska, L.; Demlova, R.; Pavelka, Z.; Zitterbart, K.; et al. Metronomic chemotherapy with the COMBAT regimen in advanced pediatric malignancies: A multicenter experience. *Oncology* **2012**, *82*, 249–260. [CrossRef] [PubMed]
12. Stempak, D.; Gammon, J.; Halton, J.; Moghrabi, A.; Koren, G.; Baruchel, S. A Pilot Pharmacokinetic and antiangiogenic Biomarker Study of Celecoxib and Low-dose Metronomic Vinblastine or Cyclophosphamide in Pediatric Recurrent Solid Tumors. *J. Pediatr. Hematol. Oncol.* **2006**, *28*, 720–728. [CrossRef] [PubMed]
13. André, N.; Rome, A.; Coze, C.; Padovani, L.; Pasquier, E.; Camoin, L.; Gentet, J.C. Metronomic etoposide/cyclophosphamide/celecoxib regimen given to children and adolescents with refractory cancer: A preliminary monocentric study. *Clin. Ther.* **2008**, *30*, 1336–1340. [CrossRef]
14. Bouche, G.; André, N.; Banavali, S.; Berthold, F.; Berruti, A.; Bocci, G.; Brandi, G.; Cavallaro, U.; Cinieri, S.; Colleoni, M.; et al. Lessons from the Fourth Metronomic and Anti-angiogenic Therapy Meeting, 24–25 June 2014, Milan. *Ecancermedicalscience* **2014**, *8*, 463.
15. Schmiegelow, K.; Nielsen, S.N.; Frandsen, T.L.; Nersting, J. Mercaptopurine/Methotrexate maintenance therapy of childhood acute lymphoblastic leukemia: Clinical facts and fiction. *J. Pediatr. Hematol. Oncol.* **2014**, *36*, 503–517. [CrossRef]
16. Klingebiel, T.; Boos, J.; Beske, F.; Hallmen, E.; Int-Veen, C.; Dantonello, T.; Treuner, J.; Gadner, H.; Marky, I.; Kazanowska, B.; et al. Treatment of children with metastatic soft tissue sarcoma with oral maintenance compared to high dose chemotherapy: Report of the HD CWS-96. *Trial. Pediatr. Blood Cancer* **2008**, *50*, 739–745. [CrossRef]

17. Bisogno, G.; De Salvo, G.L.; Bergeron, C.; Gallego Melcón, S.; Merks, J.H.; Kelsey, A.; Martelli, H.; Minard-Colin, V.; Orbach, D.; Glosli, H.; et al. Vinorelbine and continuous low-dose cyclophosphamide as maintenance chemotherapy in patients with high-risk rhabdomyosarcoma (RMS 2005): A multicenter, open-label, randomized, phase 3 trial. *Lancet Oncol.* **2019**, *20*, 1566–1575. [CrossRef]
18. Shimada, H.; Ambros, I.M.; Dehner, L.P.; Hata, J.; Joshi, V.V.; Roald, B.; Stram, D.O.; Gerbing, R.B.; Lukens, J.N.; Matthay, K.K.; et al. The International Neuroblastoma Pathology Classification (the Shimada System). *Cancer* **1999**, *86*, 364–372. [CrossRef]
19. Brodeur, G.M.; Pritchard, J.; Berthold, F.; Carlsen, N.L.; Castel, V.; Castelberry, R.P.; De Bernardi, B.; Evans, A.E.; Favrot, M.; Hedborg, F.; et al. Revisions of the International Criteria for Neuroblastoma Diagnosis, Staging, and Response to Treatment. *J. Clin. Oncol.* **1993**, *11*, 1466–1477. [CrossRef]
20. Su, Y.; Ma, X.L.; Wang, H.M.; Qin, H.; Qin, M.Q.; Zhang, F.Q.; Jin, M.; Zhang, D.W.; Chen, C.H.; Zeng, Q.; et al. Clinical characteristics and prognostic analysis of 458 children with high-risk neuroblastoma in a single center. *Chin. J. Pediatr.* **2020**, *58*, 796–801.
21. Simon, T.; Hero, B.; Faldum, A.; Handgretinger, R.; Schrappe, M.; Klingebiel, T.; Berthold, F. Long term outcome of high-risk neuroblastoma patients after immunotherapy with antibody ch14.18 or oral metronomic chemotherapy. *BMC Cancer* **2011**, *11*, 21. [CrossRef]
22. Kushner, B.H.; Ostrovnaya, I.; Cheung, I.; Kuk, D.; Modak, S.; Kramer, K.; Roberts, S.S.; Basu, E.M.; Yataghene, K.; Cheung, N.K. Lack of survival advantage with autologous stem-cell transplantation in high-risk neuroblastoma consolidated by anti-GD2 immunotherapy and isotretinoin. *Oncotarget* **2016**, *7*, 4155–4166. [CrossRef]
23. Robison, N.J.; Campigotto, F.; Chi, S.N.; Manley, P.E.; Turner, C.D.; Zimmerman, M.A.; Chordas, C.A.; Werger, A.M.; Allen, J.C.; Goldman, S.; et al. A phase II trial of a multi-agent oral antiangiogenic (metronomic) regimen in children with recurrent or progressive cancer. *Pediatr. Blood Cancer* **2014**, *61*, 636–642. [CrossRef] [PubMed]
24. Lee, J.W.; Son, M.H.; Cho, H.W.; Ma, Y.E.; Yoo, K.H.; Sung, K.W.; Koo, H.H. Clinical significance of *MYCN* amplification in patients with high-risk neuroblastoma. *Pediatr. Blood Cancer* **2018**, *65*, e27257. [CrossRef]
25. Campbell, K.; Gastier-Foster, J.M.; Mann, M.; Naranjo, A.H.; Van Ryn, C.; Bagatell, R.; Matthay, K.K.; London, W.B.; Irwin, M.S.; Shimada, H.; et al. Association of *MYCN* Copy Number with Clinical Features, Tumor Biology, and Outcomes in Neuroblastoma: A Report from the Children's Oncology Group. *Cancer* **2017**, *123*, 4224–4235. [CrossRef]
26. Aygun, N. Biological and Genetic Features of Neuroblastoma and Their Clinical Importance. *Curr. Pediatric Rev.* **2018**, *14*, 73–90. [CrossRef] [PubMed]
27. Campbell, K.; Shyr, D.; Bagatell, R.; Fischer, M.; Nakagawara, A.; Nieto, A.C.; Brodeur, G.M.; Matthay, K.K.; London, W.B.; DuBois, S.G. Comprehensive evaluation of context dependence of the prognostic impact of MYCN amplification in neuroblastoma: A report from the International Neuroblastoma Risk Group (INRG) project. *Pediatr. Blood Cancer* **2019**, *66*, e27819. [CrossRef] [PubMed]
28. Canete, A.; Gerrard, M.; Rubie, H.; Castel, V.; Di Cataldo, A.; Munzer, C.; Ladenstein, R.; Brichard, B.; Bermúdez, J.D.; Couturier, J.; et al. Poor survival for infants with MYCN-amplified metastatic neuroblastoma despite intensified treatment: The International Society of Paediatric Oncology European Neuroblastoma Experience. *J. Clin. Oncol.* **2009**, *27*, 1014–1019. [CrossRef] [PubMed]
29. Yamazaki, F.; Yamasaki, K.; Kiyotani, C.; Hashii, Y.; Shioda, Y.; Hara, J.; Matsumoto, K. Thiotepa–melphalan myeloablative therapy for high-risk neuroblastoma. *Pediatr. Blood Cancer* **2021**, *68*, e28896. [CrossRef]
30. Pinto, N.R.; Applebaum, M.A.; Volchenboum, S.L.; Matthay, K.K.; London, W.B.; Ambros, P.F.; Nakagawara, A.; Berthold, F.; Schleiermacher, G.; Park, J.R.; et al. Advances in Risk Classification and Treatment Strategies for Neuroblastoma. *J. Clin. Oncol.* **2015**, *33*, 3008–3017. [CrossRef]
31. London, W.B.; Castel, V.; Monclair, T.; Ambros, P.F.; Pearson, A.D.; Cohn, S.L.; Berthold, F.; Nakagawara, A.; Ladenstein, R.L.; Iehara, T.; et al. Clinical and biologic features predictive of survival after relapse of neuroblastoma: A report from the International Neuroblastoma Risk Group project. *J. Clin. Oncol.* **2011**, *29*, 3286–3292. [CrossRef] [PubMed]
32. London, W.B.; Bagatell, R.; Weigel, B.J.; Fox, E.; Guo, D.; Van Ryn, C.; Naranjo, A.; Park, J.R. Historical time to disease progression and progression—free survival in patients with recurrent/refractory neuroblastoma treated in the modern era on Children's Oncology Group early-phase trials. *Cancer* **2017**, *123*, 4914–4923. [CrossRef] [PubMed]
33. Moreno, L.; Rubie, H.; Varo, A.; Le Deley, M.C.; Amoroso, L.; Chevance, A.; Garaventa, A.; Gambart, M.; Bautista, F.; Valteau-Couanet, D.; et al. Outcome of children with relapsed or refractory neuroblastoma: A meta-analysis of ITCC/SIOPEN European phase II clinical trials. *Pediatr. Blood Cancer* **2017**, *64*, 25–31. [CrossRef] [PubMed]

cancers

MDPI

Article

Identification of an RNA-Binding-Protein-Based Prognostic Model for Ewing Sarcoma

Yi Chen [1,2,*], Huafang Su [3], Yanhong Su [1], Yifan Zhang [1,2], Yingbo Lin [1] and Felix Haglund [1,2]

1 Department of Oncology-Pathology, Karolinska Institutet, Solna, 17176 Stockholm, Sweden; yanhong.su@ki.se (Y.S.); yifan.zhang@ki.se (Y.Z.); yingbo.lin@ki.se (Y.L.); felix.haglund@ki.se (F.H.)
2 Clinical Pathology and Cancer Diagnostics, Karolinska University Hospital, Solna, 17176 Stockholm, Sweden
3 Department of Radiation and Medical Oncology, The First Affiliated Hospital of Wenzhou Medical University, Wenzhou 325000, China; suhuafang@wzhospital.cn
* Correspondence: yi.chen@ki.se

Simple Summary: Ewing sarcoma (ES) is an aggressive childhood tumor for which response to chemotherapy is central to long-term prognosis, but few prognostic markers have been identified. RNA-binding proteins (RBPs) are strong regulators of cell behavior, working, for example, through post-translational modifications of mRNA. In this study, we investigated whether patterns in the RBP levels were related to outcomes in ES patients. A total of three distinct patterns were recognized, and additional modelling suggested that 10 RPBs had predictive value, suggesting that this model could be used in a clinical setting to identify patients with a higher risk of mortality.

Abstract: RNA-binding proteins (RBPs) are important transcriptomic regulators and may be important in tumorigenesis. Here, we sought to investigate the clinical impact of RBPs for patients with Ewing sarcoma (ES). ES transcriptome signatures were characterized from four previously published cohorts and grouped into new training and validation cohorts. A total of three distinct subtypes were identified and compared for differences in patient prognosis and RBP signatures. Next, univariate Cox and Lasso regression models were used to identify hub prognosis-related RBPs and construct a prognostic risk model, and prediction capacity was assessed through time-dependent receiver operating characteristics (ROCs), Kaplan–Meier curves, and nomograms. Across the three RBP subtypes, 29 significant prognostic-associated RBP genes were identified, of which 10 were used to build and validate an RBP-associated prognostic risk model (RPRM) that had a stable predictive value and could be considered valuable for clinical risk-stratification of ES. A comparison with immunohistochemistry validation showed a significant association between overall survival and NSUN7 immunoreactivity, which was an independent favorable prognostic marker. The association of RBP signatures with ES clinical prognosis provides a strong rationale for further investigation into RBPs molecular mechanisms.

Keywords: Ewing sarcoma; RNA-binding proteins; regulation network; prognosis prediction; risk model

Citation: Chen, Y.; Su, H.; Su, Y.; Zhang, Y.; Lin, Y.; Haglund, F. Identification of an RNA-Binding-Protein-Based Prognostic Model for Ewing Sarcoma. *Cancers* **2021**, *13*, 3736. https://doi.org/10.3390/cancers13153736

Academic Editors: Saurabh Agarwal and Jianhua Yang

Received: 28 May 2021
Accepted: 23 July 2021
Published: 25 July 2021

Publisher's Note: MDPI stays neutral with regard to jurisdictional claims in published maps and institutional affiliations.

1. Introduction

Ewing sarcoma (ES) represents the second-most common primary bone malignancy affecting children and adolescents, with an incidence of 2.9 per million/year [1–3]. It is an aggressive tumor typically characterized by a fusion of the Ewing sarcoma breakpoint region 1 (EWSR1) with an erythroblast transformation specific (ETS) transcription factor gene, most frequently (>95%) the friend leukemia virus integration 1 (FLI1) gene [4,5]. Most types of ES harbors the t (11;22) (q24;q12) chromosomal translocation leading to the EWS–FLI1 fusion transcript. Genome-wide association studies identifying molecular features and genetic profiles suggest a generally low mutational burden for ES [6–8]. The STAG2 gene is the most frequently mutated (17%) compared to normal paired cases, followed by CDKN2A (12%), TP53 (7%), and EZH2 (2.7%). Genetic lesions of STAG2 and TP53 and deletions of CDKN2A are mutually exclusive and associated with the worst ES prognosis [6,9,10]. Recently, the Euro-Ewing 99 clinical trial revealed 3-year and 8-year overall survival (OS) rates of 72–78% and 56–65%, respectively [11]. Patients with relapsed and advanced ES have a dismal prognosis, especially when the disease has metastasized, and a short survival time is expected [12]. Therefore, prognostic biomarkers and novel intervention targets are urgently needed.

RNA-binding proteins (RBPs) are a class of proteins that interact with a series of RNAs: messenger (mRNAs), ribosomal (rRNAs), non-coding (ncRNAs), micro (miRNAs), transfer (tRNAs), small nuclear (snRNAs), and small nucleolar (snoRNAs). They exert their functions by associating with their RNA targets and generating ribonucleoprotein complexes that regulate transcription, RNA maturation, translation, and metabolism and maintain post-transcriptional genome integrity [13–17]. Currently, more than 1500 RBPs have been identified by large-scale quantitative methods accounting for about 7.5% of all protein-coding genes in the human genome [18,19]. EWS is a multifaceted RBP that has been implicated in modulating transcription, pre-mRNA splicing, and, importantly, in causing epigenetic remodeling in the tumorigenesis of ES [20,21]. Another example is IGF2BP3, a novel post-transcriptional regulator responsible for poor survival probability [22,23]. Our group previously reported that a higher expression of RBPs related to rRNA metabolism and mRNA splicing were significantly overrepresented in ES patients who had first-line treatment failure [22]. Thus, we hypothesized that a systematic study to explore prognosis-related RBPs would increase our knowledge of treatment resistance and facilitate risk-stratification.

In this study, we obtained ES gene expression data from previously published GEO datasets and investigated the RBP landscape in ES associated with patient outcomes. Next, the prognostic value of a 10-RBP expression signature was validated in an independent cohort. We also identified *NSUN7* as a novel independent prognostic factor that could be a diagnostic and therapeutic target.

2. Material and Methods

2.1. ES Data Processing

Both the ES gene expression microarray data and clinical data were derived from the NCBI Gene Expression Omnibus (GEO) database: GSE63155 (46 ES cases, follow-up time: median 1876 days, range 155–3987 days) [24], GSE63166 (39 ES cases, follow-up time: median 2085 days, range 286–4500 days) [24], GSE17679 (32 ES cases, follow-up time: median 1159 days, range 138–5766 days) [25], and GSE34620 (38 ES cases, follow-up time: median 1549 days, range 270–4017 days) [26]. The patient characteristics involved in this study are shown in Table 1. In these four datasets, Affymetrix HuEx1.0 (GPL5175) and Affymetrix HG-U133 Plus 2.0 (GPL570) platform data were merged into one integrated dataset. Batch effects were removed using the ComBat function of the R package surrogate variable analysis (sva) [27], and expression values were quantile-normalized across different samples. A total of 85patients from GSE63155 and GSE63166 were involved in the training cohort, while 70 from GSE17679 and GSE34620 were included in the validation cohort. We obtained a comprehensive RBP catalog consisting of 1542 genes and 318 transcriptional factors (TFs) for further analysis [19,28]. Unsupervised clustering was performed using the R package ConsesusClusterPlus to classify patients into distinct subtypes according to RBP expression.

Table 1. GEO datasets and patient characteristics of ES.

GEO ID	No. of ES Cases Included	Platform	Age		Sex		Outcomes	
			<18	≥18	Male	Female	Dead	Alive
GSE63155	46	HuEx1.0 (GPL5175)	41	5	27	19	14	32
GSE63166	39	HuEx1.0 (GPL5175)	31	8	19	20	11	28
GSE17679	32	U133Plus2.0 (GPL570)	16	16	22	10	20	12
GSE34620	38	U133Plus2.0 (GPL570)	27	11	20	18	21	17

2.2. Identification of Differentially Expressed RBPs

The R package linear models for microarray data (LIMMA) [29] was used to perform differentially expressed RBP (DERBP) analysis in the training cohort by comparing clusters of patient subtypes that were significantly associated with overall survival. A false discovery rate (FDR) of <0.05 or *p*-value of <0.05 was set as the cut-off criterion. A Venn diagram was used to determine the overlapping DERBPs between two sets of comparisons. Additionally, the DETFs and DEHallmarks were generated based on the same criteria.

2.3. Protein–Protein Interaction (PPI) Network Construction and Functional Enrichment Analyses

To further explore the potential molecular functions, the DERBPs were submitted into the Search Tool for the Retrieval of Interacting Genes (STRING) database for building the protein–protein interaction (PPI) network [30]. The PPIs with combined scores ≥ 0.7 were selected, after which the network was constructed and visualized using Cytoscape software (version 3.8.2) (National Institute of General Medical Sciences, Bethesda, MD, USA) [31]. Any genes with a connectivity of ≥1 (node/edge) were screened as hub genes for downstream analyses. The top 50 central nodes were ranked by The MCC (maximum clique centrality) algorithm of the CytoHubba plug-in [32] in Cytoscape. The functional enrichment analyses were performed using the ClueGO and CluePedia plug-ins in Cytoscape [33,34]. The main parameters of the constructing network with ClueGO were as follows: marker list, homo sapiens; ontologies/pathways, GO (biological process) and REACTOME pathways; GO tree interval, level = 6; GO term/pathway selection, min # genes = 6 and % genes = 6.000; and GO term/pathway network connectivity (kappa score) = 0.45. Only GO terms/pathways with $p < 0.01$ were selected.

2.4. Prognosis-Related RBPs and the Interaction with TFs

After combining the expression levels of the hub genes with the survival status and follow-up time, a univariate Cox proportional regression analysis was used to determine the hub genes related to overall survival in the training cohort. The association between the expression level of significant survival-related genes and DETFs was assessed by Pearson correlation analysis. A correlation coefficient of ≥0.4 and *p*-value of ≤0.05 were set as the cut-off criteria. The Cytoscape ClusterViz plug-in [35] was employed to identify biological network modules between RBPs and TFs.

2.5. Establishment and Validation of an RBP-Associated Prognostic Risk Model (RPRM)

The survival-related genes were subjected to penalized multivariate Cox proportional hazard survival modeling by an algorithm for variable selection based on L1-penalized Least Absolute Shrinkage and Selection Operator (LASSO) regression [36]. The process of prognostic model construction was repeated for the combination of parameter values for 1000 iterations. Subsequently, the resulting models were combined through cross-validation. The risk score formula for each sample was calculated as follows:

$$\text{Risk score} = \beta 1 * \text{Exp1} + \beta 2 * \text{Exp2} + \beta i * \text{Expi}$$

where β represents the coefficient value, and Exp represents the gene expression level. Subsequently, A time-dependent receiver operating characteristic (ROC) curve was performed to evaluate the model performance by calculating the area under the curve (AUC). As a result, patients were classified into high- or low-risk groups according to the median risk score. The RBP-associated prognostic risk model (RPRM) was externally validated by the validation cohort (GSE17679 and GSE34620) to access its generalizability. Afterwards, we developed a nomogram-combing gene expression to improve risk stratification and quantify the risk assessment for survival probability at one year, three years, and five years in individual patients.

2.6. Assessment of Gene Expression Level and Prognostic Significance in RPRM

Our group had previously identified several pathways that were significantly associated with first-line treatment failure in ES [22]. Here, a single sample gene set enrichment analysis (ssGSEA) was used to define the estimated enrichment score of the gene signature for each sample using gene set variation analysis (GSVA) in R package [37]. Hallmark gene sets were downloaded from the Molecular Signatures Database v7.4. Gene expression levels in RPRM were assessed between dead and live patients. Significant genes were cross-validated by the validation cohort and our ES RNA-seq data.

2.7. Immunohistochemistry

We investigated the expression of NSUN7 in ES tumor tissue from patients (n = 24) who underwent standard treatment protocols (ISG/SSG III or EuroEwing99/2012 protocols) [38]. These were therapy-naive preoperative biopsies (n = 9), but if these were unavailable for analysis, viable tumor cells from a resection specimen (n = 11) or metastatic specimen (n = 4) were used.

Immunohistochemical staining for *NSUN7* was done using a manual protocol with a polyclonal antibody (anti-*NSUN7*, Product #PA5-54257 from ThermoFisher Scientific, Waltham, MA, USA) at 1:50 dilution. Formalin-fixed paraffin embedded (FFPE) tissue was sectioned in 4 um thick sections and deparaffinized. Antigen retrieval was performed for 10 min in low pH (citrate). Slides were incubated with the primary antibody at 4 °C overnight, followed by incubation with a biotinylated secondary antibody and signal generation using the VECTASTAIN® Elite ABC-HRP Kit (Vector Laboratories, Burlingame, CA, USA).

Two clinical pathologists evaluated the staining, and after agreeing on a scoring method, the slides were scored independently in a blinded fashion. The agreement was substantial (82.6%, Cohen's K 0.620). There was disagreement in four cases, but consensus was reached after a second review. The slides fit into three scoring categories:

- Negative ("Most of the tumor cells were completely negative"): >75% completely negative, <25% with very weak cytoplasmic staining);
- Weak ("Cases exhibited a diffuse weak staining with small areas with negative or stronger immunoreactivity"): >75% of tumor cells exhibited weak immunoreactivity); and
- Moderate ("Cases showed stronger cytoplasmic immunoreactivity in most of the tumor cells, but smaller areas with either weaker or negative immunoreactivity were sometimes observed"): >75% of the tumor cells showed moderate cytoplasmic reactivity.

Since the number of patients was limited, the staining groups were compared in two different fashions: all groups (negative vs. weak vs. moderate) or simplified (negative vs. positive).

2.8. Statistical Analysis

Normally distributed variables were analyzed with an unpaired two-tailed Student's t-test, whereas a Mann–Whitney U test or a Wilcoxon rank-sum test was used for non-normally distributed variables. The relationship between the parameters was evaluated using the Spearman rank correlation coefficient. A log-rank test Kaplan–Meier curve and a Cox proportional hazard regression were performed for survival analysis using the R package "survminer" and "survival". Additionally, L1-penalized LASSO regression and ROC curves were conducted using the R package "glmnet" and "pROC", respectively [39–41]. R version 4.0.3 (R Foundation for Statistical Computing, Vienna, Austria) was used to execute all statistical tests and plots.

3. Results

3.1. Data Preprocessing of the ES Dataset

Data from 155 ES patients with complete follow-up and expression data from GSE61355, GSE63156, GSE34620, and GSE17679 were bioinformatically pretreated to remove batch effects. The 15,016 overlapping genes across the four GEO datasets are depicted in the Venn diagrams in Figure S1A. Data before and after normalization were carefully inspected by principal component analysis (PCA), suggesting the batch effect was successfully removed using ComBat (Figure S1B,C). In addition, the gene expression profiles were visualized in box plots to show the impact on batch effect removal (Figure S1D,E).

3.2. Identification of DERBPs in ES Patients and Transcriptional Subtypes

To investigate how RBPs affect the prognostic significance in ES, we compared the gene expression between dead and alive patients using the Wilcoxon rank-sum test ($p < 0.05$) in the training cohort. This analysis identified 22 DERBPs, and the consensus k-mean clustering on the 22 RBP signatures clearly divided ES patients into three main subtypes with clustering stability decreasing for k = 2 to 6 (Figures 1A and S2). They were clustered into RS1 (n = 32), RS2 (n = 42), and RS3 (n = 11) with distinct prognostic differences. Interestingly, patients in Cluster 1 had the best clinical prognosis, whereas patients in Cluster 3 had the poorest (log-ranktest, $p < 0.05$) (Figure 1B). Following this, a heatmap visualized the expression levels of 22 RBPs in three RBP subtypes with survival status, where most were RS3s dominated by down-regulated transcripts (Figure 1C). Among 1542 RBPs quantified in 85 ES patients with two comparisons (RS1 vs. RS2, RS1 vs. RS3), 377 (24.4%) were significantly down-regulated, and 374 (24.3%) were significantly up-regulated in RS3 compared to RS1; 150 (9.7%) were significantly down-regulated, and 137 (8.9%) were significantly up-regulated in RS2 compared to RS1 (Wilcoxon rank-sum test, FDR < 0.05; Figure 1D). A total of 90 overlapping up-regulated and 125 overlapping down-regulated DERBPs were identified by the two comparisons and selected for further analysis (Figure 1E).

3.3. Functional Enrichment Analysis and PPI Network of DERBPs

Among the DERBPs, we wanted to identify the key components using co-variation analysis. Functional enrichment analysis was performed for the 218 overlapping DERBPs: 90 up-regulated, 125 down-regulated, and 3 with distinct expression profiles. Gene ontology (GO) biological process and REACTOME pathway analyses were applied to clarify the DERBP gene function. As shown in Table S1, most of the DERBPs were enriched in the GO categories of RNA metabolism (i.e., RNA processing, RNA metabolic process, and RNA catabolic process) and protein metabolism (i.e., translation, ribonucleoprotein complex assembly). In addition, enrichment analysis of the REACTOME pathways revealed that RNA metabolism and mRNA splicing and translation were mostly enriched for DERBPs (Table S2). The GO term/pathway network is shown in Figure 2A,B ($p < 0.01$). Next, we built a high-confidence PPI network that contained 171 nodes (genes) and 1234 edges, which were selected for further analysis. To identify the top 50 hub nodes, the cytoHubba plug-in was applied, and key genes were selected from the PPI network according to the MCC score (Figure 2C). Detailed gene descriptions and connectivity degrees of the top 15 hub genes are shown in Table S3. EIF4A3 was identified as having the largest number of edges interacting with 54 other genes, followed by POLR2F and SRSF1.

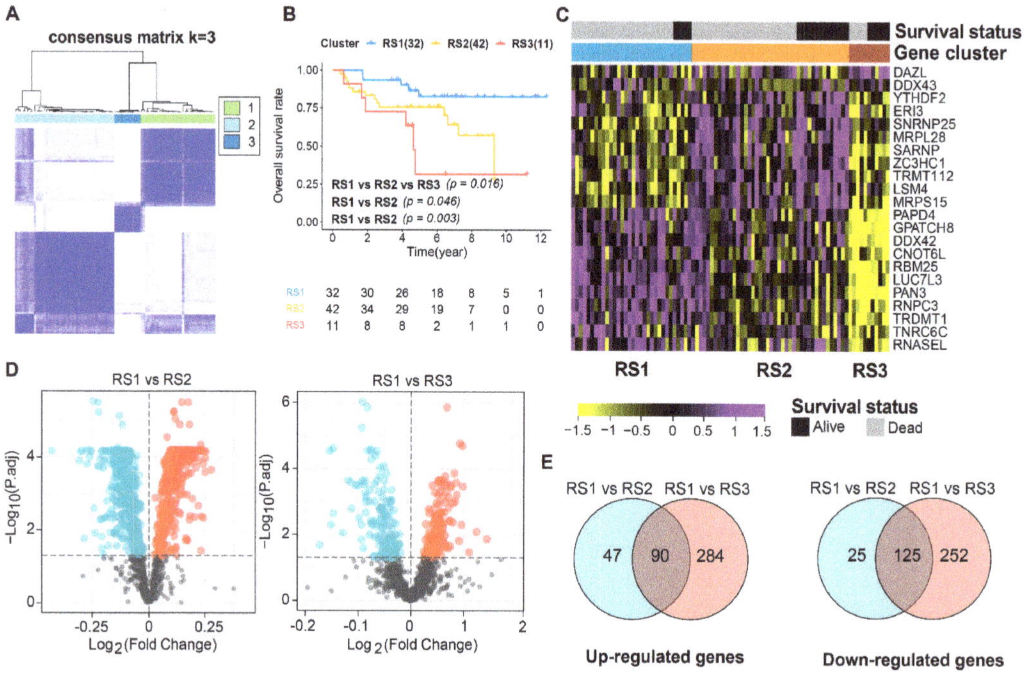

Figure 1. DERBP identification and the transcriptional subtypes of ES (**A**). Consensus matrices of identified clusters (k = 3). (**B**). Kaplan-Meier curves show the overall survival in ES patients among the three subtypes ($p = 0.016$, $p = 0.045$, $p = 0.003$ for RS1 vs. RS2 vs. RS3, RS1 vs. RS2 and RS1 vs. RS3, respectively). (**C**). Abundances of 22 DERBPs (identified in dead ES patients) in the three RSs. (**D**). Expression profile of RBPs between RS1 versus RS2 and RS1 versus RS3, respectively. Red and blue dots represent upregulated and downregulated RBPs in RS2 or RS3, respectively (FDR < 0.05). (**E**). Overlap up-regulated and down-regulated DEREPs between RS1 versus RS2 and RS1 versus RS3.

3.4. Prognosis-Related RPBs and the Regulatory Network

To investigate the prognostic significance of these 171 RBPs involved in the PPI network, a univariate Cox regression analysis was performed from which 29 prognostic-associated hub RBP genes were obtained: 12 having favorable factors with a hazard ratio (HR) of <1, and 17 having risk factors (HR > 1) (Figure 3A).

Next, we identified DETFs across the three RSs to explore the regulatory mechanisms of these RPBs. We found 144 TFs significantly expressed in RS3 compared to RS1, and 28 TFs significantly expressed in RS2 compared to RS1 (FDR < 0.05). The overlapping 18 DETFs in the two comparisons were then selected (Figure 3B). A heatmap was constructed to indicate their expression levels in three subtypes with survival status (Figure 3C). We consequently created a regulatory network based on these 18 TFs and our 29 prognosis-related RPBs. A correlation score of more than 0.4 and $p < 0.01$ were set as the cut-off values. The TF-based regulatory network topology was grouped into three clusters through the ClusterViz plug-in in Cytoscape. Notably, the schematic clearly illustrated the regulatory relationships among these TFs and RBPs (Figure 3D, Table S4).

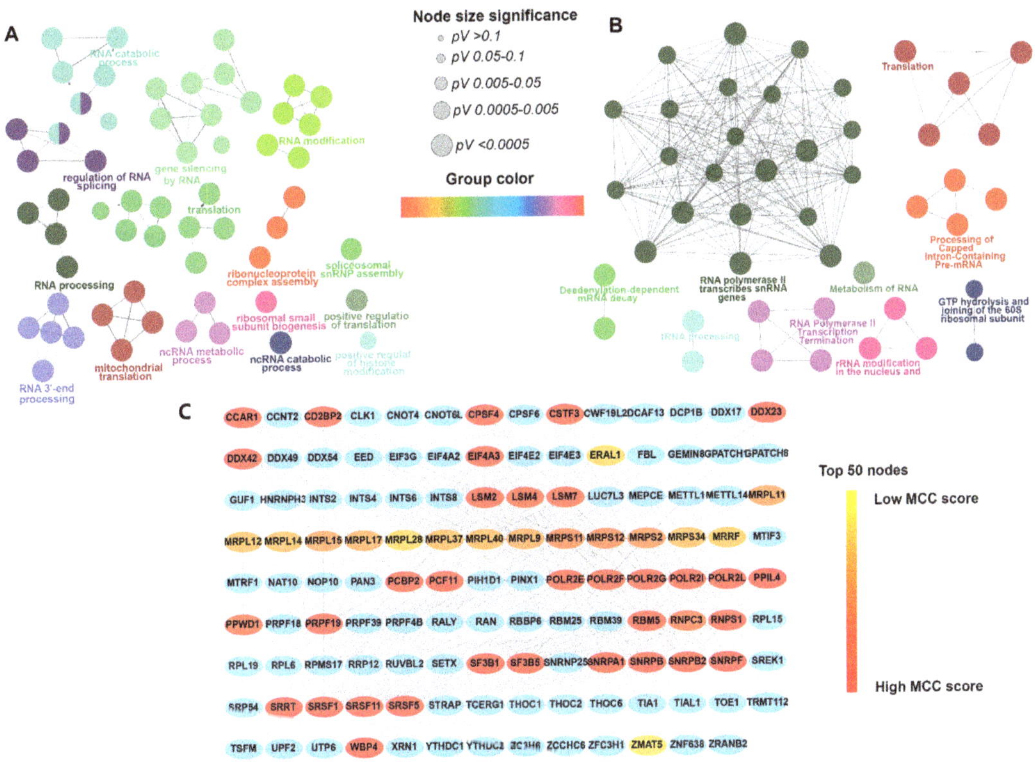

Figure 2. Functional enrichment analysis of DERBPs and PPI Network (**A**). GO terms (biological process) grouped network (kappa score levels ≥ 0.45, $p < 0.01$). Ellipse represents the GO terms. The node size represents the significance of the term enrichment (p-value), and the colors represent different functional groups. (**B**). The REACTOME pathway grouped network (kappa score levels ≥ 0.45, $p < 0.01$). Ellipse represents the GO terms. The node size represents the significance of the term enrichment (p-value), and the colors represent different functional groups. (**C**). The MCC score of top 50 genes in the PPI network of DERBPs (combined score ≥ 0.7) is represented by a red to yellow gradient.

3.5. Construction and Validation of the RPBs-Associated Prognostic Risk Model (RPRM)

After identifying the 29 prognosis-related RBPs, 1000 iterations of LASSO-penalized multivariate modeling were constructed, which led to 10 features with a non-zero, coefficient-based-risk model called RPRM (Figure 3E,F). We then computed the RPRM for each patient based on the risk score formula. The ROC curve exhibited significant prognostic performances for the AUC (1-year, 3-year, and 5-year OS predictions were 0.960, 0.915, and 0.817 in the training cohort and 0.745, 0.700, and 0.792 in the validation cohort, respectively (Figures 3G and 4D). The samples in both the training and validation cohorts were subsequently separated into high- and low-risk groups according to the median risk score. Assessments of the survival of the two groups by Kaplan–Meier estimates showed that high-risk patients had a significantly worse overall survival than the low-risk patients in both cohorts ($p < 0.001$) (Figure 4A,E). Moreover, the distribution of risk score, survival time, and survival status of each patient is illustrated in Figure 4B. Among the 10 significant genes in the risk model, *NSUN7*, *RPL15*, *ZCCHC6*, *DCP1B*, and *GPATCH8* were associated with a favorable prognosis, while *DDX23*, *STRAP*, *PRDX1*, *RPL6*, and *DCAF13* were considered to be risk factors (Figure 4C). Subsequently, a nomogram combining these genes was generated, and each patient was assigned a series of scores corresponding to all of the involved genes. Then the 1-, 3-, and 5-year survival probabilities were projected to the final sum of the scores (Figure 4E).

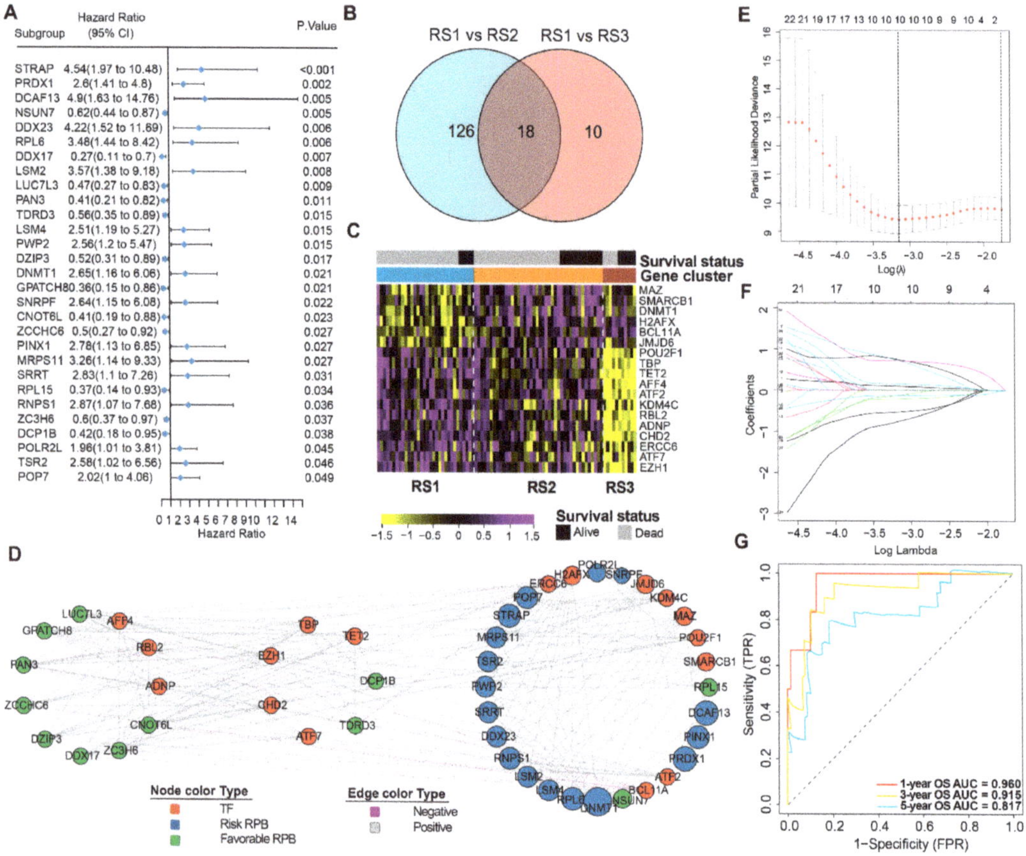

Figure 3. RBP regulatory network and prognostic risk model (**A**). Univariate analyses of 29 significant DERBPs with overall survival ($p < 0.05$). (**B**) Abundances of 18 DETFs (identified in dead ES patients) in the three RSs. (**C**) Overlap of DETFs between RS1 versus RS2 and RS1 versus RS3. (**D**). DERBPs network in ES. Circles represents all genes. Red indicates TF; blue indicates high risk RBP; and green indicates low risk RBP. The size of each circle indicates the degree of correlation. Grey lines indicate positive correlations, and purple lines indicate negative correlations ($r > 0.4$, $p < 0.01$). (**E**). Partial likelihood deviance under each log (lambda) was drawn in a LASSO Cox regression model. (**F**). The change trajectory of each independent variable in the model. (**G**). ROC curve of the prognostic values of the RPRM risk model in training group in 1-, 3- and 5-year OS.

Since our group previously revealed that a series of pathways (e.g., apoptotic process, PI3K pathway, RNA splicing, rRNA metabolic process, glycolysis) were significantly associated with first-line treatment failure in ES patients, we performed an ssGSEA analysis in 50 hallmark gene sets to determine the relationship between pathway-enrichment and risk scores (Table S5). As shown in Figure 4G, 23 hallmark gene sets were significantly enriched between high- and low-risk patients, in which only protein secretion and bile-acid metabolism was positively associated with better prognosis. A total of ten overrepresenting gene sets are illustrated in Figure S3A ($-\log_{10}$(FDR) > 3). Notably, positive correlations between the risk and ssGSEA scores in high-risk patients were shown in 16 gene sets ($p < 0.05$) (Figure 4H, Table S6), where the reactive oxygen species pathway, increased UV response, and glycolysis represent the top three highest correlation coefficients ($p < 0.001$; r = 0.67, 0.61, and 0.60, respectively) (Figure S3B–D).

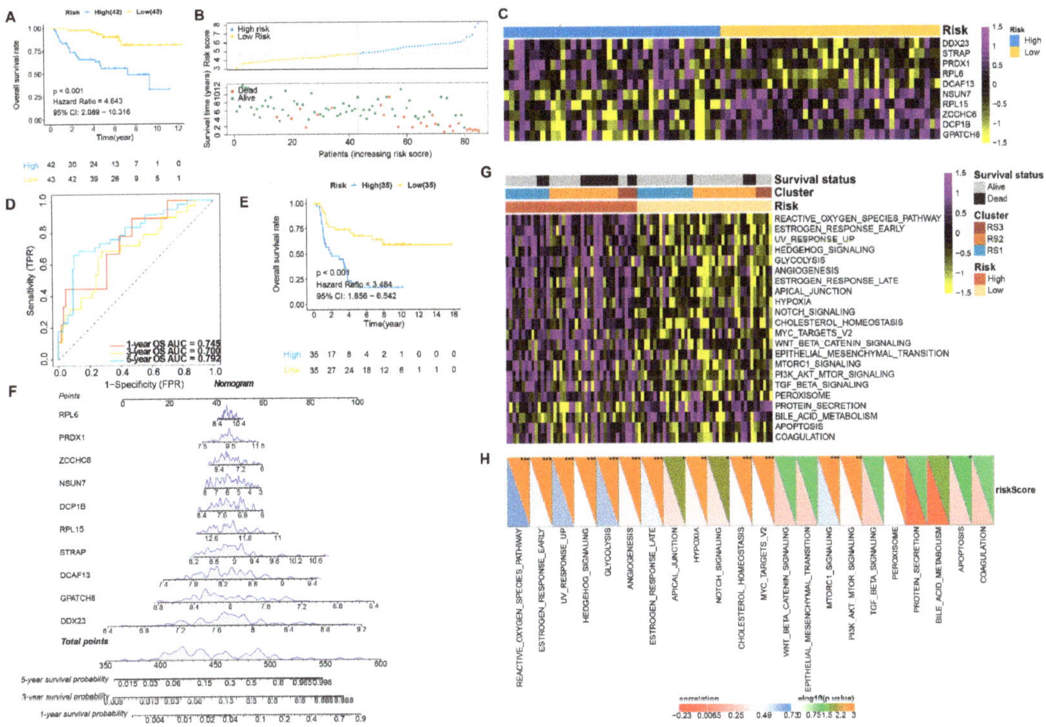

Figure 4. Validation and assessment of RPRM (**A**). Kaplan–Meier curves show overall survival between high-risk and low-risk patients in the training group ($p < 0.001$). (**B**). Distribution of risk score and survival time of patients in the training group. The patients were divided into high-risk and low-risk subgroups based on the median value of the risk score. Blue and yellow dots represent high- and low-risk patients, respectively. In the plot below, red and green dots indicate dead and live patients, respectively. (**C**). Abundances of 10 significant RBPs (involved in RPRM) in the training group. (**D**). ROC curve of the prognostic values of RPRM risk model in 1-, 3- and 5-year validation groups. (**E**). Kaplan–Meier curves show overall survival in validation groups between high- and low-risk patients ($p < 0.001$). (**F**). Nomogram for predicting 1-, 3-, and 5-year survival probability of ES patients in the training group. The total score of these genes for each patient is on the total points axis, which corresponds to the survival probabilities plotted on the three axes below. (**G**). Abundances of 23 DEHallmarks between high- and low-risk patients in the three RSs. (**H**). The correlation between LDA score and enrichment scores of 23 DEHallmarks. * $p < 0.05$; ** $p < 0.01$, *** $p < 0.001$.

3.6. Validation of the Prognostic Value and Expression of the RBPs Involved in RPRM

To further assess the independent prognostic value of each key RBPs in ES, the Kaplan–Meier estimates were used to evaluate the relationship between these key RBPs and OS. As a result, six gene expressions (*DCP18*, *DDX23*, *GRATCH8*, *NSUN7*, *RPL6*, and *ZCCHC6*) were identified as being significantly associated with OS in the training cohort: high (≥median) and low (<median). A total of four of these were beneficial to prognosis, and two were associated with poor outcomes ($p < 0.05$) (Figures 5A and S5). The Kaplan–Meier plots of the other remaining four genes are shown in Figure S4 ($p > 0.05$). Additionally, we found that the expression level of *DDX23*, *RPL6*, and *NSUN7* were significantly correlated with survival status (Wilcoxon rank-sum test, $p < 0.05$, Figure 5B), but only *NSUN7* could be verified (log-rank test, Wilcoxon rank-sum test, $p < 0.05$, Figure 5C,D).

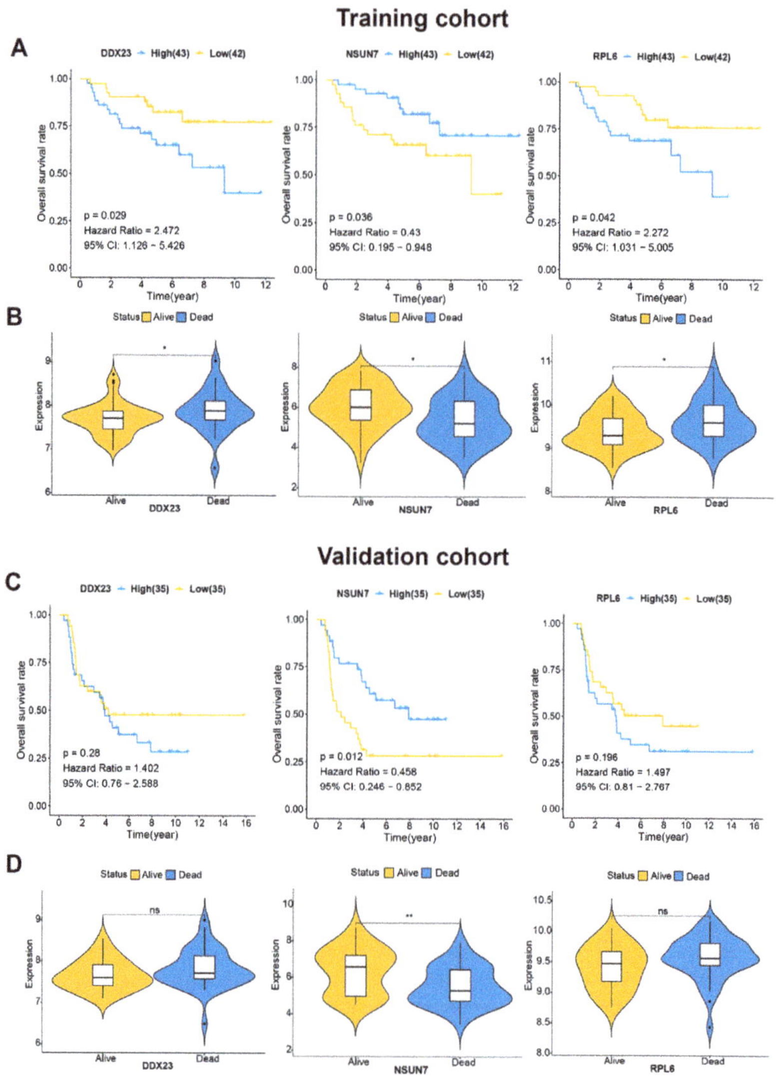

Figure 5. Validation of the prognostic and expression value of the RBPs involved in RPRM in training cohort (**A**). Kaplan–Meier curves show the overall survival in training group with high-risk and low-risk subgroups by *DDX23*, *NSUN7*, and *RPL6*, based on the median value of these genes, respectively. ($p < 0.05$). (**B**). The expression levels of three significant genes in the training group by survival status. The Wilcoxon rank sum test was used to compare the differences between groups. * $p < 0.05$. (**C**). Kaplan–Meier curves show overall survival in validation group with high-risk and low-risk subgroups by *DDX23*, *NSUN7*, and *RPL6* based on the median value of these genes, respectively. (**D**). The expression levels of *DDX23*, *RPL6*, *NSUN7* in validation group by survival status. The Wilcoxon rank sum test was used to compare the differences between groups. ** $p < 0.01$, ns: no significant.

The immunoreactivity of *NSUN7* was scored as negative ($n = 8$), weak ($n = 6$) or moderate ($n = 10$) in 24 patients (Table S7). The proportion of negative cases was higher in pre-treated tissue (7/15) than in therapy-naïve (1/9) biopsies, which may indicate that *NSUN7*-negative cells are more resistant to treatment. In the therapy-naive biopsies, 5/9 cases were classified as good responders according to the treatment protocol (<10% viable tumor cells after treatment), and in the post-therapy specimens, only 2/11 were good responders. We found a significant association between *NSUN7* immunoreactivity and overall survival (Fisher's exact test, $p = 0.0069$, Figure 6A) and that *NSUN7* negative cases had a shorter overall survival compared to positive ones (log-rank test, $p = 0.044$, Figure 6B). From the preoperative biopsies, only a single patient died (scored as negative for *NSUN7*). The prognostic association across three comparisons (negative vs. weak vs. moderate) are shown in Figure S6 and Figure 6C (chi-square test, $p = 0.0084$; log-rank test, $p = 0.101$).

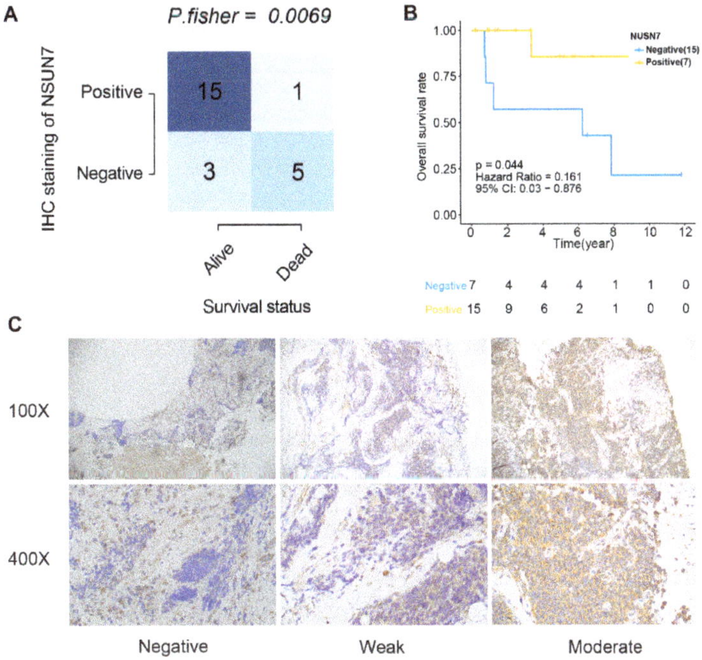

Figure 6. Examples of histology imaging and follow-up in patients with ES. (**A**). Comparison of the survival status and the *NSUN7* IHC scores in 24 ES patients (Fisher's exact test). (**B**). Kaplan–Meier curve shows the overall survival of *NSUN7* positive and negative in the subgroup of IHC staining. (**C**). Comparison between *NSUN7* immunoreactivity in negative, weak, and moderate scoring at 100× and 400× magnification.

These results strongly suggest that *NSUN7* may be an important biomarker for ES therapy resistance and that the absence of immunoreactivity may be a prognostic marker.

4. Discussion

Our group had previously reported an association between RBP expression and treatment resistance in Ewing sarcoma [22]. Here, we sought to evaluate the putative relationship between RBPs and patient outcome in four previously published transcriptome datasets. We identified three RBP-related transcriptional subtypes that were significantly associated with overall patient survival. We constructed and validated a 10-gene prognostic risk model that showed good clinical applicability as a prognostic marker for ES patients. Finally, one individual transcript in the model (NSUN7) was validated on a protein level in a separate cohort.

It is well known that RBPs are critical for various biological processes. Perhaps not surprisingly, we identified 171 prognostic-associated RBPs that were mostly enriched in RNA (mRNA, ncRNA, rRNA) metabolism, splicing, and translation, and these were similar to our previous findings from the RNA-sequencing of a single Ewing sarcoma cohort. Differential expression between the three RBP-related subtypes also identified 18 transcription factors that significantly co-varied with the 171 RBPs, potentially identifying the regulatory network that define the three RBP subtypes. While many of the identified transcription factors were described as central to cancer development (including SMARCB1, MAZ, EZH1 and H2AFX), none had an established role in ES, likely due to the limited number of functional studies of this tumor type.

Current chemotherapy protocols can cure a significant proportion of ES patients, but progression under chemotherapy is likely to drive resistance mechanisms. It may therefore be valuable to identify patients with a high likelihood of limited treatment response to offer alternative therapies in clinical trials. Since our 10-gene prognostic model was based on patients with current treatment regimes, our model could be used to identify patients who are at a high risk of death and who derive little benefit from current chemotherapy protocols.

When considering each gene in the model, only six were significantly associated with patient outcomes in the training cohort. The expression levels of three (*DDX23*, *RPL6*, and *NSUN7*) were related to survival status, but only *NSUN7* was validated as an individual marker, which may be related to the semi-quantitative nature of immunohistochemistry protocols. While this may sound disheartening, it should be interpreted as meaning that multiple gene analyses are required to create a functional prognostic panel for ES. For comparison, validated and implemented prognostic panels for breast cancer contain ~50 genes [42].

Consistent with our previous findings, we also found that cancer-related pathways were mostly activated in the ES samples with high-risk scores. Of these, high expression of hypoxia-inducible factor 1 α (HIF1α) emerged as an novel hallmark pathway, which was described to be correlated with a hypoxic tumor microenvironment in ES [43]. These findings are in line with the well-known prognostic value of serum lactate dehydrogenase (LDH) in ES patients [44].

5. Conclusions

In summary, this study identified three distinct RBPs signatures in ES. Through DERBP stratification, an RPRM risk model was constructed and validated to be a potential prognostic factor for ES patients. Major limitations of this study should be noted, including a relatively small sample size (155 patients) and the lack of detailed clinical characteristics for many of the cases. However, our results provide a strong rationale for the continued investigation of RBPs since they seem central to chemotherapy resistance and ES patient outcomes.

Supplementary Materials: The following are available online at https://www.mdpi.com/article/10.3390/cancers13153736/s1, Figure S1. GEO data processing and normalization (A). A Venn diagram illustrates the overlap of the genes across the four involved GEO datasets of ES. (B,C). The PCA plots show the overall GEO data profiles of GSE17679, GSE34620, GSE61355, and GSE61366 before and after removing batch effect, respectively. (D,E). The box plots show the expression profiles of GSE17679, GSE34620, GSE61355, and GSE61366 before and after removing batch effect, respectively. Figure S2. Consensus matrices of identified clusters (k = 2, 4–6); Figure S3. (A). The distribution of significant gene sets enriched in high-risk patients at cutoff with $-\log10$ (FDR) > 3. (B–D). Violin plots show the correlation between risk score and enrichment scores of reactive oxygen species pathway, UV response up, and glycolysis, respectively. Figure S4. Kaplan-Meier curves show the overall survival in training group with high-risk and low-risk subgroups by DCF13, PRDX1, RPL15, and STRAP, based on the median value of these gene expression levels, respectively. ($p < 0.05$). Figure S5. (A). Kaplan-Meier curves show the overall survival in training group with high-risk and low-risk subgroups by DCP1B, GPATCH8, and ZCCHC6, based on the median value of these gene expression levels, respectively. ($p < 0.05$). (B). The expression levels of DCP1B, GPATCH8, and ZCCHC6 in training group by survival status. The Wilcoxon rank sum test was used to compare the differences between groups. * $p < 0.05$. Figure S6. (A). Comparison of the survival status and the NSUN7 IHC scores (negative vs. weak vs. moderate) in 24 Ewing sarcoma patients (Fisher's exact test). (B). Kaplan-Meier curve show the overall survival of NSUN7 (negative vs. weak vs. moderate) in the subgroup of IHC staining. Table S1. The top 15 GO biological processes of DERBPs in ES. Table S2. The top 15 REACTOME pathways of DERBPs in ES. Table S3. The top 15 hub genes with the highest degree of connectivity in the PPI network. Table S4. Correlations between RBPs and TFs. Table S5. ssGSEA scores of 50 Hallmarks and risk score in training cohort. Table S6. The correlations between risk score and 23 significant hallmarks. Table S7. IHC scoring and corresponding clinical characteristics in 24 Ewing sarcoma patients' cohort.

Author Contributions: Conceptualization, Y.C., H.S., and F.H.; methodology, Y.C., H.S., and F.H.; software, Y.C.; validation, Y.C., Y.Z., Y.L., and F.H.; formal analysis, Y.C., H.S., Y.Z., Y.S., Y.L., and F.H.; investigation, Y.C., H.S., Y.S., Y.Z., Y.L., and F.H.; resources, Y.C. and F.H.; data curation, all authors; writing—original draft preparation, Y.C.; writing—review and editing, Y.C., H.S., Y.S., Y.Z., Y.L., and F.H.; visualization, Y.C., H.S., and F.H.; supervision, F.H.; project administration, Y.C. and F.H.; funding acquisition, F.H. All authors have read and agreed to the published version of the manuscript.

Funding: This work is supported by the Swedish Cancer Society, the Swedish Childhood Cancer Foundation, the Cancer Society in Stockholm, the Stockholm County Council, and the Karolinska Institutet.

Institutional Review Board Statement: The study was conducted according to the guidelines of the Declaration of Helsinki and approved by the Institutional Review Board (2017-1114).

Informed Consent Statement: Informed consent was obtained from all subjects involved in the study.

Data Availability Statement: The data presented in this study are available on request from the corresponding author.

Conflicts of Interest: The authors declare no conflict of interest.

References

1. Arndt, C.A.; Rose, P.S.; Folpe, A.L.; Laack, N.N. Common musculoskeletal tumors of childhood and adolescence. *Mayo Clin. Proc.* **2012**, *87*, 475–487. [CrossRef] [PubMed]
2. Esiashvili, N.; Goodman, M.; Marcus, R.B., Jr. Changes in incidence and survival of Ewing sarcoma patients over the past 3 decades: Surveillance Epidemiology and End Results data. *J. Pediatr. Hematol. Oncol.* **2008**, *30*, 425–430. [CrossRef] [PubMed]
3. Randall, R.L.; Lessnick, S.L.; Jones, K.B.; Gouw, L.G.; Cummings, J.E.; Cannon-Albright, L.; Schiffman, J.D. Is There a Predisposition Gene for Ewing's Sarcoma? *J. Oncol.* **2010**, *2010*, 397632. [CrossRef]
4. Delattre, O.; Zucman, J.; Plougastel, B.; Desmaze, C.; Melot, T.; Peter, M.; Kovar, H.; Joubert, I.; de Jong, P.; Rouleau, G.; et al. Gene fusion with an ETS DNA-binding domain caused by chromosome translocation in human tumours. *Nature* **1992**, *359*, 162–165. [CrossRef] [PubMed]
5. Sorensen, P.H.; Lessnick, S.L.; Lopez-Terrada, D.; Liu, X.F.; Triche, T.J.; Denny, C.T. A second Ewing's sarcoma translocation, t (21;22), fuses the EWS gene to another ETS-family transcription factor, ERG. *Nat. Genet.* **1994**, *6*, 146–151. [CrossRef] [PubMed]
6. Brohl, A.S.; Solomon, D.A.; Chang, W.; Wang, J.; Song, Y.; Sindiri, S.; Patidar, R.; Hurd, L.; Chen, L.; Shern, J.F.; et al. The genomic landscape of the Ewing Sarcoma family of tumors reveals recurrent STAG2 mutation. *PLoS Genet.* **2014**, *10*, e1004475. [CrossRef]
7. Crompton, B.D.; Stewart, C.; Taylor-Weiner, A.; Alexe, G.; Kurek, K.C.; Calicchio, M.L.; Kiezun, A.; Carter, S.L.; Shukla, S.A.; Mehta, S.S.; et al. The genomic landscape of pediatric Ewing sarcoma. *Cancer Discov.* **2014**, *4*, 1326–1341. [CrossRef]
8. Shukla, N.; Schiffman, J.; Reed, D.; Davis, I.J.; Womer, R.B.; Lessnick, S.L.; Lawlor, E.R.; The COG Ewing Sarcoma Biology Committee. Biomarkers in Ewing Sarcoma: The Promise and Challenge of Personalized Medicine. A Report from the Children's Oncology Group. *Front. Oncol.* **2013**, *3*, 141. [CrossRef]
9. Huang, H.Y.; Illei, P.B.; Zhao, Z.; Mazumdar, M.; Huvos, A.G.; Healey, J.H.; Wexler, L.H.; Gorlick, R.; Meyers, P.; Ladanyi, M. Ewing sarcomas with p53 mutation or p16/p14ARF homozygous deletion: A highly lethal subset associated with poor chemoresponse. *J. Clin. Oncol.* **2005**, *23*, 548–558. [CrossRef]
10. Kovar, H.; Auinger, A.; Jug, G.; Aryee, D.; Zoubek, A.; Salzer-Kuntschik, M.; Gadner, H. Narrow spectrum of infrequent p53 mutations and absence of MDM2 amplification in Ewing tumours. *Oncogene* **1993**, *8*, 2683–2690.
11. Whelan, J.; Le Deley, M.C.; Dirksen, U.; Le Teuff, G.; Brennan, B.; Gaspar, N.; Hawkins, D.S.; Amler, S.; Bauer, S.; Bielack, S.; et al. High-Dose Chemotherapy and Blood Autologous Stem-Cell Rescue Compared With Standard Chemotherapy in Localized High-Risk Ewing Sarcoma: Results of Euro-E.W.I.N.G.99 and Ewing-2008. *J. Clin. Oncol.* **2018**, *36*, 3110. [CrossRef]

12. Lawlor, E.R.; Sorensen, P.H. Twenty Years on: What Do We Really Know about Ewing Sarcoma and What Is the Path Forward? *Crit. Rev. Oncog.* **2015**, *20*, 155–171. [CrossRef]
13. Fu, X.D.; Ares, M., Jr. Context-dependent control of alternative splicing by RNA-binding proteins. *Nat. Rev. Genet.* **2014**, *15*, 689–701. [CrossRef] [PubMed]
14. Nishida, K.; Kuwano, Y.; Nishikawa, T.; Masuda, K.; Rokutan, K. RNA Binding Proteins and Genome Integrity. *Int. J. Mol. Sci.* **2017**, *18*, 1341. [CrossRef]
15. Martin, K.C.; Ephrussi, A. mRNA localization: Gene expression in the spatial dimension. *Cell* **2009**, *136*, 719–730. [CrossRef] [PubMed]
16. Moore, M.J.; Proudfoot, N.J. Pre-mRNA processing reaches back to transcription and ahead to translation. *Cell* **2009**, *136*, 688–700. [CrossRef]
17. Sonenberg, N.; Hinnebusch, A.G. Regulation of translation initiation in eukaryotes: Mechanisms and biological targets. *Cell* **2009**, *136*, 731–745. [CrossRef]
18. Baltz, A.G.; Munschauer, M.; Schwanhausser, B.; Vasile, A.; Murakawa, Y.; Schueler, M.; Youngs, N.; Penfold-Brown, D.; Drew, K.; Milek, M.; et al. The mRNA-bound proteome and its global occupancy profile on protein-coding transcripts. *Mol. Cell* **2012**, *46*, 674–690. [CrossRef]
19. Gerstberger, S.; Hafner, M.; Tuschl, T. A census of human RNA-binding proteins. *Nat. Rev. Genet.* **2014**, *15*, 829–845. [CrossRef]
20. Araya, N.; Hirota, K.; Shimamoto, Y.; Miyagishi, M.; Yoshida, E.; Ishida, J.; Kaneko, S.; Kaneko, M.; Nakajima, T.; Fukamizu, A. Cooperative interaction of EWS with CREB-binding protein selectively activates hepatocyte nuclear factor 4-mediated transcription. *J. Biol. Chem.* **2003**, *278*, 5427–5432. [CrossRef] [PubMed]
21. Rossow, K.L.; Janknecht, R. The Ewing's sarcoma gene product functions as a transcriptional activator. *Cancer Res.* **2001**, *61*, 2690–2695.
22. Chen, Y.; Hesla, A.C.; Lin, Y.; Ghaderi, M.; Liu, M.; Yang, C.; Zhang, Y.; Tsagkozis, P.; Larsson, O.; Haglund, F. Transcriptome profiling of Ewing sarcomas—Treatment resistance pathways and IGF-dependency. *Mol. Oncol.* **2020**, *14*, 1101–1117. [CrossRef]
23. Mancarella, C.; Pasello, M.; Ventura, S.; Grilli, A.; Calzolari, L.; Toracchio, L.; Lollini, P.L.; Donati, D.M.; Picci, P.; Ferrari, S.; et al. Insulin-Like Growth Factor 2 mRNA-Binding Protein 3 is a Novel Post-Transcriptional Regulator of Ewing Sarcoma Malignancy. *Clin. Cancer Res.* **2018**, *24*, 3704–3716. [CrossRef]
24. Volchenboum, S.L.; Andrade, J.; Huang, L.; Barkauskas, D.A.; Krailo, M.; Womer, R.B.; Ranft, A.; Potratz, J.; Dirksen, U.; Triche, T.J.; et al. Gene Expression Profiling of Ewing Sarcoma Tumors Reveals the Prognostic Importance of Tumor-Stromal Interactions: A Report from the Children's Oncology Group. *J. Pathol. Clin. Res.* **2015**, *1*, 83–94. [CrossRef]
25. Savola, S.; Klami, A.; Myllykangas, S.; Manara, C.; Scotlandi, K.; Picci, P.; Knuutila, S.; Vakkila, J. High Expression of Complement Component 5 (C5) at Tumor Site Associates with Superior Survival in Ewing's Sarcoma Family of Tumour Patients. *ISRN Oncol.* **2011**, *2011*, 168712. [CrossRef]
26. Postel-Vinay, S.; Veron, A.S.; Tirode, F.; Pierron, G.; Reynaud, S.; Kovar, H.; Oberlin, O.; Lapouble, E.; Ballet, S.; Lucchesi, C.; et al. Common variants near TARDBP and EGR2 are associated with susceptibility to Ewing sarcoma. *Nat. Genet.* **2012**, *44*, 323–327. [CrossRef]
27. Leek, J.T.; Johnson, W.E.; Parker, H.S.; Jaffe, A.E.; Storey, J.D. The sva package for removing batch effects and other unwanted variation in high-throughput experiments. *Bioinformatics* **2012**, *28*, 882–883. [CrossRef] [PubMed]
28. Mei, S.; Meyer, C.A.; Zheng, R.; Qin, Q.; Wu, Q.; Jiang, P.; Li, B.; Shi, X.; Wang, B.; Fan, J.; et al. Cistrome Cancer: A Web Resource for Integrative Gene Regulation Modeling in Cancer. *Cancer Res.* **2017**, *77*, e19–e22. [CrossRef] [PubMed]
29. Ritchie, M.E.; Phipson, B.; Wu, D.; Hu, Y.; Law, C.W.; Shi, W.; Smyth, G.K. Limma powers differential expression analyses for RNA-sequencing and microarray studies. *Nucleic Acids Res.* **2015**, *43*, e47. [CrossRef]
30. Szklarczyk, D.; Gable, A.L.; Lyon, D.; Junge, A.; Wyder, S.; Huerta-Cepas, J.; Simonovic, M.; Doncheva, N.T.; Morris, J.H.; Bork, P.; et al. STRING v11: Protein-protein association networks with increased coverage, supporting functional discovery in genome-wide experimental datasets. *Nucleic Acids Res.* **2019**, *47*, D607–D613. [CrossRef] [PubMed]
31. Shannon, P.; Markiel, A.; Ozier, O.; Baliga, N.S.; Wang, J.T.; Ramage, D.; Amin, N.; Schwikowski, B.; Ideker, T. Cytoscape: A software environment for integrated models of biomolecular interaction networks. *Genome Res.* **2003**, *13*, 2498–2504. [CrossRef]
32. Chin, C.H.; Chen, S.H.; Wu, H.H.; Ho, C.W.; Ko, M.T.; Lin, C.Y. cytoHubba: Identifying hub objects and sub-networks from complex interactome. *BMC Syst. Biol.* **2014**, *8* (Suppl. 4), S11. [CrossRef]
33. Bindea, G.; Galon, J.; Mlecnik, B. CluePedia Cytoscape plugin: Pathway insights using integrated experimental and in silico data. *Bioinformatics* **2013**, *29*, 661–663. [CrossRef]
34. Bindea, G.; Mlecnik, B.; Hackl, H.; Charoentong, P.; Tosolini, M.; Kirilovsky, A.; Fridman, W.H.; Pages, F.; Trajanoski, Z.; Galon, J. ClueGO: A Cytoscape plug-in to decipher functionally grouped gene ontology and pathway annotation networks. *Bioinformatics* **2009**, *25*, 1091–1093. [CrossRef]
35. Wang, J.; Zhong, J.; Chen, G.; Li, M.; Wu, F.X.; Pan, Y. ClusterViz: A Cytoscape APP for Cluster Analysis of Biological Network. *IEEE/ACM Trans. Comput. Biol. Bioinform.* **2015**, *12*, 815–822. [CrossRef]
36. Goeman, J.J. L1 penalized estimation in the Cox proportional hazards model. *Biom. J.* **2010**, *52*, 70–84. [PubMed]
37. Hanzelmann, S.; Castelo, R.; Guinney, J. GSVA: Gene set variation analysis for microarray and RNA-seq data. *BMC Bioinform.* **2013**, *14*, 7. [CrossRef] [PubMed]

38. Biswas, B.; Bakhshi, S. Management of Ewing sarcoma family of tumors: Current scenario and unmet need. *World J. Orthop.* **2016**, *7*, 527–538. [CrossRef] [PubMed]
39. Simon, N.; Friedman, J.; Hastie, T.; Tibshirani, R. Regularization Paths for Cox's Proportional Hazards Model via Coordinate Descent. *J. Stat. Softw.* **2011**, *39*, 1–13. [CrossRef] [PubMed]
40. Gerds, T.A.; Kattan, M.W.; Schumacher, M.; Yu, C. Estimating a time-dependent concordance index for survival prediction models with covariate dependent censoring. *Stat. Med.* **2013**, *32*, 2173–2184. [CrossRef]
41. Vickers, A.J.; Cronin, A.M.; Elkin, E.B.; Gonen, M. Extensions to decision curve analysis, a novel method for evaluating diagnostic tests, prediction models and molecular markers. *BMC Med. Inform. Decis. Mak.* **2008**, *8*, 53. [CrossRef] [PubMed]
42. Nielsen, T.; Wallden, B.; Schaper, C.; Ferree, S.; Liu, S.; Gao, D.; Barry, G.; Dowidar, N.; Maysuria, M.; Storhoff, J. Analytical validation of the PAM50-based Prosigna Breast Cancer Prognostic Gene Signature Assay and nCounter Analysis System using formalin-fixed paraffin-embedded breast tumor specimens. *BMC Cancer* **2014**, *14*, 177. [CrossRef] [PubMed]
43. Stahl, D.; Gentles, A.J.; Thiele, R.; Gutgemann, I. Prognostic profiling of the immune cell microenvironment in Ewing s Sarcoma Family of Tumors. *Oncoimmunology* **2019**, *8*, e1674113. [CrossRef] [PubMed]
44. Li, S.; Yang, Q.; Wang, H.; Wang, Z.; Zuo, D.; Cai, Z.; Hua, Y. Prognostic significance of serum lactate dehydrogenase levels in Ewing's sarcoma: A meta-analysis. *Mol. Clin. Oncol.* **2016**, *5*, 832–838. [CrossRef] [PubMed]

cancers

Systematic Review

The Role of Chemotherapy in Management of Inoperable, Metastatic and/or Recurrent Melanotic Neuroectodermal Tumor of Infancy—Own Experience and Systematic Review

Małgorzata Styczewska [1,*,†], Małgorzata A. Krawczyk [2,†], Ines B. Brecht [3], Konrad Haug [3], Ewa Iżycka-Świeszewska [4], Jan Godziński [5,6], Anna Raciborska [7], Marek Ussowicz [8], Wojciech Kukwa [9], Natalia Cwalina [1], Emil Lundstrom [1] and Ewa Bień [2,*]

1 The English Division Pediatric Oncology Scientific Circle, Medical University of Gdansk, 80-210 Gdansk, Poland; ncwalina@gmail.com (N.C.); emil.lundstrom@vgregion.se (E.L.)
2 Department of Pediatrics, Hematology and Oncology, Medical University of Gdansk, 80-210 Gdansk, Poland; mkrawczyk@gumed.edu.pl
3 Department of Pediatric Hematology and Oncology, University of Tübingen, 72076 Tübingen, Germany; ines.brecht@med.uni-tuebingen.de (I.B.B.); konrad.haug@med.uni-tuebingen.de (K.H.)
4 Department of Pathology and Neuropathology, Medical University of Gdansk, 80-210 Gdansk, Poland; ewa.izycka-swieszewska@gumed.edu.pl
5 Department of Pediatric Surgery, Marciniak Hospital, 54-049 Wroclaw, Poland; jgodzin@wp.pl
6 Department of Pediatric Traumatology and Emergency Medicine, Wroclaw Medical University, 50-345 Wroclaw, Poland
7 Department of Oncology and Surgical Oncology for Children and Youth, Institute of Mother and Child, 01-211 Warsaw, Poland; anna.raciborska@hoga.pl
8 Department of Pediatric Hematology, Oncology and Bone Marrow Transplantation, Wroclaw Medical University, 50-556 Wroclaw, Poland; marek.ussowicz@umed.wroc.pl
9 Department of Otorhinolaryngology, Faculty of Medicine and Dentistry, Medical University of Warsaw, 00-739 Warsaw, Poland; wojciechkukwa@gmail.com
* Correspondence: mstyczewska@gumed.edu.pl (M.S.); ebien@gumed.edu.pl (E.B.); Tel.: +48-58-349-28-80 (M.S. & E.B.)
† The authors M.S. and M.A.K. contributed equally to the study.

Citation: Styczewska, M.; Krawczyk, M.A..; Brecht, I.B..; Haug, K.; Iżycka-Świeszewska, E.; Godziński, J.; Raciborska, A.; Ussowicz, M.; Kukwa, W.; Cwalina, N.; et al. The Role of Chemotherapy in Management of Inoperable, Metastatic and/or Recurrent Melanotic Neuroectodermal Tumor of Infancy—Own Experience and Systematic Review. *Cancers* **2021**, *13*, 3872. https://doi.org/10.3390/cancers13153872

Academic Editors: Saurabh Agarwal and Jianhua Yang

Received: 31 May 2021
Accepted: 28 July 2021
Published: 31 July 2021

Simple Summary: Melanotic Neuroectodermal Tumor of Infancy (MNTI) is a very rare neoplasm that most commonly develops within maxilla in infants. It usually has a benign clinical course and is treated with only surgery. However, patients with large, inoperable, metastatic or multiply recurring MNTI may require systemic treatment. The role of pre- and post-surgery chemotherapy (CHT) in the management of MNTI is unclear. Here, we have presented the disease courses and outcomes of four infants treated with multidrug CHT due to inoperable/recurrent MNTI. Additionally, a systematic literature review was performed which revealed 38 similar cases in the last 42 years. Most children with primarily inoperable MNTI responded to CHT, which allowed physicians to perform complete, non-mutilating delayed surgery. However, it is still uncertain whether CHT administered after incomplete resection of MNTI prevents recurrence. This study aimed to contribute to the establishment of standards of management in patients with inoperable, metastatic or persistently recurring MNTIs, which are currently lacking.

Abstract: Melanotic Neuroectodermal Tumor of Infancy (MNTI) is a very rare pediatric neoplasm of neural crest origin. In most cases, it develops in infants as a localized tumor of the maxilla, and surgery is usually curative. In less than 10% of patients with inoperable, metastatic or persistently recurring MNTI, chemotherapy (CHT) may be considered; however, its role is still unclear. The aim of our study was to assess the efficacy of CHT in children with large, inoperable, metastatic and/or recurrent MNTI. Four such infants, treated with CHT in Polish and German centers of pediatric oncology, were presented. Additionally, a systematic literature search of the PubMed/MEDLINE, Scopus and Web of Science databases was performed, yielding 38 similar cases within the last 42 years. Neoadjuvant CHT, based mainly on the protocols for neuroblastoma, was often effective, allowing for complete delayed surgery in most cases. However, the role of adjuvant CHT in preventing recurrences after incomplete resection of MNTI remains unclear. Disseminated inoperable MNTI was almost universally associated with poor response to CHT and unfavorable outcome. Further investigations to elaborate standards of management in patients with inoperable, metastatic or persistently recurring MNTIs are necessary to improve outcomes.

Keywords: melanotic neuroectodermal tumor of infancy; inoperable; metastatic; recurrent; chemotherapy; adjuvant; neoadjuvant; systemic treatment

1. Introduction

Melanotic neuroectodermal tumor of infancy (MNTI) is a very rare neoplasm mainly affecting children in the first year of life, with a median age at presentation of 4.5 months [1]. The clinical manifestation of MNTI is usually a fast-growing, painless, non-ulcerated solid mass with blue or black discoloration [2]. The tumor is localized most often in the head and neck region, with maxilla (62.2%), skull (15.6%) and mandible (7.8%) being the most common anatomical sites [1]. Localizations of MNTI within epididymis, extremities, ovary and endolymphatic sac have also been reported [3–10].

For many years, the origin of the tumor remained unclear and different theories were proposed, including odontogenic, retinal, neuroepithelial and neural crest derivation of the neoplasm [11]. This uncertainty is reflected in the many different names that have been used to describe this entity, including melanotic progonoma, congenital melanocarcinoma, pigmented epulis, melanoameloblastoma, retinal anlage tumor, pigmented tumor of the jaw of infants and others. Histochemical, biochemical and ultrastructural studies performed in recent decades have confirmed that MNTI belongs to the group of neoplasms originating from neural crest cells [12–14].

MNTI consists of two different populations of cells, often arranged in nests, cords, sheets and pseudo-glandular structures, embedded in fibro-collagenous stroma. One of the cell types are relatively large epithelioid cells resembling melanocytes, usually staining positive for both epithelioid and selected melanocytic markers. The second are small, round blue undifferentiated cells with neuroepithelial phenotype [2,15–17].

WHO 2017 "Head & neck tumors classification" defines MNTI as a locally aggressive, rapidly growing tumor [18,19]. In localized disease, surgery is a mainstay of treatment and, if performed radically, it is usually curative. In case of a local recurrence, which occurs in 15–20% of cases, the recommended treatment strategy is the repeated complete excision of the tumor [20].

Several cases of malignant MNTI with aggressive clinical course and presence of distant metastases have been described in the literature [21–24]. In such cases, as well as in primarily inoperable or persistently recurring MNTI, the optimal treatment strategy is uncertain. Since radical excision without severe mutilation is usually not possible, and radiotherapy (RTX) should not be proposed to young infants, a systemic treatment has to be considered. However, due to the rarity of the disease, only a few guidelines regarding the clinical management of children with inoperable, metastatic, or recurrent MNTI have been established to date, and the role of chemotherapy (CHT) as a neoadjuvant or adjuvant treatment has yet to be fully determined [20].

In this work, we presented four infants with inoperable and/or recurrent MNTI requiring systemic treatment. A systematic review of the literature regarding all cases of MNTI treated with CHT was also performed to evaluate the efficacy of CHT in MNTI.

2. Case Reports

2.1. Patient 1

A 2-month-old female infant was admitted due to a fast-growing tumor of the right side of the maxilla, disfiguring the right cheek, nose and upper lip. The first symptoms occurred when the girl was 6 weeks old, when a tooth within the hard palate surrounded by a small, hard, non-painful bulge was found by the mother (Figure 1). It was diagnosed by a GP as a neonatal tooth and a decision was made to wait and see. However, during the next two weeks, the mass rapidly increased in size. Magnetic resonance (MR) displayed a pathological, well defined, solid mass measuring 23 × 22 × 18 mm, located within the alveolar process on the right side of the maxilla, reaching the midline. The bud of the right incisor was visible within the mass. The tumor bulged outside in the region of the nasal alar and invaginated into the mouth. No evidence of destruction of the hard palate, penetration into the orbital or nasal cavities and no evidence of enlarged lymph nodes (LN) were found (Figure 2).

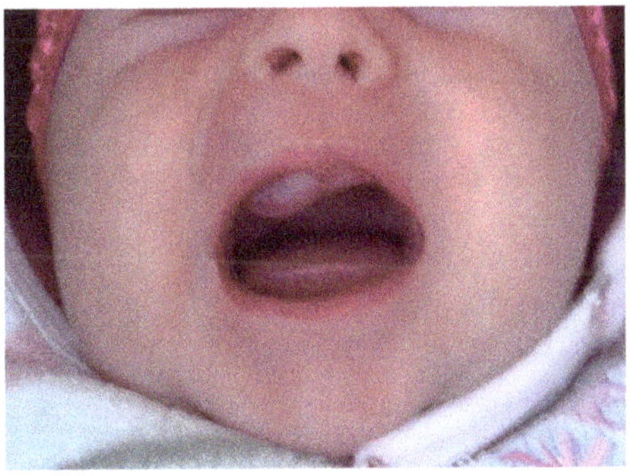

Figure 1. Patient 1: an intraoral tumor originating from the right side of the maxilla at first admission to the clinic.

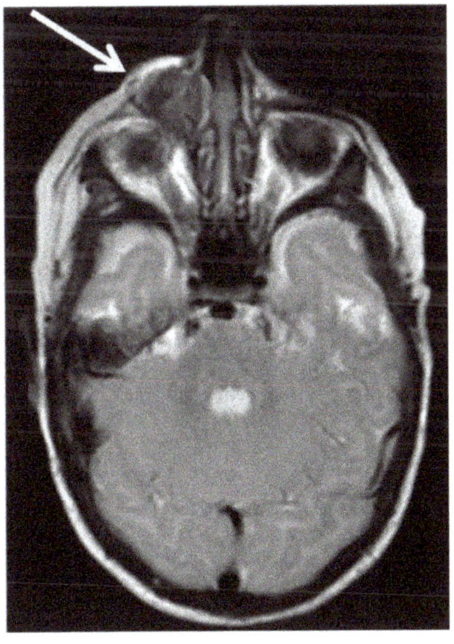

Figure 2. Patient 1: an axial T2-weighted MR scan shows well-defined pathological solid mass originating from the alveolar process of the right maxilla.

The tumor was resected in the otolaryngology department; however, clear margins were not achieved. The histopathological examination revealed MNTI with typical morphology and immunophenotype (Figure 3). Three weeks after the surgery, a local recurrence was observed and the patient was referred to the department of Pediatric Oncology. MR revealed a mass measuring 16.5 × 16.5 × 22 mm visible in the postoperative area (Figure 4).

The mass was invading the nasal concha and reaching the medial angle of the right eye; however, it did not penetrate into the orbit. Because the tumor was deemed unresectable without severe mutilation, a neoadjuvant CHT was introduced to reduce its size and extent. The patient received several courses of CHT according to the protocols for neuroblastoma (NBL) and sarcoma. The responses to particular courses of CHT varied (Table 1) but eventually, after eight courses, a partial regression of the tumor was achieved which enabled its complete non-mutilating resection. No recurrence was observed within 4.5 years post treatment.

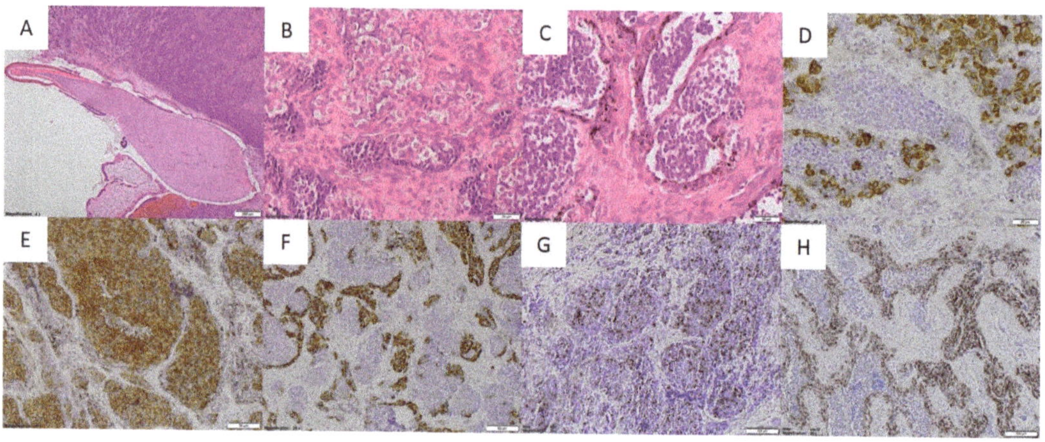

Figure 3. Patient 1: histopathological examination of MNTI. (**A**) Neoplastic tissue surrounding developing tooth with a root resorption (HE, 40×); (**B**) Tumor composed of small hyperchromatic and bigger epithelioid pale cells' islands embedded in mesenchymal stroma (HE, 400×); (**C**) Nests of small neuroectodermal cells rimmed with brown melanocytic cells (HE, 400×); (**D**) Pan-cytokeratin expression in epithelial population (CKAE1.AE3, 400×); (**E**) Synaptophysin staining with different intensity (synaptophysin, 200×); (**F**) Melanocytic marker HMB-45 within the pigmented cells (HMB-45, 200×); (**G**) Proliferative index up to 15% (Ki-67, 200×); (**H**) SOX10 nuclear expression within the tumor (SOX10, 200×).

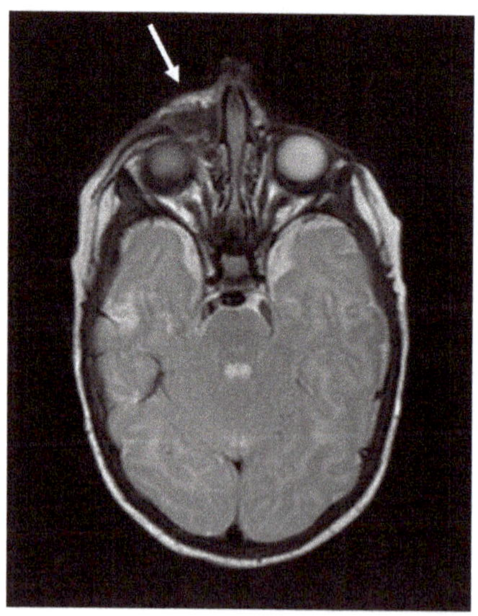

Figure 4. Patient 1: an axial T2-weighted MR scan shows recurrence of MNTI in the postoperative area.

Table 1. Chemotherapy in patient 1.

Type of Treatment	Cytostatic Drugs	Number of Courses	Response
neoadjuvant	CADO (cyclophosphamide, vincristine, doxorubicin)	1	stable disease (tumor growth stopped)
neoadjuvant	VP-16 + Carbo (etoposide + carboplatin)	2	minor partial response → slow tumor progression
neoadjuvant	I2VAdr (ifosfamide, vincristine, doxorubicin with dose reduction to 3/4)	1	stable disease
neoadjuvant	I2VE (ifosfamide with dose reduction to 2/3, vincristine, etoposide)	1	stable disease
neoadjuvant	CADO (cyclophosphamide, vincristine, doxorubicin)	1	stable disease
neoadjuvant	VP-16 + Carbo (etoposide + carboplatin)	1	minor partial response
neoadjuvant	CO (vincristine, cyclophosphamide)	1	minor partial response

2.2. Patient 2

A 6-month-old girl was admitted due to visible facial asymmetry involving the left nasal region. The diagnostic imagings revealed the tumor within the left maxilla, which was then resected subtotally in a regional hospital. Based on histopathological examination, the diagnosis of MNTI was made. After two months, a local recurrence was diagnosed and the girl was referred to the department of Pediatric Oncology. The surgical excision of the relapse was performed but, once again, it proved incomplete. This time the histopathological assessment revealed malignant histological features of MNTI. One month after the surgery, the girl presented with the second local recurrence. She was qualified for treatment according to the protocol for skeletal Ewing sarcoma (ES) (Table 2). After six courses of VIDE, the tumor regressed substantially, which allowed for the delayed resection of the remaining mass. Unfortunately, the surgery did not ensure clear margins. Therefore, the child was given adjuvant CHT (8 courses of VAC) postoperatively. She finished therapy in complete remission of MNTI. However, seven months later, the girl was diagnosed with a second malignancy: acute lymphoblastic leukemia (ALL). She was treated according to the ALL IC-BFM 2009 protocol and was qualified for allogeneic hematopoietic stem cell transplantation (allo-HSCT). Currently, she is alive without evidence of disease with a follow-up of over 4 years after the end of ALL therapy. Genetic testing for germline pathogenic variants predisposing to cancer was not performed.

Table 2. Chemotherapy in patient 2.

Type of the Treatment	Cytostatic Drugs	Number of Courses	Response
neoadjuvant	VIDE (vincristine, ifosfamide, doxorubicin, etoposide)	6	partial response
	microscopically incomplete (R1) resection of the tumor		
adjuvant	VAC (vincristine, dactinomycin cyclophosphamide)	8	no recurrence
ALL treatment (chemotherapy + allo-HSCT)	vincristine, daunorubicin, L-asparaginase, cyclophosphamide, ifosfamid, cytarabine, inthrathecal cytarabine, intrathecal metothrexate, doxorubicin, 6-mercaptopurin, 6-tioguanine, vindesine, etopozide	>2 years of treatment	no recurrence of ALL and MNTI

ALL = acute lymphoblastic leukemia; allo-HSCT = allogeneic hematopoietic stem cell transplantation.

2.3. Patient 3

A 3-month-old female was admitted to the clinic due to an enlarging occipital tumor. The lesion was visible at birth, and the parents noticed that its size increased within the following weeks. During a routine checkup, a muscular weakness of the lower extremities was observed in the child, which led to CT and MR of the central nervous system (CNS). The imaging studies showed a suboccipital mass measuring 40 × 40 × 31 mm with epidural infiltration (Figure 5). It was compressing the spinal cord (C1-C3) and was located directly adjacent to the vertebral arteries. A biopsy followed by histopathological examination was performed, revealing the diagnosis of MNTI. Because of the rapid tumor growth, neoadjuvant CHT was started according to NBL protocol NB2004 (Table 3). In the second course of CHT, an accidental overdosage of doxorubicin happened and the next planned course of CHT was cancelled. Instead, a watch and wait policy was applied. The volume of the tumor was stable for three months, but then a slow progression occurred. Due to the high local tumor extension, only a partial resection was possible. The remaining mass was still compressing the spinal cord; thus, three further courses of CHT, not including anthracyclines, were administered. A follow-up echocardiography performed a year after treatment completion revealed anthracyclin-induced cardiomyopathy. The remaining tumor mass did not change in size for the next three years, but then a slow progression of the tumor was noted in the MR. Therefore, the current indication for RTX is under discussion.

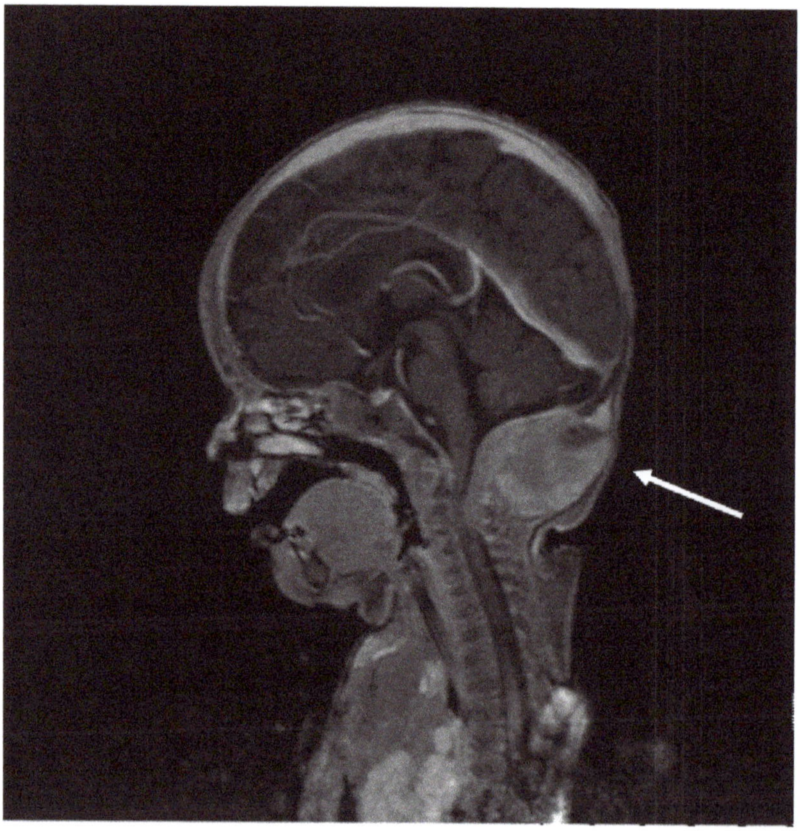

Figure 5. Patient 3: a sagittal T1-weighted MR scan shows large, suboccipital, mainly extracranial pathological solid mass compressing the spinal cord (C1–C3).

Table 3. Chemotherapy in patient 3.

Type of the Treatment	Cytostatic Drugs	Number of Courses	Response
neoadjuvant	N4 (vincristine, cyclophosphamide, doxorubicin)	1	stable disease
neoadjuvant	N4 (vincristine, cyclophosphamide, doxorubicin—accidental overdosage)	1	stable disease→watch&wait→slow progression
	partial resection of the tumor		
adjuvant	N5 (cisplatin, etoposide, vindesine)	1	not assessed
adjuvant	N6 (vincristine, dacarbazin, ifosfamide, without doxorubicin)	1	stable disease
adjuvant	N5 (cisplatin, etoposide, vindesine)	1	stable disease→follow-up →progession

2.4. Patient 4

A 4-month-old female was admitted due to a fast-growing tumor of the right maxilla, dislocating the nasal septum. An MR was performed, showing a solid 10 × 10 mm, well defined, slightly lobulated mass of the frontal medial palate with an embedded tooth. No infiltration of surrounding structures was detected. The diagnosis of MNTI was made based on the histopathological examination of the material from tumor biopsy. Subsequently, the resection of the tumor was performed, which was found microscopically incomplete. Three months later, the tumor recurrence was diagnosed. It was not feasible for resection, so CHT according to the NBL protocol NB2004 was started. After four courses of treatment with vincristine, cyclophosphamide and doxorubicin, the tumor size stabilized (Table 4). A second resection was performed; however, again, the tumor was not excised completely. Since then, no disease progression has been stated for over 1.5 years.

Table 4. Chemotherapy in patient 4.

Type of the Treatment	Cytostatic Drugs	Number of Courses	Response
neoadjuvant	N4 (vincristine, cyclophosphamide, doxorubicin)	4	stable disease

3. Review of the Literature

3.1. Methodology

The systematic literature review was performed according to the Preferred Reporting Items for Systematic reviews and Meta-Analyses (PRISMA) 2020 guidelines by searching the publications listed in the PubMed/MEDLINE, Scopus and Web of Science databases up to 26 April 2021 (Figure 6). The systematic search was performed by one reviewer (M.S.). Independently, the publications were searched by N.C. and E.L. (searched up until 2016) and M.K. (searched 2017–2021). Subsequently, all papers were verified by senior author (E.B.). Duplicated publications were removed with the use of automatic tool (Mendeley). Keyword search terms included: "melanotic neuroectodermal tumor of infancy", "retinal anlage tumor" and "melanotic progonoma" combined with terms "chemotherapy" and "systemic treatment". No additional search filters were applied. Additionally, the cases of children with MNTI treated with CHT were selected manually from the four main reviews [1,25–27]. The reference lists inserted in all publications were also searched for additional cases. Reports published in a language other than English were excluded. The data retrieved from each article included: age and sex of the patient, primary tumor site and size, presence of metastases, course of treatment, details of CHT treatment, follow-up and outcome. The demographic data of the patients were carefully analyzed to avoid duplication of cases. The series from literature was completed by four cases registered in the German Pediatric Rare Tumor Registry (STEP-Registry) and the Polish Pediatric Rare Tumors Database. Written informed consent was received from all parents at the time of registration.

Neoadjuvant CHT was defined as a CHT introduced (either in the first-line treatment or in the treatment of relapse) with the aim of diminishing an inoperable tumor and enabling delayed surgery. Adjuvant CHT was defined as a CHT introduced after tumor resection to treat residual tumor mass/avoid recurrence.

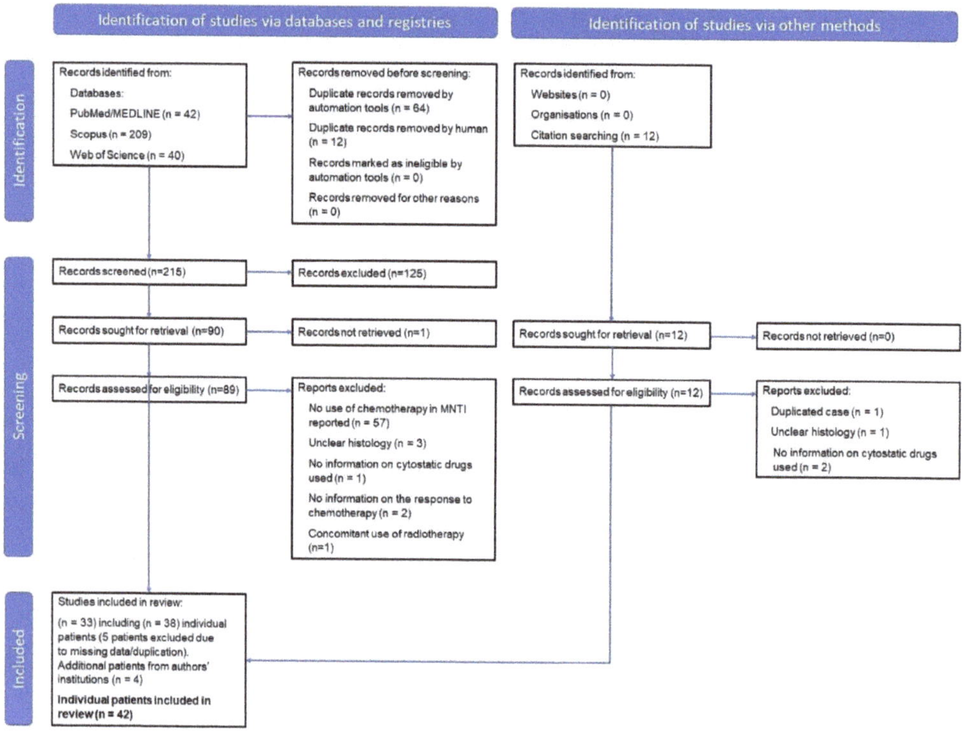

Figure 6. Flowchart of the study selection process (according to PRISMA 2020 guidelines).

3.2. *Results*

The literature search yielded 54 children with MNTI treated with CHT. Among them, two cases were duplicates [16,28] and four cases were excluded due to unclear histology [29–32]. Eight cases were excluded due to lack of information on the cytostatic drugs administered to patients [16,33–36] or no data on the response to CHT [37,38]. In two patients, CHT was combined with RTX; in those cases, therefore, the response to CHT was impossible to define [23,39]. Eventually, 42 cases were analyzed, including 38 cases previously published in the literature and four cases of our own. CHT was given in all cases, either as a first-line treatment (n = 24) or as a treatment after relapse (n = 18). In each patient, the particular neoadjuvant/adjuvant CHT was analyzed only if data regarding the response to CHT were available.

The demographic data of patients showed that the median age at diagnosis was 4 months. The anatomical locations of the tumors were: maxilla (n = 19; 45.2%), skull (n = 9; 21.4%), mandible (n = 5; 11.9%), femur (n = 2; 4.8%), CNS (n = 2; 4.8%), orbit (n = 2; 4.8%), epididymis (n = 1; 2.4%), suboccipital area wih an infiltration of the spinal canal (n = 1; 2.4%) and soft tissue of forearm (n = 1; 2.4%). Among 39 cases with known gender, a clear male predilection was found (the male to female ratio was 27:12; 69.2%:30.8%) (Table 5).

Table 5. Demographic and clinical data of the analyzed patients.

All Patients	*n* = 42 (100%)
own cases	n = 4 (9.5%)
literature reports	n = 38 (90.5%)
Sex	
female	n = 12 (28.6%)
male	n = 27 (64.3%)
unknown	n = 3 (7.1%)
Median age at diagnosis (mo)	4 (range: 0–48)
Tumor site	
maxilla	n = 19 (45.2%)
skull	n = 9 (21.4%)
mandible	n = 5 (11.9%)
femur	n = 2 (4.8%)
CNS	n = 2 (4.8%)
orbit	n = 2 (4.8%)
epididymis	n = 1 (2.4%)
suboccipital area	n = 1 (2.4%)
forearm	n = 1 (2.4%)
Median tumor size at diagnosis (cm)	4 (range: 1–20.5)
Metastases at diagnosis	
no	n = 12 (28.6%)
yes	n = 4 (9.5%)
lymph nodes	n = 3 (7.1%)
distant metastases	n = 1 (2.4%)
unknown	n = 26 (61.9%)

CHT = chemotherapy; cm = centimeters; CNS = central nervous system; mo = months.

3.2.1. Chemotherapy in the First-Line Treatment of MNTI

Chemotherapy was used as a component of the first-line treatment in 24 patients with MNTI, including one of our four patients (Table 6, Supplementary Materials Table S1). This subgroup consisted of 14 males and 7 females (in 3 patients, no data on gender was reported) aged between 1 day and 4 years (median age 4 months). The maximal tumor diameter ranged from 2.3 to 20.5 cm with a median size of 6 cm. The most commonly used CHT regimes included cyclophosphamide, doxorubicin, vincristine, ifosfamide, etoposide and cisplatin (Supplementary Materials Table S1).

Table 6. Summary data of the patients receiving CHT in the first-line treatment of MNTI (n = 24).

Type of CHT	Number of Patients	Response to CHT	Outcome
adjuvant		10 (41.7%) *	
after macroscopically incomplete (R2) surgery of the primary tumor	5	CR: 2	NED: 2
		PR: 1	NED: 1
		SD: 2	AWD: 2*
after macroscopically incomplete (R2) surgery of the metastatic lesion	1	PR: 1	DOD: 1
after microscopically incomplete (R1) surgery of the primary tumor	2	NA: 2	NED: 2
after complete (R0) surgery of the primary tumor and involved LNs	2	NA: 2	NED: 2
neoadjuvant		12 (50%) *	
after biopsy of primary tumor	11	PR: 7	NED: 4 D of complications: 1 LFU: 2
		SD: 3	NED: 1 AWD: 1 * LFU: 1
		PD: 1	NED: 1
after biopsy of involved LN	1	PR: 1	DOD: 1
the only treatment		3 (12,5%) *	
after biopsy	3	PR: 2	AWD: 2 *
		SD: 1	LFU: 1
together	24 (100%)	CR: 2 PR: 12 SD: 4 SD; SD: 1 ** PD: 1 NA: 4	NED: 13 AWD: 4 * DOD: 2 D of complications: 1 LFU: 4

AWD = alive with disease; CHT = chemotherapy; CR = complete response; D = died; DOD = died of disease; LFU = lost to follow-up; LNs = lymph nodes; NA = not assessable; NED = no evidence of disease; PD = progressive disease; PR = partial response; SD = stable disease; * the numbers/percentage scores do not sum up to the summarized numbers/100% due to the overlap of the patients (details provided in the supplementary Table S1); ** patient who underwent both neoadjuvant and adjuvant CHT.

Adjuvant CHT in the First-Line Treatment of MNTI

Ten patients with MNTI received adjuvant CHT postoperatively. In eight patients, the preceding surgery was incomplete. In seven, it was a resection of a localized tumor and in one patient, it was a resection of a metastatic focus [21,40–45]. Adjuvant CHT was also given to two children with regional disease after complete resections of the primary tumors and involved lymph nodes [22,46].

Among eight patients who received CHT after incomplete surgery, response to CHT was assessable in six. Complete response was noted in two patients, both with MNTI of the skull. Their CHT regimes consisted of either alternate courses of cyclophosphamide and doxorubicin [42] or the 8-in-1 protocol for medulloblastoma [41]. In two children, partial responses were obtained. In a child with localized MNTI of the skull, it allowed for the second-look complete surgery [43], while in a patient with metastatic disease, the partial response to CHT only lasted for a short time and was followed by PD [21]. Two children, including one of our patients (number 3), achieved stabilization of disease; however, progression occurred thereafter [40].

Two children in whom CHT followed complete resection of the tumor and involved lymph nodes and two children who received adjuvant CHT after R1 resection are alive without recurrences [22,44–46].

Altogether, seven of ten patients treated with adjuvant CHT did not experience recurrence or PD, with a follow-up ranging from 10 months to 4.5 years.

Neoadjuvant CHT in the First-Line Treatment of MNTI

In 12 patients with inoperable and/or metastatic MNTI at diagnosis, the neoadjuvant CHT was introduced. The diagnosis was set based on a biopsy of the primary tumor (11 patients) or of an involved lymph node (one patient).

In eight patients, partial responses to CHT were achieved, followed by delayed tumor resections [4,43,47–52]. The surgery was complete in five patients, of whom three are alive; one was lost to follow-up after late recurrence and one died due to acute cardiomyopathy. Incomplete delayed resections were performed in two children, of whom one survived. In one patient, the completeness of resection was not defined and the follow-up was not provided [52].

Four of 12 patients did not respond well to neoadjuvant CHT (stable disease in three, PD in one patient), however, in all of those cases, delayed surgeries were attempted [3,53,54]. Three patients survived, and one is currently being treated for late tumor recurrence.

In total, 6 of 11 patients treated with neoadjuvant CHT due to inoperable and/or metastatic MNTI at diagnosis were disease-free at the time of publication [4,47,48,50,53,54]. One patient is alive with disease (our patient, number 3), two patients died (including one due to acute therapy-related complications) [49,51] and three patients were lost to follow-up [3,43,52].

CHT as the Only First-Line Treatment of MNTI

In three patients with inoperable MNTI, no tumor resection was performed. In two of them, the disease was successfully controlled by CHT only. In one child, a subtotal calcification of the tumor was observed during 10 months of multidrug CHT. One year after CHT cessation, no progression occurred [55]. In another child, CHT—consisting of etoposide and carboplatin—produced partial response with no regrowth within the next 16 months [16]. Conversely, in one patient with a huge inoperable MNTI of the maxilla, the response to one course of CHT composed of vincristine, cyclophosphamide and dactinomycin was poor (10% regression). The parents refused further treatment and the patient was lost to follow-up [56].

3.2.2. Chemotherapy in the Treatment of Recurrence of MNTI

There were 18 patients with recurrence of MNTI who were treated with CHT (Table 7, Supplementary Materials Table S1). This group included three of our patients (numbers 1, 2 and 4). In all children, the recurrence developed after the first-line therapy composed of surgical tumor excision. In most publications, the completeness of resection was not specified; however, at least seven patients initially had only curettage or partial resection. The median time from initial surgery to the first relapse was 1.5 months (range: 2 weeks–18 months), with 16 local, one regional and one metastatic recurrence. Chemotherapy was administered as a treatment of the first, second and third relapses in 13, three and two patients, respectively.

The CHT regimes to treat MNTI recurrence comprised mainly cyclophosphamide, doxorubicin, vincristine, etoposide and carboplatin (Supplementary Materials Table S2).

Table 7. Summary data of the patients receiving CHT in the treatment of recurrent MNTI (n = 18).

Type of CHT	Number of Patients	Response to CHT	Outcome
adjuvant	12 (66.7%) *		
after macroscopically incomplete (R2) surgery of the recurrent tumor	4	CR: 1 PR: 2 SD: 1	NED: 1 * AWD: 2 AWD: 1
after microscopically incomplete (R1) surgery of the recurrent tumor	5	NA: 5	NED: 5*
after unspecified surgery of the recurrent tumor	3	PD: 2 NA: 1	DOD: 1 LFU: 1 (in progression) * NED: 1*
neoadjuvant	8 (44.4%) *		
after local recurrence	6	CR: 1 PR: 3 SD: 1 PD: 1	NED: 1* NED: 3* AWD: 1 NED: 1*
after regional/metastatic recurrence	2	PR: 1 PD: 1	LFU: 1 (in progression) * DOD: 1
together	18 (100%)	CR: 2 PR: 2 PR; PD: 1 ** PR; NA: 1 ** SD: 2 PD: 4 NA: 6	NED: 11 * AWD: 4 DOD: 2 LFU: 1*

AWD = alive with disease; CHT = chemotherapy; CR = complete response; DOD = died of disease; LFU: lost to follow-up; LN = lymph nodes; NA = not assessable; NED = no evidence of disease; PD = progressive disease; SD = stable disease; * the numbers/percentage scores do not sum up to the summarized numbers/100% due to the overlap of the patients (details provided in the supplementary Table S2); ** patients who underwent both neoadjuvant and adjuvant CHT.

Adjuvant CHT in the Treatment of Recurrence of MNTI

Adjuvant CHT was used in 12 patients with recurrent MNTI; in nine cases, after incomplete tumor excision (R1 in five, R2 in four), and in three patients with unspecified completeness of excision of the recurrence. The response to CHT was assessable in six children. One of them achieved long-term complete response [57], two had partial response after five and six months of CHT, respectively [24,43] and one patient achieved stabilization of disease [39]. In two children, disease progression was stated during CHT [53,58].

In six patients, including our patient number 2, the efficacy of CHT was impossible to evaluate as the CHT followed R1 or subtotal resections with no residual mass visible in imaging studies. However, in all of them, no recurrence was stated and the children have no evidence of disease with follow-ups ranging from 1 to 17 years [36,59,60].

CHT was given after resection of the first or second recurrence in nine and three children, respectively. All but one of the patients who received adjuvant CHT in the treatment of the first relapse had favorable outcomes [36,39,43,57,59,60]. At the time of publication, six of these children (four R1, one R2, one not specified) were alive and free of disease, one patient with residual disease remained stable and one child with partial response was still in treatment. One patient who developed metastatic relapse did not respond to CHT and was lost to follow-up with disseminated and progressive tumor [53].

Among three patients who developed two relapses each and received adjuvant CHT after resection of the second recurrence, one did not respond to CHT and died of PD [58]. Another patient had a partial response; however, progression of disease was stated 5 months after cessation of CHT. After further CHT and RTX, at the time of publication, this child was alive with disease [24]. The third patient (our patient number 2), however, was successfully cured of the second recurrence of maxillary MNTI. She received neoadjuvant CHT before the resection of the recurrence followed by adjuvant CHT. However, due to the second malignancy (ALL) which was diagnosed seven months after the completed MNTI treatment, she underwent another oncological therapy consisting of CHT and allo-HSCT. She has remained disease free with a follow-up of 6.5 years.

Neoadjuvant CHT in the Treatment of Recurrence of MNTI

Eight patients were reported to start therapy of relapse with CHT. It was the first, second and third recurrence in five, one and two patients, respectively. Six patients presented with local, one with regional and one with metastatic relapse.

Among six children with local relapses, one patient achieved complete response to VECI protocol adapted from NBL treatment [61]. The partial response to CHT was stated in three children, including two presented in our study. All of them underwent successful tumor resection and remained disease free [17]. In one child (our patient number 4), CHT followed by incomplete tumor resection led to long-term disease stabilization. One patient developed PD during neoadjuvant CHT. Nevertheless, complete resection of the tumor was possible [62]. All these children were alive at the time of publication, with follow-up periods ranging from 6 months to 14 years.

Among two patients with regional/metastatic relapses, one child did not respond to CHT. The other achieved temporary remission after CHT, but both eventually died due to PD [53,63].

4. Discussion

The optimal management of children with repeatedly relapsing, inoperable and/or disseminated MNTI has not yet been determined, due to the extreme rarity of this subgroup of MNTI. The Children's Cancer & Leukaemia Group (CCLG) Guidelines published in 2004 suggested that, in such cases, systemic CHT would be an option [20]. In this paper, based on literature review and our own experiences, we aimed to determine the incidence of MNTIs requiring systemic treatment and analyze the efficacy of the most frequently used CHT regimes.

We found 54 reported cases of MNTI treated with CHT. For 38 patients, sufficient data on cytostatic drugs, response to CHT and disease course was available. Four cases of MNTI treated in oncological centers in Poland and Germany were added, resulting in 42 cases feasible for analysis. Three out of our four cases developed in the maxilla; however, nearly 55% of the studied group had extramaxillary tumors. This was significantly more frequent in comparison to children with localized MNTI, treated with surgery and/or RTX, as reported by Rachidi et al. (34%). It seems that aggressive MNTIs tend to develop in sites other than the maxilla. Additionally, the significant male predilection observed in patients treated with CHT (M:F 69.2%:30.8%) stands in contrary to the almost equal gender distribution reported in patients with localized MNTIs (M:F 57%:43%) [1].

In 2004, the CCLG proposed two CHT protocols for inoperable MNTI. Their regimen 1 consists of cyclophosphamide and vincristine and was proposed for management of tumor recurrence occurring despite two surgical interventions. The more aggressive regimen 2 is intended for therapy of inoperable or metastatic disease at diagnosis or in case of no response to regimen 1. It was adapted from the protocol OPEC/OJEC for stage 4 NBL and includes vincristine (O), cisplatin (P), etoposide (E), cyclophosphamide (C) and carboplatin (J) [20]. However, these protocols have not been widely used. Few authors reported successful use of regimen 1 [36,44] and only two cases of MNTI treated with adjuvant OPEC/OJEC protocol after incomplete resection have been published since 2004 [36]. Although both latter patients survived, the effectiveness of these regimens is difficult to assess. Various other CHT protocols have been used for inoperable/metastatic MNTI in the last 15 years [22,24,39,45,50,51,54].

Including the aforementioned patients treated according to CCLG guidelines, 12 out of 39 reported cases of inoperable/recurrent and/or metastatic MNTI were treated with protocols directly adapted from NBL treatment [36,44,45,54,60–62]. Within this group are three of our patients, among whom one (patient number 1) responded to a regime consisting of vincristine, cyclophosphamide and carboplatin with etoposide, successfully used in infant NBL. Two other children (patients 3 and 4) achieved stabilization of the disease following NB2004 regimen.

The reason for choosing regimes adapted from NBL treatment is the neural crest origin of both MNTI and NBL. Neven at al. reported a 1p deletion and gain of chromosome 7q in one child with MNTI, indicating the genetic analogy to NBL [61]. In a subset of children with MNTI, elevated urine concentrations of vanillylmandelic acid (VMA) and homovanillic acid (HVA) have been observed, similarly to NBL [21,22,51]. Moreover, the component of neuroblast-like cells within MNTI has been suggested to be responsible for the aggressiveness of the tumor. In some cases of MNTI, the small, round blue component significantly predominated over epithelioid cells in metastatic lesions, whereas in the primary tumor, the two components were balanced. [46,58]. In cases evaluated histologically both before and after CHT, the CHT resulted in almost complete reduction of the small round blue cell component of MNTI; however, the reduction of epithelioid cells was poor [4,62].

Accordingly, the efficacy of NBL-derived protocols in MNTI treatment seems to be satisfactory. Notably, all six patients treated with adjuvant CHT after R1 resections (two in first-line treatment, four in relapsed tumor) experienced no recurrence [36,44,45,60]. The neoadjuvant CHT adapted from NBL treatment was used in the treatment of primary tumor and relapses in two and four patients, respectively. The responses varied; however, all patients were alive at the time of publication (all but one underwent delayed surgery) [54,61,62].

In our patient number 2, a CHT protocol for ES was successfully used as both neoadjuvant and adjuvant treatment of recurrent maxillary MNTI. In five other children, the name of the regimen was not mentioned, but the chemotherapeutic combinations were probably also adapted from ES treatment [22,50,55–57]. The cellular origin of ES has not been fully determined. Some authors have suggested neural crest derivation similar to MNTI [64]. Consequently, several MNTI cases were found to express CD99, a marker used in diagnosis of ES group tumors [3,45,65,66]. In a series of children with MNTI, CD99 was expressed in one out of eight tumors and was associated with malignant behavior [15].

In our analysis, the efficacy of MNTI treatment based on ES regimes was difficult to assess due to the small number of cases. Excluding our aforementioned patient, two other children were treated with adjuvant CHT. One of them had no recurrence after R0 resection of primary tumor and metastatic LNs [22]. Another child treated with CHT after R2 surgery of recurrent MNTI demonstrated complete response [57]. Among three patients who underwent neoadjuvant CHT adapted from ES protocol in the first-line treatment, one achieved PR and two had stable disease [50,55,56]. Therefore, in general, the efficacy of regimes adapted from ES treatment seems to be acceptable in MNTI.

Some patients with MNTI were treated according to the protocols for other pediatric solid tumors. Creytens et al. reported the successful use of neoadjuvant CHT according to the RMS 2005 protocol, which resulted in a significant response allowing for delayed complete tumor resection [4]. Another patient had a partial response to a protocol for high-risk soft tissue sarcomas; however, progression occurred 5 months after the end of treatment [24]. Our patient number 1 also received two courses of CHT based on the Cooperativen Weichteilsarkom Studiengruppe (CWS)-2006 protocol for soft tissue sarcomas (I2VAdr and I2VE), which produced stabilization of tumor size. Two other children with MNTI of the pineal region and skull were treated according to the 8-in-1 protocol used in the treatment of malignant intracranial tumors. In one patient, after subtotal surgery, a complete response and no recurrence after 10 months were observed [41]. In another, the CHT followed partial resection and resulted in stabilization of the tumor, lasting for nine months after cessation of treatment [40]. Response of MNTI to CHT meant for other pediatric solid tumors may result from the use of similar cytostatic drugs (e.g., vincristine, doxorubicin) in these neoplasms.

In order to optimize therapeutic approaches to MNTI, several attempts were made to find genetic connections of MNTI to other malignancies. Gomes et al. revealed BRAFV600E mutation in one of three analyzed cases of MNTI which may indicate further options for targeted therapy [67]. Moreover, a case of fibular MNTI with a germline CDKN2A loss-of-function mutation was also described. Since inherited CDKN2A mutations are associated with familial melanoma syndromes and the epithelioid MNTI cells are supposed to be immature melanocytes, it was suggested that the genetic rearrangement might be partially responsible for the development of MNTI in this particular patient [66]. However, none of these discoveries has yet been translated into clinical management of the tumor.

Most children with inoperable/recurrent and/or metastatic MNTI analyzed in this study were treated with CHT administered preoperatively (neoadjuvant CHT) and/or after surgical attempts (adjuvant CHT).

In children with inoperable or incompletely resected primary MNTI or MNTI relapse, neoadjuvant CHT seems to be promising. In eight out of 12 children who underwent neoadjuvant CHT in the first-line treatment, partial tumor shrinkage was stated [4,43,47–52]. This enabled physicians to perform delayed surgery, which was complete in 5 out of 8. Moreover, two children were successfully treated for MNTI with CHT only. In these patients, the CHT was probably intended to be neoadjuvant but, eventually, the delayed surgeries were not performed [16,55]. Both patients were alive at the time of publication without further progression or recurrence. It seems, therefore, that in the subset of children with initially inoperable MNTI in whom complete or good partial response to CHT is observed, long-term remission may be achieved even without performing delayed resection. This approach may reduce possible mutilation; however, the risk of progression/recurrence after treatment with CHT only is difficult to determine.

Among eight children in whom neoadjuvant CHT was used as a first treatment of relapse, one patient achieved complete response to NBL VECI protocol [61], while four others (including two of our patients) achieved partial response. This allowed for delayed non-mutilating surgeries [17,53].

However, the final outcome did not always correlate with response to neoadjuvant CHT. After complete delayed surgery following good response to CHT of primary MNTI, 2 out of 5 patients died due to recurrence or complications [43,49]. Conversely, 2 out of 4 children who did not respond to CHT survived after delayed surgery, independent of its completeness [53,54]. In relapsed MNTI, complete delayed resection was achieved in a patient who did not respond to CHT [62].

Adjuvant CHT is usually the only available therapeutic option in patients with macroscopic tumor masses or metastases remaining after surgery, especially when reoperation is not possible. However, it seems to be beneficial only in a subset of cases. Adjuvant CHT following initial partial (R2) resection of MNTI was effective in 3 out of 6 patients [42,43,68]. Among four patients who underwent CHT after partial (R2) resection of relapse, one achieved complete response, one partial response, and one was still in treatment at the time of publication [24,43,57].

Most authors agree that adjuvant CHT might prevent recurrence in patients with microscopically positive surgical margins (R1), high mitotic rate, aneuploidy and unfavorable location of MNTI [21,44,45,49,60]. In our study, however, only seven such cases were reported (two in the first-line treatment and five after recurrence) [36,44,45,60]. Lack of visible tumor in imaging studies makes it impossible to assess the response to CHT. However, in all the children undergoing adjuvant CHT, no recurrence was noted after follow-ups of 1 to 17 years. On the other hand, the recurrence rate of MNTI does not clearly correlate with the radicality of surgery or the use of adjuvant chemotherapy [1,16]. There are numerous reports on uneventful courses of MNTI after R1 or subtotal surgery with no adjuvant treatment [16,36]. Moreover, several cases of spontaneous regression or calcification of the macroscopic residual MNTI were noted [48,54,69]. However, a recently published systematic review including 429 MNTIs of maxilla and mandible, suggests that the risk of recurrence increases when only curettage of the tumor is performed [18]. Therefore, it would be of great value to identify the subset of patients with high risk of relapse of MNTI after R1 surgery, which would benefit from adjuvant CHT. One significant factor predicting the risk of relapse is the patient's age at diagnosis. In children diagnosed within the first 2 months of life, the recurrence rate is the highest, while in those beyond 4, 5 months of age it is minimal [1]. Most recurrences occur within the first 4 weeks after surgery and the risk of regrowth decreases with time, with only anecdotal reports of recurrences developing more than six months after completion of treatment [25]. In our study group, among 18 patients who were treated with CHT due to relapse of MNTI, the median age at diagnosis was 3 months, as compared to 4 months in the whole study group.

The localization of the primary MNTI may also be a risk factor for relapse. The recurrence rate was found to be higher in MNTIs affecting the mandible (33.3%) and the skull (31.8%), than in maxillary (19.3%) and intracranial (12.5%) tumors [25]. However, this tendency was not observed in the group of patients treated with CHT for recurrent MNTI, as the relapsed tumors were localized predominately in the maxilla (n = 13; 72.2%), skull (n = 3; 16.7%), mandible (n = 1; 5.6%) and orbit (n = 1; 5.6%).

No clear histological or biological markers of malignancy or recurrence in MNTI have been reported to date. Aneuploidy was suggested to be a negative prognostic factor, since two such cases reported by Pettinato et al. recurred within 1 month. However, the recurrences were observed in the diploid tumors as well [18,70]. Some authors implied that the presence of necrosis and mitotic figures may be associated with a malignant course of disease [53]. Still, these features have also been found in benign tumors [65]. Based on a single case, Barrett et al. suggested that a high Ki-67 proliferation marker and the presence of CD99 expression could indicate aggressive clinical behavior of MNTI [15]. However, the Ki-67 status assessed in 36 cases of maxillary/mandible MNTIs did not correlate with recurrence rate [18] and the CD99 expression was also reported in typical MNTIs, with benign histologies and outcomes [2,65].

Thus, in spite of various attempts, it is currently impossible to reliably assess the risk of recurrence based on clinical characteristics of MNTI. In our opinion, adjuvant CHT may be considered in children in whom the tumor has been removed incompletely, particularly in those within the first two months of life, when the recurrence risk is the highest and the use of radiotherapy is avoided. However, the potential short- and long-term adverse effects of cytostatic drugs in young infants should be carefully taken under consideration. Choi et al. reported the death of a patient with MNTI due to cardiomyopathy during treatment including doxorubicin [49]. Therefore, the risk-to-benefit ratio should always be carefully evaluated for each patient with MNTI before adjuvant CHT is given.

There seems to be a group of patients with MNTI who respond poorly to CHT, regardless of which cytostatic drugs and protocols are used. Two out of three patients in whom distant organ metastases were present at diagnosis or were detected during the course of treatment succumbed to the disease or were lost to follow up in PD [21,53]. The only patient with leptomeningeal dissemination of recurrent MNTI who was alive with disease three years following treatment responded to salvage RTX treatment [24]. Among four children with metastases limited to LN, the outcome was good in two patients who underwent complete excision of the main mass and involved LN, followed by adjuvant CHT [22,46]. However, two other patients in whom radical surgery was not feasible, died of disease despite long, multimodal treatment [51,63]. It suggests that the presence of inoperable disseminated disease is associated with poor response to CHT and, subsequently, with unfavorable outcomes.

The optimal follow-up regimen for patients with MNTI after surgical resection has not yet been established. In patients with maxillary or mandibular tumors, almost all local and regional recurrences were reported to occur within 6 months after resection. Therefore, it has been suggested to perform regular and frequent clinical and ultrasound examinations [1]: weekly in in the first month after surgery, every second week for the next

3 months and once a month for the following 6–12 months. It would also be important to perform MR of the resected tumor area every 3 months during this time period.

In patients with MNTI localized within the skull or CNS, relapses may occur later, even years after primary surgery [25]. Moreover, in the analyzed group, all three patients with distant metastases of MNTI had primary tumors located in the skull or CNS and developed intracranial metastases (in one case, disseminated via ventriculoperitoneal shunt) [21,24,53]. This may indicate a greater (but still very small) risk of dissemination in children with MNTI in these locations. Therefore, it may be suggested that patients with CNS/skull MNTI should undergo regular surveillance with MR of the head for at least 5 years after treatment. As no other locations of distant metastases of MNTI have been described in recent decades, it seems that routine follow-up imaging examinations of the chest, abdominal cavity and pelvis are not necessary.

In children with inoperable, metastatic and/or multiply relapsing MNTI who were treated with CHT—either neoadjuvant or adjuvant—the follow-up examinations should additionally include regular clinical, laboratory and imaging assessment of the renal, liver, eye, hearing and cardiac function. The risk of second malignancy should also be considered and properly addressed.

5. Conclusions

There are no established guidelines for optimal management in children with inoperable, metastatic or repeatedly recurrent MNTI. In such cases, CHT may be administered on a neoadjuvant and/or adjuvant basis. Considering the biological similarity of neuroblast-like components of MNTI and NBL, CHT adapted from NBL treatment protocols has been used most frequently with beneficial outcomes. Neoadjuvant CHT is usually effective, allowing physicians to perform delayed complete and non-mutilating surgery in a large subset of patients. Our review and own experiences indicate that CHT may be considered in a child with MNTI not feasible for initial tumor resection; however, significant awareness of severe complications is needed.

The role of adjuvant CHT in preventing recurrences after subtotal excision of MNTI remains unclear. The presence of inoperable disseminated disease is almost universally associated with poor response to CHT and unfavorable outcomes. Further investigations on the histological and molecular features of MNTI are required to adjust the treatment to the biology of the tumor and search for new targeted therapies. International collaborative studies with unified standards of management of children with inoperable, metastatic or persistently recurring MNTI and prospective enrollment in national or international databases are mandatory if we are to improve the knowledge and the prognosis of children with these rare types of MNTI.

6. Limitations of the Study

This retrospective study had several limitations. First, the analyzed group of patients is relatively small and heterogenous, so it is difficult to undoubtedly determine the exact efficacy of neoadjuvant and adjuvant CHT in the treatment of MNTI. Moreover, the definitions of response to CHT (partial response, stable disease, progression) have not been consistent and clearly defined between reports.

Additionally, some of the studies included in our analysis were published a few decades ago. Thus, we cannot exclude that, in a subset of analyzed patients, the tumor would nowadays be pathologically classified as a neoplasm other than MNTI.

Supplementary Materials: The following are available online at https://www.mdpi.com/article/10.3390/cancers13153872/s1, Table S1: Detailed data of the patients treated with chemotherapy in the first-line treatment of MNTI; Table S2: Detailed data of the patients treated with chemotherapy in the recurrence of MNTI; Supplementary material S3: Systematic literature search protocol.

Author Contributions: Conceptualization, M.S., M.A.K. and E.B.; methodology, M.S., N.C., E.L. and E.B.; software, M.S.; validation, E.I.-Ś., I.B.B. and E.B.; formal analysis, M.S., M.A.K. and E.B.; investigation, M.S., M.A.K., E.I.-Ś., N.C., E.L.; resources, M.A.K., I.B.B., A.R., M.U., W.K., J.G.; data curation, M.S.; writing—original draft preparation, M.S., M.A.K., K.H.; writing—review and editing, I.B.B., E.I.-Ś., E.B.; supervision, E.B.; project administration, E.B. All authors have read and agreed to the published version of the manuscript.

Funding: This research received no external funding.

Institutional Review Board Statement: Ethical review and approval were waived for this study, as it included retrospective case report descriptions and literature-based systematic review only.

Informed Consent Statement: Written informed consent was obtained from the parents of the patients to publish this paper.

Data Availability Statement: The data presented in this study (additional information on four own patients) are available on request from the corresponding author. The data are not publicly available due to privacy of the patients. The other data presented in this study (cases of patients with MNTI found in reviewed articles) are openly available in public databases (PubMED/MEDLINE, Scopus, Web of Science).

Conflicts of Interest: The authors declare no conflict of interest. The funders had no role in the design of the study; in the collection, analyses, or interpretation of data; in the writing of the manuscript, or in the decision to publish the results.

References

1. Rachidi, S.; Sood, A.J.; Patel, K.G.; Nguyen, S.A.; Hamilton, H.; Neville, B.W.; Day, T.A. Melanotic neuroectodermal tumor of infancy: A systematic review. *J. Oral Maxillofac. Surg.* **2015**, *73*, 1946–1956. [CrossRef]
2. Chaudhary, A.; Wakhlu, A.; Mittal, N.; Misra, S.; Mehrotra, D.; Wakhlu, A.K. Melanotic Neuroectodermal Tumor of Infancy: 2 Decades of Clinical Experience With 18 Patients. *J. Oral Maxillofac. Surg.* **2009**, *67*, 47–51. [CrossRef]
3. Rekhi, B.; Suryavanshi, P.; Desai, S.; Gulia, A.; Desai, S.; Juvekar, S.L.; Puri, A.; Jambhekar, N.A. Melanotic neuroectodermal tumor of infancy in thigh of an infant—A rare case report with diagnostic implications. *Skelet. Radiol.* **2011**, *40*, 1079–1084. [CrossRef]
4. Creytens, D.; Ferdinande, L.; Lecoutere, E.; Van Dorpe, J. Melanotic neuroectodermal tumour of infancy presenting as an undifferentiated round cell tumour in the soft tissue of the forearm. *Pathology* **2017**, *49*, 87–90. [CrossRef]
5. Ghersin, Z.J.; Kuo, D.J. Melanotic Neuroectodermal Tumor of Infancy in the Epididymis: A Brief Report and Review of the Role of Chemotherapy in Management. *J. Pediatr. Hematol. Oncol.* **2016**, *38*, e144–e146. [CrossRef] [PubMed]
6. Burton, K.R.; Ngan, B.Y.; Navarro, O.M. Epididymal melanotic neuroectodermal tumor of infancy: A rare cause of scrotal mass in an infant. *J. Clin. Ultrasound* **2019**, *47*, 100–103. [CrossRef] [PubMed]
7. Lei, L.; Ellsworth, B.D.; Young, L.W.; Kheradpour, A.; Zuppan, C.W. Spontaneous Regression of Diffuse Periosteal Melanotic Neuroectodermal Tumor of Infancy in the Tibia, With 13-year Follow-up. *J. Pediatr. Hematol. Oncol.* **2019**, *41*, 148–151. [CrossRef] [PubMed]
8. Kirigin, M.S.; Džombeta, T.; Seiwerth, S.; Mesić, M.; Stepan Giljević, J.; Krušlin, B. Melanotic Neuroectodermal Tumor of Infancy of the Upper Arm. *Med. Princ. Pract.* **2017**, *26*, 582–585. [CrossRef]
9. Choy, J.; Abouzari, M.; Mahboubi, H.; Linskey, M.E.; Djalilian, H.R. Melanotic Neuroectodermal Tumor Presenting as Endolymphatic Sac Tumor. *Ear Nose Throat J.* **2019**, *98*, 537–539. [CrossRef] [PubMed]
10. Liu, Z.; Li, M.; Tang, X.; Xiao, Y.; Xiao, Z.; Li, Y. Melanotic neuroectodermal tumor of infancy in ovary: A rare case report. *Medicine* **2019**, *98*. [CrossRef] [PubMed]
11. Lurie, H.I. Congenital melanocarcinoma, melanotic adamantinoma, retinal anlage tumor, progonoma, and pigmented epulis of infancy. *Cancer* **1961**, *14*, 1090–1108. [CrossRef]
12. Nikai, H.; Ijuhin, N.; Yamasaki, A.; Nutani, K.; Imai, K. Ultrastructural Evidence for Neural Crest Origin of the Melanotic Neuroectodermal Tumor of Infancy. *J. Oral Pathol. Med.* **1977**, *6*, 221–232. [CrossRef] [PubMed]
13. Kapadia, S.B.; Frisman, D.M.; Hitchcock, C.L.; Popek, E.J. Melanotic Neuroectodermal Tumor of Infancy. Clinicopathological, Immunohistochemical, and Flow Cytometric Study. *Am. J. Surg. Pathol.* **1993**, *17*, 566–573. [CrossRef] [PubMed]
14. Soles, B.S.; Wilson, A.; Lucas, D.R.; Heider, A. Melanotic neuroectodermal tumor of infancy. *Arch. Pathol. Lab. Med.* **2018**, *142*, 1358–1363. [CrossRef]
15. Barrett, A.W.; Morgan, M.; Ramsay, A.D.; Farthing, P.M.; Newman, L.; Speight, P.M. A clinicopathologic and immunohistochemical analysis of melanotic neuroectodermal tumor of infancy. *Oral Surg. Oral Med. Oral Pathol. Oral Radiol. Endod.* **2002**, *93*, 688–698. [CrossRef] [PubMed]
16. Moreau, A.; Galmiche, L.; Minard-Colin, V.; Rachwalski, M.; Belhous, K.; Orbach, D.; Joly, A.; Picard, A.; Kadlub, N. Melanotic neuroectodermal tumor of infancy (MNTI) of the head and neck: A French multicenter study. *J. Cranio-Maxillo-Facial Surg.* **2018**, *46*, 201–206. [CrossRef] [PubMed]

17. Enriquez, A.M.; Carnate, J.M., Jr. Melanotic neuroectodermal tumor of infancy. *Philipp. J. Otolaryngol. Neck Surg.* **2011**, *26*, 51–54. [CrossRef]
18. Chrcanovic, B.R.; Gomez, R.S. Melanotic neuroectodermal tumour of infancy of the jaws: An analysis of diagnostic features and treatment. *Int. J. Oral Maxillofac. Surg.* **2019**, *48*, 1–8. [CrossRef]
19. El-Naggar, A.; Chan, J.; Grandis, J.; Takata, T.; Slootweg, P. *WHO Classification of Head and Neck Tumors*, 4th ed.; International Agency for Research on Cancer (IARC): Lyon, France, 2017; ISBN 978-92-832-2438-9.
20. Jenkinson, H.; Grundy, R. *Guidelines for the Management of Melanotic Neuroectodermal Tumour of Infancy*; Children's Cancer and Leukaemia Group, Study Group Rare Tumour Committee: Leicester, United Kingdom, 2004.
21. Yoo, I.H.; Yum, S.K.; Oh, S.-J.J.; Kim, K.-M.M.; Jeong, D.C. Melanotic neuroectodermal tumor of infancy disseminated by a ventriculoperitoneal shunt and diagnosed from the inguinal sac. *J. Pediatr. Hematol. Oncol.* **2014**, *36*, 61–64. [CrossRef]
22. Azari, A.; Petrisor, D.; Wright, J.; Ghali, G.E. Metastatic Melanotic Neuroectodermal Tumor of Infancy: Report of a Case and Review of the Literature. *J. Oral Maxillofac. Surg.* **2016**, *74*, 2431–2440. [CrossRef]
23. Furtado, S.V.; Ghosal, N.; Hegde, A.S. Calvarial malignant melanotic neuroectodermal tumour of infancy presenting with widespread intracranial metastasis. *J. Cranio-Maxillo-Facial Surg.* **2012**, *40*, e170–e173. [CrossRef]
24. Pereira, A.A.C.; de Jesus Rozante, M.M.; Bauer Doveinis, R.; Porcelli Salvarani, C.; Hissa Anegawa, T.; da Costa Souza, P.; Brat, D.J.; de Oliveira Borges, A.C. The recurrence of the melanotic neuroectodermal tumour of infancy: An unusual presentation of a rare tumour. *Ecancermedicalscience* **2020**, *14*. [CrossRef]
25. Kruse-Lösler, B.; Gaertner, C.; Bürger, H.; Seper, L.; Joos, U.; Kleinheinz, J. Melanotic neuroectodermal tumor of infancy: Systematic review of the literature and presentation of a case. *Oral Surg. Oral Med. Oral Pathol. Oral Radiol. Endodontol.* **2006**, *102*, 204–216. [CrossRef]
26. Mosby, E.L.; Lowe, M.W.; Cobb, C.M.; Ennis, R.L. Melanotic Neuroectodermal Tumor of Infancy: Review of the Literature and Report of a Case. *J. Oral Maxillofac. Surg.* **1992**, *50*, 886–894. [CrossRef]
27. Cutler, L.S.; Chaudhry, A.P.; Topazian, R. Melanotic Neuroectodermal Tumor of Infancy: An Ultrastructural Study, Literature Review, and Reevaluation. *Cancer* **1981**, *48*, 257–270. [CrossRef]
28. Block, J.C.; Waite, D.E.; Dehner, L.P.; Leonard, A.S.; Ogle, R.G.; Gatto, D.J. Pigmented neuroectodermal tumor of infancy. An example of rarely expressed malignant behavior. *Oral Surg. Oral Med. Oral Pathol.* **1980**, *49*, 279–285. [CrossRef]
29. Mirich, D.R.; Blaser, S.I.; Harwood-Nash, D.C.; Armstrong, D.C.; Becker, L.E.; Posnick, J.C. Melanotic Neuroectodermal Tumor of Infancy: Clinical, Radiologic, and Pathologic Findings in Five Cases. *Am. J. Neuroradiol.* **1991**, *12*, 689–697. [PubMed]
30. Nadarajah, J.; Garg, A.; Bohara, S.; Garg, K.; Devaranjan Sebastian, L.J.; Suri, V.; Bakhshi, S.; Singh, M. Calvarial Melanotic Neuroectodermal Tumor of Infancy with Rhabdomyosarcomatous differentiation—A Rare Case. *World Neurosurg.* **2021**, *145*, 134–141. [CrossRef] [PubMed]
31. Ahmed, J.; Chuckwulobelu, R.; Sebire, N.J.; Hartley, B.E.J.; Dunaway, D.J. Hemimandibulectomy and autologous costochondral rib graft reconstruction for a case of melanotic neuroectodermal tumour of infancy arising within the mandible. *Int. J. Pediatr. Otorhinolaryngol. Extra* **2007**, *2*, 189–193. [CrossRef]
32. Sobel, N.; Carcangiu, M.L. Primary Pigmented Neuroectodermal Tumor of the Uterine Cervix. *Int. J. Surg. Pathol.* **1994**, *2*, 31–36. [CrossRef]
33. Ogata, A.; Fujioka, Y.; Nagashima, K.; Tashiro, K.; Aida, T.; Abe, H. Malignant melanotic neuroectodermal tumor arising from the pineal body. *Acta Neuropathol.* **1989**, *77*, 654–658. [CrossRef]
34. Johnson, R.E.; Scheithauer, B.W.; Dahlin, D.C. Melanotic Neuroectodermal Tumor of Infancy. A Review of Seven Cases. *Cancer* **1983**, *52*, 661–666. [CrossRef]
35. Naidoo, J.; Potgieter, L.; Wieselthaler, N.; Pillay, K. Melanotic neuroectodermal tumour of infancy, a rare cause for a bump on the head. *Child's Nerv. Syst.* **2013**, *29*, 167–172. [CrossRef] [PubMed]
36. Rickart, A.J.; Drummond-Hay, V.; Suchak, A.; Sadiq, Z.; Sebire, N.J.; Slater, O.; Mills, C. Melanotic neuroectodermal tumour of infancy: Refining the surgical approach. *Int. J. Oral Maxillofac. Surg.* **2019**, *48*, 1307–1312. [CrossRef]
37. Hojan-Jezierska, D.; Chomiak, A.; Czopor, A.; Matthews-Kozanecka, M.; Majewska, A.; Urbaniak-Olejnik, M.; Matthews-Brzozowska, T. Ototoxicity after platinum-based chemotherapy in the treatment of melanotic neuroectodermal tumour of infancy. *Oncol. Lett.* **2020**, *19*, 3411–3416. [CrossRef]
38. Ebel, F.; Thieringer, F.M.; Kunz, C.; Klein-Franke, A.; Scheinemann, K.; Guzman, R.; Soleman, J. Melanotic neuroectodermal tumor of infancy to the skull: Case-based review. *Child's Nerv. Syst.* **2020**, *36*, 679–688. [CrossRef]
39. Haque, S.; McCarville, M.B.; Sebire, N.; McHugh, K. Melanotic neuroectodermal tumour of infancy: CT and MR findings. *Pediatr. Radiol.* **2012**, *42*, 699–705. [CrossRef]
40. Cohen, B.H.; Handler, M.S.; DeVivo, D.C.; Garvin, J.H.; Hays, A.P.; Carmel, P. Central nervous system melanotic neuroectodermal tumor of infancy: Value of chemotherapy in management. *Neurology* **1988**, *38*, 163–164. [CrossRef] [PubMed]
41. Hoshino, S.; Takahashi, H.; Shimura, T.; Nakazawa, S.; Naito, Z.; Asano, G. Melanotic neuroectodermal tumor of infancy in the skull associated with high serum levels of catecholamine. *J. Neurosurg.* **1994**, *80*, 919–924. [CrossRef] [PubMed]
42. Patankar, T.; Prasad, S.; Goel, A.; Perumpillichira, J.; Desai, A.P. Malignant melanotic neuroectodermal tumour of infancy affecting the occipital squama. *J. Postgrad. Med.* **1998**, *44*, 73–75.
43. Kumari, T.P.; Venugopal, M.; Mathews, A.; Kusumakumary, P. Effectiveness of chemotherapy in melanotic neurectodermal tumor of infancy. *Pediatr. Hematol. Oncol.* **2005**, *22*, 199–206. [CrossRef]

44. Murphy, C.; Pears, J.; Kearns, G.J. Melanotic neuroectodermal tumour of infancy: Surgical and chemotherapeutic management. *Ir. J. Med. Sci.* **2016**, *185*, 753–756. [CrossRef]
45. Higashi, K.; Ogawa, T.; Onuma, M.; Usubuchi, H.; Imai, Y.; Takata, I.; Hidaka, H.; Watanabe, M.; Sasahara, Y.; Koyama, S.; et al. Clinicopathological features of melanotic neuroectodermal tumor of infancy: Report of two cases. *Auris Nasus Larynx* **2016**, *43*, 451–454. [CrossRef] [PubMed]
46. De Chiara, A.; Van Tornout, J.M.; Hachitanda, Y.; Ortega, J.A.; Shimada, H. Melanotic Neuroectodermal Tumor of Infancy. A Case Report of Paratesticular Primary with Lymph Node Involvment. *Am. J. Pediatr. Hematol. Oncol.* **1992**, *14*, 356–360. [CrossRef] [PubMed]
47. Mello, R.J.V.; Vidal, A.K.L.; Fittipaldi, H.M.; Montenegro, L.T.; Calheiros, L.M.C.; Rocha, G.I. Melanotic Neuroectodermal Tumor of Infancy: Clinicopathologic Study of a Case, with Emphasis on the Chemotherapeutic Effects. *Int. J. Surg. Pathol.* **2000**, *8*, 247–251. [CrossRef] [PubMed]
48. Hered, R.W.; Smithwick IV, W.; Sandler, E.; Goldstein, J.D. Orbital melanotic neuroectodermal tumor of infancy successfully treated with chemotherapy and subtotal excision. *J. AAPOS* **2007**, *11*, 504–505. [CrossRef]
49. Choi, I.S.; Kook, H.; Han, D.K.; Baek, H.J.; Jung, S.T.; Lee, J.H.; Park, J.G.; Hwang, T.J. Melanotic Neuroectodermal Tumor of Infancy in the Femur: A Case Report and Review of the Literature. *J. Pediatr. Hematol. Oncol.* **2007**, *29*, 854–857. [CrossRef] [PubMed]
50. Maroun, C.; Khalifeh, I.; Alam, E.; Akl, P.A.; Saab, R.; Moukarbel, R.V. Mandibular melanotic neuroectodermal tumor of infancy: A role for neoadjuvant chemotherapy. *Eur. Arch. Oto-Rhino-Laryngol.* **2016**, *273*, 4629–4635. [CrossRef]
51. Nicosia, G.; Spennato, P.; Aliberti, F.; Cascone, D.; Quaglietta, L.; Errico, M.E.; Muto, M.; Ionna, F.; Cinalli, G. Giant melanotic neuroectodermal tumor of infancy (melanotic progonoma) of the head and neck: Report of a malignant case. *J. Neurosurg. Pediatr.* **2017**, *19*, 538–545. [CrossRef]
52. Khemiri, S.; Feki, J.; Khanfir, A.; Abdelmoula, M.; Frikha, M. Melanotic Neuroectodermal Tumor of Infancy: Presentation of a Case Affecting the Mandible. *Acta Med. Iran.* **2019**, *57*, 143–146. [CrossRef]
53. Pierre-Kahn, A.; Cinalli, G.; Lellouch-Tubiana, A.; Villarejo, F.J.; Sainte-Rose, C.; Pfister, A.; Couly, G. Melanotic Neuroectodermal Tumor of the Skull and Meninges in Infancy. *Pediatr. Neurosurg.* **1992**, *18*, 6–15. [CrossRef]
54. Moreau, A.; Galmiche, L.; Belhous, K.; Franchi, G.; Couloigner, V.; Nevoux, J.; Aymard, P.A.; Picard, A.; Minard-Colin, V.; Kadlub, N. Prenatal Diagnosis of a Melanotic Neuroectodermal Tumor of Infancy (MNTI): A Case Report with a Favorable Outcome after Chemotherapy Failure and Incomplete Resection. *J. Pediatr. Hematol. Oncol.* **2018**, *40*, 320–324. [CrossRef]
55. Woessmann, W.; Neugebauer, M.; Gossen, R.; Blütters-Sawatzki, R.; Reiter, A. Successful chemotherapy for melanotic neuroectodermal tumor of infancy in a baby. *Med. Pediatr. Oncol.* **2003**, *40*, 198–199. [CrossRef]
56. Sailukar, M.; Bhagwat, R.; Seth, T. Melanocytic neuroectodermal tumor of infancy. *J. Indian Assoc. Pediatr. Surg.* **2007**, *12*, 136–137. [CrossRef]
57. Blank, E.; Runckel, D.N. Case Report 119. *Skeletal Radiol.* **1980**, *5*, 179–182. [CrossRef] [PubMed]
58. Shokry, A.; Briner, J.; Makek, M. Malignant Melanotic Neuroectodermal Tumor of Infancy: A Case Report. *Pediatr. Pathol.* **1986**, *5*, 217–223. [CrossRef] [PubMed]
59. Atkinson Jr, G.O.; Davis, P.C.; Patrick, L.E.; Winn, K.J.; Ball, T.I.; Wyly, J.B. Melanotic neuroectodermal tumor of infancy. MR findings and a review of the literature. *Pediatr. Radiol.* **1989**, *20*, 20–22. [CrossRef] [PubMed]
60. Shaia, W.T.; DiNardo, L.J.; Underhill, T.E.; Cesca, C.E. Recurrent melanotic neuroectodermal tumor of infancy. *Am. J. Otolaryngol.* **2002**, *23*, 249–252. [CrossRef] [PubMed]
61. Neven, J.; Hulsbergen-van der Kaa, C.; Groot-Loonen, J.; de Wilde, P.C.M.; Merkx, M.A.W. Recurrent melanotic neuroectodermal tumor of infancy: A proposal for treatment protocol with surgery and adjuvant chemotherapy. *Oral Surg. Oral Med. Oral Pathol. Oral Radiol. Endod.* **2008**, *106*, 493–496. [CrossRef]
62. Davis, J.M.; DeBenedictis, M.; Frank, D.K.; Lessin, M.E. Melanotic Neuroectodermal Tumor of Infancy: A Wolf in Sheep's Clothing. *Ann. Otol. Rhinol. Laryngol.* **2015**, *124*, 97–101. [CrossRef] [PubMed]
63. Dehner, L.P.; Sibley, R.K.; Sauk, J.J.; Vickers, R.A.; Nesbit, M.E.; Leonard, A.S.; Waite, D.E.; Neeley, J.E.; Ophoven, J. Malignant Melanotic Neuroectodermal Tumor of Infancy. A Clinical, Pathologic, Ultrastructural and Tissue Culture Study. *Cancer* **1979**, *43*, 1389–1410. [CrossRef]
64. Riggi, N.; Suva, M.L.; Stamenkovic, I. Ewing's sarcoma origin: From duel to duality. *Expert Rev. Anticancer Ther.* **2009**, *9*, 1025–1030. [CrossRef] [PubMed]
65. Marschall, J.S.; Kushner, G.M.; Shumway, B.S. Agressive histologic features do not predict biologic behavior in melanotic neuroectodermal tumor of infancy. *J. Oral Maxillofac. Pathol.* **2018**, *22*, 260–262. [CrossRef] [PubMed]
66. Barnes, D.J.; Hookway, E.; Athanasou, N.; Kashima, T.; Oppermann, U.; Hughes, S.; Swan, D.; Lueerssen, D.; Anson, J.; Hassan, A.B. A germline mutation of CDKN2A and a novel RPLP1-C19MC fusion detected in a rare melanotic neuroectodermal tumor of infancy: A case report. *BMC Cancer* **2016**, *16*, 629. [CrossRef] [PubMed]
67. Gomes, C.C.; Diniz, M.G.; de Menezes, G.H.F.; Castro, W.H.; Gomez, R.S. BRAFV600E Mutation in Melanotic Neuroectodermal Tumor of Infancy: Toward Personalized Medicine? *Pediatrics* **2015**, *136*, e267–e269. [CrossRef]
68. Hoshina, Y.; Hamamoto, Y.; Suzuki, I.; Nakajima, T.; Ida-Yonemochi, H.; Saku, T. Melanotic neuroectodermal tumor of infancy in the mandible: Report of a case. *Oral Surg. Oral Med. Oral Pathol. Oral Radiol. Endod.* **2000**, *89*, 594–599. [CrossRef]

69. Judd, P.L.; Pedod, D.; Harrop, K.; Becker, J. Melanotic neuroectodermal tumor of infancy. *Oral Surg. Oral Med. Oral Pathol.* **1990**, *69*, 723–726. [CrossRef]
70. Pettinato, G.; Manivel, J.C.; D'Amore, E.S.G.; Jaszcz, W.; Gorlin, R.J. Melanotic neuroectodermal tumor of infancy: A reexamination of a histogenetic problem based on immunohistochemical, flow cytometric, and ultrastructural study of 10 cases. *Am. J. Surg. Pathol.* **1991**, *15*, 233–245. [CrossRef] [PubMed]

cancers

MDPI

Article

Somatic Disease in Survivors of Childhood Malignant Bone Tumors in the Nordic Countries

Camilla Pedersen [1,*], Catherine Rechnitzer [2], Elisabeth Anne Wreford Andersen [1], Line Kenborg [1], Filippa Nyboe Norsker [1], Andrea Bautz [1], Thomas Baad-Hansen [3], Laufey Tryggvadottir [4,5], Laura-Maria Madanat-Harjuoja [6], Anna Sällfors Holmqvist [7,8], Lars Hjorth [7,8], Henrik Hasle [9], Jeanette Falck Winther [1,10] and on behalf of the ALiCCS Study Group †

1 Danish Cancer Society Research Center, 2100 Copenhagen, Denmark; elian@cancer.dk (E.A.W.A.); kenborg@cancer.dk (L.K.); fnorsker@hotmail.com (F.N.N.); andrea.bautz@gmail.com (A.B.); jeanette@cancer.dk (J.F.W.)
2 Department of Pediatrics and Adolescent Medicine, Copenhagen University Hospital, 2100 Copenhagen, Denmark; catherine.rechnitzer@regionh.dk
3 Department of Orthopedic Surgery, Sarcoma Centre of Aarhus University Hospital, 8200 Aarhus, Denmark; thombaad@rm.dk
4 The Icelandic Cancer Registry, 105 Reykjavik, Iceland; laufeyt@krabb.is
5 Faculty of Medicine, University of Iceland, 102 Reykjavik, Iceland
6 The Finnish Cancer Registry, 00130 Helsinki, Finland; Laura.Madanat@cancer.fi
7 Department of Clinical Sciences Lund, Lund University, 222 41 Lund, Sweden; anna.sallfors-holmqvist@med.lu.se (A.S.H.); Lars.Hjorth@skane.se (L.H.)
8 Department of Pediatric Hematology and Oncology, Skane University Hospital, 221 85 Lund, Sweden
9 Department of Pediatrics, Aarhus University Hospital, 8200 Aarhus, Denmark; hasle@dadlnet.dk
10 Department of Clinical Medicine, Faculty of Health, Aarhus University and University Hospital, 8200 Aarhus, Denmark
* Correspondence: camped@cancer.dk
† The ALiCCS Study Group consists of all researchers, data managers and clinicians who have been involved in setting up the large Nordic cohort of childhood cancer survivors within the research program 'Adult Life after Childhood Cancer in Scandinavia' (ALiCCS). Further, the group comprises an oversight Board composed of three cancer epidemiologists and three pediatric oncologists.

Citation: Pedersen, C.; Rechnitzer, C.; Andersen, E.A.W.; Kenborg, L.; Norsker, F.N.; Bautz, A.; Baad-Hansen, T.; Tryggvadottir, L.; Madanat-Harjuoja, L.-M.; Holmqvist, A.S.; et al. Somatic Disease in Survivors of Childhood Malignant Bone Tumors in the Nordic Countries. *Cancers* **2021**, *13*, 4505. https://doi.org/10.3390/cancers13184505

Academic Editors: Saurabh Agarwal and Jianhua Yang

Received: 10 August 2021
Accepted: 3 September 2021
Published: 7 September 2021

Simple Summary: The treatment of osteosarcoma and Ewing sarcoma, the two major types of malignant bone tumors in children, has progressed considerably during the last decades, with more patients becoming long-term survivors. This improvement has resulted in an increasing number of patients with long-term adverse health consequences from the life-saving treatment. The aim of this study was to provide a detailed, comprehensive overview of somatic diseases that require hospitalization in long-term survivors of osteosarcoma and Ewing sarcoma. This study contributes new insights into the risk of somatic late effects in survivors of osteosarcoma and Ewing sarcoma which are urgently requested by pediatric oncologists, researchers, and by survivors and their families. The study provides an essential basis for the development of preventive intervention strategies and for optimal patient counseling and follow-up care, which all contribute to improving the health and quality of life in survivors.

Abstract: Survivors of malignant bone tumors in childhood are at risk of long-term adverse health effects. We comprehensively reviewed cases of somatic diseases that required a hospital contact in survivors of osteosarcoma and Ewing sarcoma. In a population-based cohort study, 620 five-year survivors of osteosarcoma (n = 440) or Ewing sarcoma (n = 180), diagnosed before the age of 20 years in Denmark, Finland, Iceland, and Sweden during 1943–2008, were followed in the national hospital registers. Overall rates of hospital contacts for any somatic disease and for 12 main diagnostic groups and 120 specific disease categories were compared with those in a matched comparison cohort (n = 3049) randomly selected from the national population registers. The rate of hospital contact for any somatic disease was 80% higher in survivors of malignant bone tumors than in comparisons and remained elevated up to 30 years after diagnosis. The rate of hospital contacts was higher after Ewing sarcoma (rate ratio (RR) 2.24; 95% confidence interval (CI) 1.76–2.85) than after osteosarcoma

(RR 1.67; 95% CI 1.41–1.98). Elevated rates were observed for 11 main diagnostic groups, including infections, second malignant neoplasms, and diseases of the skin, bones, and circulatory, digestive, endocrine, and urinary systems. Survivors of malignant bone tumors in childhood are at increased risk of somatic diseases many years after diagnosis. This comprehensive study contributes new insight into the risk of late effects in survivors of osteosarcoma and Ewing sarcoma, which is an essential basis for optimal patient counseling and follow-up care.

Keywords: childhood malignant bone tumors; survivorship; late effects; somatic disease; cohort study

1. Introduction

Osteosarcoma and Ewing sarcoma are the commonest primary malignant bone tumors in children and adolescents, accounting for approximately 6% of all childhood cancers [1,2]. The commonest site for both tumors is the extremities, but more Ewing sarcomas than osteosarcomas occur in the axial skeleton [1,2]. Chemotherapy and surgery are the primary therapy for osteosarcoma, as the tumor is not responsive to radiation at conventional doses [2]. The chemotherapeutic agents used in the treatment of osteosarcoma include cisplatin, doxorubicin, and high-dose methotrexate, and those for Ewing sarcoma are mainly vincristine, ifosfamide, doxorubicin, etoposide, actinomycin D, and cyclophosphamide [1–3]. Ewing sarcoma is controlled locally with surgery, radiation therapy, or a combination of the two [1]. Previously, most bone tumors in the extremities were managed by amputation, however, advances in high-dose chemotherapy and in surgical techniques have made limb salvage the leading surgical procedure [2].

Survival after osteosarcoma and Ewing sarcoma diagnosed in childhood improved significantly with the introduction of chemotherapy in the 1970s, from 25% and 10% before use of chemotherapy, respectively, to 60–70% [1,2]. Improved survival has resulted in increasing numbers of patients with long-term adverse health effects of the tumors and their treatments.

The majority of previous studies of somatic late effects in survivors of malignant bone tumors in childhood addressed either exclusively the overall risk of adverse health effects and health care use [4–6] or the risk of one or a few specific somatic late effects, especially secondary cancers [7–20]. Only a few studies [21–23] evaluated the risks of survivors of childhood malignant bone tumors for hospitalization for conditions in a wide range of main diagnostic groups, including infections, neoplasms, and diseases of the endocrine, circulatory, nervous, respiratory, digestive, and genitourinary systems. None of the studies reported risk estimates for subtypes of malignant bone tumors or for specific underlying disease categories.

In order to provide a detailed, comprehensive overview of somatic diseases that require hospital contact in long-term survivors of malignant bone tumors, we followed all children with a malignant bone tumor diagnosed in the Nordic countries. High-quality nationwide hospital registers and population registers allow long-term follow-up for medically verified diagnoses, with virtually no loss to follow-up. This is the first population-based cohort study that provides relative rates of hospital contacts for the full range of main diagnostic groups in survivors of osteosarcoma and Ewing sarcoma and relative and absolute rates for 120 specific disease categories in survivors of childhood malignant bone tumors.

2. Materials and Methods

2.1. Survivors of Malignant Bone Tumors and Comparison Cohort

The study was conducted within the large Nordic population-based research program Adult Life after Childhood Cancer in Scandinavia (ALiCCS) [24], which comprises all children in the Nordic countries with cancer diagnosed before the age of 20 years from the start of the national cancer registries in the 1940s and 1950s until 2008. In accordance with the International Classifications of Childhood Cancer (ICCC) [25,26], 1661 children were registered with a primary malignant bone tumor in Denmark, Finland, Iceland, or Sweden in the ALiCCS cohort. Subsequently, 101 children were excluded because the morphology code of their tumor was not for a bone tumor or was for an unverified cancer, leaving 1560 children with a malignant bone tumor.

Since the start of population registries in the Nordic countries (Denmark, 1968; Finland, 1971; Iceland, 1955; Sweden, 1968), all residents have been assigned a unique personal identification number which allows accurate linkage of information across nationwide registers of health, migration, and vital status for all residents of the Nordic countries, with virtually no loss to follow-up.

Five comparisons for each cancer survivor were selected at random from the population registers and matched to survivors by sex, age, and country (Denmark, Finland, and Iceland) or county (Sweden). The vital and migration status of survivors and comparisons was obtained from the national population registers. Comparisons had to be alive on the date of the cancer diagnosis of the corresponding survivor and without a cancer diagnosis before 20 years of age. A total of 7796 comparisons from the general population were matched to the 1560 survivors of a malignant bone tumor. After relevant exclusions, the final dataset comprised 620 five-year survivors and 3049 comparisons (Figure 1).

In order to stratify risk estimates by the major types of bone tumor, we created two sub-cohorts consisting of survivors of osteosarcoma and of Ewing sarcoma (Supplemental Table S1, online). Topographical sites were obtained from the cancer registries and grouped into the categories "extremities" and "axial skeleton".

2.2. Hospital Contacts for Somatic Diseases

By linking the cohort to the national hospital registers, we obtained a full history of inpatient admissions and outpatient visits for somatic diseases ("hospital contacts"). The hospital registers contain the dates of admission and discharge, a primary discharge diagnosis, and supplementary diagnoses coded according to the International Classification of Diseases (ICD) [27–30]. Registration by treating physicians is mandatory. To estimate the disease burden among survivors, we grouped the diagnoses into 12 main groups according to the ICD, further subdivided into 120 disease categories (Supplemental Table S2, online). We included only the primary discharge diagnosis in the analyses. Information on secondary cancers in survivors and primary cancers in comparisons was obtained from the cancer registries, as the hospital registers do not distinguish between primary and secondary cancers.

2.3. Statistical Analyses

Follow-up started 5 years after the date of cancer diagnosis for survivors and the corresponding date for comparisons and ended at the date of death, migration, or the end of the study (Denmark, 31 October 2010; Finland, 31 December 2012; Iceland, 31 December 2008; Sweden, 31 December 2009), whichever came first. As the comparisons were restricted to individuals without a childhood cancer, follow-up for cancer (second malignant neoplasm in survivors and first primary cancer in comparisons) did not start until they were 20 years of age. Only the first diagnosis in each of the 120 disease categories was retained. As the disease categories were grouped into 12 main groups, individuals could have several diagnoses in each main group.

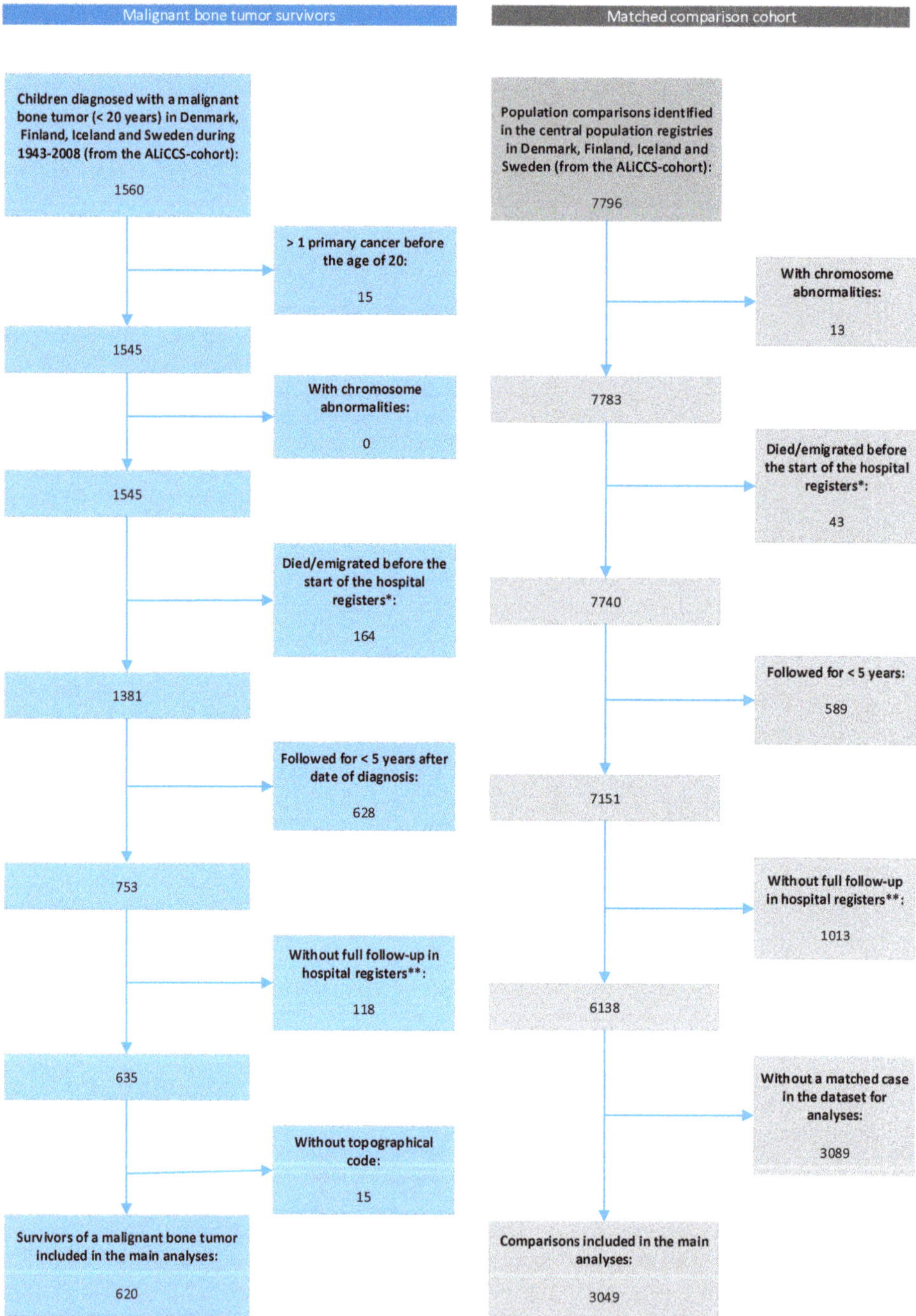

Figure 1. Flow diagram showing exclusions from the cohort of malignant bone tumor survivors and the matched comparisons cohort. * Information on inpatient admissions was available from the start of the hospital registers (Denmark, 1977; Finland, 1983; Iceland, 1999; Sweden, 1968), and outpatient visits were available for Denmark from 1995 and for Sweden from 2001. ** Individuals diagnosed more than 5 years before the start of the hospital registers were excluded to avoid a gap in information on the history of hospital contacts and to ensure accurate rates of hospital contacts.

Rate ratios (RRs) and 95% confidence intervals (CIs) for hospital contacts were calculated in marginal rates models [31] with age as the underlying time scale and allowing each individual to have several hospital contacts during follow-up. To account for recurrent hospital contacts for some individuals, a sandwich estimator was used for variance. We compared the overall rates of hospital contacts for any disease in the cohort of all survivors and in survivors of osteosarcoma and Ewing sarcoma with the rate in the matched comparisons. Rates were stratified by sex, age at diagnosis (0–14; 15–19 years), period of diagnosis (1961–1969; 1970–1979; 1980–1989; ≥1990), years since diagnosis (5–9; 10–19; ≥20), and cancer site (extremities, axial skeleton). RRs for hospital contacts were also calculated for the 12 main diagnostic groups. Estimates were calculated for all survivors of a malignant bone tumor and for osteosarcoma and Ewing sarcoma separately. RRs and rate differences for the 120 disease categories were calculated in unadjusted Poisson models for all survivors of a malignant bone tumor and matched comparisons.

The cumulative incidence of first hospital contact was estimated by accounting for death as a competing event and with year since diagnosis or entry as the underlying time scale. To compare the disease burden of survivors and of comparisons, we calculated the mean cumulative count of all first hospital contacts in the 120 disease categories. The cumulative incidence and the mean cumulative count were calculated for all survivors of a malignant bone tumor and for survivors of osteosarcoma and Ewing sarcoma separately.

Statistical analyses were performed in R version 3.6.1 and Stata version 14.2. The study was approved by the national bioethics committees, the data protection authorities, or the national institute for health and welfare in the respective countries.

3. Results

The survivor cohort consisted of 440 (71.0%) patients with osteosarcoma and 180 (29.0%) with Ewing sarcoma. Table 1 summarizes key characteristics of survivors and comparisons.

Table 1. Characteristics of the cohort of malignant bone tumor survivors overall, of the sub-cohorts of osteosarcoma and Ewing sarcoma, and of the matched comparison cohort.

	5-Year Survivors of Malignant Bone Tumors			Comparisons
	All N (%)	Osteosarcoma [a] N (%)	Ewing Sarcoma N (%)	N (%)
Overall	620 (100.0) [cb]	440 (100.0)	180 (100.0)	3049 (100.0) [c]
Country				
Denmark and Iceland	145 (23.4)	92 (20.9)	53 (29.4)	699 (22.9)
Finland	176 (28.4)	146 (33.2)	30 (16.7)	870 (28.5)
Sweden	299 (48.2)	202 (45.9)	97 (53.9)	1480 (48.5)
Period of diagnosis				
1961–1979	129 (20.8)	90 (20.5)	39 (21.7)	-
1980–1989	152 (24.5)	112 (25.5)	40 (22.2)	-
1990–2008	339 (54.7)	238 (54.1)	101 (56.1)	-
Sex				
Male	338 (54.5)	234 (53.2)	104 (57.8)	1666 (54.6)
Female	282 (45.5)	206 (46.8)	76 (42.2)	1383 (45.4)
Age at diagnosis				
Mean (SD)	13.7 (4.0)	14.2 (3.7)	12.3 (4.4)	-
0–14	347 (56.0)	226 (51.4)	121 (67.2)	-
15–19	273 (44.0)	214 (48.6)	59 (32.8)	-
Cancer site				
Extremities	485 (78.2)	382 (86.8)	103 (57.2)	-
Axial skeleton	135 (21.8)	58 (13.2)	77 (42.8)	-

SD: standard deviation; [a] Including osteosarcoma-like bone sarcomas treated as osteosarcoma; [b] The 620 five-year survivors of a malignant bone tumor diagnosed in childhood or adolescence were followed in the national hospital registers for a median of 12.5 years (range: 0–42 years), accruing 8854 person-years. [c] The 3049 comparisons were followed-up for a median of 13.8 years (range: 0–42 years) and accrued 47,478 person-years.

A total of 537 hospital contacts were observed among the 620 survivors of a childhood malignant bone tumor during follow-up, and survivors had an 80% higher rate of hospital contacts for any disease than comparisons (RR 1.80; 95% CI 1.56–2.08; Table 2). After stratification by the two major types of malignant bone tumor, a statistically significantly higher rate ($p = 0.044$) was seen for survivors of Ewing sarcoma (RR 2.24; 95% CI 1.76–2.85) than of osteosarcoma (RR 1.67; 95% CI 1.41–1.98).

The rate of hospital contacts for any disease was higher among men (RR = 2.08) than women (RR = 1.54; Table 2). The rate of hospital contacts remained elevated throughout follow-up, being highest 5–9 years and ≥20 years after the diagnosis.

The relative rates of hospital contacts for survivors of childhood malignant bone tumors were increased for 11 of the 12 main disease groups (statistically significantly in 10 of 12) (Figure 2). The rate of hospital contacts for respiratory diseases (RR 0.92; 95% CI 0.68–1.22) was similar to that in matched comparisons. The highest rates were seen for diseases of the blood and blood-forming organs (RR 5.46; 95% CI 2.20–13.59), followed by diseases of the skin and subcutaneous tissue (RR 3.57; 95% CI 2.37–5.38) and malignant neoplasms (RR 3.47; 95% CI 2.18–5.52). In survivors of osteosarcoma, the highest rate was seen for diseases of the skin and subcutaneous tissue (RR 3.88), followed by diseases of the blood and blood-forming organs (RR 3.30) and malignant neoplasms (RR 2.87). In survivors of Ewing sarcoma, the three highest rates were observed for diseases of the blood and blood-forming organs (RR 12.81), malignant neoplasms (RR 5.33), and diseases of the circulatory system (RR 4.22).

The RRs for hospital contacts for the specific disease categories with more than five hospital contacts observed among the survivors are shown in Figure 3. The highest relative rate was observed for heart failure (RR 16), which was observed, however, in only nine survivors (Supplemental Table S3 is a comprehensive list of numbers of hospital contacts, RRs, and 95% CIs). Other disease categories for which the relative rate was >5 were sepsis, erysipelas, anemias, pericardial-, myocardial-, and endocardial disease, diseases of arteries, arterioles, and capillaries, and other disorders of the skin and subcutaneous tissue. Stratification of the overall relative rate for respiratory diseases by specific disease category showed higher rates for pneumonia but lower rates for acute upper respiratory infections and other disorders of the upper respiratory tract.

Figure 3 also provides estimates of the rate difference. None of the estimates were >4 per 1000 person-years. The main diagnostic group with the highest observed rate difference was diseases of bone, joint, and soft tissue, for which survivors of a malignant bone tumor had 3.98 more first hospital contacts per 1000 person-years than the comparisons. The highest rate difference of 2.38 per 1000 person-years was seen for osteomyelitis and other diseases of bone and joint.

At 25 years after diagnosis, 49.8% (95% CI 44.6–55.0%) of all survivors of a malignant bone tumor and 39.3% (95% CI 36.9–41.6%) of the comparisons had had at least one hospital contact for any disease (Figure 4A), and the elevated risk persisted up to 30 years after diagnosis. Stratification by type of bone tumor showed that 55.5% of Ewing sarcoma survivors and 47.6% of osteosarcoma survivors had had at least one hospital contact for any disease 25 years after diagnosis (Figure 4B).

Figure 4C,D shows the mean cumulative count, including all first hospital contacts for the 120 disease categories. By 25 years after diagnosis, survivors of a malignant bone tumor had had an average of 0.98 hospital contacts, and comparisons had had 0.64.

Table 2. Rate ratios of hospital contacts (RRs) and 95% confidence intervals (CIs) for all survivors of malignant bone tumors and for the sub-cohorts of survivors of osteosarcoma and Ewing sarcoma compared with the matched comparison cohort, stratified by sex, age at diagnosis, year of diagnosis, years since diagnosis, and cancer site.

	All Survivors of Malignant Bone Tumors		*p*-Value	Osteosarcoma		Ewing Sarcoma	
	Observed Number of Hospital Contacts	RR (95% CI)		Observed Number of Hospital Contacts	RR (95% CI)	Observed Number of Hospital Contacts	RR (95% CI)
Overall	537	1.80 (1.56 to 2.08)		386	1.67 (1.41 to 1.98)	151	2.24 (1.76 to 2.85)
Test for same effect in survivors of osteosarcoma and Ewing sarcoma			0.044				
Sex							
Men	302	2.08 (1.69 to 2.56)		210	1.91 (1.50 to 2.42)	92	2.63 (1.85 to 3.73)
Women	235	1.54 (1.26 to 1.87)		176	1.46 (1.16 to 1.84)	59	1.82 (1.37 to 2.42)
Test for interaction *			0.038				
Age at diagnosis (years)							
0–14	299	1.64 (1.36 to 1.98)		198	1.51 (1.20 to 1.89)	101	1.99 (1.48 to 2.67)
15–19	238	2.05 (1.63 to 2.58)		188	1.90 (1.47 to 2.45)	50	2.93 (1.89 to 4.54)
Test for interaction *			0.138				
Period of diagnosis							
1961–1969	54	3.54 (1.84 to 6.80)		44	3.13 (1.48 to 6.59)	10	8.19 (5.37 to 12.47)
1970–1979	174	1.47 (1.15 to 1.86)		117	1.30 (0.98 to 1.71)	57	1.99 (1.34 to 2.95)
1980–1989	163	1.69 (1.31 to 2.17)		127	1.65 (1.23 to 2.21)	36	1.87 (1.30 to 2.67)
≥1990	146	2.17 (1.69 to 2.79)		98	2.02 (1.53 to 2.67)	48	2.58 (1.63 to 4.08)
Test for interaction *			0.025				
Time since diagnosis (years)							
5–9	185	2.14 (1.73 to 2.64)		127	1.94 (1.53 to 2.46)	58	2.76 (1.90 to 4.03)
10–19	160	1.44 (1.16 to 1.79)		113	1.34 (1.03 to 1.74)	47	1.74 (1.26 to 2.41)
≥20	192	1.92 (1.50 to 2.45)		146	1.80 (1.36 to 2.39)	46	2.40 (1.59 to 3.62)
Test for interaction *			0.016				
Cancer site							
Extremities	412	1.75 (1.48 to 2.08)		320	1.66(1.37 to 2.01)	92	2.16 (1.57 to 2.98)
Axial skeleton	125	1.98 (1.56 to 2.53)		66	1.73(1.27 to 2.37)	59	2.37 (1.67 to 3.36)
Test for interaction *			0.394				

* Test for common effect, whether the effect of being a survivor of a malignant bone tumor is the same in men and women, across different age groups at diagnosis, different period of diagnosis, different time intervals since diagnosis, or in different cancer sites.

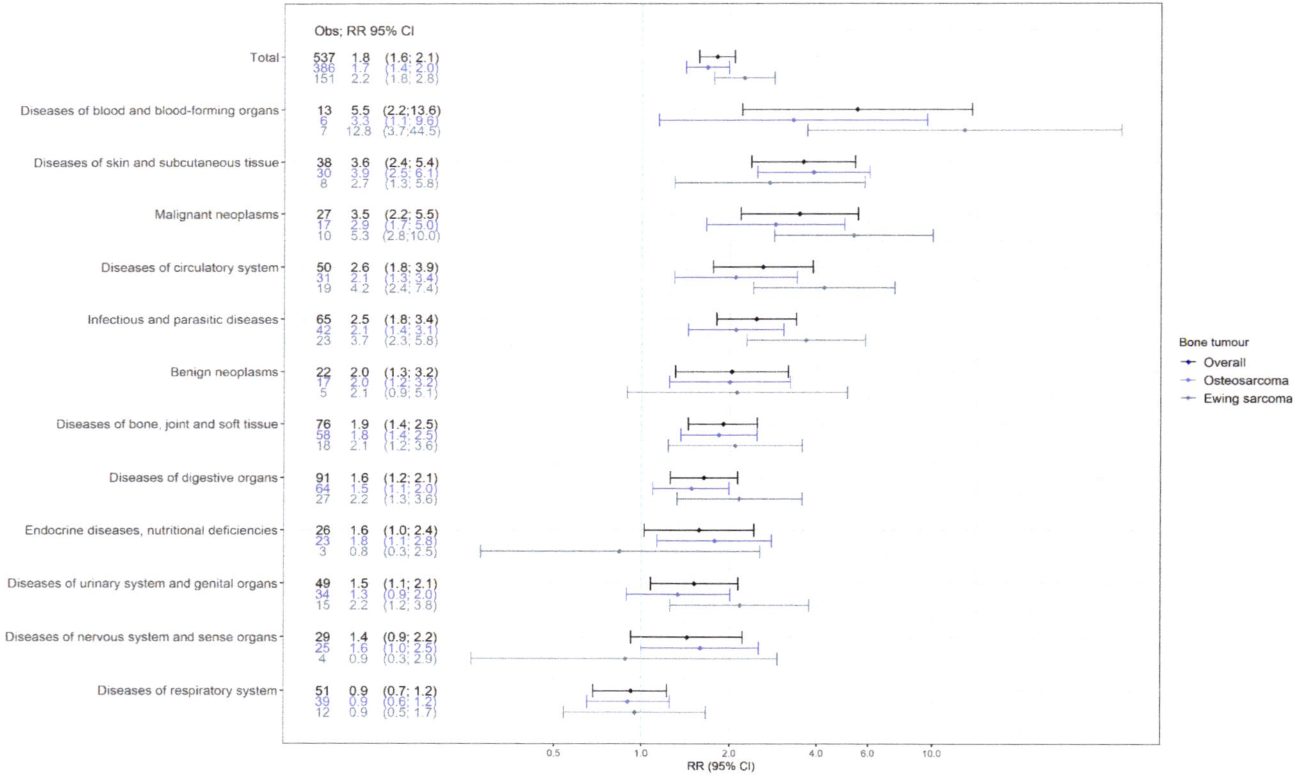

Figure 2. Rate ratios (RRs) and 95% confidence intervals (CIs) for hospital contacts for the 12 main diagnostic groups among survivors of a malignant bone tumor and stratified by osteosarcoma and Ewing sarcoma. Obs: observed number of hospital contacts by survivors.

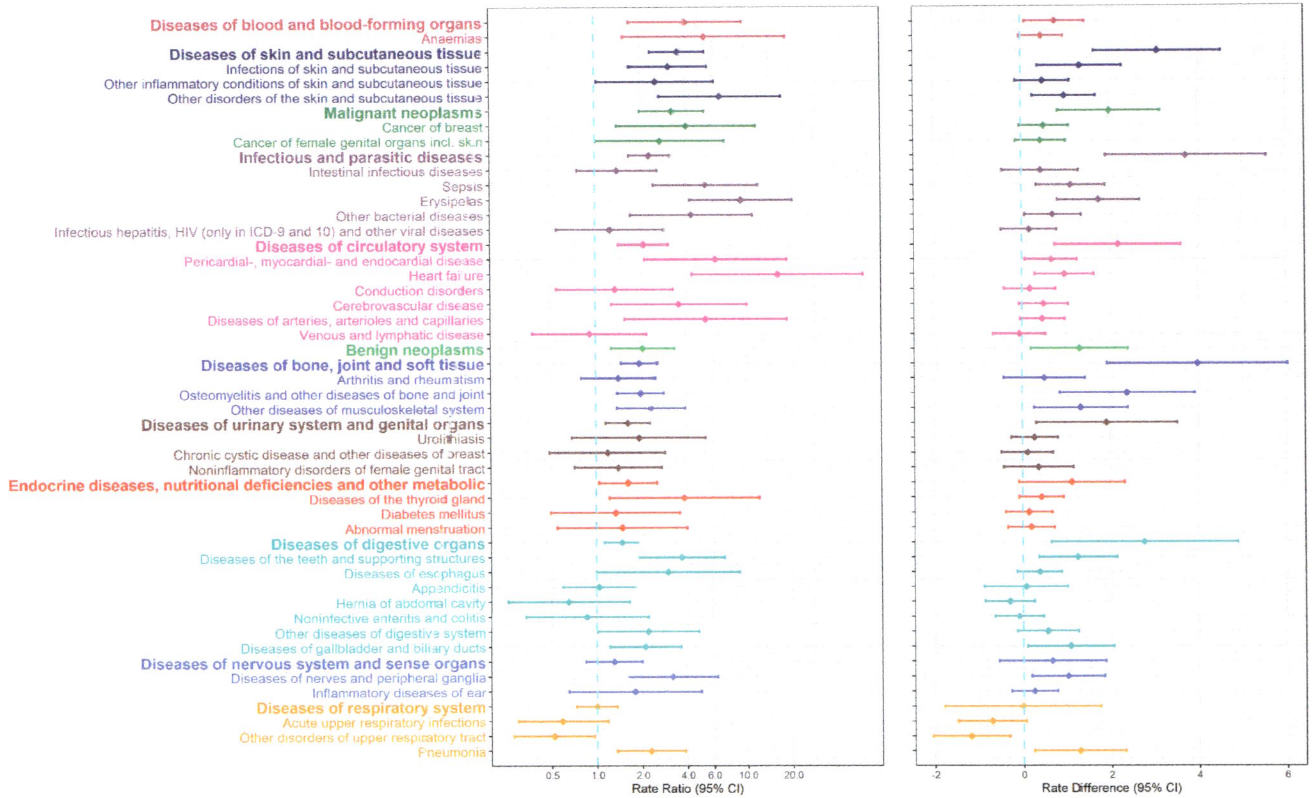

Figure 3. Rate ratios of hospital contacts, rate differences per 1000 person years, and 95% confidence intervals (CIs) for first diagnosis of each of the 120 specific disease categories for survivors compared with the matched comparisons. Only diseases for which survivors had five or more hospital contacts are shown.

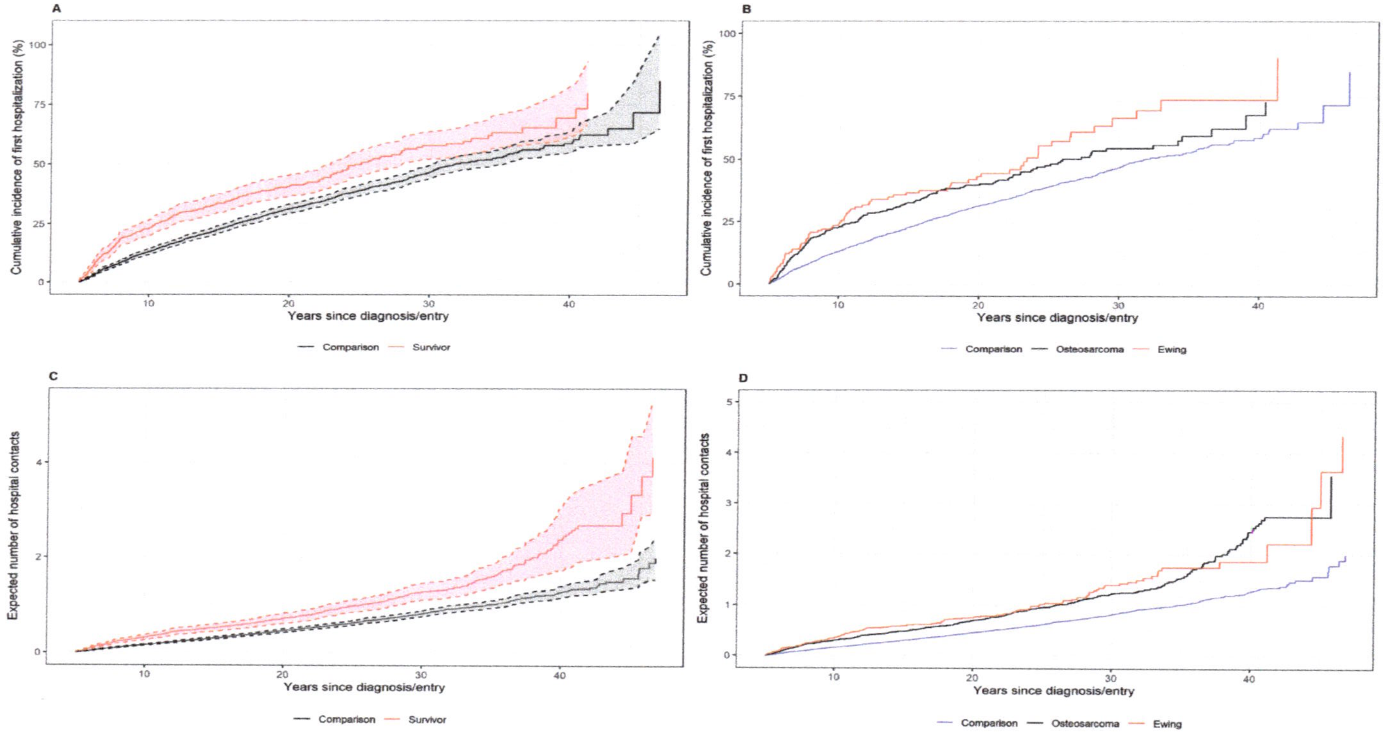

Figure 4. Cumulative incidence of first hospital contact and 95% confidence intervals (CIs) by time since diagnosis or entry for survivors and comparisons (panels **A** and **B**). Expected numbers of hospital contacts (mean cumulative count) with 95% CI for survivors and their matched comparisons by time since diagnosis or entry (panels **C** and **D**).

4. Discussion

This population-based cohort study of 620 survivors of a childhood malignant bone tumor in four countries showed an 80% higher rate of hospital contacts for somatic disease among survivors than matched comparisons. Survivors of Ewing sarcoma had higher rates of hospital contacts than survivors of osteosarcoma. The relative rates of hospital contacts by survivors were higher than those by comparisons for 11 of 12 main diagnostic groups.

Our finding of a 1.8-times increased risk for a hospital contact for any disease in survivors of malignant bone tumors is in line with the results of a study within the North American Childhood Cancer Survivor Study (CCSS) of self-reported hospitalization by survivors of childhood cancer, which gave a standardized incidence ratio (SIR) of 1.6 for bone cancer survivors (n = 854) [22]. In a cohort study of hospitalization of five-year survivors of childhood, adolescent, and young adult cancer in Scotland [21], the standardized hospitalization rate ratio (SHR) for 188 survivors of bone cancer was 3.8 for all diagnoses combined. The increased risks for hospital contacts for infections and diseases of the circulatory, endocrine, nervous, digestive, and genitourinary systems observed in our study are in accordance with the results of both these studies, although they reported markedly higher relative risks for second malignant neoplasms (SHR 10.3 and SIR 6.8) than the RR of 3.5 observed in our study. The lower rate of second malignant neoplasms seen in our study may be influenced by the inclusion of second malignant neoplasms occurring only after the age of 20 years.

To the best of our knowledge, this is the first population-based comparison of the risk of hospital contacts in survivors of osteosarcoma and Ewing sarcoma and to provide estimates of the relative and absolute rate of hospital contacts for 120 disease categories in survivors of malignant bone tumors. The rate of hospital contacts was higher among survivors of Ewing sarcoma than of osteosarcoma for most of the main diagnostic groups, including second malignant neoplasms, diseases of the circulatory system, infections, and diseases of the blood and blood-forming organs. The higher rates seen in survivors of Ewing sarcoma is probably due to the more frequent location in the axial skeleton and the use of radiation therapy for local control. Radiation of the chest is a well-established risk factor for cardiovascular disease [32] and for second malignancies such as sarcomas occurring in the radiation field [13].

Heart failure was the specific disease category with the highest relative rate, which is most likely attributable to the use of the cardiotoxic drug doxorubicin in the treatment of both osteosarcoma and Ewing sarcoma. The elevated risk for pericardial, myocardial, and endocardial disease is also likely to be related to doxorubicin and/or radiation therapy with the heart in the radiation field. Both limb-salvage surgery and amputation often require several surgical revisions over time to maintain a functional limb or in case of prosthesis-related infection, loosening, or breakage [1,2]. The highest rate difference was seen for osteomyelitis and other diseases of bone and joint, which is probably related to complications of endoprosthesis.

Cancer of the breast was the second malignant neoplasm with the highest relative rate in female survivors in our study, which could be related to both radiotherapy to the chest and chemotherapy. The majority of second malignancies reported in survivors of Ewing sarcoma are acute myeloid leukemia, myelodysplastic syndrome and sarcomas in the radiation field and secondary malignancies observed in osteosarcomas include leukemia, breast, lung, kidney, central nervous system, and colon cancer [2]. Most cases of leukemia in survivors of malignant bone tumors occur within five years from diagnosis [2], which explains why they were not captured in our study. The highest rate of hospital contacts of survivors was seen for diseases of the blood and blood-forming organs, especially anemias, which might be in line with the fact that hematological abnormalities in survivors of osteosarcoma have been previously documented even a long time after treatment [33].

The similar rates of hospital contacts of both survivors and comparisons for the main diagnostic group respiratory diseases were a result of a higher rate of pneumonia, but lower rates of acute upper respiratory infections and other disorders of the upper respiratory tract. Both the Scottish study [21] and the CCSS study [22] found elevated estimates of 2.1 and 2.9 for respiratory and pulmonary diseases but provided no information on the relative risks for specific respiratory diseases.

The major strengths of this study include the population-based design and the high-quality nationwide hospital registers in the four countries, with medically verified discharge diagnoses. Moreover, as 70% of the survivors in this study who were alive five years after diagnosis survived to 40 years of age, we were able to investigate long-term somatic late effects. The matched comparison cohort was selected at random from the Nordic population registers, providing unbiased rates of hospital contacts by the general population. In addition, the survivor cohort was restricted to those with full follow-up in the national hospital registers to prevent gaps in information and ensure accurate rates of hospital contacts.

The major limitation of the study is the lack of information on treatment, which prevented investigation of the association between specific cancer treatment and risk of somatic late effects. Further, only morbidity serious enough to require a hospital contact was included, and less serious morbidity treated by a general practitioner in primary healthcare was not captured. Some of the relative risk estimates for Ewing sarcoma had high uncertainty because of the few survivors and events in this sub-group.

5. Conclusions

Survivors of malignant bone tumors are at increased risk of hospital contacts for somatic diseases, and the elevated risk persists up to 30 years after diagnosis. The rates of hospital contacts were increased for 11 of 12 main diagnostic groups. Survivors of Ewing sarcoma had higher rates of hospital contacts than survivors of osteosarcoma. This comprehensive population-based study provides detailed information on, and new insight into, the risk of late effects in survivors of childhood malignant bone tumors, which serves as an essential basis for patient counseling and for ensuring optimal follow-up care.

Supplementary Materials: The following are available online at https://www.mdpi.com/article/10.3390/cancers13184505/s1: Table S1: Grouping of malignant bone tumors in the present study; Table S2: Definition of the 12 main diagnostic groups and 120 disease categories according to the disease codes of the International Classification of Diseases, 7th to 10th revisions (ICD-7 to ICD-10); Table S3: Rate ratios (RRs) and rate differences (RDs) of hospital contact per 1000 person years and 95% confidence intervals (CIs) for first diagnosis of each of the 120 specific disease categories for survivors compared with the matched comparisons.

Author Contributions: Conceptualization, C.P., J.F.W., C.R., H.H.; methodology, C.P., J.F.W., C.R., H.H., L.K., T.B.-H. and F.N.N.; software, A.B.; formal analysis, E.A.W.A.; investigation, C.P.; data curation, J.F.W., A.B., A.S.H., L.H., L.-M.M.-H. and L.T.; writing—original draft preparation, C.P.; writing—review and editing, all authors; visualization, C.P. and E.A.W.A.; supervision, J.F.W.; project administration, C.P.; funding acquisition, C.P., J.F.W., H.H. and L.H. All authors have read and agreed to the published version of the manuscript.

Funding: This research was funded by the Danish Childhood Cancer Foundation, grant number 2014-48, 2015-32, the Swedish Childhood Cancer Fund, grant number PR2014-0121 and by the Scientific Committee of the Danish Cancer Society, grant number A7588.

Institutional Review Board Statement: The study was approved by the national bioethics committees, the data protection authorities, or the national institute for health and welfare in the respective countries.

Informed Consent Statement: Not applicable.

Data Availability Statement: Data are not available due to Danish legislation. However, the study group welcomes collaboration with other researchers using our registry data. For further information regarding collaboration, please contact Jeanette Falck Winther, (jeanette@cancer.dk).

Conflicts of Interest: The authors declare no conflict of interest. The funders had no role in the design of the study; in the collection, analyses, or interpretation of data; in the writing of the manuscript, or in the decision to publish the results.

References

1. Heare, T.; Hensley, M.A.; Dell'Orfano, S. Bone tumors: Osteosarcoma and Ewing's sarcoma. *Curr. Opin. Pediatr.* **2009**, *21*, 365–372. [CrossRef]
2. Gorlick, R.; Perisoglou, M.; Whelan, J. Bone tumors. In *Pediatric Hematology and Oncology Scientific Principles and Clinical Practice*; Estlin, E., Gilbertson, R., Wynn, R., Eds.; Wiley-Blackwell: Hoboken, NJ, USA, 2010; pp. 234–257.
3. Gaspar, N.; Hawkins, D.S.; Dirksen, U.; Lewis, I.J.; Ferrari, S.; Le Deley, M.-C.; Kovar, H.; Grimer, R.; Whelan, J.; Claude, L.; et al. Ewing Sarcoma: Current Management and Future Approaches through Collaboration. *J. Clin. Oncol.* **2015**, *33*, 3036–3046. [CrossRef] [PubMed]
4. Fidler, M.M.; Frobisher, C.; Guha, J.; Wong, K.; Kelly, J.; Winter, D.L.; Sugden, E.; Duncan, R.; Whelan, J.; Reulen, R.C.; et al. Long-term adverse outcomes in survivors of childhood bone sarcoma: The British Childhood Cancer Survivor Study. *Br. J. Cancer* **2015**, *112*, 1857–1865. [CrossRef]
5. Kirchhoff, A.C.; Fluchel, M.N.; Wright, J.; Ying, J.; Sweeney, C.; Bodson, J.; Stroup, A.M.; Smith, K.R.; Fraser, A.; Kinney, A. Risk of hospitalization for survivors of childhood and adolescent cancer. *Cancer Epidemiol. Biomark. Prev.* **2014**, *23*, 1280–1289. [CrossRef]
6. Oeffinger, K.C.; Mertens, A.C.; Sklar, C.A.; Kawashima, T.; Hudson, M.M.; Meadows, A.T.; Friedman, D.L.; Marina, N.; Hobbie, W.; Kadan-Lottick, N.; et al. Chronic health conditions in adult survivors of childhood cancer. *N. Engl. J. Med.* **2006**, *355*, 1572–1582. [CrossRef] [PubMed]
7. Aung, L.; Gorlick, R.G.; Shi, W.; Thaler, H.; Shorter, N.A.; Healey, J.H.; Huvos, A.G.; Meyers, P.A. Second malignant neoplasms in long-term survivors of osteosarcoma: Memorial Sloan-Kettering Cancer Center Experience. *Cancer* **2002**, *95*, 1728–1734. [CrossRef] [PubMed]
8. Bacci, G.; Ferrari, C.; Longhi, A.; Ferrari, S.; Forni, C.; Bacchini, P.; Palmerini, E.; Briccoli, A.; Pignotti, E.; Balladelli, A.; et al. Second malignant neoplasm in patients with osteosarcoma of the extremities treated with adjuvant and neoadjuvant chemotherapy. *J. Pediatr. Hematol. Oncol.* **2006**, *28*, 774–780. [CrossRef] [PubMed]
9. Bhatia, S.; Krailo, M.D.; Chen, Z.; Burden, L.; Askin, F.B.; Dickman, P.S.; Grier, H.E.; Link, M.P.; Meyers, P.; Perlman, E.; et al. Therapy-related myelodysplasia and acute myeloid leukemia after Ewing sarcoma and primitive neuroectodermal tumor of bone: A report from the Children's Oncology Group. *Blood* **2007**, *109*, 46–51. [CrossRef] [PubMed]
10. Friedman, D.L.; Whitton, J.; Leisenring, W.; Mertens, A.C.; Hammond, S.; Stovall, M.; Donaldson, S.S.; Meadows, A.T.; Robison, L.L.; Neglia, J.P. Subsequent neoplasms in 5-year survivors of childhood cancer: The Childhood Cancer Survivor Study. *J. Natl. Cancer Inst.* **2010**, *102*, 1083–1095. [CrossRef]
11. Goldsby, R.; Burke, C.; Nagarajan, R.; Zhou, T.; Chen, Z.; Marina, N.; Friedman, D.; Neglia, J.; Chuba, P.; Bhatia, S. Second solid malignancies among children, adolescents, and young adults diagnosed with malignant bone tumors after 1976: Follow-up of a Children's Oncology Group cohort. *Cancer* **2008**, *113*, 2597–2604. [CrossRef]
12. Hamilton, S.N.; Carlson, R.; Hasan, H.; Rassekh, S.R.; Goddard, K. Long-term Outcomes and Complications in Pediatric Ewing Sarcoma. *Am. J. Clin. Oncol.* **2017**, *40*, 423–428. [CrossRef] [PubMed]
13. Kuttesch, J.F., Jr.; Wexler, L.H.; Marcus, R.B.; Fairclough, D.; Weaver-McClure, L.; White, M.; Mao, L.; Delaney, T.F.; Pratt, C.B.; Horowitz, M.E.; et al. Second malignancies after Ewing's sarcoma: Radiation dose-dependency of secondary sarcomas. *J. Clin. Oncol.* **1996**, *14*, 2818–2825. [CrossRef] [PubMed]
14. Longhi, A.; Ferrari, S.; Tamburini, A.; Luksch, R.; Fagioli, F.; Bacci, G.; Ferrari, C. Late effects of chemotherapy and radiotherapy in osteosarcoma and Ewing sarcoma patients: The Italian Sarcoma Group Experience (1983–2006). *Cancer* **2012**, *118*, 5050–5059. [CrossRef] [PubMed]
15. Navid, F.; Billups, C.; Liu, T.; Krasin, M.J.; Rodriguez-Galindo, C. Second cancers in patients with the Ewing sarcoma family of tumours. *Eur. J. Cancer* **2008**, *44*, 983–991. [CrossRef]
16. Novakovic, B.; Fears, T.R.; Horowitz, M.E.; Tucker, M.A.; Wexler, L.H. Late effects of therapy in survivors of Ewing's sarcoma family tumors. *J. Pediatr. Hematol. Oncol.* **1997**, *19*, 220–225. [CrossRef]
17. Olsen, J.H.; Moller, T.; Anderson, H.; Langmark, F.; Sankila, R.; Tryggvadottir, L.; Winther, J.F.; Rechnitzer, C.; Jonmundsson, G.; Christensen, J.; et al. Lifelong cancer incidence in 47,697 patients treated for childhood cancer in the Nordic countries. *J. Natl. Cancer Inst.* **2009**, *101*, 806–813. [CrossRef]
18. Nagarajan, R.; Kamruzzaman, A.; Ness, K.K.; Marchese, V.G.; Sklar, C.; Mertens, A.; Yasui, Y.; Robison, L.L.; Marina, N. Twenty years of follow-up of survivors of childhood osteosarcoma: A report from the Childhood Cancer Survivor Study. *Cancer* **2011**, *117*, 625–634. [CrossRef] [PubMed]
19. Ginsberg, J.P.; Goodman, P.; Leisenring, W.; Ness, K.K.; Meyers, P.A.; Wolden, S.L.; Smith, S.M.; Stovall, M.; Hammond, S.; Robison, L.L.; et al. Long-term survivors of childhood Ewing sarcoma: Report from the childhood cancer survivor study. *J. Natl. Cancer Inst.* **2010**, *102*, 1272–1283. [CrossRef]
20. Marina, N.M.; Liu, Q.; Donaldson, S.S.; Sklar, C.A.; Armstrong, G.T.; Oeffinger, K.C.; Leisenring, W.; Ginsberg, J.P.; Henderson, T.O.; Neglia, J.; et al. Longitudinal follow-up of adult survivors of Ewing sarcoma: A report from the Childhood Cancer Survivor Study. *Cancer* **2017**, *123*, 2551–2560. [CrossRef]

21. Brewster, D.H.; Clark, D.; Hopkins, L.; Bauer, J.; Wild, S.H.; Edgar, A.B.; Wallace, W.H. Subsequent hospitalisation experience of 5-year survivors of childhood, adolescent, and young adult cancer in Scotland: A population based, retrospective cohort study. *Br. J. Cancer* **2014**, *110*, 1342–1350. [CrossRef]
22. Kurt, B.A.; Nolan, V.G.; Ness, K.K.; Neglia, J.P.; Tersak, J.M.; Hudson, M.M.; Armstrong, G.T.; Hutchinson, R.J.; Leisenring, W.M.; Oeffinger, K.C.; et al. Hospitalization rates among survivors of childhood cancer in the Childhood Cancer Survivor Study cohort. *Pediatr. Blood Cancer* **2012**, *59*, 126–132. [CrossRef] [PubMed]
23. De Fine Licht, S.; Rugbjerg, K.; Gudmundsdottir, T.; Bonnesen, T.G.; Asdahl, P.H.; Holmqvist, A.S.; Madanat-Harjuoja, L.; Tryggvadottir, L.; Wesenberg, F.; Hasle, H.; et al. Long-term inpatient disease burden in the Adult Life after Childhood Cancer in Scandinavia (ALiCCS) study: A cohort study of 21,297 childhood cancer survivors. *PLoS Med.* **2017**, *14*, e1002296. [CrossRef] [PubMed]
24. Asdahl, P.H.; Winther, J.F.; Bonnesen, T.G.; Licht, S.D.F.; Gudmundsdottir, T.; Anderson, H.; Madanat-Harjuoja, L.; Tryggvadottir, L.; Småstuen, M.C.; Holmqvist, A.S.; et al. The Adult Life after Childhood Cancer in Scandinavia (ALiCCS) Study: Design and Characteristics. *Pediatr. Blood Cancer* **2015**, *62*, 2204–2210. [CrossRef] [PubMed]
25. Birch, J.M.; Marsden, H.B. A classification scheme for childhood cancer. *Int. J. Cancer* **1987**, *40*, 620–624. [CrossRef]
26. Steliarova-Foucher, E.; Stiller, C.; Lacour, B.; Kaatsch, P. International Classification of Childhood Cancer, third edition. *Cancer* **2005**, *103*, 1457–1467. [CrossRef]
27. Ludvigsson, J.F.; Andersson, E.; Ekbom, A.; Feychting, M.; Kim, J.-L.; Reuterwall, C.; Heurgren, M.; Olausson, P.O. External review and validation of the Swedish national inpatient register. *BMC Public Health* **2011**, *11*, 450. [CrossRef]
28. Lynge, E.; Sandegaard, J.L.; Rebolj, M. The Danish National Patient Register. *Scand. J. Public Health* **2011**, *39* (Suppl. 7), 30–33. [CrossRef]
29. Schmidt, M.; Schmidt, S.A.; Sandegaard, J.L.; Ehrenstein, V.; Pedersen, L.; Sorensen, H.T. The Danish National Patient Registry: A review of content, data quality, and research potential. *Clin. Epidemiol.* **2015**, *7*, 449–490. [CrossRef]
30. Sund, R. Quality of the Finnish Hospital Discharge Register: A systematic review. *Scand. J. Public Health* **2012**, *40*, 505–515. [CrossRef]
31. Amorim, L.D.; Cai, J. Modelling recurrent events: A tutorial for analysis in epidemiology. *Int. J. Epidemiol.* **2015**, *44*, 324–333. [CrossRef]
32. Lipshultz, S.E.; Adams, M.J.; Colan, S.D.; Constine, L.S.; Herman, E.H.; Hsu, D.T.; Hudson, M.M.; Kremer, L.C.; Landy, D.C.; Miller, T.L.; et al. Long-term cardiovascular toxicity in children, adolescents, and young adults who receive cancer therapy: Pathophysiology, course, monitoring, management, prevention, and research directions: A scientific statement from the American Heart Association. *Circulation* **2013**, *128*, 1927–1995. [CrossRef] [PubMed]
33. Lohmann, D.J.; Hasle, H. Hematological Changes Mimicking Myelodysplastic Syndrome Following Treatment for Osteosarcoma. *J. Pediatr. Hematol. Oncol.* **2015**, *37*, 170–174. [CrossRef]

cancers

Article

Surfaceome Profiling of Rhabdomyosarcoma Reveals B7-H3 as a Mediator of Immune Evasion

Roxane R. Lavoie [1], Patricio C. Gargollo [1], Mohamed E. Ahmed [1], Yohan Kim [1], Emily Baer [1], Doris A. Phelps [2], Cristine M. Charlesworth [3], Benjamin J. Madden [3], Liguo Wang [4], Peter J. Houghton [2], John Cheville [5], Haidong Dong [1,6], Candace F. Granberg [1] and Fabrice Lucien [1,*]

1 Department of Urology, Mayo Clinic, Rochester, MN 55902, USA; r.lavoie.roxane@mayo.edu (R.R.L.); gargollo.patricio@mayo.edu (P.C.G.); mohamed.ahmed@mayo.edu (M.E.A.); kim.yohan@mayo.edu (Y.K.); Emily.E.Baer@dmu.edu (E.B.); dong.haidong@mayo.edu (H.D.); granberg.candace@mayo.edu (C.F.G.)
2 Greehey Children's Cancer Research Institute, San Antonio, TX 78229, USA; PhelpsD@uthscsa.edu (D.A.P.); houghtonp@uthscsa.edu (P.J.H.)
3 Proteomic Core, Mayo Clinic, Rochester, MN 55902, USA; charlesworth.cristine@mayo.edu (C.M.C.); madden.benjamin@mayo.edu (B.J.M.)
4 Division of Computational Biology, Mayo Clinic, Rochester, MN 55902, USA; Wang.Liguo@mayo.edu
5 Department of Anatomic Pathology, Mayo Clinic, Rochester, MN 55902, USA; Cheville.John@mayo.edu
6 Department of Immunology, Mayo Clinic, Rochester, MN 55902, USA
* Correspondence: Lucien-Matteoni.Fabrice@mayo.edu

Simple Summary: Rhabdomyosarcoma (RMS) is the most common soft-tissue sarcoma in children, and there is a critical need to develop efficacious and tolerable anticancer therapies against this aggressive disease. To uncover druggable RMS-associated tumor antigens, we analyzed the cell surface protein repertoire in RMS tumor cells and normal tissue. We identified several surface proteins highly enriched in RMS, including the immune checkpoint molecule B7-H3. A further analysis using patient specimens showed that B7-H3 is overexpressed in most of RMS tumors and weakly or not detected in normal organs. Interestingly, we found that B7-H3 depletion was associated with higher immune cell killing activity against tumor cells. In line with this, high B7-H3 tumor expression was associated with lower CD8 T-cell density. Our study reveals novel RMS-associated proteins for the development of targeted therapies. In addition, we demonstrate that targeting B7-H3 function can pave the way for the design of new immunotherapies in the treatment of RMS.

Abstract: Novel therapeutic strategies are needed for the treatment of rhabdomyosarcoma (RMS), the most common soft-tissue sarcoma in children. By using a combination of cell surface proteomics and transcriptomic profiling of RMS and normal muscle, we generated a catalog of targetable cell surface proteins enriched in RMS tumors. Among the top candidates, we identified B7-H3 as the major immunoregulatory molecule expressed by RMS tumors. By using a large cohort of tissue specimens, we demonstrated that B7-H3 is expressed in a majority of RMS tumors while not detected in normal human tissues. Through a deconvolution analysis of the RMS tumor RNA-seq data, we showed that B7-H3-rich tumors are enriched in macrophages M1, NK cells, and depleted in $CD8^+$-T cells. Furthermore, in vitro functional assays showed that B7-H3 knockout in RMS tumor cells increases T-cell mediated cytotoxicity. Altogether, our study uncovers new potential targets for the treatment of RMS and provides the first biological insights into the role of B7-H3 in RMS biology, paving the way for the development of next-generation immunotherapies.

Keywords: rhabdomyosarcoma; targeted therapies; cell surface proteomics; B7-H3

Citation: Lavoie, R.R.; Gargollo, P.C.; Ahmed, M.E.; Kim, Y.; Baer, E.; Phelps, D.A.; Charlesworth, C.M.; Madden, B.J.; Wang, L.; Houghton, P.J.; et al. Surfaceome Profiling of Rhabdomyosarcoma Reveals B7-H3 as a Mediator of Immune Evasion. *Cancers* **2021**, *13*, 4528. https://doi.org/10.3390/cancers13184528

Academic Editor: Saurabh Agarwal

Received: 29 June 2021
Accepted: 17 August 2021
Published: 9 September 2021

1. Introduction

Rhabdomyosarcoma (RMS) is the most common childhood soft tissue sarcoma, with nearly 20% of patients presenting with locally aggressive and/or metastatic disease [1].

RMS is classified into two molecular subtypes based on the expression of the fusion gene *PAX3/7 FOXO1* [1,2]. Patients with fusion-positive RMS usually have a poorer outcome compared to fusion-negative RMS [3]. The *PAX3/7-FOXO1* fusion protein drives an oncogenic transcriptional program, including the upregulation of genes involved in invasion, proliferation, and survival [4]. In contrast, fusion-negative RMS does not harbor *PAX3/7-FOXO1* fusion but a variety of mutations on oncogenes, including *RAS*, *PIK3CA*, *FGFR4*, and *TP53* [5]. Multimodal therapies, including surgery, radiotherapy, and chemotherapy, have been the standard care for the treatment of RMS. The current chemotherapy regimens consist of combinations of Vincristine, Dactinomycin, and Cyclophosphamide and have shown to improve the survival rates by 60–90% in patients with localized disease [6–8]. In contrast, 90% of RMS-related deaths occur within two years following diagnosis and are mostly related to disease recurrence. Importantly, long-term effects and life-threatening complications often occur in the lifetime of RMS survivors [9,10]. This underscores the need for effective and more tolerable therapies for the treatment of RMS.

Antibody-based therapies, including immune checkpoint inhibitors, bispecific antibodies, antibody–drug conjugates, and CAR-T therapy, have transformed the therapeutic landscape of adult cancers [11]. Antibody-based therapies have a common denominator with the specific binding of a protein expressed on the surface of tumor cells. Cell surface proteins are ideal targets for anticancer therapies as their accessible extracellular domain makes pharmacological interventions more effective. To date, cell surface proteins account for ~60% of all FDA-approved drugs [11]. Furthermore, it provides an alternative strategy to treat tumors with undruggable intracellular oncogenic alterations that drive transcriptional and translational programs, resulting in the aberrant expression of druggable cell surface proteins [12–14]. The surface proteome is composed of cell adhesion molecules; nutrients; metabolite transporters; immunoregulatory molecules; and growth factor receptors that control cancer cell behaviors, aggressiveness, and their response to therapy. The identification of tumor-associated surface proteins may provide important insights into tumor biology and be further exploited as therapeutic vulnerabilities.

In this study, we employed cell surface proteomics (surfaceomics) to establish a comprehensive map of cell surface proteins expressed in RMS. By integrating the transcriptomic data and proteomic data of RMS tumors and normal tissue, we revealed an RMS-specific cell surface protein signature. Among the top upregulated proteins, we identified B7-H3 as the major immunoregulatory molecule expressed in RMS. We found that B7-H3 mediates tumor immune evasion through functional assays and transcriptomic characterization of the tumor immune landscape of RMS tumors. Our findings support B7-H3 as a therapeutic target for antibody-based therapies and provide new biological insights on its immunomodulatory role in RMS. Finally, this study demonstrates the potential of surfaceomics for cancer research and drug development.

2. Materials and Methods

2.1. Cells and Reagents

Human pediatric rhabdomyosarcoma cell lines SJCRH30 (RH30) (alveolar rhabdomyosarcoma (aRMS) fusion-positive, cat# CRL-2061) and RD (embryonal RMS (eRMS), fusion-negative, cat# CCL-136) were purchased from ATCC (Manassas, Virginia, USA). Two normal human skeletal muscle cell lines were purchased from Lonza (Basel Switzerland) (cat#PCS-950-010) and ATCC (cat#CC-2561). The RMS cell lines RH36 (fusion-negative) and RH18 (fusion-positive) were obtained from Dr. Peter Houghton. The RH30 cell line was grown in RPMI-1640 with 10% FBS (Gibco, Waltham, MA, USA) and 1-mg/mL Pen–Strep (Gibco), the RD cell line was grown in DMEM with 10% FBS (Gibco) and 1-mg/mL Pen–Strep (Gibco), and the normal muscle cell lines were cultured in HSkMC Growth Medium (Cell Applications, cat#151-500, San Diego, CA, USA). The RMS-MC02 primary cells were isolated from a resected embryonal (fusion-negative) RMS tumor (IRB #16-006956) and cultured in DMEM-F12 (Gibco) with 10% FBS (Gibco) and 1-mg/mL Pen–Strep (Gibco). The B7-H3 knockout RH30 cell line was generated using CRISPR/Cas-9 (Santa

Cruz Biotechnology, sc-402032, Dallas, TX, USA), and negative selection was performed by flow cytometry using anti-B7H3 PE (DCN.70 clone, BioLegend, cat#331606, San Diego, CA, USA). All cell lines were maintained in an incubator with a humidified atmosphere and 5% CO_2 at 37 °C.

2.2. Cell Surface Biotinylation and Mass Spectrometry

A total of 5×10^7 cells from each cell line were subjected to cell surface biotinylation. EZ-Link-Sulfo-NHS-SS-biotin (Thermo Scientific, cat#21331, San Diego, CA, USA) was added to cultured cells for 30 min at 4 °C. The cells were washed with 50 mM of glycine to quench the unbound biotin. The cells were lysed in NP-40 lysis buffer, and the biotinylated cell surface proteins were affinity-purified on streptavidin magnetic beads (Thermo Scientific, cat#88816). After stripping off the nonspecifically bound proteins by several rounds of washing with the lysis buffer, the labeled proteins were reduced with 10-mM TCEP (Thermo Scientific, cat#77720) for 30 min at 50 °C and followed by alkylation with iodoacetamide (Thermo Scientific, 90034) for 30 min in the dark at room temperature. The proteins were run in a SDS-PAGE electrophoresis gel and submitted for mass spectrometry (additional information in the Supplementary Materials) [15–17].

2.3. Bioinformatic Annotation of Cell Surface Proteins

Genes encoding for the cell surface proteins were assembled from Gene Ontology GO:0005886 (plasma membrane) and UniProt using the keywords "homo sapiens" and "single-pass transmembrane domain" and "multi-pass transmembrane domain".

2.4. Expression Analysis of RMS-Enriched Cell Surface Proteins in RMS Tumors and Normal Tissue

The RNA sequencing data was processed and analyzed as previously reported [5] (GEO: GSE108022). The gene-level raw read counts matrix file was downloaded from the GEO database, and differential expression analyses were performed using DESeq2 [18]. Specifically, the DESeq2 default parameters/methods were used to estimate the size factors, estimate the dispersion, and perform the statistical tests. The cell surface proteins differently expressed between RMS (n = 101) and normal muscles (n = 5), fusion-negative (n = 66) and normal muscles, and fusion-positive (n = 35) and normal muscles with a fold change $\geq$ two and adjusted p-value < 0.05 were undertaken for further analysis. To analyze the normal tissue expression of the RMS-enriched cell surface proteins, tissue-specific transcriptomes and proteomes were obtained from Jiang, L et al. that consisted of transcriptomic and proteomic analyses of 201 samples from 32 tissue types of 14 normal individuals [19]. The median relative protein and RNA abundances for each tissue type were transformed into absolute values (reversed log2) for further analysis. To identify the best therapeutic candidates, a composite ranking of the cell surface proteins was generated, where the cell surface proteins were ranked with equal weights based on their RNA and protein expression in the RMS tumors and normal tissue. For tumor expression, the genes and proteins were ranked from the most expressed to the least expressed. For normal expression, the genes and proteins were ranked from the least expressed to the most expressed.

2.5. Deconvolution Analysis of Bulk RNA-seq

From the GSE108022 RNA sequencing data, the genes with an average RPKM value across all samples less than 0.5 were removed, and then, the fraction of cell types (B cells, M1 macrophages, M2 macrophages, monocytes neutrophils, NK cells, CD4 T cells, CD8 T cells, Tregs, and dendritic cells) were identified and estimated using the QuanTIseq pipeline [20]. The RMS tumors were stratified by B7-H3 expression (25% highest and 25% lowest), and the immune cell abundance was compared between both groups.

2.6. T-Cell Cytotoxicity Assay

Adherent RH30 wild-type and B7-H3KO cells were plated in a 96-well plate and incubated 30 min at 37 °C with calcein-AM (Biolegend, 425201). The cells were then washed twice with Live Cell Imaging Solution (Gibco, A14291DJ) and kept in RPMI-1640 (no phenol red) supplemented with 10% HI-FBS. PBMC were isolated from the normal donor blood apheresis cones using a Ficoll gradient and activated overnight with 5 µg/mL of phytohemagglutinin (PHA-L; Millipore, cat#431784, Burlington, MA, USA). PBMC were added to each well at a ratio 1:10 and 1:20 (Target:Effector). Each condition was performed in triplicate. The calcein fluorescence was recorded every 10 min for 16 h using the EVOS FL Auto (Thermo) system equipped with the onstage incubator set at 37 °C and 5% CO_2. The total number of calcein-positive cells and fluorescence intensities of intracellular calcein were calculated for each condition. The tumor cell viability was calculated at 5 h post-incubation. The survival index was obtained by the mean calcein fluorescence in treated cells divided by the mean fluorescence in the control (tumor cells only). An image analysis was performed using Fiji software (National Institute of Health, Bethesda, MD, USA).

2.7. *Statistical Analysis*

The normality of distribution was assessed using the Shapiro-Wilk normality test. The Student's *t*-test (parametric) and Mann–Whitney *U* test (nonparametric) were employed to compare the two groups. A paired Wilcoxon test (Wilcoxon's signed rank test) was performed to analyze the data presented in Figure 6C. The results were considered significant for *p*-values < 0.05. The *p*-values were either specified in the figure or denoted as asterisk: * $p < 0.05$, ** $p < 0.01$, and *** $p < 0.001$. All data were analyzed and plotted in GraphPad Prism 9.0.1 (San Diego, CA, USA).

Additional materials and methods for the Western blot, immunohistochemistry, and flow cytometry can be found in the Supplementary Materials.

3. Results

3.1. *Identification of Cell Surface Protein Repertoires in RMS and Normal Muscle*

In order to define a set of cell surface proteins with an extracellular domain that can serve as potential therapeutic targets in RMS, we used a combination of cell surface capture and proteomic profiling on five RMS cell lines (three fusion-negative and two fusion-positive) and two normal skeletal muscle cell lines (Figure 1A). The proteins expressed in all fusion-negative and/or fusion-positive RMS cell lines by more than two-fold with a false discovery rate <0.05 compared to normal muscle were selected as target candidates. While the biotinylation method enriches for cell surface proteins, mass spectrometry can still reveal the cytosolic proteins interacting with plasma membrane proteins that are considered as "false" positives. To validate the subcellular location of the proteins identified, several publicly available databases were interrogated (Figure S1A). Gene Ontology GO:0005886 encompassed other annotation databases by providing the most exhaustive list of cell surface proteins. By using GO:0005886 for filtering data, we ensured not losing the true positives. An independent transcriptomic dataset (GSE108022) of 101 RMS (66 fusion-negative and 35 fusion-positive) and five normal skeletal muscle tissues was used to determine the expression profiles of the genes coding for cell surface proteins identified by mass spectrometry [5]. The publicly transcriptomic dataset Genotype-Tissue Expression (GTEx) containing gene expression profiles of 54 different human tissues from 948 donors was used to determine the expression of the target candidates in normal tissues. Finally, we also included the expression of target proteins obtained from a recent large-scale proteomic profiling of normal organs [19]. RNA and protein expression profiles in RMS and normal tissue were used to create a composite rank and identify targetable RMS-specific/enriched cell surface proteins.

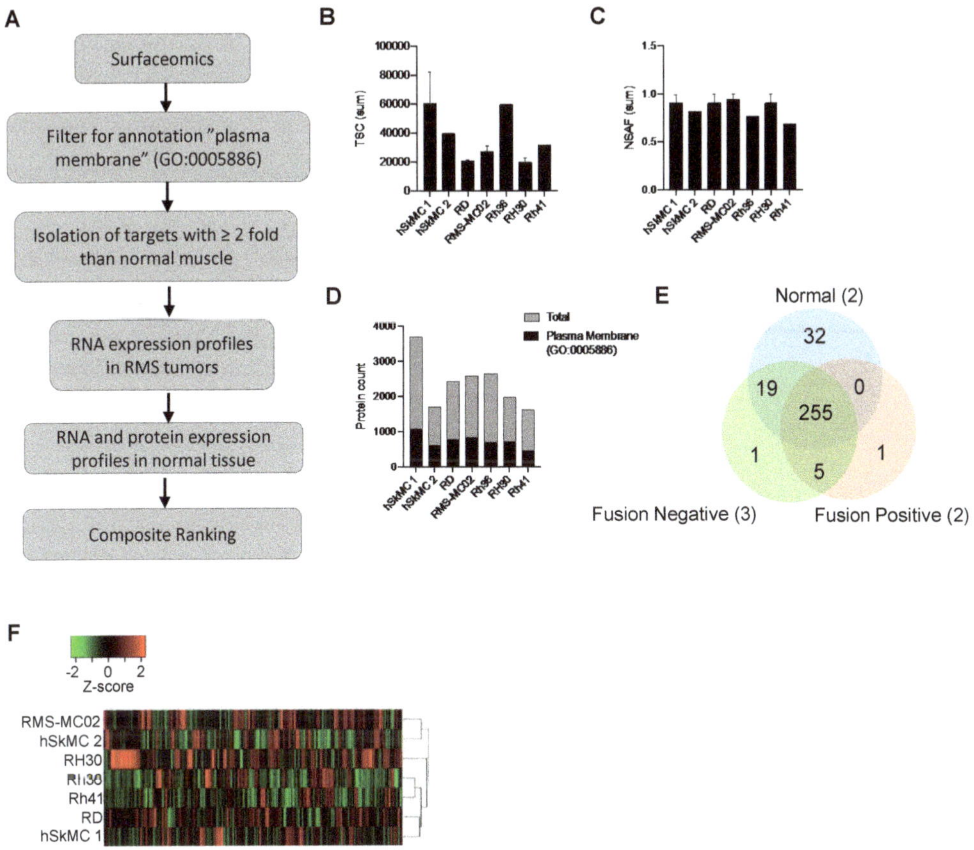

Figure 1. Distinct cell-surface protein signatures are identified in RMS subtypes and normal muscle. (**A**) Overall representation of the bioinformatics strategy to identify highly confident RMS-enriched cell surface proteins by combined proteomic and transcriptomic analysis of RMS. (**B**) Sum of TSC (total spectral count) and (**C**) NSAF (normalized spectral abundance factor) for each cell line analyzed. (**D**) Total protein count for each cell line. Count of proteins with the annotation plasma membrane is superimposed on the total protein count (black bars). (**E**) Venn diagram showing number of proteins identified in all cell lines of each subtype. (**F**) Heatmap of hierarchical clustering for cell surface protein data set filtered with the annotation "plasma membrane" (GO:0005886), Average linkage, Euclidean distance.

Upon an analysis of the cell surface proteome by label-free mass spectrometry, significant differences in the number of peptides (total spectral counts, TSC) were detected between individual samples (Figure 1B). To normalize our data, we employed a normalized spectral count by the protein length (NSAF), which improves the quantification of protein abundance with label-free proteomics. Upon applying NSAF, the total spectral counts were normalized, and no statistically significant differences the between samples were observed (Figure 1C). A total of 5061 different proteins were found in combined cell lines (Figure 1D). Protein identification revealed an average of 2676 proteins detected per cell line, with hSkMC1 showing the highest number of detected proteins (3694 proteins) (Table S1). From the pool of proteins identified by mass spectrometry, an average of 1455 proteins were located at the plasma membrane (Table S2). When the filtering step was applied with Gene Ontology GO:0005886 to tease out the proteins with incorrect subcellular locations, we found an average of 31.6% of the proteins annotated as located at the plasma membrane. The protein repertoires were heterogeneous within types of cell lines

(Figures S1B–D and 1D). Despite sharing similarities, several proteins were found exclusively in one cell line compared to the other cell lines of the same group. When diseased and normal muscle-specific protein repertoires were compared, we identified a total of 255 proteins commonly expressed by the two normal skeletal muscle cell lines and the five RMS cell lines (Figure 1E). We found seven and 32 cell surface proteins specifically expressed by RMS cells and normal muscle cells, respectively. When the RMS cell lines were grouped based on their fusion status, only one protein was exclusively expressed in each RMS subtype. The unsupervised hierarchical clustering of the cell lines based on the protein repertoires showed four clusters (Figure 1F). The fusion-negative RMS cell lines RMS-MC02 and hSkMC1 segregated together, while the fusion-negative RMS cell lines RD and hSkMC2 clustered together. Fusion-positive RH41 and fusion-negative RH36 segregated together, and fusion-positive RH30 was separated from all other cell lines. The proteomic profiling of the RMS cell lines and normal skeletal muscle revealed cell type-specific surfaceome signatures, but it did not differentiate the RMS from normal muscle and fusion-positive from fusion-negative RMS cells.

3.2. Validation of RMS-Enriched Cell Surface Proteins by Combined Proteomic and Transcriptomic Analysis

The abundance of the cell surface proteins was compared between the normal muscle, fusion-negative, and fusion-positive RMS cell lines. A high degree of correlation in the protein expression was observed for the two normal skeletal muscle cell lines (Pearson's $r = 0.7994$ $p < 0.0001$). Among the fusion-negative and fusion-positive cell lines, positive associations were also observed, but the correlation coefficients varied from 0.2829 (fusion-positive) to 0.4736 (fusion-negative) (Figure S1E). Among the cell surface proteins commonly expressed in fusion-negative and fusion-positive cell lines, 57 proteins were upregulated by more than two-fold in RMS compared to normal muscle (Table S3). Similarly, 85 and 99 proteins were upregulated by more than two-fold in fusion-negative and fusion-positive RMS, respectively (Tables S4 and S5). To identify high-confidence cell surface targets in RMS, we interrogated a transcriptomic dataset (GSE108022) that consists of 66 fusion-negative RMS tumors, 35 fusion-positive RMS tumors, and 5 normal muscle tissues to analyze the gene expressions of our target candidates [5]. By using a two-fold increase and a p-value of 0.05 as the cut-off, we found that 42.1% (24 out of 57) of the targets overexpressed in all RMS cell lines at the protein level were also increased at the transcriptional level (Figure 2A and Table S6). When the samples were sub-grouped based on their fusion status, 35.3% (30 out of 85) and 38.4% (38 out of 99) of the targets were upregulated in RMS compared to normal muscle at the RNA level (Tables S7 and S8).

One major barrier to the development of effective antibody-based therapies is the expression of tumor targets within normal tissue causing on-target, off-tumor toxicity [21]. To improve the safety and minimize the off-tumor toxicity, tumor-enriched cell surface proteins must have a limited expression in normal organs. To determine which RMS-enriched cell surface proteins may be suitable as targets for antibody-based therapies, we analyzed their expression in normal tissues using publicly available GTEx transcriptomic and proteomic datasets [19,22] (Figure 2B–D). We also included Mesothelin (MSLN), CEACAM5, and HER2 (ERBB2), three common targets for CAR-T therapy in solid tumors demonstrating an acceptable safety profile in humans [23,24]. Among the three selected CAR-T targets, ERBB2 was the most expressed in normal tissue at the RNA level, followed by CEACAM5 and MSLN (Figure 2B and Table S9). Many (80.3%) (41/51) of the RMS-enriched cell surface proteins had lower normal gene expression than ERBB2. The proteomic data showed that ERBB2 was also the most expressed protein among the three CAR-T targets (Figure 2C and Table S10), and 49% (25/51) of the RMS-enriched cell surface proteins had a lower relative protein expression compared to ERBB2. The least and most abundant proteins in normal tissue were EFNA5 and MARCKSL1, respectively. A correlation analysis of the protein and RNA relative expression in normal tissues showed that most RMS-associated surface antigen candidates have a lower RNA expression and similar protein expression than the three CAR-T targets (Figure 2D).

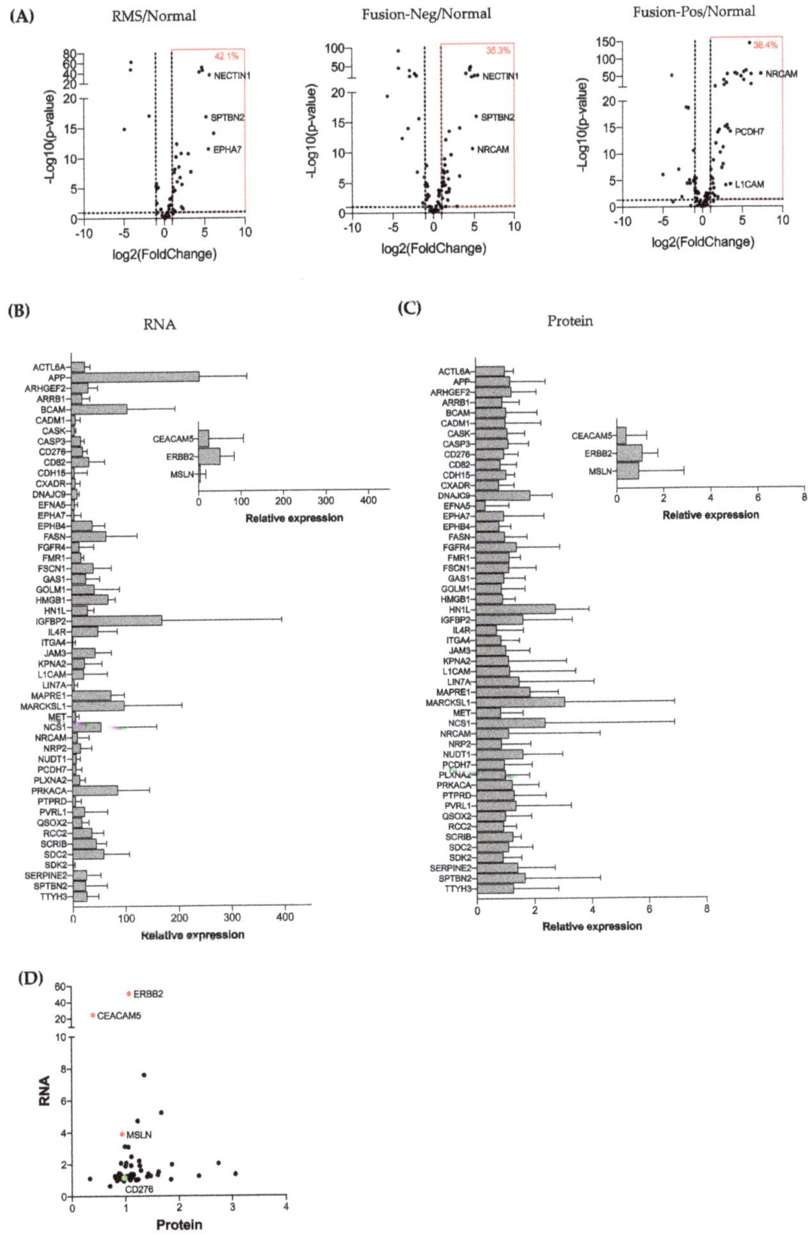

Figure 2. Validation of RMS-enriched cell-surface proteins by combined proteomic and transcriptomic analyses. (**A**) Volcano plot reporting *p*-values against fold changes for transcriptomic analysis of RMS tumors and normal muscle. The red box represents significantly up-regulated genes. Relative expression of (**B**) RNA and (**C**) protein of RMS-enriched cell-surface antigens in normal tissues including three CAR-T targets. (**D**) Correlation analysis between RNA and protein relative expression of RMS-enriched cell-surface antigens in normal tissues. Selected CAR-T targets and CD276 (B7-H3) are represented in red and green, respectively.

Based on the RNA and protein expression on the tumor and normal tissues of cell surface antigens, we generated a composite rank by assigning equal weights to transcriptomic and proteomic ranks of overexpressed genes and proteins in RMS (Tables 1–3). For tumor expression, the gene and protein with the highest expression was assigned rank 1. For normal expression, the gene and protein with the lowest expression was assigned rank 1. A gene enrichment analysis revealed the enrichment of the proteins associated to neurogenesis, axon guidance, and cell adhesion (Figure S2).

Table 1. Composite ranking of RMS-enriched cell-surface proteins based on combined transcriptomic and proteomic analyses.

Entry	Gene Names	Gene	Proteomic RMS	Transcriptomic RMS	Proteomic Normal	Transcriptomic Normal	Composite Sum	Composite Rank
P55283	CDH4	CDH4	1	12	1	1	15	1
P52803	EFNA5 EPLG7 LERK7	EFNA5	1	9	2	4	16	2
Q92823	NRCAM KIAA0343	NRCAM	4	1	13	6	24	3
O60245	PCDH7 BHPCDH	PCDH7	9	8	5	5	27	4
Q15375	EPHA7 EHK3 HEK11	EPHA7	18	3	7	2	30	5
Q15223	NECTIN1 HVEC PRR1 PVRL1	NECTIN1	1	2	19	14	36	6
O14936	CASK LIN2	CASK	5	21	12	3	41	7
O60462	NRP2 VEGF165R2	NRP2	7	22	3	9	41	7
P54826	GAS1	GAS1	8	14	6	17	45	9
Q06787	FMR1	FMR1	6	15	16	11	48	10
Q16658	FSCN1 FAN1 HSN SNL	FSCN1	12	5	14	19	50	11
Q6ZRP7	QSOX2 QSCN6L1 SOXN	QSOX2	20	11	10	10	51	12
Q9C0H2	TTYH3 KIAA1691	TTYH3	10	7	18	16	51	12
O15020	SPTBN2 KIAA0302 SCA5	SPTBN2	14	4	20	15	53	14
P32004	L1CAM CAML1 MIC5	L1CAM	15	10	15	13	53	14
Q5ZPR3	CD276 B7H3 PSEC0249 UNQ309/PRO352	CD276	23	13	8	12	56	16
O75051	PLXNA2 KIAA0463 OCT PLXN2 UNQ209/PRO235	PLXNA2	24	16	9	8	57	17
P09429	HMGB1 HMG1	HMGB1	11	24	4	22	61	18
Q8WXX5	DNAJC9	DNAJC9	17	18	22	7	64	19
P49006	MARCKSL1 MLP MRP	MARCKSL1	16	6	24	24	70	20
Q9H910	JPT2 C16orf34 HN1L L11	HN1L	13	19	23	18	73	21
P34741	SDC2 HSPG1	SDC2	22	20	11	21	74	22
Q15691	MAPRE1	MAPRE1	19	17	21	23	80	23
Q14160	SCRIB CRIB1 KIAA0147 LAP4 SCRB1 VARTUL	SCRIB	21	23	17	20	81	24

Table 2. Composite ranking of fusion-negative RMS-enriched cell-surface proteins based on combined transcriptomic and proteomic analyses.

Entry	Gene Names	Gene	Surfaceomic RMS	Transcriptomic RMS	Proteomic Normal	Transcriptomic Normal	Composite Sum	Composite Rank
P52803	EFNA5 EPLG7 LERK7	EFNA5	1	10	1	3	15	1
Q92823	NRCAM KIAA0343	NRCAM	8	4	12	7	31	2
Q15223	NECTIN1 HVEC PRR1 PVRL1	NECTIN1	1	1	19	13	34	3
O60245	PCDH7 BHPCDH	PCDH7	15	11	5	4	35	4
Q9BY67	CADM1 IGSF4 IGSF4A NECL2 SYNCAM TSLC1	CADM1	1	19	9	6	35	4
P27701	CD82 KAI1 SAR2 ST6 TSPAN27	CD82	1	14	2	20	37	6
O14910	LIN7A MALS1 VELI1	LIN7A	1	17	22	1	41	7
P22455	FGFR4 JTK2 TKF	FGFR4	13	3	20	9	45	8
O14936	CASK LIN2	CASK	7	26	11	2	46	9
Q8NBJ4	GOLM1 C9orf155 GOLPH2 PSEC0242	GOLM1	10	13	3	22	48	10
P54826	GAS1	GAS1	12	16	6	18	52	11
Q06787	FMR1	FMR1	9	18	15	12	54	12
Q9C0H2	TTYH3 KIAA1691	TTYH3	16	7	18	17	58	13
O15020	SPTBN2 KIAA0302 SCA5	SPTBN2	18	2	25	14	59	14
P18065	IGFBP2 BP2 IBP2	IGFBP2	1	5	24	29	59	14
P36639	NUDT1 MTH1	NUDT1	25	12	23	5	65	16
Q16658	FSCN1 FAN1 HSN SNL	FSCN1	24	8	14	21	67	17
Q6ZRP7	QSOX2 QSCN6L1 SOXN	QSOX2	26	24	7	10	67	17
P42574	CASP3 CPP32	CASP3	14	30	13	11	68	19
O96019	ACTL6A BAF53 BAF53A INO80K	ACTL6A	30	23	8	16	77	20
P07093	SERPINE2 PI7 PN1	SERPINE2	27	15	21	15	78	21
P09429	HMGB1 HMG1	HMGB1	20	28	4	26	78	21
P62166	NCS1 FLUP FREQ	NCS1	17	9	28	24	78	21
Q8WXX5	DNAJC9	DNAJC9	21	22	27	8	78	21
P34741	SDC2 HSPG1	SDC2	28	21	10	25	84	25
Q14160	SCRIB CRIB1 KIAA0147 LAP4 SCRB1 VARTUL	SCRIB	19	25	17	23	84	26
P05067	APP A4 AD1	APP	11	29	16	30	86	27
P49006	MARCKSL1 MLP MRP	MARCKSL1	23	6	30	28	87	28
Q9H910	JPT2 C16orf34 HN1L L11	HN1L	22	27	29	19	97	28
Q15691	MAPRE1	MAPRE1	29	20	26	27	102	30

Table 3. Composite ranking of fusion-positive RMS-enriched cell-surface proteins based on combined transcriptomic and proteomic analyses.

Entry	Gene Names	Gene	Surfaceomic RMS	Transcriptomic RMS	Proteomic Normal	Transcriptomic Normal	Composite Sum	Composite Rank
P55283	CDH4	CDH4	1	17	1	1	20	1
Q58EX2	SDK2 KIAA1514	SDK2	1	6	11	4	22	2
Q96KG7	MEGF10 KIAA1780	MEGF10	15	9	1	1	26	3
O00762	UBE2C UBCH10	UBE2C	23	2	1	1	27	4
P52803	EFNA5 EPLG7 LERK7	EFNA5	1	23	4	10	38	5
P49407	ARRB1 ARR1	ARRB1	1	13	13	18	45	6
P78310	CXADR CAR	CXADR	1	26	6	13	46	7
Q15375	EPHA7 EHK3 HEK11	EPHA7	22	3	17	6	48	8
Q92823	NRCAM KIAA0343	NRCAM	8	1	26	14	49	9
O60245	PCDH7 BHPCDH	PCDH7	13	12	15	11	51	10
P24394	IL4R IL4RA 582J2.1	IL4R	1	21	5	31	58	11
Q15223	NECTIN1 HVEC PRR1 PVRL1	NECTIN1	1	4	35	21	61	12
P23468	PTPRD	PTPRD	18	5	34	7	64	13
O60462	NRP2 VEGF165R2	NRP2	11	30	9	15	65	14
P08581	MET	MET	16	36	8	12	72	15
P13612	ITGA4 CD49D	ITGA4	38	22	10	5	75	16
O14936	CASK LIN2	CASK	9	33	25	9	76	17
P32004	L1CAM CAML1 MIC5	L1CAM	17	11	29	20	77	18
P52292	KPNA2 RCH1 SRP1	KPNA2	14	16	27	22	79	19
Q9P258	RCC2 KIAA1470 TD60	RCC2	26	14	12	27	79	19
Q06787	FMR1	FMR1	10	25	30	17	82	21
P54826	GAS1	GAS1	21	24	16	24	85	22
Q9C0H2	TTYH3 KIAA1691	TTYH3	19	10	33	23	85	22
Q6ZRP7	QSOX2 QSCN6L1 SOXN	QSOX2	36	15	19	16	86	24
P55291	CDH15 CDH14 CDH3	CDH15	37	19	23	8	87	25
Q92974	ARHGEF2 KIAA0651 LFP40	ARHGEF2	12	20	32	26	90	26
Q16658	FSCN1 FAN1 HSN SNL	FSCN1	28	7	28	29	92	27
Q5ZPR3	CD276 B7H3 PSEC0249 UNQ309/PRO352	CD276	29	27	18	19	93	28
P54760	EPHB4 HTK MYK1 TYRO11	EPHB4	31	32	7	28	98	29
Q9BX67	JAM3 UNQ859/PRO1868	JAM3	33	18	21	30	102	30
P09429	HMGB1 HMG1	HMGB1	20	38	14	34	106	31
P49006	MARCKSL1 MLP MRP	MARCKSL1	34	8	38	37	117	32
P50895	BCAM LU MSK19	BCAM	27	31	22	38	118	33
P49327	FASN FAS	FASN	32	34	20	33	119	34
Q9H910	JPT2 C16orf34 HN1L L11	HN1L	30	29	37	25	121	35
Q15691	MAPRE1	MAPRE1	25	28	36	35	124	36
P17612	PRKACA PKACA	PKACA	24	35	31	36	126	37
P34741	SDC2 HSPG1	SDC2	35	37	24	32	128	38

Using combined proteomic and transcriptomic analyses of RMS tumors and normal tissue, we identified several cell surface proteins enriched in RMS and with limited expression in normal tissue, suggesting they are potential RMS target candidates for antibody-based therapies.

3.3. The Immune Checkpoint Molecule B7-H3 Is Upregulated in Both Fusion-Negative and Fusion-Positive RMS

Among the cell surface proteins that were upregulated at the RNA and protein levels in RMS compared to the normal muscle and limited expression in normal tissue, we identified the immune checkpoint molecule B7-H3 (Tables 1–3 and Figure 2D). B7-H3 (CD276) is a member of the B7 family that consists of 10 immune checkpoint molecules, including the well-known immunotherapeutic target B7-H1 (PD-L1) [25]. B7-H3 represents a promising target for the treatment of RMS, because it not only appears as one of the most upregulated cell surface antigens in tumors but, also, its immunoregulatory function may provide novel insights on the mechanisms associated with tumor immune evasion for the development of new immune checkpoint inhibitors.

The Gene expression profiles of the B7 molecules showed B7-H3 as the most expressed B7 molecule in RMS, and the tumor expression was significantly higher in 100% of both fusion-negative RMS (9.45-fold) and fusion-positive RMS (7.03-fold) (Figure 3A). The B7-H6 gene expression was significantly increased in fusion-negative RMS. B7-H4 was upregulated in both RMS subtypes but at a greater extent in fusion-positive RMS ($p = 0.06$). Strikingly, both PD-L1 (B7-H1) and PD-L2 (B7-DC) were significantly downregulated in RMS compared to normal muscle. The cell surface protein expression of the B7 molecules was also assessed by flow cytometry in normal skeletal muscle cells, fusion-negative (RD), and fusion-positive RMS (RH30) (Figure 3B). Similar findings were obtained with B7-H3 being the most expressed B7 member, with a 3.42-fold increase in RMS cells. In contrast, PD-L1 and PD-L2 were downregulated in RMS, which corroborates with transcriptomic data. While B7-H4 and B7-H7 were not detected on the surfaces of normal and cancer cells, B7-H6 were weakly expressed in normal muscle and downregulated in RMS cell lines. In the proteomic data, the B7-H6 protein was expressed at low levels in three RMS cell lines (Table S1). A Western blot of the cell lines showed a similar increase of the total B7-H3 proteins in the RMS cell lines compared to the normal muscle (Figure 3C). We analyzed the tissue specificity score of B7-H3 in normal tissue as described previously [19]. The tissue specificity (TS) score defines the enrichment of B7H3 across normal tissues. A score superior of 2.5 means a protein is tissue-enriched, and a score superior of 4 defines a tissue-specific protein. An expression analysis in normal tissue showed a positive correlation between the RNA and protein expression of B7-H3 (Spearman correlation $\rho = 0.84$), and no tissue specificity was observed (Figure 3D).

To validate the differential expression of B7-H3 in normal tissue and RMS tumors, we first evaluated the sensitivity and specificity of four commercial antibodies using tumor xenografts of RMS wild-type and knockout cells for B7-H3 (Figure S3). While all antibodies showed positive signals and well-defined membrane staining in RH30 wild-type cells, no unspecific signal was detected with the AF1027 and EPNCIR122 antibodies (Figure S3C). The AF1027 clone was selected for tissue staining.

We performed staining on a normal human tissue microarray and RMS tumor sections. The tissue microarray consisted of three specimens for 32 normal tissues. No positive detection of B7-H3 was observed in normal human tissues (Figure 4A). We also analyzed B7-H3 tissue expression in a cohort of 132 RMS specimens ($N = 97$ patients) obtained from our institutional tissue registry (Table S11). B7-H3 expression was detected in 122/132 specimens (91.5%) with only 10 samples (8.5%) with negative expression. The RMS tumor specimens showed various levels of B7-H3 expression and intratumoral heterogeneity (Figure 4B). The staining of RMS tumor cells with Myogenin and MyoD1 revealed that B7-H3 is expressed by tumor cells and not detected in the stroma or tumor-infiltrating immune cells. The quantification of B7-H3 expression did not show any statistical differences between FN-RMS and FP-RMS. The median H-score was 60 and 80 for FN-RMS and FP-RMS, respectively.

Furthermore, no association was observed between the B7-H3 expression and histological or molecular subtype (Table S11). Altogether, these findings showed that B7-H3 is not expressed by normal tissue at both the RNA and protein levels, while it is strongly expressed in most RMS tumors, regardless of the molecular subtype.

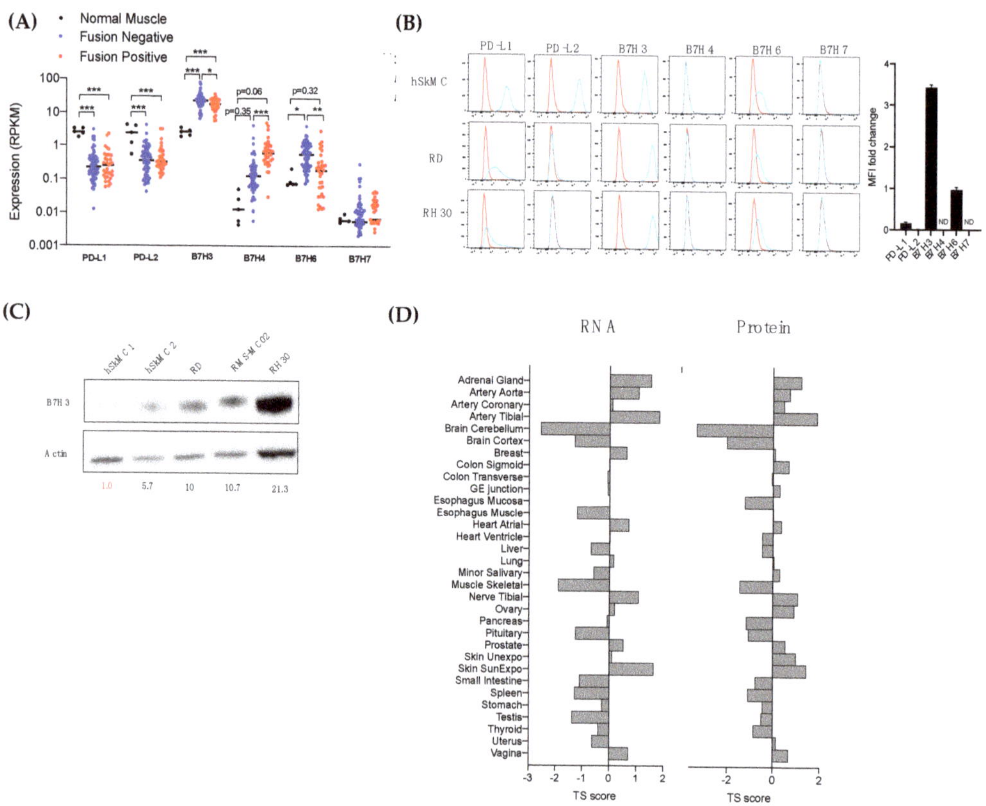

Figure 3. Expression of the immune checkpoint molecule B7-H3 in RMS and normal tissue. (**A**) Normalized expression (RPKM) of B7 molecules from RNA sequencing of normal muscle (N = 5), fusion negative (N = 66), and fusion positive (N = 35) RMS samples. Each data point represents an individual sample. The median is noted by a horizontal bar and significance denoted by asterisks (*** $p < 0.001$; ** $p < 0.01$; * $p < 0.05$; n.s., not significant, Student's *t*-test). (**B**) Histograms of flow cytometric analysis of B7 molecule expression in RMS and normal muscle cell lines. Bar graph shows mean fluorescence intensity (MFI) fold change in RMS cell lines compared to normal muscle. ND indicates "not detectable" expression (**C**) Western blot of B7H3 expression in 5 different cell lines. Band intensity for B7-H3 normalized on actin is indicated for each cell line. (**D**) Tissue Specific (TS) score of protein and RNA expression of B7H3 in each normal tissue (obtained from Jiang et al. [15]).

3.4. *Tumor B7-H3 Overexpression Is Associated with Low Infiltration of CD8⁺-T Cells in RMS Tumors and Impaired Antitumor Immune Response*

B7-H3 has shown paradoxical roles in tumor immunity [26]. Described originally as a costimulatory molecule, B7-H3 has also been linked to inhibition of the antitumor immune response and immune evasion [27,28]. The immunomodulatory role of B7-H3 and its relationship with tumor-infiltrating immune cells in RMS remains unknown. To address this unmet need, we investigated the immune landscape of fusion-positive and fusion-negative RMS by a deconvolution analysis of bulk RNA-seq data. We employed QuantiSeq, a computational pipeline, for the characterization and quantification of 10 tumor-infiltrating immune cell subsets [20]. An additional population called "Other" includes nonimmune cells such as malignant cells and fibroblasts. The relative abundance of different immune cell subsets was determined for normal muscle, fusion-negative, and fusion-positive RMS (Figure 5A,B). Higher proportions of monocytes, B-cells, macrophages M2, NK cells, CD4-T cells, CD8-T cells, and regulatory T cells were observed in both fusion-negative and fusion-positive RMS compared to normal muscle. In contrast, a higher content in neutrophils was found in normal muscle. Monocytes, neutrophils, and CD4-T cells represented ~70% of the immune cell content in RMS tumors, and no difference was observed between fusion-negative and fusion-positive RMS.

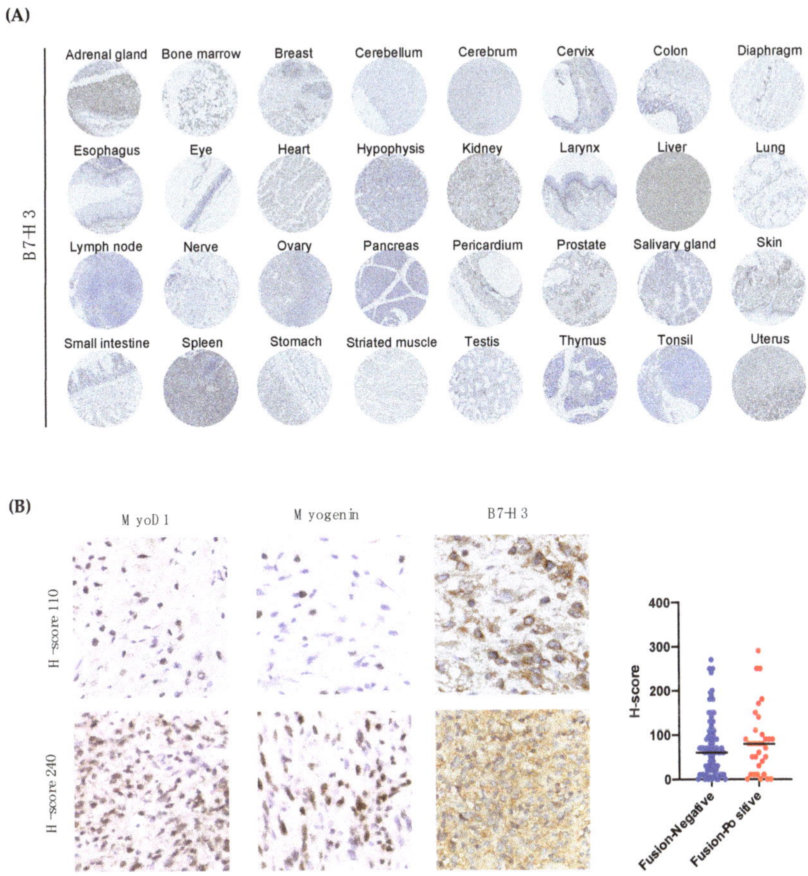

Figure 4. B7-H3 tissue expression in normal organs and RMS tumor specimens. (**A**) Immunohistochemical staining of B7H3 in a microarray of 32 normal tissues. Representative picture of triplicates for each tissue. (**B**) Representative picture of immunohistochemistry for MyoD1, Myogenin and B7H3 in RMS tumors. Scatter plot shows quantification of B7-H3 expression using H-scoring system in fusion-negative and fusion-positive RMS tumors (N = 132).

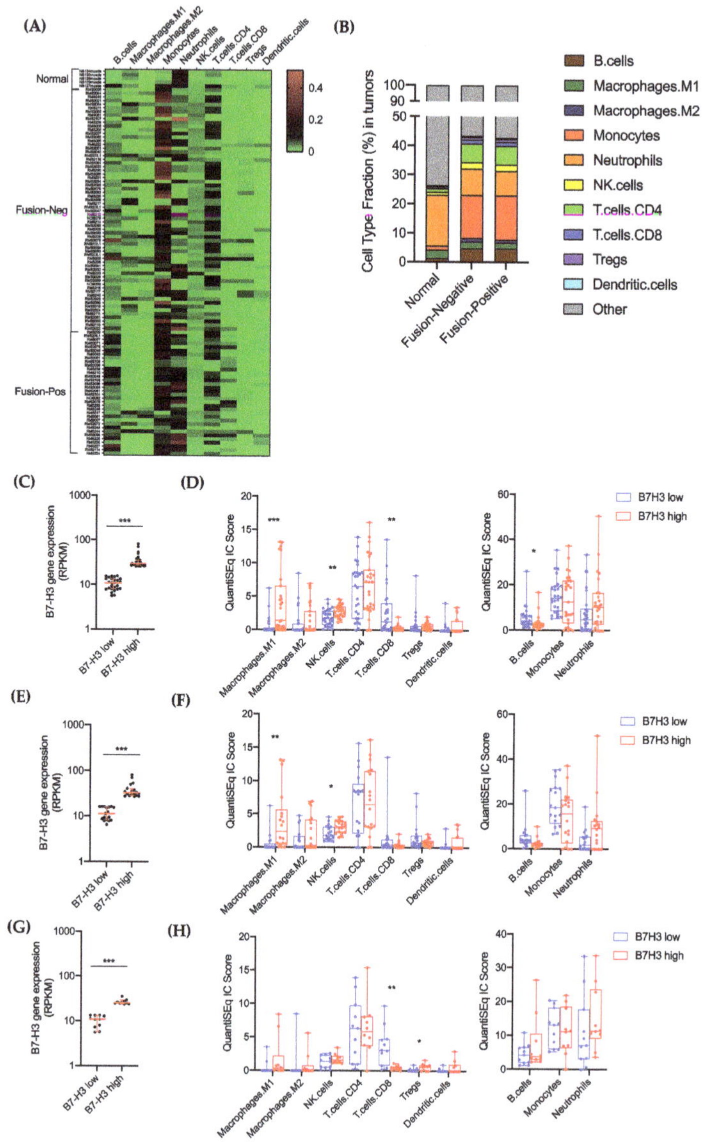

Figure 5. Relationship of B7-H3 expression and tumor-infiltrating immune cells in RMS. (**A**) Heatmap of immune cell abundance estimated by QuantiSeq analysis in normal muscle, fusion-negative and fusion-positive RMS. (**B**) Mean percentages of different immune cell subsets in normal muscle, fusion-negative and fusion-positive RMS. (**C**,**E**,**G**) B7-H3 RNA expression in quartile subgroups (25% highest, 25% lowest) in RMS tumors (**C**), FN-RMS (**E**) and FP-RMS (**G**) used for immune cell abundance analysis. (**D**,**F**,**H**) Bar graphs showing immune cell abundance between B7-H3-low and B7-H3-rich RMS tumors (**D**), FN-RMS only (**F**) and FP-RMS only (**H**). Significance is denoted by asterisks (*** $p < 0.001$; ** $p < 0.01$; * $p < 0.05$; Student's t-test).

To determine the relationship between B7-H3 expression and tumor-infiltrating immune cells, we measured the immune cell abundance in B7-H3-low and B7-H3-high tumors using quartile subgroups (25% highest and 25% lowest). Regardless of the fusion status,

B7-H3-high tumors were enriched in M1 macrophages and NK cells and depleted in CD8-T cells and B cells (Figure 5C,D). In fusion-negative RMS only, M1 macrophages and NK cells were significantly higher in B7-H3-rich tumors. Despite a trend toward a decrease of CD8-T-cell abundance in B7-H3-rich tumors, it was not statistically significant (Figure 5E,F). In fusion-positive RMS, only the infiltration of regulatory T cells and CD8-T cells were significantly affected by B7-H3 expression, where B7-H3-rich tumors have a higher infiltration of Tregs and depletion of CD8-T cells (Figure 5G,H). Correlation matrices were used to determine the degree of relationships between B7-H3 expression and the abundance of different immune cell subsets (Table S12). In fusion-negative RMS, a significant positive correlation was found between B7-H3 and M1 macrophages and neutrophils. A moderate negative correlation was also observed between B7-H3 and M2 macrophages and monocytes. Only CD8-T cells and B7-H3 showed a significant negative correlation in both fusion-negative and fusion-positive RMS.

To confirm a potential association between B7-H3 and CD8-T cells, we performed a T-cell cytotoxicity assay by coculturing activated PBMC isolated from 10 healthy donors with RH30 wild-type or knockout cells for B7-H3 and monitored the tumor cell survival over time. No significant difference was observed in B7-H3 knockout RH30 cells compared to the wild type (Figure 6A). For eight out of 10 normal donors, the loss of B7-H3 was associated with an increase in tumor cell killing (Figure 6B,C). A decrease of 34% and 45% in tumor cell survival was observed with B7-H3 knockout tumor cells at a ratio of 1:10 and 1:20 RH30:PBMC, respectively.

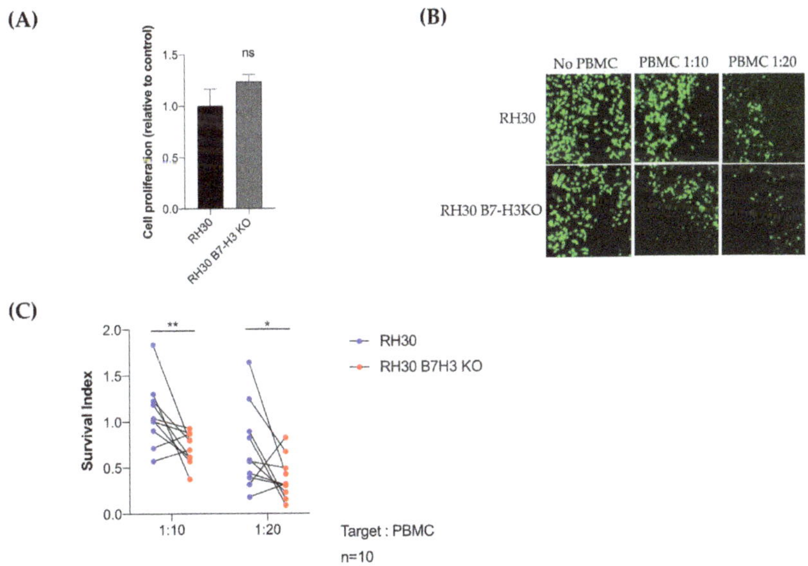

Figure 6. Loss of B7-H3 expression in RMS tumor cells is associated with higher T-cell cytotoxicity. (**A**) Histogram showing relative proliferation of B7-H3 knockout RH30 cells compared to wild-type cells. (**B**) Representative pictures of calcein-labeled wild-type and B7-H3 knockout RH30 cells after co-incubation with activated PBMCs for 4 hours. (**C**) Survival index of wild-type and B7-H3 knockout RH30 cells co-cultured with activated PBMC (N = 10 normal donors) at a ratio target: PBMC of 1:10 or 1:20. (** $p < 0.01$; * $p < 0.05$, Paired *t*-test).

Altogether, these data show that B7-H3 tumor expression is associated with the distinct immune composition of RMS tumors rich in M1 macrophages, M2 macrophages, and neutrophils and depleted in T cells. In addition, B7-H3 expression in RMS is associated with the inhibition of T-cell cytotoxic functions.

4. Discussion

In this study, we conducted the first comprehensive characterization of the cell surface proteome (surfaceome) of RMS tumors using cell surface capture and mass spectrometry-based proteomics to identify new therapeutic targets. As a result of the technological advances and reduced costs, conventional proteomics has become an attractive tool for target discovery and therapeutic development. However, it suffers from low sensitivity for cell surface proteins, which are significantly less abundant and soluble than cytosolic proteins [29]. Although surfaceomic approaches have been shown to circumvent this challenge and enhance target discovery, it has not been widely adopted, as several challenges have limited its use [30–32]. It requires a large number of starting living cells that may complicate the use of primary cells that have a limited number of passages in 2D cultures. Alongside the established RMS cell lines, we successfully analyzed the surfaceome of one primary cell line cultured from a patient-derived RMS (RMS-MC02), while three other primary cell lines did not reach the number of cells desired. Although a limited number of cell lines have been used for surfaceomic profiling, we included a bioinformatic analysis of publicly available transcriptomic and proteomic datasets of RMS specimens and normal tissues to reveal high-confidence therapeutic targets for the treatment of RMS. We benefited from the RNA sequencing and proteomic profiling of normal organs to determine the basal expression of RMS-enriched surface proteins in human organs. The success of antibody-based and cell-based therapies relies not only on the recognition of an antigen highly expressed on tumor cells but, also, on the minimal on-target off-tumor toxicities caused by the expression of the same antigen on normal cells [23]. By analyzing the gene and protein expressions in RMS and normal tissue, we uncovered a repertoire of surface antigens targetable with targeted therapies and immunotherapies.

Among the cell surface proteins enriched in FN-RMS and FP-RMS, we rediscovered several molecules previously reported by other groups [5]. For instance, we identified the FGFR4 receptor overexpressed in FN-RMS, which is in line with the prior observations of FGFR4 mutation and amplification in this RMS subtype [5]. Several well-known *PAX3/7-FOXO1* target genes were also reported among the top cell surface proteins in FP-RMS, including MET, IL4R, FMR1, and NRCAM [33]. Interestingly, FMR and NRCAM were also upregulated in FN-RMS, suggesting similarities in the protein repertoires regardless of the fusion status. An outstanding work from Shern et al. showed that fusion-negative and fusion-positive RMS display common altered pathways, including the RAS/PIK3CA axis [5]. The hierarchical clustering of RMS surfaceome signatures does not separate fusion-negative and fusion-positive RMS, which corroborates with the previous findings. Gene overexpression either through *PAX3/7-FOXO1* activity or mutation-associated amplification in FN-RMS may explain the high similarities in the cell surface protein repertoire of both molecular subtypes. This is of utmost interest, as it suggests that therapeutic strategies can be designed to target proteins commonly enriched in both RMS subtypes.

Our surfaceomic analysis also revealed new targetable RMS-enriched cell surface proteins. Among the commonly overexpressed proteins in both RMS subtypes, we identified the cell adhesion molecule CDH4 and ephrin receptors EFNA5 and EPHA7. CDH4 (R-Cadherin) has been previously found amplified in 43.6% of osteosarcomas, and its overexpression was associated with metastasis and a poor prognosis [34]. Similarly, CDH4 is an important driver of metastasis in glioblastoma [35]. In skeletal muscle, CDH4 has been reported to block myogenesis process and induce myoblast transformation, suggesting a potential oncogenic role in rhabdomyosarcoma [36]. Ephrin A5 and A7 are tyrosine kinase receptors that both bind the ephrin A5 ligand [37]. Ephrin receptors play a large role in embryonic and neural development [38]. In disease, Ephrin receptors and ligands can mediate the metastatic potential of cancer cells [39]. Little is known about the exact roles of Ephrin A5 and A7 in cancer progression. A few studies point towards a paradoxical role with pro- and antimetastatic functions, suggesting disease-specific activity [40–42]. The identification of new RMS-enriched tumor antigens, such as CDH4, Ephrin A5, and A7, support the clinical relevance of using surfaceomics for target discovery and drug development. Moreover, it provides novel biological insights into RMS pathology. Further studies are warranted to elucidate the roles of CDH4 and Ephrin A5/A7 in RMS, paving the way for the development of new treatments preventing disease recurrence and metastasis.

Another interesting target identified in our study is the immune checkpoint molecule B7-H3. B7-H3 is a member of the B7 family that contains ten members, including the well-known immune checkpoint PD-L1, the target of FDA-approved immunotherapies [25]. We identified B7-H3 as the major B7 immune molecule expressed on the surface of RMS cells. In our cohort of RMS specimens, only 8.55% of samples tested were negative for B7-H3, and no difference was observed between FN-RMS or FP-RMS. Interestingly, PD-L1 was poorly expressed in RMS cells, and it was significantly lower than in normal muscle. This is in accordance with the previous reports of a minimal or negative expression of PD-L1 in rhabdomyosarcoma [43,44]. B7-H3 has a multifaceted role in cancer, including immunological and nonimmunological functions [45,46]. Herein, we provided the first biological insights of the B7-H3 role in RMS. We analyzed the impact of B7-H3 expression in the immune composition of RMS tumors by deconvolution of the RNA-seq data. Interestingly, we found that RMS tumors rich in B7-H3 are depleted in CD8-T cells. Furthermore, B7-H3 knockout in RMS cells was associated with greater T-cell-mediated cytotoxicity. Altogether, this suggests that B7-H3 acts as an immune-inhibitory molecule in RMS. The underlying molecular and cellular mechanisms of B7-H3-mediated antitumor immunity remain to be elucidated, and unlike PD-L1, the B7-H3 receptor on immune cells has not been identified yet. Interestingly, we also found a positive correlation between B7-H3 expression and an abundance of neutrophils and M1 macrophages in FN-RMS. In colorectal cancer and hepatocellular carcinoma, B7-H3 tumor expression was associated with a higher infiltration of $CD68^+$ macrophages and the polarization of M1 to M2 macrophages [47,48]. While it was not statistically significant, we also observed higher infiltration of M2 macrophages in B7-H3-rich FN-tumors. This suggests that B7-H3 may drive the polarization of M1 macrophages towards the tumor-promoting M2 phenotype. Finally, the overexpression of B7-H3 in tumors compared to normal muscle was associated with a higher monocyte infiltration. Tumor-associated monocytes can differentiate into myeloid suppressor-derived cells, which are an important contributor of the immunosuppressive tumor microenvironment. While we cannot exclude a role of B7-H3 in MDSC differentiation, B7-H3 can also be expressed by MDSCs and contribute to CD8-T-cell inhibition and tumor progression [49–52].

While clinical trials evaluating PD-L1/PD-1 inhibitors in pediatric sarcoma have been unsuccessful, B7-H3 has become a popular target for new, targeted therapies [53]. Antibody–drug conjugates and CAR-T therapy are currently evaluated in clinical trials and are poised to positively change the therapeutic landscape of childhood cancers [54–56]. Our work provides novel mechanistic insights into the role of B7-H3 in tumor immune evasion and RMS progression. A complete characterization of B7-H3 function and regulation will pave the way for developing new B7-H3-based immunotherapies for the treatment of RMS.

5. Conclusions

While clinical trials evaluating PD-L1/PD-1 inhibitors in pediatric sarcoma have been unsuccessful, B7-H3 has become a popular target for new, targeted therapies [49]. Antibody–drug conjugates and CAR-T therapy are currently being evaluated in clinical trials and are poised to positively change the therapeutic landscape of childhood cancers [50–52]. Our work provides novel mechanistic insights on the role of B7-H3 in tumor immune evasion and RMS progression. A complete characterization of the B7-H3 functions and regulations will pave the way for developing new B7-H3-based immunotherapies for the treatment of RMS.

Supplementary Materials: The following are available online at https://www.mdpi.com/article/10.3390/cancers13184528/s1, Table S1. Normalized protein abundance (expressed in NSAF values) in RMS and normal muscle cell lines; Table S2. List of proteins identified as cell-surface proteins (GO:00005886); Table S3. List of cell-surface proteins upregulated by 2-fold in RMS compared to normal muscle; Table S4. List of cell-surface proteins upregulated by 2-fold in fusion-negative

RMS compared to normal muscle; Table S5. List of cell-surface proteins upregulated by 2-fold in fusion-positive RMS compared to normal muscle; Table S6. List of cell-surface proteins from Table S4 upregulated by 2-fold in RMS at the RNA level; Table S7. List of cell-surface proteins from Table S5 upregulated by 2-fold in fusion-negative RMS at the RNA level; Table S8. List of cell-surface proteins from Table S6 upregulated by 2-fold in fusion-positive RMS at the RNA level; Table S9. Normal tissue median RNA expression of RMS-enriched cell-surface proteins; Table S10. Normal tissue median protein expression of RMS-enriched cell-surface proteins; Table S11. B7-H3 tissue expression in RMS specimens and association with clinicopathological features; Table S12. Correlation analysis of B7-H3 expression with immune cell abundance in RMS tumors; Figure S1. Comparison of cell-surface protein repertoires in RMS and normal muscle; Figure S2. Gene enrichment analysis for RMS-enriched cell-surface proteins; Figure S3. Validation of antibody specificity for B7-H3 tissue staining.

Author Contributions: R.R.L., Y.K., and E.B., performed all the experiments. R.R.L. and F.L. designed the experiments and interpreted the data. D.A.P. and P.J.H. performed the cell surface capturing for the RH36, RH18, and RH41 cells. C.M.C. and B.J.M. performed the mass spectrometry and proteomic data analyses. L.W. performed the RNA-seq analysis. P.C.G., M.E.A., and C.F.G. provided the clinical input, data abstraction, and access to the tumor tissue specimens. J.C. reviewed the pathology cases and performed the B7-H3 H-scoring. H.D. provided technical support for the T-cell cytotoxic assays and immune cell analysis. All the authors reviewed and edited the manuscript. F.L. conceptualized and directed the study. All authors have read and agreed to the published version of the manuscript.

Funding: This research was funded by Hyundai on Wheels (C.G. and P.G.); Rein in Sarcoma (F.L., P.G., and H.D.); a departmental startup grant (F.L.); and generous benefactors.

Institutional Review Board Statement: This study was conducted according to the guidelines of the Declaration of Helsinki and approved by the Institutional Review Board of Mayo Clinic (16-006956, October 2016; 19-012426, January 2020).

Informed Consent Statement: Informed consent for use of the biospecimens and data was obtained from all the subjects involved in this study.

Data Availability Statement: All data presented here are available in the Supplementary Materials.

Conflicts of Interest: The authors declare no potential conflict of interest.

References

1. Davis, R.J.; D'Cruz, C.M.; Lovell, A.M.; Biegel, A.J.; Barr, F.G. Fusion of PAX7 to FKHR by the Variant t(1;13)(p36;q14) Translocation in Alveolar Rhabdomyosarcoma. *Cancer Res.* **1994**, *54*, 2869–2872.
2. Barr, F.G.; Galili, N.; Holick, J.; Biegel, J.A.; Rovera, G.; Emanuel, B.S. Rearrangement of the PAX3 Paired Box Gene in the Paediatric Solid Tumour Alveolar Rhabdomyosarcoma. *Nat. Genet.* **1993**, *3*, 113–117. [CrossRef]
3. Skapek, S.X.; Anderson, J.; Barr, F.G.; Bridge, J.A.; Gastier-Foster, J.M.; Parham, D.M.; Rudzinski, E.R.; Triche, T.; Hawkins, D.S. PAX-FOXO1 Fusion Status Drives Unfavorable Outcome for Children with Rhabdomyosarcoma: A Children's Oncology Group Report. *Pediatr. Blood Cancer* **2013**, *60*, 1411–1417. [CrossRef]
4. Cao, L.; Yu, Y.; Bilke, S.; Walker, R.L.; Mayeenuddin, L.H.; Azorsa, D.O.; Yang, F.; Pineda, M.; Helman, L.J.; Meltzer, P.S. Genome-Wide Identification of PAX3-FKHR Binding Sites in Rhabdomyosarcoma Reveals Candidate Target Genes Important for Development and Cancer. *Cancer Res.* **2010**, *70*, 6497–6508. [CrossRef]
5. Shern, J.F.; Chen, L.; Badgett, T.; Getz, G.; Chmielecki, J.; Mora, J.; Anderson, J.R.; Skapek, S.X.; Barr, F.G.; Meyerson, M.; et al. Comprehensive Genomic Analysis of Rhabdomyosarcoma Reveals a Landscape of Alterations Affecting a Common Genetic Axis in Fusion-Positive and Fusion-Negative Tumors. *Cancer Discov.* **2014**, *4*, 216–231. [CrossRef] [PubMed]
6. Crist, W.; Gehan, A.E.; Ragab, A.H.; Dickman, P.S.; Donaldson, S.S.; Fryer, C.; Hammond, D.; Hays, D.M.; Herrmann, J.; Heyn, R. The Third Intergroup Rhabdomyosarcoma Study. *J. Clin. Oncol.* **1995**, *13*, 610–630. [CrossRef]
7. Maurer, H.M.; Beltangady, M.; Gehan, E.A.; Crist, W.; Hammond, D.; Hays, D.M.; Heyn, R.; Lawrence, W.; Newton, W.; Ortega, J.; et al. The Intergroup Rhabdomyosarcoma Study-I. A Final Report. *Cancer* **1988**, *61*, 209–220. [CrossRef]
8. Maurer, H.M.; Gehan, E.A.; Beltangady, M.; Crist, W.; Dickman, P.S.; Donaldson, S.S.; Fryer, C.; Hammond, D.; Hays, D.M.; Herrmann, J.; et al. The Intergroup Rhabdomyosarcoma Study-II. *Cancer* **1993**, *71*, 1904–1922. [CrossRef]
9. Punyko, J.A.; Mertens, A.C.; Gurney, J.G.; Yasui, Y.; Donaldson, S.S.; Rodeberg, D.A.; Raney, R.B.; Stovall, M.; Sklar, C.A.; Robison, L.L.; et al. Long-Term Medical Effects of Childhood and Adolescent Rhabdomyosarcoma: A Report from the Childhood Cancer Survivor Study. *Pediatr. Blood Cancer* **2005**, *44*, 643–653. [CrossRef] [PubMed]

10. Reulen, R.C.; Winter, D.L.; Frobisher, C.; Lancashire, E.R.; Stiller, C.A.; Jenney, M.E.; Skinner, R.; Stevens, M.C.; Hawkins, M.M.; British Childhood Cancer Survivor Study Steering Group. Long-term Cause-Specific Mortality among Survivors of Childhood Cancer. *JAMA* **2010**, *304*, 172–179. [CrossRef] [PubMed]
11. Lu, R.-M.; Hwang, Y.-C.; Liu, I.-J.; Lee, C.-C.; Tsai, H.-Z.; Li, H.-J.; Wu, H.-C. Development of Therapeutic Antibodies for the Treatment of Diseases. *J. Biomed. Sci.* **2020**, *27*, 1–30. [CrossRef]
12. Leung, K.K.; Wilson, G.M.; Kirkemo, L.L.; Riley, N.M.; Coon, J.J.; Wells, J.A. Broad and Thematic Remodeling of the Surfaceome and Glycoproteome on Isogenic Cells Transformed with Driving Proliferative Oncogenes. *Proc. Natl. Acad. Sci. USA* **2020**, *117*, 7764–7775. [CrossRef]
13. Chen, W.; Mou, K.Y.; Solomon, P.; Aggarwal, R.; Leung, K.K.; Wells, J.A. Large Remodeling of the Myc-Induced Cell Surface Proteome in B Cells and Prostate Cells Creates New Opportunities for Immunotherapy. *Proc. Natl. Acad. Sci. USA* **2021**, *118*. [CrossRef]
14. Martinko, A.J.; Truillet, C.; Julien, O.; Diaz, E.J.; Horlbeck, A.M.; Whiteley, G.; Blonder, J.; Weissman, J.S.; Bandyopadhyay, S.; Evans, M.J.; et al. Targeting RAS-Driven Human Cancer Cells with Antibodies to Upregulated and Essential Cell-Surface Proteins. *eLife* **2018**, *7*. [CrossRef]
15. Nesvizhskii, A.I.; Keller, A.; Kolker, E.; Aebersold, R. A Statistical Model for Identifying Proteins by Tandem Mass Spectrometry. *Anal. Chem.* **2003**, *75*, 4646–4658. [CrossRef] [PubMed]
16. Deng, N.; Li, Z.; Pan, C.; Duan, H. freeQuant: A Mass Spectrometry Label-Free Quantification Software Tool for Complex Proteome Analysis. *Sci. World J.* **2015**, *2015*, 1–11. [CrossRef]
17. Lowerison, M.; Huang, C.; Lucien, F.; Chen, S.; Song, P. Ultrasound Localization Microscopy of Renal Tumor Xenografts in Chicken Embryo is Correlated to Hypoxia. *Sci. Rep.* **2020**, *10*, 1–13. [CrossRef] [PubMed]
18. Jiang, L.; Wang, M.; Lin, S.; Jian, R.; Li, X.; Chan, J.; Dong, G.; Fang, H.; Robinson, A.E.; Snyder, M.P.; et al. A Quantitative Proteome Map of the Human Body. *Cell* **2020**, *183*, 269–283.e19. [CrossRef]
19. Bonifant, C.; Jackson, H.J.; Brentjens, R.J.; Curran, K.J. Toxicity and Management in CAR T-Cell Therapy. *Mol. Ther.-Oncol.* **2016**, *3*, 16011. [CrossRef] [PubMed]
20. Uhlén, M.; Hallström, B.M.; Lindskog, C.; Mardinoglu, A.; Pontén, F.; Nielsen, J. Transcriptomics Resources of Human Tissues and Organs. *Mol. Syst. Biol.* **2016**, *12*, 862. [CrossRef]
21. Mackay, M.; Afshinnekoo, E.; Rub, J.; Hassan, C.; Khunte, M.; Baskaran, N.; Owens, B.; Liu, L.; Roboz, G.J.; Guzman, M.L.; et al. The Therapeutic Landscape for Cells Engineered with Chimeric Antigen Receptors. *Nat. Biotechnol.* **2020**, *38*, 233–244. [CrossRef] [PubMed]
22. Hegde, M.; Joseph, S.K.; Pashankar, F.; DeRenzo, C.; Sanber, K.; Navai, S.; Byrd, T.T.; Hicks, J.; Xu, M.L.; Gerken, C.; et al. Tumor Response and Endogenous Immune Reactivity after Administration of HER2 CAR T Cells in a Child with Metastatic Rhabdomyosarcoma. *Nat. Commun.* **2020**, *11*, 1–15. [CrossRef] [PubMed]
23. Ni, L.; Dong, C. New B7 Family Checkpoints in Human Cancers. *Mol. Cancer Ther.* **2017**, *16*, 1203–1211. [CrossRef] [PubMed]
24. Kontos, F.; Michelakos, T.; Kurokawa, T.; Sadagopan, A.; Schwab, J.H.; Ferrone, C.R.; Ferrone, S. B7-H3: An Attractive Target for Antibody-based Immunotherapy. *Clin. Cancer Res.* **2020**, *27*, 1227–1235. [CrossRef] [PubMed]
25. Chapoval, A.I.; Ni, J.; Lau, J.S.; Wilcox, R.A.; Flies, D.B.; Liu, D.; Dong, H.; Sica, G.L.; Zhu, G.; Tamada, K.; et al. B7-H3: A Costimulatory Molecule for T Cell Activation and IFN-γ Production. *Nat. Immunol.* **2001**, *2*, 269–274. [CrossRef]
26. Lee, Y.-H.; Martin-Orozco, N.; Zheng, P.; Li, J.; Zhang, P.; Tan, H.; Park, H.J.; Jeong, M.; Chang, S.H.; Kim, B.-S.; et al. Inhibition of the B7-H3 Immune Checkpoint Limits Tumor Growth by Enhancing Cytotoxic Lymphocyte Function. *Cell Res.* **2017**, *27*, 1034–1045. [CrossRef]
27. Finotello, F.; Mayer, C.; Plattner, C.; Laschober, G.; Rieder, D.; Hackl, H.; Krogsdam, A.; Loncova, Z.; Posch, W.; Wilflingseder, D.; et al. Molecular and Pharmacological Modulators of the Tumor Immune Contexture Revealed by Deconvolution of RNA-seq Data. *Genome Med.* **2019**, *11*, 1–20. [CrossRef]
28. Wollscheid, B.; Bausch-Fluck, D.; Henderson, C.; O'Brien, R.; Bibel, M.; Schiess, R.; Aebersold, R.; Watts, J.D. Mass-Spectrometric Identification and Relative Quantification of N-Linked Cell Surface Glycoproteins. *Nat. Biotechnol.* **2009**, *27*, 378–386. [CrossRef]
29. Nix, M.A.; Mandal, K.; Geng, H.; Paranjape, N.; Lin, Y.-H.T.; Rivera, J.M.; Marcoulis, M.; White, K.L.; Whitman, J.D.; Bapat, S.P.; et al. Surface Proteomics Reveals CD72 as a Target for In Vitro–Evolved Nanobody-Based CAR-T Cells in KMT2A/MLL1-Rearranged B-ALL. *Cancer Discov.* **2021**. [CrossRef]
30. Heider, M.; Eichner, R.; Stroh, J.; Morath, V.; Kuisl, A.; Zecha, J.; Lawatscheck, J.; Baek, K.; Garz, A.-K.; Rudelius, M.; et al. The IMiD Target CRBN Determines HSP90 Activity Toward Transmembrane Proteins Essential in Multiple Myeloma. *Mol. Cell* **2021**, *81*, 1170–1186.e10. [CrossRef]
31. Lee, J.; Bangayan, N.J.; Chai, T.; Smith, B.A.; Pariva, T.E.; Yun, S.; Vashisht, A.; Zhang, Q.; Park, J.W.; Corey, E.; et al. Systemic Surfaceome Profiling Identifies Target Antigens for Immune-Based Therapy in Subtypes of Advanced Prostate Cancer. *Proc. Natl. Acad. Sci. USA* **2018**, *115*, E4473–E4482. [CrossRef]
32. Davicioni, E.; Finckenstein, F.G.; Shahbazian, V.; Buckley, J.D.; Triche, T.J.; Anderson, M.J. Identification of a PAX-FKHR Gene Expression Signature that Defines Molecular Classes and Determines the Prognosis of Alveolar Rhabdomyosarcomas. *Cancer Res.* **2006**, *66*, 6936–6946. [CrossRef]
33. Tang, Q.; Lu, J.; Zou, C.; Shao, Y.; Chen, Y.; Narala, S.; Fang, H.; Xu, H.; Wang, J.; Shen, J.; et al. CDH4 is a Novel Determinant of Osteosarcoma Tumorigenesis and Metastasis. *Oncogene* **2018**, *37*, 3617–3630. [CrossRef] [PubMed]

34. Ceresa, D.; Alessandrini, F.; Bosio, L.; Marubbi, D.; Reverberi, D.; Malatesta, P.; Appolloni, I. Cdh4 Down-Regulation Impairs in Vivo Infiltration and Malignancy in Patients Derived Glioblastoma Cells. *Int. J. Mol. Sci.* **2019**, *20*, 4028. [CrossRef]
35. Kucharczak, J.; Charrasse, S.; Comunale, F.; Zappulla, J.; Robert, B.; Teulon-Navarro, I.; Pèlegrin, A.; Gauthier-Rouviere, C. R-Cadherin Expression Inhibits Myogenesis and Induces Myoblast Transformation via Rac1 GTPase. *Cancer Res.* **2008**, *68*, 6559–6568. [CrossRef] [PubMed]
36. Boyd, A.W.; Bartlett, P.F.; Lackmann, M. Therapeutic Targeting of EPH Receptors and Their Ligands. *Nat. Rev. Drug Discov.* **2013**, *13*, 39–62. [CrossRef] [PubMed]
37. Klein, R. Eph/Ephrin Signalling during Development. *Development* **2012**, *139*, 4105–4109. [CrossRef] [PubMed]
38. Wang, B. Cancer Cells Exploit the Eph-Ephrin System to Promote Invasion and Metastasis: Tales of Unwitting Partners. *Sci. Signal.* **2011**, *4*, pe28. [CrossRef]
39. Giaginis, C.; Tsoukalas, N.; Bournakis, E.; Alexandrou, P.; Kavantzas, N.; Patsouris, E.; Theocharis, S. Ephrin (Eph) receptor A1, A4, A5 and A7 Expression in Human Non-Small Cell Lung Carcinoma: Associations with Clinicopathological Parameters, Tumor Proliferative Capacity and Patients' Survival. *BMC Clin. Pathol.* **2014**, *14*, 8. [CrossRef]
40. Li, R.; Jin, M.; Sun, Y.; Jiang, A.; Wu, Y.; Li, C.; Yan, H.; Jin, H. Knockdown of Ephrin Receptor A7 Suppresses the Proliferation and Metastasis of A549 Human Lung Cancer Cells. *Mol. Med. Rep.* **2016**, *13*, 3190–3196. [CrossRef]
41. Di, W.; Weinan, X.; Xin, L.; Zhiwei, Y.; Xinyue, G.; Jinxue, T.; Mingqi, L. Long Noncoding RNA SNHG14 Facilitates Colorectal Cancer Metastasis through Targeting EZH2-Regulated EPHA7. *Cell Death Dis.* **2019**, *10*, 1–13. [CrossRef] [PubMed]
42. Gabrych, A.; Pęksa, R.; Kunc, M.; Krawczyk, M.; Izycka-Swieszewska, E.; Biernat, W.; Bień, E. The PD-L1/PD-1 Axis Expression on Tumor-Infiltrating Immune Cells and Tumor Cells in Pediatric Rhabdomyosarcoma. *Pathol. Res. Pract.* **2019**, *215*, 152700. [CrossRef] [PubMed]
43. Bertolini, G.; Bergamaschi, L.; Ferrari, A.; Renne, S.L.; Collini, P.; Gardelli, C.; Barisella, M.; Centonze, G.; Chiaravalli, S.; Paolino, C.; et al. PD-L1 Assessment in Pediatric Rhabdomyosarcoma: A Pilot Study. *BMC Cancer* **2018**, *18*, 652. [CrossRef] [PubMed]
44. Flem-Karlsen, K.; Fodstad, O.; Tan, M.; Nunes-Xavier, C.E. B7-H3 in Cancer—Beyond Immune Regulation. *Trends Cancer* **2018**, *4*, 401–404. [CrossRef] [PubMed]
45. Li, G.; Quan, Y.; Che, F.; Wang, L. B7-H3 in Tumors: Friend or Foe for Tumor Immunity? *Cancer Chemother. Pharmacol.* **2018**, *81*, 245–253. [CrossRef]
46. Mao, Y.; Chen, L.; Wang, F.; Zhu, D.; Ge, X.; Hua, D.; Sun, J. Cancer Cell-Expressed B7-H3 Regulates the Differentiation of Tumor-Associated Macrophages in Human Colorectal Carcinoma. *Oncol. Lett.* **2017**, *14*, 6177–6183. [CrossRef]
47. Kang, F.-B.; Wang, L.; Li, N.; Zhang, Y.-G.; Sun, D.-X. Hepatocellular Carcinomas Promote Tumor-Associated Macrophage M2-Polarization via Increased B7-H3 Expression. *Oncol. Rep.* **2014**, *33*, 274–282. [CrossRef]
48. Yim, J.; Koh, J.; Kim, S.; Song, S.G.; Ahn, H.K.; Kim, A.Y.; Jeon, Y.K.; Chung, D.H. Effects of B7-H3 Expression on Tumour-Infiltrating Immune Cells and Clinicopathological Characteristics in Non–Small-Cell Lung Cancer. *Eur. J. Cancer* **2020**, *133*, 74–85. [CrossRef]
49. Zhang, G.; Huang, H.; Zhu, Y.; Yu, G.; Gao, X.; Xu, Y.; Liu, C.; Hou, J.; Zhang, X. A Novel Subset of B7-H3+CD14+HLA-DR−/Lowmyeloid-Derived Suppressor Cells are Associated with Progression of Human NSCLC. *OncoImmunology* **2015**, *4*, e977164. [CrossRef]
50. Baumann, T.; Dunkel, A.; Schmid, C.; Schmitt, S.; Hiltensperger, M.; Lohr, K.; Laketa, V.; Donakonda, S.; Ahting, U.; Lorenz-Depiereux, B.; et al. Regulatory Myeloid Cells Paralyze T Cells through Cell–Cell Transfer of the Metabolite Methylglyoxal. *Nat. Immunol.* **2020**, *21*, 555–566. [CrossRef]
51. Si, Y.; Merz, S.F.; Jansen, P.; Wang, B.; Bruderek, K.; Altenhoff, P.; Mattheis, S.; Lang, S.; Gunzer, M.; Klode, J.; et al. Multidimensional Imaging Provides Evidence for Down-Regulation of T Cell Effector Function by MDSC in Human Cancer Tissue. *Sci. Immunol.* **2019**, *4*, eaaw9159. [CrossRef] [PubMed]
52. Yang, S.; Wei, W.; Zhao, Q. B7-H3, a Checkpoint Molecule, as a Target for Cancer Immunotherapy. *Int. J. Biol. Sci.* **2020**, *16*, 1767–1773. [CrossRef]
53. Majzner, R.G.; Theruvath, J.L.; Nellan, A.; Heitzeneder, S.; Cui, Y.; Mount, C.W.; Rietberg, S.P.; Linde, M.H.; Xu, P.; Rota, C.; et al. CAR T Cells Targeting B7-H3, a Pan-Cancer Antigen, Demonstrate Potent Preclinical Activity Against Pediatric Solid Tumors and Brain Tumors. *Clin. Cancer Res.* **2019**, *25*, 2560–2574. [CrossRef] [PubMed]
54. Theruvath, J.; Sotillo, E.; Mount, C.W.; Graef, C.M.; Delaidelli, A.; Heitzeneder, S.; Labanieh, L.; Dhingra, S.; Leruste, A.; Majzner, R.G.; et al. Locoregionally Administered B7-H3-Targeted CAR T Cells for Treatment of Atypical Teratoid/Rhabdoid Tumors. *Nat. Med.* **2020**, *26*, 712–719. [CrossRef] [PubMed]
55. Scribner, J.A.; Brown, J.G.; Son, T.; Chiechi, M.; Li, P.; Sharma, S.; Li, H.; De Costa, A.; Li, Y.; Chen, Y.; et al. Preclinical Development of MGC018, a Duocarmycin-Based Antibody–Drug Conjugate Targeting B7-H3 for Solid Cancer. *Mol. Cancer Ther.* **2020**, *19*, 2235–2244. [CrossRef]
56. Love, I.M.; Huber, W.; Anders, S. Moderated Estimation of Fold Change and Dispersion for RNA-seq Data with DESeq2. *Genome Biol.* **2014**, *15*, 550. [CrossRef]

cancers MDPI

Review

The Psychoneuroimmunology of Stress Regulation in Pediatric Cancer Patients

Gillian E. White [1], Jessica E. Caterini [1], Victoria McCann [2], Kate Rendall [1], Paul C. Nathan [3], Shawn G. Rhind [4,5], Heather Jones [3] and Greg D. Wells [1,*]

1 Translational Medicine, The Hospital for Sick Children, Toronto, ON M5G 1X8, Canada; gillianwhite018@gmail.com (G.E.W.); jcaterini@qmed.ca (J.E.C.); kate.rendall@mail.utoronto.ca (K.R.)
2 School of Medicine, Queen's University, Kingston, ON K7L 3N6, Canada; vmccann@qmed.ca
3 Division of Hematology/Oncology, The Hospital for Sick Children, Toronto, ON M5G 1X8, Canada; paul.nathan@sickkids.ca (P.C.N.); heather.jones@sickkids.ca (H.J.)
4 Defence Research and Development Canada, Toronto Research Centre, Toronto, ON M3K 2C9, Canada; Shawn.Rhind@drdc-rddc.gc.ca
5 Faculty of Kinesiology & Physical Education, University of Toronto, Toronto, ON M5S 2W6, Canada
* Correspondence: greg.wells@sickkids.ca; Tel.: +1-416-710-4618

Simple Summary: There are many commonalities between children with cancer and other populations that experience early-life stress. Thus, it is important to review the existing research surrounding the stress response in the pediatric cancer population. In this review, we describe the psychoneuroimmunology behind stress regulation and the differences observed in stress regulatory pathways in childhood cancer patients. Our objective is to provide a clinically relevant summary of the stress pathways contributing to, and exacerbating, childhood illness and outline some potential interventions.

Citation: White, G.E.; Caterini, J.E.; McCann, V.; Rendall, K.; Nathan, P.C.; Rhind, S.G.; Jones, H.; Wells, G.D. The Psychoneuroimmunology of Stress Regulation in Pediatric Cancer Patients. *Cancers* **2021**, *13*, 4684. https://doi.org/10.3390/cancers13184684

Academic Editors: Saurabh Agarwal and Jianhua Yang

Received: 30 July 2021
Accepted: 17 September 2021
Published: 18 September 2021

Abstract: Stress is a ubiquitous experience that can be adaptive or maladaptive. Physiological stress regulation, or allostasis, can be disrupted at any point along the regulatory pathway resulting in adverse effects for the individual. Children with cancer exhibit significant changes to these pathways in line with stress dysregulation and long-term effects similar to those observed in other early-life stress populations, which are thought to be, in part, a result of cytotoxic cancer treatments. Children with cancer may have disruption to several steps in the stress-regulatory pathway including cognitive-affective function, neurological disruption to stress regulatory brain regions, altered adrenal and endocrine function, and disrupted tissue integrity, as well as lower engagement in positive coping behaviours such as physical activity and pro-social habits. To date, there has been minimal study of stress reactivity patterns in childhood illness populations. Nor has the role of stress regulation in long-term health and function been elucidated. We conclude that consideration of stress regulation in childhood cancer may be crucial in understanding and treating the disease.

Keywords: stress; allostasis; stress regulatory pathways; stress dysregulation; stress reactivity; early life stress; childhood cancer; cytotoxic cancer treatment; interventions

1. Introduction

Stress is a ubiquitous experience with significant impacts on health and function. Stress refers to the physiological state of the body in response to a stressor, whereas the stressors themselves are challenges, threats, demands and constraints that provide barriers to the normal daily functioning of the individual, thereby inducing stress and its related consequences on the body [1]. Stress responses function acutely to defend homeostasis during changing internal and external demands [2,3] and transduce external stimuli into physiological signals to support effective navigation of the environment and encode relevant information for future events [4]. Depending on the nature of the stressor and the interpretation of it, stress can be adaptive ("eustress"), broadening one's ability to cope with environmental conditions and challenging experiences effectively, or maladaptive ("distress"), straining one's ability to cope and contributing to general dysfunction [2,5,6].

Significant or chronic exposure to stress activation can cause long-term changes to how the stress-regulatory system responds to future stressors [6–8]. This is particularly true of significant stress exposure during times of developmental plasticity, such as childhood and adolescence [9–13]. Several chronic diseases in adults and children, including cardiovascular disease, metabolic conditions, cancer and immunologic conditions have been attributed to chronic stress and/or dysfunctional stress regulation, in addition to a number of psychological and cognitive conditions and nonclinical functional outcomes following significant or chronic stress exposure [5,14–16]. Risk and prevalence of these stress-related adverse outcomes have been reported in populations of children experiencing significant stress or trauma [17] and are thought to be a result of alterations in stress regulatory pathways which can be detected by examining patterns of physiological signalling in response to an acute stress exposure [5,14–16].

While the adverse outcomes of childhood trauma and chronic stress have been well documented, current definitions of Adverse Childhood Events (ACEs) and early life stress models from which this research draws [4,9,10,13,15,17,18] do not include childhood illness. Stress dysregulation may provide a valuable paradigm through which to understand long-term health and wellbeing in populations experiencing childhood illness, given psychological strain and physically aggressive medical treatments.

Approximately 1 in 300 children will develop cancer between birth and 20 years of age [19] which involves significant psychological burden, physical distress and barriers to engaging in positive stress coping behaviour such as pro-social and physical activity habits. Children with cancer and survivors show evidence of neurological [20–27], physiological [28–34] and psychosocial [24,25,27,35,36] changes potentially attributable to stress dysregulation (see Figure 1). Further, they report higher risk of late effects similar to those reported in other early life stress populations, suggesting that stress dysregulation may be worth investigating as a putative pathway, or as a modulator of the damage caused by cancer treatment. Psychoneuroimmunology, a field of study that is inherently implicated in stress regulation, examines the relationships between human behaviour, the nervous and endocrine systems, and the immune system [37]. This review describes the psychoneuroimmunology of stress regulation and support for potential clinical relevance in childhood cancer patients and survivors [38–40].

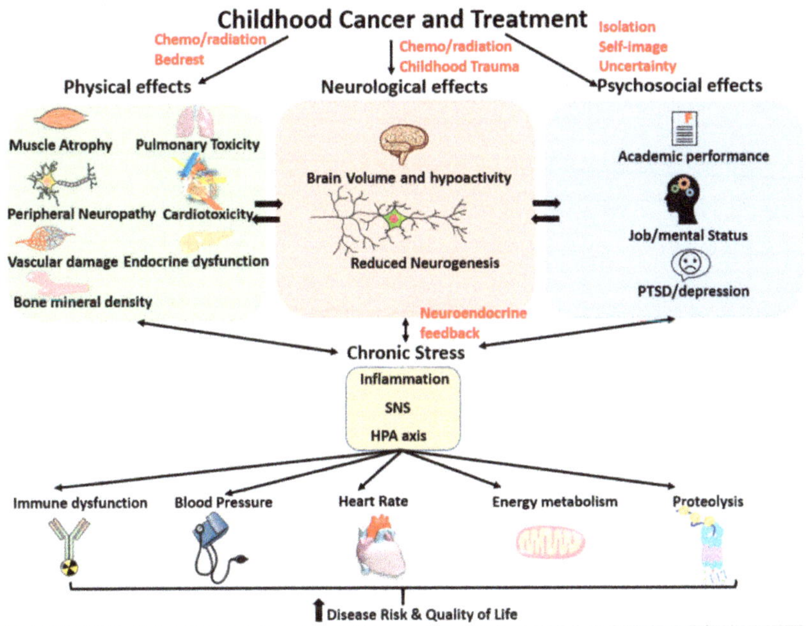

Figure 1. Illustration representation of the interrelationships between childhood cancer and its treatment and the physical, neurological and psychosocial effects that lead to chronic stress. A bidirectional relationship exists between physical, neurological and psychosocial effects, as well as between chronic stress and dysfunction in these systems. The major health related effects of chronic stress leading to disease risk and lower quality of life.

2. Stress Regulation & Allostasis

Selye defines stress as the body's adaptive response to a "noxious agent", involving a deviation from resting state, or homeostasis [2]. Homeostasis refers to the maintenance physiological variables essential for human life, such as internal body temperature, or pH, within a precise range [3]. Physiological stress regulation, also termed 'allostasis', is an adaptive process aimed at keeping the body's systems in physiological ranges, or maintaining homeostasis, despite changing internal and external environmental conditions [3]. Stressors can be both biological or psychological, and thus can be real or perceived, past, current, anticipated or recalled [41]. Since biological stressors are typically internal, they evoke a direct physiological response without the engagement of higher order cognitive processing [42]. Psychological stimuli, however, must undergo interpretation, and thus elicits a physiological stress response indirectly [14,43–45]. Therefore, a psychological response is mounted in response to the perception of stress, rather than the stressor itself. For the purposes of this review, we will only consider psychological stressors and the allostatic processes that they evoke.

The allostatic processes are an integration of subjective, often subconscious appraisal of a physical or psychological stimulus, which is converted into neural and endocrine signalling in turn activating target organs that elicit an allostatic response as shown in Figure 2. Effective allostatic responses are characterised by rapid upregulation of signaling hormones and effector tissue activation, commensurate with the level of threat encountered, and rapid termination upon cessation of stimulus—producing maximal physiological response to effectively defend homeostasis during stress experience with minimal exposure to signalling biomediators and associated energy costs [46].

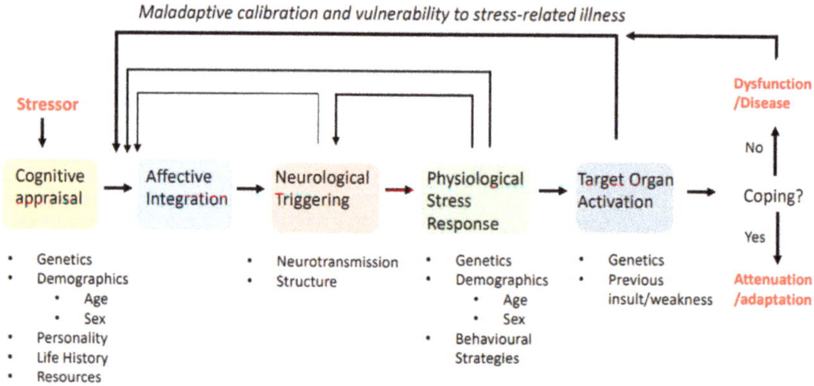

Figure 2. A systems approach to stress responses. Some stimulus—internal or external, perceived or real, past, present or future—is interpreted as threatening through cognitive appraisal and integrated with affective feedback. Together, the cognitive-affective appraisal determines the saliency of the threat. If sufficient, this will cause neurological triggering of stress activation pathways in the brain, resulting in increased biochemical signaling of these pathways and their subsequent physiological activation of target organs. If a stable state cannot be achieved or maintained to meet the demands of the stressor, then excessive activation and associated wear on target tissues may result and possibly lead to dysfunctional signalling with excessive feedback by mediators on earlier pathway steps. Adapted with permission from ref. [14]. Copyright 2012 Springer.

Allostasis is considered to be a mechanism by which the body can maintain homeostasis in the face of variable environments—stability through change [47]. To this end it can be considered from an energy conservation perspective in which the physiological responses are predictive and aim to maximise energy efficiency in coping with environmental demands [48]. It can also be considered a calibrating mechanism in which the mediators of the response feedback to the earlier steps to adapt optimally to environmental contexts [4]. The hypothesis of allostasis as a mechanism in long-term health and functional outcomes in childhood cancer patients and survivors is rooted in the data indicating altered regulation of the cognitive, psychological, physiological pathways that regulate physiological responses to stress. We suggest that understanding and mitigating this dysregulation of physiological responses to stress is integral to improving care in this population. Therefore, for the purposes of this paper, we take the view of allostasis as the process by which the body responds physiologically to stressors in order to regain homeostasis.

Acutely, allostasis is adaptive and functions to maintain homeostasis in face of a prospective disruption by mediating a fight or flight response [8]. Biomediators of allostasis, specifically cortisol, also provide feedback to brain regions responsible for triggering allostatic responses in a self-regulating process that tunes the response to the stressor presented and terminates the acute activation of allostasis [6–8]. This same biochemical feedback calibrates the system to future stress exposures through both neurological encoding and affective working memory [4,6,8].

There are important functional consequences of chronic allostatic stress. The catabolic and energy costly processes needed to mount allostatic responses contribute to wear and tear on target organs, potentially exacerbating underlying weakness or dysfunction in these tissues. Further, the chronic feedback of biomediators on stress regulatory corticolimbic brain centres, particularly the hippocampus, alter their structure and function, ultimately impacting the overall regulation and responsivity of the system [49–54]. Due to the self-regulatory nature of stress regulation, dysregulation can occur with disruption to any point in the pathway shown in Figure 2. Typically, dysregulation that contributes to stress-related health and dysfunction is characterised by one of four patterns of stress reactivity: (1). Overly frequent activation; (2). No habituation to familiar stimuli; (3). Inefficient termination; or (4). Hypoactivation of one pathway resulting in compensatory overactivation of the other [3].

While stress can contribute to the development or acceleration of illness, it does not cause illness or dysfunction per se. Rather, the adaptive function of allostasis requires systemic upregulation. Often these are catabolic processes, resulting in wear and tear over time or "allostatic load" [3,10,18,55,56], which describes the cost that this prolonged upregulation and activation of compensatory effectors has on the body [57]. Allostatic load exacerbates existing tissue weaknesses, whether hereditary genetic predispositions or susceptibility from a previous insult, or a concomitant one such as cancer therapy. The aggregate of this subclinical dysfunction can have significant health and functional implications and has been associated with all-cause morbidity and mortality [3,55].

Much research to date has focused on individuals who experienced significant childhood adversity or ACE's (such as physical or emotional child abuse, neglect, parental substance abuse, household and family turmoil, etc.) but leaves out childhood illness [4,7,10,11,17,18,58]. Many commonalities exist between populations experiencing psychosocial early life stress and children experiencing childhood illness [13,17,18,28,58–60]. This presents the possibility that children treated for illness many be susceptible to long-term health and dysfunction associated with altered stress regulation. Childhood cancer patients experience significant stress prolonged over the duration of treatment, as well as significant physical and psychosocial effects of treatment and disruptions to their normal developmental opportunities [23,61]. As a disruption to any step in the pathway can result in dysfunctional regulation, we will review the role of each step and evidence of prospective dysfunction in childhood cancer patients and survivors.

3. Cognitive-Affective Appraisal

Allostatic processes are regulated by the integration of inputs to corticolimbic brain regions including the prefrontal cortex (PFC), hippocampus, amygdala and brainstem. These regions are implicated in many other functions besides stress circuitry such as decision making and higher cognition, learning and memory, emotional processing and the judgement of salience, among others [62]. Neurocognitive and neuropsychological function development corresponds with the timing of brain development through childhood and adolescence. In brief, functions associated with emotional reactivity develop ahead of those necessary for cognitive reappraisal and self-regulation during childhood [12,63], in such a way that the time-lag in development in conjunction with social contexts of adolescence is often attributed to the behavioural trends (impulsivity, risk taking) and greater stress reactivity seen during adolescence [12,64].

For evolutionary reasons, stimuli that involve social judgement, are goal oriented and include unpredictability, novelty and uncertainty are most salient and reliably evoke a physiological response [65,66]. The subjective nature of cognitive-affective appraisal presents significant inter-individual variability and is influenced by lived experience, disposition, cognitive and psychological strategies, as well as positive coping behaviours such as physical activity and social connections. Human and animal models of chronic stress have shown impairments to memory and executive functions, and these also appear to be long-term deficits reported in children and adults who have experienced early life stress [20,49].

Studies investigating the neurocognitive consequences of chemotherapy during childhood (see Figure 1) have also found impairments to working memory and executive function [20,67–69], as well as processing speed, task efficiency, attention, memory and learning [70–72]. An important association has been made between these functions and self-regulatory behaviour, such as effective coping [73]. As noted by Campbell et al. (2007), the consequences of these stress-induced impairments are present not only in school settings, but in many other domains of life such as social relationships, emotional control, coping skills, the workplace and overall quality of life [70]. In line with this, Krull et al. (2013) found that over 10 years post-treatment, 28–59% of childhood cancer survivors reported neurocognitive and neuropsychological impairments, the greatest being reduced attention and executive functions, which were most closely associated with treatment with dexamethasone, a synthetic glucocorticoid, when excluding those who received cranial radiation therapy [35]. They further found that survivors reported an approximately 5% annual increase in self-reported behavioural issues related to self-regulation, which impacted functioning in academic and occupational settings [35].

Additional risk to effective cognitive-affective stress appraisal may be conferred by lower engagement in stress buffering behaviours such as socialisation and physical activity that promote adaptive coping [74–76]. Indeed, in a recent study of childhood Acute Lymphoblastic Leukemia (ALL) survivors, correlations were found between low levels of physical activity and inattention [77]. Further, disruption to corticolimbic brain regions central to these interpretive processes has been reported in several childhood cancer populations and is thought to be a result of cytotoxic treatments including chemotherapies that are neurotoxic and/or cross the blood brain barrier.

4. Neurological Triggering

During threat appraisal, corticotrophin-releasing hormone (CRH) is released from the hypothalamus and activates the hypothalamic-pituitary-adrenal (HPA) axis, while norepinephrine from the Locus-Coeruleus activates the autonomic nervous system (ANS), and ANS input directly activates target organs, and the sympatho-adreno-medullary (SAM) pathway [78]. The glucocorticoid cascade hypothesis posits that significant or enduring stress exposure will result in excessive cortisol exposure leading to altered hippocampal functional control over HPA scaling and termination, which in turn leads to further cortisol exposure and propagation of dysregulated stress signalling, ultimately leading to adverse health and functional effects [79]. Critically, the biomediators released by neurological triggering (cortisol and catecholamines) feed back to the brain, influencing both the cognitive-affective experience of stress, as well as continued neurological triggering. Cortisol feedback in particular is thought to be critical to tuning and terminating the activation of the HPA axis, by influencing neurotransmission of the hippocampus. Cortisol-mediated neuroplastic and neurotransmission changes to the hippocampus are commonly thought to be integral to the development of stress dysregulation. This may be of particular interest to populations treated with pharmacological glucocorticoids, such as prednisone and dexamethasone, both of which have CNS penetrance and are used for CNS prophylaxis for pediatric leukemia patients [23].

The protracted nature of brain development is such that subcortical and limbic structures (amygdala, hippocampus, brain stem) development precedes prefrontal and frontal cortex development [63,80,81]. Even though most adult hippocampal networks are apparent during childhood, their connections to lateral lobes increase throughout childhood (ages 4–10) [81] and PFC development is not fully developed until early adulthood [82].

Significant evidence in both human and animal models have found changes to neurological structures responsible for cognitive-affective processing and neurological triggering to be affected by exposure to chronic stress or early life stress [18,49,83–88]. Chronic stress in animal models has shown reduced hippocampal volume [83], reduced neurogenesis in the hippocampus [84–86], PFC atrophy [49] and amygdalar hypertrophy [49], concomitant with altered cognitive functions associated with those regions [49,83–86]. Human studies have found similar changes in altered hippocampal structure and function [18,87], PFC impairment [18,88] and other neurological impairments [88] in populations experiencing early life stress. Some inconsistencies exist in neurological studies of children, which have been attributed to the protracted nature of neurological development, such that the full extent of impact of early life stress on regional brain structure and function does not become apparent until the third decade of life [12,81,89]. It follows that the nature of brain changes is also sensitive developmentally, such that the regions that are in development when the stress occurs are most likely to experience long-term dysfunction [61,90].

Disruption to corticolimbic brain regions critical to stress regulation have been reported in childhood cancer patients treated with chemotherapy and/or cranial radiation. It is important to note that most studies examining the relationship between chemotherapy and brain structure and function changes have been performed mostly using cross-sectional study designs, as detailed in Table 1. Many of the neurological consequences associated with chemotherapy, including lower white matter volume [22,91–93], altered hippocampal microstructure [84,86,94] and altered PFC microstructure [73], are likely to impact neurological triggering and feedback effects on cognitive-affective appraisal. Both human [67,94] and animal [84–86] studies have reported lower hippocampal volumes and impaired neurogenesis related to various chemotherapeutic agents, similar to findings in other early life stress studies [18] (see Table 1). Amygdala changes have also been reported in adult cancer populations [26,94,95] and have been related to adverse psychological effects [26,95], and recently, reduced amygdala and dorsal striatum brain matter volume has been found in pediatric cancer populations [96]. Chemotherapy-treated survivors of childhood cancer displayed lower cerebellar volumes, versus healthy controls, which was associated with both poorer performance on neurocognitive testing and exposure to dexamethasone [97]. Despite some evidence of altered limbic and subcortical structure and function following chemotherapy, the majority of studies report differences in frontal and pre-frontal brain structures, as well as impairments in the function of these structures [20,21,35,67,70,98–100]. This may be due to the importance of these regions in global intellect and other neurocognitive deficits that have taken priority in this research space. The functional implications of these brain changes have not been studied in the context of stress, however, adrenal insufficiency and HPA dysfunction has been reported in children receiving cranial radiation suggesting that disruption to corticolimbic brain regions can have downstream effects on physiological stress signalling.

Table 1. Stress regulation consequences of cancer treatment on paediatric populations.

Author, Year	Purpose/Aim	Design/Methods	Participants/Sample	Findings	Limitations
Neurocognitive outcomes and cognitive-affective appraisal					
Brinkman (2012) [20]	To investigate the relationship between white matter and cognitive processes in adult survivors of childhood medulloblastoma.	Cross-sectional study Participants underwent neurocognitive testing and MRI (diffusion-tensor imaging)	n = 20 Participants were survivors of MB treated at St. Jude's	Neurocognitive impairment was common in many domains of function. Reduced white matter integrity was associated with worse performance on tasks of executive function.	Lack of control group mall sample size Patients were treated with outdated methods for MB treatment
Brown (1996) [21]	To investigate the intellectual and academic functioning of children with leukemia treated with intrathecal chemotherapy when compared to cancer patients not treated with CNS prophylaxis.	Prospective cohort study Intellectual assessments and academic achievement tests conducted at baseline and at years 1, 2 and 3 following diagnosis.	n = 38 children with leukemia (receiving CNS prophylactic chemotherapy) n = 25 children with other cancers (not treated with CNS prophylaxis) Participants were diagnosed at the Women's and Children's Hospital	3 years after diagnosis, the CNS treated children performed worse on academic tests of reading, spelling and arithmetic.	Lack of healthy control group Small sample size and high attrition rate
Davidson (2000) [22]	To investigate neurological changes in cancer patients treated with high-dose MTX compared to healthy controls.	Cross-sectional study Cognitive assessment and proton magnetic resonance spectroscopy/MRI in cancer patients and healthy controls.	n = 11 children with ALL, non-Hodgkin lymphoma, or osteosarcoma, undergoing MTX treatment. n = 17 healthy controls	The choline/water ratio was lower in patients than controls. Abnormal white matter was observed in 3 cancer patients (and potentially a 4th).	Due to the small sample size, any differences in metabolites would need to be large to be detected

Table 1. *Cont.*

Author, Year	Purpose/Aim	Design/Methods	Participants/Sample	Findings	Limitations
Kadan-Lottick (2009) [23]	To evaluate differences in neurocognitive and academic performance in children with ALL treated with DX versus prednisone.	Cross-sectional study Patients were previously randomised to corticosteroid treatment. Patients underwent a half-day neurocognitive assessment.	n = 51 DX-treated patients n = 41 prednisone-treated patients Participants were previously enrolled in Children's Cancer Group 1991 trial.	The only significant difference between groups was on a test of reading (DX-treated scored worse)	Lack of control group
Matsuoka (2003) [26]	To assess structural differences in the amygdala of cancer survivors with/without intrusive recollections.	Cross-sectional study MRI and volumetric analysis of the amygdala was performed in both groups.	n = 35 breast cancer survivors with intrusive recollections n = 41 breast cancer survivors with no intrusive recollections Participants recruited from the outpatient clinic of the Division of Breast Surgery, National Cancer Center Hospital East	Amygdala volume was lower in participants with intrusive recollections even after controlling for age, height and depression diagnoses.	Lack of healthy control group Risk of recall bias Early life stressors were not evaluated
Krull (2013) [35]	To investigate predictors, patterns and rates of neurocognitive impairment in adult survivors of childhood ALL decades after treatment.	Cross-sectional study Participants underwent neurocognitive testing and completed a self-rating questionnaire.	n = 567 Participants were ALL survivors from the St. Jude lifetime cohort study.	Participants treated with chemotherapy exhibited impairment across all neurocognitive domains. Risk for executive function problems increased with survival time (cranial radiation therapy dose-dependent).	Lack of control group Did not adjust for SES Dose variability now may not reflect that of decades ago
Reddick (2014) [67]	To prospectively validate reduced white matter volume, its influential factors and neurocognitive impairments in childhood cancer survivors	Cross-sectional study MRI, volumetrics and neurocognitive testing on participants.	n = 383 childhood cancer survivors (199 ALL, 184 brain tumor) n = 67 healthy siblings	Brain tumor survivors had lower white matter volume than ALL survivors, who were lower than controls, this was associated with treatment parameters. Childhood cancer survivors performed worse than controls neurocognitive tests.	Limited area in which white matter volume was quantified, was used to assess total tissue volume for a specific anatomical region Cross-sectional design limits temporal data
Wolfe (2013) [68]	To assess the relationship between cardiorespiratory fitness and executive functioning in radiation-treated pediatric cancer survivors.	Cross-sectional study Participants underwent fMRI imaging while completing an n-back test and cardiorespiratory fitness testing on a cycle ergometer.	n = 9 childhood posterior fossa tumor survivors Participants were recruited from the neuro-oncology clinic at the Children's Hospital of Alabama	Higher cardiorespiratory fitness was associated with increased working memory and efficiency of neural functioning in pediatric cancer survivors.	Small sample size and lack of controls: limited power and generalisability Correlational data limits the inference of causality Cerebellar activation was not assessed due to heterogeneity of lesions in participants
Stefancin (2020) [69]	To explore the association between chemotherapy and working memory function in childhood cancer survivors and healthy controls.	Cross-sectional study fMRI was acquired while participants performed a visual n-back test.	n = 15 pediatric cancer survivors, patients at the Stony Brook Children's Hospital n = 15 healthy controls	Working memory impairment was present in pediatric cancer survivors when compared to controls. In survivors, correct responses generated a decreased BOLD response in the posterior cingulate, incorrect responses generated a greater BOLD response in the angular gyrus, and no response generate a greater BOLD response in the superior parietal lobule.	Cross-sectional design Increased BOLD signal could either indicate increased activation, or decreased neuronal efficiency Heterogeneity of sample in terms of cancer type and treatment

Table 1. *Cont.*

Author, Year	Purpose/Aim	Design/Methods	Participants/Sample	Findings	Limitations
van der Plas (2021) [71]	To investigate the prevalence of neurocognitive impairments in survivors of childhood ALL and if age at diagnosis, chemotherapy, and chronic conditions correlate with risk of impairment.	Cross-sectional study Participants completed the Childhood Cancer Survivor Study Neurocognitive Questionnaire. Neurocognitive impairment associations with treatment exposures and chronic conditions were examined.	*n* = 1207 survivors of ALL *n* = 2273 siblings Participants were enrolled in the Childhood Cancer Survivor Study	ALL survivors reported increased impairments in memory and task efficiency when compared to healthy controls. In male survivors, impairments in memory were associated with increased dosage of MTX and DX exposure, while impairments in task efficiency were associated with neurologic and pulmonary conditions. In female survivors, endocrine conditions were associated with higher risk of impairments in memory and task efficiency.	Self-report of cognitive impairment Risk of participation bias Many other factors that were not studied might be associated with neurocognitive impairment
Williams (2020) [72]	To examine if childhood cancer survivors with injuries to the brain are at a higher risk for chronic health conditions and if this is associated with neurocognitive impairment later in life.	Cross-sectional study All participants completed neurocognitive testing and a clinical examination.	*n* = 2859 adult survivors of childhood cancer: 1598 had CNS therapy Participants were treated at St Jude's and enrolled in the St Jude Lifetime Cohort Study	Participants that were CNS-treated performed worse than those that were not CNS-treated on neurocognitive testing and had more global neurocognitive impairments. There was a dose-dependent association between severity/burden of treatment and global impairment in CNS-treated participants. Chronic health conditions such as cardiovascular and pulmonary conditions were associated with impairments in memory, processing speed and attention in CNS-treated participants with neurological conditions.	Lack of healthy control group Lack of neurocognitive data immediately following treatment in survivors
Lesnik (1998) [73]	To assess frontal-cerebellar morphological characteristics and function in survivors of childhood ALL that were treated with intrathecal MTX, while using an effect size model to increase validity in a small sample.	Cross-sectional study Neuropsychological testing and MRI of cerebellar lobuli (I-V and VI-VII) and prefrontal cortices was assessed in participants.	*n* = 10 childhood survivors of ALL *n* = 10 age, sex, and socioeconomic status matched healthy controls	There were deficits in neuropsychological testing and morphometric and functional characteristics of cerebellar lobuli and prefrontal cortices in MTX-treated childhood ALL survivors. Evidence supported the involvement of the hypothesised subsystem; the cerebellar-frontal system.	Correlational data limits the inference of causality Small sample size
Peng (2021) [77]	To investigate behavioural and neurocognitive functioning in survivors of childhood ALL and evaluate the associated clinical and socio-environmental factors.	Prospective, cross-sectional Participants completed neurocognitive testing and self-reported emotional, behavioural and socio-environmental variables via questionnaires and checklists. Chronic health conditions and clinical variables were pulled from patient charts.	152 survivors of childhood ALL: 32 received cranial radiation therapy 120 received chemotherapy Participants were patients at the Long-term Follow-up Clinic of the Prince of Wales Hospital	Lower levels of self-reported physical activity were correlated with inattention and sluggish cognitive tempo. A minority of survivors had impairments in motor-processing, and attention, and developed treatment related chronic conditions.	Small sample size of cranial radiation therapy group may have eliminated differences in neurocognition between groups Risk of sampling bias Lack of healthy control group Self-report data Incomplete patient records

Table 1. *Cont.*

Author, Year	Purpose/Aim	Design/Methods	Participants/Sample	Findings	Limitations
Harila-Saari (1998) [91]	To evaluate changes in MRI scans of the brain of childhood-treated ALL survivors and correlate the observed abnormalities with neuropsychological impairments.	Prospective cohort study MRI immediately after the cessation of treatment and 5 years later, as well as neuropsychological testing (various tests) was conducted.	$n = 32$ Participants were childhood survivors of ALL: 15 chemotherapy-treated 17 combined chemotherapy and cranial radiation-treated Participants were patients at the Department of Pediatrics at the University of Oulu	Abnormalities in MRI were heterogeneous and infrequent among participants and did not correlate with neuropsychological function. Most participants did have neuropsychological impairments, however.	Small sample size, limited statistical power
Iuvone (2002) [92]	To investigate correlations between cognitive measures and abnormalities in MRI and computerized tomography scans of childhood ALL survivors.	Prospective cohort study Cognitive testing (various tests) and prospective MRI and computerized tomography imaging were conducted once a year for 4 years.	$n = 21$ Participants were children with ALL who received CNS prophylaxis (cranial irradiation and intrathecal MTX). Participants were patients at the Division of Pediatric Oncology, Catholic University	Abnormalities in white matter were associated with poor performance on a task of visual motor integration in approximately half of the participants. Intracerebral calcifications were correlated with MTX doses, and impaired cognitive testing. Females were more vulnerable to the treatment effects.	Small sample size
Monje (2013) [94]	To explore the correlates of dysfunctional episodic memory in CNS prophylaxis-treated survivors of childhood ALL.	Cross-sectional study Participants episodically encoded visual scenes and underwent fMRI while completing a memory paradigm.	$n = 10$ CNS prophylaxis and chemotherapy-treated adult survivors of childhood ALL, patients at the Dana Farber Cancer Institute $n = 10$ age matched controls	Survivors of childhood ALL demonstrated altered BOLD signal and atrophy in the hippocampus, and poor recognition memory when compared to controls. Unsuccessful encoding in ALL survivors showed increased hippocampal BOLD signal. Differences in memory among ALL survivors was related to the magnitude of BOLD response in areas responsible for successful encoding.	Small sample size
Spitzhüttl (2021) [96]	To investigate gray and white matter volume in childhood cancer survivors and the relationship to cognitive processes.	Cross-sectional study MRI T1 weighted images were acquired for voxel-based morphometry and cognitive and fine motor coordination assessments were completed by all participants.	$n = 43$ childhood cancer survivors (non-CNS cancer), treated at the University Children's Hospital Bern or the University Children's Hospital Zurich $n = 43$ healthy controls	Amygdala and dorsal striatum white and gray matter volume were lower in cancer survivors. Fine motor coordination of the right hand and executive function was poorer in survivors, although still within the normal range.	Cross-sectional design, risk of cohort effects Correlations were performed for each ROI and variable separately
Phillips (2020) [97]	To examine the association between glucocorticoid and MTX treatment and disruptions to the cerebello-thalamo-cortical network and antioxidant system in the brain of survivors of childhood ALL.	Cross-sectional study Brain volumes, neurocognitive testing, functional and effective connectivity, and the association between MTX and DX treatment and neurocognitive outcomes were assessed in childhood ALL survivors and healthy controls.	$n = 176$ childhood ALL survivors, recruited from the St Jude Children's Research Hospital Total Therapy Study XV $n = 82$ age and SES matched healthy controls from the community	Survivors had decreased cerebellar volumes compared to controls, which was associated with DX exposure. In females, effective connectivity disruption was associated with poorer executive function.	Controls did not complete neurocognitive testing Biomarkers were not available to assess oxidative injury pre-treatment associated with genetics or disease Risk of confounding effects of cytarabine on brain volume Not a representative population of all ALL chemotherapy treated patients

Table 1. *Cont.*

Author, Year	Purpose/Aim	Design/Methods	Participants/Sample	Findings	Limitations
Carey (2008) [98]	To evaluate differences in white and gray matter between ALL survivors and healthy controls.	Cross-sectional study T1 weighted MRI images, and subsequent voxel based morphometry, and neuropsychological evaluations were acquired from participants.	n = 9 long term ALL survivors treated with chemotherapy, patients at the University of Arizona Pediatric Hematology/Oncology Late Effects Clinic n = 14 healthy controls	ALL survivors had reduced white matter in the right frontal lobes and performed worse on tests of math, attention, visual-construction skills and mental flexibility when compared to controls. Neurocognitive impairments were associated with regional decreases in white matter volume.	Small sample size Risk of confounding factors
Reddick (2006) [100]	To assess differences in neurocognitive functioning and its relationship with white matter volume in survivors of childhood ALL when compared to healthy controls.	Cross-sectional study MRI imaging, and subsequent voxel based morphometry, as well as neurocognitive tests of academics, intelligence and attention, were performed on ALL survivors and controls.	n = 112 ALL survivors n = 33 healthy siblings	Survivors of ALL performed significantly worse on tests of attention and had decreased white matter volume when compared to controls. Decreased white-matter volume was associated with impaired academics, intelligence and attention.	Limited area in which white matter volume was quantified was used to assess total tissue volume for a specific anatomical region Cross-sectional design limits temporal data
Stefanski (2020) [101]	To examine neurocognitive and psychosocial outcomes in adult survivors of childhood leukemia that were treated with bone marrow transplantation or intensive chemotherapy.	Cross-sectional study Participants completed questionnaires on emotional distress, neurocognitive problems, social attainment and health-related quality of life.	n = 482 adult survivors of AML: 183 bone marrow transplantation-treated 299 intensive chemotherapy-treated n = 3190 siblings Participants were enrolled in the Childhood Cancer Survivor Study	Survivors had greater impairments in health-related quality of life, emotional distress and neurocognitive functioning than siblings. Survivors had greater risk for unemployment, lower education and income, and not having a partner	Lack of differences in treatment groups may have been due to sample sizes and limited power Risk of participation bias Self-report Siblings may not be representative of the general population
Biological/cellular aging and inflammatory outcomes					
Kennedy (2004) [35]	To assess the effects of ALL treatment in children on antioxidant status and the association between antioxidant stress, oxidative stress and complications.	Prospective cohort study At baseline (diagnosis), 3 months and 6 months, antioxidant plasma concentrations, total antioxidant capacity and DNA oxidised base 8-oxodeoxyguanosine were assessed.	n = 103 newly diagnosed children and adolescents being treated for ALL	Plasma vitamin A, antioxidants, total antioxidant capacity and DNA oxidised base 8-oxodeoxyguanosinconcentrations changed over 6 months. Beneficial associations were found between higher concentrations and various treatment dose parameters. Adverse relationships were also found.	Criteria for deficiency states may be limited, i.e., children with leukemia might have higher requirements
Mazur (2004) [32]	To evaluate serum levels of cytokines in children after treatment for ALL was finished.	Cross-sectional study Serum concentrations of cytokines measured using an enzyme linked immunosorbent assay.	n = 30 healthy controls 5 groups of 30 ALL patients: 1, 3-, 6-, 9- and 12-months post-treatment (n = 150 total), treated at the Department of Pediatric Hematology and Chemotherapy, Zabrze	There were significant differences in interleukin-8, tumor necrosis factor-alpha and interleukin-2 serum concentrations between ALL patients and healthy controls.	Cross-sectional design limits the inference of causality

Table 1. *Cont.*

Author, Year	Purpose/Aim	Design/Methods	Participants/Sample	Findings	Limitations
Papageorgiou (2005) [102]	To compare TAC and corrected TAC between cancer free children and children with malignancy at the time of diagnosis and during chemotherapy.	Cross-sectional study All children were under a free diet during the study. TAC and corrected TAC levels were evaluated from blood samples.	*n* = 20 children with malignancy, recruited from the University Hospital of Heraklion *n* = 80 control participants	TAC and corrected TAC decrease progressively during cycles of chemotherapy in children with malignancy.	Small sample size Different treatment regimes used among patients Variability in patient diet and other potential confounding variables
Hasan (2020) [103]	To assess differences in serum TOS, TAC and the OSI of ALL and AML patients compared to healthy controls.	Cross-sectional study Erel's methods were utilised to assess TOS and TAC, and OSI was calculated in leukemia patients and controls.	*n* = 60 leukemia patients, patients at the Hereditary Hematology Center *n* = 70 age and gender matched healthy controls	TOS and OSI were significantly higher in leukemia patients when compared to controls, and antioxidant levels were significantly lower. Oxidative stress was present in both ALL and AML.	
Vatanen (2017) [104]	To analyse the prevalence of frailty and physical health limitations among long-term survivors of high-risk neuroblastoma and to investigate whether frail health is associated with markers of inflammation and telomere length.	Cross-sectional study Frailty is defined as 3 or more of the following: low lean muscle mass, low energy expenditure, slowness, weakness, exhaustion. Cardiovascular function and telomere length analysis were also performed.	*n* = 19 cancer survivors *n* = 20 healthy controls	Prevalence of frailty was significantly higher in survivors versus controls (47% vs. 0%). 68% or survivors reported limitations in vigorous activity versus 0% of controls. Survivors had significantly shorter telomeres and significantly higher CRP levels.	Small sample size Definition of frailty not wholly comprehensive
Song (2020) [105]	To analyse and compare leukocyte telomere length and age-related attrition between childhood cancer survivors and non-cancer controls. Leukocyte telomere lengths were also analysed for association with treatment exposures, chronic health conditions and health behaviours among survivors.	Retrospective cohort study with prospective clinical follow-up. Leukocyte telomere length was measured using whole genome sequencing. Common non-neoplastic health conditions and subsequent malignant neoplasms were clinically assessed.	*n* = 2427 childhood cancer survivors, recruited from St. Jude's Children's Hospital *n* = 293 non-cancer controls	Leukocyte telomere length was significantly shorter in childhood cancer survivors compared to non-cancer controls. Shorter leukocyte telomere length was correlated with specific treatments including chest and abdominal irradiation, glucocorticoid and vincristine chemotherapies.	Risk of confounding bias Correlational data
Qin (2021) [106]	To evaluate EAA and its association to chronic health conditions, health behaviour and treatment exposures in survivors of childhood cancer.	Cross-sectional study Methylation data was generated from cancer survivors and controls. EAA was calculated as residuals from a linear regression of epigenetic age and chronological age. EAA adjusted least square mean was compared across health behaviours and treatment exposures. The associations between EAA and 20 different chronic health conditions was assessed.	*n* = 2139 childhood cancer survivors *n* = 282 frequency matched controls Participants were enrolled in the St Jude Lifetime Cohort Study	EAA was greater in childhood cancer survivors than in controls. Among survivors, higher EAA was observed in patients that were previously various cancer treatments. Associations between several chronic health conditions hypertension, myocardial infarction, obstructive pulmonary deficit, peripheral motor and sensory neuropathy, and pulmonary diffusion deficits and EAA were observed.	Limited power due to few participants having specific chronic health conditions and non-matching controls Only treatments within 5 years of diagnosis were considered Results cannot be generalised, as all participants were of European descent Temporal associations of health behaviours and EAA were unavailable

Table 1. *Cont.*

Author, Year	Purpose/Aim	Design/Methods	Participants/Sample	Findings	Limitations
Ravera (2021) [107]	To investigate the molecular and metabolic markers of early aging in survivors of childhood cancer.	Cross-sectional study Mononuclear cells were isolated from the blood of childhood cancer survivors and healthy controls, and assessed using biochemical, proteomic and molecular biology analyses.	n = 196 childhood cancer survivors n = 154 healthy controls	Survivors had an increased biological age by decades compared to their chronological age. Survivors had inefficient oxidative phosphorylation which was associated with decreased energy and the switch to lactate fermentation, increased lipid peroxidation and decreased expression of genes/proteins involved in metabolism and mitochondrial biogenesis.	Cross-sectional study design
Hayek (2020) [108]	To investigate the prevalence of frailty in survivors of childhood cancer, and its association with cancer treatment and other factors.	Retrospective cohort study Participants completed a baseline and follow-up questionnaires. A generalised linear model evaluated associations between frailty, treatment and other variables.	n = 10,899 childhood cancer survivors n = 2097 siblings Participants were enrolled in the Childhood Cancer Survivor Study	Survivors had increased frailty compared to siblings. Radiation treatment and lung surgery were associated with increased risk of frailty for survivors.	Lack of participation from all eligible subjects may inflate/deflate prevalence estimates Risk of recall bias Risk of survival bias

Abbreviations: MRI, magnetic resonance imaging; MB, medulloblastoma; CNS, central nervous system; ALL, acute lymphoblastic leukemia; MTX, methotrexate; DX, dexamethasone; AML, acute myelogenous leukemia; fMRI, functional magnetic resonance imaging; BOLD, blood-oxygen-level-dependent; DNA, Deoxyribonucleic Acid; TAC, total antioxidant capacity; TOS, total oxidant status; OSI, oxidative stress index; CRP, c-reactive protein; EAA, epigenetic age acceleration. Table 1 inclusion criteria: A scoping review of the literature was completed examining: (1). Primary research articles examining structural and functional neurological impairments in pediatric cancer patients; (2). Primary research articles examining inflammatory consequences and biological or cellular aging in pediatric cancer patients; (3). Pediatric cancer patients or survivors of childhood cancer (pediatric and adult).

Whether these neurological changes are a direct cause of anti-neoplastic treatment (i.e., glucocorticoids, intrathecal methotrexate, cranial radiation) or endogenous stress processes, all can disrupt effective stress regulation and therefore, may be relevant to supporting long-term health and function in childhood cancer patients and survivors.

5. Physiological Stress Response

The signalling response evoked by neurological triggering is comprised of three distinct but overlapping systems. Brainstem activation of the ANS occurs almost immediately via neural inputs to visceral target organs activated for allostasis as well as through adrenal stimulation of catecholamines into circulation (SAM), while the hypothalamus activates the HPA axis via stimulation of the pituitary to release ACTH. These two arms of allostatic control, SAM signalling via catecholamines and HPA signalling via glucocorticoids, are responsible for the physiological and affective experience of stress.

Secondary signalling by inflammatory factors is also implicated in short and long-term effects and experiences of stress. The biomediators of inflammation, cytokines, can be produced by neutrophil demargination and activation of immune cells [109]. In the short term, this primes the immune system to protect the body from impending injury but in the long-term can contribute to reduced immune function, worse tissue healing and chronic inflammation and associated physiological and psychological disturbance [79,110].

Basal HPA and HPA-reactivity increases with age, with a marked increase around puberty (—13 for girls, 15 for boys) [64,111,112]. This is likely due to the effects of changing environmental demands, developing neurological structures that enable relevant neurocognitive and neuropsychological functions (i.e., goal-oriented behaviour is governed by the PFC and is thought to be an essential component of psychological stress provocation), and hormonal changes associated with puberty [88,111,113].Stress response patterns can be used to infer how the whole system is functioning and whether an individual appears to have a resilient or vulnerable stress phenotype. These acute patterns are thought to be demonstrative of longitudinal stress regulation and prospective allostatic load and associated consequences for health and wellbeing. In a resilient and optimised system, the response pattern should show an immediate increase with stressor onset followed by a rapid termination upon resolution or cessation of stress exposure. Unnecessary activation of stress responses overexposes corticolimbic brain regions to the neuroplastic and neurotransmission effects of biomediators. Children who have experienced significant adversity show divergence in their stress reactivity patterns [3]. Some have higher basal activation and hypoactivity to acute stress exposure in one or multiple signalling systems, while others show hyperactivity and impaired termination. Elevated inflammation is also a common finding that is a purported mechanism of long-term adverse health and functional outcomes in chronic stress populations [32], the degree of which is different to that of an inflammatory response to infection. Acute stress reactivity to psychological stimuli has been used in a variety of pediatric populations to predict stress-related risks for health and wellbeing [64,111,114–118] and it is generally considered that any deviation (hypoactivation or hyperactivation), is likely indicative of dysfunctional signalling and regulation [16,58].

Few studies have examined stress reactivity patterns in childhood illness populations or considered the role of stress regulation in long-term health and function, despite the significant psychological distress in addition to the direct physiological effects of medical treatments. This is even more surprising considering that many childhood illnesses require the use of synthetic glucocorticoids for treatment. Thus, little evidence for the chronic effects of stress exposure exists in childhood cancer, although some studies have reported that elevated inflammation and oxidative stress persists post treatment [31,102]. Kennedy (2005) found total antioxidant capacity to be lower in ALL patients 6 months post-treatment, and this was correlated with better clinical outcomes including lower rate of infection and hospitalisation, higher quality of life and better treatment tolerance [31]. Similarly, Mazur et al. (2004) found elevated circulating cytokines, TNFα, IL-2 and IL-8, in ALL patients 3, 6 and 12-months post-treatment [32]. Importantly, studies of adult patients have shown a relationship between cytokine status and neurocognitive function [35] providing support for a link between physiological activation and cognitive-affective capacities. However, no research has explicitly investigated physiological stress response profiles, which might provide clues into mechanisms of adverse long-term and late effects in this population. As altered acute stress profiles and biochemical signalling is considered the mechanism linking stressful experiences to long-term health and dysfunction in other populations experiencing significant stress, this is an important area for further research in childhood cancer patient and survivor populations that may provide insight into late effects of childhood cancer [119].

6. Target Organ Activation

The culmination of stress responses is in allostatic processes preparing the body for a threat and defending homeostasis in the face of changing environmental conditions. However, what was an adaptive response for physical stressors evolutionarily may not be effective for psychological stressors of contemporary lives. The inappropriate activation of stress systems is thought to contribute to a wide range of illnesses reflecting allostatic load [55].

Allostatic load, the aggregation of these subclinical issues or progression of a subclinical issue into a clinical issue over time, is associated with significant risk of morbidity and mortality [3,120]. Children who experience significant life stress have higher risk of heart disease, diabetes, cancer, chronic lung disease, skeletal fractures, liver disease, mental distress disability and overall worse health ratings [17,121]. At the tissue level, evidence of oxidative stress including cellular aging [122–124] and shortened telomeres [125,126] indicate systemic tissue disruption by early life stress.

The consequences of early life stress and chronic stress is further demonstrated by the higher rates of morbidity, as cellular vulnerabilities result in dysfunction, ultimately increasing risk of all-cause mortality in these populations [5]. These same issues are common side-effects observed in children's cancer treatment, especially those treated with synthetic glucocorticoids [28,60,103].

In line with other populations of children who have experienced significant stress, childhood cancer survivors have also been reported to have shortened telomeres, and this has been associated with higher chronic inflammation [104] and higher incidence of late effects during survivorship [105]. Pediatric cancer survivors differ from age-related controls in terms of activation of the adaptive immune system, chronic, low-grade inflammation, as well as immune tolerance resulting from the synthesis of immunomodulators via the tryptophan-kynurenine metabolic pathway [127]. These changes resemble an aging phenotype observed in older populations [128] and are indicative of allostatic load [127]. Some research shows that pediatric cancer survivors have increased biological age relative to their chronological age, as indicated by shortened telomeres [104,105], epigenetic age acceleration [106] and biochemical and molecular markers such as inefficient oxidative phosphorylation, increased lipid peroxidation and decreased expression of metabolic proteins and those involved in mitochondrial biogenesis [107]. Childhood cancer survivors also report an increased incidence of premature frailty associated with radiation treatment [108]. The cytotoxic nature of childhood cancer treatments can cause significant damage and disruption to developing organ systems, which may present more inherent vulnerability to stress exacerbation than the general population. Childhood cancer patients have a higher risk of many chronic illnesses including cardiovascular disease, secondary cancer, metabolic conditions, depression and anxiety, as well as subclinical but lower reported health related quality of life [24,28,60,101]. Dysrhythmias and other indicators of cardiac dysfunction and conductive symptoms are reported in adult survivors of childhood cancers [59], which is most commonly attributed to anthracycline exposure used for anti-neoplastic treatment [129], or radiation to a field that involves the heart. In a population of children and youth who survived mixed types of cancer, 28.2% were reported to exhibit hypertensive or pre-hypertensive signs [130]. Similarly, Cardous Ubbink (2010) found increased risk of hypertension in adult survivors of childhood cancer, related to BMI, cyclophosphamide, cisplatin or abdominal radiation [33]. In a population of adult survivors of childhood cancer, stress and distress were associated with adverse cardiovascular health conditions such as hypertension, dysrhythmia, dyslipidemia and metabolic syndrome [131]. Even though stress reactivity and function of the systems involved in stress responses have not been investigated in childhood cancer patients or survivors, many of the conditions experienced by this population over their lifetime are those that can be developed or worsened by dysfunctional stress regulation [3,28,60].

It is not possible to differentiate between the contribution of stress dysregulation of target organs over and above that caused by direct cytotoxicity of anti-neoplastic treatment. However, it is still relevant to reducing the burden of illness, as stress dysregulation effects on tissue integrity and function are self-propagating and many adverse health and functional outcomes worsen with time since treatment in childhood cancer survivors. Thus, it is possible that these work in concert to contribute to adverse late-effects, with direct effects of treatment producing vulnerabilities and initial weakness, and stress and other long-term pathways contributing to worsening of function with time. For example, direct disruption to corticolimbic brain regions mediating upstream stress regulation can be further exacerbated by altered stress signalling and may be important targets for intervention to promote better health and function during survivorship.

7. Considerations for Interventions

Each step in the stress regulatory pathway can alter the function of the system acutely and over time influence the potential of stress to contribute to adverse health and dysfunction. Potential intervention options to reduce the burden of childhood illness centre around stress-buffering behaviours and may include encouraging social connection [132,133], pro-social behaviour [133] and physical activity [75,134,135], while teaching effective coping strategies. Mutable individual factors related to disposition and behaviour can also have a significant impact on acute physiological activation to a stress stimulus and the effects of stress on long-term health and function. The mutable factors, such as intrapsychic coping strategies, pro-social behaviour and physical activity habits, should be considered clinically meaningful as it relates to any contribution of stress dysregulation on long-term health and function [136]. Programs promoting social connections and development can be expected to have psychosocial and psychobiological benefits through positive changes to cognitive-affective processing and neurological triggering. Expectations of negative social judgement reliably provoke physiological stress responses [65,66]. Strong social connections may reduce negative expectations during cognitive-affective appraisal, reducing physiological activation of stress systems [132]. Further, social support has a strong influence over acute stress reactivity due to the release of oxytocin in the brain, inhibiting CRH production, thereby reducing neurological triggering of the HPA axis and subsequent physiological and affective experiences of a given stressor. Outcomes of social programs in clinical populations rarely focus on clinical indicators, however, positive effects of social support are reported improve stress management [132,133,137].

Several studies have shown that parents of children with cancer have a higher incidence of post-traumatic stress disorder and related symptoms when compared to parents of healthy children [138–140], and that these symptoms were associated between the parents and their children [139]. This suggests that the parents' stress of having a child with cancer may have consequences for the children themselves. Parental stress has been found to be a significant predictor of functional impairment in childhood cancer survivors [141], and childhood cancer survivors may experience different parenting styles including parental overprotection due to stress [142,143]. However, these findings are not conclusive as some research has suggested parenting styles are not different from children without a history of serious illness [144]. These findings indicate the need for further investigation and potentially psychological interventions in childhood cancer patients and parents alike.

Programs promoting physical activity can be expected to have neurobiological and psychobiological benefits through positive changes to cognitive-affective processing, neurological triggering, as well as stress signalling and impacts on target organs [134,135]. Physical activity and fitness both have adaptive effects on stress reactivity and can influence the stress regulatory pathways at multiple steps. Acutely, activity promotes positive mood, reduces negative affect and alters dopamine-GR signalling [75]. Chronically, physical activity promotes executive function, increases neurogenesis of the hippocampus and reduces inflammation as well as promotes healthy function of many of the organ systems of allostatic responses. Together, the effects of physical activity can be expected to reduce acute activation, promote habituation to future stressors and counteract adverse effects of allostatic load on target organ systems [145,146].

Treatment for childhood cancer must prioritise the eradication of the cancer itself; however, secondary considerations must be given to reducing burden of illness during survivorship and improving quality of life and function. To this end, interventions that support stress buffering behaviours may have a beneficial impact on childhood cancer patient and survivors. While clinicians may already promote these types of support for their patients, understanding that these benefits not only improve experiences and subjective quality of life but are likely to have clinical implications for long-term health and wellbeing is critical to ensuring that they are included in holistic treatment of children's cancer.

8. Clinical Implications and Future Research

Stress regulation is integral to how we navigate dynamic environments in everyday life. Stress, homeostasis and allostasis are concepts that have been developed for decades, with recent attempts to quantify these states being rooted in the study of thermodynamics [147]. The thermodynamic entropy-based stress model proposes that adverse health states are caused by positive stress entropic load, while negative stress entropic load leads to a protective health state, leading to the idea that energy balance may be a crucial intervention for chronic disease [147]. The chronic or inappropriate activation of stress regulatory signaling or target organ activation contributes to wear and tear on critical organ systems and can contribute to adverse health and functional outcomes [3,8]. Even though neurological changes caused by biomediator feedback occur during excessive or prolonged stress signaling, disruption to the systems regulation can occur at any step in the pathway due to self-regulatory nature. There is evidence that childhood cancer patients may have disruption to several steps in the stress-regulatory pathway including cognitive-affective function, neurological disruption to stress regulatory brain regions, altered adrenal and endocrine function, and disrupted tissue integrity, as well as lower engagement in positive coping behaviours such as physical activity and pro-social habits. Childhood cancer patients experience an array of adverse late effects of their cancer that may be brought on by or exacerbated by dysfunctional stress regulation and adversely affect their physical and mental health. Stress regulation may be a valuable lens through which to examine these chronic morbidities in childhood cancer populations. Further research is needed to better understand acute stress reactivity and stress signaling, as well as the connections between different pathways (i.e., cognitive-affective function and stress outcomes). Larger cohort studies may be necessary to accommodate interindividual variability in stress impacts on individuals and the nature of dysfunction. Even though it is not possible to differentiate between direct and indirect effects of cancer treatment during childhood, a better understanding of how neurological, physiological and psychological disruptions during the experience of childhood cancer interact to produce late effects is important. As treatments continue to improve survival rates in this pediatric clinical population, an emphasis on understanding how to improve health and wellbeing during survivorship has emerged. Clinical recognition of stress as a model during treatment, understanding clinical implications of programs supporting positive coping behaviours—psychological, social, physical activity, may thus be timely.

Author Contributions: Conceptualisation, G.E.W. and G.D.W.; methodology, G.E.W. and G.D.W.; investigation, G.E.W., G.D.W., P.C.N. and H.J.; data Curation, G.E.W., J.E.C., K.R., V.M. and G.D.W.; writing—original draft preparation, G.E.W.; writing—review and editing, G.E.W., J.E.C., K.R., P.C.N., S.G.R., H.J. and G.D.W.; supervision, G.D.W.; project administration, G.D.W.; funding acquisition, G.D.W. All authors have read and agreed to the published version of the manuscript.

Funding: This research was funded by The Exercise Medicine Fund at the Hospital for Sick Children.

Conflicts of Interest: The authors declare no conflict of interest.

References

1. Wheaton, B.; Montazer, S. *A Handbook for the Study of Mental Health*; Cambridge University Press: Tennessee, TN, USA, 2012; pp. 171–199. [CrossRef]
2. Selye, H. Stress and disease. *Laryngoscope* **1955**, *65*, 514. [CrossRef]
3. McEwen, B.S. The neurobiology of stress: From serendipity to clinical relevance. *Brain Res.* **2000**, *886*, 172–189. [CrossRef]
4. Del Giudice, M.; Ellis, B.J.; Shirtcliff, E.A. The Adaptive Calibration Model of stress responsivity. *Neurosci. Biobehav. Rev.* **2011**, *35*, 1562–1592. [CrossRef]
5. McEwen, B.S.; Stellar, E. Stress and the individual. Mechanisms leading to disease. *Arch. Intern. Med.* **1993**, *153*, 2093–2101. [CrossRef] [PubMed]
6. De Kloet, E.R.; Joels, M.; Holsboer, F. Stress and the brain: From adaptation to disease. *Nat. Rev. Neurosci.* **2005**, *6*, 463–475. [CrossRef] [PubMed]
7. Cohen, S.; Janicki-Deverts, D.; Doyle, W.J.; Miller, G.E.; Frank, E.; Rabin, B.S.; Turner, R.B. Chronic stress, glucocorticoid receptor resistance, inflammation, and disease risk. *Proc. Natl. Acad. Sci. USA* **2012**, *109*, 5995–5999. [CrossRef]
8. Sapolsky, R.M. Why Stress Is Bad for Your Brain. *Science* **1996**, *273*, 749–750. [CrossRef] [PubMed]
9. Kaiser, R.H.; Clegg, R.; Goer, F.; Pechtel, P.; Beltzer, M.; Vitaliano, G.; Olson, D.P.; Teicher, M.H.; Pizzagalli, D.A. Childhood stress, grown-up brain networks: Corticolimbic correlates of threat-related early life stress and adult stress response. *Psychol. Med.* **2017**, *48*, 1157–1166. [CrossRef]

10. Shonkoff, J.P.; Garner, A.S.; Siegel, B.S.; Dobbins, M.I.; Earls, M.F.; Garner, A.S.; McGuinn, L.; Pascoe, J.; Wood, D.L. The Lifelong Effects of Early Childhood Adversity and Toxic Stress. *Pediatrics* **2011**, *129*, e232–e246. [CrossRef] [PubMed]
11. Romeo, R.D. The Teenage Brain. *Curr. Dir. Psychol. Sci.* **2013**, *22*, 140–145. [CrossRef] [PubMed]
12. Iacono, L.L.; Carola, V. The impact of adolescent stress experiences on neurobiological development. *Semin. Cell Dev. Biol.* **2018**, *77*, 93–103. [CrossRef] [PubMed]
13. Stavrou, S.; Critselis, E.; Darviri, C.; Charmandari, E.; Nicolaides, N.C.; Chrousos, G.P. Paediatric stress: From neuroendocrinology to contemporary disorders. *Eur. J. Clin. Investig.* **2017**, *47*, 262–269. [CrossRef] [PubMed]
14. Everly, G.S.; Lating, J.M. *A Clinical Guide to the Treatment of the Human Stress Response*; Springer: New York, NY, USA, 2012; pp. 53–65. [CrossRef]
15. Danese, A.; Pariante, C.M.; Caspi, A.; Taylor, A.; Poulton, R. Childhood maltreatment predicts adult inflammation in a life-course study. *Proc. Natl. Acad. Sci. USA* **2007**, *104*, 1319–1324. [CrossRef]
16. VanItallie, T.B. Stress: A risk factor for serious illness. *Metabolism* **2002**, *51*, 40–45. [CrossRef] [PubMed]
17. Felitti, V.J.; Anda, R.F.; Nordenberg, D.; Williamson, D.F.; Spitz, A.M.; Edwards, V.; Koss, M.P.; Marks, J.S. Relationship of Childhood Abuse and Household Dysfunction to Many of the Leading Causes of Death in Adults: The Adverse Childhood Experiences (ACE) Study. *Am. J. Prev. Med.* **1998**, *14*, 245–258. [CrossRef]
18. Danese, A.; McEwen, B.S. Adverse childhood experiences, allostasis, allostatic load, and age-related disease. *Physiol. Behav.* **2011**, *106*, 29–39. [CrossRef]
19. Institute of Medicine and National Research Council. *Childhood Cancer Survivorship: Improving Care and Quality of Life*; The National Academies Press: Washington, DC, USA, 2003. [CrossRef]
20. Brinkman, T.M.; Reddick, W.; Luxton, J.; Glass, J.O.; Sabin, N.D.; Srivastava, D.K.; Robison, L.L.; Hudson, M.M.; Krull, K.R. Cerebral white matter integrity and executive function in adult survivors of childhood medulloblastoma. *Neuro Oncol.* **2012**, *14*, iv25–iv36. [CrossRef]
21. Brown, R.T.; Sawyer, M.B.; Antoniou, G.; Toogood, I.; Rice, M.; Thompson, N.; Madan-Swain, A. A 3-Year Follow-Up of the Intellectual and Academic Functioning of Children Receiving Central Nervous System Prophylactic Chemotherapy for Leukemia. *J. Dev. Behav. Pediatr.* **1996**, *17*, 392–398. [CrossRef] [PubMed]
22. Davidson, A.; Payne, G.; Leach, M.O.; McVicar, D.; Britton, J.M.; Watson, M.; Tait, D.M. Proton Magnetic Resonance Spec-troscopy (1H-MRS) of the Brain Following High-dose Methotrexate Treatment for Childhood Cancer. *Med. Pediatr. Oncol.* **2000**, *35*, 28–34. [CrossRef]
23. Kadan-Lottick, N.S.; Brouwers, P.; Breiger, D.; Kaleita, T.; Dziura, J.; Liu, H.; Chen, L.; Nicoletti, M.; Stork, L.; Bostrom, B.; et al. A comparison of neurocognitive functioning in children previously randomized to dexamethasone or prednisone in the treatment of childhood acute lymphoblastic leukemia. *Blood* **2009**, *114*, 1746–1752. [CrossRef]
24. Zebrack, B.J.; Gurney, J.G.; Oeffinger, K.; Whitton, J.; Packer, R.J.; Mertens, A.; Turk, N.; Castleberry, R.; Dreyer, Z.; Robison, L.L.; et al. Psychological Outcomes in Long-Term Survivors of Childhood Brain Cancer: A Report From the Childhood Cancer Survivor Study. *J. Clin. Oncol.* **2004**, *22*, 999–1006. [CrossRef]
25. Kwak, M.; Zebrack, B.J.; Meeske, K.A.; Embry, L.; Aguilar, C.; Block, R.; Hayes-Lattin, B.; Li, Y.; Butler, M.; Cole, S. Trajectories of Psychological Distress in Adolescent and Young Adult Patients With Cancer: A 1-Year Longitudinal Study. *J. Clin. Oncol.* **2013**, *31*, 2160–2166. [CrossRef] [PubMed]
26. Matsuoka, Y.; Yamawaki, S.; Inagaki, M.; Akechi, T.; Uchitomi, Y. A volumetric study of amygdala in cancer survivors with intrusive recollections. *Biol. Psychiatry* **2003**, *54*, 736–743. [CrossRef]
27. Zeltzer, L.K.; Recklitis, C.; Buchbinder, D.; Zebrack, B.; Casillas, J.; Tsao, J.C.; Lu, Q.; Krull, K. Psychological Status in Childhood Cancer Survivors: A Report From the Childhood Cancer Survivor Study. *J. Clin. Oncol.* **2009**, *27*, 2396–2404. [CrossRef]
28. Hudson, M.M.; Ness, K.K.; Gurney, J.G.; Mulrooney, D.A.; Chemaitilly, W.; Krull, K.R.; Green, D.M.; Armstrong, G.T.; Nottage, K.A.; Jones, K.E.; et al. Clinical Ascertainment of Health Outcomes Among Adults Treated for Childhood Cancer. *JAMA* **2013**, *309*, 2371–2381. [CrossRef] [PubMed]
29. Bower, J.E.; Ganz, P.A.; Aziz, N.; Olmstead, R.; Irwin, M.; Cole, S.W. Inflammatory responses to psychological stress in fatigued breast cancer survivors: Relationship to glucocorticoids. *Brain Behav. Immun.* **2007**, *21*, 251–258. [CrossRef]
30. Patterson, B.C.; Truxillo, L.; Wasilewski-Masker, K.; Mertens, A.C.; Meacham, L.R. Adrenal function testing in pediatric cancer survivors. *Pediatr. Blood Cancer* **2009**, *53*, 1302–1307. [CrossRef]
31. Kennedy, D.D.; Ladas, E.J.; Rheingold, S.R.; Blumberg, J.; Kelly, K.M. Antioxidant status decreases in children with acute lymphoblastic leukemia during the first six months of chemotherapy treatment. *Pediatr. Blood Cancer* **2004**, *44*, 378–385. [CrossRef]
32. Mazur, B.; Mertas, A.; Sońta-Jakimczyk, D.; Szczepanski, T.; Janik-Moszant, A. Concentration of IL-2, IL-6, IL-8, IL-10 and TNF-alpha in children with acute lymphoblastic leukemia after cessation of chemotherapy. *Hematol. Oncol.* **2004**, *22*, 27–34. [CrossRef]
33. Cardous-Ubbink, M.C.; Geenen, M.M.; Schade, K.J.; Heinen, R.C.; Caron, H.N.; Kremer, L.C.M.; Leeuwen, F.E.V. Hypertension in long-term survivors of childhood cancer: A nested case-control study. *Eur. J. Cancer* **2010**, *46*, 782–790. [CrossRef]
34. Brennan, B.M.D.; Shalet, S.M. Endocrine late effects after bone marrow transplant. *Br. J. Haematol.* **2002**, *118*, 58–66. [CrossRef] [PubMed]

35. Krull, K.R.; Brinkman, T.M.; Li, C.; Armstrong, G.T.; Ness, K.K.; Srivastava, D.K.; Gurney, J.G.; Kimberg, C.; Krasin, M.J.; Pui, C.-H.; et al. Neurocognitive Outcomes Decades After Treatment for Childhood Acute Lymphoblastic Leukemia: A Report from the St Jude Lifetime Cohort Study. *J. Clin. Oncol.* **2013**, *31*, 4407–4415. [CrossRef]
36. Schultz, K.A.P.; Ness, K.K.; Whitton, J.; Recklitis, C.; Zebrack, B.; Robison, L.L.; Zeltzer, L.; Mertens, A.C. Behavioral and Social Outcomes in Adolescent Survivors of Childhood Cancer: A Report from the Childhood Cancer Survivor Study. *J. Clin. Oncol.* **2007**, *25*, 3649–3656. [CrossRef]
37. Ader, R.; Cohen, N. Psychoneuroimmunology: Conditioning and Stress. *Annu. Rev. Psychol.* **1993**, *44*, 53–85. [CrossRef]
38. Ader, R. *Psychoneuroimmunology*, 1st ed.; Academic Press: New York, NY, USA, 1 January 1981. Available online: https://www.elsevier.com/books/psychoneuroimmunology/ader/978-0-12-043780-1 (accessed on 19 July 2021).
39. Kiecolt-Glaser, J.; Glaser, R. Psychoneuroimmunology and cancer: Fact or fiction? *Eur. J. Cancer* **1999**, *35*, 1603–1607. [CrossRef]
40. Schlesinger, M.; Yodfat, Y. *Psychoneuroimmunology, Stress and Disease*; Taylor & Francis Group: London, UK. Available online: https://www.taylorfrancis.com/chapters/edit/10.1201/9780367812522-6/psychoneuroimmunology-stress-disease-michael-schlesinger-yair-yodfat (accessed on 19 July 2021).
41. Girdano, G.A.; Dusek, D.E.; Everly, G.S. *Controlling Stress and Tension*, 8th ed.; Pearson: London, UK, 2009.
42. Cannon, W.B. The Emergency Function of the Adrenal Medulla in Pain and the Major Emotions. *Am. J. Physiol. Content* **1914**, *33*, 356–372. [CrossRef]
43. Lazarus, R.S.; Folkman, S. *Stress, Appraisal, and Coping*; Springer Publishing Company: New York, NY, USA, 1984.
44. Van Eck, M.; Berkhof, H.; Nicolson, N.; Sulon, J. The Effects of Perceived Stress, Traits, Mood States, and Stressful Daily Events on Salivary Cortisol. *Psychosom. Med.* **1996**, *58*, 447–458. [CrossRef]
45. Dayas, C.V.; Buller, K.M.; Crane, J.; Xu, Y.; Day, T. Stressor categorization: Acute physical and psychological stressors elicit distinctive recruitment patterns in the amygdala and in medullary noradrenergic cell groups. *Eur. J. Neurosci.* **2001**, *14*, 1143–1152. [CrossRef]
46. McEwen, B.S.; Gianaros, P.J. Stress—And Allostasis-Induced Brain Plasticity. *Annu. Rev. Med.* **2011**, *62*, 431–445. [CrossRef]
47. Sterling, P.; Eyer, J. *Allostasis: A New Paradigm to Explain Arousal Pathology*; Fisher, S., Reason, J., Eds.; John Wiley & Sons: New York, NY, USA, 1988.
48. Sterling, P. *Allostasis, Homeostasis, and the Costs of Physiological Adaptation*; Cambridge University Press: Cambridge, UK, 2004; pp. 17–64. [CrossRef]
49. Vyas, A.; Mitra, R.; Rao, B.S.S.; Chattarji, S. Chronic Stress Induces Contrasting Patterns of Dendritic Remodeling in Hippocampal and Amygdaloid Neurons. *J. Neurosci.* **2002**, *22*, 6810–6818. [CrossRef]
50. Pu, Z.; Krugers, H.J.; Joëls, M. Corticosterone Time-Dependently Modulates β-Adrenergic Effects on Long-Term Potentiation in the Hippocampal Dentate Gyrus. *Learn. Mem.* **2007**, *14*, 359–367. [CrossRef]
51. Dannlowski, U.; Stuhrmann, A.; Beutelmann, V.; Zwanzger, P.; Lenzen, T.; Grotegerd, D.; Domschke, K.; Hohoff, C.; Ohrmann, P.; Bauer, J.; et al. Limbic Scars: Long-Term Consequences of Childhood Maltreatment Revealed by Functional and Structural Magnetic Resonance Imaging. *Biol. Psychiatry* **2011**, *71*, 286–293. [CrossRef] [PubMed]
52. Kesler, S.; Janelsins, M.; Koovakkattu, D.; Palesh, O.; Mustian, K.; Morrow, G.; Dhabhar, F.S. Reduced hippocampal volume and verbal memory performance associated with interleukin-6 and tumor necrosis factor-alpha levels in chemotherapy-treated breast cancer survivors. *Brain, Behav. Immun.* **2013**, *30*, S109–S116. [CrossRef]
53. Conrad, C.D.; Magariños, A.M.; LeDoux, J.E.; McEwen, B.S. Repeated restraint stress facilitates fear conditioning independently of causing hippocampal CA3 dendritic atrophy. *Behav. Neurosci.* **1999**, *113*, 902–913. [CrossRef]
54. Stuart, F.A.; Segal, T.Y.; Keady, S. Adverse psychological effects of corticosteroids in children and adolescents. *Arch. Dis. Child.* **2005**, *90*, 500–506. [CrossRef]
55. Karlamangla, A.S.; Singer, B.H.; McEwen, B.S.; Rowe, J.W.; Seeman, T. Allostatic load as a predictor of functional decline: MacArthur studies of successful aging. *J. Clin. Epidemiol.* **2002**, *55*, 696–710. [CrossRef]
56. Taylor, S.E. Mechanisms linking early life stress to adult health outcomes. *Proc. Natl. Acad. Sci. USA* **2010**, *107*, 8507–8512. [CrossRef]
57. Ramsay, D.S.; Woods, S.C. Clarifying the roles of homeostasis and allostasis in physiological regulation. *Psychol. Rev.* **2014**, *121*, 225–247. [CrossRef] [PubMed]
58. Lovallo, W.R. Do low levels of stress reactivity signal poor states of health? *Biol. Psychol.* **2011**, *86*, 121–128. [CrossRef] [PubMed]
59. Robison, L.L.; Hudson, M.M. Survivors of childhood and adolescent cancer: Life-long risks and responsibilities. *Nat. Rev. Cancer* **2013**, *14*, 61–70. [CrossRef]
60. Oeffinger, K.C.; Mertens, A.C.; Sklar, C.A.; Kawashima, T.; Hudson, M.M.; Meadows, A.T.; Friedman, D.L.; Marina, N.; Hobbie, W.; Kadan-Lottick, N.S.; et al. Chronic Health Conditions in Adult Survivors of Childhood Cancer. *N. Engl. J. Med.* **2006**, *355*, 1572–1582. [CrossRef]
61. Andersen, S.L.; Tomada, A.; Vincow, E.S.; Valente, E.; Polcari, A.; Teicher, M.H. Preliminary Evidence for Sensitive Periods in the Effect of Childhood Sexual Abuse on Regional Brain Development. *J. Neuropsychiatry Clin. Neurosci.* **2008**, *20*, 292–301. [CrossRef] [PubMed]
62. Kovner, R.; Oler, J.A.; Kalin, N.H. Cortico-Limbic Interactions Mediate Adaptive and Maladaptive Responses Relevant to Psychopathology. *Am. J. Psychiatry* **2019**, *176*, 987–999. [CrossRef]

63. Gogtay, N.; Giedd, J.N.; Lusk, L.; Hayashi, K.M.; Greenstein, D.; Vaituzis, A.C.; Nugent, T.F.; Herman, D.H.; Clasen, L.S.; Toga, A.W.; et al. Dynamic mapping of human cortical development during childhood through early adulthood. *Proc. Natl. Acad. Sci. USA* **2004**, *101*, 8174–8179. [CrossRef]
64. Gunnar, M.R.; Wewerka, S.; Frenn, K.; Long, J.D.; Griggs, C. Developmental changes in hypothalamus—Pituitary—Adrenal activity over the transition to adolescence: Normative changes and associations with puberty. *Dev. Psychopathol.* **2009**, *21*, 69–85. [CrossRef]
65. Mason, J.W. A Review of Psychoendocrine Research on the Sympathetic-Adrenal Medullary System. *Psychosom. Med.* **1968**, *30*, 631–653. [CrossRef]
66. Dickerson, S.S.; Kemeny, M.E. Acute Stressors and Cortisol Responses: A Theoretical Integration and Synthesis of Laboratory Research. *Psychol. Bull.* **2004**, *130*, 355–391. [CrossRef]
67. Reddick, W.E.; Taghipour, D.J.; Glass, J.O.; Ashford, J.; Xiong, X.; Wu, S.; Bonner, M.; Khan, R.B.; Conklin, H.M. Prognostic factors that increase the risk for reduced white matter volumes and deficits in attention and learning for survivors of childhood cancers. *Pediatr. Blood Cancer* **2014**, *61*, 1074–1079. [CrossRef]
68. Wolfe, K.R.; Madan-Swain, A.; Hunter, G.R.; Reddy, A.T.; Baños, J.; Kana, R.K. An fMRI investigation of working memory and its relationship with cardiorespiratory fitness in pediatric posterior fossa tumor survivors who received cranial radiation therapy. *Pediatr. Blood Cancer* **2013**, *60*, 669–675. [CrossRef] [PubMed]
69. Stefancin, P.; Cahaney, C.; Parker, R.I.; Preston, T.; Coulehan, K.; Hogan, L.; Duong, T.Q. Neural correlates of working memory function in pediatric cancer survivors treated with chemotherapy: An fMRI study. *NMR Biomed.* **2020**, *33*, e4296. [CrossRef] [PubMed]
70. Campbell, L.K.; Scaduto, M.; Sharp, W.; Dufton, L.; Slyke, D.V.; Whitlock, J.A.; Compas, B. A meta-analysis of the neurocognitive sequelae of treatment for childhood acute lymphocytic leukemia. *Pediatr. Blood Cancer* **2007**, *49*, 65–73. [CrossRef]
71. van der Plas, E.; Qiu, W.; Nieman, B.J.; Yasui, Y.; Liu, Q.; Dixon, S.B.; Kadan-Lottick, N.S.; Weldon, C.B.; Weil, B.R.; Jacola, L.M.; et al. Sex-Specific Associations Between Chemotherapy, Chronic Conditions, and Neurocognitive Impairment in Acute Lymphoblastic Leukemia Survivors: A Report From the Childhood Cancer Survivor Study. *J. Natl. Cancer Inst.* **2020**, *113*, 588–596. [CrossRef] [PubMed]
72. Williams, A.M.; Cheung, Y.T.; Hyun, G.; Liu, W.; Ness, K.K.; Ehrhardt, M.J.; Mulrooney, D.A.; Bhakta, N.; Banerjee, P.; Brinkman, T.M.; et al. Childhood Neurotoxicity and Brain Resilience to Adverse Events during Adulthood. *Ann. Neurol.* **2020**, *89*, 534–545. [CrossRef] [PubMed]
73. Lesnik, P.G.; Ciesielski, K.T.; Hart, B.L.; Benzel, E.C.; Sanders, J.A. Evidence for Cerebellar-Frontal Subsystem Changes in Children Treated With Intrathecal Chemotherapy for Leukemia: Enhanced Data Analysis Using an Effect Size Model. *Arch. Neurol.* **1998**, *55*, 1561–1568. [CrossRef]
74. Ness, K.K.; Leisenring, W.M.; Huang, S.; Hudson, M.M.; Gurney, J.G.; Whelan, K.; Hobbie, W.L.; Armstrong, G.T.; Robison, L.L.; Oeffinger, K.C. Predictors of inactive lifestyle among adult survivors of childhood cancer. *Cancer* **2009**, *115*, 1984–1994. [CrossRef]
75. Chen, C.; Nakagawa, S.; An, Y.; Ito, K.; Kitaichi, Y.; Kusumi, I. The exercise-glucocorticoid paradox: How exercise is beneficial to cognition, mood, and the brain while increasing glucocorticoid levels. *Front. Neuroendocr.* **2017**, *44*, 83–102. [CrossRef] [PubMed]
76. Bernstein, E.E.; McNally, R.J. Acute aerobic exercise hastens emotional recovery from a subsequent stressor. *Health Psychol.* **2017**, *36*, 560–567. [CrossRef]
77. Peng, L.; Yang, L.S.; Yam, P.; Lam, C.S.; Chan, A.S.-Y.; Li, C.K.; Cheung, Y.T. Neurocognitive and Behavioral Outcomes of Chinese Survivors of Childhood Lymphoblastic Leukemia. *Front. Oncol.* **2021**, *11*, 655669. [CrossRef]
78. Godoy, L.D.; Rossignoli, M.T.; Delfino-Pereira, P.; Garcia-Cairasco, N.; de Umeoka, E.H. A Comprehensive Overview on Stress Neurobiology: Basic Concepts and Clinical Implications. *Front. Behav. Neurosci.* **2018**, *12*, 127. [CrossRef]
79. Sapolsky, R.M.; Krey, L.C.; McEwen, B.S. The Neuroendocrinology of Stress and Aging: The Glucocorticoid Cascade Hypothesis. *Endocr. Rev.* **1986**, *7*, 284–301. [CrossRef] [PubMed]
80. Uytun, M.C. Development Period of Prefrontal Cortex. Available online: https://www.intechopen.com/chapters/63179 (accessed on 30 July 2021).
81. Blankenship, S.L.; Redcay, E.; Dougherty, L.R.; Riggins, T. Development of hippocampal functional connectivity during childhood. *Hum. Brain Mapp.* **2016**, *38*, 182–201. [CrossRef] [PubMed]
82. Lenroot, R.K.; Giedd, J.N. Brain development in children and adolescents: Insights from anatomical magnetic resonance imaging. *Neurosci. Biobehav. Rev.* **2006**, *30*, 718–729. [CrossRef] [PubMed]
83. Isgor, C.; Kabbaj, M.; Akil, H.; Watson, S.J. Delayed effects of chronic variable stress during peripubertal-juvenile period on hippocampal morphology and on cognitive and stress axis functions in rats. *Hippocampus* **2004**, *14*, 636–648. [CrossRef]
84. Seigers, R.; Schagen, S.B.; Beerling, W.; Boogerd, W.; van Tellingen, O.; van Dam, F.S.A.M.; Koolhaas, J.M.; Buwalda, B. Long-lasting suppression of hippocampal cell proliferation and impaired cognitive performance by methotrexate in the rat. *Behav. Brain Res.* **2008**, *186*, 168–175. [CrossRef] [PubMed]
85. Seigers, R.; Schagen, S.B.; Coppens, C.M.; van der Most, P.J.; van Dam, F.S.A.M.; Koolhaas, J.M.; Buwalda, B. Methotrexate decreases hippocampal cell proliferation and induces memory deficits in rats. *Behav. Brain Res.* **2009**, *201*, 279–284. [CrossRef]
86. Yang, M.; Kim, J.-S.; Song, M.-S.; Kim, S.-H.; Kang, S.S.; Bae, C.-S.; Kim, J.-C.; Wang, H.; Shin, T.; Moon, C. Cyclophosphamide impairs hippocampus-dependent learning and memory in adult mice: Possible involvement of hippocampal neurogenesis in chemotherapy-induced memory deficits. *Neurobiol. Learn. Mem.* **2010**, *93*, 487–494. [CrossRef]

87. Whittle, S.; Dennison, M.; Vijayakumar, N.; Simmons, J.G.; Yücel, M.; Lubman, D.; Pantelis, C.; Allen, N. Childhood Maltreatment and Psychopathology Affect Brain Development During Adolescence. *J. Am. Acad. Child Adolesc. Psychiatry* **2013**, *52*, 940–952.e1. [CrossRef]
88. Arnsten, A.F.T. Stress weakens prefrontal networks: Molecular insults to higher cognition. *Nat. Neurosci.* **2015**, *18*, 1376–1385. [CrossRef]
89. Dennis, E.L.; Thompson, P.M. Typical and atypical brain development: A review of neuroimaging studies. *Dialog.-Clin. Neurosci.* **2013**, *15*, 359–384. [CrossRef]
90. Pechtel, P.; Lyons-Ruth, K.; Anderson, C.M.; Teicher, M.H. Sensitive periods of amygdala development: The role of maltreatment in preadolescence. *NeuroImage* **2014**, *97*, 236–244. [CrossRef]
91. Harila-Saari, A.H.; Pääkkö, E.L.; Vainionpää, L.K.; Pyhtinen, J.; Lanning, B.M. A Longitudinal Magnetic Resonance Imaging Study of the Brain in Survivors of Childhood Acute Lymphoblastic Leukemia. *Cancer* **1998**, *83*, 2608–2617. [CrossRef]
92. Iuvone, L.; Mariotti, P.; Colosimo, C.; Guzzetta, F.; Ruggiero, A.; Riccardi, R. Long-term cognitive outcome, brain computed tomography scan, and magnetic resonance imaging in children cured for acute lymphoblastic leukemia. *Cancer* **2002**, *95*, 2562–2570. [CrossRef] [PubMed]
93. Moleski, M. Neuropsychological, Neuroanatomical, and Neurophysiological Consequences of CNS Chemotherapy for Acute Lymphoblastic Leukemia. *Arch. Clin. Neuropsychol. Off. J. Natl. Acad. Neuropsychol.* **2000**, *15*, 603–630. [CrossRef]
94. Monje, M.; Thomason, M.E.; Rigolo, L.; Wang, Y.; Waber, D.P.; Sallan, S.E.; Golby, A.J. Functional and structural differences in the hippocampus associated with memory deficits in adult survivors of acute lymphoblastic leukemia. *Pediatr. Blood Cancer* **2012**, *60*, 293–300. [CrossRef] [PubMed]
95. Yoshikawa, E.; Nishi, D.; Matsuoka, Y.J. Association between regular physical exercise and depressive symptoms mediated through social support and resilience in Japanese company workers: A cross-sectional study. *BMC Public Health* **2016**, *16*, 553. [CrossRef]
96. Spitzhüttl, J.S.; Kronbichler, M.; Kronbichler, L.; Benzing, V.; Siegwart, V.; Schmidt, M.; Pastore-Wapp, M.; Kiefer, C.; Slavova, N.; Grotzer, M.; et al. Cortical Morphometry and Its Relationship with Cognitive Functions in Children after non-CNS Cancer. *Dev. Neurorehabilit.* **2021**, *24*, 266–275. [CrossRef]
97. Phillips, N.S.; Kesler, S.R.; Scoggins, M.A.; Glass, J.O.; Cheung, Y.T.; Liu, W.; Banerjee, P.; Ogg, R.J.; Srivastava, D.; Pui, C.-H.; et al. Connectivity of the Cerebello-Thalamo-Cortical Pathway in Survivors of Childhood Leukemia Treated With Chemotherapy Only. *JAMA Netw. Open* **2020**, *3*, e2025839. [CrossRef]
98. Carey, M.E.; Haut, M.W.; Reminger, S.L.; Hutter, J.J.; Theilmann, R.; Kaemingk, K.L. Reduced Frontal White Matter Volume in Long-Term Childhood Leukemia Survivors: A Voxel-Based Morphometry Study. *Am. J. Neuroradiol.* **2008**, *29*, 792–797. [CrossRef] [PubMed]
99. Askins, M.A.; Moore, B.D. Preventing Neurocognitive Late Effects in Childhood Cancer Survivors. *J. Child Neurol.* **2008**, *23*, 1160–1171. [CrossRef]
100. Reddick, W.E.; Shan, Z.Y.; Glass, J.O.; Helton, S.; Xiong, X.; Wu, S.; Bonner, M.J.; Howard, S.C.; Christensen, R.; Khan, R.B.; et al. Smaller white-matter volumes are associated with larger deficits in attention and learning among long-term survivors of acute lymphoblastic leukemia. *Cancer* **2006**, *106*, 941–949. [CrossRef] [PubMed]
101. Stefanski, K.J.; Anixt, J.S.; Goodman, P.; Bowers, K.; Leisenring, W.; Baker, K.S.; Burns, K.; Howell, R.; Davies, S.; Robison, L.L.; et al. Long-Term Neurocognitive and Psychosocial Outcomes After Acute Myeloid Leukemia: A Childhood Cancer Survivor Study Report. *J. Natl. Cancer Inst.* **2020**, *113*, 481–495. [CrossRef] [PubMed]
102. Papageorgiou, M.; Stiakaki, E.; Dimitriou, H.; Malliaraki, N.; Notas, G.; Castanas, E.; Kalmanti, M. Cancer chemotherapy reduces plasma total antioxidant capacity in children with malignancies. *Leuk. Res.* **2005**, *29*, 11–16. [CrossRef] [PubMed]
103. Hasan, S.J.; Hasan, H.R. An evaluation of the oxidative stress index in sera of acute myeloid leukemia and acute lymphoblastic leukemia patients. *Ann. Trop. Med. Public Health* **2020**, *23*, 231–821. [CrossRef]
104. Vatanen, A.; Hou, M.; Huang, T.; Söder, O.; Jahnukainen, T.; Kurimo, M.; Ojala, T.H.; Sarkola, T.; Turanlahti, M.; Saarinen-Pihkala, U.M.; et al. Clinical and biological markers of premature aging after autologous SCT in childhood cancer. *Bone Marrow Transplant.* **2017**, *52*, 600–605. [CrossRef] [PubMed]
105. Song, N.; Li, Z.; Qin, N.; Howell, C.R.; Wilson, C.L.; Easton, J.; Mulder, H.L.; Edmonson, M.N.; Rusch, M.C.; Zhang, J.; et al. Shortened Leukocyte Telomere Length Associates with an Increased Prevalence of Chronic Health Conditions among Survivors of Childhood Cancer: A Report from the St. Jude Lifetime Cohort. *Clin. Cancer Res.* **2020**, *26*, 2362–2371. [CrossRef] [PubMed]
106. Qin, N.; Li, Z.; Song, N.; Wilson, C.L.; Easton, J.; Mulder, H.; Plyler, E.; Neale, G.; Walker, E.; Zhou, X.; et al. Epigenetic Age Acceleration and Chronic Health Conditions Among Adult Survivors of Childhood Cancer. *J. Natl. Cancer Inst.* **2020**, *113*, 597–605. [CrossRef]
107. Ravera, S.; Vigliarolo, T.; Bruno, S.; Morandi, F.; Marimpietri, D.; Sabatini, F.; Dagnino, M.; Petretto, A.; Bartolucci, M.; Mu-raca, M.; et al. Identification of Biochemical and Molecular Markers of Early Aging in Childhood Cancer Survivors. *Biorxiv* **2021**, 438017. [CrossRef]
108. Hayek, S.; Gibson, T.M.; Leisenring, W.M.; Guida, J.L.; Gramatges, M.M.; Lupo, P.J.; Howell, R.M.; Oeffinger, K.C.; Bhatia, S.; Edelstein, K.; et al. Prevalence and Predictors of Frailty in Childhood Cancer Survivors and Siblings: A Report From the Childhood Cancer Survivor Study. *J. Clin. Oncol.* **2020**, *38*, 232–247. [CrossRef]
109. Gleeson, M.; Bishop, N.; Walsh, N. *Exercise Immunology*; Routledge: London, UK, 2013. [CrossRef]

110. Dhabhar, F.S. Effects of stress on immune function: The good, the bad, and the beautiful. *Immunol. Res.* **2014**, *58*, 193–210. [CrossRef]
111. Sumter, S.R.; Bokhorst, C.L.; Miers, A.C.; Pelt, J.V.; Westenberg, P.M. Age and puberty differences in stress responses during a public speaking task: Do adolescents grow more sensitive to social evaluation? *Psychoneuroendocrinology* **2010**, *35*, 1510–1516. [CrossRef]
112. Evans, B.E.; Greaves-Lord, K.; Euser, A.S.; Tulen, J.H.M.; Franken, I.H.A.; Huizink, A.C. Determinants of Physiological and Perceived Physiological Stress Reactivity in Children and Adolescents. *PLoS ONE* **2013**, *8*, e61724. [CrossRef]
113. Dennis, M. *The Frontal Lobes*; Cambridge University Press: Cambridge, UK, 2009; pp. 128–162. [CrossRef]
114. Gordis, E.B.; Granger, D.; Susman, E.J.; Trickett, P.K. Asymmetry between salivary cortisol and α-amylase reactivity to stress: Relation to aggressive behavior in adolescents. *Psychoneuroendocrinology* **2006**, *31*, 976–987. [CrossRef]
115. Klimes-Dougan, B.; Hastings, P.D.; Granger, D.A.; Usher, B.A.; Zahn–Waxler, C. Adrenocortical activity in at-risk and normally developing adolescents: Individual differences in salivary cortisol basal levels, diurnal variation, and responses to social challenges. *Dev. Psychopathol.* **2001**, *13*, 695–719. [CrossRef]
116. Buske-Kirschbaum, A.; Jobst, S.; Wustmans, A.; Kirschbaum, C.; Rauh, W.; Hellhammer, D. Attenuated Free Cortisol Response to Psychosocial Stress in Children with Atopic Dermatitis. *Psychosom. Med.* **1997**, *59*, 419–426. [CrossRef]
117. Hlavacova, N.; Solarikova, P.; Marko, M.; Brezina, I.; Jezova, D. Blunted cortisol response to psychosocial stress in atopic patients is associated with decrease in salivary alpha-amylase and aldosterone: Focus on sex and menstrual cycle phase. *Psychoneuroendocrinology* **2017**, *78*, 31–38. [CrossRef] [PubMed]
118. Stroud, L.R.; Foster, E.; Papandonatos, G.D.; Handwerger, K.; Granger, D.A.; Kivlighan, K.T.; Niaura, R. Stress response and the adolescent transition: Performance versus peer rejection stressors. *Dev. Psychopathol.* **2009**, *21*, 47–68. [CrossRef]
119. Erdmann, F.; Frederiksen, L.E.; Bonaventure, A.; Mader, L.; Hasle, H.; Robison, L.L.; Winther, J.F. Childhood cancer: Survival, treatment modalities, late effects and improvements over time. *Cancer Epidemiol.* **2020**, *71*, 101733. [CrossRef]
120. Sapolsky, R.M.; Romero, L.M.; Munck, A.U. How Do Glucocorticoids Influence Stress Responses? Integrating Permissive, Suppressive, Stimulatory, and Preparative Actions. *Endocr. Rev.* **2000**, *21*, 55–89. [CrossRef] [PubMed]
121. Gilbert, L.K.; Breiding, M.J.; Merrick, M.T.; Thompson, W.W.; Ford, D.C.; Dhingra, S.S.; Parks, S.E. Childhood Adversity and Adult Chronic Disease An Update from Ten States and the District of Columbia. *Am. J. Prev. Med.* **2010**, *48*, 345–349. [CrossRef]
122. Rentscher, K.E.; Carroll, J.E.; Repetti, R.L.; Cole, S.W.; Reynolds, B.M.; Robles, T.F. Chronic stress exposure and daily stress appraisals relate to biological aging marker p16INK4a. *Psychoneuroendocrinology* **2018**, *102*, 139–148. [CrossRef] [PubMed]
123. Entringer, S.; Epel, E.S. The stress field ages: A close look into cellular aging processes. *Psychoneuroendocrinology* **2019**, *113*, 104537. [CrossRef] [PubMed]
124. Yegorov, Y.E.; Poznyak, A.V.; Nikiforov, N.G.; Sobenin, I.A.; Orekhov, A.N. The Link between Chronic Stress and Accelerated Aging. *Biomedicines* **2020**, *8*, 198. [CrossRef] [PubMed]
125. Tyrka, A.R.; Price, L.H.; Kao, H.-T.; Porton, B.; Marsella, S.A.; Carpenter, L.L. Childhood Maltreatment and Telomere Shortening: Preliminary Support for an Effect of Early Stress on Cellular Aging. *Biol. Psychiatry* **2010**, *67*, 531–534. [CrossRef]
126. Simon, N.M.; Smoller, J.W.; McNamara, K.L.; Maser, R.S.; Zalta, A.K.; Pollack, M.H.; Nierenberg, A.A.; Fava, M.; Wong, K.-K. Telomere Shortening and Mood Disorders: Preliminary Support for a Chronic Stress Model of Accelerated Aging. *Biol. Psychiatry* **2006**, *60*, 432–435. [CrossRef]
127. Tanaka, M.; Tóth, F.; Polyák, H.; Szabó, Á.; Mándi, Y.; Vécsei, L. Immune Influencers in Action: Metabolites and Enzymes of the Tryptophan-Kynurenine Metabolic Pathway. *Biomedicines* **2021**, *9*, 734. [CrossRef] [PubMed]
128. Sulicka-Grodzicka, J.; Surdacki, A.; Seweryn, M.; Mikołajczyk, T.; Rewiuk, K.; Guzik, T.; Grodzicki, T. Low-grade chronic inflammation and immune alterations in childhood and adolescent cancer survivors: A contribution to accelerated aging? *Cancer Med.* **2021**, *10*, 1772–1782. [CrossRef]
129. Wolf, C.M.; Reiner, B.; Kühn, A.; Hager, A.; Müller, J.; Meierhofer, C.; Oberhoffer, R.; Ewert, P.; Schmid, I.; Weil, J. Subclinical Cardiac Dysfunction in Childhood Cancer Survivors on 10-Years Follow-Up Correlates With Cumulative Anthracycline Dose and Is Best Detected by Cardiopulmonary Exercise Testing, Circulating Serum Biomarker, Speckle Tracking Echocardiography, and Tissue Doppler Imaging. *Front. Pediatr.* **2020**, *8*, 123. [CrossRef]
130. Haddy, T.B.; Mosher, R.B.; Reaman, G.H. Hypertension and prehypertension in long-term survivors of childhood and adolescent cancer. *Pediatr. Blood Cancer* **2007**, *49*, 79–83. [CrossRef]
131. Lubas, M.M.; Wang, M.; Jefferies, J.L.; Ness, K.K.; Ehrhardt, M.J.; Krull, K.R.; Mulrooney, D.A.; Srivastava, D.K.; Howell, R.M.; Robison, L.L.; et al. The Contribution of Stress and Distress to Cardiovascular Health in Adult Survivors of Childhood Cancer. *Cancer Epidemiol. Biomark. Prev.* **2020**, *30*, 286–294. [CrossRef]
132. Heinrichs, M.; Baumgartner, T.; Kirschbaum, C.; Ehlert, U. Social support and oxytocin interact to suppress cortisol and subjective responses to psychosocial stress. *Biol. Psychiatry* **2003**, *54*, 1389–1398. [CrossRef]
133. Southwick, S.M.; Vythilingam, M.; Charney, D.S. The Psychobiology of Depression and Resilience to Stress: Implications for Prevention and Treatment. *Annu. Rev. Clin. Psychol.* **2005**, *1*, 255–291. [CrossRef]
134. Vaynman, S.; Ying, Z.; Gomez-Pinilla, F. Hippocampal BDNF mediates the efficacy of exercise on synaptic plasticity and cognition. *Eur. J. Neurosci.* **2004**, *20*, 2580–2590. [CrossRef]
135. Daly, M.; McMinn, D.; Allan, J.L. A bidirectional relationship between physical activity and executive function in older adults. *Front. Hum. Neurosci.* **2015**, *8*, 1044. [CrossRef] [PubMed]

136. Roos, L.G.; Janson, J.; Sturmbauer, S.C.; Bennett, J.M.; Rohleder, N. Higher trait reappraisal predicts stronger HPA axis habituation to repeated stress. *Psychoneuroendocrinology* **2019**, *101*, 12–18. [CrossRef]
137. Swaab, D.F.; Pool, C.W.; Nijveldt, F. Immunofluorescence of vasopressin and oxytocin in the rat hypothalamo-neurohypophyseal system. *J. Neural Transm.* **1975**, *36*, 195–215. [CrossRef] [PubMed]
138. Kazak, A.E.; Barakat, L.P.; Meeske, K.; Christakis, D.; Meadows, A.T.; Casey, R.; Penati, B.; Stuber, M.L. Posttraumatic stress, family functioning, and social support in survivors of childhood leukemia and their mothers and fathers. *J. Consult. Clin. Psychol.* **1997**, *65*, 120–129. [CrossRef] [PubMed]
139. Barakat, L.P.; Kazak, A.E.; Meadows, A.T.; Casey, R.; Meeske, K.; Stuber, M.L. Families Surviving Childhood Cancerf: A Comparison of Posttraumatic Stress Symptoms with Families of Healthy Children. *J. Pediatr. Psychol.* **1997**, *22*, 843–859. [CrossRef] [PubMed]
140. Phipps, S.; Larson, S.; Long, A.; Rai, S.N. Adaptive Style and Symptoms of Posttraumatic Stress in Children with Cancer and Their Parents. *J. Pediatr. Psychol.* **2005**, *31*, 298–309. [CrossRef] [PubMed]
141. Hile, S.; Erickson, S.J.; Agee, B.; Annett, R.D. Parental stress predicts functional outcome in pediatric cancer survivors. *Psycho-Oncol.* **2014**, *23*, 1157–1164. [CrossRef] [PubMed]
142. Ernst, M.; Brähler, E.; Klein, E.M.; Jünger, C.; Wild, P.S.; Faber, J.; Schneider, A.; Beutel, M.E. Parenting in the face of serious illness: Childhood cancer survivors remember different rearing behavior than the general population. *Psycho-Oncol.* **2019**, *28*, 1663–1670. [CrossRef]
143. Hullmann, S.E.; Wolfe-Christensen, C.; Meyer, W.H.; McNall-Knapp, R.Y.; Mullins, L.L. The relationship between parental overprotection and health-related quality of life in pediatric cancer: The mediating role of perceived child vulnerability. *Qual. Life Res.* **2010**, *19*, 1373–1380. [CrossRef] [PubMed]
144. Tillery, R.; Long, A.; Phipps, S. Child Perceptions of Parental Care and Overprotection in Children with Cancer and Healthy Children. *J. Clin. Psychol. Med. Settings* **2014**, *21*, 165–172. [CrossRef]
145. Shepard, J.D.; Barron, K.W.; Myers, D. Corticosterone delivery to the amygdala increases corticotropin-releasing factor mRNA in the central amygdaloid nucleus and anxiety-like behavior. *Brain Res.* **2000**, *861*, 288–295. [CrossRef]
146. Gustafson, M.P.; Wheatley-Guy, C.M.; Rosenthal, A.C.; Gastineau, D.; Katsanis, E.; Johnson, B.D.; Simpson, R.J. Exercise and the immune system: Taking steps to improve responses to cancer immunotherapy. *J. Immunother. Cancer* **2021**, *9*, e001872. [CrossRef] [PubMed]
147. Tanaka, M.; Vécsei, L. Monitoring the kynurenine system: Concentrations, ratios or what else? *Adv. Clin. Exp. Med.* **2021**, *30*, 775–778. [CrossRef]

cancers

Review

Molecular Stratification of Childhood Ependymomas as a Basis for Personalized Diagnostics and Treatment

Margarita Zaytseva [1,*], Ludmila Papusha [1], Galina Novichkova [1] and Alexander Druy [1,2]

[1] Dmitry Rogachev National Medical Research Center of Pediatric Hematology, Oncology and Immunology, 117997 Moscow, Russia; ludmila.mur@mail.ru (L.P.); Galina.Novichkova@fccho-moscow.ru (G.N.); dr-drui@yandex.ru (A.D.)
[2] Research Institute of Medical Cell Technologies, 620026 Yekaterinburg, Russia
* Correspondence: astice@list.ru

Simple Summary: The current trend in neuropathology directs to the integrated histo-molecular approach. The traditional concept of histological grade should be complemented by comprehensive diagnostics with the mandatory use of molecular genetic markers. As a consequence, basic types of CNS tumors fall into multiple nosological entities that can be morphologically similar while having fundamentally different pathogenesis and clinical presentation. This trend is particularly evident for ependymal tumors, which harbor molecular markers of decisive importance for the prognosis. This minireview emphasizes recent achievements in ependymoma biology research closely connected with state-of-the-art diagnostics.

Abstract: Ependymomas are among the most enigmatic tumors of the central nervous system, posing enormous challenges for pathologists and clinicians. Despite the efforts made, the treatment options are still limited to surgical resection and radiation therapy, while none of conventional chemotherapies is beneficial. While being histologically similar, ependymomas show considerable clinical and molecular diversity. Their histopathological evaluation alone is not sufficient for reliable diagnostics, prognosis, and choice of treatment strategy. The importance of integrated diagnosis for ependymomas is underscored in the recommendations of Consortium to Inform Molecular and Practical Approaches to CNS Tumor Taxonomy. These updated recommendations were adopted and implemented by WHO experts. This minireview highlights recent advances in comprehensive molecular-genetic characterization of ependymomas. Strong emphasis is made on the use of molecular approaches for verification and specification of histological diagnoses, as well as identification of prognostic markers for ependymomas in children.

Keywords: ependymoma; risk stratification; molecular group; prognosis

Citation: Zaytseva, M.; Papusha, L.; Novichkova, G.; Druy, A. Molecular Stratification of Childhood Ependymomas as a Basis for Personalized Diagnostics and Treatment. *Cancers* **2021**, *13*, 4954. https://doi.org/10.3390/cancers13194954

Academic Editors: Saurabh Agarwal and Jianhua Yang

Received: 7 September 2021
Accepted: 29 September 2021
Published: 1 October 2021

1. Introduction

Ependymal tumors (ependymomas, EPNs), a common type of malignant neoplasms of the central nervous system (CNS), constitute about 10% of all intracranial tumors and about 20% of spinal cord tumors. EPNs rank third in the prevalence of pediatric CNS tumors (after glial and embryonal tumors) [1]. Despite the use of advanced protocols that include maximal safe surgical resection followed by localized radiotherapy, the mortality remains high due to frequent relapses explained by the strong metastatic potential of EPNs complemented by an efficient spread of metastases with cerebrospinal fluid.

Adverse predictors for EPNs are early age at onset, residual tumor tissue after resection, and metastatic lesions in CNS [2–5]; however, the detailed prognosis for EPNs is often hampered by (1) clinical and morphological diversity of the tumors and (2) complex relations of histopathological grades with the prognosis [6,7]. In line with the modern trends in neuropathology, the aggressiveness of a tumor and, accordingly, the prognosis is mainly determined by molecular-genetic aberrations, whereas the conventional, histologically defined grade becomes subsidiary [8,9]. For pediatric EPNs, the relevance of molecular stratification is especially obvious.

Over recent years, the diagnostics of CNS malignancies has been significantly reconsidered. The accent has been shifted from pathomorphology to molecular profiling and the search for clinically informative markers that would justify the selection of a particular therapy. Molecular framework-based stratification schemes have been developed and introduced into clinical practice for a number of CNS tumors; examples include *IDH1/2* mutations and 1p/19q codeletions for gliomas and oligodendrogliomas [10]; *KIAA1549-BRAF* fusions, *MYB/MYBL* rearrangements, recurrent pathogenic mutations in *BRAF* and *H3F3A* for pediatric astrocytomas [11,12]; and four molecular groups with the account of *MYC/MYCN* amplification for medulloblastomas [13].

EPNs of different molecular etiologies occupy distinct anatomical compartments within CNS. Recurrent genetic or epigenetic alterations found in EPNs are invariably linked to tumor localization. Molecular subgrouping of EPNs is superior to histopathological grading based on the WHO criteria [14]. Gene expression signatures and related subgrouping have shown the highest prognostic value among other studied molecular criteria. A tumor retains its affiliation to a particular subgroup indefinitely (it cannot be switched during progression and/or relapse of the disease), which increases its clinical significance [5,14,15]. An advanced EPN classification has been recently proposed by the Consortium to Inform Molecular and Practical Approaches to CNS Tumor Taxonomy (cIMPACT-NOW) update 7, aimed at connecting localization-dependent molecular groups with tumor progression modes and outcomes [16]. This view has been supported by WHO experts and reflected in the summary of the upcoming fifth edition of the WHO Classification of Tumors of the Central Nervous System (WHO CNS5) [9]. According to the newest CNS tumor nomenclature, ependymomas are subdivided into supratentorial (ST-EPNs), infratentorial (a.k.a. posterior fossa ependymomas, PF-EPNs), and spinal (Sp-EPNs) by localization of the primary tumor; these groups are further stratified by (epi)genetic features.

2. Molecular Profiles of ST-EPNs

ST-EPNs are fairly rare and show considerable genetic heterogeneity. ST-EPNs have been recently stratified into two major groups: supratentorial ependymoma, *ZFTA* fusion-positive (ST-EPN-ZFTA) and supratentorial ependymoma, *YAP1* fusion-positive (ST-EPN-YAP1) [9] consistently with gene expression and/or DNA methylation signatures revealed by transcriptomic methods and/or whole-genome DNA methylation profiling, respectively.

2.1. ST-EPN-ZFTA Group

Gain-of-function rearrangements in *ZFTA* or *YAP1* are specific for ST-EPNs. At that, ST-EPN-ZFTA tumors are prevalent (50–75% and 25% of ST-EPNs in children and adults, respectively [7,14,17–20]), while ST-EPN-YAP1 tumors are rare (3–10% in different cohorts [7,14,18,19,21,22]). The archetypal chimeric transcript harbored by *ZFTA*-rearranged ependymomas is *ZFTA–RELA*, hence the ST-EPN-RELA is a traditional designation for this group [14]. Alternative *ZFTA* fusions (non-*RELA*, e.g., *ZFTA-NCOA1*, *ZFTA-NCOA2*, *ZFTA-MAML2* [23–28], and *MN1-ZFTA* [28]) are less common.

Recurrent *ZFTA–RELA* fusion is a unique molecular hallmark of *ZFTA*-positive EPNs not found in other CNS tumors. Nine different transcript variants have been described, differing by breakpoints in *RELA* and its partner gene; the prevalent isoform comprises *ZFTA* exon 2 spliced to *RELA* exon 2 [17,22,29]. Formation of the *ZFTA–RELA* intrachromosomal gene fusion results from multiple double-strand breaks in 11q13.1 with subsequent random reassociation (typical for chromotrypsis); hence the diversity of fusion points for such transcripts. The oncogenic impact of classical *ZFTA–RELA* fusions was elucidated in recent studies [30,31].

RELA encodes the RelA (p65) subunit of the dimeric nuclear factor-κB (NF-κB), most known as a master regulator of immune responses and inflammation. NF-κB promotes apoptosis inhibition, cell growth, and pro-angiogenic signaling—the basic components of oncogenesis and tumor progression. Expression of *RELA* chimeras results in constitutive activation of NF-κB signaling pathway [17] and associated resistance of the tumor to chemo and radiation therapies [32]. ZFTA chimeric proteins accumulate in the nucleus. A zinc finger domain in the truncated ZFTA protein endows the chimeras with extraordinary high affinity to DNA. The oncoprotein interferes with chromatin structure at ST-EPN-associated loci, enabling the RELA transactivation domain to induce their transcription [30]. Moreover, apart from the canonical NF-κB pathway activation, *ZFTA–RELA* fusions may trigger other gene expression programs through recruitment of transcriptional co-activators BRD4, EP300, and CBP, which participate in chromatin-related pathways and represent potential druggable targets [31,33].

EPNs with the *ZFTA* gene fused with a non-*RELA* partner gene are considerably less common. These tumors have variable histological structures and, apart from the ependymomal component, may additionally involve pleomorphic xanthoastrocytoma-like, astroblastoma-like, malignant teratoma-like, embryonal tumor-like, or sarcoma-like patterns. Despite the heterogeneous morphology, these tumors are (epi)genetically similar and tend to resemble the classic *ZFTA–RELA*-fused EPNs, as revealed by methylome assay. A detailed analysis of DNA methylation profiles allows subdivision of these tumors into two clusters, one of them comprising tumors with histological features of astroblastomas and xanthoastrocytomas, harboring *ZFTA–MAML2* and *MN1–ZFTA* rearrangements; the second cluster comprises tumors histologically resembling small-cell sarcomatoid carcinomas and undifferentiated sarcomas, harboring *ZFTA–NCOA1* and *ZFTA–NCOA2* rearrangements [24–28].

The presence of recurrent *ZFTA–RELA* fusions has been repeatedly implicated as an adverse prognostic factor [5,14,18]. Five-year rates of event-free survival (EFS) and overall survival (OS) for ST-EPN-ZFTA tumors never exceed 29% and 75%, respectively [14]. Within the ST-EPN-ZFTA group, additional risks of relapse have been associated with 1q gains [5]. Interestingly, the St Jude Young Children 07 (SJYC07) study (encompassing ependymal tumors diagnosed in <3-year-olds) identified similar 4-year EFS rates for ST-EPN-ZFTA, ST-EPN-YAP1, and PF-EPN group A [7]. Consistently, two other studies conducted independently by the Italian Association of Pediatric Hematology and Oncology AIEOP and the Children's Oncology Group (trial ACNS0121) revealed no difference in survival rates for ST-EPNs with and without *ZFTA–RELA* fusion [4,34].

ZFTA (non-*RELA*)-fused EPNs have an especially dismal prognosis, with EFS rates significantly lower compared with classical *ZFTA–RELA*-fused EPNs, while the corresponding OS rates are comparable [26]. However, these findings are preliminary, given the small number of cases reported so far. In the context of ST-EPN heterogeneity, it might be useful to consider *ZFTA* fusions with atypical (non-*RELA*) partners as a distinguishing feature for a separate group, the prognostic and clinical relevance of which is yet to be specified.

2.2. ST-EPN-YAP1 Group

ST-EPN-YAP1 tumors show an aberrant activity of transcription co-activator YAP1 (Yes-associated protein 1) related to its abnormal accumulation in the nucleus. With YAP1 being a direct regulator of TEAD and SMAD transcription factors, its escape from Hippo-dependent sequestration through accumulation in the nucleus results in sustained proliferative signaling via WNT and Hedgehog. More accurately, the nucleus accumulates the oncogenic fusion protein YAP1–MAMLD1 transferred from the cytoplasm to the nucleus independently of its YAP1-Ser127 phosphorylation status that limits the nuclear import of YAP1 in normal cells [35]. Apart from the prevalent *YAP1–MAMLD1* fusions, ST-EPN-YAP1 may harbor relatively rare structural variants, e.g., *YAP1–FAM118B* [14,36]. In some cases, the formation of *YAP1* fusions involves focal copy number alterations mapping to the 3′ portion of the gene (11q22.1–11q21.2) [37]. Unlike *ZFTA*-positive ependymomas, ST-EPN-YAP1 tumors have balanced genomes with local aberrations in the *YAP1* locus and no evidence of chromothripsis.

Andreiuolo et al. (2019) reported a multicenter retrospective study on what is so far the largest cohort of patients with *YAP1*-positive EPNs ($n = 14$). Overall survival for these patients (median observation time of 4.8 years within the range of 0.6–16 years) constituted 100%. It is important to note that the boy-to-girl ratio for the studied cohort was 1:6.5, and only three of the patients were over three years old at the time of diagnosis (the median age at diagnosis constituted 8.2 months) [37]. The best survival rates for the ST-EPN-YAP1 group among other EPNs were also reported by other authors [7,14]. Careful de-escalation of conventional EPN treatment protocols specifically for ST-EPN-YAP1 patients is currently under scrutiny. An opportunity to exclude (delay or dismiss) radiation therapy alleviates the risks of severe cognitive dysfunctions, endocrinopathies, and secondary tumors [38].

2.3. *Non-ZFTA/Non-YAP1 ST-EPNs*

The molecular diversity of ST-EPNs exceeds the currently established ZFTA-YAP stereotype. Tumors with neither *ZFTA* nor *YAP1* alterations are considered as a separate group, and recent findings emphasize the need for a finer specification. A distinct entity is formed by *PLAGL1* rearranged EPNs, harboring *EWSR1–PLAGL1* and less commonly *PLAGL1–FOXO1* or *PLAGL1–EP300* fusions [39], which echoes molecular landscapes of soft tissue sarcomas and a group of rare mesenchymal (non-meningothelial) and glioneuronal CNS tumors with *EWSR1*–non-ETS fusions [40,41]. Nevertheless, for the vast majority of ST-EPNs lacking recurrent chromosomal rearrangements, the oncogenic driver events remain elusive. Several reports reveal the presence of fusion genes *MAML2–ASCL2, MARK2–ADCY3* [19], *PTEN–TAS2R1* [14], *PATZ1–MN1, MYH9–SEC14L2, MTMR3–NCOA3* [24], *TMEFF2–FOXO1, PCGF1–CREBBP* [20], *FOXO1–STK24*, as well as *EP300–BCORL1* in such tumors [21]. Olsen et al. (2015) described two cases of hemispheric infantile EPN-like gliomas with *ALK* fusions (*CCDC88A–ALK* and *KTN1–ALK*), both of them morphologically ambiguous: the tumors showed glial phenotypes and resembled glioblastomas [42]. In the summary of the upcoming WHO CNS5, such tumors have been reclassified and renamed as infant-type hemispheric gliomas harboring receptor tyrosine kinase gene rearrangements [9]. Torre et al. (2020) reported in-frame fusions *AGK–BRAF* and *MYO5A–NTRK3* as potential targets for therapeutic inhibition [20]. In sum, these observations indicate the absence of a single driver mechanism for this group of tumors while underscoring the importance of their distinction from other CNS neoplasms.

3. Molecular Profiles of PF-EPNs

PF-EPNs, more prevalent in children than adults, constitute about 2/3 of intracranial ependymal tumors of childhood. Based on high-throughput molecular techniques, PF-EPNs are subdivided into two molecular groups: PF-EPN group A (PF-EPN-A), and PF-EPN group B (PF-EPN-B) [9].

3.1. *PF-EPN-A Group*

This highly heterogeneous group comprises 85–90% of infratentorial EPNs. PF-EPN-A tumors are often located laterally within the posterior fossa and occur predominantly in infants and young children, twice more frequently in boys than in girls; the average age at diagnosis constitutes 3.5 years [14]. The patients are at high risk of relapse, even under multimodal therapy and in the absence of extra adverse prognostic markers [5]. Identification with PF-EPN-A represents a strong independent prognostic factor associated with the worst rates of survival. According to Zapotocky et al. (2019), 5-year and 10-year EFS for PF-EPN-A constitute, respectively, 43% and 37% [3]. Relapses are typical for PF-EPN-A but not PF-EPN-B and, consistently, 10-year OS rates for PF-EPN-A are significantly lower than for PF-EPN-B (56–58% vs. 88–100%) [4,14,15]. Additional negative clinical predictors for PF-EPN-A are the presence of residual tumor tissue (incomplete resection) and adjuvant radiation therapy refusal [3]. Retrospective evaluation of outcomes for the patients receiving treatment under HIT-2000 protocol implicated residual tumors, 1q gains, and high mitotic activity of tumor cells (>10 mitotic figures per 10 fields of view) as independent adverse predictors for PF-EPNs in general and PF-EPN-A tumors in particular [2]. Cytogenetic prognostic factors for PF-EPNs include 1q gains and 6q losses [43,44]. These cytogenetic abnormalities are detected in 18.9% and 8.6% of PF-EPNs, respectively. At that, the 1q gained PF-EPNs harbor 6q losses at an increased frequency of 17.7% [43]. Both types of copy number variations have been qualified as adverse predictors. Five-year progression-free survival rates were 50% for EPNs without 1q gain and 6q loss, as opposed to 32% for 1q gain only, 7.3% for 6q loss only, and 0% for both 1q gain/6q loss tumors [43]. The ultra-high risks conferred by the co-occurrence of cytogenetic markers in PF-EPN-A patients should be taken into account for the treatment regimen optimization.

PF-EPN-A tumors reveal characteristic aberrant patterns of DNA methylation, the so-called CpG-island methylator phenotype (CIMP) with extensive hypermethylation of CpG-islands in promoter regions of multiple genes. This effect critically interferes with the function of PRC2 (Polycomb repressive complex 2) [45,46]—a transcription repressor protein complex that facilitates methylation of nucleosome histone H3 at amino acid residues H3K27 and H3K9 thus inhibiting the expression of key regulatory genes responsible for cell fate determination and differentiation. Bayliss et al. (2016) revealed the deficiency or complete loss of H3K27me3 in PF-EPN-A tumors [46]. This finding complements the earlier hypotheses on the central role of epigenetic mechanisms in PF-EPN pathogenesis inferred from the absence of presumably pathogenic mutations in chromatin remodeling genes and enzymes that catalyze post-translational modifications (e.g., methylation) of histones in whole-genome sequencing data [17,45,46]. The recruitment of Polycomb group (PcG) transcription repressors to chromatin requires the presence of non-methylated CpGs; accordingly, the loss of H3K27me3 methylation has been associated with dense hypermethylation of CpG-islands preventing the recruitment of PcG proteins to chromatin by steric hindrance.

An advanced investigation of molecular mechanisms responsible for the observed epigenetic malfunctioning revealed a plausible association of the H3K27me3 deficiency with elevated expression levels of accessory proteins encoded by *EZHIP* (formerly *CXorf67*) and *EPOP* (formerly *C17orf96*) [47–49]. As demonstrated by Hübner et al. (2019), EZHIP is a competitive inhibitor of PRC2. A conservative stretch of amino acids in the C-terminal portion of EZHIP mimics the K27 methylation target in histone H3, albeit with K27M substitution. The binding of methionine M27 (instead of lysine K27) to the active center in the histone-lysine N-methyltransferase subunit of PRC2 blocks its catalytic activity [50]. Somatic missense mutations in *EZHIP* are detected in a small proportion of PF-EPN-A tumors (<10%) [48]. Jain et al. (2019) demonstrated that such mutations have no influence on H3K27me3 levels thus disproving their functional significance [51]. Noteworthy, no loss-of-function mutations in *EZHIP* (nonsense substitutions or frameshift indels) have been reported. Elevated expression of *EZHIP* in tumors may be caused by mutations in *cis*-regulatory elements; the same effect may be conferred by the formation of fusion genes involving *EZHIP* locus (for instance, *MBTD1–EZHIP* fusion described for low-grade endometrial stromal sarcoma [52]). However, no fusions comprising *EZHIP* or PRC2 subunit-encoding genes (e.g., *EED*, *SUZ12*) have been described for EPNs.

Related signatures of disrupted epigenetic regulation have been associated with *H3* K27M mutations typical for diffuse midline gliomas (DMGs) but rarely found in PF-EPN-A (<5% of the cases). Noteworthy, in EPNs such mutations are harbored by canonical histone-encoding genes *HIST1H3C* and *HIST1H3B*, whereas in DMGs they are predominantly found in a replacement histone gene *H3F3A* (90% of the cases) [48,53–55]. Given the mutually agonistic roles of the onco-histone H3 K27M and EZHIP, it would be natural to expect similar patterns of disease progression and therapy outcomes for *H3* K27M-mutant DMG and EZHIP^{high} PF-EPN-A. Indeed, in DMG, disruption of H3K27me3-mediated epigenetic regulation is associated with an extremely aggressive course of the disease, typically presenting with sustained tumor growth and polychemotherapy resistance [56–58]. Similarly, effective chemotherapy regimens for PF-EPN-A are missing [59] and therapeutic options for relapses are extremely limited [5,60–63].

Despite the uniformity of methylation profiles within PF-EPN-A, tumors of this group show considerable molecular heterogeneity and can be additionally classified into two major subgroups A1 and A2 (and ultimately into nine minor subtypes by using additional markers: gains 1q, deletions 22q, 6q, and 10q, and OTX2 protein expression). PF-EPN-A1 tumors are distinguished by pronounced expression of the homeotic HOX genes (*HOXA1/2/3/4*, *HOXB2/3/4*, *HOXC4*, and *HOXD4*) which define the segmental (rhombomeric) organization of the hindbrain in early embryogenesis. PF-EPN-A2 tumors hyperexpress *EN2*, *CNPY1*, and *IRX3*—a group of genes involved in the rhombomere differentiation. Expression of A1- and A2-specific genes within the developing hindbrain shows distinct zonality—increased expression of A2 markers is characteristic of the rostral portion at the border with the midbrain, while expression of HOX genes is more pronounced in caudal segments of the brainstem and spinal cord. Differential expression of the spatial patterning genes in A1 and A2 tumors apparently reflects their origin from different hindbrain structures. However, the practical relevance of the advanced A1/A2 subgrouping is questionable. Stratification by clinical factors (gender, age at diagnosis, tumor resection volume, and received therapy) revealed no significant differences between A1 and A2 tumors, except the patterns of relapse (PF-EPN-A1 tumors more often produce local than distant relapses, and vice versa) [48].

3.2. PF-EPN-B Group

In contrast to PF-EPN-A tumors which predominantly affect children, PF-EPN-B tumors are more common in adults. In adolescents (aged 10–17), about 45% of newly diagnosed EPNs fall into this group. The prognosis for PF-EPN-B tumors is favorable: 10-year OS rates for the patients after subtotal and gross total resections reach 66.7% and 96.1%, respectively [3,14,15,64]. Thus, the prognosis for this group strongly depends on the extent of surgical resection. The occurrence of delayed relapses (10 years after the onset) underscores the importance of long-term follow-up [64]. Patients with R0 may benefit from chemo- and radiation-sparing strategies; such possibility is being considered [38]. The observed difference in patterns of recurrence between PF-EPN-A and PF-EPN-B adds to the relevance of comprehensive molecular characterization of a tumor as early as possible.

By now, recurrent mutations or fusion genes in PF-EPN-B tumors are missing, and no clear drivers for this group have been identified. Ciliogenesis and microtubule assembly are deregulated only in PF-EPN-B tumors, while several canonical cancer-associated pathways operate in the PF-EPN-A group (VEGF, PDGF, EGFR, RAS signaling, etc.) [14]. PF-EPN-B tumors harbor major cytogenetic aberrations including gains 1q, monosomies 6, 10, and 17, trisomies 5, 8, and 18, and deletions 22q [64]. The diversity of cytogenetic profiles revealed for PF-EPN-B indicate inherent genomic instability and suggest that these tumors emerge from multiple driving events. Similarly with PF-EPN-A, the PF-EPN-B group shows significant heterogeneity, with distinct molecular subtypes of different demographics, copy number alterations, and gene expression signatures. By contrast with PF-EPN-A, gains 1q pose no extra risks for PF-EPN-B tumors. Losses 13q may represent a more reliable negative prognostic marker than gains 1q; however, this assumption requires further substantiation, particularly as the basis for de-escalation of therapy regimens. However, the extent of resection remains the strongest predictor of poor outcomes for this group. Given the patient data scarcity, advanced stratification within PF-EPN-B remains clinically irrelevant [64].

3.3. ST-EPN-ZFTA-like PF-EPNs

Unique cases of *ZFTA–MAML2*, *ZFTA–RELA*, and *ZFTA–NCOA2* fusion in PF-EPN were reported recently. These tumors revealed characteristic *ZFTA*-mediated gene expression and whole-genome DNA methylation signatures corresponding to the ST-EPN-ZFTA group; accordingly, they were classified as "ST-EPN-ZFTA" despite the infratentorial localization [65].

A summary of the intracranial EPN classification is given in Figure 1.

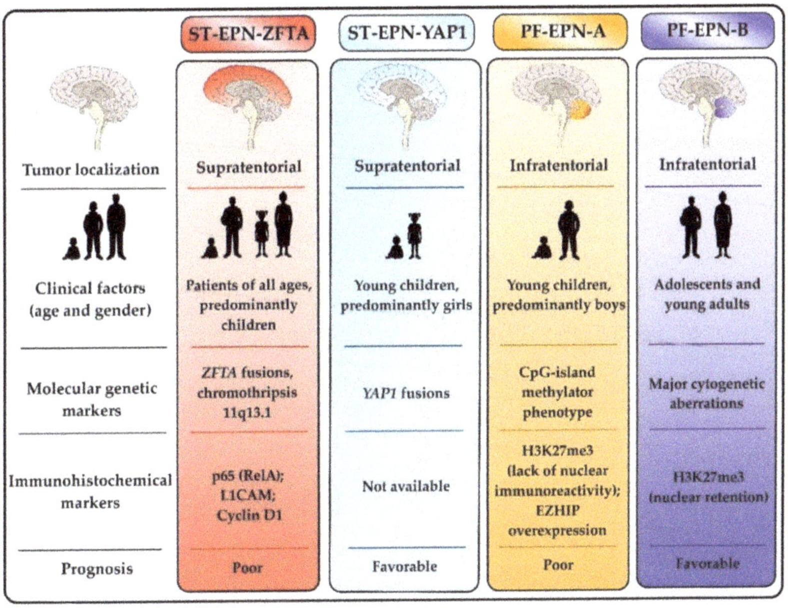

Figure 1. Basic classification of intracranial ependymal tumors.

4. Molecular Profiles of Sp-EPNs

Sp-EPNs constitute a heterogeneous group with a generally favorable prognosis. These tumors mostly occur in adult patients and are rare in children. Clinical outcomes for Sp-EPNs are better than for intracranial EPNs, with 5-year OS rates within the range of 60–90% [14]. Three molecular groups of Sp-EPNs were originally identified, including subependymomas, myxopapillary Sp-EPNs (SP-MPE), and Sp-EPNs per se; the molecular subgrouping shows excellent concordance with corresponding histopathological subtypes [14]. Our knowledge on the molecular pathogenesis of Sp-EPN tumors is limited. The groups reveal characteristic somatic copy number aberrations; most Sp-EPNs harbor 22q deletions involving neurofibromin 2 (*NF2*) tumor suppressor gene, whereas SP-MPEs show chromosomal instability.

SP-MPEs, the most prevalent type of pediatric spinal cord EPNs, predominantly arise in the conus medullaris, cauda equina, and filum terminale regions [66]. Despite their low mitotic index and slow-growing nature, SP-MPEs generally have far more aggressive behavior than other low-grade CNS tumors. Furthermore, pediatric SP-MPEs are especially aggressive, with much higher rates of local recurrence and secondary seeding to distant craniospinal sites or local spinal sites (64% cf. 32% in adults) [67]. As demonstrated by Ahmad et al. (2021), pediatric SP-MPEs exhibit aberrant activity of the mitochondrial metabolic pathways [68]. The only recurrent focal amplification identified for this group involves *HOXB* gene cluster mapping to 17q. *HOXB13* amplification represents a candidate diagnostic marker for SP-MPEs. The elevated expression of *HOXB13* enhances tumor cell proliferation and dissemination, playing a critical role in the development of metastasis [68]. Due to the high propensity for local recurrence and distant neural axis dissemination, the summary of the upcoming WHO CNS5 identifies SP-MPEs with grade 2 (rather than grade 1). SP-MPEs have distinctive histopathological features such as well-organized papillary architecture, with vascular cores and abundant mucinous extracellular matrix. Histological examination of tumor tissue is necessary and sufficient for the diagnosis of SP-MPE, whereas genetic testing is accessory [9].

A rare subtype of Sp-EPN in adult patients has been described recently, presenting highly aggressive clinical behavior with early metastasis, diffuse leptomeningeal spread throughout CNS, and resistance to standard treatment protocols. All of them harbored *MYCN* amplification and no other recurrent pathogenic events [69–71]. Importantly, these tumors formed a distinct methylation cluster of their own, and none of them clustered with any of the previously identified nine EPN groups. Recognizing the importance of clinical and molecular data on such tumors, the summary of the upcoming WHO CNS5 reports a novel nosological entity of *MYCN*-driven Sp-EPNs with dismal outcomes [9].

Genetic alterations found in particular in EPN groups are summarized in Table 1. Subgroup-specific diagnostic and candidate genes in pediatric EPNs are contained in Table 2.

Table 1. Summary of major molecular markers of ependymal tumors.

Localization	Molecular Group	Major Molecular Markers	Prevalence within the Group	Pathogenic Impact
Supratentorial	ST-EPN-ZFTA	*ZFTA–RELA* fusion, chromothripsis 11q13.1	90–95%	NF-kB pathway activation
		ZFTA–MAML2 *ZFTA–NCOA1* *ZFTA–NCOA2*	5–10%	EP300/CREBBP gene expression pathway
	ST-EPN-YAP1	*YAP1–MAMLD1*	95%	Hippo pathway activation
		YAP1–FAM118B	5%	
Infratentorial	PF-EPN-A	*EZHIP* overexpression	95%	CpG-island methylator phenotype
		HIST1H3C, HIST1H3B or *H3F3A* K27M substitution	<5%	
	PF-EPN-B	Major cytogenetic aberrations	Up to 100%	Ciliogenesis deregulation
Spinal	Sp-MPE	HOXB cluster genes amplification	Up to 100%	Mitochondrial metabolism pathways activation
	SP-EPN-MYCN	*MYCN* amplification	100%	Proliferative signaling

Table 2. Subgroup-specific diagnostic marker and candidate genes involved in recurrent chromosomal abnormalities in pediatric EPNs.

Molecular Group	Implicated Gene *	Gene Name	Cytogenetic Band	Pathogenic Impact	Evidence Level **	Evidence-Based Categorization ***	Hallmark of Cancer ****	
							Promotes	Suppresses
ST-EPN-ZFTA	*ZFTA*	Zinc finger translocation associated	11q13.1	5′-partner gene in *ZFTA–RELA* fusion	I	Tier I, level A	Genome instability	
	RELA	V-Rel avian reticu-loendotheliosis viral oncogene homolog A	11q13.1	3′-partner gene in *ZFTA–RELA* fusion	I	Tier I, level A	Escaping programmed cell death; tumor promoting inflammation	
	MAML2	Mastermind-like transcriptional coactivator 2	11q21	3′-partner gene in *ZFT–MAML2* fusion	III	Tier II, level C	Proliferative signaling; angiogenesis	
	NCOA1	Nuclear receptor coactivator 1	2p23.3	3′-partner gene in *ZFTA–NCOA1* fusion	III	Tier II, level C	Proliferative signaling; change of cellular energetics	
	NCOA2	Nuclear receptor coactivator 2	8q13.3	3′-partner gene in *ZFTA–NCOA2* fusion	III	Tier II, level C	Proliferative signaling; change of cellular energetics; escaping programmed cell death	
ST-EPN-YAP1	*YAP1*	Yes1-associated transcriptional regulator	11q22.1	5′-partner gene in *YAP1–MAMLD1* fusion	II	Tier I, level A	Proliferative signaling; escaping programmed cell death; invasion and metastasis	Escaping programmed cell death
	MAMLD1	Mastermind-like domain-containing 1	Xq28	3′-partner gene in *YAP1–MAMLD1* fusion	II	Tier I, level A	Proliferative signaling; angiogenesis	Escaping programmed cell death
	FAM118B	Family with sequence similarity 118 member B	11q24.2	3′-partner gene in *YAP1–FAM118B* fusion	IV	Tier II, level D	Unknown	
Non-ZFTA/Non-YAP1 ST-EPNs	*PLAGL1*	PLAG1-like zinc finger 1	6q24.2	3′-partner gene in *EWSR1-PLAGL1* fusion; 5′-partner gene in *PLAGL1–FOXO1* or *FLAGL1–EP300* fusion	IV	Tier II, level D	Suppression of growth	Escaping immunic response to cancer; tumor promoting inflammation; invasion and metastasis; angiogenesis

Table 2. *Cont.*

Molecular Group	Implicated Gene *	Gene Name	Cytogenetic Band	Pathogenic Impact	Evidence Level **	Evidence-Based Categorization ***	Hallmark of Cancer ****	
							Promotes	Suppresses
	EWSR1	EWS RNA binding protein 1	22q12.2	5′-partner gene in *EWSR1–PLAGL1* or *EWSR1–PATZ1* fusion	IV	Tier II, level D	Proliferative signaling; escaping programmed cell death; angiogenesis; invasion and metastasis	Genome instability and mutations
	FOXO1	Forkhead box O1	13q14.11	3′-partner gene in *PLAGL1–FOXO1* fusion	IV	Tier II, level D	Change of cellular energetics	Escaping programmed cell death
	EP300	E1A binding protein P300	22q13.2	3′-partner gene in *PLAGL1–EP300* fusion	IV	Tier II, level D	Suppression of growth	Escaping programmed cell death
	PATZ1	POZ/BTB and AT hook-containing zinc finger 1	22q12.2	3′-partner gene in *EWSR1–PATZ1* or *MN1–PATZ1* fusion	IV	Tier II, level D	Proliferative signaling; escaping programmed cell death	
	MN1	MN1 proto-oncogene, transcriptional regulator	22q12.1	5′-partner gene in *MN1-PATZ1* fusion	IV	Tier II, level D	Suppression of growth	Escaping programmed cell death
PF-EPN-A	*EZHIP*	EZH inhibitory protein	Xp11.22	Overexpression	IV	Tier II, level D		EZH1/EZH2-mediated trimethylation of H3K27
	EPOP	Elongin BC and polycomb repressive complex 2-associated protein	17q12	Overexpression	IV	Tier II, level D		EZH2-mediated trimethylation of H3K27
	HIST1H3C	H3 clustered histone 3	6p22.2	Somatic mutation	IV	Tier II, level D		EZH2-mediated trimethylation of H3K27
	HIST1H3B	H3 clustered histone 2	6p22.2	Somatic mutation	IV	Tier II, level D		EZH2-mediated trimethylation of H3K27
	H3F3A	H3.3 histone A	1q42.12	Somatic mutation	IV	Tier II, level D		EZH2-mediated trimethylation of H3K27
	BCL9	BCL9 transcription coactivator	1q21.2	Oncogene, involved in 1q gain	V	NA	Proliferative signaling; invasion and metastasis; angiogenesis	

Table 2. *Cont.*

Molecular Group	Implicated Gene *	Gene Name	Cytogenetic Band	Pathogenic Impact	Evidence Level **	Evidence-Based Categorization ***	Hallmark of Cancer ****	
							Promotes	Suppresses
	ARNT	Aryl hydrocarbon receptor nuclear translocator	1q21.3	Oncogene, involved in 1q gain	V	NA	Angiogenesis; change of cellular energetics	Invasion and metastasis
	SETDB1	SET domain bifurcated histone lysine methyltransferase 1	1q21.3	Oncogene, involved in 1q gain	V	NA	Epigenetic transcriptional repression by recruiting HP1 (CBX1, CBX3 and/or CBX5) proteins to methylated histones	
	NTRK1	Neurotrophic receptor tyrosine kinase 1	1q23.1	Oncogene, involved in 1q gain	V	NA	Proliferative signaling; escaping programmed cell death; angiogenesis	
	FCRL4	Fc receptor-like 4	1q23.1	Oncogene, involved in 1q gain	V	NA	Escaping immunic response to cancer	
	FCGR2B	Fc fragment of IgG receptor IIb	1q23.3	Oncogene, involved in 1q gain	V	NA	Suppression of growth	Escaping programmed cell death
	DDR2	Discoidin domain receptor tyrosine kinase 2	1q23.3	Oncogene, involved in 1q gain	V	NA	Invasion and metastasis	
	PBX1	PBX homeobox 1	1q23.3	Oncogene, involved in 1q gain	V	NA	Angiogenesis; escaping programmed cell death; change of cellular energetics	
	ABL2	ABL proto-oncogene 2, non-receptor tyrosine kinase	1q25.2	Oncogene, involved in 1q gain	V	NA	Proliferative signaling; invasion and metastasis; angiogenesis; genome instability and mutations; change of cellular energetics	Escaping programmed cell death
	MDM4	MDM4 regulator of P53	1q32.1	Oncogene, involved in 1q gain	V	NA	Proliferative signaling; invasion and metastasis; angiogenesis; escaping programmed cell death	Suppression of growth

Table 2. *Cont.*

Molecular Group	Implicated Gene *	Gene Name	Cytogenetic Band	Pathogenic Impact	Evidence Level **	Evidence-Based Categorization ***	Hallmark of Cancer ****	
							Promotes	Suppresses
	ELK4	ETS transcription factor ELK4	1q32.1	Oncogene, involved in 1q gain	V	NA	Proliferative signaling; escaping programmed cell death	
	RGS7	Regulator of G protein signaling 7	1q43	Oncogene, involved in 1q gain	V	NA	Change of cellular energetics	
	AKT3	AKT serine/threonine Kinase 3	1q43-q44	Oncogene, involved in 1q gain	V	NA	Proliferative signaling; suppression of growth; invasion and metastasis; angiogenesis; escaping programmed cell death; change of cellular energetics	Invasion and metastasis; angiogenesis; genome instability and mutations
	EPHA7	EPH receptor A7	6q16.1	Tumor suppressor gene, involved in 6q loss	V	NA		Escaping programmed cell death
	CCNC	Cyclin C	6q16.2	Tumor suppressor gene, involved in 6q loss	V	NA		Proliferative signaling
	PRDM1	PR/SET domain 1	6q21	Tumor suppressor gene, involved in 6q loss	V	NA	Suppression of growth	Escaping immunic response to cancer
	FOXO3	Forkhead box O3	6q21	Tumor suppressor gene, involved in 6q loss	V	NA	Change of cellular energetics	Escaping programmed cell death
	PTPRK	Protein tyrosine phosphatase receptor type K	6q22.33	Tumor suppressor gene, involved in 6q loss	V	NA	Escaping immunic response to cancer	Proliferative signaling
	BCLAF1	BCL2-associated transcription factor 1	6q23.3	Tumor suppressor gene, involved in 6q loss	V	NA		Escaping programmed cell death
	TNFAIP3	TNF alpha-induced protein 3	6q23.3	Tumor suppressor gene, involved in 6q loss	V	NA		Escaping immunic response to cancer; tumor promoting inflammation

Table 2. *Cont.*

Molecular Group	Implicated Gene *	Gene Name	Cytogenetic Band	Pathogenic Impact	Evidence Level **	Evidence-Based Categorization ***	Hallmark of Cancer ****	
							Promotes	Suppresses
	LATS1	Large tumor suppressor kinase 1	6q25.1	Tumor suppressor gene, involved in 6q loss	V	NA	Suppression of growth	Genome instability and mutations; escaping programmed cell death
	ESR1	Estrogen receptor 1	6q25.1	Tumor suppressor gene, involved in 6q loss	V	NA	Proliferative signaling; suppression of growth; escaping immunic response to cancer; invasion and metastasis	Invasion and metastasis
	ARID1B	AT-rich interaction domain 1B	6q25.3	Tumor suppressor gene, involved in 6q loss	V	NA	Suppression of growth; cell replicative immortality	Cell replicative immortality; invasion and metastasis; genome instability and mutations; escaping programmed cell death
	QKI	QKI, KH domain-containing RNA binding	6q26	Tumor suppressor gene, involved in 6q loss	V	NA	Suppression of growth; escaping programmed cell death	Escaping programmed cell death
PF-EPN-B	*LATS2*	Large tumor suppressor kinase 2	13q12.11	Tumor suppressor gene, involved in 13 q loss	V	NA	Suppression of growth; invasion and metastasis	Invasion and metastasis; genome instability and mutations; escaping programmed cell death
	CDX2	Caudal type homeobox 2	13q12.2	Tumor suppressor gene, involved in 13 q loss	V	NA		Proliferative signaling
	BRCA2	BRCA2 DNA repair associated	13q13.1	Tumor suppressor gene, involved in 13 q loss	V	NA		Genome instability and mutations; escaping programmed cell death
	RB1	RB transcriptional corepressor 1	13q14.2	Tumor suppressor gene, involved in 13 q loss	V	NA	Suppression of growth; escaping programmed cell death; change of cellular energetics	Escaping immunic response to cancer; invasion and metastasis; genome instability and mutations; escaping programmed cell death

Table 2. *Cont.*

Molecular Group	Implicated Gene *	Gene Name	Cytogenetic Band	Pathogenic Impact	Evidence Level **	Evidence-Based Categorization ***	Hallmark of Cancer ****	
							Promotes	Suppresses
	GPC5	Glypican 5	13q31.3	Tumor suppressor gene, involved in 13 q loss	V	NA	Suppression of growth; invasion and metastasis	
	SOX21	SRY-box transcription factor 21	13q32.1	Tumor suppressor gene, involved in 13 q loss	V	NA	Suppression of growth	Proliferative signaling
	ERCC5	ERCC excision repair 5, endonuclease	13q33.1	Tumor suppressor gene, involved in 13 q loss	V	NA	Genome instability and mutations; escaping programmed cell death	Genome instability and mutations
SP-MPE	*HOXB13*	Homeobox B13	17q21.32	Amplification	III	Tier II, level C	Change of cellular energetics	Escaping programmed cell death
SP-EPN-MYCN	*MYCN*	MYCN proto-oncogene, BHLH transcription factor	2p24.3	Amplification	II	Tier I, level A	Proliferative signaling; escaping immunic response to cancer; angiogenesis; genome instability and mutations; change of cellular energetics	Cell replicative immortality; invasion and metastasis; escaping programmed cell death

* The list of genes is selected from the Catalogue of Somatic Mutations in Cancer (COSMIC) Cancer Gene Census (https://cancer.sanger.ac.uk/census, accessed on 20 September 2021). Oncogenes and tumor suppressor genes are viewed as candidates for recurrent aberrations resulting in gain-of-function (1q gains) and loss-of-function (6q losses, 13q losses), respectively. ** Strength of evidence for gene diagnostic value based on Strength-of-evidence rating scheme of the Centre for Evidence-Based Medicine (https://www.cebm.net, accessed on 20 September 2021). *** Evidence-based variant (nucleotide substitution, copy-number variation, and fusion) of listed genes categorization based on the Joint Consensus Recommendation of the Association for Molecular Pathology, American Society of Clinical Oncology, and College of American Pathologists (AMP/ASCO/CAP recommendations). **** Potential roles of the cancer hallmark genes are annotated using COSMIC Cancer Gene Census (https://cancer.sanger.ac.uk/census, accessed on 20 September 2021), GeneCards: The Human Gene Database (https://www.genecards.org/, accessed on 20 September 2021), and KEGG: Kyoto Encyclopedia of Genes and Genomes (https://www.kegg.jp/kegg/, accessed on 20 September 2021). NA—nonapplicable.

5. Molecular Profiles of Subependymomas

Subependymomas are exceptionally rare slow-growing benign neoplasms, accounting for only 0.07–0.7% of all intracranial tumors [14,72]. These tumors typically arise in adults (aged 22–76 years) without strict predilection to a particular site in CNS. The most typical locations, the fourth and lateral ventricles, are encountered in up to 85% of the cases, followed by septum pellucidum, brainstem, and spine [73]. DNA methylation profiles of subependymomas differ depending on tumor localization, which provides certain grounds for molecular subgrouping [14]. However, all subependymomas have a favorable prognosis independently of localization. According to the cIMPACT-NOW group recommendations and the summary of the upcoming WHO CNS5, morphological examination provides adequate means for the diagnosis and prognostication; the integrative histo-molecular approach for subependymomas is accessory [9,16]. Recent findings suggest that the process of clonal evolution in subependymomas may give rise to more aggressive tumor clones enriched with pure EPN phenotypes, chromosome 6 losses, and *TERT* mutations. These markers, associated with increased risks of recurrence, should be considered as an indication for more intensive therapies, especially under conditions of subtotal tumor resection [44].

6. Laboratory Approaches for EPN Diagnostics

From a histological perspective, EPNs show moderate cellularity and variable mitotic activity; they consist of monomorphic rounded or oval cells with scant cytoplasm and vesicular nuclei containing granular (salt-and-pepper) chromatin. Key histological features of EPNs are perivascular (pseudo) rosettes and ependymal rosettes. EPNs are positive for glial fibrillary acidic protein (GFAP), S100, vimentin, rarely cytokeratin, and epithelial membrane antigen (EMA, positive along the luminal surface of ependymal rosettes or as characteristic dot-like or ring-like intracytoplasmic patterns) and negative for most neuronal antigens [6].

No unified standards for molecular diagnostics of EPNs have been introduced so far. The molecular group for each particular case is determined by whole-genome DNA methylation profiling as a golden standard. Comprehensive analysis of DNA methylation signatures is the method of choice for differential diagnosis within a broad spectrum of glial and embryonal tumors including CNS neuroblastoma with *FOXR2* activation, CNS Ewing sarcoma family tumor with *CIC* alteration, CNS high-grade neuroepithelial tumor with *MN1* alteration, and CNS high-grade neuroepithelial tumor with *BCOR* alteration [24,74]. The advantage of using this technique is the suitability of formalin-fixed paraffin-embedded tumor tissue for the analysis (methylated sites in genomic DNA are preserved during fixation, dehydration, etc.). However, the use of DNA methylation assays in routine laboratory practice has serious limitations, as the data processing is complex and the method itself is expensive, sophisticated, and labor-consuming. Alternative diagnostic algorithms may be based on a combination of economically justified methods including histological examination, immunohistochemistry (IHC) testing, reverse transcription-polymerase chain reaction (RT-PCR), Sanger sequencing, fluorescence in situ hybridization (FISH), and probably also the NanoString nCounter® platform (Figure 2).

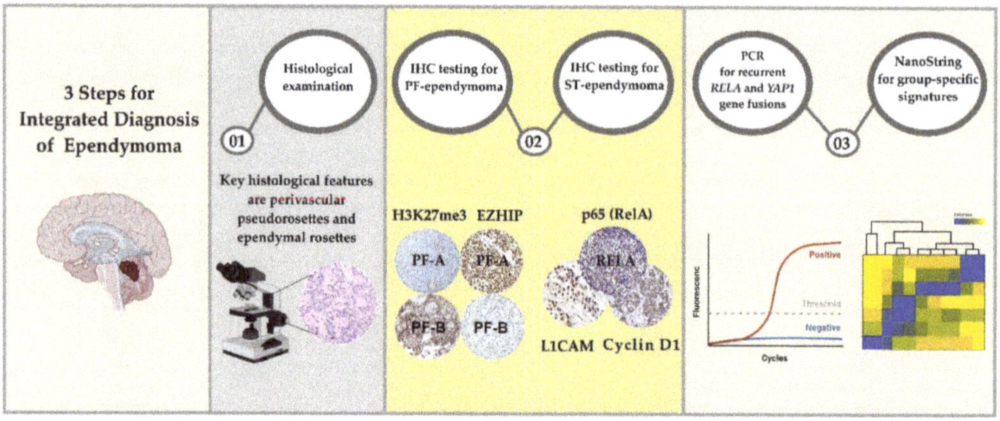

Figure 2. The integrated approach for molecular group determination applied to ependymal tumors.

6.1. *Differential Diagnosis of ST-EPNs*

In many cases, unambiguous determination of the molecular group of a tumor can be afforded by a certain combination of routine morphological and genetic approaches [75]. For instance, nuclear accumulation of p65 (RelA) and the presence of *ZFTA–RELA* fusion are sufficient for the identification of the sample with the core ST-EPN-ZFTA group harboring the classical *ZFTA–RELA* transcript. The diversity of fusion transcripts, which results from the variability of breakpoints in *RELA* and its partner genes, significantly complicates the identification of particular rearrangements and their use as markers. Conventional RT-PCR tests are targeted at the two most frequent fusions *ZFTA* (exon 2)–*RELA* (exon 2) and *ZFTA* (exon 3)–*RELA* (exon 2) [19,75,76].

A highly efficient way of screening for *ZFTA* and *RELA* rearrangements is provided by FISH with break-apart probes for one of the fusion partner genes. This approach allows detecting rearrangement of the gene of interest without accurate determination of the breakpoint [19,21]. Its excellent concordance with other methods including nuclear expression of RelA and DNA methylation profiling should be noted [19].

ZFTA gene rearrangements induce the hyperexpression of L1 cell adhesion molecule (L1CAM) and/or cyclin D1 may be considered as a surrogate marker of the ST-EPN-ZFTA group [18,20,22,23]. However, IHC is not sufficient on its own: the results require confirmation by independent alternative methods since neither of the antibodies (anti-p65, anti-L1CAM, anti-cyclin D1) has enough sensitivity and/or specificity to reliably verify the ST-EPN-ZFTA group [20,22]. The diagnosis of ST-EPN-ZFTA with alternative (non-*RELA*) fusion gene can be suspected if tumor cells show immunopositivity for L1CAM and negativity for p65 [26].

Identification of a tumor with the ST-EPN-YAP1 group is based on detection of either *YAP1* rearrangement by FISH or one of the *YAP1*-fusions (*YAP1–MAMLD1, YAP1–FAM118B*) by RT-PCR, while IHC with anti-YAP1 and/or anti-Claudin-1 is noninformative [19,37].

6.2. *Differential Diagnosis of PF-EPNs*

Panwalkar et al. (2017) proposed IHC tests for H3K27me3 as a straightforward and affordable method for PF-EPN stratification. The authors demonstrate that PF-EPN-B tumors are completely positive for H3K27me3, whereas the presence of H3K27me3 signal in less than 80% of tumor cell nuclei (against the total positivity of endothelial cells used as an internal control) is indicative of PF-EPN-A. Moreover, groupings based on whole-genome DNA methylation profiling vs. IHC tests for H3K27me3 were 99.1% concordant; the outlier was H3K27me3-negative tumor from a 12-year-old female patient, classified as PF-EPN-B on the basis of DNA methylation analysis [77]. The methodology was further enhanced by Fukuoka et al. (2018) who attempted at finer quantitation of the IHC data. In about 62% of prediagnosed PF-EPN-A tumors examined by the authors, the H3K27me3 reactivity was shown by 5–50% of the tumor cell nuclei; for the rest of PF-EPN-A tumors (about 40%), the proportion was less than 5%. In contrast, the vast majority of PF-EPN-B tumors showed intact expression of H3K27me3, with 90–95% of tumor cell nuclei stained positive, except two cases of reduced positivity (10–60%) classified as PF-EPN-B by DNA methylation profiling. Overall, the cut-off level of 80% H3K27me3-positive tumor cells allowed distinguishing PF-EPN-B from PF-EPN-A with 100% specificity and 86.7% sensitivity [21].

Antin et al. (2020) suggested complementing the routine IHC panels for CNS tumor diagnostics with anti-EZHIP staining. The intensive diffuse nuclear staining with >90% tumor cells immunopositive was obtained for PF-EPN-A (with the exception of rare *H3* K27M-mutant tumors), diffuse midline gliomas with wild-type (non-mutant) *H3F3A*, and germinomas [78]. Nambirajan et al. (2021) evaluated an extended panel of IHC markers (H3K27me3, acetyl-H3K27, H3K27M, ATRX, EZHIP, EPOP, and Tenascin-C) for the use in differential diagnostics of PF-EPNs. The authors demonstrated that a combination of EZHIP- and H3K27me3-specific antibodies is sufficient for reliable verification of PF-EPN-A; this finding illustrates the consistency between genetic and IHC profiles of a tumor [79].

Given the existence of rare EPN subtypes e.g., PF-EPN with *ZFTA* rearrangement [65] or ST-EPN harboring *ZFTA* fusions with atypical (non-*RELA*) partners, correct identification of a tumor with a particular group requires an extended diagnostic algorithm accounting for gene expression signatures. The creation of a universal diagnostic tool for the determination of molecular subtypes of EPN can be based on the analysis of gene expression signatures using the NanoString nCounter® platform. The assay must be carried out with customized panels of markers, corresponding to EPN group-specific signatures of upregulated genes, designed specifically for this purpose. Lastowska et al. (2021) identified group-specific marker genes for *RELA*-fused EPN (*RELA*, *ELL3*, *FBP2*, *PCP4L1*, and *MYO3A*), *YAP1*-fused EPN (*MRAP*, *IGF1*, *CAPS*, and *WWC1*), PF-EPN-A (*LAMA2*, *ALDH1L1*, *SLC6A13*, *IGSF1*, and *CXorf67*), and PF-EPN-B (*NELL2*, *DNAH1*, *CEP83*, *C9orf72*, and *NXNL2*) tumors. The NanoString-based approach clearly separated PF-EPNs into two clusters, corresponding to the defined PF-EPN groups, based on the expression of selected genes. Extended cluster analysis allowed subdivision of PF-EPN-A tumors into PF-EPN-A1 and PF-EPN-A2 subgroups using four marker genes (*SKAP2*, *WIF1*, and *EN2*, *CNPY1*, respectively). The described gene panel for *RELA*-positive EPN did not allow differentiation of EPN with *ZFTA–MAML2* fusion [80]. The potential of the NanoString method for diagnosis and advanced classification of other CNS tumors has been confirmed in several studies [81–83]. The approach provides reliable stratification of medulloblastomas into four groups (designated WNT, SHH, group 3, and group 4) based on the evaluation of 22 transcripts expressed differentially among the groups and contributing to characteristic group-specific gene expression signatures [81,82]. Primitive neuroectodermal tumors of the CNS (CNS-PNETs) have been already defined as four new molecularly defined entities (CNS neuroblastoma with *FOXR2* activation, CNS Ewing sarcoma family tumor with *CIC* alteration, CNS high-grade neuroepithelial tumor with *MN1* alteration, and CNS high-grade neuroepithelial tumor with *BCOR* alteration) using a single multi-gene tumor-specific signature [83]. However, differential diagnostics of EPNs in mixed series with other CNS tumors by the NanoString approach is impossible. For correct molecular stratification of ependymal tumors by the NanoString approach, all samples intended for comparison must be morphologically identified as EPN prior to the analysis.

7. Therapeutic Targeting of EPNs

Despite the profound insights into ependymal tumor biology, consensus recommendations for the management of patients with EPNs with regard to molecular diagnosis are missing. As tumors are resistant to conventional chemotherapy, the search for druggable targets is highly relevant. However, no candidate targets have been identified despite the extensive efforts in molecular profiling. Clinical studies on general cohorts showed no significant objective response to ERBB1/2 (lapatinib) and VEGF inhibitors (bevacizumab, sunitinib), despite the evidence on overexpression of ERBB2, ERBB4, or VEGF in EPNs [84–87].

In the molecular era, cancer-specific somatic aberrations should be taken into account when choosing a treatment strategy. For PF-EPNs, the pathogenesis of which involves epigenetic mechanisms, the possibility of using pharmaceuticals targeted at epigenetic modifications, including abnormalities of DNA methylation and histone modifications, seems most rational. Despite the promising results of pre-clinical research [88,89], histone deacetylase (HDAC) inhibitors showed no therapeutic activity in patients, apparently due to the intricacies of subcellular localization of HDACs and inability of the inhibitors to accumulate inside the brain tissue in concentrations sufficient for therapeutic response [90]. Nevertheless, the search for new HDAC inhibitors with appropriate brain-penetrating capacities and safety profiles may provide a useful treatment strategy [91]. The possibility of using BET-bromodomain inhibitors as anticancer therapeutics is being investigated in pediatric EPN stem cell models [92]. The upregulation of EZHIP could have important implications for therapeutic approaches. As reported by Han et al. (2020), EZHIP has a conservative PALB2-binding domain which enables its functioning as a competitive inhibitor of BRCA2. Elevated levels of EZHIP prevent the formation of BRCA1-PALB2-BRCA2 complexes thus inhibiting the homologous recombination-mediated DNA repair pathway. The results indicate the potential of PARP inhibitors as targeted therapeutics in PF-EPN-A, especially when combined with radiotherapy [61].

NF-kappa B inhibitors are considered potential therapeutic agents for the treatment of ST-EPN-ZFTA with constitutive activation of the NF-κB signaling pathway. A trial currently in phase II is evaluating the effectiveness of marizomib—a second-generation irreversible proteasome inhibitor, enrolling patients with histologically proven spinal or intracranial EPN, including but not limited to ST-EPN-ZFTA (ClinicalTrials.gov Identifier: NCT03727841, accessed on 20.09.2021). At the moment, the enrollment is complete, and the study is in progress. Specific expression of PD-L1 on both tumor and myeloid cells in ST-EPN-RELA has been demonstrated, accompanied by high levels of PD-1 expressed by tumor-infiltrating T cells (both CD4 and CD8) [93,94]. In the context of immunotherapy, ST-EPN-RELA progression may be controlled with PD-1 inhibitors, such as pembroluzimab or nivolumab [95].

Despite the principal shift in the EPN diagnostics and molecular stratification, its immediate impact on the existing treatment regimens is low. The correction would require preclinical and clinical trials for EPNs with due consideration of the molecular subgrouping.

8. Conclusions

As demonstrated by advanced studies of the last decade, ependymomas constitute a heterogeneous group of tumors and differ by molecular etiology. This minireview underscores the importance of comprehensive molecular profiling for ependymal tumors aimed at identifying specific expression signatures and/or (epi)genetic variants. Molecular identification of an ependymal tumor with a particular molecular group should follow its anatomical and histopathological assessment. Advanced stratification of patients into risk groups provides a framework for personalized management, e.g., allows de-escalation of the therapy in patients with low-risk tumors (supratentorial ependymomas group YAP1 and infratentorial ependymomas group B). Detailed understanding of causative molecular abnormalities for particular tumors is pivotal for the development of novel therapeutic options.

Author Contributions: Conceptualization, M.Z.; writing—original draft preparation, M.Z., L.P. and A.D.; writing—review and editing M.Z., L.P., G.N. and A.D.; supervision G.N. and A.D. All authors have read and agreed to the published version of the manuscript.

Funding: The study was supported by Foundation for support and development in the field of pediatric hematology, oncology and immunology "Science for Children".

Conflicts of Interest: The authors declare that they have no conflict of interest.

References

1. Ostrom, Q.T.; Patil, N.; Cioffi, G.; Waite, K.; Kruchko, C.; Barnholtz-Sloan, J.S. CBTRUS Statistical Report: Primary Brain and Other Central Nervous System Tumors Diagnosed in the United States in 2013–2017. *Neuro-Oncol.* **2020**, *22* (Suppl. 1), iv1–iv96. [CrossRef] [PubMed]
2. Jünger, S.T.; Mynarek, M.; Wohlers, I.; Dörner, E.; Mühlen, A.Z.; Velez-Char, N.; von Hoff, K.; Rutkowski, S.; Warmuth-Metz, M.; Kortmann, R.D.; et al. Improved Risk-Stratification for Posterior Fossa Ependymoma of Childhood Considering Clinical, Histological and Genetic Features—A Retrospective Analysis of the HIT Ependymoma Trial Cohort. *Acta Neuropathol. Commun.* **2019**, *7*, 181. [CrossRef] [PubMed]
3. Zapotocky, M.; Beera, K.; Adamski, J.; Laperierre, N.; Guger, S.; Janzen, L.; Lassaletta, A.; Figueiredo Nobre, L.; Bartels, U.; Tabori, U.; et al. Survival and functional outcomes of molecularly defined childhood posterior fossa ependymoma: Cure at a cost. *Cancer* **2019**, *125*, 1867–1876. [CrossRef] [PubMed]
4. Massimino, M.; Barretta, F.; Modena, P.; Witt, H.; Minasi, S.; Pfister, S.M.; Pajtler, K.W.; Antonelli, M.; Gandola, L.; Luisa Garrè, M.; et al. Second series by the Italian Association of Pediatric Hematology and Oncology of children and adolescents with intracranial ependymoma: An integrated molecular and clinical characterization with a long-term follow-up. *Neuro Oncol.* **2021**, *23*, 848–857. [CrossRef]
5. Ritzmann, T.A.; Rogers, H.A.; Paine, S.M.L.; Storer, L.C.D.; Jacques, T.S.; Chapman, R.J.; Ellison, D.; Donson, A.M.; Foreman, N.K.; Grundy, R.G. A Retrospective Analysis of Recurrent Pediatric Ependymoma Reveals Extremely Poor Survival and Ineffectiveness of Current Treatments across Central Nervous System Locations and Molecular Subgroups. *Pediatr. Blood Cancer* **2020**, *67*, e28426. [CrossRef]
6. Ellison, D.W.; Kocak, M.; Figarella-Branger, D.; Felice, G.; Catherine, G.; Pietsch, T.; Frappaz, D.; Massimino, M.; Grill, J.; Boyett, J.M.; et al. Histopathological Grading of Pediatric Ependymoma: Reproducibility and Clinical Relevance in European Trial Cohorts. *J. Negat. Results BioMed* **2011**, *10*, 7. [CrossRef]
7. Upadhyaya, S.A.; Robinson, G.W.; Onar-Thomas, A.; Orr, B.A.; Billups, C.A.; Bowers, D.C.; Bendel, A.E.; Hassall, T.; Crawford, J.R.; Partap, S.; et al. Molecular Grouping and Outcomes of Young Children with Newly Diagnosed Ependymoma Treated on the Multi-Institutional SJYC07 Trial. *Neuro-Oncology* **2019**, *21*, 1319–1330. [CrossRef]
8. Louis, D.N.; Wesseling, P.; Aldape, K.; Brat, D.J.; Capper, D.; Cree, I.A.; Eberhart, C.; Figarella-Branger, D.; Fouladi, M.; Fuller, G.N.; et al. cIMPACT-NOW update 6: New entity and diagnostic principle recommendations of the cIMPACT-Utrecht meeting on future CNS tumor classification and grading. *Brain Pathol.* **2020**, *30*, 844–856. [CrossRef]
9. Louis, D.N.; Perry, A.; Wesseling, P.; Brat, D.J.; Cree, I.A.; Figarella-Branger, D.; Hawkins, C.; Ng, H.K.; Pfister, S.M.; Reifenberger, G.; et al. The 2021 WHO Classification of Tumors of the Central Nervous System: A summary. *Neuro Oncol.* **2021**, *23*, 1231–1251. [CrossRef]
10. Leu, S.; von Felten, S.; Frank, S.; Boulay, J.L.; Mariani, L. IDH Mutation Is Associated with Higher Risk of Malignant Transformation in Low-Grade Glioma. *J. Neurooncol.* **2016**, *127*, 363–372. [CrossRef]
11. Jairam, V.; Kann, B.H.; Park, H.S.; Miccio, J.A.; Beckta, J.M.; Yu, J.B.; Prabhu, R.S.; Gao, S.J.; Mehta, M.P.; Curran, W.J.; et al. Defining an Intermediate-Risk Group for Low-Grade Glioma: A National Cancer Database Analysis. *Anticancer Res.* **2019**, *39*, 2911–2918. [CrossRef]
12. Yang, R.R.; Aibaidula, A.; Wang, W.W.; Chan, A.K.; Shi, Z.F.; Zhang, Z.Y.; Chan, D.; Poon, W.S.; Liu, X.Z.; Li, W.C.; et al. Pediatric low-grade gliomas can be molecularly stratified for risk. *Acta Neuropathol.* **2018**, *136*, 641–655. [CrossRef]
13. Schwalbe, E.C.; Lindsey, J.C.; Nakjang, S.; Crosier, S.; Smith, A.J.; Hicks, D.; Rafiee, G.; Hill, R.M.; Iliasova, A.; Stone, T.; et al. Novel Molecular Subgroups for Clinical Classification and Outcome Prediction in Childhood Medulloblastoma: A Cohort Study. *Lancet Oncol.* **2017**, *18*, 958–971. [CrossRef]
14. Pajtler, K.W.; Witt, H.; Sill, M.; Jones, D.T.W.; Hovestadt, V.; Kratochwil, F.; Wani, K.; Tatevossian, R.; Punchihewa, C.; Johann, P.; et al. Molecular Classification of Ependymal Tumors across All CNS Compartments, Histopathological Grades, and Age Groups. *Cancer Cell* **2015**, *27*, 728–743. [CrossRef]
15. Ramaswamy, V.; Hielscher, T.; Mack, S.C.; Lassaletta, A.; Lin, T.; Pajtler, K.W.; Jones, D.T.W.; Luu, B.; Cavalli, F.M.G.; Aldape, K.; et al. Therapeutic Impact of Cytoreductive Surgery and Irradiation of Posterior Fossa Ependymoma in the Molecular Era: A Retrospective Multicohort Analysis. *J. Clin. Oncol.* **2016**, *34*, 2468–2477. [CrossRef]
16. Ellison, D.W.; Aldape, K.D.; Capper, D.; Fouladi, M.; Gilbert, M.R.; Gilbertson, R.J.; Hawkins, C.; Merchant, T.E.; Pajtler, K.; Venneti, S.; et al. CIMPACT-NOW Update 7: Advancing the Molecular Classification of Ependymal Tumors. *Brain Pathol.* **2020**, *30*, 863–866.
17. Parker, M.; Mohankumar, K.M.; Punchihewa, C.; Weinlich, R.; Dalton, J.D.; Li, Y.; Lee, R.; Tatevossian, R.G.; Phoenix, T.N.; Thiruvenkatam, R.; et al. C11orf95–RELA Fusions Drive Oncogenic NF-KB Signalling in Ependymoma. *Nature* **2014**, *506*, 451–455. [CrossRef]

18. Malgulwar, P.B.; Nambirajan, A.; Pathak, P.; Faruq, M.; Rajeshwari, M.; Singh, M.; Suri, V.; Sarkar, C.; Sharma, M.C. C11orf95-RELA Fusions and Upregulated NF-KB Signalling Characterise a Subset of Aggressive Supratentorial Ependymomas That Express L1CAM and Nestin. *J. Neurooncol.* **2018**, *138*, 29–39. [CrossRef]
19. Pagès, M.; Pajtler, K.W.; Puget, S.; Castel, D.; Boddaert, N.; Tauziède-Espariat, A.; Picot, S.; Debily, M.; Kool, M.; Capper, D.; et al. Diagnostics of Pediatric Supratentorial RELA Ependymomas: Integration of Information from Histopathology, Genetics, DNA Methylation and Imaging. *Brain Pathol.* **2019**, *29*, 325–335. [CrossRef]
20. Torre, M.; Alexandrescu, S.; Dubuc, A.M.; Ligon, A.H.; Hornick, J.L.; Meredith, D.M. Characterization of Molecular Signatures of Supratentorial Ependymomas. *Mod. Pathol.* **2020**, *33*, 47–56. [CrossRef]
21. Fukuoka, K.; Kanemura, Y.; Shofuda, T.; Fukushima, S.; Yamashita, S.; Narushima, D.; Kato, M.; Honda-Kitahara, M.; Ichikawa, H.; Kohno, T.; et al. Significance of Molecular Classification of Ependymomas: C11orf95-RELA Fusion-Negative Supratentorial Ependymomas Are a Heterogeneous Group of Tumors. *Acta Neuropathol. Commun.* **2018**, *6*, 134. [CrossRef]
22. Gessi, M.; Giagnacovo, M.; Modena, P.; Elefante, G.; Gianno, F.; Buttarelli, F.R.; Arcella, A.; Donofrio, V.; Diomedi Camassei, F.; Nozza, P.; et al. Role of Immunohistochemistry in the Identification of Supratentorial C11ORF95-RELA Fused Ependymoma in Routine Neuropathology. *Am. J. Surg. Pathol.* **2019**, *43*, 56–63. [CrossRef]
23. Pietsch, T.; Wohlers, I.; Goschzik, T.; Dreschmann, V.; Denkhaus, D.; Dörner, E.; Rahmann, S.; Klein-Hitpass, L. Supratentorial Ependymomas of Childhood Carry C11orf95–RELA Fusions Leading to Pathological Activation of the NF-KB Signaling Pathway. *Acta Neuropathol.* **2014**, *127*, 609–611. [CrossRef]
24. Zschernack, V.; Jünger, S.T.; Mynarek, M.; Rutkowski, S.; Garre, M.L.; Ebinger, M.; Neu, M.; Faber, J.; Erdlenbruch, B.; Claviez, A.; et al. Supratentorial Ependymoma in Childhood: More than Just RELA or YAP. *Acta Neuropathol.* **2021**, *141*, 455–466. [CrossRef]
25. Tamai, S.; Nakano, Y.; Kinoshita, M.; Sabit, H.; Nobusawa, S.; Arai, Y.; Hama, N.; Totoki, Y.; Shibata, T.; Ichimura, K.; et al. Ependymoma with C11orf95-MAML2 Fusion: Presenting with Granular Cell and Ganglion Cell Features. *Brain Tumor Pathol.* **2021**, *38*, 64–70. [CrossRef]
26. Tauziède-Espariat, A.; Siegfried, A.; Nicaise, Y.; Kergrohen, T.; Sievers, P.; Vasiljevic, A.; Roux, A.; Dezamis, E.; Benevello, C.; Machet, M.C.; et al. Supratentorial non-RELA, ZFTA-fused ependymomas: A comprehensive phenotype genotype correlation highlighting the number of zinc fingers in ZFTA-NCOA1/2 fusions. *Acta Neuropathol. Commun.* **2021**, *9*, 135. [CrossRef]
27. Zheng, T.; Ghasemi, D.R.; Okonechnikov, K.; Korshunov, A.; Sill, M.; Maass, K.K.; Goncalves, B.; da Silva, P.; Ryzhova, M.; Gojo, J.; et al. Cross-species genomics reveals oncogenic dependencies in ZFTA/C11orf95 fusion-positive supratentorial ependymomas. *Cancer Discov.* **2021**, *11*, 2230–2247. [CrossRef]
28. Tomomasa, R.; Arai, Y.; Kawabata-Iwakawa, R.; Fukuoka, K.; Nakano, Y.; Hama, N.; Nakata, S.; Suzuki, N.; Ishi, Y.; Tanaka, S.; et al. Ependymoma-like tumor with mesenchymal differentiation harboring C11orf95-NCOA1/2 or -RELA fusion: A hitherto unclassified tumor related to ependymoma. *Brain Pathol.* **2021**, *31*, e12943. [CrossRef]
29. de Sousa, G.R.; Marie, S.; Oba-Shinjo, S.M.; Ramalho, L.; Tone, L.G.; Valera, E.T. A novel type of C11orf95-LOC-RELA fusion in a grade II supratentorial ependymoma: Report of a case with literature review. *Childs Nerv. Syst.* **2019**, *35*, 689–694. [CrossRef]
30. Zhu, J.J.; Jillette, N.; Li, X.N.; Cheng, A.W.; Lau, C.C. C11orf95-RELA reprograms 3D epigenome in supratentorial ependymoma. *Acta Neuropathol.* **2020**, *140*, 951–960. [CrossRef]
31. Arabzade, A.; Zhao, Y.; Varadharajan, S.; Chen, H.-C.; Jessa, S.; Rivas, B.; Stuckert, A.J.; Solis, M.; Kardian, A.; Tlais, D.; et al. ZFTA-RELA Dictates Oncogenic Transcriptional Programs to Drive Aggressive Supratentorial Ependymoma. *Cancer Discov.* **2021**, *11*, 2200–2215. [CrossRef] [PubMed]
32. Godwin, P.; Baird, A.M.; Heavey, S.; Barr, M.P.; O'Byrne, K.J.; Gately, K. Targeting Nuclear Factor-Kappa B to Overcome Resistance to Chemotherapy. *Front. Oncol.* **2013**, *3*, 120. [CrossRef] [PubMed]
33. Ozawa, T.; Kaneko, S.; Szulzewsky, F.; Qiao, Z.; Takadera, M.; Narita, Y.; Kondo, T.; Holland, E.C.; Hamamoto, R.; Ichimura, K. C11orf95-RELA Fusion Drives Aberrant Gene Expression through the Unique Epigenetic Regulation for Ependymoma Formation. *Acta Neuropathol. Commun.* **2021**, *9*, 36. [CrossRef] [PubMed]
34. Merchant, T.E.; Bendel, A.E.; Sabin, N.D.; Burger, P.C.; Shaw, D.W.; Chang, E.; Wu, S.; Zhou, T.; Eisenstat, D.D.; Foreman, N.K.; et al. Conformal Radiation Therapy for Pediatric Ependymoma, Chemotherapy for Incompletely Resected Ependymoma, and Observation for Completely Resected, Supratentorial Ependymoma. *J. Clin. Oncol.* **2019**, *37*, 974–983. [CrossRef]
35. Pajtler, K.W.; Wei, Y.; Okonechnikov, K.; Silva, P.B.G.; Vouri, M.; Zhang, L.; Brabetz, S.; Sieber, L.; Gulley, M.; Mauermann, M.; et al. YAP1 Subgroup Supratentorial Ependymoma Requires TEAD and Nuclear Factor I-Mediated Transcriptional Programmes for Tumorigenesis. *Nat. Commun.* **2019**, *10*, 3914. [CrossRef]
36. Wang, J.; Wang, L.; Fu, L.; Li, Q.C.; Qiu, X.S.; Wang, E.H.; Yu, J.H. Supratentorial ependymoma with YAP1:FAM118B fusion: A case report. *Neuropathology* **2021**, *41*, 133–138. [CrossRef]
37. Andreiuolo, F.; Varlet, P.; Tauziède-Espariat, A.; Jünger, S.T.; Dörner, E.; Dreschmann, V.; Kuchelmeister, K.; Waha, A.; Haberler, C.; Slavc, I.; et al. Childhood Supratentorial Ependymomas with *YAP1-MAMLD1* Fusion: An Entity with Characteristic Clinical, Radiological, Cytogenetic and Histopathological Features: YAP1-MAMLD1 Ependymoma. *Brain Pathol.* **2019**, *29*, 205–216. [CrossRef]
38. Pajtler, K.W.; Mack, S.C.; Ramaswamy, V.; Smith, C.A.; Witt, H.; Smith, A.; Hansford, J.R.; von Hoff, K.; Wright, K.D.; Hwang, E.; et al. The Current Consensus on the Clinical Management of Intracranial Ependymoma and Its Distinct Molecular Variants. *Acta Neuropathol.* **2017**, *133*, 5–12. [CrossRef]

39. Sievers, P.; Henneken, S.C.; Blume, C.; Sill, M.; Schrimpf, D.; Stichel, D.; Okonechnikov, K.; Reuss, D.E.; Benzel, J.; Maaß, K.K.; et al. Recurrent fusions in PLAGL1 define a distinct subset of pediatric-type supratentorial neuroepithelial tumors. *Acta Neuropathol.* **2021**. [CrossRef]
40. Lopez-Nunez, O.; Cafferata, B.; Santi, M.; Ranganathan, S.; Pearce, T.M.; Kulich, S.M.; Bailey, K.M.; Broniscer, A.; Rossi, S.; Zin, A.; et al. The spectrum of rare central nervous system (CNS) tumors with EWSR1-non-ETS fusions: Experience from three pediatric institutions with review of the literature. *Brain Pathol.* **2021**, *31*, 70–83. [CrossRef]
41. Siegfried, A.; Rousseau, A.; Maurage, C.A.; Pericart, S.; Nicaise, Y.; Escudie, F.; Grand, D.; Delrieu, A.; Gomez-Brouchet, A.; Le Guellec, S.; et al. EWSR1-PATZ1 gene fusion may define a new glioneuronal tumor entity. *Brain Pathol.* **2019**, *29*, 53–62. [CrossRef]
42. Olsen, T.K.; Panagopoulos, I.; Meling, T.R.; Micci, F.; Gorunova, L.; Thorsen, J.; Due-Tønnessen, B.; Scheie, D.; Lund-Iversen, M.; Krossnes, B.; et al. Fusion Genes with *ALK* as Recurrent Partner in Ependymoma-like Gliomas: A New Brain Tumor Entity? *Neuro Oncol.* **2015**, *17*, 1365–1373. [CrossRef]
43. Baroni, L.V.; Sundaresan, L.; Heled, A.; Coltin, H.; Pajtler, K.W.; Lin, T.; Merchant, T.E.; McLendon, R.; Faria, C.; Buntine, M.; et al. Ultra high-risk PFA ependymoma is characterized by loss of chromosome 6q. *Neuro Oncol.* **2021**, *23*, 1360–1370. [CrossRef]
44. Thomas, C.; Thierfelder, F.; Träger, M.; Soschinski, P.; Müther, M.; Edelmann, D.; Förster, A.; Geiler, C.; Kim, H.Y.; Filipski, K.; et al. TERT promoter mutation and chromosome 6 loss define a high-risk subtype of ependymoma evolving from posterior fossa subependymoma. *Acta Neuropathol.* **2021**, *141*, 959–970. [CrossRef]
45. Mack, S.C.; Witt, H.; Piro, R.M.; Gu, L.; Zuyderduyn, S.; Stütz, A.M.; Wang, X.; Gallo, M.; Garzia, L.; Zayne, K.; et al. Epigenomic Alterations Define Lethal CIMP-Positive Ependymomas of Infancy. *Nature* **2014**, *506*, 445–450. [CrossRef]
46. Bayliss, J.; Mukherjee, P.; Lu, C.; Jain, S.U.; Chung, C.; Martinez, D.; Sabari, B.; Margol, A.S.; Panwalkar, P.; Parolia, A.; et al. Lowered H3K27me3 and DNA Hypomethylation Define Poorly Prognostic Pediatric Posterior Fossa Ependymomas. *Sci. Transl. Med.* **2016**, *8*, 366ra161. [CrossRef]
47. Beringer, M.; Pisano, P.; Di Carlo, V.; Blanco, E.; Chammas, P.; Vizán, P.; Gutiérrez, A.; Aranda, S.; Payer, B.; Wierer, M.; et al. EPOP Functionally Links Elongin and Polycomb in Pluripotent Stem Cells. *Mol. Cell* **2016**, *64*, 645–658. [CrossRef]
48. Pajtler, K.W.; Wen, J.; Sill, M.; Lin, T.; Orisme, W.; Tang, B.; Hübner, J.-M.; Ramaswamy, V.; Jia, S.; Dalton, J.D.; et al. Molecular Heterogeneity and CXorf67 Alterations in Posterior Fossa Group A (PFA) Ependymomas. *Acta Neuropathol.* **2018**, *136*, 211–226. [CrossRef]
49. Michealraj, K.A.; Kumar, S.A.; Kim, L.; Cavalli, F.; Przelicki, D.; Wojcik, J.B.; Delaidelli, A.; Bajic, A.; Saulnier, O.; MacLeod, G.; et al. Metabolic Regulation of the Epigenome Drives Lethal Infantile Ependymoma. *Cell* **2020**, *181*, 1329–1345.e24. [CrossRef]
50. Hübner, J.M.; Müller, T.; Papageorgiou, D.N.; Mauermann, M.; Krijgsveld, J.; Russell, R.B.; Ellison, D.W.; Pfister, S.M.; Pajtler, K.W.; Kool, M. EZHIP/CXorf67 mimics K27M mutated oncohistones and functions as an intrinsic inhibitor of PRC2 function in aggressive posterior fossa ependymoma. *Neuro Oncol.* **2019**, *21*, 878–889. [CrossRef]
51. Jain, S.U.; Do, T.J.; Lund, P.J.; Rashoff, A.Q.; Diehl, K.L.; Cieslik, M.; Bajic, A.; Juretic, N.; Deshmukh, S.; Venneti, S.; et al. PFA Ependymoma-Associated Protein EZHIP Inhibits PRC2 Activity through a H3 K27M-like Mechanism. *Nat. Commun.* **2019**, *10*, 2146. [CrossRef]
52. Dewaele, B.; Przybyl, J.; Quattrone, A.; Finalet Ferreiro, J.; Vanspauwen, V.; Geerdens, E.; Gianfelici, V.; Kalender, Z.; Wozniak, A.; Moerman, P.; et al. Identification of a Novel, Recurrent *MBTD1-CXorf67* Fusion in Low-grade Endometrial Stromal Sarcoma. *Int. J. Cancer* **2014**, *134*, 1112–1122. [CrossRef]
53. Khuong-Quang, D.A.; Buczkowicz, P.; Rakopoulos, P.; Liu, X.Y.; Fontebasso, A.M.; Bouffet, E.; Bartels, U.; Albrecht, S.; Schwartzentruber, J.; Letourneau, L.; et al. K27M Mutation in Histone H3.3 Defines Clinically and Biologically Distinct Subgroups of Pediatric Diffuse Intrinsic Pontine Gliomas. *Acta Neuropathol.* **2012**, *124*, 439–447. [CrossRef]
54. Karremann, M.; Gielen, G.H.; Hoffmann, M.; Wiese, M.; Colditz, N.; Warmuth-Metz, M.; Bison, B.; Claviez, A.; van Vuurden, D.G.; von Bueren, A.O.; et al. Diffuse High-Grade Gliomas with H3 K27M Mutations Carry a Dismal Prognosis Independent of Tumor Location. *Neuro Oncol.* **2018**, *20*, 123–131. [CrossRef]
55. Zaytseva, M.A.; Shekhtman, A.P.; Papusha, L.I.; Valiakhmetova, E.F.; Yasko, L.A.; Druy, A.E. Analysis of genetic aberrations in pediatric high-grade gliomas. *Advances in Molecular Oncology* **2020**, *7*, 37–47. (In Russian) [CrossRef]
56. Infinger, L.; Stevenson, C. Re-Examining the Need for Tissue Diagnosis in Pediatric Diffuse Intrinsic Pontine Gliomas: A Review. *Curr. Neuropharmacol.* **2016**, *15*, 129–133. [CrossRef]
57. Lu, V.M.; Alvi, M.A.; McDonald, K.L.; Daniels, D.J. Impact of the H3K27M mutation on survival in pediatric high-grade glioma: A systematic review and meta-analysis. *J. Neurosurg. Pediatr.* **2018**, *23*, 308–316. [CrossRef]
58. Mosaab, A.; El-Ayadi, M.; Khorshed, E.N.; Amer, N.; Refaat, A.; El-Beltagy, M.; Hassan, Z.; Soror, S.H.; Zaghloul, M.S.; El-Naggar, S. Histone H3K27M Mutation Overrides Histological Grading in Pediatric Gliomas. *Sci. Rep.* **2020**, *10*, 8368. [CrossRef]
59. Rudà, R.; Reifenberger, G.; Frappaz, D.; Pfister, S.M.; Laprie, A.; Santarius, T.; Roth, P.; Tonn, J.C.; Soffietti, R.; Weller, M.; et al. EANO guidelines for the diagnosis and treatment of ependymal tumors. *Neuro Oncol.* **2018**, *20*, 445–456. [CrossRef]
60. Merchant, T.E.; Boop, F.A.; Kun, L.E.; Sanford, R.A. A retrospective study of surgery and reirradiation for recurrent ependymoma. *Int. J. Radiat. Oncol. Biol. Phys.* **2008**, *71*, 87–97. [CrossRef]
61. Han, J.; Yu, M.; Bai, Y.; Yu, J.; Jin, F.; Li, C.; Zeng, R.; Peng, J.; Li, A.; Song, X.; et al. Elevated CXorf67 Expression in PFA Ependymomas Suppresses DNA Repair and Sensitizes to PARP Inhibitors. *Cancer Cell* **2020**, *38*, 844–856.e7. [CrossRef] [PubMed]

62. Tsai, J.W.; Manoharan, N.; Alexandrescu, S.; Zimmerman, M.A.; Scully, J.; Chordas, C.; Clymer, J.; Wright, K.D.; Filbin, M.; Ullrich, N.J.; et al. Outcomes after first relapse of childhood intracranial ependymoma. *Pediatr. Blood Cancer* **2021**, *68*, e28930. [CrossRef] [PubMed]
63. Adolph, J.E.; Fleischhack, G.; Mikasch, R.; Zeller, J.; Warmuth-Metz, M.; Bison, B.; Mynarek, M.; Rutkowski, S.; Schüller, U.; von Hoff, K.; et al. Local and systemic therapy of recurrent ependymoma in children and adolescents: Short- and long-term results of the E-HIT-REZ 2005 study. *Neuro Oncol.* **2021**, *23*, 1012–1023. [CrossRef] [PubMed]
64. Cavalli, F.M.G.; Hübner, J.M.; Sharma, T.; Luu, B.; Sill, M.; Zapotocky, M.; Mack, S.C.; Witt, H.; Lin, T.; Shih, D.J.H.; et al. Heterogeneity within the PF-B ependymoma subgroup. *Acta Neuropathol.* **2018**, *136*, 227–237. [CrossRef]
65. Keenan, C.; Graham, R.T.; Harreld, J.H.; Lucas, J.T., Jr.; Finkelstein, D.; Wheeler, D.; Li, X.; Dalton, J.; Upadhyaya, S.A.; Raimondi, S.C.; et al. Infratentorial C11orf95-fused gliomas share histologic, immunophenotypic, and molecular characteristics of supratentorial RELA-fused ependymoma. *Acta Neuropathol.* **2020**, *140*, 963–965. [CrossRef]
66. Abdallah, A.; Emel, E.; Gündüz, H.B.; Sofuoğlu, Ö.E.; Asiltürk, M.; Abdallah, B.G. Long-Term Surgical Resection Outcomes of Pediatric Myxopapillary Ependymoma: Experience of Two Centers and Brief Literature Review. *World Neurosurg.* **2020**, *136*, e245–e261. [CrossRef]
67. Bagley, C.A.; Wilson, S.; Kothbauer, K.F.; Bookland, M.J.; Epstein, F.; Jallo, G.I. Long term outcomes following surgical resection of myxopapillary ependymomas. *Neurosurg. Rev.* **2009**, *32*, 321–334. [CrossRef]
68. Ahmad, O.; Chapman, R.; Storer, L.C.; Luo, L.; Heath, P.R.; Resar, L.; Cohen, K.J.; Grundy, R.G.; Lourdusamy, A. Integrative molecular characterization of pediatric spinal ependymoma: The UK Children's Cancer and Leukaemia Group study. *Neurooncol. Adv.* **2021**, *3*, vdab043.
69. Ghasemi, D.R.; Sill, M.; Okonechnikov, K.; Korshunov, A.; Yip, S.; Schutz, P.W.; Scheie, D.; Kruse, A.; Harter, P.N.; Kastelan, M.; et al. MYCN amplification drives an aggressive form of spinal ependymoma. *Acta Neuropathol.* **2019**, *138*, 1075–1089. [CrossRef]
70. Swanson, A.A.; Raghunathan, A.; Jenkins, R.B.; Messing-Jünger, M.; Pietsch, T.; Clarke, M.J.; Kaufmann, T.J.; Giannini, C. Spinal Cord Ependymomas With MYCN Amplification Show Aggressive Clinical Behavior. *J. Neuropathol. Exp. Neurol.* **2019**, *78*, 791–797. [CrossRef]
71. Raffeld, M.; Abdullaev, Z.; Pack, S.D.; Xi, L.; Nagaraj, S.; Briceno, N.; Vera, E.; Pittaluga, S.; Lopes Abath Neto, O.; Quezado, M.; et al. High level MYCN amplification and distinct methylation signature define an aggressive subtype of spinal cord ependymoma. *Acta Neuropathol. Commun.* **2020**, *8*, 101. [CrossRef]
72. D'Amico, R.S.; Praver, M.; Zanazzi, G.J.; Englander, Z.K.; Sims, J.S.; Samanamud, J.L.; Ogden, A.T.; McCormick, P.C.; Feldstein, N.A.; McKhann, G.M.; et al. Subependymomas Are Low-Grade Heterogeneous Glial Neoplasms Defined by Subventricular Zone Lineage Markers. *World Neurosurg.* **2017**, *107*, 451–463. [CrossRef]
73. Kweh, B.; Rosenfeld, J.V.; Hunn, M.; Tee, J.W. Tumor characteristics and surgical outcomes of intracranial subependymomas: A systematic review and meta-analysis. *J. Neurosurg.* **2021**, *1*, 1–13. [CrossRef]
74. Korshunov, A.; Okonechnikov, K.; Schmitt-Hoffner, F.; Ryzhova, M.; Sahm, F.; Stichel, D.; Schrimpf, D.; Reuss, D.E.; Sievers, P.; Suwala, A.K.; et al. Molecular analysis of pediatric CNS-PNET revealed nosologic heterogeneity and potent diagnostic markers for CNS neuroblastoma with FOXR2-activation. *Acta Neuropathol. Commun.* **2021**, *9*, 20. [CrossRef]
75. de Sousa, G.R.; Lira, R.C.P.; de Almeida Magalhães, T.; da Silva, K.R.; Nagano, L.F.P.; Saggioro, F.P.; Baroni, M.; Marie, S.K.N.; Oba-Shinjo, S.M.; Brandelise, S.; et al. A coordinated approach for the assessment of molecular subgroups in pediatric ependymomas using low-cost methods. *J. Mol. Med.* **2021**, *99*, 1101–1113. [CrossRef]
76. Matsumoto, Y.; Ichikawa, T.; Kurozumi, K.; Otani, Y.; Date, I. Clinicopathological and Genetic Features of Supratentorial Cortical Ependymomas. *World Neurosurg.* **2019**, *129*, e417–e428. [CrossRef]
77. Panwalkar, P.; Clark, J.; Ramaswamy, V.; Hawes, D.; Yang, F.; Dunham, C.; Yip, S.; Hukin, J.; Sun, Y.; Schipper, M.J.; et al. Immunohistochemical analysis of H3K27me3 demonstrates global reduction in group-A childhood posterior fossa ependymoma and is a powerful predictor of outcome. *Acta Neuropathol.* **2017**, *134*, 705–714. [CrossRef]
78. Antin, C.; Tauziède-Espariat, A.; Debily, M.A.; Castel, D.; Grill, J.; Pagès, M.; Ayrault, O.; Chrétien, F.; Gareton, A.; Andreiuolo, F.; et al. EZHIP is a specific diagnostic biomarker for posterior fossa ependymomas, group PFA and diffuse midline gliomas H3-WT with EZHIP overexpression. *Acta Neuropathol. Commun.* **2020**, *8*, 183. [CrossRef]
79. Nambirajan, A.; Sharma, A.; Rajeshwari, M.; Boorgula, M.T.; Doddamani, R.; Garg, A.; Suri, V.; Sarkar, C.; Sharma, M.C. EZH2 inhibitory protein (EZHIP/Cxorf67) expression correlates strongly with H3K27me3 loss in posterior fossa ependymomas and is mutually exclusive with H3K27M mutations. *Brain Tumor Pathol.* **2021**, *38*, 30–40. [CrossRef]
80. Łastowska, M.; Matyja, E.; Sobocińska, A.; Wojtaś, B.; Niemira, M.; Szałkowska, A.; Krętowski, A.; Karkucińska-Więckowska, A.; Kaleta, M.; Ejmont, M.; et al. Transcriptional profiling of paediatric ependymomas identifies prognostically significant groups. *J. Pathol. Clin. Res.* **2021**. online ahead of print. [CrossRef]
81. Northcott, P.A.; Shih, D.J.H.; Remke, M.; Cho, Y.J.; Kool, M.; Hawkins, C.; Eberhart, C.G.; Dubuc, A.; Guettouche, T.; Cardentey, Y.; et al. Rapid, reliable, and reproducible molecular sub-grouping of clinical medulloblastoma samples. *Acta Neuropathol.* **2012**, *123*, 615–626. [CrossRef]
82. Druy, A.E.; Yasko, L.A.; Konovalov, D.M.; Ektova, A.P.; Valiakhmetova, E.F.; Olshanskaya, Y.V.; Maschan, A.A.; Novichkova, G.A.; Papusha, L.I. Identification of medulloblastoma molecular subgroups by gene expression profiling. *Pediatr. Hematol./Oncol. Immunopathol.* **2017**, *16*, 85–89. (In Russian) [CrossRef]

83. Łastowska, M.; Trubicka, J.; Sobocińska, A.; Wojtas, B.; Niemira, M.; Szałkowska, A.; Krętowski, A.; Karkucińska-Więckowska, A.; Kaleta, M.; Ejmontet, M.; et al. Molecular identification of CNS NB-FOXR2, CNS EFT-CIC, CNS HGNET-MN1 and CNS HGNET-BCOR pediatric brain tumors using tumor-specific signature genes. *Acta Neuropathol. Commun.* **2020**, *8*, 105. [CrossRef]
84. Gururangan, S.; Fangusaro, J.; Young Poussaint, T.; Onar-Thomas, A.; Gilbertson, R.J.; Vajapeyam, S.; Gajjar, A.; Goldman, S.; Friedman, H.S.; Packer, R.J.; et al. Lack of efficacy of bevacizumab + irinotecan in cases of pediatric recurrent ependymoma–a Pediatric Brain Tumor Consortium study. *Neuro Oncol.* **2012**, *14*, 1404–1412. [CrossRef]
85. DeWire, M.; Fouladi, M.; Turner, D.C.; Wetmore, C.; Hawkins, C.; Jacobs, C.; Yuan, Y.; Liu, D.; Goldman, S.; Fisher, P.; et al. An open-label, two-stage, phase II study of bevacizumab and lapatinib in children with recurrent or refractory ependymoma: A collaborative ependymoma research network study (CERN). *J. Neurooncol.* **2015**, *123*, 85–91. [CrossRef]
86. Wetmore, C.; Daryani, V.M.; Billups, C.A.; Boyett, J.M.; Leary, S.; Tanos, R.; Goldsmith, K.C.; Stewart, C.F.; Blaney, S.M.; Gajjar, A. Phase II evaluation of sunitinib in the treatment of recurrent or refractory high-grade glioma or ependymoma in children: A children's Oncology Group Study ACNS1021. *Cancer Med.* **2016**, *5*, 1416–1424. [CrossRef]
87. Fouladi, M.; Stewart, C.F.; Blaney, S.M.; Onar-Thomas, A.; Schaiquevich, P.; Packer, R.J.; Goldman, S.; Geyer, J.R.; Gajjar, A.; Kun, L.E.; et al. A molecular biology and phase II trial of lapatinib in children with refractory CNS malignancies: A pediatric brain tumor consortium study. *J. Neurooncol.* **2013**, *114*, 173–179. [CrossRef]
88. Milde, T.; Kleber, S.; Korshunov, A.; Witt, H.; Hielscher, T.; Koch, P.; Kopp, H.G.; Jugold, M.; Deubzer, H.E.; Oehme, I.; et al. A novel human high-risk ependymoma stem cell model reveals the differentiation-inducing potential of the histone deacetylase inhibitor Vorinostat. *Acta Neuropathol.* **2011**, *122*, 637–650. [CrossRef]
89. Rahman, R.; Osteso-Ibanez, T.; Hirst, R.A.; Levesley, J.; Kilday, J.P.; Quinn, S.; Peet, A.; O'Callaghan, C.; Coyle, B.; Grundy, R.G. Histone deacetylase inhibition attenuates cell growth with associated telomerase inhibition in high-grade childhood brain tumor cells. *Mol. Cancer Ther.* **2010**, *9*, 2568–2581. [CrossRef]
90. Bukowinski, A.; Chang, B.; Reid, J.M.; Liu, X.; Minard, C.G.; Trepel, J.B.; Lee, M.J.; Fox, E.; Weigel, B.J. A phase 1 study of entinostat in children and adolescents with recurrent or refractory solid tumors, including CNS tumors: Trial ADVL1513, Pediatric Early Phase-Clinical Trial Network (PEP-CTN). *Pediatr. Blood Cancer* **2021**, *68*, e28892. [CrossRef]
91. Antonelli, R.; Jiménez, C.; Riley, M.; Servidei, T.; Riccardi, R.; Soriano, A.; Roma, J.; Martínez-Saez, E.; Martini, M.; Ruggiero, A.; et al. CN133, a Novel Brain-Penetrating Histone Deacetylase Inhibitor, Hampers Tumor Growth in Patient-Derived Pediatric Posterior Fossa Ependymoma Models. *Cancers* **2020**, *12*, 1922. [CrossRef] [PubMed]
92. Servidei, T.; Meco, D.; Martini, M.; Battaglia, A.; Granitto, A.; Buzzonetti, A.; Babini, G.; Massimi, L.; Tamburrini, G.; Scambia, G.; et al. The BET Inhibitor OTX015 Exhibits In Vitro and In Vivo Antitumor Activity in Pediatric Ependymoma Stem Cell Models. *Int. J. Mol. Sci.* **2021**, *22*, 1877. [CrossRef] [PubMed]
93. Witt, D.A.; Donson, A.M.; Amani, V.; Moreira, D.C.; Sanford, B.; Hoffman, L.M.; Handler, M.H.; Levy, J.; Jones, K.L.; Nellan, A.; et al. Specific expression of PD-L1 in RELA-fusion supratentorial ependymoma: Implications for PD-1-targeted therapy. *Pediatr. Blood Cancer* **2018**, *65*, e26960. [CrossRef] [PubMed]
94. Nambirajan, A.; Malgulwar, P.B.; Sharma, A.; Boorgula, M.T.; Doddamani, R.; Singh, M.; Suri, V.; Sarkar, C.; Sharma, M.C. Clinicopathological evaluation of PD-L1 expression and cytotoxic T-lymphocyte infiltrates across intracranial molecular subgroups of ependymomas: Are these tumors potential candidates for immune check-point blockade? *Brain Tumor Pathol.* **2019**, *36*, 152–161. [CrossRef]
95. Perruccio, K.; Mastronuzzi, A.; Lupattelli, M.; Arcioni, F.; Capolsini, I.; Cerri, C.; Gurdo, G.M.I.; Massei, M.S.; Mastrodicasa, E.; Caniglia, M. Targeted Therapy with Sirolimus and Nivolumab in a Child with Refractory Multifocal Anaplastic Ependymoma. *Reports* **2021**, *4*, 12. [CrossRef]

cancers

Article

Characteristics of Nephroblastoma/Nephroblastomatosis in Children with a Clinically Reported Underlying Malformation or Cancer Predisposition Syndrome

Nils Welter [1], Angelo Wagner [1], Rhoikos Furtwängler [1], Patrick Melchior [2], Leo Kager [3], Christian Vokuhl [4], Jens-Peter Schenk [5], Clemens Magnus Meier [6], Stefan Siemer [7], Manfred Gessler [8] and Norbert Graf [1,*]

1 Department of Pediatric Oncology and Hematology, Saarland University, 66421 Homburg, Germany; nils.welter@uks.eu (N.W.); s9aowagn@stud.uni-saarland.de (A.W.); rhoikos.furtwaengler@uks.eu (R.F.)
2 Department of Radiation Oncology, Saarland University, 66421 Homburg, Germany; patrick.melchior@uks.eu
3 St. Anna Kinderspital, Department of Pediatrics, Medical University Vienna, Kinderspitalgasse 6, 1090 Vienna, Austria; leo.kager@stanna.at
4 Section of Pediatric Pathology, University of Bonn, Venusberg-Campus 1, 53127 Bonn, Germany; Christian.vokuhl@ukbonn.de
5 Division of Pediatric Radiology, Clinic for Diagnostic and Interventional Radiology, University of Heidelberg, Im Neuenheimer Feld 430, 69120 Heidelberg, Germany; jens-peter.schenk@med.uni-heidelberg.de
6 Department of General, Visceral, Vascular and Pediatric Surgery, Saarland University, 66421 Homburg, Germany; clemens-magnus.meier@uks.eu
7 Department of Urology and Pediatric Urology, Saarland University, 66421 Homburg, Germany; stefan.siemer@uks.eu
8 Developmental Biochemistry and Comprehensive Cancer Center Mainfranken, Theodor-Boveri-Institute/Biocenter, University of Würzburg, 97074 Würzburg, Germany; gessler@biozentrum.uni-wuerzburg.de
* Correspondence: norbert.graf@uks.eu; Tel.: +49-(0)6841-1628397

Citation: Welter, N.; Wagner, A.; Furtwängler, R.; Melchior, P.; Kager, L.; Vokuhl, C.; Schenk, J.-P.; Meier, C.M.; Siemer, S.; Gessler, M.; et al. Characteristics of Nephroblastoma/ Nephroblastomatosis in Children with a Clinically Reported Underlying Malformation or Cancer Predisposition Syndrome. *Cancers* **2021**, *13*, 5016. https://doi.org/10.3390/cancers13195016

Academic Editors: Saurabh Agarwal and Jianhua Yang

Received: 7 August 2021
Accepted: 29 September 2021
Published: 7 October 2021
Corrected: 16 November 2021

Simple Summary: It is well known that different cancer predisposition syndromes are associated with characteristic WT-features. The following findings from our retrospective analysis of patients with nephroblastoma treated according to the SIOP/GPOH trials between 1989 and 2017 are relevant: (1) The outcome of patients with a cancer predisposition syndrome is not always favorable despite early diagnosis, small tumors and less metastatic disease. This finding is partly depending on complications related to the underlying syndrome. (2) Predisposition syndromes seem to be underdiagnosed as several clinical and pathological features of Wilms tumor being clearly linked to a cancer predisposition syndrome did not lead to genetic counseling before and after WT diagnosis. As a conclusion, in children with a nephroblastoma and specific clinical and pathological features that are in line with a nephroblastoma cancer predisposition syndrome such a syndrome should always be considered and ruled out if unknown at the time of tumor diagnosis.

Abstract: (1) Background: about 10% of Wilms Tumor (WT) patients have a malformation or cancer predisposition syndrome (CPS) with causative germline genetic or epigenetic variants. Knowledge on CPS is essential for genetic counselling. (2) Methods: this retrospective analysis focused on 2927 consecutive patients with WTs registered between 1989 and 2017 in the SIOP/GPOH studies. (3) Results: Genitourinary malformations (GU, $N = 66$, 2.3%), Beckwith-Wiedemann spectrum (BWS, $N = 32$, 1.1%), isolated hemihypertrophy (IHH, $N = 29$, 1.0%), Denys-Drash syndrome (DDS, $N = 24$, 0.8%) and WAGR syndrome ($N = 20$, 0.7%) were reported most frequently. Compared to others, these patients were younger at WT diagnosis (median age 24.5 months vs. 39.0 months), had smaller tumors (349.4 mL vs. 487.5 mL), less often metastasis (8.2% vs. 18%), but more often nephroblastomatosis (12.9% vs. 1.9%). WT with IHH was associated with blastemal WT and DDS with stromal subtype. Bilateral WTs were common in WAGR (30%), DDS (29%) and BWS (31%). Chemotherapy induced reduction in tumor volume was poor in DDS (0.4% increase) and favorable in BWS (86.9% reduction). The event-free survival (EFS) of patients with BWS was significantly ($p = 0.002$) worse than in others. (4) Conclusions: CPS should be considered in WTs with specific clinical features resulting in referral to a geneticist. Their outcome was not always favorable.

Keywords: nephroblastoma; clinical malformations; cancer predisposition syndromes; tumor surveillance; outcome

1. Introduction

Nephroblastoma or Wilms tumor (WT), the most common kidney tumor in childhood [1], can be cured in more than 90% today [2–4]. According to SIOP clinical studies and trials patients are diagnosed by imaging studies alone and preoperatively treated with AV (actinomycin and vincristine for 4 weeks) with localized or with (AV plus doxorubicin for 6 weeks) with metastatic tumors. During the registration process of patients, data on the kind of a cancer predisposition syndromes (CPS) or a malformation are provided by the treating hospital without further specifying malformations. In addition, participation in a surveillance protocol before the diagnosis in case of a CPS is registered in the database.

With 8 to 17% overall and up to 24% in bilateral WTs it has one of the highest association rates with congenital anomalies of all childhood cancers [5,6]. Such malformations and CPS related to the development of WTs are characterized by genetic or epigenetic alterations. For example, the WAGR syndrome, is clinically defined by a variable occurrence of WT in combination with aniridia, genitourinary malformations and a range of developmental delays [7–9]. It is caused by chromosome 11p13 deletions, including WT1 and neighboring genes, whereas Denys-Drash syndrome (DDS) is due to a dominant-negative WT1 mutation. DDS is characterized by the triad of WT, nephropathy and, if applicable, male pseudohermaphroditism [10–12]. In addition, genitourinary malformations (GU) have been linked to WT1 mutations [13,14]. Isolated hemihypertrophy (IHH) [15,16] and the Beckwith-Wiedemann spectrum (BWS) are overgrowth syndromes with elevated risk to develop WT. BWS shows a high variability of macroglossia, abdominal wall defects, visceromegaly, gigantism and hypoglycaemia caused by genetic and epigenetic alterations at 11p15.3 [12,17,18]. Other WT associated syndromes such as Perlman syndrome or Simpson-Golabi-Behmel syndrome are much rarer and have a different genetic background. In this paper we focus on the five most frequent WT malformations or CPS, namely WAGR, DDS, GU, IHH and BWS, to compare their clinical, pathological and outcome data with data from WTs without a known CPS.

2. Materials and Methods

We conducted a retrospective investigation on data of 2927 patients with WT and/or nephroblastomatosis from Germany, Austria and Switzerland enrolled in the SIOP/GPOH 9, 93-01 and 2001 studies between 1989 and 2017. Details of their treatment protocols have been reviewed previously [19]. Ethical approval was obtained from the Ärztekammer des Saarlandes (No: 136/01 from 20 September 2002 and, 248/13 from 13 January 2014). All parents or legal guardians of the affected children gave informed consent for study participation.

Pseudonymized data of all patients were stored in a central and encrypted SQL database. All patients identified in the database with a clinically documented malformation or CPS were reviewed by NW and NG, and details on these patients—including presentation, treatments and outcome—were collected from the SIOP-RTSG/GPOH database and, retrospectively, from status report forms, radiology, pathology and surgery reports, progress letters and telephone notes available at the data center. The identification of patients with malformations or CPS was based solely on clinical data provided by the registration CRF where associated congenital malformations or a syndrome were asked specifically for Aniridia, WAGR, genitourinary malformations, Denys Drash syndrome, BWS, IHH, Perlman syndrome. Free text could specify other malformations or syndromes that are not listed. This information is based on clinical characteristics. No information is provided if the syndrome was confirmed by genetic analysis. Patients with WAGR were also included in the paper by Hol et al. [8]. Tumor volume was calculated from imaging studies using the ellipsoid formula in those patients were CT or MRI of the tumor was available. Nephroblastomatosis was based on reference histology defined as multiple or diffuse nephrogenic rests but not further specified as perilobar, intralobar or both, as this information was not available for all patients with nephroblastomatosis. For statistical analysis all data were anonymized. IBM SPSS Statistics, version 25 and 27, was used for descriptive analyses (histograms, boxplots, pie charts, frequency charts and bar charts) and statistical comparisons (T-test for independent samples, Levene test, Chi-square test, multivariate analysis and Kaplan-Meier survival curves with Log Rank). p-values below 0.05 were considered as statistically significant. Overall survival (OS) included the time period between diagnosis and death of any reason, and event free survival (EFS) between diagnosis and any event, including recurrence of WT or nephroblastomatosis, death or loss to follow up.

3. Results

3.1. Characteristics of Study Population

An underlying malformation or syndrome was recorded in 198 out of all 2927 (6.8%) patients (Table 1). Bilateral disease occurred in 253 (8.6%) of patients and 29 patients with CPS or malformation were included in a surveillance program before diagnosis of a WT and/or nephroblastomatosis. In 137/2927 (4.6%) patients nephroblastomatosis was diagnosed, either isolated (73; 2.4%) or in conjunction with WT (64; 2.1%). This investigation highlights specifically a sub-cohort of 171 patients, who presented with one of the five most frequent malformations or syndromes (Table 1), that is GU (N = 66), BWS (N = 32), IHH (N = 29), DDS (N = 24) and WAGR syndrome (N = 20). In addition, 27 patients were diagnosed with a variety of other malformations or syndromes (Table S1). Interestingly, there was no increase of the percentage of patients with CPS or GU over time. Up to year 2000, 54.9% of CPS or GU were diagnosed and 45.1% in the following years.

Table 1. Frequency of malformations and CPS in WT/nephroblastomatosis (NBL).

	Frequency								
	All Patients with WT and/or NBL			Only Bilateral WT and/or NBL			Only Screened Patients with CPS/Malformation and WT and/or NBL		
All WT	2927	-	100%	253	8.6%	**	29 ***	****	**
WAGR	20		0.7%	6	2.4%	30.0%	8	27.6%	40.0%
GU	66		2.3%	8	3.2%	12.1%	1	3.4%	1.5%
DDS	24	$\sum = 171$	0.8%	7	2.8%	29.2%	3	10.3%	12.5%
BWS	32		1.1%	10	4.0%	31.3%	10	34.5%	31.3%
IHH	29		1.0%	4	1.6%	13.8%	4	13.8%	13.8%
Other *	27	-	0.9%	7	2.8%	25.9%	3	10.3%	11.1%
All	198	-	6.8%	42	16.6%	21.2%	29	100%	14.6%

* see Table S1; ** % related to the specific malformations or CPS, *** screening in 29 of 198 patients with malformations or CPS, **** % related to the 29 screened patients with malformations or CPS. WT: Wilms tumor; CPS: cancer predisposition syndrome; NBL: nephroblastomatosis.

3.2. Ultrasound Surveillance Every 3 Months

Altogether 29 (14.6%) patients had been screened by ultrasound every 3 months after the diagnosis of a CPS. The highest screening frequency resulting in the diagnosis of WT/nephroblastomatosis was reported in patients with WAGR (40%) and BWS (31.3%). (Table 1). No data were available on why not all children with an underlying syndrome were included in a screening program.

3.3. Gender Distribution and Age at Diagnosis of WT/Nephroblastomatosis

Gender distribution in the whole cohort of patients with syndrome-associated WT is similar to the total group of patients with WT/nephroblastomatosis with a predominance of females (51.4% females vs. 48.1% males, 0.5% gender not known) with the exception of GU cases. 43 males (3.2% of 1353) and 23 females (1.5% of 1562) were affected by GU malformations. In DDS a slight male predominance has also been observed, but without statistical significance.

With a median age of 24.5 months (mean: 40.3 +/− 36.7 months), patients with associated malformations or syndromes were significantly (*t*-test: $p < 0.001$) younger at diagnosis of WT/nephroblastomatosis than patients without a malformation or syndrome (median age: 39.0 months; mean age: 50.0 +/− 51.6 months). Patients with WAGR (median age: 21 months; mean age: 23.8 +/− 9.0 months) and DDS (median age: 16.0 months; mean age: 16.7 +/− 12.2 months) were even significantly younger at diagnosis than patients with other syndromes (Figure 1).

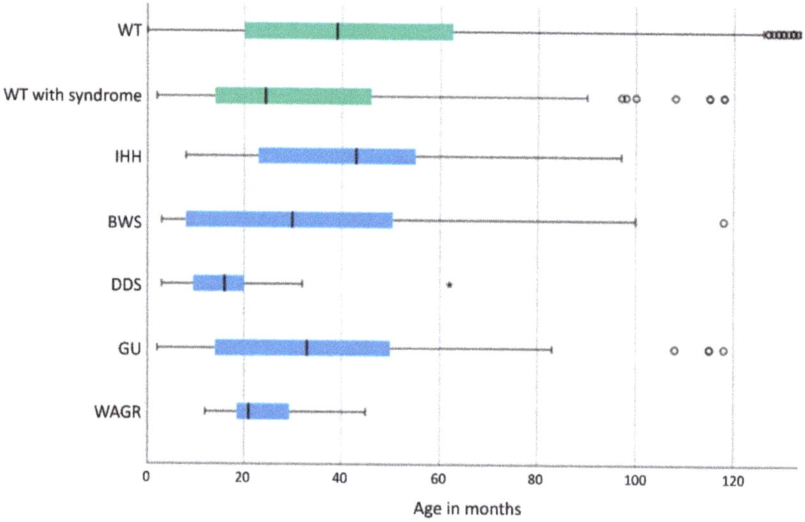

Figure 1. Age of WT/nephroblastomatosis at diagnosis of syndrome-associated WTs displayed as a boxplot (The line in the bar shows the median, the left end of the bar the lower quartile and the right end of the bar the upper quartile, the horizontal line ranges from the minimum to the maximum of data with dots and stars as outliers). DDS and WAGR show a significant lower age at diagnosis compared to all other syndromes ($p <$ =0.001). In 3 patients with CPS WT was diagnosed beyond 10 years of age (IHH (173 months), GU (146 months, 325 months)) and in 143 WT without a syndrome beyond 120 months, the last one at 628 months (data not shown). WT: Wilms tumor; GU: Genitourinary malformations; BWS: Beckwith-Wiedemann spectrum; IHH: isolated hemihypertrophy; DDS: Denys-Drash syndrome; WAGR: Wilms tumor, aniridia, genitourinary abnormalities, range of developmental delays.

3.4. *Bilaterality*

There was a statistically significant higher incidence of bilaterality in patients with (21.2%) than without a syndrome (7.4%) ($p < 0.05$), especially in patients with BWS (31.3%), WAGR (30.0%) and DDS (29.2%) compared to other patients with WT and/or nephroblastomatosis (Table 1).

3.5. *Metastatic Disease in Patients with CPS or Malformations*

In the whole group of patients 18% (529/2927) had a primary metastatic disease (stage IV) while in the cohort of patients with the five most common WT associated syndromes, primary metastatic disease was observed in less than 8.0% (14/171) ($p < 0.001$) with the exception of IHH (17.2%, $p = 0.014$) (see Table S3).

3.6. *Histology*

According to the SIOP studies and trials histology of WT is classified in low, intermediate and high risk, depending on the availability of diffuse anaplasia, the percentage of necrosis and the percentage of blastema, epithelia and stroma in the vital tumor part after preoperative chemotherapy. Blastemal type WT and diffuse anaplasia are high risk tumors. Stromal type is mainly associated with WT1 mutations and is not responding on preoperative chemotherapy despite the fact that patients with stromal type WT have an excellent outcome in case of a localized tumor. The histological risk group together with the local and overall stage defines postoperative treatment. In addition, information about nephrogenic rests or nephroblastomatosis are provided. Mixed type, an intermediate risk tumor, is, with the exception of WAGR syndrome, the most common histological subtype for all WTs with or without syndromes. Patients with CPS are significantly more likely to have isolated nephroblastomatosis. In particular, a significantly increased proportion of isolated nephroblastomatosis is observed in WAGR, BWS and IHH (Table 2).

Table 2. Association between nephroblastomatosis (NBL) and CPS or malformation in the whole cohort of patients and bilateral disease; * chi-square $p \leq 0.001$. WT: Wilms tumor; CPS: cancer predisposition syndrome; NBL: nephroblastomatosis; GU: Genitourinary malformations; BWS: Beckwith-Wiedemann spectrum; IHH: isolated hemihypertrophy; DDS: Denys-Drash syndrome; WAGR: Wilms tumor, aniridia, genitourinary abnormalities, range of developmental delays.

	Isolated NBL		**WT + NBL**		**WT Only**		**Total**	
Total	73	2.5%	64	2.2%	2790	95.3%	2927	100%
Bilateral disease	31	12.3%	61	24.1%	161	63.6%	253	100%
Patients with CPS or GU	22 *	12.9% *	11	6.4%	138	80.7%	171	100%
Patients without CPS or GU	51	1.9%	53	1.9%	2652	96.2%	2756	100%
WAGR	7 *	35.0% *	2	10.0%	11	55.0%	20	100%
BWS	7 *	21.9% *	3	9.4%	22	68.8%	32	100%
IHH	5 *	17.2% *	0	0.0%	24	82.8%	29	100%
DDS	1	4.2%	2	8.3%	21	87.5%	24	100%
GU	2	3.0%	4	6.1%	60	90.9%	66	100%

There was also a statistically significant association of IHH with the blastemal subtype after preoperative chemotherapy ($p = 0.040$) and of DDS with stromal subtype ($p < 0.001$) (Table 3).

Table 3. Association between histological subtypes and CPS or malformation; * chi-square: $p \leq 0.040$. WT: Wilms tumor; GU: Genitourinary malformations; BWS: Beckwith-Wiedemann spectrum; IHH: isolated hemihypertrophy; DDS: Denys-Drash syndrome; WAGR: Wilms tumor, aniridia, genitourinary abnormalities, range of developmental delays.

	Stromal Subtype		**Blastemal Subtype after Preoperative Chemotherapy**		**Other Histological Subtypes**	
All WT	270	9.2%	215	7.3%	2442	83.4%
WAGR	0	0.0%	1	5.0%	19	95.0%
GU	4	6.1%	5	7.6%	57	86.3%
DDS	9 *	37.5% *	0	0.0%	15	62.5%
BWS	0	0.0%	3	9.4%	29	90.6%
IHH	0	0.0%	5 *	17.2% *	24	82.8%

3.7. *Tumor Volume*

Tumor volume (TV) at diagnosis and after preoperative chemotherapy was available in 1798 of 2927 (61.1%) patients (in 1698 patients without and in 91 with CPS or GU) (Table 4). In children with WT and CPS or GU TV at diagnosis was significantly lower than in patients without (349.4 mL vs. 487.5 mL; $p < 0.001$) (Table 4). Furthermore, with the exception of DDS a significant TV reduction can be achieved by preoperative chemotherapy in WTs with CPS or GU with the largest effect of 86.9% in patients with BWS showing an average TV after preoperative chemotherapy of only 38.3 mL (Table 4). In contrast, in DDS no real change of TV under preoperative chemotherapy was observed (Table 4).

Table 4. Tumor volume (TV) at diagnosis and volume reduction achieved by preoperative (preop.) chemotherapy, * *t*-test: $p < 0.001$ for lower initial TV in patients with CPS. TV is not available for all patients. Standard deviation (SD); CPS: cancer predisposition syndrome; GU: Genitourinary malformations; BWS: Beckwith-Wiedemann spectrum; IHH: isolated hemihypertrophy; DDS: Denys-Drash syndrome; WAGR: Wilms tumor, aniridia, genitourinary abnormalities, range of developmental delays.

	Mean Tumor Volume (TV) and [SD]						
	at Diagnosis		**after Preop. Chemo**		**Volume Reduction**		
Patients without CPS or GU (*N* = 1698)	487.5 mL *	[383.0]	228.0 mL	[279.8]	259.5 mL	[326.7]	53.2%
Patients with CPS or GU (*N* = 91)	349.4 mL *	[381.7]	189.6 mL	[255.7]	159.8 mL	[315.3]	45.7%
WAGR (*N* = 10)	104.9 mL	[179.0]	84.4 mL	[156.8]	20.4 mL	[90.1]	19.4%
GU (*N* = 32)	464.0 mL	[329.3]	254.0 mL	[213.2]	210.0 mL	[281.2]	45.3%
DDS (*N* = 15)	379.3 mL	[256.3]	380.7 mL	[375.0]	−1.4 mL	[194.8]	−0.4%
BWS (*N* = 17)	292.9 mL	[539.1]	38.3 mL	[54.2]	254.5 mL	[514.3]	86.9%
IHH (*N* = 17)	307.7 mL	[416.7]	112.7 mL	[244.7]	195.0 mL	[226.7]	63.4%
Patients with CPS or GU undergoing surveillance (*N* = 11)	62.7 mL	[112.0]	55.4 mL	[142.9]	7.3 mL	[36.9]	11.6%

Tumor volume at diagnosis in the 11 patients with CPS undergoing surveillance and treated with preoperative chemotherapy is significantly smaller (TV in CPS patients with surveillance: mean 62.7 mL, median: 21.3 mL. TV in CPS patients without surveillance: mean: 388.8 mL, median: 321.5 mL) (Figure 2).

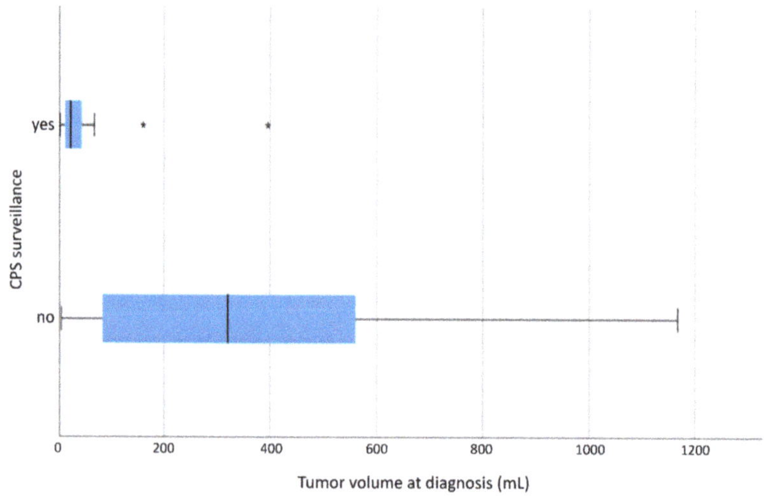

Figure 2. TV at diagnosis of CPS patients as a function of CPS surveillance displayed as a boxplot. TV at diagnosis of 17 patients with CPS surveillance and of 101 patients without CPS surveillance. 2 outliers in patients with CPS surveillance at 160.0 mL and 396.0 mL and 3 outliers in patients without CPS surveillance at 1631.0 mL, 1632.0 mL and 2051.0 mL. CPS patients with CPS surveillance show significantly smaller TV at diagnosis ($p < 0.001$). "*": outliers; CPS: cancer predisposition syndrome.

3.8. Outcome

There was no statistically significant influence on EFS for the whole group of patients with a CPS (Figure 3A). However, patients with BWS showed a significantly worse EFS (Figure 3B) and a higher relapse rate (34.4%) compared to other patients with WT and/or nephroblastomatosis (13.7%). Out of 22 patients with BWS and only unilateral disease at diagnosis 5 patients relapsed of whom 3 showed metachronic disease (3, 4.5 and 6 years after initial diagnosis). One of these 3 patients developed also lung and liver metastasis and died 6 years after diagnosis. Of the other two relapsed patients one patient developed a local relapse in the same kidney and the other one developed lung metastasis without local or metachronic relapse and both survived. Apart from these three patients, metachronic disease occurred only in two further patients with CPS, one with WAGR syndrome and one with IHH. The contralateral kidney tumors were diagnosed in these patients 7 years (WAGR) and 10 months (IHH) after initial diagnosis, respectively. Further analysis suggested that EFS tends to be worse in patients with nephroblastomatosis and a syndrome than in patients without nephroblastomatosis (Figure 3C), particularly if they had developed WT in addition (Figure 3D). Table S2 gives an overview of outcome data. The only significant difference was seen in BWS for 5- and 10y EFS.

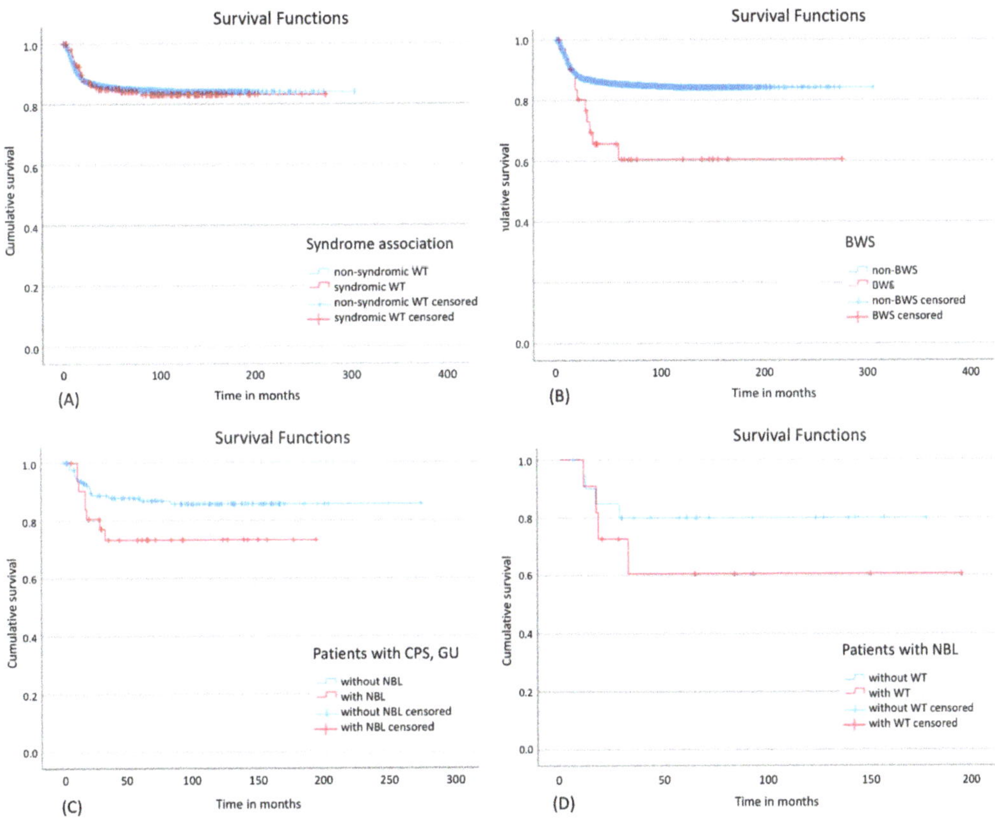

Figure 3. Event-free survival in different subgroups: (**A**) Influence of a syndrome on EFS in patients with WT and/or nephroblastomatosis; Log Rank: $p = 0.890$; (**B**) Influence of BWS on EFS in patients with WT and/or nephroblastomatosis; Log Rank: $p = 0.002$; (**C**) Influence of nephroblastomatosis on EFS in patients with a CPS; Log Rank: $p = 0.086$; (**D**) Influence of WT on EFS in patients with nephroblastomatosis; Log Rank: $p = 0.315$. CPS: cancer predisposition syndrome; BWS: Beckwith-Wiedemann spectrum; NBL: nephroblastomatosis, GU: Genitourinary malformations.

In a multivariate analysis of all patients, nephroblastomatosis and bilaterality had a significant influence on the risk of relapse and death. If only CPS and GU WT patients were considered, such risk was only found for relapse but not for death (Table 5).

Table 5. Multivariate analysis in patients with a WT and/or nephroblastomatosis and only in patients with a syndrome; *: $p < 0.05$. CPS: cancer predisposition syndrome.

	WT and/or Nephroblastomatosis						**CPS Patients**					
Factor	**Relapse**			**Death**			**Relapse**			**Death**		
values	*p*-Value	Hazard Ratio	EFS (%)	*p*-Value	Hazard Ratio	OS (%)	*p*-Value	Hazard Ratio	EFS (%)	*p*-Value	Hazard Ratio	OS (%)
CPS patients	0.594	0.931	83.2	0.139	1.422	88.1						
bilaterality	0.000	1.579 *	73.4	0.030	1.976 *	88.2	0.003	3.013 *	65.4	0.639	1.861	85.6
nephro-blastomatosis	0.005	1.220	72.1	0.074	0.266 *	96.2	0.032	1.264	73.3	0.167	0.225	96.7

4. Discussion

In our retrospective analysis we found that 5.8% of patients with WT and/or nephroblastomatosis are associated with the top five syndromes (WAGR, BWS, DDS, IHH and GU) in agreement with previous literature. With the exception of patients with GU and DDS, female patients are more frequently affected. Patients with syndromes show smaller TVs both at diagnosis and after preoperative chemotherapy, which might be due to the inclusion in a screening program [20]. The statistically significant lower frequency of metastatic disease at diagnosis in patients with a syndrome does not translates into a better EFS. Therefore, other factors such as nephroblastomatosis and comorbidities must be considered to explain their EFS, especially in patients with BWS. (Table 6).

Table 6. Summary of results in the top 5 syndromes associated with WT. * significant results ($p < 0.05$); ** isolated nephroblastomatosis, Standard deviation (SD), Standard error (SE). BWS: Beckwith-Wiedemann spectrum; IHH: isolated hemihypertrophy; DDS: Denys-Drash syndrome; WAGR: Wilms tumor, aniridia, genitourinary abnormalities, range of developmental delays.

	Prevalence(%)	**Median Age at Diagnosis, [SD] (Month)**		**Gender**	**Characteristic Histology**	**Bilaterality (%)**	**Average Volume Reduction by Preoperative Chemotherapy**	**5y-EFS (%), {SE}**		**Confirmed by**
WAGR	0.7	21.0 *	[9.0]	m < f	NBL **	30.0 *	19.4 %	87.5	{0.1}	[9,21,22]
GU	2.3	33.0	[47.3]	m > f *	-	12.1	45.3 %	87.6	{0.4}	[23]
DDS	0.8	16.0 *	[12.2]	m > f	stromal type	29.2 *	−0.4 %	94.7	{0.5}	[24–27]
BWS	1.1	30.0	[29.7]	m < f	NBL **	31.3 *	86.9 %	60.6 *	{0.1}	[28,29]
IHH	1.0	43.0	[34.2]	m < f	NBL **/ blastemal type after preop. chemo	13.8	63.4 %	84.6	{0.1}	[6,15,30,31]

4.1. Prevalence and Surveillance

The prevalence of syndromes in patients with WT is lower in our series compared to the 8–17% in the literature [5,6]. This may be due to an underreporting in our retrospective multicenter study where standardized reporting was carried out at diagnosis, hence early in life with a probably incompletely symptomatology. BWS, for example, a syndrome with variable features, is described with a prevalence of 1 up to 8% in other studies [12,18,29]. Therefore, not all patients with a WT CPS are included in ongoing ultrasound screening programs. In our data the screening rate is depending on the clinical symptomatology and highest in WAGR with 40% (Table 1). MacFarland et al. reported 12 patients diagnosed with BWS after a WT was already known [32]. This may explain our low prevalence of 1.1% and also undiagnosed BWS in other studies. Clinicians need to recognize subtle manifestations of syndromes in WT patients to not overlook them. For clinical diagnosis of BWS a new consensus statement has been published in this respect [33]. Knowledge about specific associations between different syndromes and WT will allow an earlier diagnosis of such WT, with CPSs demanding genetic testing, counselling and subsequently screening programs. In this respect, it is important to separate clearly between the WT1 associated CPS (DDS and WAGR) and the imprinting disorders (BWS and IHH) showing differences in WT characteristics, e.g., age at diagnosis or the response to preoperative chemotherapy as shown in our analysis.

4.2. Age at Diagnosis of WT/Nephroblastomatosis

Of all patients with syndromes, those with DDS are diagnosed the earliest followed by WAGR [9]. Patients with GU and BWS also tend to be diagnosed earlier than those without syndromes. The late median diagnosis at 45 months in IHH compared to other syndromes suggests that IHH manifests itself clinically rather subtly [12,30] which is why no surveillance for WT was carried out. Therefore, our results suggest that early age at diagnosis of a WT without a syndrome should always raise awareness of a CPS. In addition, in patients with the diagnosis of such a syndrome screening for WT needs to start early and regularly up to the age when the manifestation of a WT becomes more unlikely [20]. According to our data, for patients with DDS and WAGR such a screening may stop already at the age of 4 or 5 years as also recommended by Hol et al. [8], whereas in the other syndromic patients screening should continue at least up to the age of 7 years as also consented for BWS [20] (Figure 1). In the work of Diller et al. the average age of diagnosis of patients with GU was 13 months and thus significantly earlier compared to our data (42.9 months). This difference in median age may be related to a different approach as they analysed blood samples from 201 patients with a history of WT for constitutional WT1 mutations, which was not carried out in our cohort of patients [34]. As a result of their work, this underscores the need for a regular screening in patients with GU.

4.3. Tumor Volume and Response to Preoperative Chemotherapy

Tumor response to preoperative chemotherapy varies significantly between the different syndromes and depends on the presence of a stromal subtype or WT1 mutation. Thus, we confirm a poor response in patients with DDS [26,27]. These patients have comparably larger initial TVs that can even increase after preoperative chemotherapy (Table 4). A WT1 mutation/deletion path driven propensity for stromal components or even stromal predominance is a likely reason [35,36]. The TV reduction after preoperative chemotherapy is also poor in WAGR patients with WT1 deletion despite of missing stromal type WT (Table 4). In contrast, an excellent response on preoperative chemotherapy is achieved in patients with BWS. As a consequence of a poor TV reduction, a stromal subtype or WT1 aberrations with an underlying syndrome may be possible. This is especially true for patients with bilateral disease.

4.4. Bilaterality

The significantly increased frequency of bilaterality in patients with WAGR, GU, DDS is consistent with previous work [9,23,24]. In patients with BWS we found a higher percentage of bilaterality (31.3%) than Breslow et al. (21%) [9]. Consequently, patients with unilateral disease and WAGR, DDS or BWS should always be regarded as predisposed for bilaterality [4] and in cases of bilateral disease these three syndromes need to be kept in mind, if not diagnosed yet.

4.5. Metastatic Disease

Metastases are a significantly less frequent event in patients with syndromes as compared to non-syndromic WT. This may be due to an early diagnosis of WT or nephroblastomatosis in patients with syndromes and underscores the importance of a screening program [20].

4.6. Histology

The majority of patients with syndromes have intermediate-risk WT. As described earlier, patients with DDS are significantly more often affected by WT with a stromal subtype and never with high risk tumors [24,25]. There is a significant association with blastemal subtype after preoperative chemotherapy in patients with IHH and a trend in BWS due to the frequently IGF2 driven biology. In contrast to the work of Green et al., focal and diffuse anaplasia do not occur as first histology in our data [30]. However after nephroblastomatosis the development of diffuse anaplasia in case of a secondary WT is a relatively frequent event [37]. We confirm the association between WAGR, BWS and IHH and nephroblastomatosis found by others [9,21,31,33] in contrast to patients with DDS and GU [23]. As nephroblastomatosis is found more often in patients with the above mentioned three syndromes this raises the question whether all patients with nephroblastomatosis need to be examined for CPS if not yet known [33].

4.7. Outcome

With the exception of patients with BWS showing a significantly worse EFS and increased risk of relapse, CPS or GU in general have no impact on EFS. Breslow et al. found no difference in WT with BWS neither in OS nor in EFS [9]. However, if nephroblastomatosis is present in our data, EFS tends to be worse, especially if these patients develop a WT as already shown by Furtwängler et al. [37,38]. Therefore, patients with nephroblastomatosis independent of a predisposition syndrome must be followed in regular intervals after the end of treatment for longer periods of time to diagnose a relapse early in order to keep their overall survival as high as for other patients [39].

5. Conclusions

Diagnosis of WT at an early age, bilateral tumors or nephroblastomatosis in patients without a known CPS should always raise suspicion of an underlying CPS and genetic testing and counselling should be offered to these patients and families. Screening for WT in patients with a syndrome may stop earlier after the age of 4 to 5 years in patients with DDS and WAGR as also recommended by Hol et al. [8], whereas for the other syndromes this should last up to the age of 7 years and needs to continue in cases of nephroblastomatosis even in CR after the end of first line treatment.

Supplementary Materials: The following are available online at https://www.mdpi.com/article/10.3390/cancers13195016/s1, Table S1: Further WT associated syndromes and malformations (number of involved patients in brackets), Table S2: Data on outcome of different patient groups. (NBL: Nephroblastomatosis); * $p < 0.05$, Table S3: Frequencies of metastatic disease in patients with or without malformations or CPS.

Author Contributions: Conceptualization, N.G.; methodology, N.W. and N.G.; project administration, N.G.; validation, N.W., R.F. and N.G.; formal analysis, N.W. and N.G.; investigation, N.W., R.F. and N.G.; data curation, R.F. and N.G.; visualization, N.W.; writing—original draft preparation, N.W. and N.G.; writing—review and editing, N.W., A.W., R.F., P.M., L.K., C.V., J.-P.S., C.M.M., S.S., M.G. and N.G.; supervision, N.G.; funding acquisition, N.G., R.F. All authors have read and agreed to the published version of the manuscript.

Funding: This work was partly funded by the German Cancer Aid (Deutsche Krebshilfe, grant No: 70-1899 and 50-2709-GR2) and the Elterninitiative krebskranker Kinder im Saarland e.V.

Institutional Review Board Statement: The study was conducted according to the guidelines of the Declaration of Helsinki and approved by the Institutional Ethics Committee of the Ärztekammer des Saarlandes (No: 136/01 from 20.09.2002 and, 248/13 from 13.01.2014).

Informed Consent Statement: Informed consent was obtained from all subjects, their parents or legal guardians involved in the study.

Data Availability Statement: The data presented in this study are available on request from the corresponding author. The data are not publicly available due to ongoing analysis.

Acknowledgments: We thank all patients and their families for taking part in the corresponding clinical studies and also to the participating hospitals. We acknowledge support by the Deutsche Forschungsgemeinschaft (DFG, German Research Foundation) and Saarland University within the funding programme Open Access Publishing.

Conflicts of Interest: The authors declare no conflict of interest.

References

1. Stiller, C.A.; Parkint, D.M. International variations in the incidence of childhood soft-tissue sarcomas. *Paediatr. Perinat. Epidemiol.* **1994**, *8*, 107–119. [CrossRef]
2. Pritchard-Jones, K.; Moroz, V.; Vujanić, G.; Powis, M.; Walker, J.; Messahel, B.; Hobson, R.; Levitt, G.; Kelsey, A.; Mitchell, C. Treatment and outcome of Wilms' tumour patients: An analysis of all cases registered in the UKW3 trial. *Ann. Oncol.* **2012**, *23*, 2457–2463. [CrossRef]
3. Dome, J.S.; Graf, N.; Geller, J.I.; Fernandez, C.V.; Mullen, E.A.; Spreafico, F.; Van Den Heuvel-Eibrink, M.; Pritchard-Jones, K. Advances in wilms tumor treatment and biology: Progress through international collaboration. *J. Clin. Oncol.* **2015**, *33*, 2999–3007. [CrossRef]
4. van den Heuvel-Eibrink, M.M.; Hol, J.A.; Pritchard-Jones, K.; Tinteren, H.V.; Furtwängler, R.; Verschuur, A.C.; Vujanic, G.M.; Leuschner, I.; Brok, J.; Rübe, C.; et al. Position Paper: Rationale for the treatment of Wilms tumour in the UMBRELLA SIOP-RTSG 2016 protocol. *Nat. Rev. Urol.* **2017**, *14*, 743–752. [CrossRef]
5. Narod, S.A.; Hawkins, M.M.; Robertson, C.M.; Stiller, C.A. Congenital anomalies and childhood cancer in Great Britain. *Am.J.Hum.Genet.* **1997**, *60*, 474–485.
6. Merks, J.H.M.; Caron, H.N.; Hennekam, R.C.M. High Incidence of Malformation Syndromes in a Series of 1, *073 Child. Cancer* **2005**, *143*, 132–143.
7. Shannon, R.S.; Mann, J.R.; Harper, E.; Harnden, D.G.; Morten, J.E.N.; Herbert, A. Wilms's tumour and aniridia: Clinical and cytogenetic features. *Arch. Dis. Child.* **1982**, *57*, 685–690.
8. Hol, J.A.; Jongmans, M.C.J.; Sudour-Bonnange, H.; Ramírez-Villar, G.L.; Chowdhury, T.; Rechnitzer, C.; Pal, N.; Schleiermacher, G.; Karow, A.; Kuiper, R.P.; et al. Clinical characteristics and outcomes of children with WAGR syndrome and Wilms tumor and/or nephroblastomatosis: The 30-year SIOP-RTSG experience. *Cancer* **2021**, *127*, 628–638. [CrossRef]
9. Breslow, N.E.; Norris, R.; Norkool, P.A.; Kang, T.; Beckwith, J.B.; Perlman, E.J.; Ritchey, M.L.; Green, D.M.; Nichols, K.E. Characteristics and outcomes of children with the Wilms tumor-aniridia syndrome: A report from the National Wilms Tumor Study Group. *J. Clin. Oncol.* **2003**, *21*, 4579–4585. [CrossRef]
10. Drash, A.; Sherman, F.; Hartmann, W.H.; Blizzard, R.M. A syndrome of pseudohermaphroditism, Wilms' tumor, hypertension, and degenerative renal disease. *J. Pediatr.* **1970**, *76*, 585–593. [CrossRef]
11. Mueller, R.F. the month The Denys-Drash syndrome. *J. Med Genet.* **1994**, *36*, 471–477. [CrossRef]
12. Scott, R.H.; Stiller, C.A.; Walker, L.; Rahman, N. Syndromes and constitutional chromosomal abnormalities associated with Wilms tumour. *J. Med. Genet.* **2006**, *43*, 705–715. [CrossRef] [PubMed]
13. Pritchard-Jones, K.; Fleming, S.; Davidson, D.; Bickmore, W.; Porteous, D.; Gosden, C.; Bard, J.; Buckler, A.; Pelletier, J.; Housman, D.; et al. The candidate Wilms' tumour gene is involved in genitourinary development. *Nature* **1990**, *346*, 194–197. [CrossRef]
14. Royer-Pokora, B.; Ragg, S.; Heckl-Östreicher, B.; Held, M.; Loos, U.; Call, K.; Glaser, T.; Housman, D.; Saunders, G.; Zabel, B.; et al. Direct pulsed field gel electrophoresis of Wilms' tumors shows that dna deletions in 11 p 13 are rare. *Genes Chromosom. Cancer* **1991**, *3*, 89–100. [CrossRef] [PubMed]
15. Hoyme, H.E.; Seaver, L.H.; Jones, K.L.; Procopio, F.; Crooks, W.; Feingold, M. Isolated hemihyperplasia (hemihypertrophy): Report of a prospective multicenter study of the incidence of neoplasia and review. *Am. J. Med. Genet.* **1998**, *79*, 274–278. [CrossRef]
16. Grundy, P.E.; Feinberg, A.P.; Niemitz, E.L.; Brandenburg, S.A.; DeBaun, M.R. Children with Idiopathic Hemihypertrophy and Beckwith-Wiedemann Syndrome Have Different Constitutional Epigenotypes Associated with Wilms Tumor. *Am. J. Hum. Genet.* **2005**, *77*, 887–891.
17. Pettenati, M.J.; Haines, J.L.; Higgins, R.R.; Wappner, R.S.; Palmer, C.G.; Weaver, D.D. Wiedemann-Beckwith syndrome: Presentation of clinical and cytogenetic data on 22 new cases and review of the literature. *Hum. Genet.* **1986**, *74*, 143–154. [CrossRef]
18. Brioude, F.; Toutain, A.; Giabicani, E.; Cottereau, E.; Cormier-Daire, V.; Netchine, I. Overgrowth syndromes—Clinical and molecular aspects and tumour risk. *Nat. Rev. Endocrinol.* **2019**, *15*, 299–311. [CrossRef]
19. Dome, J.S.; Perlman, E.J.; Graf, N. Risk Stratification for Wilms Tumor: Current Approach and Future Directions. *Am. Soc. Clin. Oncol. Educ. B.* **2014**, *34*, 215–223. [CrossRef]
20. Mussa, A.; Duffy, K.A.; Carli, D.; Griff, J.R.; Fagiano, R.; Kupa, J.; Brodeur, G.M.; Ferrero, G.B.; Kalish, J.M. The effectiveness of Wilms tumor screening in Beckwith–Wiedemann spectrum. *J. Cancer Res. Clin. Oncol.* **2019**, *145*, 3115–3123. [CrossRef]
21. Beckwith, J.B. Nephrogenic rests and the pathogenesis of Wilms tumor: Developmental and clinical considerations. *Am. J. Med. Genet.* **1998**, *79*, 268–273. [CrossRef]

22. Bonaïti-Pellié, C.; Chompret, A.; Tournade, M.-F.; Hochez, J.; Moutou, C.; Zucker, J.-M.; Steschenko, D.; Brunat-Mentigny, M.; Roché, H.; Tron, P.; et al. Genetics and epidemiology of Wilms' tumor: The French Wilms' tumor study. *Med. Pediatr. Oncol.* **1992**, *20*, 284–291. [CrossRef]
23. Diller, L.; Ghahremani, M.; Morgan, J.; Grundy, P.; Reeves, C.; Breslow, N.; Green, D.; Neuberg, D.; Pelletier, J.; Li, F.P. Constitutional WT1 mutations in Wilms' tumor patients. *J. Clin. Oncol.* **1998**, *16*, 3634–3640. [CrossRef]
24. Auber, F.; Jeanpierre, C.; Denamur, E.; Jaubert, F.; Schleiermacher, G.; Patte, C.; Cabrol, S.; Leverger, G.; Nihoul-Fékété, C.; Sarnacki, S. Management of Wilms tumors in Drash and Frasier syndromes. *Pediatr. Blood Cancer* **2009**, *52*, 55–59. [CrossRef] [PubMed]
25. Little, S.E.; Hanks, S.P.; King-Underwood, L.; Jones, C.; Rapley, E.A.; Rahman, N.; Pritchard-Jones, K. Frequency and Heritability of WT1 Mutations in Nonsyndromic Wilms' Tumor Patients: A UK Children's Cancer Study Group Study. *J. Clin. Oncol.* **2004**, *22*, 4140–4146. [CrossRef]
26. Pelletier, J.; Bruening, W.; Kashtan, C.E.; Mauer, S.M.; Manivel, J.C.; Striegel, J.E.; Houghton, D.C.; Junien, C.; Habib, R.; Fouser, L. Germline mutations in the Wilms' tumor suppressor gene are associated with abnormal urogenital development in Denys-Drash syndrome. *Cell* **1991**, *67*, 437–447. [CrossRef]
27. Weirich, A.; von Harrach, M.; Royer-Pokora, B.; Schneider, D.; Leuschner, I.; Schumacher, V.; Graf, N.; Autschbach, F.; Uschkereit, C.; Beier, M. Clinical relevance of mutations in the Wilms tumor suppressor 1 gene WT1 and the cadherin-associated protein β1 gene CTNNB1 for patients with Wilms tumors. *Cancer* **2008**, *113*, 1080–1089.
28. Sotelo-Avila, C.; Gonzalez-Crussi, F.; Fowler, J.W. Complete and incomplete forms of Beckwith-Wiedemann syndrome: Their oncogenic potential. *J. Pediatr.* **1980**, *96*, 47–50. [CrossRef]
29. DeBaun, M.R.; Tucker, M.A. Risk of cancer during the first four years of life in children from The Beckwith-Wiedemann Syndrome Registry. *J. Pediatr.* **1998**, *132*, 398–400. [CrossRef]
30. Green, D.M.; Breslow, N.E.; Beckwith, J.B.; Norkool, P. Screening of children with hemihypertrophy, aniridia, and Beckwith-Wiedemann syndrome in patients with wilms tumor: A report from the national Wilms tumor study. *Med. Pediatr. Oncol.* **1993**, *21*, 188–192. [CrossRef]
31. Beckwith, J.B.; Kiviat, N.B.; Bonadio, J.F. Nephrogenic rests, nephroblastomatosis, and the pathogenesis of Wilms' tumor. *Pediatr. Pathol.* **1990**, *10*, 1–36. [CrossRef]
32. MacFarland, S.P.; Duffy, K.A.; Bhatti, T.R.; Bagatell, R.; Balamuth, N.J.; Brodeur, G.M.; Ganguly, A.; Mattei, P.A.; Surrey, L.F.; Balis, F.M.; et al. Diagnosis of Beckwith-Wiedemann syndrome in children presenting with Wilms tumor. *Pediatr. Blood Cancer* **2018**, e27296. [CrossRef]
33. Brioude, F.; Kalish, J.M.; Mussa, A.; Foster, A.C.; Bliek, J.; Ferrero, G.B.; Boonen, S.E.; Cole, T.; Baker, R.; Bertoletti, M.; et al. Expert consensus document: Clinical and molecular diagnosis, screening and management of Beckwith–Wiedemann syndrome: An international consensus statement. *Nat. Rev. Endocrinol.* **2018**, *14*, 229–249. [CrossRef] [PubMed]
34. Ripperger, T.; Bielack, S.S.; Borkhardt, A.; Brecht, I.B.; Burkhardt, B.; Calaminus, G.; Debatin, K.-M.; Deubzer, H.; Dirksen, U.; Eckert, C.; et al. Childhood cancer predisposition syndromes-A concise review and recommendations by the Cancer Predisposition Working Group of the Society for Pediatric Oncology and Hematology. *Am. J. Med. Genet. Part A* **2017**, *173*, 1017–1037. [CrossRef]
35. Royer-Pokora, B.; Weirich, A.; Schumacher, V.; Uschkereit, C.; Beier, M.; Leuschner, I.; Graf, N.; Autschbach, F.; Schneider, D.; von Harrach, M. Clinical relevance of mutations in the Wilms tumor suppressor 1 gene WT1 and the cadherin-associated protein beta1 gene CTNNB1 for patients with Wilms tumors: Results of long-term surveillance of 71 patients from International Society of Pediatric Oncolog. *Cancer* **2008**, *113*, 1080–1089. [CrossRef] [PubMed]
36. Verschuur, A.C.; Vujanic, G.M.; Van Tinteren, H.; Jones, K.P.; de Kraker, J.; Sandstedt, B. Stromal and epithelial predominant Wilms tumours have an excellent outcome: The SIOP 93 01 experience. *Pediatr. Blood Cancer* **2010**, *55*, 233–238. [CrossRef]
37. Furtwängler, R.; Schmolze, M.; Gräber, S.; Leuschner, I.; Amann, G.; Schenk, J.P.; Niggli, F.; Kager, L.; Schweinitz, D.V.; Graf, N. Pretreatment for bilateral nephroblastomatosis is an independent risk factor for progressive disease in patients with stage V nephroblastoma. *Klin. Padiatr.* **2014**, *226*, 175–181. [CrossRef]
38. Furtwängler, R.; Nourkami, N.; Alkassar, M.; von Schweinitz, D.; Stehr, M.; Graf, N. Sydromes and syndrome-like features in bilateral Wilms Tumor are associated with inferior outcome. *Pediatr. Blood Cancer* **2010**, *55*, 885.
39. Furtwängler, R.; Müller, M.; Nourkami-Tutdibi1, N.; Warmann, S.; Hubertus, J.; Vokuhl, C.; Leuschner, I.; Schenk, J.-P.; Kager, L.; Graf, N. Treatment of Nephroblastomatosis: The GPOH Experience 1993–2014. *Pediatr. Blood Cancer* **2016**, *63*, S34.

cancers

MDPI

Correction

Correction: Welter et al. Characteristics of Nephroblastoma/Nephroblastomatosis in Children with a Clinically Reported Underlying Malformation or Cancer Predisposition Syndrome. *Cancers* 2021, *13*, 5016

Nils Welter [1], Angelo Wagner [1], Rhoikos Furtwängler [1], Patrick Melchior [2], Leo Kager [3], Christian Vokuhl [4], Jens-Peter Schenk [5], Clemens Magnus Meier [6], Stefan Siemer [7], Manfred Gessler [8] and Norbert Graf [1,*]

1 Department of Pediatric Oncology and Hematology, Saarland University, 66421 Homburg, Germany; nils.welter@uks.eu (N.W.); s9aowagn@stud.uni-saarland.de (A.W.); rhoikos.furtwaengler@uks.eu (R.F.)
2 Department of Radiation Oncology, Saarland University, 66421 Homburg, Germany; patrick.melchior@uks.eu
3 St. Anna Kinderspital, Department of Pediatrics, Medical University Vienna, Kinderspitalgasse 6, 1090 Vienna, Austria; leo.kager@stanna.at
4 Section of Pediatric Pathology, University of Bonn, Venusberg-Campus 1, 53127 Bonn, Germany; Christian.vokuhl@ukbonn.de
5 Division of Pediatric Radiology, Clinic for Diagnostic and Interventional Radiology, University of Heidelberg, Im Neuenheimer Feld 430, 69120 Heidelberg, Germany; jens-peter.schenk@med.uni-heidelberg.de
6 Department of General, Visceral, Vascular and Pediatric Surgery, Saarland University, 66421 Homburg, Germany; clemens-magnus.meier@uks.eu
7 Department of Urology and Pediatric Urology, Saarland University, 66421 Homburg, Germany; stefan.siemer@uks.eu
8 Developmental Biochemistry and Comprehensive Cancer Center Mainfranken, Theodor-Boveri-Institute/Biocenter, University of Würzburg, 97074 Würzburg, Germany; gessler@biozentrum.uni-wuerzburg.de
* Correspondence: norbert.graf@uks.eu; Tel.: +0049-(0)6841-1628397

Citation: Welter, N.; Wagner, A.; Furtwängler, R.; Melchior, P.; Kager, L.; Vokuhl, C.; Schenk, J.-P.; Meier, C.M.; Siemer, S.; Gessler, M.; et al. Correction: Welter et al. Characteristics of Nephroblastoma/Nephroblastomatosis in Children with a Clinically Reported Underlying Malformation or Cancer Predisposition Syndrome. *Cancers* 2021, *13*, 5016. *Cancers* **2021**, *13*, 5743. https://doi.org/10.3390/cancers13225743

Received: 1 November 2021
Accepted: 12 November 2021
Published: 16 November 2021

In the original article [1] there was a mistake in Table 2 as published. Table 2 contains wrong percentages in lines *Bilateral disease* and *Patients with CPS or GU*. For this reason the table should be replaced with the correct one as shown below.

Table 2. Association between nephroblastomatosis (NBL) and CPS or malformation in the whole cohort of patients and bilateral disease; * chi-square $p \leq 0.001$. WT: Wilms tumor; CPS: cancer predisposition syndrome; NBL: nephroblastomatosis; GU: Genitourinary malformations; BWS: Beckwith-Wiedemann spectrum; IHH: isolated hemihypertrophy; DDS: Denys-Drash syndrome; WAGR: Wilms tumor, aniridia, genitourinary abnormalities, range of developmental delays.

	Isolated NBL		WT + NBL		WT Only		Total	
Total	73	2.5%	64	2.2%	2790	95.3%	2927	100%
Bilateral disease	31	12.3%	61	24.1%	161	63.6%	253	100%
Patients with CPS or GU	22 *	12.9% *	11	6.4%	138	80.7%	171	100%
Patients without CPS or GU	51	1.9%	53	1.9%	2652	96.2%	2756	100%
WAGR	7 *	35.0% *	2	10.0%	11	55.0%	20	100%
BWS	7 *	21.9% *	3	9.4%	22	68.8%	32	100%
IHH	5 *	17.2% *	0	0.0%	24	82.8%	29	100%
DDS	1	4.2%	2	8.3%	21	87.5%	24	100%
GU	2	3.0%	4	6.1%	60	90.9%	66	100%

The authors apologize for any inconvenience caused and state that the scientific conclusions are unaffected. The original article has been updated.

Reference

1. Welter, N.; Wagner, A.; Furtwängler, R.; Melchior, P.; Kager, L.; Vokuhl, C.; Schenk, J.-P.; Meier, C.M.; Siemer, S.; Gessler, M.; et al. Characteristics of Nephroblastoma/Nephroblastomatosis in Children with a Clinically Reported Underlying Malformation or Cancer Predisposition Syndrome. *Cancers* **2021**, *13*, 5016. [CrossRef] [PubMed]

cancers

MDPI

Article

EWSR1-WT1 Target Genes and Therapeutic Options Identified in a Novel DSRCT In Vitro Model

Margit Bleijs [1,2], Corine Pleijte [1,2], Sem Engels [1,2], Femke Ringnalda [1,2], Friederike Meyer-Wentrup [1], Marc van de Wetering [1] and Hans Clevers [1,2,3,*]

1 Princess Máxima Center for Pediatric Oncology, 3584 CT Utrecht, The Netherlands; m.w.bleijs-2@prinsesmaximacentrum.nl (M.B.); cpleijte@umcutrecht.nl (C.P.); s.a.g.engels-3@prinsesmaximacentrum.nl (S.E.); f.c.a.ringnalda@prinsesmaximacentrum.nl (F.R.); f.meyer-wentrup@prinsesmaximacentrum.nl (F.M.-W.); m.l.vandewetering@prinsesmaximacentrum.nl (M.v.d.W.)
2 Oncode Institute, 3521 AL Utrecht, The Netherlands
3 Hubrecht Institute, Royal Netherlands Academy of Arts and Sciences and University Medical Center, 3584 CT Utrecht, The Netherlands
* Correspondence: h.clevers@hubrecht.eu

Simple Summary: Desmoplastic small round cell tumor (DSRCT) is an extremely rare soft tissue sarcoma arising in the abdomen of adolescents and young adults. This sarcoma is driven by a single genomic rearrangement, resulting in the expression of the *EWSR1-WT1* fusion gene. No effective treatment exists for DSRCT patients, which highlights the need for preclinical models to study disease progression and drug sensitivity. The aim of this study is to develop a pre-clinical DSRCT in vitro model, which enables investigating the molecular target genes of the *EWSR1-WT1* fusion gene and allows for medium-throughput drug screening to discover new therapeutic options.

Abstract: Desmoplastic small round cell tumor (DSRCT) is a rare and aggressive soft tissue sarcoma with a lack of effective treatment options and a poor prognosis. DSRCT is characterized by a chromosomal translocation, resulting in the *EWSR1-WT1* gene fusion. The molecular mechanisms driving DSRCT are poorly understood, and a paucity of preclinical models hampers DSRCT research. Here, we establish a novel primary patient-derived DSRCT in vitro model, recapitulating the original tumor. We find that *EWSR1-WT1* expression affects cell shape and cell survival, and we identify downstream target genes of the EWSR1-WT1 fusion. Additionally, this preclinical in vitro model allows for medium-throughput drug screening. We discover sensitivity to several drugs, including compounds targeting RTKs. MERTK, which has been described as a therapeutic target for several malignancies, correlates with *EWSR1-WT1* expression. Inhibition of MERTK with the small-molecule inhibitor UNC2025 results in reduced proliferation of DSRCT cells in vitro, suggesting MERTK as a therapeutic target in DSRCT. This study underscores the usefulness of preclinical in vitro models for studying molecular mechanisms and potential therapeutic options.

Keywords: DSRCT; EWSR1-WT1; preclinical model; in vitro model; pediatric cancer; sarcoma

Citation: Bleijs, M.; Pleijte, C.; Engels, S.; Ringnalda, F.; Meyer-Wentrup, F.; van de Wetering, M.; Clevers, H. EWSR1-WT1 Target Genes and Therapeutic Options Identified in a Novel DSRCT In Vitro Model. *Cancers* **2021**, *13*, 6072. https://doi.org/10.3390/cancers13236072

Academic Editors: David Wong, Saurabh Agarwal and Jianhua Yang

Received: 24 October 2021
Accepted: 28 November 2021
Published: 2 December 2021

1. Introduction

Desmoplastic small round cell tumor (DSRCT) is a highly aggressive and rare soft tissue sarcoma [1–3]. This sarcoma was first characterized as a separate entity by Gerald and Rosai in 1989, who described the histologic appearance of DSRCT as nests of small round blue cells, separated by desmoplastic stroma [4]. DSRCT exhibits features of multi-phenotypic differentiation, since mesenchymal, epithelial, and neural markers are expressed within this sarcoma [3,4]. Genetically, DSRCT is characterized by a chromosomal translocation, resulting in a gene fusion involving Ewing sarcoma region 1 (*EWSR1*) and Wilms tumor 1 (*WT1*). The chimeric protein contains the N-terminal domain of EWSR1 fused to the DNA-binding domain of WT1, including zinc fingers 2–4 [5]. The N-terminal domain of EWSR1 is thought to act as a transcriptional activator, while the zinc finger domains of WT1 bind DNA of regulatory elements in its target genes [6]. Previously, it has been shown that two different isoforms of *EWSR1-WT1* are expressed, distinguished by the presence or absence of a three amino acid KTS domain, which induce distinct transcriptional profiles [5,7,8].

Recently, a very comprehensive study classified molecular subtypes of sarcomas, including DSRCT [9]. Several downstream targets of its *EWSR1-WT1* fusion gene were identified. Kang and colleagues showed that EWSR1-WT1 directly binds to the promoter of neural reprogramming factor *ASCL1*, activating neural gene expression [7]. Gedminas and colleagues identified EWSR1-WT1 target genes using an siRNA knock-down model of the fusion in the patient-derived JN-DSRCT-1 and patient-derived xenograft (PDX)-derived BER cell lines [10–12]. They discovered several downstream target pathways that are commonly deregulated in fusion-positive sarcomas and observed an overlap between EWSR1-WT1 and EWSR1-FLI1 (the fusion gene driving Ewing sarcoma) gene signatures. Hingorani and colleagues performed bulk RNA sequencing of patient-derived DSRCT specimens and identified CD200 and CD276 as potentially targetable immune checkpoint molecules, whose expression are independent of *EWSR1-WT1* expression. Additionally, they performed WT1 ChIP-sequencing and established an shRNA knock-down model of *EWSR1-WT1* in the JN-DSRCT-1 cell line and identified IGF2 and FGFR4 as potential therapeutic targets in DSRCT patients [13]. While these studies give important insights on the downstream targets of the *EWSR1-WT1* fusion gene, molecular mechanisms triggered by the gene fusion, which account for the aggressive proliferation of DSRCT, have remain elusive.

Preclinical models, such as cell lines, organoids, and patient-derived xenografts (PDXs), are key for investigating tumor progression and molecular mechanisms [14]. While several PDX models have been established from DSRCT specimens by transplantation into the flank of the mouse [11], only two in vitro models of DSRCT have been published: one primary patient-derived and one PDX-derived cell line [10–12]. The paucity of preclinical DSRCT models hampers the research on mechanisms driving DSRCT and discovery of therapeutic options. Here, we establish a novel primary patient-derived DSRCT model in vitro, which recapitulates morphological and transcriptomic features of the originating tumor. We use a shRNA knock-down model to identify molecular target genes of the EWSR1-WT1 fusion. Additionally, this preclinical in vitro model allows for medium-throughput drug screening to discover drug sensitivity in DSRCT cells. Finally, we discover promising therapeutic targets, including an EWSR1-WT1-driven receptor tyrosine kinase (RTK).

2. Materials and Methods

2.1. Patient-Derived DSRCT Specimen

Surgically resected tissue was obtained from a DSRCT patient at the Amsterdam UMC with informed consent. Tumor material was washed with Advanced Dulbecco's Modified Eagle's Medium F12 (AdDMEM-F12) (Gibco™, Thermo Fisher Scientific, #12634010, Waltham, MA, USA), supplemented with 1% pen/strep (Gibco™, Thermo Fisher Scientific, #15140122, Life Technologies, Merelbeke, Belgium), 1% Glutamax (Gibco™, Thermo Fisher Scientific, #35050038, Life Technologies, Merelbeke, Belgium), 1% Hepes (Gibco™, Thermo Fisher Scientific, #15630056, Life Technologies, Merelbeke, Belgium), and minced into tumor pieces using scalpels, as well as mechanically disrupted by pipetting up and down before plating into suspension cell culture plates.

2.2. Cell Culture

The patient-derived OV-054 DSRCT cell line was maintained in DSRCT medium, containing Advanced Dulbecco's Modified Eagle's Medium F12 (AdDMEM-F12) (Gibco™, Thermo Fisher Scientific, #12634010, Waltham, MA, USA), supplemented with 1% pen/strep (Gibco™, Thermo Fisher Scientific, #15140122, Life Technologies, Merelbeke, Belgium), 1% Glutamax (Thermo Fisher Scientific, #35050038), 1% N2 (Gibco, #17502048), 1% Hepes (Gibco™, Thermo Fisher Scientific, #15630056, Life Technologies, Merelbeke, Belgium), 2% B27 supplement minus vitamin A, 50x (Gibco™, Thermo Fisher Scientific, #12587010, Life Technologies, Merelbeke, Belgium), 0.25% N-Acetyl-L-cysteine (NAC) (Sigma Aldrich, #A9165, St. Louis, MO, USA), 50 ng/mL FGF-basic (154 a.a.) (PeproTech, #100-18B, Rocky Hill, NJ, USA), 50 ng/mL EGF (PeproTech, #AF-100-15, Rocky Hill, NJ, USA), 10 ng/mL IGF1 (PeproTech, #100-11, Rocky Hill, NJ, USA), 10 ng/mL Rho Kinase Inhibitor (AbMole BioScience, #M1817, Houston, TX, USA), 10 ng/mL BMP4 (PeproTech, #120-05ET, Rocky Hill, NJ, USA), and 0.1% BME (Cultrex Reduced Growth Factor Basement Membrane Extract, Type 2, #3533-005-2, Bio-Techne, Minneapolis, MN, USA). Cells were incubated at 37 °C with 5% CO_2 and passaged 1:5 every 10 days using TrypLE™ Express Enzyme (Gibco™, Thermo Fisher Scientific, #12605010, Life Technologies, Merelbeke, Belgium).

HEK293T cells were maintained in Dulbecco's Modified Eagle's Medium (Gibco™, Thermo Fisher Scientific, #31966-047, Life Technologies, Merelbeke, Belgium), supplemented with 10% Fetal Calf Serum (FCS) (Gibco™, Thermo Fisher Scientific, #A4766801, Life Technologies, Merelbeke, Belgium), 1% pen/strep (Gibco™, Thermo Fisher Scientific, #15140122, Life Technologies, Merelbeke, Belgium), and 1% Hepes (Gibco™, Thermo Fisher Scientific, #15630056, Life Technologies, Merelbeke, Belgium). Cells were incubated at 37 °C with 5% CO_2 and passaged 1:20 every 3–4 days using TrypLE™ Express Enzyme (Gibco™, Thermo Fisher Scientific, #12605010, Life Technologies, Merelbeke, Belgium).

JN-DSRCT-1 cells were maintained in Dulbecco's Modified Eagle's Medium F12 (DMEM-F12) (Gibco™, Thermo Fisher Scientific, #11320033, Waltham, MA, USA), supplemented with 1% pen/strep (Gibco™, Thermo Fisher Scientific, #15140122, Life Technologies, Merelbeke, Belgium) and 10% Fetal Calf Serum (FCS) (Gibco™, Thermo Fisher Scientific, #A4766801, Life Technologies, Merelbeke, Belgium). Cells were incubated at 37 °C with 5% CO_2 and passaged 1:10 every 3–4 days using TrypLE™ Express Enzyme (Gibco™, Thermo Fisher Scientific, #12605010, Life Technologies, Merelbeke, Belgium).

2.3. Single-Cell RNA Sequencing

OV-054 DSRCT cells were digested using TrypLE™ Express Enzyme (Gibco™, Thermo Fisher Scientific, #12605010, Life Technologies, Merelbeke, Belgium) and mechanically dissociated by pipetting up and down. The 2D and 3D OV-054 DSRCT cells were subsequently sorted for DAPI negative cells using the FACSJazz (BD bioscience, Franklin Lakes, NJ, USA) and SH800S Cell Sorter (SONY Europe B.V., Weybridge, Surrey, UK), respectively. Single DSRCT cells were sorted into a 384-well plate for SORT-seq performed by Single Cell Discoveries B.V. [15]. Gene counts per gene per well are provided in Tables S1–S4. Cluster analysis was performed using RaceID3 [16]. WT1 reads represent *EWSR1-WT1* fusion expression, since wild-type WT1 is not expressed in OV-054 DSRCT cells.

2.4. Comparison R2 Dataset

The DSRCT patient described in this paper is included in the INFORM study [17] and available microarray data were analyzed in R2 (www.r2.amc.nl, accessed on 14 January 2020). Two gene sets were created from the gene expression list of the original DSRCT patient: high-expressed genes (Z-score > 1) and low-expressed genes (Z-score < −1). Average read counts of the genes in these gene sets were collected from the here generated scRNAseq data of the 2D and 3D cultured DSRCT cells.

2.5. Introduction of shRNAs into OV-054 DSRCT Cells

shRNAs targeting the *EWSR1-WT1* breakpoint or *WT1* exon 8–10 and a non-targeting shRNA were cloned into pLKO-Tet-On Vector (Addgene, #21915) (Table S5). shRNAs were introduced in the patient-derived OV-054 DSRCT cells using lentiviral transduction. Cells were selected by addition of 0.5 μg/mL puromycin (InvivoGen, #ant-pr-1, San Diego, CA, USA) to the culture medium. Transcription of shRNAs was induced with 1.0 μg/mL doxycycline hyclate (Sigma Aldrich, #D9891, St. Louis, MO, USA). For shRNA 3, we harvested RNA 8, 24, and 32 h post-doxicyclin induction. For shRNA 2, we harvested RNA 40 h post-doxicyclin induction.

2.6. RNA Isolation, PCR, and qPCR

RNA isolation was performed using Direct-zol RNA MicroPrep Kit (Zymo Research, R2062, Orange, CA, USA), according to the manufacturer's protocol. cDNA synthesis was performed using SuperScript III reverse transcriptase (Thermo Fisher Scientific, #12574026, Life Technologies, Bleiswijk, The Netherlands) and random primers (Promega, #C1181, Madison, WI, USA). q-PCR was performed using iQ^{TM} SYBR Green Supermix (BIO-RAD Laboratories, #1708882, Hercules, CA, USA). *EWSR1-WT1* was amplified from cDNA by the primers FW: 3′-TCCTACAGCCAAGCTCAAGTC-5′ and REV: 3′-ACCTTCGTTCACAGTGGTTG-5′. *MERTK* was amplified from cDNA by the primers FW: 3′-GTGTCCAAGGGAGTGCAG-5′ and REV: 3′-CTCAGCGGATCAGCTTCC-5′. Cq-values were normalized to *GAPDH* and amplified by the primers: FW: 3′-CACATCGCTCA GACACCATG-5′ and REV: 3′-TGACGGTGCCATGGAATTTG-5′. For PCR and sequencing of the *EWSR1-WT1* breakpoint, the same primers were used, as mentioned above, for the q-PCR. The reverse primer was used for sequencing of the fusion breakpoint by Macrogen Europe BV.

2.7. RNA Sequencing

RNA samples of OV-054 DSRCT cells, upon induction of shRNA 2, 3, and NT, were paired-end sequenced by Macrogen Europe BV, using Illumina TruSeq stranded total RNA library construction with Ribo-Zero Gold and Novaseq6000 S4 2 × 150 bp. To allow for distinguishing between reads that derived from *EWSR1-WT1* fusion transcripts, wild-type *EWSR1* transcripts, or wild-type *WT1* transcripts, the hg38 reference transcriptome was obtained from Ensembl and edited. Therefore, the full-length *EWSR1* was split into 5′*EWSR1* (exon 1–7) and 3′*EWSR1* (exon 8–17) and full-length *WT1* was split into 5′*WT1* (exon 1–7) and 3′*WT1* (exon 8–10). Paired-end sequencing reads were aligned to this edited reference transcriptome using the Burrows–Wheeler aligner software package (BWA-0.7.17) [18]. RPKM-normalized read counts are provided in Tables S6 and S7. Transcript count analysis and figures were made using R (http://www.r-project.org, accessed on 30 April 2020).

2.8. Time-Lapse Imaging and Particle Analysis

OV-054 DSRCT cells with shRNA 3 and NT shRNA were induced with DOX, 24 h prior to live imaging. Live imaging was performed using the Leica DMi8, while incubated at 37 °C and 5% CO_2. Images were taken every 2 min from representative areas in each well for 16 h in total (24–40 h post-DOX). Cell counts and cell area from pictures of DOX-induced and non-induced shRNAs in OV-054 DSRCT cells were analyzed using particle analysis in ImageJ (National Institutes of Health and the Laboratory for Optical and Comutational Instrumentation (LOCI, University of Wisconsin), Madison, WI, USA). The number of segments represent the number of cells, and the segment areas represent the area of each cell. Cell migration and cell adhesion genes were identified using the DAVID gene annotation tool [19,20]. Cell-matrix interactions were quantified using the cell counter tool in ImageJ (National Institutes of Health and the Laboratory for Optical and Comutational Instrumentation (LOCI, University of Wisconsin), Madison, WI, USA).

2.9. Medium-Throughput Compound Screen

Two days prior to the addition of compounds, 5×10^3 OV-054 DSRCT cells were seeded, per well, in 384-well plates. A panel of 201 drugs and DMSO controls were added to the 384-well plates in different concentrations using the Beckman Coulter Biomek i7 Hybrid liquid handling workstation. After 5 days of exposure of the compounds, cell viability was measured with CellTiter-Glo® assay (Promega, #G9681, Madison, WI, USA) using the Spectramax i3x luminescence microplate reader. Values were normalized to DMSO. AUC Z-scores were calculated by $(AUC\text{-}AUC_{avg})/st.dev$ for OV-054 DSRCT cultures and two Ewing cultures (ES-046 and ES-041). Relevant plasma concentrations from previous phase I and II trials (Table S8) were plotted to compare the viability curves.

2.10. UNC2025 Screen

OV-054 DSRCT cells, JN-DSRCT-1 cells, and control human small intestinal (huSI) organoids were incubated with different concentrations of MERTK inhibitor UNC2025 (Selleckchem, #S7576, Houston, TX, USA): 25, 50, 100, 200, or 400 nM was added to the culture medium. Seven days after the addition of UNC2025, representative pictures were taken, and cells were treated with TryplE™ Express Enzyme (Gibco™, Thermo Fisher Scientific, #12605010, Life Technologies, Merelbeke, Belgium) and counted. Viability of huSI organoids were measured by organoid size using representative pictures and the area measurement tool in ImageJ (National Institutes of Health and the Laboratory for Optical and Comutational Instrumentation (LOCI, University of Wisconsin), Madison, WI, USA).

3. Results

3.1. Establishment and Characterization of a DSRCT In Vitro Preclinical Model

Surgically resected tissue was obtained from a DSRCT patient and successfully cultured in vitro under various conditions, including as spheroids in 3D basement membrane extract type 2 (BME) droplets, as a 2D layer in 0.1% BME, and in suspension without BME (Figure 1A). The DSRCT in vitro cultures were established and maintained for over two years in serum-free basal medium (AdDMEM/F12+), supplemented with the specific growth factors EGF, FGF2, IGF1, RKI, and BMP4. While these growth factors were essential for initial growth of DSRCT cells, the established OV-054 DSRCT cultures were able to be maintained without the additional growth factors EGF, FGF2, IGF1, RKI, and BMP4 for over 10 passages. The addition of the growth factors independently did not affect DSRCT cell expansion in vitro, compared to the control medium without the additional growth factors (Figure S1), suggesting that these OV-054 DSRCT cells are able to adapt fast and are able to propagate independently of their microenvironment. For consistency, all further experiments were conducted using the complete medium, in which the OV-054 DSRCT model was established. H&E staining on OV-054 DSRCT cultures revealed nests of small round cells with large nuclei (Figure 1A), which is typically seen in DSRCT tissue [4]. Generally, patient-derived tumor tissue grown in 3D BME droplets can form organized structures, so-called organoids, that can capture disease heterogeneity [21]. To investigate whether our 2D and 3D DSRCT in vitro models also consist of distinct cell populations, single-cell RNA sequencing (scRNAseq) was performed. scRNAseq analysis of 2D and 3D OV-054 DSRCT cultures revealed a homogeneous, tight population of DSRCT cells (Figure 1B,C). The slight difference between clustering of 2D versus 3D cultured cells was likely due to sequencing depth variation (Figure 1D). Approximately 0–2 reads, per million (RPM) of *WT1*, were picked up per cell in the scRNAseq analysis (Figure 1G), indicating low *EWSR1-WT1* expression. Importantly, the expression of *EWSR1-WT1* was similar between the 2D and 3D cultures (Figure 1E,G).

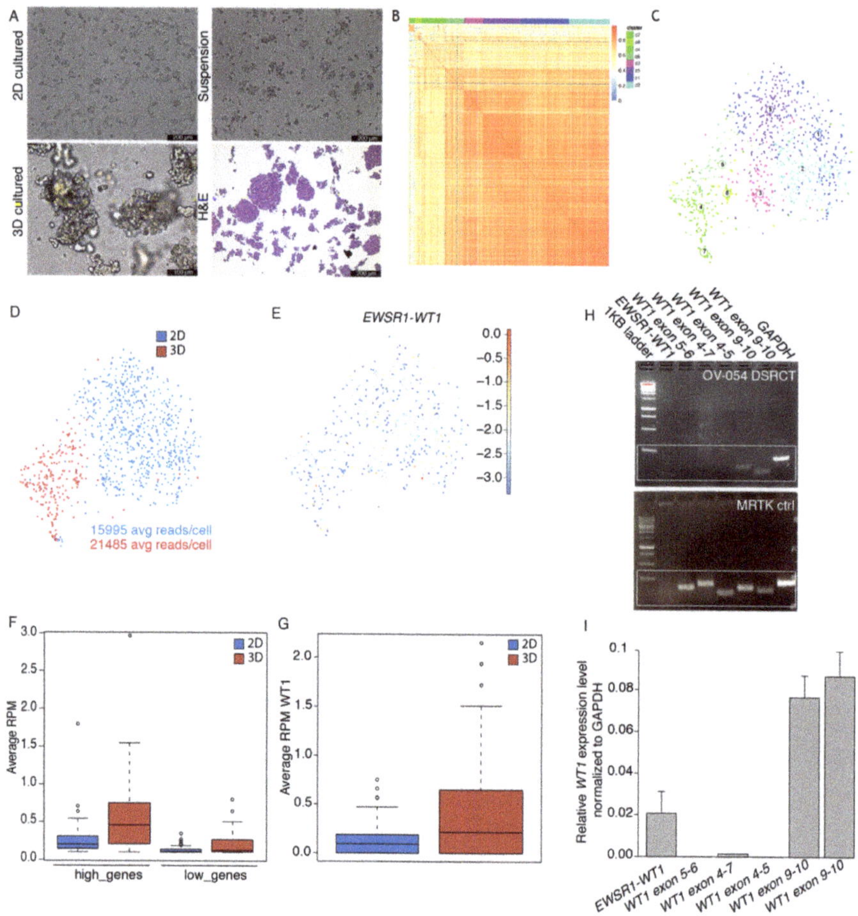

Figure 1. Establishment of a patient-derived DSRCT model in vitro. (**A**) Pictures of OV-054 DSRCT cells in different growth conditions: 2D on 0.1% BME, 3D in BME droplets and in suspension without BME. H&E staining is performed on cells in suspension. Scale bar: 100 µm (3D) and 200 µm (2D, suspension, and H&E). (**B**) Heatmap depicting cell clustering, based on differentially expressed genes. (**C**) tSNE plot depicting clusters of OV-054 DSRCT cells. (**D**) tSNE plot depicting clustering of 2D (blue) and 3D (red) cultured OV-054 DSRCT cells. (**E**) tSNE plot depicting *EWSR1-WT1* expression in 2D and 3D OV-054 DSRCT cells. (**F**) Box plot depicting the average RPM in 2D (blue) and 3D (red) cultured DSCRT cells for highly expressed and lowly expressed gene sets. (**G**) Box plot depicting the average *WT1* RPM in 2D (blue) and 3D (red) cultured OV-054 DSRCT cells. (**H**) Picture depicting gel electrophoresis of PCR amplicons of the *EWSR1-WT1* fusion breakpoint and different regions of *WT1* from cDNA of OV-054 DSRCT cells and MRTK control cells. (**I**) Bar plot depicting relative expression of *EWSR1-WT1* and different regions of *WT1* by qPCR, normalized to *GAPDH*.

The DSRCT patient, described in this study, was enrolled in the INFORM (individualized therapy for relapsed malignancies in childhood) pilot study [17]. The INFORM registry applied comprehensive molecular profiling to provide information on actionable gene variants, which may be used for subsequent therapeutic approaches. Microarray data of the original tumor of this DSRCT patient were obtained from the INFORM study and used to create two gene sets from the gene expression list: high (Z-score > 1) and low genes (Z-score < −1). Average read counts of the two gene sets were collected from the scRNAseq data of the 2D and 3D cultured DSRCT cells, and a similar trend was found between expression of the gene sets in the in vitro cultures, compared to the original tumor (Figure 1F). This indicated that both the 2D and 3D cultured DSRCT cells closely resemble features of the original tumor. For consistency, subsequent experiments were conducted with 2D cultured OV-054 DSRCT cells. To check whether wild-type *WT1* is expressed in OV-054 DSRCT cells, the *EWSR1-WT1* fusion breakpoint from cDNA, as well as several regions of *WT1*, were amplified. In our in vitro OV-054 DSRCT model, only the 3′-most exons of *WT1* were expressed (Figure 1H,I), while the expression of *WT1* exon 4–7 was absent. This indicated that exons 9–10 from *WT1* were derived from the *EWSR1-WT1* fusion cDNA and that wild-type *WT1* was not expressed in these cells (Figure 1H,I). As a control, cDNA of a *WT1*-expressing malignant rhabdoid tumor of the kidney (MRTK) model was taken along, which, indeed, showed expression of all regions of the *WT1* gene (Figure 1H).

3.2. *shRNAs Targeting the EWSR1-WT Breakpoint Create an Effective Knock-Down of the Fusion mRNA*

The OV-054 DSRCT in vitro model enabled us to study the molecular target genes of the fusion driving DSRCT. By generating a knock-down model of the *EWSR1-WT1* fusion mRNA, target genes could be identified that were affected by the expression level of *EWSR1-WT1*. To this end, we sequenced the breakpoint of *EWSR1-WT1* cDNA, in order to identify the exact location of the gene-fusion and found that *EWSR1* exon 7 was fused to *WT1* exon 8. This enabled us to design short hairpin RNAs (shRNAs), targeting the exact *EWSR1-WT1* breakpoint (shRNA 3, 5) or exon 8–10 of *WT1* (shRNA 2, 4, 6). With these, we created a DOX-inducible knock-down of the *EWSR1-WT1* fusion mRNA (Figure 2A). As a control, a non-targeting (NT), DOX-inducible scrambled shRNA sequence was taken along. An effective knock-down of the *EWSR1-WT1* fusion was observed with shRNA 2, 3, and 5 after 24 h of DOX induction (Figure 2B). When cells were harvested 8, 16, 24, 32, and 40 h after DOX induction, a gradual decrease of *EWSR1-WT1* transcripts was observed (Figure 2C), where shRNA 3 resulted in the most efficient knock-down. To identify genes that are affected upon *EWSR1-WT1* knock-down, RNA sequencing was performed. RNA sequencing reads were mapped to a reference transcriptome, in which *EWSR1* and *WT1* were split into a 3′end and a 5′end, in order to distinguish between the *EWSR1-WT1* fusion and wild-type *EWSR1* and *WT1* transcripts. 5′*WT1* was not expressed (Figure 2D), again confirming that wild-type *WT1* was not expressed in these DSRCT cells; 5′*EWSR1* and 3′*WT1* showed a gradual decrease upon DOX induction (Figure 2D), indicative of efficient knock-down of the *EWSR1-WT1* fusion transcripts. A slight decrease was seen in 3′*EWSR1* transcripts (Figure 2D), suggesting that shRNA 3 could also target wild-type *EWSR1*. However, the RPKM did not drop below the 3′*EWSR1* levels in the NT shRNA control, indicating that this effect was minimal. Sequencing reads were picked-up from both +KTS and −KTS isoforms, which were decreased to a similar extent after induction of shRNA 3 (Figure 2E). Hence, knock-down of the *EWSR1-WT1* fusion with shRNA 3 did not result in an isoform bias. Together, these data showed that induction of shRNA 3 (targeting the *EWSR1-WT1* breakpoint in DSRCT cells) resulted in an efficient knock-down of the fusion gene. Wild-type *EWSR1* and *WT1* were not affected, and no bias was observed for the +KTS or −KTS isoform. This *EWSR1-WT1* knock-down model was used to further investigate genes affected by levels of *EWSR1-WT1* expression.

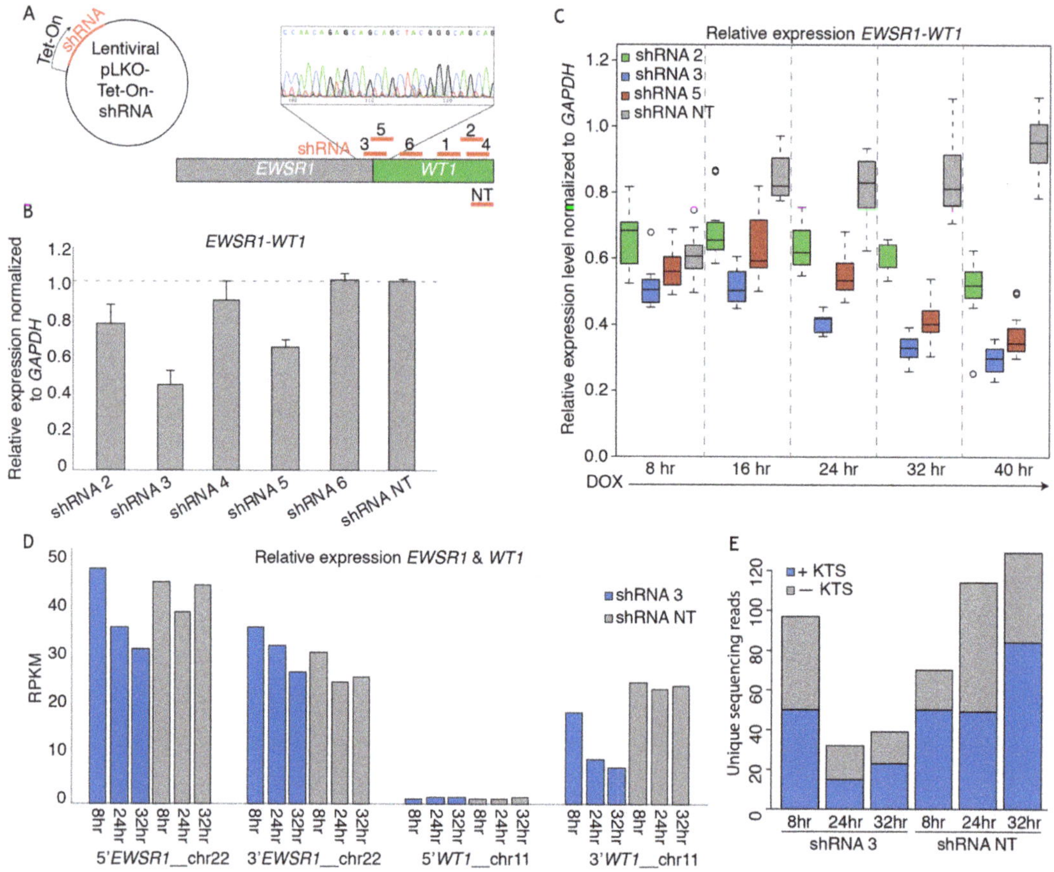

Figure 2. shRNA-mediated *EWSR1-WT1* knock-down in the DSRCT in vitro model. (**A**) Animation depicting the design of shRNAs targeting the breakpoint of *EWSR1-WT1*, 3′*WT1*, and a non-targeting (NT) shRNA. shRNAs were cloned into a lentiviral Tet-On backbone. (**B**) Bar plot depicting relative *EWSR1-WT1* expression levels 24 h post-DOX-induction of shRNAs normalized to *GAPDH*, as determined by qPCR. (**C**) Box plot depicting relative *EWSR1-WT1* expression levels at 8, 16, 24, 32, and 40 h post-DOX-induction of shRNA 2, 3, 5, and NT normalized to *GAPDH*, as determined by qPCR. (**D**) Bar plot depicting RPKM counts of 5′*EWSR1*, 3′*EWSR1*, 5′*WT1*, and 3′*WT1* at 8, 24, and 32 h post-DOX-induction of shRNA 3 and NT, as determined by mRNA paired-end sequencing. (**E**) Bar plot depicting number of +KTS (blue) and −KTS (gray) isoform reads, as determined by mRNA paired-end sequencing.

3.3. *EWSR1-WT1 Expression Affects Cell Shape, Cell Propagation, and the Transcriptome*

To investigate the phenotypic effects of EWSR1-WT1 in OV-054 DSRCT cells, time-lapse imaging was performed. Knock-down of the *EWSR1-WT1* fusion mRNA, in a time lapse between 24 h and 40 h after DOX induction, resulted in a decrease of cell–cell adhesion and increase of cell-matrix adhesion (Figures 3A and S2). Additionally, after 4 days of DOX-induction of shRNA 3, OV-054 DSRCT cells appeared more stretched, creating more cell-matrix adhesion, while non-induced cells, as well as cells with DOX-induction of the NT shRNA control, remained in clusters of inter-adhesive cells (Figure 3B). Upon knock-down of *EWSR1-WT1*, the number of cells decreased (Figure 3C), while the area of the cells increased (Figure 3D), showing that decreased expression of *EWSR1-WT1* quickly affects shape and expansion of DSRCT cells.

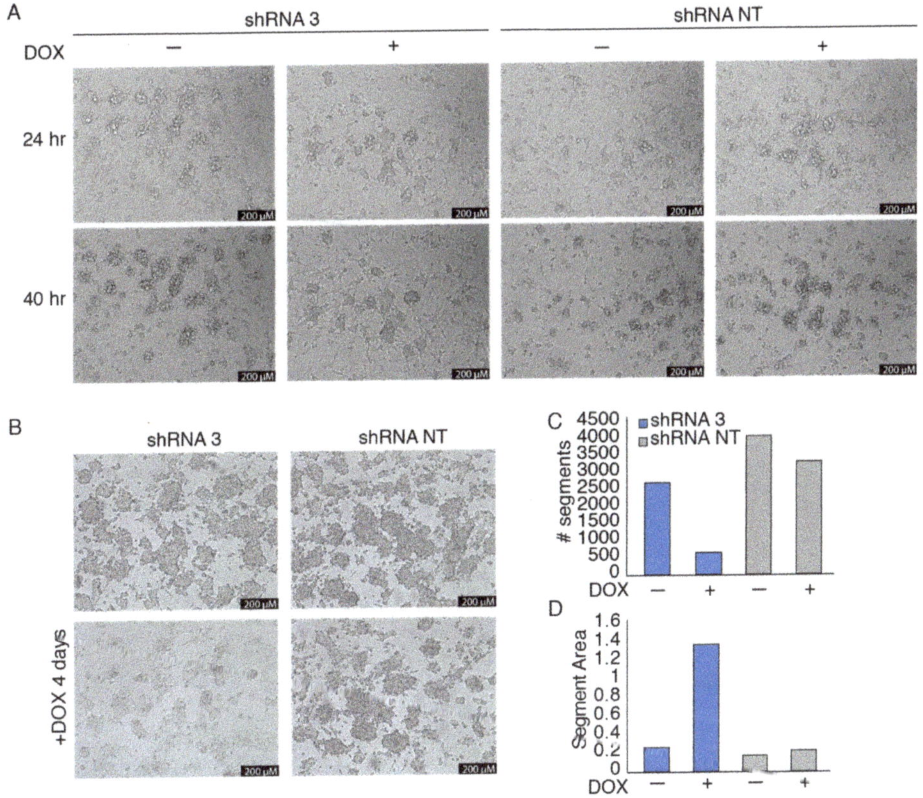

Figure 3. *EWSR1-WT1* expression affects OV-054 DSRCT cell shape and propagation. (**A**) Time-lapse images of OV-054 DSRCT cells upon *EWSR1-WT1* knock-down 24 and 40 h post-DOX-induction; scale bar: 200 µm. (**B**) Representative pictures taken 4 days post-DOX-induction of shRNA 3 and shRNA NT in DSRCT cells; scale bar: 200 µm. (**C**) Bar plot depicting number of segments, representing OV-054 DSRCT cells 4 days post-DOX-induction of shRNA 3 and shRNA NT, using particle analysis in ImageJ. (**D**) Bar plot depicting segment area, representing OV-054 DSRCT cells found 4 days post-DOX-induction of shRNA 3 and shRNA NT, using particle analysis in ImageJ.

To investigate the effect of *EWSR1-WT1* expression on the transcriptome, RNA sequencing was performed 8, 24, and 32 h after induction of shRNA 2, 3, and NT. To control for off-target effects, both shRNA 2 and shRNA 3 were used, which target *WT1* exon 10 and the *EWSR1-WT1* breakpoint, respectively. For both shRNA 2 and shRNA 3, genes that were at least 25% up- or downregulated were selected, compared to the NT shRNA control. A cut-off was set at 5 RPKM, in order to extract truly expressed genes. Genes that were affected by both shRNA 2- and shRNA 3-mediated knock-down of the *EWSR1-WT1* fusion were called true target genes of EWSR1-WT1 (Figure 4A). After 32 h of DOX-induction, we identified 75 genes that were downregulated upon *EWSR1-WT1* knock-down and 174 genes that were upregulated, upon *EWSR1-WT1* knock-down (Figure 4B). The expression levels of the 50 most upregulated genes and most downregulated genes, upon *EWSR1-WT1* knock-down, are shown in heatmaps (Figure 4C,D). The phenotypic effects that we observed were accompanied by an increased expression of several cell migration and cell adhesion genes (Figure S3C). While phenotypically the DSRCT cells appeared more mesenchymal upon knock-down of the *EWSR1-WT1* fusion, no evidence was found for epithelial to mesenchymal transition (EMT) (Figure S3D) [22]. Together, these data showed that cell migration and adhesion genes are upregulated upon knock-down of *EWSR1-WT1*, resulting in phenotypic changes, including a decreased cell expansion and increase in cell surface area. Subsequently, OV-054 DSRCT cells, with decreased *EWSR1-WT1* levels, create more cell-matrix interactions (Figure S3A,B).

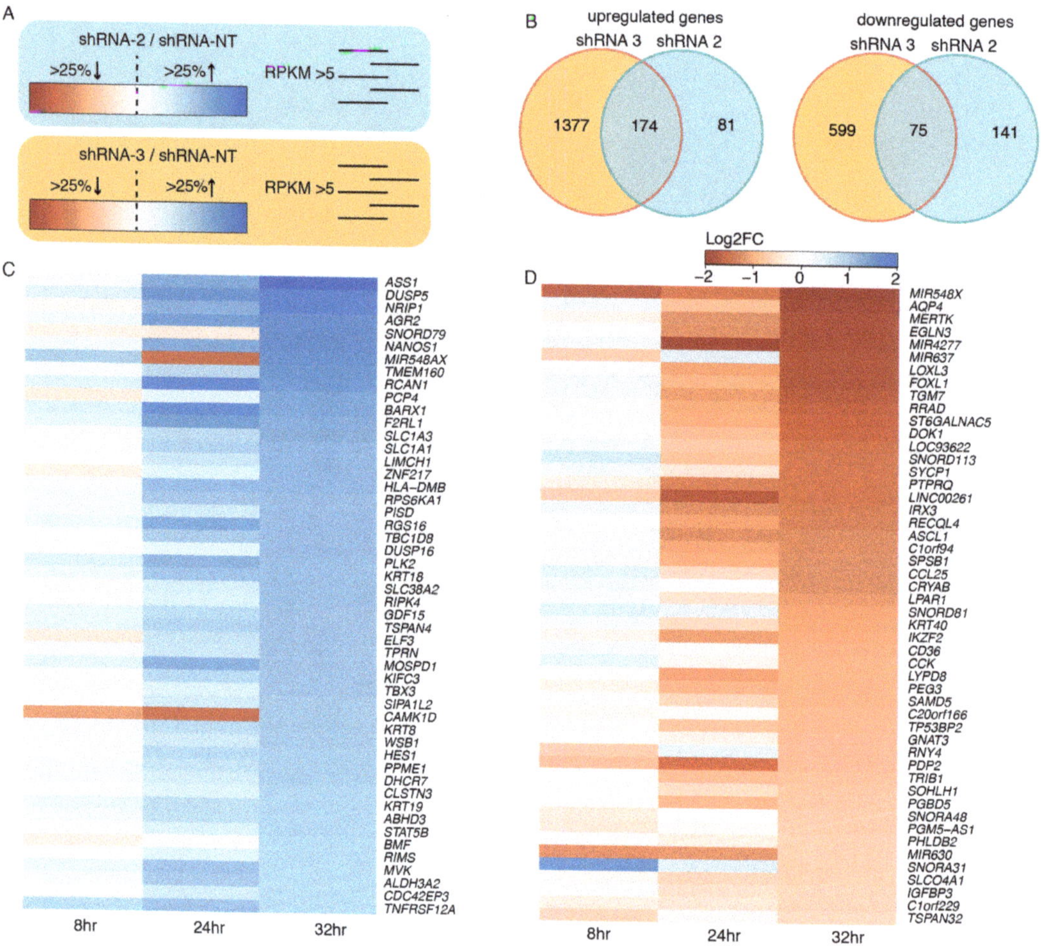

Figure 4. Target genes of EWSR1-WT1. (**A**) Animation of the approach to find EWSR1-WT1 target genes: select genes that are at least 25% up- or downregulated and have an RPKM value >5 for both shRNA 2 and shRNA 3. (**B**) Venn diagram depicting upregulated and downregulated genes, upon knock-down of *EWSR1-WT1* with shRNA 2 and shRNA 3. (**C**) Heatmap depicting the 50 highest upregulated genes, upon *EWSR1-WT1* knock-down, colors represent log2FC, compared to shRNA NT. (**D**) Heatmap depicting 50 most downregulated genes, upon *EWSR1-WT1* knock-down, colors represent log2FC, compared to shRNA NT.

The genes that were affected upon knock-down of *EWSR1-WT1* were next compared to a recent study, in which a siRNA-mediated knock-down of the *EWSR1-WT1* fusion gene was performed in the JN-DSRCT-1 and BER cell lines [10]. From the 174 upregulated genes upon *EWS-WT1* knock-down in our OV-054 DSRCT cells, 6% (11/174) showed an overlap with both the JN-DSRCT-1 and BER lines: *CST1, FILIP1L, CHST1, ALCAM, EPHA4, KRT8, ELF3, KRT18, SLC1A1, PCP4,* and *ASS1* (Figure 5A). We found that 9% (7/75) of the downregulated genes, upon *EWSR1-WT1* knock-down, in our OV-054 DSRCT cells were shared with the other two in vitro models: *LOXL3, FOXL1, RRAD, ST6GALNAC5, CCL25, IGFBP3,* and *C1QL4* (Figure 5B). ShRNA3 resulted in a more efficient knock-down of

EWSR1-WT1, compared to shRNA 2-mediated knock-down. When only the genes affected upon shRNA 3-mediated knock-down were compared with the JN-DSRCT-1 and BER cell lines, we indeed discovered additional genes to overlap between the three in vitro models (Figure S4). The overlap of the transcriptomic effects, upon knock-down of *EWSR1-WT1* in three independent DSRCT cell lines, confirmed that these genes are direct or indirect targets of the EWSR1-WT1 fusion and potentially play an important role in tumor development and progression.

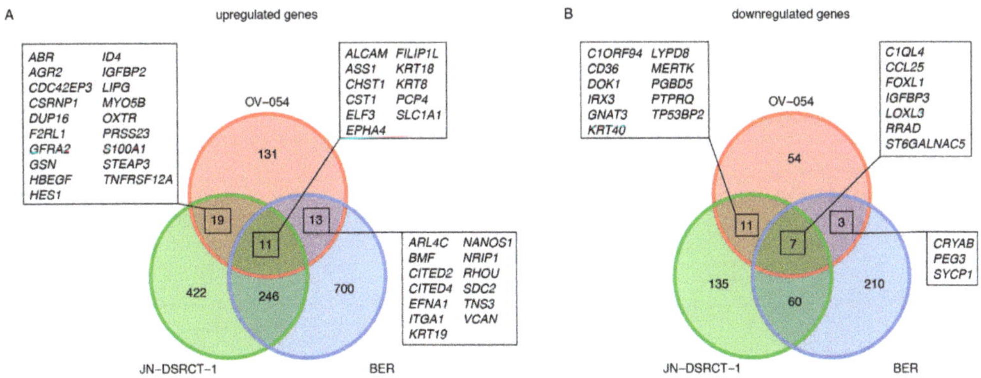

Figure 5. Comparison of *EWSR1-WT1* target genes in OV-054, JN-DSRCT-1, and BER cell lines [10]. (**A**) Venn diagram depicting overlapping upregulated genes, upon knock-down of *EWSR1-WT1*. (**B**) Venn diagram depicting overlapping downregulated genes, upon knock-down of *EWSR1-WT1*.

3.4. Drug Screen on OV-054 Cells Reveals Effective Compounds Targeting RTKs and Downstream Pathways

The established preclinical OV-054 DSRCT in vitro model enabled us to perform a medium-throughput drug screen for DSRCT. The screen involved a panel of 201 different compounds. Several of these compounds affected DSRCT cell viability. The area under the curve (AUC) Z-scores of OV-054 DSRCT cells were compared to the Z-scores of two Ewing sarcoma in vitro models, ES-041 and ES-046, to unravel compounds that showed a higher sensitivity in DSRCT cells (Figure 6A). As DSRCT is closely related to Ewing sarcoma, the compounds showing AUC Z-score possibly present sarcoma specific sensitivity. From these drugs, the compounds were selected that presented with an IC50 value, which was below or around the relevant plasma concentrations (Figure 6B). Several of the compounds that showed an effective response were targeting apoptosis regulators (such as XIAP and BCL2L), including AZD5582, Birinapant, and Navitoclax (Figure 6B), suggesting that the regulation of apoptosis is relevant in DSRCT cells. Interestingly, many other compounds that showed an effective response, including regorafenib, lapatinib, entrectinib, linsitinib, crizotinib, dovitinib, sorafenib, vandetanib, and brigatinib (Figure 6B), target RTKs. Additionally, compounds targeting downstream targets of RTK signaling pathways (Figure 6B), such as the PI3K-AKT and the mTOR signaling pathways, showed effective cell killing, including AT7519 and AZD8055. Together, the drug screening data suggested that the regulation of apoptosis- and RTK-driven signaling pathways are important for DSRCT tumor cell survival.

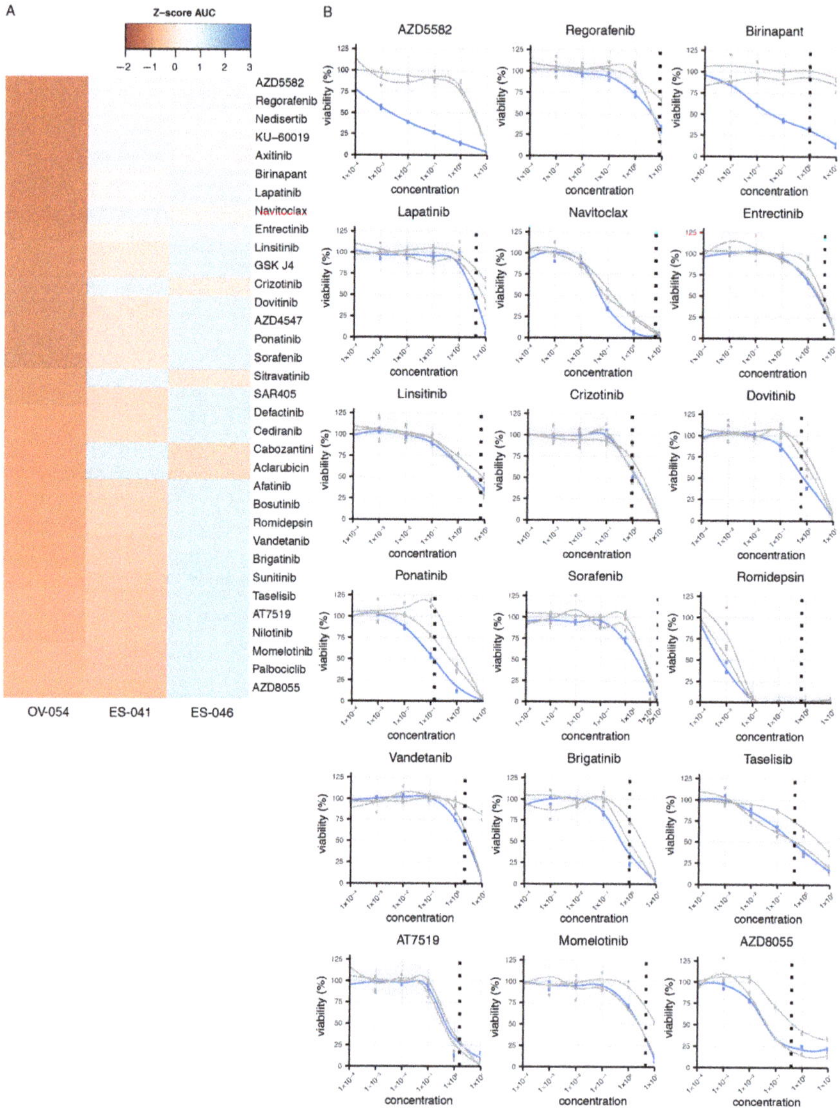

Figure 6. Compound screen on OV-054 DSRCT cells reveals compounds that affect cell viability. (**A**) Heatmap depicting the most sensitive compounds for OV-054 DSRCT, compared to two Ewing sarcoma models (ES-041 and ES-046); colors represent AUC Z-scores. (**B**) Graphs depicting cell viability of OV-054 DSRCT (blue) and the two Ewing sarcoma models (gray), upon a 5-day incubation with different drug concentrations (µM). Known plasma concentrations are shown with the vertical dashed line [23–33].

3.5. *MERTK, Regulated by EWSR1-WT1, Is a Potential Therapeutic Target in DSRCT*

Since RTKs and downstream pathways of RTKs appeared to be important for DSRCT tumor progression, we looked for RTKs that are regulated by *EWSR1-WT1* expression. *MERTK* is one of the most downregulated genes, upon knock-down of *EWSR1-WT1*, in OV-054 DSRCT cells (Figure 4C) and the JN-DSRCT-1 cell line (Figure 5B), suggesting that high expression of *MERTK* (Figure 7A) is regulated by the EWSR1-WT1 fusion. Interestingly, MERTK has been described as a therapeutic target in several cancers, including melanoma, leukemia, glioblastoma, and gastric cancer [34–36]. To investigate the function of MERTK in DSRCT cells, different concentrations of a MERTK/FLT3 small-molecule inhibitor UNC2025 were added to the culture medium, and live cells were counted after 4 and 7 days. Inhibition of MERTK resulted in reduced cell expansion of OV-054 DSRCT cells in vitro in a dose dependent manner (Figure 7B,D), and a similar effect was found in the JN-DSRCT-1 cell line (Figure 7B,D). In both DSRCT cell lines, the IC50 of UNC2025 was around 104 nM (Figure 7C), in line with a previous study of UNC2025-mediated MERTK inhibition in leukemia [34]. Normal human small intestinal (huSI) organoids were exposed to UNC2025 as a control and were only affected by UNC2025 at the highest concentration of 400 nM (Figure 7B,D). This shows that the effect of UNC2025, seen in both DSRCT in vitro models, is a targeted effect, rather than a non-specific cytotoxic effect. Together, this suggested that MERTK is an important driver for cell proliferation in DSRCT and a potential therapeutic target.

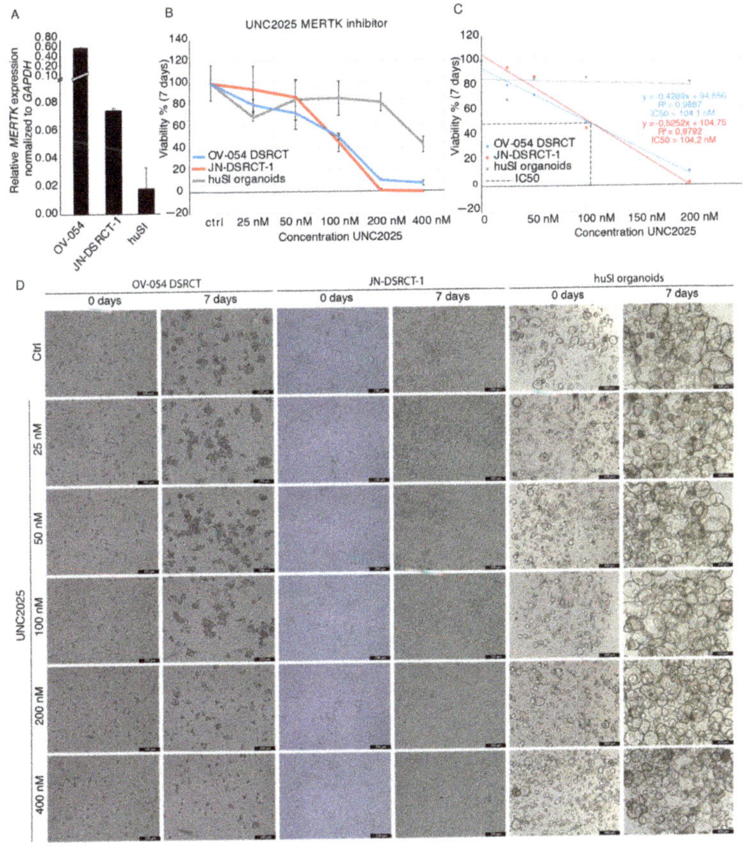

Figure 7. MERTK inhibitor UNC2025 affects DSRCT tumor expansion in vitro. (**A**) Barplot depicting qPCR results of relative *MERTK* expression in OV-054, JN-DSRCT-1, and huSI organoids, normalized to *GAPDH*. (**B**) Graph depicting cell viability of OV-054 DSRCT, JN-DSRCT-1, and huSI organoids, upon administration of 0, 25, 50, 100, 200, and 400 nM UNC2025. (**C**) Graph depicting IC50 of UNC2025 on both OV-054 DSRCT and JN-DSRCT-1 cells (**D**) Representative pictures of OV-054 DSRCT, JN-DSRCT-1, and huSI organoids in vitro, days 0 and 7, after administration of 0, 25, 50, 100, 200, and 400 nM UNC2025 (10× objective).

4. Discussion

DSRCT is a highly aggressive and rare soft tissue sarcoma, characterized by a chromosomal translocation, resulting in the *EWSR1-WT1* gene fusion. Expression of *EWSR1-WT1* presumably regulates the genetic targets responsible for oncogenesis in DSRCT. The 5-year, event-free survival rate is 18% [37], showing the urgence of novel therapies to improve the outcomes for DSRCT patients.

OV-054 DSRCT cells in vitro grow as nests of small round cells with large nuclei, typically also seen in DSRCT tissue [4]. Both 2D and 3D culture conditions of OV-054 DSRCT consist of a homogeneous cell population, exhibiting similarities to the transcriptional profile of the original tumor. The *EWSR1-WT1* fusion is expressed similarly under 2D and 3D culture conditions. This novel DSRCT in vitro model can be used to investigate molecular pathways driving this rare sarcoma type and further explore therapeutic options, which is urgently needed to improve the poor prognosis of this sarcoma.

To explore the molecular mechanisms of *EWSR1-WT1*, we used an shRNA knock-down approach on our primary DSRCT in vitro model. When the genes that were affected upon shRNA-mediated knock-down of *EWSR1-WT1* were compared with similar gene sets previously described for the JN-DSRCT-1 and BER cell lines [10], we indeed found overlap in up- and downregulated genes. Of note, *ASCL1* was downregulated upon *EWSR1-WT1* knock-down, which was previously described as a direct target of the gene fusion [7]. This confirms that these genes are (in)direct targets of the EWSR1-WT1 fusion and likely play a role in tumor development and progression.

Gedminas and colleagues show similarities between the molecular mechanisms of EWSR1-FLI1 in Ewing sarcoma and EWSR1-WT1 in DSRCT, despite their different DNA-binding domains [10]. These common features include the dysregulation of the DNA damage response, an alteration in the E2F transcription factor family members and modulation of other pathways, including TGFb and IGF/mTOR signaling. While the mechanism behind these common features is not clear, DSRCT cells show a striking dependence on ERG expression. This is a close family member of FLI1, and it is upregulated by EWSR1-WT1 in DSRCT [10]. Interestingly, Franzetti and colleagues found that a knock-down of *EWSR1-FLI1* expression in Ewing sarcoma affected cell dynamics [38]. Major changes were observed in the dynamics of the actin cytoskeleton and cell-to-cell adhesions shifted to cell-matrix adhesion, associated with an increase of cell migration and invasion potential in vivo. The dynamical changes of the actin cytoskeleton and a shift from cell–cell adhesion to cell-matrix adhesion, shown in the Ewing sarcoma model, are similar to the observations we found in DSRCT, upon knock-down of *EWSR1-WT1* in vitro. In our study, knock-down of *EWSR1-WT1* in DSRCT cells affected cell shape and propagation. We found several cell migration genes and cell–cell adhesion genes to be affected upon *EWSR1-WT1* knock-down. Thus, despite the differences in DNA binding motifs of EWSR1-FLI1 and EWSR1-WT1, the underlying mechanisms driving Ewing sarcoma and DSRCT are possibly similar.

Because of the rarity of DSRCT, this sarcoma is often excluded from clinical trials. Currently, there are 10 clinical trials ongoing or completed that involve DSRCT patients and just one of these clinical trials included an RTK-targeting compound, i.e., sorafenib (ClinicalTrials.gov identifier: NCT01946529). Unfortunately, the interim analysis of this clinical trial determined that the therapy did not meet the anticipated response; therefore, the trial was stopped. To discover novel therapeutic entities, we performed a medium-throughput drug screen on the established preclinical OV-054 DSRCT in vitro model. A total of 201 different compounds were included in the drug screen for DSRCT cell viability. The compounds to which the OV-054 DSRCT cells were sensitive included several drugs targeting the regulators of apoptosis and many drugs targeting RTKs and downstream pathways of RTKs, including the PI3K-AKT and mTOR signaling pathways. Our data suggests that the regulation of apoptosis is important in DSRCT cells and that RTK-driven pathways are key players in DSRCT tumor progression. Therefore, other compounds that target RTKs or downstream pathways of RTKs might show a more effective response, either alone or in combination with current standard chemotherapy.

MERTK levels in OV-054 DSRCT are decreased upon knock-down of *EWSR1-WT1*, showing that this gene is likely regulated by the fusion protein. Interestingly, *MERTK* expression is higher in the OV-054 DSRCT cells, compared to the JN-DSRCT-1 cells. This inconsistency can be the result of the different fusion breakpoints of the *EWSR1-WT1* fusion genes in the OV-054 DSRCT and JN-DSRCT1 cells. However, we cannot exclude the possibility that this effect is a result of differences in the culture media. Despite these different expression levels of *MERTK* in the different DSRCT in vitro models, the effect of small-molecule inhibitor UNC2025 was similar between the OV-054 DSRCT model and the JN-DSRCT-1 cell line. MERTK has several downstream signaling pathways, including MAPK/ERK, PI3K/AKT, and JAK/STAT, regulating multiple biological processes [35,39,40]. MERTK is involved in multiple malignancies, including leukemia, glioma, melanoma, and rhabdomyosarcoma [35,36], while it has been described as novel therapeutic target in several of these malignancies [36]. Here, we show that MERTK inhibition, with small-molecule inhibitor UNC2025, affects propagation of OV-054 DSRCT cells in vitro, providing a rationale for investigating MERTK as a therapeutic target in DSRCT by small molecule inhibitors, such as UNC2025.

5. Conclusions

Using a primary DSRCT patient-derived 2D and 3D cell culture system, we were able to characterize the molecular mechanisms that are driven by the DSRCT-specific EWSR1-WT1 fusion protein. The preclinical DSRCT in vitro model also enables us to perform a medium-throughput drug screen. This screen reveals compounds that affect the cellular pathways that are important for DSRCT cell viability, including RTK-driven signaling pathways. Interestingly, we show that expression the RTK family member *MERTK* correlates with the expression of the *EWSR1-WT1* fusion gene. To our knowledge, this is the first study that reveals effective therapeutic compounds that likely target EWSR1-WT1-driven mechanisms.

Supplementary Materials: The following are available online at https://www.mdpi.com/article/10.3390/cancers13236072/s1. Figure S1: effect growth factors on DSRCT cultures, Figure S2: time lapse DSRCT, upon EWSR1-WT1 knock-down, Figure S3: changes in cell-matrix interactions, Figure S4: comparison target genes to JN-DSRCT-1 and BER cell line, Table S1: scRNAseq_OV-054_2D_TM335.transcripts, Table S2: scRNAseq_OV-054_2D_TM336.transcripts, Table S3: scRNAseq_ OV-054_3D_TM93.transcripts, Table S4: scRNAseq_OV-054_3D_TM94.transcripts, Table S5: shRNA sequences, Table S6: RNAseq_EWS-WT1_KD_shR3-NT_rpkm, Table S7: RNAseq_EWS-WT1_KD_shR2-NT_rpkm, Table S8: plasma concentrations compounds.

Author Contributions: Conceptualization, M.v.d.W., M.B.; methodology, M.B., M.v.d.W.; software, M.B.; validation, M.B., C.P., S.E., F.R., M.v.d.W.; formal analysis, M.B.; investigation, M.B., C.P., S.E., F.R.; re-sources, M.v.d.W., F.M.-W.; data curation, M.B., F.R.; writing—original draft preparation, M.B.; writing—review and editing, M.B., C.P., S.E., F.R., M.v.d.W., H.C.; visualization, M.B.; supervision, M.v.d.W., H.C.; project administration, M.B.; funding acquisition, H.C., M.v.d.W., F.M.-W. All authors have read and agreed to the published version of the manuscript.

Funding: This research was funded by Stichting Kinderen Kankervrij, grant number 262.

Institutional Review Board Statement: Experiments with human material were approved (MEC-2016-739) by the medical ethical committee of the Erasmus Medical Center (Rotterdam, the Netherlands) and Princess Máxima Center for Pediatric Oncology (Utrecht, The Netherlands).

Informed Consent Statement: Informed consent was obtained from all subjects involved in the study.

Data Availability Statement: Single-cell RNA sequencing gene counts are provided in Tables S1–S4. RPKM-normalized RNA sequencing data are provided in Tables S6 and S7.

Acknowledgments: The authors thank Reinier van den Linden for help with flow cytometry; the Máxima Single Cell Genomics Facility for help with the scRNA data processing; Kimberley Ober, Selma Eising, and Jan Molenaar from the Máxima Drug Screening Facility for their screening services.

Conflicts of Interest: The authors declare no conflict of interest.

References

1. *Soft Tissue and Bone Tumours: WHO Classification of Tumours*, 5th ed.; IARC: Lyon, France, 2020.
2. Bulbul, A.; Fahy, B.N.; Xiu, J.; Rashad, S.; Mustafa, A.; Husain, H.; Hayes-Jordan, A. Desmoplastic small round blue cell tumor: A review of treatment and potential therapeutic genomic alterations. *Sarcoma* **2017**, *2017*, 1278268. [CrossRef] [PubMed]
3. Hayes-Jordan, A.; LaQuaglia, M.P.; Modak, S. Management of desmoplastic small round cell tumor. *Semin. Pediatr. Surg.* **2016**, *25*, 299–304. [CrossRef] [PubMed]
4. Gerald, W.L.; Rosai, J. Case 2. Desmoplastic small cell tumor with divergent differentiation. *Pediatr. Pathol.* **1989**, *9*, 177–183. [CrossRef] [PubMed]
5. Karnieli, E.; Werner, H.; Rauscher, F.J., 3rd; Benjamin, L.E.; LeRoith, D. The IGF-I receptor gene promoter is a molecular target for the Ewing's sarcoma-Wilms' tumor 1 fusion protein. *J. Biol. Chem.* **1996**, *271*, 19304–19309. [CrossRef] [PubMed]
6. Gerald, W.L.; Haber, D.A. The EWS-WT1 gene fusion in desmoplastic small round cell tumor. *Semin. Cancer Biol.* **2005**, *15*, 197–205. [CrossRef]
7. Kang, H.J.; Park, J.H.; Chen, W.; Kang, S.I.; Moroz, K.; Ladanyi, M.; Lee, S.B. EWS-WT1 oncoprotein activates neuronal reprogramming factor ASCL1 and promotes neural differentiation. *Cancer Res.* **2014**, *74*, 4526–4535. [CrossRef]
8. Kim, J.; Lee, K.; Pelletier, J. The desmoplastic small round cell tumor t(11;22) translocation produces EWS/WT1 isoforms with differing oncogenic properties. *Oncogene* **1998**, *16*, 1973–1979. [CrossRef]
9. Watson, S.; Perrin, V.; Guillemot, D.; Reynaud, S.; Coindre, J.M.; Karanian, M.; Guinebretiere, J.M.; Freneaux, P.; Le Loarer, F.; Bouvet, M.; et al. Transcriptomic definition of molecular subgroups of small round cell sarcomas. *J. Pathol.* **2018**, *245*, 29–40. [CrossRef] [PubMed]
10. Gedminas, J.M.; Chasse, M.H.; McBrairty, M.; Beddows, I.; Kitchen-Goosen, S.M.; Grohar, P.J. Desmoplastic small round cell tumor is dependent on the EWS-WT1 transcription factor. *Oncogenesis* **2020**, *9*, 41. [CrossRef] [PubMed]
11. Markides, C.S.; Coil, D.R.; Luong, L.H.; Mendoza, J.; Kozielski, T.; Vardeman, D.; Giovanella, B.C. Desmoplastic small round cell tumor (DSRCT) xenografts and tissue culture lines: Establishment and initial characterization. *Oncol. Lett.* **2013**, *5*, 1453–1456. [CrossRef] [PubMed]
12. Nishio, J.; Iwasaki, H.; Ishiguro, M.; Ohjimi, Y.; Fujita, C.; Yanai, F.; Nibu, K.; Mitsudome, A.; Kaneko, Y.; Kikuchi, M. Establishment and characterization of a novel human desmoplastic small round cell tumor cell line, JN-DSRCT-1. *Lab. Investig.* **2002**, *82*, 1175–1182. [CrossRef]
13. Hingorani, P.; Dinu, V.; Zhang, X.; Lei, H.; Shern, J.F.; Park, J.; Steel, J.; Rauf, F.; Parham, D.; Gastier-Foster, J.; et al. Transcriptome analysis of desmoplastic small round cell tumors identifies actionable therapeutic targets: A report from the Children's Oncology Group. *Sci. Rep.* **2020**, *10*, 12318. [CrossRef] [PubMed]
14. Bleijs, M.; van de Wetering, M.; Clevers, H.; Drost, J. Xenograft and organoid model systems in cancer research. *EMBO J.* **2019**, *38*, e101654. [CrossRef] [PubMed]
15. Muraro, M.J.; Dharmadhikari, G.; Grun, D.; Groen, N.; Dielen, T.; Jansen, E.; van Gurp, L.; Engelse, M.A.; Carlotti, F.; de Koning, E.J.; et al. A single-cell transcriptome atlas of the human pancreas. *Cell Syst.* **2016**, *3*, 385–394.e3. [CrossRef]
16. Herman, J.S.; Sagar; Grun, D. FateID infers cell fate bias in multipotent progenitors from single-cell RNA-seq data. *Nat. Methods* **2018**, *15*, 379–386. [CrossRef] [PubMed]
17. Van Tilburg, C.M.; Pfaff, E.; Pajtler, K.W.; Langenberg, K.P.S.; Fiesel, P.; Jones, B.C.; Balasubramanian, G.P.; Stark, S.; Johann, P.D.; Blattner-Johnson, M.; et al. The pediatric precision oncology inform registry: Clinical outcome and benefit for patients with very high-evidence targets. *Cancer Discov.* **2021**, *11*, 2764–2779. [CrossRef] [PubMed]
18. Li, H.; Durbin, R. Fast and accurate short read alignment with Burrows-Wheeler transform. *Bioinformatics* **2009**, *25*, 1754–1760. [CrossRef] [PubMed]
19. Da huang, W.; Sherman, B.T.; Lempicki, R.A. Bioinformatics enrichment tools: Paths toward the comprehensive functional analysis of large gene lists. *Nucleic Acids. Res.* **2009**, *37*, 1–13. [CrossRef] [PubMed]
20. Da Huang, W.; Sherman, B.T.; Lempicki, R.A. Systematic and integrative analysis of large gene lists using DAVID bioinformatics resources. *Nat. Protoc.* **2009**, *4*, 44–57. [CrossRef]
21. Calandrini, C.; Schutgens, F.; Oka, R.; Margaritis, T.; Candelli, T.; Mathijsen, L.; Ammerlaan, C.; van Ineveld, R.L.; Derakhshan, S.; de Haan, S.; et al. An organoid biobank for childhood kidney cancers that captures disease and tissue heterogeneity. *Nat. Commun.* **2020**, *11*, 1310. [CrossRef]
22. Pastushenko, I.; Blanpain, C. EMT transition states during tumor progression and metastasis. *Trends Cell Biol.* **2019**, *29*, 212–226. [CrossRef] [PubMed]
23. Bedi, S.; Khan, S.A.; AbuKhader, M.M.; Alam, P.; Siddiqui, N.A.; Husain, A. A comprehensive review on Brigatinib—A wonder drug for targeted cancer therapy in non-small cell lung cancer. *Saudi Pharm. J.* **2018**, *26*, 755–763. [CrossRef]
24. Chen, E.X.; Hotte, S.; Hirte, H.; Siu, L.L.; Lyons, J.; Squires, M.; Lovell, S.; Turner, S.; McIntosh, L.; Seymour, L. A Phase I study of cyclin-dependent kinase inhibitor, AT7519, in patients with advanced cancer: NCIC Clinical Trials Group IND 177. *Br. J. Cancer* **2014**, *111*, 2262–2267. [CrossRef] [PubMed]
25. Juric, D.; Krop, I.; Ramanathan, R.K.; Wilson, T.R.; Ware, J.A.; Sanabria Bohorquez, S.M.; Savage, H.M.; Sampath, D.; Salphati, L.; Lin, R.S.; et al. Phase I dose-escalation study of taselisib, an oral PI3K inhibitor, in patients with advanced solid tumors. *Cancer Discov.* **2017**, *7*, 704–715. [CrossRef]

26. Kang, Y.K.; Yoo, C.; Ryoo, B.Y.; Lee, J.J.; Tan, E.; Park, I.; Park, J.H.; Choi, Y.J.; Jo, J.; Ryu, J.S.; et al. Phase II study of dovitinib in patients with metastatic and/or unresectable gastrointestinal stromal tumours after failure of imatinib and sunitinib. *Br. J. Cancer* **2013**, *109*, 2309–2315. [CrossRef] [PubMed]
27. Liston, D.R.; Davis, M. Clinically relevant concentrations of anticancer drugs: A guide for nonclinical studies. *Clin. Cancer Res.* **2017**, *23*, 3489–3498. [CrossRef] [PubMed]
28. Macaulay, V.M.; Middleton, M.R.; Eckhardt, S.G.; Rudin, C.M.; Juergens, R.A.; Gedrich, R.; Gogov, S.; McCarthy, S.; Poondru, S.; Stephens, A.W.; et al. Phase I dose-escalation study of linsitinib (OSI-906) and erlotinib in patients with advanced solid tumors. *Clin. Cancer Res.* **2016**, *22*, 2897–2907. [CrossRef] [PubMed]
29. Meneses-Lorente, G.; Bentley, D.; Guerini, E.; Kowalski, K.; Chow-Maneval, E.; Yu, L.; Brink, A.; Djebli, N.; Mercier, F.; Buchheit, V.; et al. Characterization of the pharmacokinetics of entrectinib and its active M5 metabolite in healthy volunteers and patients with solid tumors. *Investig. New Drugs* **2021**, *39*, 803–811. [CrossRef]
30. Naing, A.; Aghajanian, C.; Raymond, E.; Olmos, D.; Schwartz, G.; Oelmann, E.; Grinsted, L.; Burke, W.; Taylor, R.; Kaye, S.; et al. Safety, tolerability, pharmacokinetics and pharmacodynamics of AZD8055 in advanced solid tumours and lymphoma. *Br. J. Cancer* **2012**, *107*, 1093–1099. [CrossRef]
31. Wilson, W.H.; O'Connor, O.A.; Czuczman, M.S.; LaCasce, A.S.; Gerecitano, J.F.; Leonard, J.P.; Tulpule, A.; Dunleavy, K.; Xiong, H.; Chiu, Y.L.; et al. Navitoclax, a targeted high-affinity inhibitor of BCL-2, in lymphoid malignancies: A phase 1 dose-escalation study of safety, pharmacokinetics, pharmacodynamics, and antitumour activity. *Lancet Oncol.* **2010**, *11*, 1149–1159. [CrossRef]
32. Zheng, J.; Xin, Y.; Zhang, J.; Subramanian, R.; Murray, B.P.; Whitney, J.A.; Warr, M.R.; Ling, J.; Moorehead, L.; Kwan, E.; et al. Pharmacokinetics and Disposition of Momelotinib Revealed a Disproportionate Human Metabolite-Resolution for Clinical Development. *Drug Metab. Dispos.* **2018**, *46*, 237–247. [CrossRef]
33. Zhu, X.; Trueman, S.; Straubinger, R.M.; Jusko, W.J. Physiologically-based pharmacokinetic and pharmacodynamic models for gemcitabine and birinapant in pancreatic cancer xenografts. *J. Pharmacokinet. Pharmacodyn.* **2018**, *45*, 733–746. [CrossRef]
34. DeRyckere, D.; Lee-Sherick, A.B.; Huey, M.G.; Hill, A.A.; Tyner, J.W.; Jacobsen, K.M.; Page, L.S.; Kirkpatrick, G.G.; Eryildiz, F.; Montgomery, S.A.; et al. UNC2025, a mertk small-molecule inhibitor, is therapeutically effective alone and in combination with methotrexate in leukemia models. *Clin. Cancer Res.* **2017**, *23*, 1481–1492. [CrossRef]
35. Graham, D.K.; DeRyckere, D.; Davies, K.D.; Earp, H.S. The TAM family: Phosphatidylserine sensing receptor tyrosine kinases gone awry in cancer. *Nat. Rev. Cancer* **2014**, *14*, 769–785. [CrossRef] [PubMed]
36. Schlegel, J.; Sambade, M.J.; Sather, S.; Moschos, S.J.; Tan, A.C.; Winges, A.; DeRyckere, D.; Carson, C.C.; Trembath, D.G.; Tentler, J.J.; et al. MERTK receptor tyrosine kinase is a therapeutic target in melanoma. *J. Clin. Investig.* **2013**, *123*, 2257–2267. [CrossRef] [PubMed]
37. Zhang, S.; Zhang, Y.; Yu, Y.H.; Li, J. Results of multimodal treatment for desmoplastic small round cell tumor of the abdomen and pelvis. *Int. J. Clin. Exp. Med.* **2015**, *8*, 9658–9666. [PubMed]
38. Franzetti, G.A.; Laud-Duval, K.; van der Ent, W.; Brisac, A.; Irondelle, M.; Aubert, S.; Dirksen, U.; Bouvier, C.; de Pinieux, G.; Snaar-Jagalska, E.; et al. Cell-to-cell heterogeneity of EWSR1-FLI1 activity determines proliferation/migration choices in Ewing sarcoma cells. *Oncogene* **2017**, *36*, 3505–3514. [CrossRef] [PubMed]
39. Brandao, L.N.; Winges, A.; Christoph, S.; Sather, S.; Migdall-Wilson, J.; Schlegel, J.; McGranahan, A.; Gao, D.; Liang, X.; Deryckere, D.; et al. Inhibition of MerTK increases chemosensitivity and decreases oncogenic potential in T-cell acute lymphoblastic leukemia. *Blood Cancer J.* **2013**, *3*, e101. [CrossRef]
40. Lee-Sherick, A.B.; Eisenman, K.M.; Sather, S.; McGranahan, A.; Armistead, P.M.; McGary, C.S.; Hunsucker, S.A.; Schlegel, J.; Martinson, H.; Cannon, C.; et al. Aberrant Mer receptor tyrosine kinase expression contributes to leukemogenesis in acute myeloid leukemia. *Oncogene* **2013**, *32*, 5359–5368. [CrossRef]

cancers

Review

Childhood Malignant Brain Tumors: Balancing the Bench and Bedside

Colin Thorbinson [1] and John-Paul Kilday [1,2,*]

[1] Children's Brain Tumour Research Centre, Department of Paediatric Oncology, Royal Manchester Children's Hospital, Manchester University NHS Foundation Trust, Manchester M13 9WL, UK; Colin.Thorbinson@mft.nhs.uk
[2] The Centre for Paediatric, Teenage and Young Adult Cancer, Institute of Cancer Sciences, The University of Manchester, Manchester M20 4BX, UK
* Correspondence: John-Paul.Kilday@mft.nhs.uk; Tel.: +44-(0)161-701-8418

Simple Summary: Brain tumors remain the most common childhood solid tumors, accounting for approximately 25% of all pediatric cancers. They also represent the most common cause of cancer-related illness and death in this age group. Recent years have witnessed an evolution in our understanding of the biological underpinnings of many childhood brain tumors, potentially improving survival through both improved risk group allocation for patients to provide appropriate treatment intensity, and novel therapeutic breakthroughs. This review aims to summarize the molecular landscape, current trial-based standards of care, novel treatments being explored and future challenges for the three most common childhood malignant brain tumors—medulloblastomas, high-grade gliomas and ependymomas.

Abstract: Brain tumors are the leading cause of childhood cancer deaths in developed countries. They also represent the most common solid tumor in this age group, accounting for approximately one-quarter of all pediatric cancers. Developments in neuro-imaging, neurosurgical techniques, adjuvant therapy and supportive care have improved survival rates for certain tumors, allowing a future focus on optimizing cure, whilst minimizing long-term adverse effects. Recent times have witnessed a rapid evolution in the molecular characterization of several of the common pediatric brain tumors, allowing unique clinical and biological patient subgroups to be identified. However, a resulting paradigm shift in both translational therapy and subsequent survival for many of these tumors remains elusive, while recurrence remains a great clinical challenge. This review will provide an insight into the key molecular developments and global co-operative trial results for the most common malignant pediatric brain tumors (medulloblastoma, high-grade gliomas and ependymoma), highlighting potential future directions for management, including novel therapeutic options, and critical challenges that remain unsolved.

Keywords: pediatric; brain; tumor; medulloblastoma; glioma; ependymoma

Citation: Thorbinson, C.; Kilday, J.-P. Childhood Malignant Brain Tumors: Balancing the Bench and Bedside. *Cancers* **2021**, *13*, 6099. https://doi.org/10.3390/cancers13236099

Academic Editors: Saurabh Agarwal and Jianhua Yang

Received: 8 November 2021
Accepted: 29 November 2021
Published: 3 December 2021

1. Introduction

Brain tumors are the most common solid tumors of childhood, accounting for approximately 25% of all pediatric malignancies, and represent the leading cause of cancer-induced morbidity and mortality in this age group [1]. With an incidence of approximately 6 per 100,000 children in industrialized society [2], these tumors represent a spectrum of clinically, pathologically and biologically diverse subtypes which can pose significant challenges in conducting research and clinical trials, necessitating international collaboration.

Over recent decades, cure rates for selected pediatric brain tumors (most notably medulloblastoma) have improved [3], predominantly as a consequence of advances in multiparametric neuro-imaging, neurosurgical techniques, radiation therapy and multi-agent chemotherapy, together with improved supportive care. However, such survival advances are typically offset by a therapy-induced toxicity burden for the patient, with wide-reaching consequences for the child, their family and society. Moreover, for the majority of brain tumors, prognosis has remained static for over 30 years despite these technological improvements.

To overcome this impasse, the pediatric neuro-oncology community has shifted focus to develop risk-stratified treatment protocols that aim to reduce iatrogenic morbidity while maintaining outcomes for favorable-risk lesions, and improve cure rates for tumors refractory to conventional therapy, either through intensification or novel agents. This strategy has been supplemented by an evolution in our understanding of the molecular pathogenesis of almost all pediatric brain tumors.

Such molecular advances have identified potential cells of origin, and led to the identification of multiple biologically distinct subgroups within most brain tumor entities, therein allowing accurate risk stratification for affected children when incorporated with clinical, histological and survival data. In addition, oncogenic biological pathways amenable to manipulation using novel targeted agents have been identified.

This article will provide a summary of the most common malignant pediatric brain tumors (medulloblastoma, high-grade gliomas and ependymoma) with particular focus on inherent molecular advancements and potential future directions for management, including novel therapeutic options.

2. Medulloblastoma

2.1. Background

Medulloblastoma (MB) represents the most common malignant brain tumor in children, accounting for approximately 20% of all central nervous system (CNS) tumors [2,4]. It also comprises over 60% of intracranial embryonal tumors, a recently characterized entity consisting of atypical teratoid rhabdoid tumors (ATRTs), embryonal tumors with multilayer rosettes (ETMRs), CNS neuroblastoma with *FOX2* alteration and malignant neuroepithelial tumors with *BCOR* alteration [5].

Arising within the cerebellum, MBs are observed across all age categories but are most frequently identified at a median age of five years [6]. Demographic, histological and prognostic heterogeneity embody MB, while it represents the first brain tumor where revolutionary global initiatives (such as the Medulloblastoma Advanced Genomics International Consortium (MAGIC)) have transformed our understanding of the molecular underpinnings of MB pathogenesis, enabling improved patient risk stratification to potentially influence clinical outcome [7].

2.2. Histopathology

MBs share a primitive embryonal phenotype comprising malignant cells of stereotypic histological patterns, dominated by neuronal antigen expression [8]. World Health Organization (WHO) pathological classification systems have historically divided MB into a classic subtype accounting for 72% of all cases, a desmoplastic/nodular variant of which medulloblastoma with extensive nodularity (MBEN) is a subgroup and a large cell/anaplastic variant which has historically been assigned an adverse prognostic association [5,9].

2.3. Molecular Classification

In the past decade, seminal transcriptomic MB analyses led to a global consensus establishing the identification of four discrete molecular subgroups, likely arising from distinct cells of origin—wingless-activated (WNT), sonic hedgehog (SHH), Group 3 and Group 4 MB [10,11]. Further molecular scrutiny of these four groups has now identified somatic mutations targeting chromatin modification as the leading driver for MB heterogeneity via epigenetic dysregulation [12]; further subdivisions have now been established [13–16] (Figure 1).

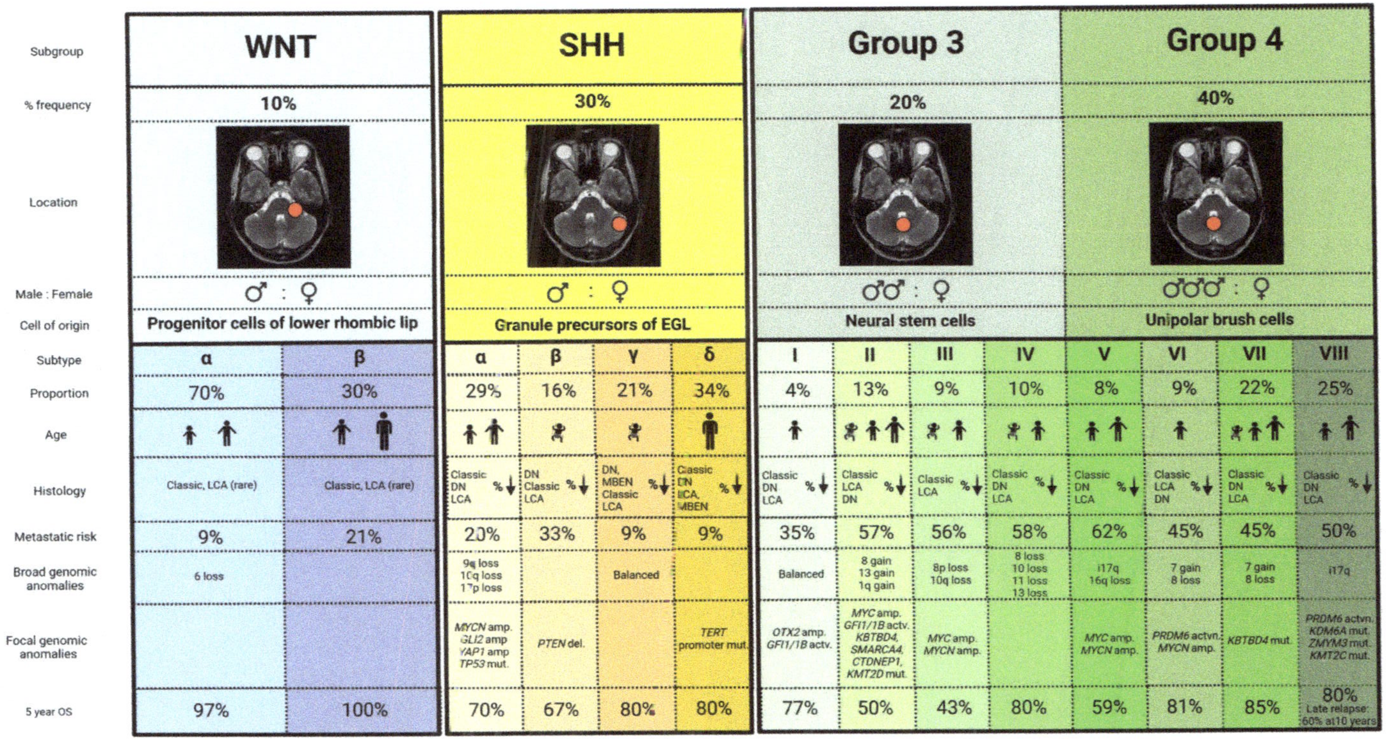

Figure 1. Molecular subgroups and in-group subtypes of medulloblastoma; the four globally recognized molecular subgroups of medulloblastoma (WNT, SHH, Group 3 and Group 4) are shown, together with the current subtypes within WNT and SHH subgroups, as per [13], and Groups 3 and 4, in accordance with [14,15]. Two WNT-activated subtypes are reported, alongside 4 SHH subtypes. Groups 3 and 4 are likely now best considered as a spectrum of 8 different subtypes, each with biological and clinical characteristics. Age-related cartoons depict infant, young child (2–5 years), child (5–12 years), adolescent and older (12+ years). Key: OS = overall survival, DN = desmoplastic/nodular histology, LCA = large cell anaplastic histology, MBEN = medulloblastoma with extensive nodularity, amp. = amplification, mut. = mutation, del. = deletion, and actvn. = activation.

2.3.1. WNT Activated (WNT)

WNT MBs account for approximately 10% of all MBs, and often arise in older children with equal gender distribution [11]. Typically occurring in the midline, they frequently invade the lateral recess of the brainstem through the foramen of Luschka, due to a lower rhombic lip cell of origin [17,18]. They rarely metastasize and morphology is typically of the classic variant [8].

Somatic activating mutations in exon 3 of *CTTNB1*, which encodes B-catenin, are found in 80–90% of WNT MB, with 85–90% displaying monosomy 6 [19–23]. Mutations in the adenomatous polyposis coli (*APC*) gene are common in WNT tumors lacking *CTTNB1* mutations [15,24]. Less frequently occurring mutations include *TP53*, *SMARCA4*, *KMT2D* and *DDX3X* [11,15,25,26]. *TP53* mutation occurs only in a minority of WNT MB, and is not prognostic, unlike the SHH subtype [27].

2.3.2. Sonic Hedgehog-Activated-Activated (SHH)

SHH MB represents approximately 30% of all cases, presenting predominantly in a bimodal age distribution; below three years and in young adults [5,8,10]. Originating from granule progenitor cells SHH MBs localize almost exclusively within cerebellar hemispheres [17,28]. All nodular desmoplastic MBs belong to the SHH subgroup, although other histologies can be observed [21,29]. They are most commonly localized at diagnosis and morphology frequently correlates with underlying genetic abnormalities.

SHH MBs are characterized by activation of the SHH pathway as a result of somatic or germline mutations in a number of genes including *SMO*, *PTCH1* and *SUFU* [30]. While *PTCH1* mutations are seen across 30–50% of SHH MBs, *SUFU* and *SMO* mutations are typically seen in infant and adult SHH MBs, respectively [30]. *TP53* mutations typically arise in childhood SHH MBs [27]. Recent epigenomic profiling has identified a further four clinically distinct granular molecular subclasses of SHH MB, alpha, beta, gamma and delta [13]. SHH-alpha MBs predominate in children, whereas infants are most commonly associated with SHH-beta and SHH-gamma, and SHH-delta is typically observed in adult patients [8].

2.3.3. Group 3

Group 3 tumors account for 25% of all MB cases, predominate in males and occur most frequently in younger children between the ages of 2 and 5 years [8]. Thought to arise from neural stem cell origin [31], Group 3 MBs have a short symptom interval and are frequently metastatic at diagnosis with small primary tumors [11,28,32].

As with Group 4 MB, Group 3 tumors are not characterized by a signature oncogenic pathway. Nevertheless, Group 3 MBs can be associated with activation of GABAergic and photoreceptor pathways [33,34]. Broad genomic aberrations are a feature, while recurrent somatic nucleotide variants are infrequent [7,12,26]. *MYC* amplification is the most common finding (in approximately 17% of cases) commonly occurring within a complex chromosomal rearrangement at the 8q24 locus, resulting in *MYC–PVT1* fusion [7,12,13,34]. The presence of isochromosome 17q, activation of growth factor proto-oncogenes *GFI1* and *GFI1B*, and amplification of transcription factor *OTX2* are also observed [13,15,35].

2.3.4. Group 4

Group 4 tumors represent 35% of all MBs, have a male predisposition and are the dominant molecular subgroup in children of 3 to 16 years of age [8,36]. Similarly to Group 3 MB, they arise in the fourth ventricle and are frequently metastatic at diagnosis, but have a longer symptom interval [11,32].

Genetic abnormalities seen in Group 4 tumors include inactivating mutations of the histone demethylase *KDMS6A* and histone modulator *PRDM6*, tandem duplications of *SNCAIP* and amplifications of *CDK6* and *MYCN* [7,12,25,26,33]. Chromosomal copy number variations include deletion of chromosome 8, 11 or 18p, gain of chromosome 1 or 17q and isochromosome 17q, the most common cytogenetic abnormality in the subgroup [37].

2.4. Prognostic Factors

Typical risk-stratification systems for MB incorporate age, extent of tumor resection, and metastatic status to define standard and high-risk cohorts, in turn determining therapy administered. Standard-risk patients are older than 3 years, have undergone gross or near total excision (below 1.5 cm^2 of residual tumor) with localized disease while remaining patients are classified as high risk. However, these and historical prognostic markers (such as anaplastic morphology) may indeed be surrogates for the underlying MB molecular subgroup, suggesting future stratifications require further refinement.

Pediatric patients with standard-risk WNT-activated MB have an excellent prognosis with a 5 year progression-free survival above 90% following standard therapy. SHH MB demonstrates a range of outcomes. Infant SHH MBs beta and gamma have disparate outcomes, with beta conferring a poor prognosis, and gamma good outcomes [38,39]. *TP53* germline positive SHH MBs confer a poor prognosis with a post-therapy 5 year survival of just 30–40%, particularly when associated with *MYCN* and *GLI2* amplification [40], whereas wildtype SHH MB are associated with a favorable outcome with a 5 year survival of approximately 80% [8,27,30].

Group 3 and 4 MBs also demonstrate variable outcomes, influenced by inherent molecular heterogeneity spanning both groups [14]. For example, Group 3 MB generally carry a poor prognosis, particularly *MYC* amplified cases which are often refractory to conventional therapy [41–43], while Group 4 MBs demonstrate a variable prognosis, incorporating favorable-risk MBs harboring chromosome 11 loss or chromosome 17 gain [14]. Infantile Group 4 MBs are infrequent but carry a poor prognosis [44].

2.5. Current Management/Clinical Trials

The sequential trial-based addition of adjuvant craniospinal radiotherapy and combination chemotherapy to maximal safe tumor resection has improved survival rates for standard-risk patients immeasurably over the last 50 years and is now the accepted standard of care (Table 1). However, such improved cure rates are achieved at a significant burden to the survivor, with most experiencing chronic neurocognitive and neuroendocrine morbidities [45,46]. While standard-risk patients have benefited from a trial-validated reduction in craniospinal radiotherapy intensity [47] (Table 1), high-risk patients continue to require high-dose radiotherapy (36 Gy) and intensified chemotherapy regimens to maintain a 5 year progression-free survival (PFS) of up to 70% [48,49] (Table 1).

Current trial designs utilize refined patient risk stratifications which incorporate the additional knowledge of molecular MB subgroups. Open standard-risk studies including the Children's Oncology Group (COG) ACNS1422 (NCT02724579), the North American SJMB12 (NCT01878617) and the European SIOP PNET5 trial (NCT02066220) are assessing whether treatment intensity can be reduced without compromising survival rates for favorable-risk MBs (particularly WNT-activated MBs).

Table 1. Multinational collaborative clinical trials in pediatric medulloblastoma, high-grade gliomas and ependymoma, published since 2000.

Year	Trial	Treatment Strategy	Inclusion Criteria	No. Patients	Results
Medulloblastoma					
1992–2000	SIOP PNET III [50]	Randomization Arm 1: RT alone (35 Gy CSI + 20 Gy PF boost) Arm 2: 4 cycles alternating Carbo/VP16 and Cyclo/VP16 followed by RT	Age 3–16 yrs Standard-risk MB	179	5 yr EFS 59.8% vs. 74.2% RT + chemotherapy superior
1996–2000	COG A9961 [3]	Radiotherapy: 23.4 Gy CSI + 32.4 Gy PF boost + weekly VCR Continuation chemotherapy randomization: Arm 1: CCNU/Cis/VCR Arm 2: Cis/Cyclo/VCR	Age 3–21 yrs Standard-risk MB	421	10 yr EFS 74% vs. 78% None superior
2001–2006	HIT-SIOP PNET-4 [51]	Radiotherapy randomization Arm 1: HFRT (36 Gy CSI, 24 Gy PF boost, 8 Gy TB boost) Arm 2: STRT (23.4 Gy CSI, 30 Gy PF boost) Continuation chemotherapy 8 cycles Cis/CCNU/VCR	Age 4–<22 years Standard-risk MB	340	5 yr EFS 77% vs. 78 None superior
2004–2016	COG ACNS0331 [52]	Radiotherapy *Children aged 3–7 years randomized:* Randomization 1: CSI: Low-dose (LDCSI) 18 Gy vs. Standard dose (SDCSI) 23.4 Gy Randomization 2: Involved field RT boost vs. Standard volume boost *Children $\geq$ 8 yrs receive CSI 23.4 Gy, then randomized:* Randomization 3: Involved field RT boost (IFRT) vs. Arm 2: Standard volume boost (PFRT) Continuation chemotherapy 9 cycles (6 $\times$ CCNU/Cis/VCR, 3 $\times$ Cytoxan/VCR)	Age 3–<21 yrs Standard-risk MB	513	5 yr EFS/OS LDCSI 72.1%/78.1% SDCSI 82.6%/85.9% LDCSI higher event rates and worse survival PFRT 80.8%/85.2% IFRT 82.2%/84.1% None superior
1990–1996	POG 9031 [49]	Arm 1: 3 cycles Cis/VP16, followed by RT (CSI 35.2–44.0 Gy, PF dose 53.2–54.4 Gy) then 7 cycles VCR/Cyclo continuation chemotherapy Arm 2: RT (CSI 35.2–44.0 Gy, PF dose 53.2–54.4 Gy) followed by 3 cycles Cis/VP16 and 7 cycles VCR/Cyclo continuation chemotherapy	Age 3–18 yrs High-risk MB	224	5 yr EFS/OS: 66%/73.1% vs. 70%/76.1% None superior

Table 1. *Cont.*

Year	Trial	Treatment Strategy	Inclusion Criteria	No. Patients	Results
Medulloblastoma					
1996–2007	SJMB96 [48]	Radiotherapy Risk Stratified: SR: 23.4 Gy, 36 Gy PF dose and 55.8 Gy TB dose; HR: 36–39.6 Gy and 55.8 Gy TB dose (50.4 Gy dose to metastatic sites) Chemotherapy 4 × Cis/Cyclo/VCR with stem cell rescue	Age 3–20 yrs Standard and High-risk MB	134	5 yr EFS/OS: SR 83%/85% HR 70%/70%
2007–2017	SJYC07 [38]	Induction chemotherapy LR and IR: MTX/VCR/Cis/Cyclo HR: MTX/VCR/Cis/Cyclo + Vinblastine Consolidation therapy LR: 2 cycles Carbo/Cyclo/VP16 IR ≥ 12 mths old: Focal RT (54 Gy TB dose); IR < 12 months old: 2 × cycles Carbo/Cyclo/VP16 HR < 3 years old: Topo/Cyclo (8 weeks); HR ≥3 years old: could opt for CSI (23.4–39.6 Gy) Continuation chemotherapy All Groups: 6 cycles oral Cyclo/Topo/Erlotinib	Age < 3 yrs newly diagnosed MB OR Age 3–5 yrs -non-metastatic -no high-risk features	81	LR: 1 yr EFS 78.3%, (accrual suspended as EFS below stopping rule). 5 yr EFS/OS: LR 55.3%/85.9% IR: 24.6%/52.8% HR: 16.7%/41%
2013–2016	ACNS1221 [39]	Induction chemotherapy 3 cycles Cyclo/VCR/MTX/VP16/Carbo Reassessment CR/CCR: No further treatment PRD: Second look surgery + 2 cycles Cyclo/VCR/Carbo/VP16	Age < 4 yrs Localized ND or MBEN	25	2 yr PFS/OS 52%/92% Failed to achieve 2 yr PFS target of 90%; study closed early
2007–2018	ACNS0332 [53]	Randomization Arm 1: Standard treatment (CSI 36 Gy, PF 55.8 Gy + 6 cycles Cis/Cyclo/VCR maintenance) Arm 2: Standard treatment + RT with Carbo Arm 3: Standard treatment + isotretinoin during maintenance Arm 4: Standard treatment + RT with Carbo + isotretinoin during maintenance	3–21 yrs High-risk MB	261	Survival advantage for Grp 3 MB receiving RT with carboplatin. 5 yr EFS/OS: 73.2%/82.3% vs. 53.7%/63.7% Isotretinoin therapy futile

Table 1. *Cont.*

Year	Trial	Treatment Strategy	Inclusion Criteria	No. Patients	Results
High-Grade Gliomas					
2004–2005	ACNS0126 [54]	RT (HGG 54 Gy, DIPG 59.4 Gy) + concomitant low-dose TMZ, followed by 10 cycles of higher dose TMZ continuation therapy	Age 3–≤22 yrs	HGG = 107 DIPG = 63	1 yr EFS/OS 14%/40% No improvement vs. historical controls
2005–2007	ACNS0423 [55]	RT (GTR 54 Gy, STR 59.4 Gy, spinal cord lesions 50.4–54 Gy) + concomitant low-dose TMZ, followed by up to 6 cycles of higher dose TMZ + CCNU continuation	Age 3–≤22 yrs	108	3 yr EFS/OS 22%/19% Improved vs. ACNS0126
2007–2008	ACNS0222 [56]	RT (54 Gy) with motexafin-gadolinium as a potent radiosensitizer	Age ≤ 21 yrs Unifocal DIPG	60	1 yr EFS/OS 18%/53% No Improvement
2011–2015	HERBY [57]	Randomization Arm 1: RT (54 Gy) + low-dose TMZ, continuation high-dose TMZ 12 months Arm 2: RT (54 Gy) + low-dose TMZ + Bev, continuation high-dose TMZ + Bev 12 mnths	Age ≥ 3–≤18 yrs Non–brainstem	116	1 yr median EFS 11.8 vs. 8.2 mnths No improvement
2014–2020	BIOMEDE 1 [58]	Randomization Arm 1: RT + Everolimus Arm 2: RT + Dasatinib Arm 3: RT + Erlotinib	Age 6 mths–25 yrs DIPG	193	Median OS Arms 1, 2, 3 10.9, 9.5 and 9 mnths No improvement
Ependymoma					
2003–2007	ACNS0121 [59]	Stratum 1: Completely resected differentiated, ST ependymoma undergo observation Stratum 2: Incompletely resected ependymoma undergo chemotherapy, second surgery and RT Stratum 3: Near-total or macroscopic GTR undergo conformal RT Stratum 4: Microscopic GTR undergo conformal RT, excluding differentiated, ST lesions	Age 1–21 yrs	356	5 yr EFS/OS Strata 1: 61%/100% Strata 2: 37.2%/70.2% Strata 3: 67%/83.3% Strata 4: 70%/88.3%
2010–2017	ACNS0831 [60]	PF tumours gross/near total resection: randomization Arm 1: RT alone Arm 2: RT + 4 cycles VCR/Cis/Cyclo/VP16	Age 1–21 yrs	451	3 yr EFS 71% vs. 80% ? chemotherapy superior

RT: radiotherapy; CSI: craniospinal irradiation; PF: posterior fossa; Carbo: carboplatin; VP16: etoposide; Cyclo: cyclophosphamide; MB; medulloblastoma; EFS: event-free survival; VCR; vincristine; CCNU: lomustine; Cis: cisplatin; HFRT: hyper-fractionated radiotherapy; STRT: standard radiotherapy; TB: tumor bed; OS: overall survival; SR: standard risk; HR: high risk; MTX: methotrexate; LR: low risk; IR: intermediate risk; Topo: topotecan; CR: complete response; CCR: continuous complete response; PRD: persistent residual disease; Ifos: ifosfamide; GTR: gross total resection; DIPG: diffuse intrinsic pontine glioma; yrs: years; mnths: months.

Caution regarding de-escalation of therapy for WNT-activated MBs is evident from the premature termination of trial NCT02212574 which abandoned craniospinal irradiation for these patients, and a recent retrospective analysis of 93 WNT-activated MBs where relapse was associated with a reduction in the cumulative dosing of maintenance chemotherapy [61].

The PNET5 trial is also assessing the radio-sensitizing effect of carboplatin for non-WNT MB, while SJMB12 is the addition of targeted drug therapy in conjunction with conventional agents for specific molecular subgroups (SHH and high-risk Group 3 and 4 MBs). A European biomarker-driven phase III trial for newly stratified high-risk MB opened to recruitment in 2021. Of interest, post-operative residual tumor is not considered a high-risk feature in this study. The trial incorporates a double-randomized design, comparing the efficacy of hyper-fractionated radiotherapy and additional high-dose chemotherapy against standard radiotherapy, followed by a comparison of multimodal continuation chemotherapy versus single agent temozolomide (EudraCT Number: 2018-004250-17).

Infant MB represents a distinct, intensive chemotherapy-only treatment group [29]. Outcomes for infants with nodular desmoplastic SHH MB can be excellent, although it appears that this requires the inclusion of intrathecal methotrexate in addition to systemic therapy [38,39,62,63]. The COG ACNS0334 study of non-nodular desmoplastic MBs, incorporating both induction and high-dose tandem consolidation cycles of chemotherapy reported 100% survival for metastatic SHH MBs and a survival advantage for the incorporation of methotrexate at induction in Group 3MBs [64].

2.6. Novel Therapies

Advances in molecular understanding of MB pathogenesis have also provided the opportunity for the application of subgroup-specific novel targeted therapeutics, notably for SHH MBs. Vismodegib and sonidegib are *SMO* inhibitors that have shown objective responses in pediatric recurrent SHH MB [65–71]. For most patients, such responses were not sustained, as a result of mutations downstream from *SMO* re-activating the pathway [30]. Another important consideration of this therapy is the association with premature growth plate fusions which has led to modification of the current SJMB12 study [70–72]. Agents such as silmitasertib, targeting *SMO* downstream mutations in the SHH pathway, are under evaluation in relapsed SHH MB (NCT03904862). *GLI* inhibition by arsenic trioxide is another area of drug development in SHH MB and early phase pediatric tumor trial data are awaited (NCT00024258).

For non-SHH tumors, the aforementioned SJMB12 study is evaluating the addition of pemetrexed and gemcitabine to conventional chemotherapeutic agents for high-risk Group 3 and 4 MBs (large cell anaplastic histology, metastatic disease or *MYC/MYCN* upregulation) after promising high throughput in vitro drug assay analysis [73]. The CDK4/6-cyclin D-Rb pathway was identified as a potential therapeutic target in xenograft models for non-WNT MB [74]. Other proposed approaches include *HDAC* inhibitors, *PI3K* inhibition and BET-bromodomain inhibition to downregulate *MYC* expression in Group 3 MBs, and *LSD1* inhibition of *GFI1/GFI1B* overexpression when present in Group 3 and 4 MBs [75–78].

Finally, despite the challenge posed by the lack of immunogenic targets in CNS tumors, immunotherapy has been proposed as a potential treatment option in relapsed/refractory MB [79]. Anti-EPHA2, HER2 and IL-13Rα2 chimeric antigen receptor T-cell (CAR-T) therapy has been shown to successfully treat murine Group 3 MBs [80] and early phase trials in children have commenced (NCT03500991, NCT04661384).

3. High-Grade Gliomas

3.1. Background

This group encapsulates all malignant lesions of glial origin. Alongside embryonal tumors, pediatric high-grade gliomas (pHGGs) are one of the most common malignant tumor groups of the childhood central nervous system, with a collective incidence of 1.1 per

100,000 children [2]. Despite a paradigm shift in our understanding of pHGG molecular subgrouping being distinct from adult counterparts, and some therapeutic successes for particular entities (such as infant HGG), little progress has been made over recent decades to improve the dismal prognosis; pHGGs account for over 40% of all childhood brain tumor deaths [81]. As a result, they remain the focus of several experimental therapeutic research teams.

3.2. Histopathology

The vast majority of pHGGs can be classified as anaplastic astrocytomas (WHO Grade III), or glioblastoma (Grade IV). Historically, a minority of diffuse intrinsic pontine gliomas (DIPGs) were morphologically consistent with diffuse astrocytoma (Grade II), likely resulting from sampling bias. However, the identification of pathognomonic oncogenic mutations in DIPG (particularly in histones 3.1 and 3.3), together with established malignant clinical characteristics, resulted in an amendment to current WHO nomenclature, with DIPGs now classified as diffuse midline gliomas with *H3K27* mutation (Grade IV) [5].

3.3. Molecular Classification

Clear biological distinctions between pHGGs and adult counterparts are now established [82,83], providing a rationale for the failure of many novel therapies derived from adult tumor research. Molecular heterogeneity within pHGGs is also well described [84–91]. The largest molecular meta-analysis of pHGGs published to date, incorporating genomic, epigenomic and transcriptomic profiling has now identified at least nine pHGG subgroups with inherent biological and/or clinical characteristics such as age, tumor location and prognosis [90]. These subgroups express recurrent signature aberrations, which may lead to further refinement of subdivisions in the future (Figure 2).

The predominant pHGG subgroups express mutations of histones *HIST1H3B* (*H3.1*) at position K27, *H3.2* (rarely) and *H3F3A* (*H3.3*) at positions K27 and G34 [90,92]. H3K27M pHGGs are characterized biologically by aberrant expression resulting from loss of trimethylation at lysine 27 on Histone 3 [93,94], and clinically by their midline location (pons, midbrain, thalamus, spina cord) and younger patient age [90,91]. H3.3 G34 subgroup pHGGs are typically located in hemispheric locations, impacting adolescent and older age groups [90,91,95]. The midline location may contribute to the significantly poorer prognosis reported in K27 pHGGs versus G34 counterparts [85,90,91,95], although the mutations alone have been reported as independent prognostic markers in multivariate analysis [90]. Secondary aberrations within the pHGG histone subgroups have also been identified. *TOP3A*, *CCND2*, *PDGFRA*, *PPM1D*, *TP53* and *FGFR1* mutations are more frequently identified in H3.3K27 pHGGs, while H3.1K27 tumors often demonstrate *PI3K* and *ACVR1* mutations and H3.3 G34 pHGGs typically contain *TP53* and *ATRX* mutations [90].

Other subgroups include the IDH mutant pHGGs, associated with a frontal location, an adolescent age range and improved prognosis, hypermutant pHGGs as seen in DNA replication repair deficiency disorders, infant HGGs characterized by *NTRK* mutations and pleomorphic xanthoastrocytoma-like pHGGs and *BRAF* mutated pHGGs, which may represent low-grade lesions that have undergone malignant transformation [90,91,95]. The latter two subgroups may be amenable to novel targeted inhibitor agents and often demonstrate good responses to therapy and improved survival outcomes. A final 'wild-type' subgroup comprises pHGGs harboring mutations in genes such as *NF1*, *MYCN*, *EGFR*, and *CDK6* [90].

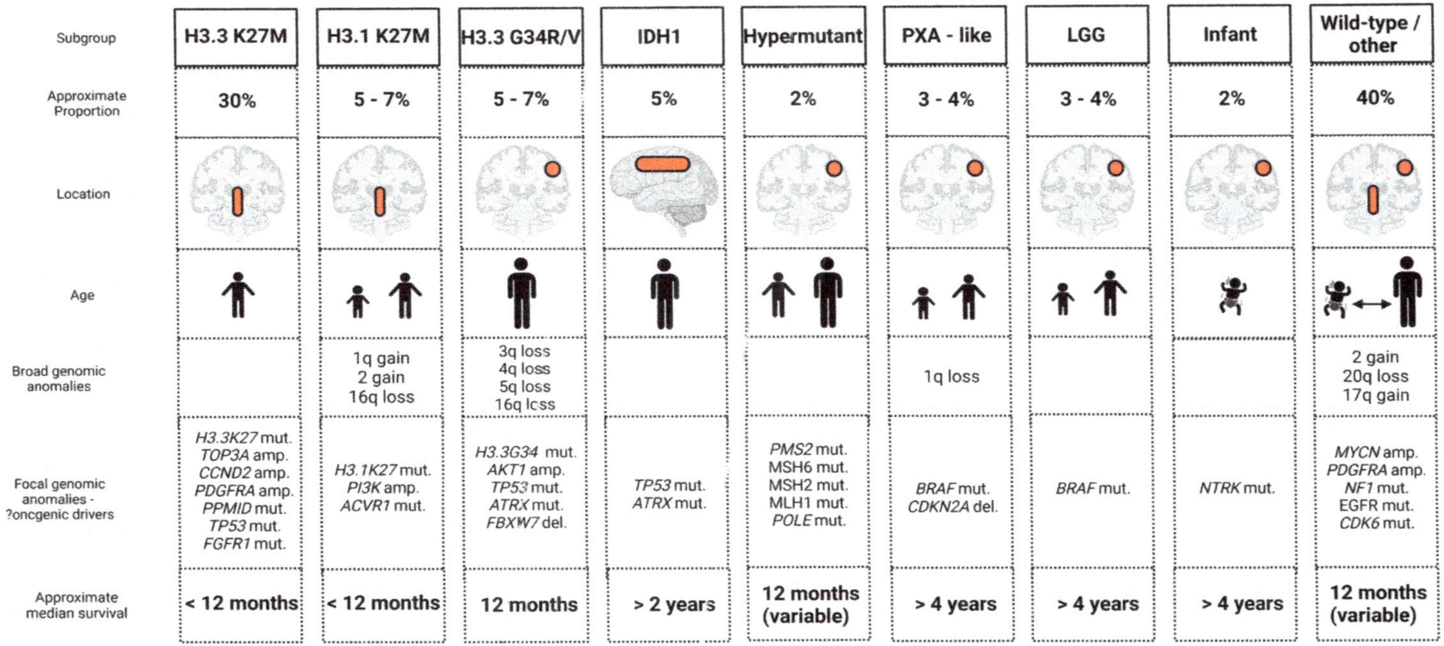

Figure 2. Molecular subgroups of pediatric high-grade glioma. At least nine subgroups are thought to exist, with biological and clinical features highlighted in accordance with [84,90,91]. Age-related cartoons depict infant, young child (2–5 years), child (5–12 years), and adolescent/adult (12+ years). Key: amp. = amplification, mut. = mutation, del. = deletion, IDH = isocitrate dehydrogenase, and PXA, pleomorphic xanthoastrocytoma.

3.4. Prognostic Factors

Prior to the advent of molecular subclassification as described above, the two leading clinical prognostic factors were the extent of surgical resection and tumor histological grade with incomplete resection and Grade IV HGGs conferring a dismal prognosis [96,97]; this continues to be the case today but is supplemented by molecular stratification also. Some studies have also reported a prognostic influence of methylguanine-DNA-methyltransferase (MGMT) expression in the efficacy of temozolomide therapy and patient outcome [54,98].

3.5. Current Management/Clinical Trials

The global standard of care for pHGG, the Stupp regimen, stems from adult glioblastoma trial work, which demonstrated that the addition of the alkylating agent temozolomide alongside and after focal radiotherapy, improved progression-free and overall patient survival [99]. Given the molecular disparity between adult HGG and their childhood counterparts, it is therefore unsurprising that temozolomide in a Children's Oncology Group (COG) pHGG trial analysis (ACNS0126) did not improve outcome compared with previous trials using varied adjuvant chemotherapies [54] (Table 1). However, it remains the standard of care because of the relatively low toxicity profile in comparison to alternative regimens.

The COG ACNS0423 trial noted a marginal outcome benefit for the addition of lomustine with temozolomide [55]; however, it was unclear if this was specific to certain molecular subgroups, while the myelosuppressive toxicity of the regime often proved restrictive. The German Hirntumor (HIT) co-operative group have also reported an improved survival rate for a subset of children with glioblastoma achieving gross total resection compared to historical controls, using an intensified chemotherapy regime alongside and after RT [100].

No definitive therapeutic breakthrough has been made in the treatment of DIPG (now diffuse midline glioma, H3K27 mutant), such that the standard therapy remains radiotherapy alone (Table 1). Modern, multinational collaborative trials, such as the Innovative Therapies for Children with Cancer (ITCC) BIOMEDE study, are developing a more nuanced approach alongside focal RT, utilizing novel inhibitor therapy to target corresponding molecular aberrations present in the lesion (dasatanib, everolimus, and erlotinib) (NCT02233049). Interim overall survival analysis of 193/250 randomized patients concluded that a preferential agent was unlikely to be demonstrated, with survival rates comparable with RT alone, albeit everolimus had the most favorable toxicity profile [58].

3.6. Novel Therapies

The paradigm shift in understanding of the molecular heterogeneity of pHGG, together with the failure of conventional therapeutics to significantly improve outcomes for several years, has shifted focus towards developing novel agents that manipulate the epigenetic and genomic aberrations inherent in pHGG molecular subgroups, immunotherapies, and the development of alternative drug administration routes to penetrate the blood–brain barrier such as convection enhanced delivery for diffuse midline glioma H3K27 mutant/DIPG [92,101–105].

Success of mutational target inhibition in specific pHGG subgroups gives credence to this new therapeutic standpoint. For pHGGs with *BRAF V600E* mutations, BRAF inhibitor (BRAFi) activity has been demonstrated as salvage therapy [106–109]; international co-operative studies are recruiting (NCT03919071). Similar findings of efficacy have been made with neurotrophic tyrosine receptor kinase inhibitor agents for infant HGGs [110,111] and immune checkpoint inhibition in hypermutant pHGGs resulting from replication repair deficiency disorders [112–115]. Follow-up co-operative early phase trials are now open (NCT04267146, NCT04323046 and NCT04655404).

With respect to the other main subgroups, targeting histone modification is a therapeutic research focus for the H3.1–3.3 pHGG subgroup. Histone deacetylase inhibitors (HDACi) such as panobinostat, vorinostat and valproic acid have been postulated to improve the therapeutic landscape for this subgroup following successful HDACi in vitro pHGG studies, but translational results to date have proved disappointing [116–119]. Other agents being looked at for this subgroup include ACVR1/ALK inhibitors [120,121] and the imipridones incorporating agents such as ONC201 [122–124].

For the IDH mutant pHGG subgroup, blood–brain barrier penetrant IDH inhibitors have been developed for glioma trials (NCT02273739, NCT03343197, NCT02073994 and NCT04056910). These may be specific to IDH-1 (ivosedinib), IDH-2 (enasidenib) or both (vorasidenib). In addition, the use of PARP (poly-adenosine 50-diphosphate-ribose) inhibitors alongside temozolomide as a radiosensitizer is being explored [125].

Immunotherapeutic strategies other than checkpoint blockade are also being evaluated in pHGGs, including cancer peptide vaccine therapy with antigens such as Ephrin A2 (EphA2), interleukin 13 receptor alpha 2 (IL13Ra2), survivin and HLA-A2 (NCT01130077) [126–128], autologous dendritic cell vaccine therapy [129], and chimeric antigen receptor (CAR)-T therapy where studies are recruiting (anti-IL13aR2; NCT02208362, anti-GD2; NCT04196413, anti-B7 H3; NCT04185038).

4. Ependymoma

4.1. Background

Ependymoma is the second most common malignant brain tumor entity in children, after medulloblastoma, representing approximately 10% of all childhood CNS tumors [130]. Most cases present in patients aged below five years and have a male predominance (male: female ratio 0.23: 0.17) [130,131]. Although able to arise anywhere in the neuraxis, over 90% of pediatric ependymomas are intracranial (IC) in origin. Of these, two-thirds occur in the posterior fossa (PF), with the remaining one-third located in the supratentorial (ST) compartment [132]. Leptomeningeal metastasis is uncommon, reported in 2–20% of cases [133,134].

No inherited disorders are consistently reported to predispose to IC pediatric ependymomas. Neurofibromatosis type 2 appears to be associated with the development of spinal ependymomas but typically in the adult population [135].

4.2. Histopathology

Current histological classification of ependymoma remains according to the current WHO grading scheme, resulting in four main histological subgroups: subependymoma and myxopapillary ependymoma (grade I), classic (grade II) and anaplastic (grade III) [120]. Subependymoma typically arise in the ventricles of adults, while myxopapillary ependymoma occur exclusively in the spine [5,136]. Consequently, classic and anaplastic variants typically account for all pediatric IC ependymomas. Morphologically they are both characterized by the tumor cell formation into true rosettes (around a canal) or pseudorosettes (around a blood vessel) while anaplasia is signified by increased mitotic figures, necrosis, microvascular proliferation, and an increased an increased cellular nucleus/cytoplasmic ratio [5]. Common immunohistochemical findings include positive staining for glial fibrillary acid protein (GFAP), expression of EMA, S100 and vimentin [5,137].

The utilization of histological grading as a prognostic marker has failed to consistently be of value, in part due to the subjective nature of grade assignation and tumor heterogeneity. These factors, alongside improved understanding of the genomic landscape of pediatric ependymoma, has led to the Consortium to Inform Molecular and Practical Approaches to CNS Tumor Taxonomy (cIMPACT) to recommend that the WHO adopt a new, integrated histological/biological classification system for ependymomas [138].

4.3. Molecular Classification

Genomic and methylomic profiling of ependymoma has revealed nine distinct molecular subtypes, four of which account for most pediatric IC ependymoma across the PF (PF-A and PF-B) and ST (ST-ZFTA and ST-YAP) compartments [139] (Figure 3).

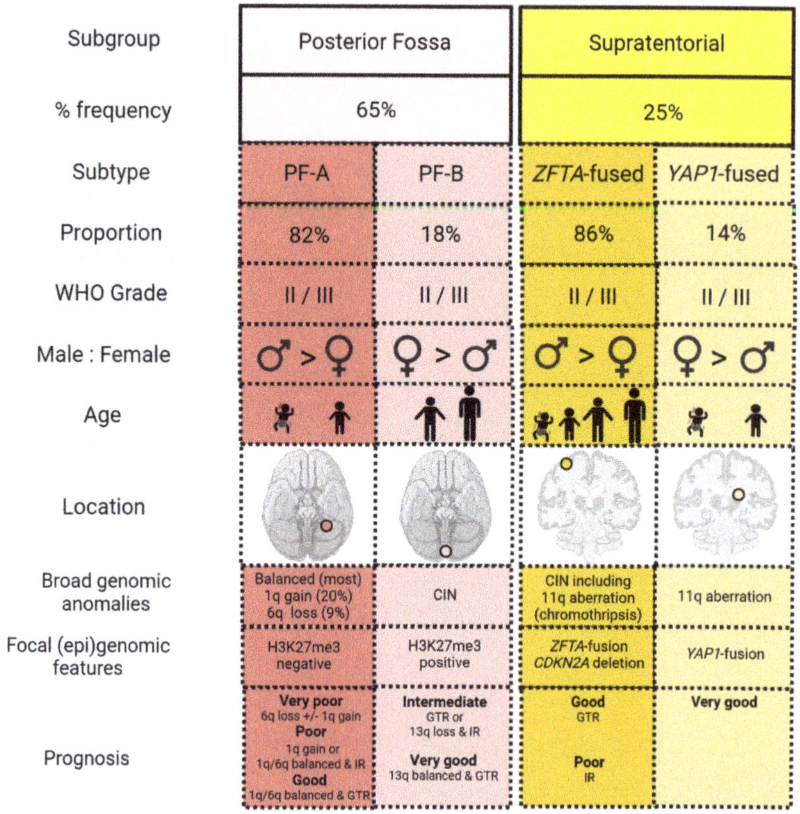

Figure 3. Predominant molecular subtypes of pediatric intracranial ependymoma. Posterior fossa and supratentorial childhood ependymomas are shown, further categorized into four in-group subtypes; PF-A, PF-B, *ZFTA*-fused and *YAP1*-fused. The clinical and biological characteristics of these subtypes are shown, in accordance with [59,120,139–153]. Nine molecular subtypes of ependymoma are reported but the remaining subtypes occur in either the spinal cord (spinal subependymoma, spinal myxopapillary ependymoma, spinal ependymoma) or the adult brain (subependymoma: PF and ST) so are not depicted in this figure. Age-related cartoons depict infant, young child (2–5 years), child (5–12 years), adolescent/adult (12+ years). Key: WHO = World Health Organization, CIN = chromosomal instability, GTR = gross total resection, IR = incomplete resection.

PF-A ependymomas are biologically characterized by epigenetic dysregulation of DNA methylation and histone modification, often accompanying lack of *H3K27* trimethylation [147,149,150]. With the exception of some genomic imbalances, namely 1q gain and 6q loss, they typically demonstrate a balanced genome [139,147,149]. They are most common in infants and young children, have a tendency towards infiltration, dissemination and consequent poor prognosis [153]. Due to their predominant lateral location and inherent invasiveness, gross total resection (GTR) is often difficult to achieve and therefore relapse rates are high [154]. PF-B ependymomas are characteristically enriched with numerous cytogenetic abnormalities and are more common in adolescents and young adults [139,152,155]. They originate in the midline yet are often amenable to surgical resection, have a low metastatic potential and therefore have a superior outcome to PF-A tumors [139,152,155]. Recent methylation profiling work to further categories these two PF subgroups have reported two major subgroups, nine minor PF-A subtypes and five PF-B subgroups displaying variable clinical and genetic heterogeneity [140,156].

Greater than 70% ST ependymomas contain a zinc finger translocation associated (ZFTA, previously C11Orf95) gene fusion, most commonly RELA-ZFTA and are termed ST-RELA or, more recently ST-ZFTA [139,143,151]. This subtype is found in children and adults, but rarely infants and is often located in frontal or parietal lobes, often with intratumoral hemorrhage, cysts or necrosis [157]. ST-YAP is the remaining molecular subgroup, characterized by the fusion of the *YAP1* oncogene with *MAMLD1* [142]. ST-YAP tumors typically arise in ventricular or periventricular locations among infants [142]. Up to 15% of ST ependymoma may not harbor a *RELA* or *YAP1* fusion [158].

4.4. Prognostic Factors

Interest remains in identifying prognostic markers to aid patient risk stratification for future ependymoma trial design to improve upon the relative poor long-term outcomes that exist. Akin to medulloblastoma, several clinical and histological putative markers (location, age, tumor grade) have been rendered obsolete by the identification of molecular subgrouping.

The most consistent clinical marker is the extent of surgical resection, with some studies reporting a 60% difference in survival between cases of complete and incomplete tumor resection [59,120,132,159–162]. The positive prognostic effect of complete excision is maintained across molecular subgroups [120,139].

The infiltrative nature, localization and predisposition to metastasis suggests PF-A ependymomas should exhibit a poorer prognosis when compared with PF-B counterparts, an assumption supported by a retrospective analysis 820 patients with PF ependymoma across four independent cohorts [161]. The recent prospective Children's Oncology Group (COG) ACN0121 clinical trial, however, found no difference between PF-A and PF-B patient survival, although likely reflecting a paucity of PF-B cases [59]. The study did identify an adverse association with 1q gain in PF-A cases, with survival as low as 30% despite tumor resection and radiotherapy administration [59]. As stated above, tumor gain of chromosome 1q and loss of chromosome 6q are the most commonly observed chromosomal imbalances in ependymoma and appear adverse prognostic factors [59,120,139,141,144–146,148,152]. A recent retrospective molecular profiling study of 212 primary PF-B ependymomas identified loss of 13q as a potential novel adverse marker [140].

A retrospective cohort study of 122 ST ependymomas identified *ZFTA/RELA* fusion as a poor prognostic marker, regardless of the attainment of resection status, with 10 year PFS and overall survival (OS) of approximately 20% and 50%, respectively [139]. The same study conversely identified excellent ST-YAP1 survival rates of 100% [139]. Nevertheless, data from the ACNS0121 clinical trial failed to show any adverse prognostic implication for ST molecular subgroups, again potentially influenced by the case numbers involved [59].

4.5. Current Management/Clinical Trials

The globally accepted standard for pediatric IC ependymomas is maximal, safe surgical resection followed by involved field adjuvant radiotherapy (RT), dosed at 54–59.4 Gy, founded from a 2009 St Jude's Children's Research Hospital single-center study of 107 children, demonstrating a 7 year PFS of 77% and OS of 85% [160]. Exceptions to this are in metastatic cases where craniospinal radiotherapy is typically utilized for older children, and infant IC ependymomas, where a chemotherapy only strategy is reserved in order to avoid or delay radiotherapy to the developing brain, with eligibility thresholds of 12 to 18 months for PF tumors and up to 3 years for ST tumors.

Concerns regarding radiotherapy-induced neurotoxicity in young children have resulted in IC ependymoma being the most common pediatric tumor treated with proton beam radiotherapy. By reducing radiation exposure to healthy tissue while delivering therapeutic doses, this modality delivers comparable disease control to modern photon radiotherapy without unexpected toxicity [163–165]. Data continue to be collated on latent toxicity [164].

Recent, large international co-operative IC ependymoma trials have been designed to validate the findings of the 2009 St. Jude's study, evaluate the utility of an aggressive surgical approach, and verify a therapeutic role for chemotherapy either pre or post-radiotherapy (Table 1), since historical data have proven contradictory and inconclusive. The North American CCG-9924 study reported a PFS benefit from immediate post-operative chemotherapy prior to radiotherapy in patients where over 90% of the tumor has been resected [166]; however, this approach has been rebutted by other trial groups [167]. Similarly, outcomes from chemotherapeutic, radiation-sparing strategies for infants have been inconsistent and ultimately disappointing for the majority of children, with only a minority ultimately sparing radiation [168–171].

The COG ACNS0121 trial confirmed the efficacy of an aggressive surgical approach followed by immediate post-operative radiotherapy, even for children below 3 years of age when compared to historical controls [59]. Long-term follow-up of these younger patients is eagerly awaited. The impact of post-operative chemotherapy to facilitate second-look surgery could not be determined. The COG ACNS0831 study followed on from ACNS0121, with the randomized addition of continuation chemotherapy (vincristine, cisplatin, cyclophosphamide and etoposide) for children treated with adjuvant focal RT following a complete or near total resection [60]. An interim "as treated" analysis of patients was undertaken due to significant non-compliance in patients randomized to receive chemotherapy. This reported a survival advantage for patients receiving chemotherapy (3 year EFS 80% vs. 71%; 1-sided p-value = 0.0121) [60].

The open phase II/III SIOP-Europe Ependymoma II trial (NCT02265770) has design similarities with the COG studies, making compliance with post-irradiation chemotherapy randomization imperative to validate the findings from ACNS 0831. Through patient allocation to three strata, the trial also attempts to evaluate the value of pre-radiotherapy chemotherapy and a 8 Gy radiotherapy boost in cases of incomplete resection, and the addition of a of a histone de-acetylase (HDAC) inhibitor, sodium valproate, for infants receiving one year of conventional multiagent chemotherapy.

4.6. *Novel Therapies*

Several biological models and patient derived xenografts have been developed to recapitulate ependymoma subgroups in order to identify new therapeutic targets and test novel therapies [172–174]. High throughput drug screening in murine models of *ZFTA* fusion-negative supratentorial ependymoma, characterized by the *Ephb2* oncogene identified 5-fluoracil (5-FU) as a potential active drug against this subtype [174,175]. Fibroblast growth factor receptor inhibitors have also been shown to have activity against patient derived ependymoma cell models and demonstrate efficacy in the clinic [176]. As detailed above, the use of histone deacetylase inhibitors as differentiation therapy is currently under evaluation in the current SIOP-Europe trial, following in vitro analyses [177,178]. Similarly, the phase I/Ib COZMOS trial is evaluating the DNA methyltransferase inhibitor 5′Azacitidine in combination with carboplatin, on the premise that inhibition of aberrant DNA methylation will have therapeutic benefit (NCT03206021). Other novel therapies being explored include chimeric antigen receptor T-Cells (HER2; NCT03500991), based on encouraging pre-clinical murine work [80] and metronomic antiangiogenic therapy [179,180].

5. Conclusions

This review exposes the need for the pediatric neuro-oncology community to address the disparity that has developed between advances at the bench compared to the bedside. The potential for an era of biology driven patient care clearly exists yet, at present, international clinical trials struggle to keep pace with the scientific progress made to date. Indeed, many are being rendered outdated before they open to recruitment when evaluated against current molecular advances. This challenge is not unsurmountable and indeed should be embraced as recent years have demonstrated a paradigm shift in our understanding of the molecular pathogenesis across principal malignant brain tumor groups, therein serving as the foundation for developing both risk stratification systems and novel agents as part of the next generation of clinical trials. Nevertheless, results from the review highlight that the statistical design, regulatory infrastructure and ultimately funding of such studies will need urgent consideration to achieve these objectives.

5.1. Clinical Trials and Therapeutic Protocols

We have shown that for pediatric medulloblastoma, the four established intrinsic molecular subgroups have now been superseded by the identification of up to 14 subtypes, each demonstrating a disparate corresponding clinical profile. In contrast, most treatment protocols over the past 20 years have continued to treat MBs with the historical backbone of craniospinal radiotherapy and multiagent chemotherapy, only recently tailoring therapy intensity according to WNT/non-WNT subgrouping, without particular focus on the three other subgroups. Encouragingly, open international trials are now attempting to stratify patients and adapt therapy according to molecular diversity. For example, the SIOP-Europe PNET5 study is following a risk-adapted treatment stratification according to low and high-risk WNT subgroups, the SHH-alpha MB subtype (which demonstrate *TP53* mutations), standard-risk biological profiles (including *MYCN* amplified Group 4 MB) and children with a germline mutational profile (NCT02066220). The SJMB12 trial, in addition to evaluating treatment de-escalation for WNT-subgroup patients, is assessing the addition of smoothened inhibitor Vismodegib for SHH MB, and the incorporation of gemcitabine and pemetrexed for high-risk Group 3 and 4 MB patients (NCT01878617). Finally, the SIOP-Europe high-risk medulloblastoma trial is using molecular screening to identify appropriate cases for increased-intensity treatments, including *MYC*/*MYCN* amplification (excluding *MYCN* amplified Group 4 MB) and SHH-alpha MB (EudraCT Number: 2018-004250-17).

Attempts to integrate molecular pathogenesis to inform on therapeutic stratification for most childhood high-grade gliomas or pediatric intracranial ependymoma unfortunately lag significantly behind the progress observed with medulloblastoma. As shown in this review, there is now compelling evidence that molecular subgrouping alone is an independent survival marker for childhood ependymoma, while prognostic adversity is further conferred by the presence of genomic aberrations including chromosome 1q gain and 6q loss in PF-A ependymomas, and potentially 13q loss in PF-B ependymomas. Despite this, international ependymoma clinical trials continue to risk stratify children according to the clinical parameters of patient age and resection status alone; an omission that will require addressing in future clinical trial strategies. With the exception of BIOMEDE 1, large-scale international pediatric HGG trials have also not incorporated biologically derived therapeutic stratification systems, principally because the finding that HGGs encompass an array of discrete subtypes is a relatively recent discovery.

As with medulloblastoma, the observation of up to 14 discrete molecular subtypes of PF ependymoma, at least 3 subtypes of ST ependymoma and up to 10 pediatric HGG subtypes clearly presents a challenge for future trial design. As can be seen from Table 1 of this article, the duration of an international pediatric brain tumor trial can take up to 10 years to complete patient accrual, and even longer to publish data. In order to tailor therapeutic intensity or introduce novel agents against the array of specific tumor subtypes now published in this review, future trials will require novel statistical designs that embrace truly global collaboration to generate timely, rigorous results as increasing molecular subcategorization will lead to significantly smaller patient subpopulations from which statistically sound conclusions must be drawn. Such collaborative efforts may also support less affluent countries to provide equity in diagnostic and therapeutic approaches. Duration of follow-up for specific patient populations will also need to be considered, as evidenced by the high proportion of late relapses in Group 3/4, subtype VIII MB and some non PF-A subgroups of ependymoma.

5.2. Conventional and Novel Therapies

While advances in adjuvant therapy have undoubtedly improved the survival of children with malignant brain tumors, the 'one-therapy-fits-all' paradigm fails to reflect and tailor to the diverse molecular landscape now apparent. As highlighted by this review, integrating clinical and biological data to generate risk-adapted treatment stratifications can potentially modify conventional therapy intensity and enable the introduction of novel agents.

De-escalation of radiotherapy dosing is being evaluated in several of the current international medulloblastoma clinical trials highlighted in the review. However, such an approach could also be considered for other molecularly-defined tumor entities including Group 4 (often subtype IV) medulloblastomas with chromosome 11 loss, completely resected ST-YAP1 ependymomas, completely resected PF-B ependymomas without 13q loss, and 'infant' or 'LGG-like' pediatric HGGs. Clearly, any de-escalation of therapy must be approached with extreme caution, as evidenced by the failure of trial NCT02212574 for WNT-activated MB, where a post-operative chemotherapy only strategy led to unacceptable relapse rates.

For some unfavorable-risk tumors, the option of increasing treatment intensity is a possibility as evidenced by current high-risk medulloblastoma trial strategies; however, any trial adopting this approach should consider incorporating disability or health status outcome measures, as they will help determine the quality of potential survivorship afforded [181]. The efficacy of chemotherapy in pediatric ependymoma remains contentious but a potential option to explore for escalation of therapy in certain cases (for instance PF-A tumors with chromosome 1q gain or 6q loss). The interim analysis results of the COG ACNS0831 trial suggested a potential survival advantage for children receiving continuation chemotherapy following tumor excision and post-operative irradiation, yet this requires validation ideally by the open phase II/III SIOP-Europe Ependymoma II trial. The administration of conventional chemotherapy agents and novel agents by alternative means, such as convection enhanced delivery to overcome the blood–brain barrier in diffuse midline glioma, H3K27M pediatric HGGs is also under consideration.

Parallel to modifying the intensity or administration of conventional therapies for childhood malignant brain tumors, much hope rests on establishing novel agents to target aberrant molecular aberrations underpinning tumorigenesis. This review highlights many of the developments in this field across medulloblastoma, pediatric high-grade gliomas and ependymomas. International trial outcomes are awaited for medulloblastoma subgroup-targeted therapy in SJMB12 and combination HDACi therapy across infants in the SIOP Ependymoma II study, while the success of BRAFi and NTRKi in certain pHGG subtypes and the evolving array of targeted primary treatment options for pediatric low-grade glioma give cause for optimism.

While encouraging, challenges nevertheless remain. As described in this review, novel agents against malignant brain tumors are being evaluated in early-phase pediatric studies, yet few successful candidates targeting the spectrum of molecular subtypes that now exist have been identified. One explanation for this is that many early-phase neuro-oncology trials in children assess novel agents in the relapse setting, rather than as primary therapy. In turn, this could potentially generate misleading results on drug efficacy, as evidenced by pre-clinical relapsed medulloblastoma work implicating clonal selection as a potential cause for the disappearance of targetable aberrations between patient-matched primary and relapsed tumors [182,183]. However, the paucity of effective novel agents also reflects the ongoing need for improved pre-clinical models that accurately replicate the specific human disease subtype interrogated, including appropriate immunocompetent murine models to test potential immunotherapies. A further explanation is that many pediatric malignant brain tumors appear driven by epigenetic dysregulation such that tumors rarely harbor immediately actionable mutations, or display significant molecular heterogeneity making resistance to single agent targeted therapy anticipated, as is described for SHH-activated medulloblastoma [30]. Consequently, it is presumed that combination therapy, utilizing novel agents alongside conventional modalities, will better enable local and disseminated disease control rather than a single agent approach in future studies.

5.3. Future Challenges

This review highlights the molecular heterogeneity across the most common pediatric malignant brain tumors, together with its relevance to current diagnostic and therapeutic protocols, and strategies to correct the consequent imbalance that arises from bench to bedside. The tumor groups discussed in this review have key clinical challenges that now warrant focus, including intensification or novel combination therapy for unfavorable-risk tumors, de-escalation of intensity for favorable-risk lesions, the treatment of relapse, and a reduction in morbidity, disability and late effects (Table 2). It is now incumbent on the neuro-oncology community to meet and overcome these challenges; in an age of digital technology and social media, where the latest global scientific breakthroughs are acknowledged promptly in the public domain, the families of our patients are demanding this of us.

Table 2. Future clinical challenges for pediatric malignant brain tumors.

Tumor Group	Future Clinical Challenge
ALL	• Modernize trial risk stratification according to biology • Improve trial design to allow timely conclusions across smaller patient populations • Enable multinational trial collaboration, including less affluent countries • Discovery of novel agents with rapid pre-clinical to clinical translation • Improved understanding of, and therapies for, recurrence (need for repeat tissue analysis via surgery, etc.) • Awareness of neuro-disability, quality of survival and protracted follow-up in trial designs
Medulloblastoma	
WNT	• Non-metastatic; de-escalation of therapy
SHH	• Metastatic/MYCN amplified/TP53 mutant; therapy intensification or novel agent(s)
Group 3	• MYC amplified and/or metastatic; therapy intensification or novel agent(s)
Group 4	• Non-metastatic and chromosome 11 loss; de-escalation of therapy • Metastatic; intensification or novel agent(s)
High-grade gliomas	• Mandating tissue analysis of brainstem lesions for trial entry • International collaborative efforts to test novel agents for specific molecular subgroups • Consideration of alternative drug delivery methods, e.g., convection enhanced delivery
Ependymoma	
PF-A	• Chromosome 1q gain +/− 6q loss; novel agents(s) or techniques including increased radiosensitization
PF-B	• Chromosome 13q balanced; de-escalation of therapy
ST-ZFTA	• Stratification of therapy dependent on extent of surgical resection
ST-YAP1	• De-escalation of therapy

Author Contributions: Conceptualization, J.-P.K., C.T.; Writing-original draft preparation, J.-P.K., C.T.; Writing-review and editing, J.-P.K., C.T. All authors have read and agreed to the published version of the manuscript.

Funding: This research received no active funding.

Conflicts of Interest: The authors declare no conflict of interest.

References

1. Heath, J.A.; Zacharoulis, S.; Kieran, M.W. Pediatric Neuro Oncol.: Current status and future directions. *Asia Pac. J. Clin. Oncol.* **2012**, *8*, 223–231. [CrossRef] [PubMed]
2. Ostrom, Q.T.; Cioffi, G.; Waite, K.; Kruchko, C.; Barnholtz-Sloan, J.S. CBTRUS Statistical Report: Primary Brain and Other Central Nervous System Tumors Diagnosed in the United States in 2014–2018. *Neuro Oncol.* **2021**, *23*, iii1–iii105. [CrossRef] [PubMed]

3. Packer, R.J.; Gajjar, A.; Vezina, G.; Rorke-Adams, L.; Burger, P.C.; Robertson, P.L.; Bayer, L.; LaFond, D.; Donahue, B.R.; Marymont, M.H.; et al. Phase III study of craniospinal radiation therapy followed by adjuvant chemotherapy for newly diagnosed average-risk medulloblastoma. *J. Clin. Oncol.* **2006**, *24*, 4202–4208. [CrossRef]
4. Khanna, V.; Achey, R.L.; Ostrom, Q.T.; Block-Beach, H.; Kruchko, C.; Barnholtz-Sloan, J.S.; de Blank, P.M. Incidence and survival trends for medulloblastomas in the United States from 2001 to 2013. *J. Neurooncol.* **2017**, *135*, 433–441. [CrossRef] [PubMed]
5. Louis, D.N.; Perry, A.; Reifenberger, G.; von Deimling, A.; Figarella-Branger, D.; Cavenee, W.K.; Ohgaki, H.; Wiestler, O.D.; Kleihues, P.; Ellison, D.W. The 2016 World Health Organization Classification of Tumors of the Central Nervous System: A summary. *Acta Neuropathol.* **2016**, *131*, 803–820. [CrossRef] [PubMed]
6. Pizer, B.L.; Clifford, S.C. The potential impact of tumour biology on improved clinical practice for medulloblastoma: Progress towards biologically driven clinical trials. *Br. J. Neurosurg.* **2009**, *23*, 364–375. [CrossRef]
7. Northcott, P.A.; Shih, D.J.; Peacock, J.; Garzia, L.; Morrissy, A.S.; Zichner, T.; Stutz, A.M.; Korshunov, A.; Reimand, J.; Schumacher, S.E.; et al. Subgroup-specific structural variation across 1000 medulloblastoma genomes. *Nature* **2012**, *488*, 49–56. [CrossRef] [PubMed]
8. Orr, B.A. Pathology, diagnostics, and classification of medulloblastoma. *Brain Pathol.* **2020**, *30*, 664–678. [CrossRef] [PubMed]
9. Eberhart, C.G.; Kepner, J.L.; Goldthwaite, P.T.; Kun, L.E.; Duffner, P.K.; Friedman, H.S.; Strother, D.R.; Burger, P.C. Histopathologic grading of medulloblastomas: A Pediatric Oncology Group study. *Cancer* **2002**, *94*, 552–560. [CrossRef]
10. Ramaswamy, V.; Remke, M.; Bouffet, E.; Bailey, S.; Clifford, S.C.; Doz, F.; Kool, M.; Dufour, C.; Vassal, G.; Milde, T.; et al. Risk stratification of childhood medulloblastoma in the molecular era: The current consensus. *Acta Neuropathol.* **2016**, *131*, 821–831. [CrossRef] [PubMed]
11. Taylor, M.D.; Northcott, P.A.; Korshunov, A.; Remke, M.; Cho, Y.J.; Clifford, S.C.; Eberhart, C.G.; Parsons, D.W.; Rutkowski, S.; Gajjar, A.; et al. Molecular subgroups of medulloblastoma: The current consensus. *Acta Neuropathol.* **2012**, *123*, 465–472. [CrossRef] [PubMed]
12. Northcott, P.A.; Jones, D.T.; Kool, M.; Robinson, G.W.; Gilbertson, R.J.; Cho, Y.J.; Pomeroy, S.L.; Korshunov, A.; Lichter, P.; Taylor, M.D.; et al. Medulloblastomics: The end of the beginning. *Nat. Rev. Cancer* **2012**, *12*, 818–834. [CrossRef]
13. Cavalli, F.M.G.; Remke, M.; Rampasek, L.; Peacock, J.; Shih, D.J.H.; Luu, B.; Garzia, L.; Torchia, J.; Nor, C.; Morrissy, A.S.; et al. Intertumoral Heterogeneity within Medulloblastoma Subgroups. *Cancer Cell* **2017**, *31*, 737–754.e736. [CrossRef]
14. Sharma, T.; Schwalbe, E.C.; Williamson, D.; Sill, M.; Hovestadt, V.; Mynarek, M.; Rutkowski, S.; Robinson, G.W.; Gajjar, A.; Cavalli, F.; et al. Second-generation molecular subgrouping of medulloblastoma: An international meta-analysis of Group 3 and Group 4 subtypes. *Acta Neuropathol.* **2019**, *138*, 309–326. [CrossRef]
15. Northcott, P.A.; Buchhalter, I.; Morrissy, A.S.; Hovestadt, V.; Weischenfeldt, J.; Ehrenberger, T.; Grobner, S.; Segura-Wang, M.; Zichner, T.; Rudneva, V.A.; et al. The whole-genome landscape of medulloblastoma subtypes. *Nature* **2017**, *547*, 311–317. [CrossRef] [PubMed]
16. Schwalbe, E.C.; Lindsey, J.C.; Nakjang, S.; Crosier, S.; Smith, A.J.; Hicks, D.; Rafiee, G.; Hill, R.M.; Iliasova, A.; Stone, T.; et al. Novel molecular subgroups for clinical classification and outcome prediction in childhood medulloblastoma: A cohort study. *Lancet Oncol.* **2017**, *18*, 958–971. [CrossRef]
17. Gibson, P.; Tong, Y.; Robinson, G.; Thompson, M.C.; Currle, D.S.; Eden, C.; Kranenburg, T.A.; Hogg, T.; Poppleton, H.; Martin, J.; et al. Subtypes of medulloblastoma have distinct developmental origins. *Nature* **2010**, *468*, 1095–1099. [CrossRef]
18. Patay, Z.; DeSain, L.A.; Hwang, S.N.; Coan, A.; Li, Y.; Ellison, D.W. MR Imaging Characteristics of Wingless-Type-Subgroup Pediatric Medulloblastoma. *AJNR Am. J. Neuroradiol.* **2015**, *36*, 2386–2393. [CrossRef]
19. Clifford, S.C.; Lusher, M.E.; Lindsey, J.C.; Langdon, J.A.; Gilbertson, R.J.; Straughton, D.; Ellison, D.W. Wnt/Wingless pathway activation and chromosome 6 loss characterize a distinct molecular sub-group of medulloblastomas associated with a favorable prognosis. *Cell Cycle* **2006**, *5*, 2666–2670. [CrossRef]
20. Jones, D.T.; Jager, N.; Kool, M.; Zichner, T.; Hutter, B.; Sultan, M.; Cho, Y.J.; Pugh, T.J.; Hovestadt, V.; Stutz, A.M.; et al. Dissecting the genomic complexity underlying medulloblastoma. *Nature* **2012**, *488*, 100–105. [CrossRef]
21. Kool, M.; Korshunov, A.; Remke, M.; Jones, D.T.; Schlanstein, M.; Northcott, P.A.; Cho, Y.J.; Koster, J.; Schouten-van Meeteren, A.; van Vuurden, D.; et al. Molecular subgroups of medulloblastoma: An international meta-analysis of transcriptome, genetic aberrations, and clinical data of WNT, SHH, Group 3, and Group 4 medulloblastomas. *Acta Neuropathol.* **2012**, *123*, 473–484. [CrossRef] [PubMed]
22. Korshunov, A.; Sahm, F.; Zheludkova, O.; Golanov, A.; Stichel, D.; Schrimpf, D.; Ryzhova, M.; Potapov, A.; Habel, A.; Meyer, J.; et al. DNA methylation profiling is a method of choice for molecular verification of pediatric WNT-activated medulloblastomas. *Neuro Oncol.* **2019**, *21*, 214–221. [CrossRef] [PubMed]
23. Northcott, P.A.; Robinson, G.W.; Kratz, C.P.; Mabbott, D.J.; Pomeroy, S.L.; Clifford, S.C.; Rutkowski, S.; Ellison, D.W.; Malkin, D.; Taylor, M.D.; et al. Medulloblastoma. *Nat. Rev. Dis. Primers* **2019**, *5*, 11. [CrossRef]
24. Waszak, S.M.; Northcott, P.A.; Buchhalter, I.; Robinson, G.W.; Sutter, C.; Groebner, S.; Grund, K.B.; Brugieres, L.; Jones, D.T.W.; Pajtler, K.W.; et al. Spectrum and prevalence of genetic predisposition in medulloblastoma: A retrospective genetic study and prospective validation in a clinical trial cohort. *Lancet Oncol.* **2018**, *19*, 785–798. [CrossRef]
25. Pugh, T.J.; Weeraratne, S.D.; Archer, T.C.; Pomeranz Krummel, D.A.; Auclair, D.; Bochicchio, J.; Carneiro, M.O.; Carter, S.L.; Cibulskis, K.; Erlich, R.L.; et al. Medulloblastoma exome sequencing uncovers subtype-specific somatic mutations. *Nature* **2012**, *488*, 106–110. [CrossRef]

26. Robinson, G.; Parker, M.; Kranenburg, T.A.; Lu, C.; Chen, X.; Ding, L.; Phoenix, T.N.; Hedlund, E.; Wei, L.; Zhu, X.; et al. Novel mutations target distinct subgroups of medulloblastoma. *Nature* **2012**, *488*, 43–48. [CrossRef]
27. Zhukova, N.; Ramaswamy, V.; Remke, M.; Pfaff, E.; Shih, D.J.; Martin, D.C.; Castelo-Branco, P.; Baskin, B.; Ray, P.N.; Bouffet, E.; et al. Subgroup-specific prognostic implications of TP53 mutation in medulloblastoma. *J. Clin. Oncol.* **2013**, *31*, 2927–2935. [CrossRef] [PubMed]
28. Raybaud, C.; Ramaswamy, V.; Taylor, M.D.; Laughlin, S. Posterior fossa tumors in children: Developmental anatomy and diagnostic imaging. *Childs Nerv. Syst.* **2015**, *31*, 1661–1676. [CrossRef]
29. Lafay-Cousin, L.; Smith, A.; Chi, S.N.; Wells, E.; Madden, J.; Margol, A.; Ramaswamy, V.; Finlay, J.; Taylor, M.D.; Dhall, G.; et al. Clinical, Pathological, and Molecular Characterization of Infant Medulloblastomas Treated with Sequential High-Dose Chemotherapy. *Pediatr. Blood Cancer* **2016**, *63*, 1527–1534. [CrossRef]
30. Kool, M.; Jones, D.T.; Jager, N.; Northcott, P.A.; Pugh, T.J.; Hovestadt, V.; Piro, R.M.; Esparza, L.A.; Markant, S.L.; Remke, M.; et al. Genome sequencing of SHH medulloblastoma predicts genotype-related response to smoothened inhibition. *Cancer Cell* **2014**, *25*, 393–405. [CrossRef]
31. Huang, G.H.; Xu, Q.F.; Cui, Y.H.; Li, N.; Bian, X.W.; Lv, S.Q. Medulloblastoma stem cells: Promising targets in medulloblastoma therapy. *Cancer Sci.* **2016**, *107*, 583–589. [CrossRef]
32. Ramaswamy, V.; Remke, M.; Shih, D.; Wang, X.; Northcott, P.A.; Faria, C.C.; Raybaud, C.; Tabori, U.; Hawkins, C.; Rutka, J.; et al. Duration of the pre-diagnostic interval in medulloblastoma is subgroup dependent. *Pediatr. Blood Cancer* **2014**, *61*, 1190–1194. [CrossRef]
33. Cho, Y.J.; Tsherniak, A.; Tamayo, P.; Santagata, S.; Ligon, A.; Greulich, H.; Berhoukim, R.; Amani, V.; Goumnerova, L.; Eberhart, C.G.; et al. Integrative genomic analysis of medulloblastoma identifies a molecular subgroup that drives poor clinical outcome. *J. Clin. Oncol.* **2011**, *29*, 1424–1430. [CrossRef]
34. Northcott, P.A.; Korshunov, A.; Witt, H.; Hielscher, T.; Eberhart, C.G.; Mack, S.; Bouffet, E.; Clifford, S.C.; Hawkins, C.E.; French, P.; et al. Medulloblastoma comprises four distinct molecular variants. *J. Clin. Oncol.* **2011**, *29*, 1408–1414. [CrossRef]
35. Northcott, P.A.; Lee, C.; Zichner, T.; Stutz, A.M.; Erkek, S.; Kawauchi, D.; Shih, D.J.; Hovestadt, V.; Zapatka, M.; Sturm, D.; et al. Enhancer hijacking activates GFI1 family oncogenes in medulloblastoma. *Nature* **2014**, *511*, 428–434. [CrossRef] [PubMed]
36. Maier, H.; Dalianis, T.; Kostopoulou, O.N. New Approaches in Targeted Therapy for Medulloblastoma in Children. *Anticancer. Res.* **2021**, *41*, 1715–1726. [CrossRef]
37. Szalontay, L.; Khakoo, Y. Medulloblastoma: An Old Diagnosis with New Promises. *Curr. Oncol. Rep.* **2020**, *22*, 90. [CrossRef] [PubMed]
38. Robinson, G.W.; Rudneva, V.A.; Buchhalter, I.; Billups, C.A.; Waszak, S.M.; Smith, K.S.; Bowers, D.C.; Bendel, A.; Fisher, P.G.; Partap, S.; et al. Risk-adapted therapy for young children with medulloblastoma (SJYC07): Therapeutic and molecular outcomes from a multicentre, phase 2 trial. *Lancet Oncol.* **2018**, *19*, 768–784. [CrossRef]
39. Lafay-Cousin, L.; Bouffet, E.; Strother, D.; Rudneva, V.; Hawkins, C.; Eberhart, C.; Horbinski, C.; Heier, L.; Souweidane, M.; Williams-Hughes, C.; et al. Phase II Study of Nonmetastatic Desmoplastic Medulloblastoma in Children Younger Than 4 Years of Age: A Report of the Children's Oncology Group (ACNS1221). *J. Clin. Oncol.* **2020**, *38*, 223–231. [CrossRef] [PubMed]
40. Korshunov, A.; Remke, M.; Kool, M.; Hielscher, T.; Northcott, P.A.; Williamson, D.; Pfaff, E.; Witt, H.; Jones, D.T.; Ryzhova, M.; et al. Biological and clinical heterogeneity of MYCN-amplified medulloblastoma. *Acta Neuropathol.* **2012**, *123*, 515–527. [CrossRef] [PubMed]
41. Pietsch, T.; Schmidt, R.; Remke, M.; Korshunov, A.; Hovestadt, V.; Jones, D.T.; Felsberg, J.; Kaulich, K.; Goschzik, T.; Kool, M.; et al. Prognostic significance of clinical, histopathological, and molecular characteristics of medulloblastomas in the prospective HIT2000 multicenter clinical trial cohort. *Acta Neuropathol.* **2014**, *128*, 137–149. [CrossRef]
42. von Bueren, A.O.; Kortmann, R.D.; von Hoff, K.; Friedrich, C.; Mynarek, M.; Muller, K.; Goschzik, T.; Zur Muhlen, A.; Gerber, N.; Warmuth-Metz, M.; et al. Treatment of Children and Adolescents with Metastatic Medulloblastoma and Prognostic Relevance of Clinical and Biologic Parameters. *J. Clin. Oncol.* **2016**, *34*, 4151–4160. [CrossRef]
43. Ramaswamy, V.; Remke, M.; Adamski, J.; Bartels, U.; Tabori, U.; Wang, X.; Huang, A.; Hawkins, C.; Mabbott, D.; Laperriere, N.; et al. Medulloblastoma subgroup-specific outcomes in irradiated children: Who are the true high-risk patients? *Neuro Oncol.* **2016**, *18*, 291–297. [CrossRef] [PubMed]
44. Shih, D.J.; Northcott, P.A.; Remke, M.; Korshunov, A.; Ramaswamy, V.; Kool, M.; Luu, B.; Yao, Y.; Wang, X.; Dubuc, A.M.; et al. Cytogenetic prognostication within medulloblastoma subgroups. *J. Clin. Oncol.* **2014**, *32*, 886–896. [CrossRef]
45. Laughton, S.J.; Merchant, T.E.; Sklar, C.A.; Kun, L.E.; Fouladi, M.; Broniscer, A.; Morris, E.B.; Sanders, R.P.; Krasin, M.J.; Shelso, J.; et al. Endocrine outcomes for children with embryonal brain tumors after risk-adapted craniospinal and conformal primary-site irradiation and high-dose chemotherapy with stem-cell rescue on the SJMB-96 trial. *J. Clin. Oncol.* **2008**, *26*, 1112–1118. [CrossRef] [PubMed]
46. Mulhern, R.K.; Merchant, T.E.; Gajjar, A.; Reddick, W.E.; Kun, L.E. Late neurocognitive sequelae in survivors of brain tumours in childhood. *Lancet Oncol.* **2004**, *5*, 399–408. [CrossRef]
47. Packer, R.J.; Goldwein, J.; Nicholson, H.S.; Vezina, L.G.; Allen, J.C.; Ris, M.D.; Muraszko, K.; Rorke, L.B.; Wara, W.M.; Cohen, B.H.; et al. Treatment of children with medulloblastomas with reduced-dose craniospinal radiation therapy and adjuvant chemotherapy: A Children's Cancer Group Study. *J. Clin. Oncol.* **1999**, *17*, 2127–2136. [CrossRef] [PubMed]

48. Gajjar, A.; Chintagumpala, M.; Ashley, D.; Kellie, S.; Kun, L.E.; Merchant, T.E.; Woo, S.; Wheeler, G.; Ahern, V.; Krasin, M.J.; et al. Risk-adapted craniospinal radiotherapy followed by high-dose chemotherapy and stem-cell rescue in children with newly diagnosed medulloblastoma (St Jude Medulloblastoma-96): Long-term results from a prospective, multicentre trial. *Lancet Oncol.* **2006**, *7*, 813–820. [CrossRef]
49. Tarbell, N.J.; Friedman, H.; Polkinghorn, W.R.; Yock, T.; Zhou, T.; Chen, Z.; Burger, P.; Barnes, P.; Kun, L. High-risk medulloblastoma: A pediatric oncology group randomized trial of chemotherapy before or after radiation therapy (POG 9031). *J. Clin. Oncol.* **2013**, *31*, 2936–2941. [CrossRef]
50. Taylor, R.E.; Bailey, C.C.; Robinson, K.; Weston, C.L.; Ellison, D.; Ironside, J.; Lucraft, H.; Gilbertson, R.; Tait, D.M.; Walker, D.A.; et al. Results of a randomized study of preradiation chemotherapy versus radiotherapy alone for nonmetastatic medulloblastoma: The International Society of Paediatric Oncology/United Kingdom Children's Cancer Study Group PNET-3 Study. *J. Clin. Oncol.* **2003**, *21*, 1581–1591. [CrossRef]
51. Lannering, B.; Rutkowski, S.; Doz, F.; Pizer, B.; Gustafsson, G.; Navajas, A.; Massimino, M.; Reddingius, R.; Benesch, M.; Carrie, C.; et al. Hyperfractionated versus conventional radiotherapy followed by chemotherapy in standard-risk medulloblastoma: Results from the randomized multicenter HIT-SIOP PNET 4 trial. *J. Clin. Oncol.* **2012**, *30*, 3187–3193. [CrossRef] [PubMed]
52. Michalski, J.M.; Janss, A.; Vezina, G.; Gajjar, A.; Pollack, I.; Merchant, T.E.; Fitzgerald, T.J.; Booth, T.; Tarbell, N.J.; Li, Y.; et al. Results of COG ACNS0331: A Phase III Trial of Involved-Field Radiotherapy (IFRT) and Low Dose Craniospinal Irradiation (LD-CSI) with Chemotherapy in Average-Risk Medulloblastoma: A Report from the Children's Oncology Group. *Int. J. Radiat. Oncol. Biol. Phys.* **2016**, *96*, 937–938. [CrossRef]
53. Leary, S.E.S.; Packer, R.J.; Li, Y.; Billups, C.A.; Smith, K.S.; Jaju, A.; Heier, L.; Burger, P.; Walsh, K.; Han, Y.; et al. Efficacy of Carboplatin and Isotretinoin in Children with High-risk Medulloblastoma: A Randomized Clinical Trial from the Children's Oncology Group. *JAMA Oncol.* **2021**, *7*, 1313–1321. [CrossRef]
54. Cohen, K.J.; Heideman, R.L.; Zhou, T.; Holmes, E.J.; Lavey, R.S.; Bouffet, E.; Pollack, I.F. Temozolomide in the treatment of children with newly diagnosed diffuse intrinsic pontine gliomas: A report from the Children's Oncology Group. *Neuro Oncol.* **2011**, *13*, 410–416. [CrossRef] [PubMed]
55. Jakacki, R.I.; Cohen, K.J.; Buxton, A.; Krailo, M.D.; Burger, P.C.; Rosenblum, M.K.; Brat, D.J.; Hamilton, R.L.; Eckel, S.P.; Zhou, T.; et al. Phase 2 study of concurrent radiotherapy and temozolomide followed by temozolomide and lomustine in the treatment of children with high-grade glioma: A report of the Children's Oncology Group ACNS0423 study. *Neuro Oncol.* **2016**, *18*, 1442–1450. [CrossRef] [PubMed]
56. Bradley, K.A.; Zhou, T.; McNall-Knapp, R.Y.; Jakacki, R.I.; Levy, A.S.; Vezina, G.; Pollack, I.F. Motexafin-gadolinium and involved field radiation therapy for intrinsic pontine glioma of childhood: A children's oncology group phase 2 study. *Int. J. Radiat. Oncol. Biol. Phys.* **2013**, *85*, e55–e60. [CrossRef]
57. Grill, J.; Massimino, M.; Bouffet, E.; Azizi, A.A.; McCowage, G.; Canete, A.; Saran, F.; Le Deley, M.C.; Varlet, P.; Morgan, P.S.; et al. Phase II, Open-Label, Randomized, Multicenter Trial (HERBY) of Bevacizumab in Pediatric Patients with Newly Diagnosed High-Grade Glioma. *J. Clin. Oncol.* **2018**, *36*, 951–958. [CrossRef] [PubMed]
58. Grill, J.; Le Teuff, G.; Nysom, K.; Blomgren, K.; Hargrave, D.; MacCowage, G.; Bautista, F.; Van Vuurden, D.; Dangouloff-Ros, V.; Puget, S.; et al. Biological medicine for diffuse intrinsic pontine glioma (DIPG) eradication: Results of the three arm biomarker-driven randomized BIOMEDE 1.0 trial. *Neuro Oncol.* **2020**, *22* (Suppl. S3), iii293. [CrossRef]
59. Merchant, T.E.; Bendel, A.E.; Sabin, N.D.; Burger, P.C.; Shaw, D.W.; Chang, E.; Wu, S.; Zhou, T.; Eisenstat, D.D.; Foreman, N.K.; et al. Conformal Radiation Therapy for Pediatric Ependymoma, Chemotherapy for Incompletely Resected Ependymoma, and Observation for Completely Resected, Supratentorial Ependymoma. *J. Clin. Oncol.* **2019**, *37*, 974–983. [CrossRef]
60. Smith, A.; Onar-Thomas, A.; Ellison, D.; Owens-Pickle, E.; Wu, S.; Leary, S.; Fouladi, M. ACNS0831, Phase III Rrandomized trial of post-radiation chemotherapy in patients with newly diagnosed ependymoma ages 1 to 21 years. *Neuro Oncol.* **2021**, *22*, iii318. [CrossRef]
61. Nobre, L.; Zapotocky, M.; Khan, S.; Fukuoka, K.; Fonseca, A.; McKeown, T.; Sumerauer, D.; Vicha, A.; Grajkowska, W.A.; Trubicka, J.; et al. Pattern of Relapse and Treatment Response in WNT-Activated Medulloblastoma. *Cell Rep. Med.* **2020**, *1*, 100038. [CrossRef] [PubMed]
62. Rutkowski, S.; Bode, U.; Deinlein, F.; Ottensmeier, H.; Warmuth-Metz, M.; Soerensen, N.; Graf, N.; Emser, A.; Pietsch, T.; Wolff, J.E.; et al. Treatment of early childhood medulloblastoma by postoperative chemotherapy alone. *N. Engl. J. Med.* **2005**, *352*, 978–986. [CrossRef]
63. von Bueren, A.O.; von Hoff, K.; Pietsch, T.; Gerber, N.U.; Warmuth-Metz, M.; Deinlein, F.; Zwiener, I.; Faldum, A.; Fleischhack, G.; Benesch, M.; et al. Treatment of young children with localized medulloblastoma by chemotherapy alone: Results of the prospective, multicenter trial HIT 2000 confirming the prognostic impact of histology. *Neuro Oncol.* **2011**, *13*, 669–679. [CrossRef] [PubMed]
64. Mazewski, C.; Kang, G.; Kellie, S.; Gossett, J.; Leary, S.; Li, B.; Arigides, P.; Hayes, L.; Reddy, A.; Shaw, D.; et al. Efficacy of methotrexate (MTX) according to molecular sub-type in young children with medulloblastoma (MB): A report from Children's Oncology Group Phase III Trial ACNS0334. *Neuro Oncol.* **2020**, *22*, iii396. [CrossRef]
65. Bautista, F.; Fioravantti, V.; de Rojas, T.; Carceller, F.; Madero, L.; Lassaletta, A.; Moreno, L. Medulloblastoma in children and adolescents: A systematic review of contemporary phase I and II clinical trials and biology update. *Cancer Med.* **2017**, *6*, 2606–2624. [CrossRef] [PubMed]

66. LoRusso, P.M.; Rudin, C.M.; Reddy, J.C.; Tibes, R.; Weiss, G.J.; Borad, M.J.; Hann, C.L.; Brahmer, J.R.; Chang, I.; Darbonne, W.C.; et al. Phase I trial of hedgehog pathway inhibitor vismodegib (GDC-0449) in patients with refractory, locally advanced or metastatic solid tumors. *Clin. Cancer Res.* **2011**, *17*, 2502–2511. [CrossRef]
67. Robinson, G.W.; Orr, B.A.; Wu, G.; Gururangan, S.; Lin, T.; Qaddoumi, I.; Packer, R.J.; Goldman, S.; Prados, M.D.; Desjardins, A.; et al. Vismodegib Exerts Targeted Efficacy Against Recurrent Sonic Hedgehog-Subgroup Medulloblastoma: Results from Phase II Pediatric Brain Tumor Consortium Studies PBTC-025B and PBTC-032. *J. Clin. Oncol.* **2015**, *33*, 2646–2654. [CrossRef]
68. Li, Y.; Song, Q.; Day, B.W. Phase I and phase II sonidegib and vismodegib clinical trials for the treatment of paediatric and adult MB patients: A systemic review and meta-analysis. *Acta Neuropathol. Commun.* **2019**, *7*, 123. [CrossRef]
69. Gajjar, A.; Packer, R.J.; Foreman, N.K.; Cohen, K.; Haas-Kogan, D.; Merchant, T.E.; Committee, C.O.G.B.T. Children's Oncology Group's 2013 blueprint for research: Central nervous system tumors. *Pediatr. Blood Cancer* **2013**, *60*, 1022–1026. [CrossRef]
70. Gajjar, A.; Stewart, C.F.; Ellison, D.W.; Kaste, S.; Kun, L.E.; Packer, R.J.; Goldman, S.; Chintagumpala, M.; Wallace, D.; Takebe, N.; et al. Phase I study of vismodegib in children with recurrent or refractory medulloblastoma: A pediatric brain tumor consortium study. *Clin. Cancer Res.* **2013**, *19*, 6305–6312. [CrossRef]
71. Kieran, M.W.; Chisholm, J.; Casanova, M.; Brandes, A.A.; Aerts, I.; Bouffet, E.; Bailey, S.; Leary, S.; MacDonald, T.J.; Mechinaud, F.; et al. Phase I study of oral sonidegib (LDE225) in pediatric brain and solid tumors and a phase II study in children and adults with relapsed medulloblastoma. *Neuro Oncol.* **2017**, *19*, 1542–1552. [CrossRef] [PubMed]
72. Robinson, G.W.; Kaste, S.C.; Chemaitilly, W.; Bowers, D.C.; Laughton, S.; Smith, A.; Gottardo, N.G.; Partap, S.; Bendel, A.; Wright, K.D.; et al. Irreversible growth plate fusions in children with medulloblastoma treated with a targeted hedgehog pathway inhibitor. *Oncotarget* **2017**, *8*, 69295–69302. [CrossRef]
73. Morfouace, M.; Shelat, A.; Jacus, M.; Freeman, B.B.; Turner, D.; Robinson, S.; Zindy, F.; Wang, Y.D.; Finkelstein, D.; Ayrault, O.; et al. Pemetrexed and gemcitabine as combination therapy for the treatment of Group3 medulloblastoma. *Cancer Cell* **2014**, *25*, 516–529. [CrossRef]
74. Sangar, M.L.C.; Genovesi, L.A.; Nakamoto, M.W.; Davis, M.J.; Knobluagh, S.E.; Ji, P.; Millar, A.; Wainwright, B.J.; Olson, J.M. Inhibition of CDK4/6 by Palbociclib Significantly Extends Survival in Medulloblastoma Patient-Derived Xenograft Mouse Models. *Clin. Cancer Res.* **2017**, *23*, 5802–5813. [CrossRef] [PubMed]
75. Bandopadhayay, P.; Bergthold, G.; Nguyen, B.; Schubert, S.; Gholamin, S.; Tang, Y.; Bolin, S.; Schumacher, S.E.; Zeid, R.; Masoud, S.; et al. BET bromodomain inhibition of MYC-amplified medulloblastoma. *Clin. Cancer Res.* **2014**, *20*, 912–925. [CrossRef] [PubMed]
76. Lee, C.; Rudneva, V.A.; Erkek, S.; Zapatka, M.; Chau, L.Q.; Tacheva-Grigorova, S.K.; Garancher, A.; Rusert, J.M.; Aksoy, O.; Lea, R.; et al. Lsd1 as a therapeutic target in Gfi1-activated medulloblastoma. *Nat. Commun.* **2019**, *10*, 332. [CrossRef] [PubMed]
77. Pei, Y.; Liu, K.W.; Wang, J.; Garancher, A.; Tao, R.; Esparza, L.A.; Maier, D.L.; Udaka, Y.T.; Murad, N.; Morrissy, S.; et al. HDAC and PI3K Antagonists Cooperate to Inhibit Growth of MYC-Driven Medulloblastoma. *Cancer Cell* **2016**, *29*, 311–323. [CrossRef] [PubMed]
78. Tang, Y.; Gholamin, S.; Schubert, S.; Willardson, M.I.; Lee, A.; Bandopadhayay, P.; Bergthold, G.; Masoud, S.; Nguyen, B.; Vue, N.; et al. Epigenetic targeting of Hedgehog pathway transcriptional output through BET bromodomain inhibition. *Nat. Med.* **2014**, *20*, 732–740. [CrossRef]
79. Sayour, E.J.; Mitchell, D.A. Immunotherapy for Pediatric Brain Tumors. *Brain Sci.* **2017**, *7*, 137. [CrossRef]
80. Donovan, L.K.; Delaidelli, A.; Joseph, S.K.; Bielamowicz, K.; Fousek, K.; Holgado, B.L.; Manno, A.; Srikanthan, D.; Gad, A.Z.; Van Ommeren, R.; et al. Locoregional delivery of CAR T cells to the cerebrospinal fluid for treatment of metastatic medulloblastoma and ependymoma. *Nat. Med.* **2020**, *26*, 720–731. [CrossRef]
81. Ostrom, Q.T.; de Blank, P.M.; Kruchko, C.; Petersen, C.M.; Liao, P.; Finlay, J.L.; Stearns, D.S.; Wolff, J.E.; Wolinsky, Y., Letterio, J.J.; et al. Alex's Lemonade Stand Foundation Infant and Childhood Primary Brain and Central Nervous System Tumors Diagnosed in the United States in 2007–2011. *Neuro Oncol.* **2015**, *16* (Suppl. S10), x1–x36. [CrossRef]
82. Diaz, A.K.; Baker, S.J. The genetic signatures of pediatric high-grade glioma: No longer a one-act play. *Semin. Radiat. Oncol.* **2014**, *24*, 240–247. [CrossRef] [PubMed]
83. Paugh, B.S.; Qu, C.; Jones, C.; Liu, Z.; Adamowicz-Brice, M.; Zhang, J.; Bax, D.A.; Coyle, B.; Barrow, J.; Hargrave, D.; et al. Integrated molecular genetic profiling of pediatric high-grade gliomas reveals key differences with the adult disease. *J. Clin. Oncol.* **2010**, *28*, 3061–3068. [CrossRef]
84. Schwartzentruber, J.; Korshunov, A.; Liu, X.Y.; Jones, D.T.; Pfaff, E.; Jacob, K.; Sturm, D.; Fontebasso, A.M.; Quang, D.A.; Tonjes, M.; et al. Driver mutations in histone H3.3 and chromatin remodelling genes in paediatric glioblastoma. *Nature* **2012**, *482*, 226–231. [CrossRef]
85. Sturm, D.; Witt, H.; Hovestadt, V.; Khuong-Quang, D.A.; Jones, D.T.; Konermann, C.; Pfaff, E.; Tonjes, M.; Sill, M.; Bender, S.; et al. Hotspot mutations in H3F3A and IDH1 define distinct epigenetic and biological subgroups of glioblastoma. *Cancer Cell* **2012**, *22*, 425–437. [CrossRef]
86. Fontebasso, A.M.; Papillon-Cavanagh, S.; Schwartzentruber, J.; Nikbakht, H.; Gerges, N.; Fiset, P.O.; Bechet, D.; Faury, D.; De Jay, N.; Ramkissoon, L.A.; et al. Recurrent somatic mutations in ACVR1 in pediatric midline high-grade astrocytoma. *Nat. Genet.* **2014**, *46*, 462–466. [CrossRef] [PubMed]

87. Wu, G.; Diaz, A.K.; Paugh, B.S.; Rankin, S.L.; Ju, B.; Li, Y.; Zhu, X.; Qu, C.; Chen, X.; Zhang, J.; et al. The genomic landscape of diffuse intrinsic pontine glioma and pediatric non-brainstem high-grade glioma. *Nat. Genet.* **2014**, *46*, 444–450. [CrossRef] [PubMed]
88. Buczkowicz, P.; Hoeman, C.; Rakopoulos, P.; Pajovic, S.; Letourneau, L.; Dzamba, M.; Morrison, A.; Lewis, P.; Bouffet, E.; Bartels, U.; et al. Genomic analysis of diffuse intrinsic pontine gliomas identifies three molecular subgroups and recurrent activating ACVR1 mutations. *Nat. Genet.* **2014**, *46*, 451–456. [CrossRef]
89. Taylor, K.R.; Mackay, A.; Truffaux, N.; Butterfield, Y.; Morozova, O.; Philippe, C.; Castel, D.; Grasso, C.S.; Vinci, M.; Carvalho, D.; et al. Recurrent activating ACVR1 mutations in diffuse intrinsic pontine glioma. *Nat. Genet.* **2014**, *46*, 457–461. [CrossRef]
90. Mackay, A.; Burford, A.; Carvalho, D.; Izquierdo, E.; Fazal-Salom, J.; Taylor, K.R.; Bjerke, L.; Clarke, M.; Vinci, M.; Nandhabalan, M.; et al. Integrated Molecular Meta-Analysis of 1000 Pediatric High-Grade and Diffuse Intrinsic Pontine Glioma. *Cancer Cell* **2017**, *32*, 520–537.e525. [CrossRef] [PubMed]
91. Sturm, D.; Bender, S.; Jones, D.T.; Lichter, P.; Grill, J.; Becher, O.; Hawkins, C.; Majewski, J.; Jones, C.; Costello, J.F.; et al. Paediatric and adult glioblastoma: Multiform (epi)genomic culprits emerge. *Nat. Rev. Cancer* **2014**, *14*, 92–107. [CrossRef] [PubMed]
92. Chatwin, H.V.; Cruz Cruz, J.; Green, A.L. Pediatric high-grade glioma: Moving toward subtype-specific multimodal therapy. *FEBS J.* **2021**, *288*, 6127–6141. [CrossRef]
93. Castel, D.; Philippe, C.; Calmon, R.; Le Dret, L.; Truffaux, N.; Boddaert, N.; Pages, M.; Taylor, K.R.; Saulnier, P.; Lacroix, L.; et al. Histone H3F3A and HIST1H3B K27M mutations define two subgroups of diffuse intrinsic pontine gliomas with different prognosis and phenotypes. *Acta Neuropathol.* **2015**, *130*, 815–827. [CrossRef]
94. Lewis, P.W.; Muller, M.M.; Koletsky, M.S.; Cordero, F.; Lin, S.; Banaszynski, L.A.; Garcia, B.A.; Muir, T.W.; Becher, O.J.; Allis, C.D. Inhibition of PRC2 activity by a gain-of-function H3 mutation found in pediatric glioblastoma. *Science* **2013**, *340*, 857–861. [CrossRef]
95. Korshunov, A.; Ryzhova, M.; Hovestadt, V.; Bender, S.; Sturm, D.; Capper, D.; Meyer, J.; Schrimpf, D.; Kool, M.; Northcott, P.A.; et al. Integrated analysis of pediatric glioblastoma reveals a subset of biologically favorable tumors with associated molecular prognostic markers. *Acta Neuropathol.* **2015**, *129*, 669–678. [CrossRef]
96. Finlay, J.L.; Boyett, J.M.; Yates, A.J.; Wisoff, J.H.; Milstein, J.M.; Geyer, J.R.; Bertolone, S.J.; McGuire, P.; Cherlow, J.M.; Tefft, M.; et al. Randomized phase III trial in childhood high-grade astrocytoma comparing vincristine, lomustine, and prednisone with the eight-drugs-in-1-day regimen. Childrens Cancer Group. *J. Clin. Oncol.* **1995**, *13*, 112–123. [CrossRef]
97. Pollack, I.F.; Boyett, J.M.; Yates, A.J.; Burger, P.C.; Gilles, F.H.; Davis, R.L.; Finlay, J.L.; Children's Cancer, G. The influence of central review on outcome associations in childhood malignant gliomas: Results from the CCG-945 experience. *Neuro Oncol.* **2003**, *5*, 197–207. [CrossRef] [PubMed]
98. Pollack, I.F.; Hamilton, R.L.; Sobol, R.W.; Burnham, J.; Yates, A.J.; Holmes, E.J.; Zhou, T.; Finlay, J.L. O6-methylguanine-DNA methyltransferase expression strongly correlates with outcome in childhood malignant gliomas: Results from the CCG-945 Cohort. *J. Clin. Oncol.* **2006**, *24*, 3431–3437. [CrossRef] [PubMed]
99. Stupp, R.; Mason, W.P.; van den Bent, M.J.; Weller, M.; Fisher, B.; Taphoorn, M.J.; Belanger, K.; Brandes, A.A.; Marosi, C.; Bogdahn, U.; et al. Radiotherapy plus concomitant and adjuvant temozolomide for glioblastoma. *N. Engl. J. Med.* **2005**, *352*, 987–996. [CrossRef]
100. Wolff, J.E.; Driever, P.H.; Erdlenbruch, B.; Kortmann, R.D.; Rutkowski, S.; Pietsch, T.; Parker, C.; Metz, M.W.; Gnekow, A.; Kramm, C.M. Intensive chemotherapy improves survival in pediatric high-grade glioma after gross total resection: Results of the HIT-GBM-C protocol. *Cancer* **2010**, *116*, 705–712. [CrossRef] [PubMed]
101. Anderson, R.C.; Kennedy, B.; Yanes, C.L.; Garvin, J.; Needle, M.; Canoll, P.; Feldstein, N.A.; Bruce, J.N. Convection-enhanced delivery of topotecan into diffuse intrinsic brainstem tumors in children. *J. Neurosurg Pediatr.* **2013**, *11*, 289–295. [CrossRef] [PubMed]
102. Souweidane, M.M.; Kramer, K.; Pandit-Taskar, N.; Zhou, Z.; Haque, S.; Zanzonico, P.; Carrasquillo, J.A.; Lyashchenko, S.K.; Thakur, S.B.; Donzelli, M.; et al. Convection-enhanced delivery for diffuse intrinsic pontine glioma: A single-centre, dose-escalation, phase 1 trial. *Lancet Oncol.* **2018**, *19*, 1040–1050. [CrossRef]
103. Tosi, U.; Souweidane, M. Convection Enhanced Delivery for Diffuse Intrinsic Pontine Glioma: Review of a Single Institution Experience. *Pharmaceutics* **2020**, *12*, 660. [CrossRef]
104. Saito, R.; Kanamori, M.; Sonoda, Y.; Yamashita, Y.; Nagamatsu, K.; Murata, T.; Mugikura, S.; Kumabe, T.; Wembacher-Schroder, E.; Thomson, R.; et al. Phase I trial of convection-enhanced delivery of nimustine hydrochloride (ACNU) for brainstem recurrent glioma. *Neurooncol. Adv.* **2020**, *2*, vdaa033. [CrossRef]
105. Abedalthagafi, M.; Mobark, N.; Al-Rashed, M.; AlHarbi, M. Epigenomics and immunotherapeutic advances in pediatric brain tumors. *NPJ Precis. Oncol.* **2021**, *5*, 34. [CrossRef]
106. Robinson, G.W.; Orr, B.A.; Gajjar, A. Complete clinical regression of a BRAF V600E-mutant pediatric glioblastoma multiforme after BRAF inhibitor therapy. *BMC Cancer* **2014**, *14*, 258. [CrossRef]
107. Bautista, F.; Paci, A.; Minard-Colin, V.; Dufour, C.; Grill, J.; Lacroix, L.; Varlet, P.; Valteau-Couanet, D.; Geoerger, B. Vemurafenib in pediatric patients with BRAFV600E mutated high-grade gliomas. *Pediatr. Blood Cancer* **2014**, *61*, 1101–1103. [CrossRef]
108. Ceccon, G.; Werner, J.M.; Dunkl, V.; Tscherpel, C.; Stoffels, G.; Brunn, A.; Deckert, M.; Fink, G.R.; Galldiks, N. Dabrafenib Treatment in a Patient with an Epithelioid Glioblastoma and BRAF V600E Mutation. *Int. J. Mol. Sci.* **2018**, *19*, 1090. [CrossRef]

109. Toll, S.A.; Tran, H.N.; Cotter, J.; Judkins, A.R.; Tamrazi, B.; Biegel, J.A.; Dhall, G.; Robison, N.J.; Waters, K.; Patel, P.; et al. Sustained response of three pediatric BRAF(V600E) mutated high-grade gliomas to combined BRAF and MEK inhibitor therapy. *Oncotarget* **2019**, *10*, 551–557. [CrossRef]
110. Ziegler, D.S.; Wong, M.; Mayoh, C.; Kumar, A.; Tsoli, M.; Mould, E.; Tyrrell, V.; Khuong-Quang, D.A.; Pinese, M.; Gayevskiy, V.; et al. Brief Report: Potent clinical and radiological response to larotrectinib in TRK fusion-driven high-grade glioma. *Br. J. Cancer* **2018**, *119*, 693–696. [CrossRef]
111. Alharbi, M.; Mobark, N.A.; Balbaid, A.A.O.; Alanazi, F.A.; Aljabarat, W.A.R.; Bakhsh, E.A.; Ramkissoon, S.H.; Abedalthagafi, M. Regression of ETV6-NTRK3 Infantile Glioblastoma after First-Line Treatment with Larotrectinib. *JCO Precis. Oncol.* **2020**, *4*, 796–800. [CrossRef]
112. AlHarbi, M.; Ali Mobark, N.; AlMubarak, L.; Aljelaify, R.; AlSaeed, M.; Almutairi, A.; Alqubaishi, F.; Hussain, M.E.; Balbaid, A.A.O.; Said Marie, A.; et al. Durable Response to Nivolumab in a Pediatric Patient with Refractory Glioblastoma and Constitutional Biallelic Mismatch Repair Deficiency. *Oncologist* **2018**, *23*, 1401–1406. [CrossRef] [PubMed]
113. Bouffet, E.; Larouche, V.; Campbell, B.B.; Merico, D.; de Borja, R.; Aronson, M.; Durno, C.; Krueger, J.; Cabric, V.; Ramaswamy, V.; et al. Immune Checkpoint Inhibition for Hypermutant Glioblastoma Multiforme Resulting from Germline Biallelic Mismatch Repair Deficiency. *J. Clin. Oncol.* **2016**, *34*, 2206–2211. [CrossRef] [PubMed]
114. Larouche, V.; Atkinson, J.; Albrecht, S.; Laframboise, R.; Jabado, N.; Tabori, U.; Bouffet, E. Sustained complete response of recurrent glioblastoma to combined checkpoint inhibition in a young patient with constitutional mismatch repair deficiency. *Pediatr. Blood Cancer* **2018**, *65*, e27389. [CrossRef]
115. Le, D.T.; Durham, J.N.; Smith, K.N.; Wang, H.; Bartlett, B.R.; Aulakh, L.K.; Lu, S.; Kemberling, H.; Wilt, C.; Luber, B.S.; et al. Mismatch repair deficiency predicts response of solid tumors to PD-1 blockade. *Science* **2017**, *357*, 409–413. [CrossRef]
116. Hummel, T.R.; Wagner, L.; Ahern, C.; Fouladi, M.; Reid, J.M.; McGovern, R.M.; Ames, M.M.; Gilbertson, R.J.; Horton, T.; Ingle, A.M.; et al. A pediatric phase 1 trial of vorinostat and temozolomide in relapsed or refractory primary brain or spinal cord tumors: A Children's Oncology Group phase 1 consortium study. *Pediatr. Blood Cancer* **2013**, *60*, 1452–1457. [CrossRef] [PubMed]
117. Rasmussen, T.A.; Tolstrup, M.; Moller, H.J.; Brinkmann, C.R.; Olesen, R.; Erikstrup, C.; Laursen, A.L.; Ostergaard, L.; Sogaard, O.S. Activation of latent human immunodeficiency virus by the histone deacetylase inhibitor panobinostat: A pilot study to assess effects on the central nervous system. *Open Forum Infect. Dis.* **2015**, *2*, ofv037. [CrossRef]
118. Muscal, J.A.; Thompson, P.A.; Horton, T.M.; Ingle, A.M.; Ahern, C.H.; McGovern, R.M.; Reid, J.M.; Ames, M.M.; Espinoza-Delgado, I.; Weigel, B.J.; et al. A phase I trial of vorinostat and bortezomib in children with refractory or recurrent solid tumors: A Children's Oncology Group phase I consortium study (ADVL0916). *Pediatr. Blood Cancer* **2013**, *60*, 390–395. [CrossRef]
119. Su, J.M.; Li, X.N.; Thompson, P.; Ou, C.N.; Ingle, A.M.; Russell, H.; Lau, C.C.; Adamson, P.C.; Blaney, S.M. Phase 1 study of valproic acid in pediatric patients with refractory solid or CNS tumors: A children's oncology group report. *Clin. Cancer Res.* **2011**, *17*, 589–597. [CrossRef]
120. Pajtler, K.W.; Mack, S.C.; Ramaswamy, V.; Smith, C.A.; Witt, H.; Smith, A.; Hansford, J.R.; von Hoff, K.; Wright, K.D.; Hwang, E.; et al. The current consensus on the clinical management of intracranial ependymoma and its distinct molecular variants. *Acta Neuropathol.* **2017**, *133*, 5–12. [CrossRef]
121. Carvalho, D.; Taylor, K.R.; Olaciregui, N.G.; Molinari, V.; Clarke, M.; Mackay, A.; Ruddle, R.; Henley, A.; Valenti, M.; Hayes, A.; et al. ALK2 inhibitors display beneficial effects in preclinical models of ACVR1 mutant diffuse intrinsic pontine glioma. *Commun. Biol.* **2019**, *2*, 156. [CrossRef] [PubMed]
122. Arrillaga-Romany, I.; Chi, A.S.; Allen, J.E.; Oster, W.; Wen, P.Y.; Batchelor, T.T. A phase 2 study of the first imipridone ONC201, a selective DRD2 antagonist for oncology, administered every three weeks in recurrent glioblastoma. *Oncotarget* **2017**, *8*, 79298–79304. [CrossRef] [PubMed]
123. Chi, A.S.; Tarapore, R.S.; Hall, M.D.; Shonka, N.; Gardner, S.; Umemura, Y.; Sumrall, A.; Khatib, Z.; Mueller, S.; Kline, C.; et al. Pediatric and adult H3 K27M-mutant diffuse midline glioma treated with the selective DRD2 antagonist ONC201. *J. Neurooncol.* **2019**, *145*, 97–105. [CrossRef] [PubMed]
124. Hall, M.D.; Odia, Y.; Allen, J.E.; Tarapore, R.; Khatib, Z.; Niazi, T.N.; Daghistani, D.; Schalop, L.; Chi, A.S.; Oster, W.; et al. First clinical experience with DRD2/3 antagonist ONC201 in H3 K27M-mutant pediatric diffuse intrinsic pontine glioma: A case report. *J. Neurosurg Pediatr.* **2019**, *23*, 1–7. [CrossRef] [PubMed]
125. Lesueur, P.; Lequesne, J.; Grellard, J.M.; Dugue, A.; Coquan, E.; Brachet, P.E.; Geffrelot, J.; Kao, W.; Emery, E.; Berro, D.H.; et al. Phase I/IIa study of concomitant radiotherapy with olaparib and temozolomide in unresectable or partially resectable glioblastoma: OLA-TMZ-RTE-01 trial protocol. *BMC Cancer* **2019**, *19*, 198. [CrossRef]
126. Okada, H.; Low, K.L.; Kohanbash, G.; McDonald, H.A.; Hamilton, R.L.; Pollack, I.F. Expression of glioma-associated antigens in pediatric brain stem and non-brain stem gliomas. *J. Neurooncol.* **2008**, *88*, 245–250. [CrossRef] [PubMed]
127. Pollack, I.F.; Jakacki, R.I.; Butterfield, L.H.; Hamilton, R.L.; Panigrahy, A.; Potter, D.M.; Connelly, A.K.; Dibridge, S.A.; Whiteside, T.L.; Okada, H. Antigen-specific immune responses and clinical outcome after vaccination with glioma-associated antigen peptides and polyinosinic-polycytidylic acid stabilized by lysine and carboxymethylcellulose in children with newly diagnosed malignant brainstem and nonbrainstem gliomas. *J. Clin. Oncol.* **2014**, *32*, 2050–2058. [CrossRef]
128. Chheda, Z.S.; Kohanbash, G.; Okada, K.; Jahan, N.; Sidney, J.; Pecoraro, M.; Yang, X.; Carrera, D.A.; Downey, K.M.; Shrivastav, S.; et al. Novel and shared neoantigen derived from histone 3 variant H3.3K27M mutation for glioma T cell therapy. *J. Exp. Med.* **2018**, *215*, 141–157. [CrossRef] [PubMed]

129. Benitez-Ribas, D.; Cabezon, R.; Florez-Grau, G.; Molero, M.C.; Puerta, P.; Guillen, A.; Gonzalez-Navarro, E.A.; Paco, S.; Carcaboso, A.M.; Santa-Maria Lopez, V.; et al. Corrigendum: Immune Response Generated with the Administration of Autologous Dendritic Cells Pulsed with an Allogenic Tumoral Cell-Lines Lysate in Patients with Newly Diagnosed Diffuse Intrinsic Pontine Glioma. *Front. Oncol.* **2018**, *8*, 201. [CrossRef]
130. McGuire, C.S.; Sainani, K.L.; Fisher, P.G. Incidence patterns for ependymoma: A surveillance, epidemiology, and end results study. *J. Neurosurg.* **2009**, *110*, 725–729. [CrossRef]
131. Amirian, E.S.; Armstrong, T.S.; Aldape, K.D.; Gilbert, M.R.; Scheurer, M.E. Predictors of survival among pediatric and adult ependymoma cases: A study using Surveillance, Epidemiology, and End Results data from 1973 to 2007. *Neuroepidemiology* **2012**, *39*, 116–124. [CrossRef]
132. Kilday, J.P.; Rahman, R.; Dyer, S.; Ridley, L.; Lowe, J.; Coyle, B.; Grundy, R. Pediatric ependymoma: Biological perspectives. *Mol. Cancer Res.* **2009**, *7*, 765–786. [CrossRef] [PubMed]
133. Benesch, M.; Mynarek, M.; Witt, H.; Warmuth-Metz, M.; Pietsch, T.; Bison, B.; Pfister, S.M.; Pajtler, K.W.; Kool, M.; Schuller, U.; et al. Newly Diagnosed Metastatic Intracranial Ependymoma in Children: Frequency, Molecular Characteristics, Treatment, and Outcome in the Prospective HIT Series. *Oncologist* **2019**, *24*, e921–e929. [CrossRef] [PubMed]
134. Zacharoulis, S.; Ji, L.; Pollack, I.F.; Duffner, P.; Geyer, R.; Grill, J.; Schild, S.; Jaing, T.H.; Massimino, M.; Finlay, J.; et al. Metastatic ependymoma: A multi-institutional retrospective analysis of prognostic factors. *Pediatr. Blood Cancer* **2008**, *50*, 231–235. [CrossRef]
135. Plotkin, S.R.; O'Donnell, C.C.; Curry, W.T.; Bove, C.M.; MacCollin, M.; Nunes, F.P. Spinal ependymomas in neurofibromatosis Type 2: A retrospective analysis of 55 patients. *J. Neurosurg. Spine* **2011**, *14*, 543–547. [CrossRef]
136. Bandopadhayay, P.; Silvera, V.M.; Ciarlini, P.; Malkin, H.; Bi, W.L.; Bergthold, G.; Faisal, A.M.; Ullrich, N.J.; Marcus, K.; Scott, R.M.; et al. Myxopapillary ependymomas in children: Imaging, treatment and outcomes. *J. Neurooncol.* **2016**, *126*, 165–174. [CrossRef] [PubMed]
137. Goldwein, J.W.; Leahy, J.M.; Packer, R.J.; Sutton, L.N.; Curran, W.J.; Rorke, L.B.; Schut, L.; Littman, P.S.; D'Angio, G.J. Intracranial ependymomas in children. *Int. J. Radiat. Oncol. Biol. Phys.* **1990**, *19*, 1497–1502. [CrossRef]
138. Ellison, D.W.; Aldape, K.D.; Capper, D.; Fouladi, M.; Gilbert, M.R.; Gilbertson, R.J.; Hawkins, C.; Merchant, T.E.; Pajtler, K.; Venneti, S.; et al. cIMPACT-NOW update 7: Advancing the molecular classification of ependymal tumors. *Brain Pathol.* **2020**, *30*, 863–866. [CrossRef]
139. Pajtler, K.W.; Witt, H.; Sill, M.; Jones, D.T.; Hovestadt, V.; Kratochwil, F.; Wani, K.; Tatevossian, R.; Punchihewa, C.; Johann, P.; et al. Molecular Classification of Ependymal Tumors across All CNS Compartments, Histopathological Grades, and Age Groups. *Cancer Cell* **2015**, *27*, 728–743. [CrossRef] [PubMed]
140. Cavalli, F.M.G.; Hubner, J.M.; Sharma, T.; Luu, B.; Sill, M.; Zapotocky, M.; Mack, S.C.; Witt, H.; Lin, T.; Shih, D.J.H.; et al. Heterogeneity within the PF-EPN-B ependymoma subgroup. *Acta Neuropathol.* **2018**, *136*, 227–237. [CrossRef]
141. Andreiuolo, F.; Le Teuff, G.; Bayar, M.A.; Kilday, J.P.; Pietsch, T.; von Bueren, A.O.; Witt, H.; Korshunov, A.; Modena, P.; Pfister, S.M.; et al. Integrating Tenascin-C protein expression and 1q25 copy number status in pediatric intracranial ependymoma prognostication: A new model for risk stratification. *PLoS ONE* **2017**, *12*, e0178351. [CrossRef] [PubMed]
142. Andreiuolo, F.; Varlet, P.; Tauziede-Espariat, A.; Junger, S.T.; Dorner, E.; Dreschmann, V.; Kuchelmeister, K.; Waha, A.; Haberler, C.; Slavc, I.; et al. Childhood supratentorial ependymomas with YAP1-MAMLD1 fusion: An entity with characteristic clinical, radiological, cytogenetic and histopathological features. *Brain Pathol.* **2019**, *29*, 205–216. [CrossRef]
143. Arabzade, A.; Zhao, Y.; Varadharajan, S.; Chen, H.C.; Jessa, S.; Rivas, B.; Stuckert, A.J.; Solis, M.; Kardian, A.; Tlais, D.; et al. ZFTA-RELA Dictates Oncogenic Transcriptional Programs to Drive Aggressive Supratentorial Ependymoma. *Cancer Discov.* **2021**, *11*, 2200–2215. [CrossRef] [PubMed]
144. Araki, A.; Chocholous, M.; Gojo, J.; Dorfer, C.; Czech, T.; Heinzl, H.; Dieckmann, K.; Ambros, I.M.; Ambros, P.F.; Slavc, I.; et al. Chromosome 1q gain and tenascin-C expression are candidate markers to define different risk groups in pediatric posterior fossa ependymoma. *Acta Neuropathol. Commun.* **2016**, *4*, 88. [CrossRef] [PubMed]
145. Baroni, L.; Sundaresan, L.; Heled, A.; Coltin, H.; Pajtler, K.W.; Lin, T.; Merchant, T.E.; McLendon, R.; Faria, C.; Buntine, M.; et al. Ultra high-risk PFA ependymoma is characterized by loss of chromosome 6q. *Neuro Oncol.* **2021**, *23*, 1360–1370. [CrossRef] [PubMed]
146. Kilday, J.P.; Mitra, B.; Domerg, C.; Ward, J.; Andreiuolo, F.; Osteso-Ibanez, T.; Mauguen, A.; Varlet, P.; Le Deley, M.C.; Lowe, J.; et al. Copy number gain of 1q25 predicts poor progression-free survival for pediatric intracranial ependymomas and enables patient risk stratification: A prospective European clinical trial cohort analysis on behalf of the Children's Cancer Leukaemia Group (CCLG), Societe Francaise d'Oncologie Pediatrique (SFOP), and International Society for Pediatric Oncology (SIOP). *Clin. Cancer Res.* **2012**, *18*, 2001–2011. [CrossRef]
147. Mack, S.C.; Witt, H.; Piro, R.M.; Gu, L.; Zuyderduyn, S.; Stutz, A.M.; Wang, X.; Gallo, M.; Garzia, L.; Zayne, K.; et al. Epigenomic alterations define lethal CIMP-positive ependymomas of infancy. *Nature* **2014**, *506*, 445–450. [CrossRef]
148. Mendrzyk, F.; Korshunov, A.; Benner, A.; Toedt, G.; Pfister, S.; Radlwimmer, B.; Lichter, P. Identification of gains on 1q and epidermal growth factor receptor overexpression as independent prognostic markers in intracranial ependymoma. *Clin. Cancer Res.* **2006**, *12*, 2070–2079. [CrossRef]
149. Michealraj, K.A.; Kumar, S.A.; Kim, L.J.Y.; Cavalli, F.M.G.; Przelicki, D.; Wojcik, J.B.; Delaidelli, A.; Bajic, A.; Saulnier, O.; MacLeod, G.; et al. Metabolic Regulation of the Epigenome Drives Lethal Infantile Ependymoma. *Cell* **2020**, *181*, 1329–1345.e1324. [CrossRef] [PubMed]

150. Panwalkar, P.; Clark, J.; Ramaswamy, V.; Hawes, D.; Yang, F.; Dunham, C.; Yip, S.; Hukin, J.; Sun, Y.; Schipper, M.J.; et al. Immunohistochemical analysis of H3K27me3 demonstrates global reduction in group-A childhood posterior fossa ependymoma and is a powerful predictor of outcome. *Acta Neuropathol.* **2017**, *134*, 705–714. [CrossRef] [PubMed]
151. Parker, M.; Mohankumar, K.M.; Punchihewa, C.; Weinlich, R.; Dalton, J.D.; Li, Y.; Lee, R.; Tatevossian, R.G.; Phoenix, T.N.; Thiruvenkatam, R.; et al. C11orf95-RELA fusions drive oncogenic NF-kappaB signalling in ependymoma. *Nature* **2014**, *506*, 451–455. [CrossRef] [PubMed]
152. Witt, H.; Mack, S.C.; Ryzhova, M.; Bender, S.; Sill, M.; Isserlin, R.; Benner, A.; Hielscher, T.; Milde, T.; Remke, M.; et al. Delineation of two clinically and molecularly distinct subgroups of posterior fossa ependymoma. *Cancer Cell* **2011**, *20*, 143–157. [CrossRef]
153. Zapotocky, M.; Beera, K.; Adamski, J.; Laperierre, N.; Guger, S.; Janzen, L.; Lassaletta, A.; Figueiredo Nobre, L.; Bartels, U.; Tabori, U.; et al. Survival and functional outcomes of molecularly defined childhood posterior fossa ependymoma: Cure at a cost. *Cancer* **2019**, *125*, 1867–1876. [CrossRef] [PubMed]
154. Liu, A.K.; Foreman, N.K.; Gaspar, L.E.; Trinidad, E.; Handler, M.H. Maximally safe resection followed by hypofractionated re-irradiation for locally recurrent ependymoma in children. *Pediatr. Blood Cancer* **2009**, *52*, 804–807. [CrossRef]
155. Ramaswamy, V.; Taylor, M.D. Treatment implications of posterior fossa ependymoma subgroups. *Chin. J. Cancer* **2016**, *35*, 93. [CrossRef]
156. Pajtler, K.W.; Wen, J.; Sill, M.; Lin, T.; Orisme, W.; Tang, B.; Hubner, J.M.; Ramaswamy, V.; Jia, S.; Dalton, J.D.; et al. Molecular heterogeneity and CXorf67 alterations in posterior fossa group A (PFA) ependymomas. *Acta Neuropathol.* **2018**, *136*, 211–226. [CrossRef]
157. Nowak, J.; Junger, S.T.; Huflage, H.; Seidel, C.; Hohm, A.; Vandergrift, L.A.; von Hoff, K.; Rutkowski, S.; Pietsch, T.; Warmuth-Metz, M. MRI Phenotype of RELA-fused Pediatric Supratentorial Ependymoma. *Clin. Neuroradiol.* **2019**, *29*, 595–604. [CrossRef]
158. Fukuoka, K.; Kanemura, Y.; Shofuda, T.; Fukushima, S.; Yamashita, S.; Narushima, D.; Kato, M.; Honda-Kitahara, M.; Ichikawa, H.; Kohno, T.; et al. Significance of molecular classification of ependymomas: C11orf95-RELA fusion-negative supratentorial ependymomas are a heterogeneous group of tumors. *Acta Neuropathol. Commun.* **2018**, *6*, 134. [CrossRef]
159. Bouffet, E.; Perilongo, G.; Canete, A.; Massimino, M. Intracranial ependymomas in children: A critical review of prognostic factors and a plea for cooperation. *Med. Pediatr. Oncol.* **1998**, *30*, 319–329; discussion 329–331. [CrossRef]
160. Merchant, T.E.; Li, C.; Xiong, X.; Kun, L.E.; Boop, F.A.; Sanford, R.A. Conformal radiotherapy after surgery for paediatric ependymoma. A prospective study. *Lancet Oncol.* **2009**, *10*, 258–266. [CrossRef]
161. Ramaswamy, V.; Hielscher, T.; Mack, S.C.; Lassaletta, A.; Lin, T.; Pajtler, K.W.; Jones, D.T.; Luu, B.; Cavalli, F.M.; Aldape, K.; et al. Therapeutic Impact of Cytoreductive Surgery and Irradiation of Posterior Fossa Ependymoma in the Molecular Era: A Retrospective Multicohort Analysis. *J. Clin. Oncol.* **2016**, *34*, 2468–2477. [CrossRef] [PubMed]
162. Thompson, Y.Y.; Ramaswamy, V.; Diamandis, P.; Daniels, C.; Taylor, M.D. Posterior fossa ependymoma: Current insights. *Childs Nerv. Syst.* **2015**, *31*, 1699–1706. [CrossRef]
163. Indelicato, D.J.; Bradley, J.A.; Rotondo, R.L.; Nanda, R.H.; Logie, N.; Sandler, E.S.; Aldana, P.R.; Ranalli, N.J.; Beier, A.D.; Morris, C.G.; et al. Outcomes following proton therapy for pediatric ependymoma. *Acta Oncol.* **2018**, *57*, 644–648. [CrossRef]
164. Indelicato, D.J.; Ioakeim-Ioannidou, M.; Bradley, J.A.; Mailhot-Vega, R.B.; Morris, C.G.; Tarbell, N.J.; Yock, T.; MacDonald, S.M. Proton Therapy for Pediatric Ependymoma: Mature Results from a Bicentric Study. *Int. J. Radiat. Oncol. Biol. Phys.* **2021**, *110*, 815–820. [CrossRef] [PubMed]
165. MacDonald, S.M.; Safai, S.; Trofimov, A.; Wolfgang, J.; Fullerton, B.; Yeap, B.Y.; Bortfeld, T.; Tarbell, N.J.; Yock, T. Proton radiotherapy for childhood ependymoma: Initial clinical outcomes and dose comparisons. *Int. J. Radiat. Oncol. Biol. Phys.* **2008**, *71*, 979–986. [CrossRef] [PubMed]
166. Garvin, J.H., Jr.; Selch, M.T.; Holmes, E.; Berger, M.S.; Finlay, J.L.; Flannery, A.; Goldwein, J.W.; Packer, R.J.; Rorke-Adams, L.B.; Shiminski-Maher, T.; et al. Phase II study of pre-irradiation chemotherapy for childhood intracranial ependymoma. Children's Cancer Group protocol 9942: A report from the Children's Oncology Group. *Pediatr. Blood Cancer* **2012**, *59*, 1183–1189. [CrossRef] [PubMed]
167. von Hoff, K.; Kortmann, R.D.; Gerber, N.U.; Mynarek, M. Risk-adapted treatment for non-metastatic ependymoma: Preliminary results of the nonrandomized prospective phase II Clinical Trial Hit2000. *Neuro Oncol.* **2014**, *16*, i17.
168. Duffner, P.K.; Krischer, J.P.; Sanford, R.A.; Horowitz, M.E.; Burger, P.C.; Cohen, M.E.; Friedman, H.S.; Kun, L.E. Prognostic factors in infants and very young children with intracranial ependymomas. *Pediatr. Neurosurg.* **1998**, *28*, 215–222. [CrossRef] [PubMed]
169. Grundy, R.G.; Wilne, S.H.; Robinson, K.J.; Ironside, J.W.; Cox, T.; Chong, W.K.; Michalski, A.; Campbell, R.H.; Bailey, C.C.; Thorp, N.; et al. Primary postoperative chemotherapy without radiotherapy for treatment of brain tumours other than ependymoma in children under 3 years: Results of the first UKCCSG/SIOP CNS 9204 trial. *Eur. J. Cancer* **2010**, *46*, 120–133. [CrossRef]
170. Zacharoulis, S.; Levy, A.; Chi, S.N.; Gardner, S.; Rosenblum, M.; Miller, D.C.; Dunkel, I.; Diez, B.; Sposto, R.; Ji, L.; et al. Outcome for young children newly diagnosed with ependymoma, treated with intensive induction chemotherapy followed by myeloablative chemotherapy and autologous stem cell rescue. *Pediatr. Blood Cancer* **2007**, *49*, 34–40. [CrossRef]
171. Grill, J.; Le Deley, M.C.; Gambarelli, D.; Raquin, M.A.; Couanet, D.; Pierre-Kahn, A.; Habrand, J.L.; Doz, F.; Frappaz, D.; Gentet, J.C.; et al. Postoperative chemotherapy without irradiation for ependymoma in children under 5 years of age: A multicenter trial of the French Society of Pediatric Oncology. *J. Clin. Oncol.* **2001**, *19*, 1288–1296. [CrossRef]
172. Milde, T.; Hielscher, T.; Witt, H.; Kool, M.; Mack, S.C.; Deubzer, H.E.; Oehme, I.; Lodrini, M.; Benner, A.; Taylor, M.D.; et al. Nestin expression identifies ependymoma patients with poor outcome. *Brain Pathol.* **2012**, *22*, 848–860. [CrossRef] [PubMed]

173. Mohankumar, K.M.; Currle, D.S.; White, E.; Boulos, N.; Dapper, J.; Eden, C.; Nimmervoll, B.; Thiruvenkatam, R.; Connelly, M.; Kranenburg, T.A.; et al. An in vivo screen identifies ependymoma oncogenes and tumor-suppressor genes. *Nat. Genet.* **2015**, *47*, 878–887. [CrossRef] [PubMed]
174. Atkinson, J.M.; Shelat, A.A.; Carcaboso, A.M.; Kranenburg, T.A.; Arnold, L.A.; Boulos, N.; Wright, K.; Johnson, R.A.; Poppleton, H.; Mohankumar, K.M.; et al. An integrated in vitro and in vivo high-throughput screen identifies treatment leads for ependymoma. *Cancer Cell* **2011**, *20*, 384–399. [CrossRef]
175. Wright, K.D.; Daryani, V.M.; Turner, D.C.; Onar-Thomas, A.; Boulos, N.; Orr, B.A.; Gilbertson, R.J.; Stewart, C.F.; Gajjar, A. Phase I study of 5-fluorouracil in children and young adults with recurrent ependymoma. *Neuro Oncol.* **2015**, *17*, 1620–1627. [CrossRef] [PubMed]
176. Lotsch, D.; Kirchhofer, D.; Englinger, B.; Jiang, L.; Okonechnikov, K.; Senfter, D.; Laemmerer, A.; Gabler, L.; Pirker, C.; Donson, A.M.; et al. Targeting fibroblast growth factor receptors to combat aggressive ependymoma. *Acta Neuropathol.* **2021**, *142*, 339–360. [CrossRef]
177. Mork, S.J.; Loken, A.C. Ependymoma: A follow-up study of 101 cases. *Cancer* **1977**, *40*, 907–915. [CrossRef]
178. Rahman, R.; Osteso-Ibanez, T.; Hirst, R.A.; Levesley, J.; Kilday, J.P.; Quinn, S.; Peet, A.; O'Callaghan, C.; Coyle, B.; Grundy, R.G. Histone deacetylase inhibition attenuates cell growth with associated telomerase inhibition in high-grade childhood brain tumor cells. *Mol. Cancer Ther.* **2010**, *9*, 2568–2581. [CrossRef] [PubMed]
179. Alderete, D.; Baroni, L.; Sampor, C.; Freytes, C.; Pennella, C. Antiangiogenic metronomic therapy for children with recurrent ependymoma. *Neuro Oncol.* **2018**, *20*, i74. [CrossRef]
180. Gillan, E. Response of recurrent ependymoma to MEMMAT based metronomic antiangiogenic combination therapy utilizing tapered bevacizumab and maintenance therapy with celecoxib and fenofibrate. *Neuro Oncol.* **2020**, *22*, iii317. [CrossRef]
181. Bull, K.S.; Hornsey, S.; Kennedy, C.R.; Darlington, A.E.; Grootenhuis, M.A.; Hargrave, D.; Liossi, C.; Shepherd, J.P.; Walker, D.A.; Morris, C. Systematic review: Measurement properties of patient-reported outcome measures evaluated with childhood brain tumor survivors or other acquired brain injury. *Neurooncol. Pract.* **2020**, *7*, 277–287. [CrossRef] [PubMed]
182. Morrissy, A.S.; Garzia, L.; Shih, D.J.; Zuyderduyn, S.; Huang, X.; Skowron, P.; Remke, M.; Cavalli, F.M.; Ramaswamy, V.; Lindsay, P.E.; et al. Divergent clonal selection dominates medulloblastoma at recurrence. *Nature* **2016**, *529*, 351–357. [CrossRef] [PubMed]
183. Wu, X.; Northcott, P.A.; Dubuc, A.; Dupuy, A.J.; Shih, D.J.; Witt, H.; Croul, S.; Bouffet, E.; Fults, D.W.; Eberhart, C.G.; et al. Clonal selection drives genetic divergence of metastatic medulloblastoma. *Nature* **2012**, *482*, 529–533. [CrossRef]

cancers MDPI

Article

Altertoxin II, a Highly Effective and Specific Compound against Ewing Sarcoma

Andrew J. Robles [1,2,3], Wentao Dai [4], Saikat Haldar [4], Hongyan Ma [4], Victoria M. Anderson [4], Ross D. Overacker [5], April L. Risinger [1,2], Sandra Loesgen [5,6], Peter J. Houghton [2,3], Robert H. Cichewicz [4,*] and Susan L. Mooberry [1,2,*]

1 Department of Pharmacology, The University of Texas Health Science Center at San Antonio, San Antonio, TX 78229, USA; roblesa3@uthscsa.edu (A.J.R.); risingera@uthscsa.edu (A.L.R.)
2 Mays Cancer Center, The University of Texas Health Science Center at San Antonio, San Antonio, TX 78229, USA; houghtonp@uthscsa.edu
3 Greehey Children's Cancer Research Institute, The University of Texas Health Science Center at San Antonio, San Antonio, TX 78229, USA
4 Natural Products Discovery Group, Institute for Natural Products Applications and Research Technologies, and Department of Chemistry & Biochemistry, Stephenson Life Science Research Center, University of Oklahoma, Norman, OK 73019, USA; huidou2015@gmail.com (W.D.); saikatchembiol@gmail.com (S.H.); oucmhy@gmail.com (H.M.); vickyanderson@ou.edu (V.M.A.)
5 Department of Chemistry, Oregon State University, Corvallis, OR 97331, USA; rd.overacker@gmail.com (R.D.O.); sandra.loesgen@whitney.ufl.edu (S.L.)
6 Whitney Laboratory for Marine Bioscience, Department of Chemistry, University of Florida, St. Augustine, FL 32080, USA
* Correspondence: rhcichewicz@ou.edu (R.H.C.); mooberry@uthscsa.edu (S.L.M.); Tel.: +1-405-325-6969 (R.H.C.); +1-210-567-4788 (S.L.M.); Fax: +1-405-325-6111 (R.H.C.); +1-210-567-4300 (S.L.M.)

Citation: Robles, A.J.; Dai, W.; Haldar, S.; Ma, H.; Anderson, V.M.; Overacker, R.D.; Risinger, A.L.; Loesgen, S.; Houghton, P.J.; Cichewicz, R.H.; et al. Altertoxin II, a Highly Effective and Specific Compound against Ewing Sarcoma. *Cancers* **2021**, *13*, 6176. https://doi.org/10.3390/cancers13246176

Academic Editor: Eelco de Bree

Received: 30 October 2021
Accepted: 3 December 2021
Published: 7 December 2021

Simple Summary: Ewing sarcoma is a cancer of the bone and soft tissues that affects children and adolescents. Unfortunately, only 20–30% of patients with metastatic Ewing sarcoma survive, necessitating the need to identify new, more effective therapies. We screened natural product extracts from plants and fungal cultures to identify compounds with selective cytotoxic activity against Ewing sarcoma cells, which led to the identification of altertoxin II as a compound with highly selective activity against Ewing sarcoma cells. Mechanism of action studies showed that altertoxin II selectively induces DNA damage in Ewing sarcoma cells, but does not bind to DNA. Additionally, we found that altertoxin II has antitumor activity in a mouse model of Ewing sarcoma, suggesting it will be useful as a lead compound to help identify new molecular targets for the development of new Ewing-sarcoma-specific therapies.

Abstract: A screening program designed to identify natural products with selective cytotoxic effects against cell lines representing different types of pediatric solid tumors led to the identification of altertoxin II as a highly potent and selective cytotoxin against Ewing sarcoma cell lines. Altertoxin II, but not the related compounds altertoxin I and alteichin, was highly effective against every Ewing sarcoma cell line tested, with an average 25-fold selectivity for these cells as compared to cells representing other pediatric and adult cancers. Mechanism of action studies revealed that altertoxin II causes DNA double-strand breaks, a rapid DNA damage response, and cell cycle accumulation in the S phase. Our studies also demonstrate that the potent effects of altertoxin II are partially dependent on the progression through the cell cycle, because the G_1 arrest initiated by a CDK4/6 inhibitor decreased antiproliferative potency more than 10 times. Importantly, the cell-type-selective DNA-damaging effects of altertoxin II in Ewing sarcoma cells occur independently of its ability to bind directly to DNA. Ultimately, we found that altertoxin II has a dose-dependent in vivo antitumor efficacy against a Ewing sarcoma xenograft, suggesting that it has potential as a therapeutic drug lead and will be useful to identify novel targets for Ewing-sarcoma-specific therapies.

Keywords: natural products; Ewing sarcoma; drug discovery; fungal secondary metabolites; soil microbes

1. Introduction

Ewing sarcoma (ES) is an aggressive bone and soft tissue cancer affecting children, adolescents, and young adults. This disease is most often caused by a chromosomal translocation leading to the expression of an abnormal fusion protein that is commonly designated as EWS-FLI1 [1,2]. While the prognosis for many patients is good, with 70–80% survival for localized cancers, only 20–30% of patients with metastatic or recurrent ES survive [2,3]. There is a clear need for new effective therapies to help treat these recalcitrant cases of ES and provide disease-specific therapies that cause fewer acute and chronic side effects. Based on the knowledge that ES tumors are caused by the expression of the EWS-FLI1 fusion protein, the possibility exists for the identification of targeted fusion-protein-dependent therapies that could achieve high levels of efficacy against ES tumors while avoiding toxicity to other tissues that do not express the abnormal fusion protein. To this end, a variety of EWS-FLI1-targeting approaches have been attempted, including the inhibition of EWS-FLI1 expression, the inhibition of the transcriptional activity of EWS-FLI1, and the repression of EWS-FLI1 downstream targets [3]. Unfortunately, the identification of a selective inhibitor of EWS-FLI1 has not yet been successful, in part because EWS-FLI1 is a highly disordered protein [4].

A screening program at the National Cancer Institute evaluated natural products from the Molecular Targets Laboratory at NCI (Frederick) for small molecules that inhibit EWS-FLI1 transcription. Their screen identified multiple natural products, including mithramycin [5], trabectedin (E-743) [6], and englerin A [7]. Mithramycin advanced to clinical trials in ES but failed because the serum levels necessary for the inhibition of transcription were not obtainable [8]. While trabectedin had promising effects in a Phase I clinical trial, the Phase II trial failed to show activity, with only one of 10 patients responding [9]. A new clinical trial of trabectedin in combination with irinotecan (SARC037) is currently enrolling patients (Clinicaltrials.gov (accessed on 1 December 2021)).

Based in part on the success of the NCI in identifying multiple natural products with activity against ES, we investigated fungal-derived metabolites from the extensive natural product library at the University of Oklahoma, which contains >76,000 samples prepared from taxonomically diverse fungi, for compounds with selective cytotoxic activities against pediatric cancer cell lines. This approach has been effective in identifying multiple leads against subtypes of triple-negative breast cancer [10–15]. This unbiased, mechanism-blind screening program also led to the identification of compounds with selective cytotoxic activities against pediatric solid cancer cells lines, including ES [16,17]. Herein, we describe the identification of altertoxin II (ATXII) as a highly potent and selective cytotoxin against ES cells in vitro that also has antitumor activity in vivo against a murine xenograft model of ES.

2. Materials and Methods

2.1. General Experimental Procedures

Sulforhodamine B salt, Trizma, Dulbecco's phosphate-buffered saline (DPBS), phenylmethanesulfonyl fluoride (PMSF), dimethyl sulfoxide (DMSO), crystal violet, and Kolliphor® EL were purchased from Sigma-Aldrich (St. Louis, MO, USA). Acetic acid was purchased from Thermo Fisher Scientific (Waltham, MA, USA). Abemaciclib-mesylate was provided by Dr. Peter Houghton. The compounds used for in vitro cell treatments were dissolved in DMSO and stored at −20 °C. ATXII for in vivo studies was diluted in 50:50 DMSO:Kolliphor® EL, stored at −20 °C, and diluted 1:10 in DPBS immediately prior to use. LCMS analyses were performed on a Shimadzu UFLC system with a quadrupole mass spectrometer using a Phenomenex Kinetex C18 column (3.0 mm × 75 mm, 2.6 μm) and a CH_3CN-H_2O (0.1% HCOOH) gradient solvent system. NMR spectra were obtained on a Varian a spectrometer (500 MHz for ^{1}H and 125 MHz for ^{13}C) using acetone-d_6 (Aldrich) as the solvent. HPLC was performed on a Waters System equipped with a 1525 binary HPLC pump coupled to a 2998 PDA detector with a Phenomenex C18 column (21.2 × 250 mm or 10 × 250 mm, 5 μm).

2.2. Purification of Secondary Metabolites Altertoxin II, Altertoxin I, and Alteichin

The fungal isolate was identified as an *Alternaria* sp. (isolate code: SC5920 TV8-1) based on its ribosomal internal transcribed spacer (ITS) sequence data (GenBank accession number MW013191). The soil sample from which it was derived was obtained from South Carolina, USA, through the University of Oklahoma-s Citizen Science Soil Collection. This isolate was grown in three large mycobags (Unicorn Bags, Plano, TX, USA) charged with monolayers of Cheerios breakfast cereal supplemented with a 0.3% sucrose solution with 0.005% chloramphenicol. After four weeks, the contents of the bags were combined, homogenized, and extracted with EtOAc. The EtOAc extract (29.0 g) was subjected to silica gel vacuum liquid chromatography with elution steps of 1:1 hexanes-DCM, DCM, 10:1 DCM-MeOH, and MeOH, yielding four fractions. The 10:1 DCM/MeOH fraction (11.5 g) was then subjected to HP20SS vacuum liquid chromatography and elution performed using a step gradient of MeOH (30%, 50%, 70%, 90%, 100%) in water followed by 1:1 DCM-MeOH, yielding a total of six fractions. The bioassay analysis of the resulting fractions indicated that the selective cytotoxicity was limited to the 90% and 100% MeOH fractions. These two fractions were subjected to further bioassay-guided purification using preparative C18 HPLC (250 mm × 21.2 mm, 5 μm) with a MeOH-H_2O gradient (30:70 to 100:0), followed by isocratic semi-preparative C18 HPLC (250 mm × 10 mm, 5 μm) with MeCN-H2O (50:80) containing 0.1% formic acid to yield altertoxin II (84.0 mg) and its structural analogues altertoxin I (8.0 mg) and alteichin (54.7 mg).

2.3. Phylogenetic Analysis of Citizen Science Alternaria

ITS sequences were generated from cell lysate using ITS1 (5′-TCCGTAGGTGAACCTG CGG-3′) and ITS4 (5′ TCCTCCGCTTATTGATATGC-3′). Amplification was performed using a LightCycle 480 II (Roche) using the following conditions: 1 cycle of denaturation at 94 °C for 2 min followed by 40 cycles of denaturation at 94 °C for 1 min, annealing at 50 °C for 1 min, and extension at 72 °C for 1 min. The samples were processed with Sanger sequencing by GENEWIZ. A cohort of 198 *Alternaria* sequences were then aligned using clustalW in Mega, and within group distances were generated using the Kimura2+G algorithm. A neighbor joining tree was constructed with 500 bootstraps using the same algorithm.

2.4. Preparation of Alternaria Isolates for Metabolomic Analysis

Isolates indicated as belonging to *Alternaria* (based on BLAST comparisons to ITS data contained in the NCBI database) were cultured in duplicate on Cheerios breakfast cereal supplemented with a 0.3% sucrose solution spiked with 0.005% chloramphenicol. The cultures were extracted twice using a Tecan Freedom EVO® robot. For the extraction process, a 3 mL aliquot of ethyl acetate was added to each culture tube followed by 3 mL of water. After 4 h, the ethyl acetate layer was transferred to a deep-well 96 well-plate. To increase the recovery rates of organic metabolites, a second 3 mL aliquot of ethyl acetate was added to each culture tube, which was removed after 2 h, and the ethyl acetate samples were combined. The organic solvent was removed under vacuum and the samples were stored at −20 °C for analysis.

2.5. LCMS Detection of Altertoxin II

The ethyl acetate soluble material from each fungus was suspended in 135 μL of 90:10 methanol that had been spiked with 0.5 μM sulfadimethoxine (internal quality control standard). LCMS analyses were conducted on a Thermo Fisher Scientific Vanquish Flex Binary LC system fitted with a C18 column (Kinetex, 50 × 2.1 mm, 1.7 μM, 100 Å,

Phenomenex, Torrance, CA, USA) as an interface to a Thermo Fisher Q Exactive Plus hybrid quadrupole-orbitrap mass spectrometer. Sample elution was performed using a gradient system with increasing amounts of acetonitrile in H_2O (treated with 0.1% formic acid). For the gradient, conditions were held at 5% acetonitrile/H_2O for 1.0 min, increased to 100% acetonitrile over 8.0 min, and held for 2.0 min. The autosampler was maintained at 10 °C, while the column compartment was held at 40 °C. The samples were analyzed in random order with injection volumes of 5.0 µL. Blanks and pooled quality control samples were run after every 12 samples, alternating between a methanol blank and a media blank.

High resolution MS data were acquired in the positive ion mode using a scan range of m/z 100–1500, with a resolution of 35,000 and 17,500 for MS1 and MS2, respectively. MS2 data were acquired in a data-dependent manner: 5 MS/MS scans were acquired of the most abundant ion during each cycle. Both MS1 and MS2′s maximum injection time was 100 ms. The AGC target was 1E6 and 5E5 for MS1 and MS2, respectively. The isolation window was m/z 2. The sheath gas and auxiliary gas flow rate was set at 35 L/min and 10 L/min, respectively, at 350 °C, whereas the sweep gas flow rate was 0 L/min. The capillary temperature was maintained at 320 °C, and the spray voltage was 3.8 kV. The S-lens RF level was set to 50 V. MS2 data were collected at the apex within a window of 2–8 s and used normalized collision energy that was increased from 20% to 30% to 40%. Dynamic exclusion was used to avoid resampling ions within 10 s. Unassigned charges were excluded from the analysis.

The peak corresponding to altertoxin II was identified by the comparison of the retention time and the MS2 spectrum of the metabolite's $[M+H-H_2O]^+$ ion to an authentic sample of the metabolite. Intensity data for compound altertoxin II were plotted against the isolate number in the R software package. Isolates were classified as either high or low producers of altertoxin II based on the intensity of this ion. Isolates were characterized as high altertoxin II producers when the intensity measurement was greater than 1×10^7. This value was used, since it corresponded to roughly one order of magnitude less than the intensity of the altertoxin II ion in the *Alternaria* sp. isolate SC5920 TV8-1 ($\approx 4.30 \times 10^8$).

2.6. Cell Culture

RD-ES, SK-ES-1, A-673, D283, A204, SH-SY-5Y, SJCRH30, SK-OV-3, MDA-MB-453, MDA-MB-231, HCC1806, and HCC1937 cell lines were purchased from the American Type Culture Collection (Manassas, VA, USA). The CAL-51 cell line was purchased from Creative Bioarray (Shirley, New York, NY, USA). The BT-549 cell line was obtained from Lombardi Comprehensive Cancer Center, in Georgetown University (Washington, DC, USA), and validated by Promega (Fitchburg, WI, USA). The TC32 cell line was provided by Dr. Alexander Bishop (University of Texas' Health Science Center at San Antonio (UTHSCSA)). Cell line identities were confirmed by DNA short tandem repeat analyses (Labcorp, Burlington, NC, USA). TC32-NR0B1 and TC32-CMV cell lines were provided by Dr. Patrick Grohar (Van Andel Research Institute, Grand Rapids, MI, USA). The SK-N-BE(2)-C cell line was provided by Dr. Alexander Pertsemlidis (UTHSCSA). The Hep293TT cell line was provided by Dr. Gail Tomlinson (UTHSCSA). The SK-OV-3-MDR-1-6/6 is a single-cell clone we isolated from the SK-OV-3/MDR-1 cell line provided by Dr. Susan Kane (Division of Molecular Medicine, Beckman Research Institute of the City of Hope, Duarte, CA, USA) and cultured as previously described [18,19]. SK-ES-1, SH-SY-5Y, MDA-MB-453, MDA-MB-231, and SK-N-BE2 cells were cultured in Improved Minimum Essential Medium (Gibco/Thermo Fisher Scientific, Waltham, MA, USA) containing 25 µg/mL gentamicin (Gibco) and 10% fetal bovine serum (FBS; GE Healthcare, Little Chalfont, UK). TC32-NR0B1, TC32-CMV, RD-ES, A-673, D283, A204, SJCRH30, Hep293TT, HCC1806, BT-549, HCC1937, and Cal-51 cells were cultured in a RPMI-1640 medium (Sigma-Aldrich) containing 50 µg/mL gentamicin and 10% FBS. SK-OV-3 and SK-OV-3-M6/6 cell lines were cultured in Basal Medium Eagle (Sigma-Aldrich) containing 50 µg/mL gentamicin and 10% FBS. The cells were maintained in humidified incubators at 37 °C with 5% CO_2.

All cell lines were initially expanded and frozen as stocks in liquid nitrogen. The cells were passaged for less than 3 months after revival.

2.7. Sulforhodamine B Assay

The antiproliferative and cytotoxic activities of ATXII were evaluated using the sulforhodamine B (SRB) assay as previously described [20]. The cells were plated into 96-well plates at predetermined densities ranging from 2500 to 6500 cells/well, depending on the cell proliferation rate, and allowed to adhere overnight. The cells were treated with ATXII for 48 h prior to assessing antiproliferative and cytotoxic activities. The cell density at the time of treatment (T_0) was measured to evaluate cytotoxicity. Concentration–response curves were plotted, and the concentrations causing 50% inhibition of proliferation compared to vehicle control (GI_{50}), total growth inhibition (TGI), and 50% cell death compared to T_0 (LC_{50}) were interpolated from nonlinear regressions of the data (Prism 6; GraphPad Software, La Jolla, CA, USA).

2.8. Colony Formation Assays

A-673 ES cells were seeded at a density of 500 cells/60 mm tissue culture dish, allowed to adhere overnight, then treated with DMSO (0.5%) or the indicated concentrations of ATXII. After 4 h of treatment, the cells were washed with DPBS, and a fresh growth medium added. After 14 days of colony formation, the cells were fixed and stained with 0.5% crystal violet in 10% methanol. The colonies were counted using the GeneSnap software (PerkinElmer). The data were analyzed by one-way ANOVA with Tukey's post hoc test using Prism 6.

2.9. Whole-Cell Lysis and Immunoblotting

A-673, RD-ES, and SK-ES-1 cells were treated with DMSO (maximum 0.5%) or ATXII for the indicated time periods, harvested by scraping, and lysed in a cell extraction buffer (Thermo Fisher Scientific) containing a protease inhibitor cocktail, PMSF, and Na_3VO_4 (Sigma Aldrich). The total protein concentrations were measured with a Pierce Coomassie Plus assay kit (Life Technologies), and 10 µg of total protein was separated by SDS-PAGE on NuPAGE Bis-Tris gels (Life Technologies). The proteins were transferred to PVDF membranes (EMD Millipore, Billerica, MA, USA) and probed overnight with antibodies for Phospho-S345-Chk1, Chk1, Phospho-T68-Chk2, Chk2, Phospho-S15-p53, p53 (Cell Signaling Technology, Danvers, MA, USA), β-actin (Sigma), or FLI1 (Abcam, Cambridge, MA, USA; ab15289) diluted in an Odyssey blocking buffer in TBST (LI-COR Biosciences, Lincoln, NE, USA). The membranes were incubated with appropriate IRDye 680 or IRDye 800 secondary antibodies (LI-COR Biosciences) diluted in an Odyssey blocking buffer in TBST + 0.01% SDS. Near-infrared fluorescence signals were captured on an Odyssey FC (LI-COR Biosciences).

2.10. Flow Cytometry

The cell cycle distribution was evaluated by flow cytometry utilizing propidium iodide staining. A-673, RD-ES, and SK-ES-1 cells were treated with DMSO or the indicated concentrations of ATXII for 18 h. The cells were then harvested by scraping, washed in DPBS, and stained with Krishan's reagent [21]. The DNA content was measured using a Muse Cell Analyzer (EMD Millipore). The data were analyzed with FlowJo 10 (FlowJo LLC, Ashland, OR, USA).

2.11. RNA Sequencing

TC32 cells were treated in triplicate with 10 nM ATXII for the indicated amount of time, and total RNA was isolated using an RNeasy mini kit (Qiagen, Germantown, MD, USA). Approximately 500 ng of total RNA was used for the RNA-seq library preparation by following the Illumina TruSeq stranded mRNA sample preparation guide (Illumina, San Diego, CA, USA). RNA-seq libraries were subjected to quantification and a subsequent 50 bp single-read sequencing module with the Illumina HiSeq 3000 platform. After the sequencing run, demultiplexing with CASAVA (Illumina, San Diego, CA, USA) was employed to generate the FastQ file for each sample. An average of ~35 M reads were generated for each sample. All RNA-seq FastQ reads were aligned with the reference genome (UCSC human genome build hg19) using TopHat2 [22] default settings. The BAM files obtained after alignment were processed using HTSeq-count [23] to obtain the counts per gene in all samples. The R package DESeq [24] was used to normalize gene expression with the size factor method and to perform pairwise comparisons between groups to identify differentially expressed genes (DEGs). Genes with an FDR-adjusted p-value < 0.1 and at least a 2-fold change compared to control were considered significantly differentially expressed. Upon obtaining the differentially expressed genes from all pair-wise comparisons, we performed k-means clustering on the combined gene set using MATLAB (MathWorks). An additional quality control statistical analysis of outliers, intergroup variability, distribution levels, PCA, and hierarchical clustering analysis were performed to validate the experimental data. A pathway analysis was performed with Ingenuity Pathway Analysis (IPA, Qiagen). A gene set enrichment analysis (GSEA) was performed with the software package distributed by the Broad Institute [25].

2.12. High-Content Immunofluorescence Imaging

A-673 and SJCRH30 cells were plated into black 96-well cell carrier plates (PerkinElmer) at a density of 3500 cells/well and allowed to adhere overnight. The cells were treated in triplicate with DMSO or the indicated concentrations of ATXII for 6 or 24 h, then fixed with paraformaldehyde. After fixation, the cells were incubated in a blocking solution of 10% bovine calf serum in DPBS for 20 min at room temperature. The cells were then incubated in a primary antibody against γ-H2A.X (1:400; Cell Signaling Technology) diluted in 1% bovine serum albumin/0.3% Triton X-100 in DPBS, overnight, at 4 °C. The cells were subsequently washed with DPBS and incubated with an Alexafluor-594-conjugated secondary antibody (1:1000; Life Technologies) for 1 h at room temperature. The plates were washed with DPBS, and the nuclei stained with NucBlue live cell stain (Life Technologies) diluted in DPBS. Images were collected using an Operetta high-content imaging system (PerkinElmer) using a 20× long working distance objective and analyzed with the Columbus Image Data Storage and Analysis System (PerkinElmer). A minimum of three fields were collected per well, with all concentrations tested in triplicate.

2.13. Luciferase Reporter Assay

NR0B1 promoter activity was assessed using a luciferase reporter assay as previously described by Grohar et al. [6]. TC32-NR0B1 and TC32-CMV cells were plated in white, opaque-bottom 384-well plates (PerkinElmer) at a density of 5000 cells/well in 27 µL growth medium and allowed to adhere overnight. The cells were then treated at the indicated concentrations with ATXII diluted in 3 µL DMSO/growth medium for 6 h. Luciferase reporter activity was measured by adding 30 µL of steadylite plus a reporter gene assay reagent and measuring the luminescence on a Pherastar FS multimode plate reader (BMG Labtech). The results represent n = 3 independent experiments, with all concentrations tested in triplicate.

2.14. LLAMAS Assay

An aliquot (400 µL) of DNA solution (in the experimental group) or control buffer (in the control group) and 200 µL MeOH were mixed first prior to the addition of 10 µL alertoxin solution. The samples were passed through the ultrafiltration membrane (100 kDa cutoff) by centrifugation at 5000× g at 10 °C after 30 min incubation at room temperature. The resulting filtrates were collected for a LC-PDA-MS/MS analysis. In the dissociation step, the retained DNA-ligand complex in the upper chambers of the microcentrifuge tubes was washed with 30% MeOH in H_2O and subjected to centrifugation at 5000× g, 10 °C. After washing, the DNA-ligand complex was mixed with 600 µL of 95% MeOH in H_2O

with 1% formic acid in a new tube and then incubated with periodic vortexing at room temperature for 20 min. The solubilized contents were then transferred to microcentrifuge tubes outfitted with new ultrafiltration filters and centrifuged at 5000× *g* at 10 °C for 10 min. The filtrates were subjected to in vacuo solvent evaporation and suspended in 50 μL MeOH for the LC-MS/MS analysis.

2.15. BioLayer Interferometry

DNA binding events were detected and monitored in real time using a FortéBio Octet Red 96 Biolayer Interferometer (Molecular Devices, now Sartorius) [26]. DNA sequences d(5′-biotin-GATTTCAAGATATTAAGAAG-3′), d(5′-CTTCTTAATATCTTGAAATC-3′), d(5′-biotin-GTGCCTGGACCGCCCGACCT-3′), and d(5′-AGGTCGGGCGGTCCAGGCAC-3′) were purchased from Integrated DNA Technologies (IDT, Coralville, IA). Streptavidin biosensors and Kinetics Buffer (1× PBS, pH 7.4, 0.02% Tween-20, 0.1% albumin, and 0.05% sodium azide) were purchased directly from Molecular Devices (San Jose, CA, USA). Single-stranded DNA oligomers were annealed for 2 min at 94°C followed by cooling to room temperature over 1 h. Duplex DNA was then stored at −20 °C as a 20 μM solution in a TE buffer (10 mM Tris, 0.5 mM EDTA, 50 mM NaCl, pH 8) until use. Biotinylated, double-stranded DNA was immobilized on streptavidin (SA) sensor tips for 1600 s at 25 nM in a 1× Kinetics buffer. Compound testing was done sequentially at 125 μM in the 1× Kinetics buffer with 5% DMSO using baseline, association, and dissociation steps for 60, 1600, and 1600 s, respectively. Double reference subtraction was performed to eliminate the signal associated with atypical binding events by subtracting data obtained using a separate set of blank sensors with DNA load and a set of sensors without DNA load, both in buffer [27,28]. The data were aligned using baseline signals and the curves fitted with a 1:1 best-fit model in FortéBio's data analysis software.

2.16. Xenograft Studies in Nude Mice

Female athymic nude mice (Envigo, Indianapolis, IN) were injected s.c. with 2 × 10^6 A-673 cells and suspended in 100 μL DPBS and 100 μL Matrigel® (BD Biosciences, San Jose, CA, USA), bilaterally into each flank. Once the tumors reached a minimum volume of 150 mm^3, the mice were assigned to three different groups (n = 8 or 10 tumors/group). One group of mice received doses of 20 mg/kg on days 1, 3, 5, 8, 10, and 12, and another group of mice received doses of 40 mg/kg on days 1, 3, and 5. The third group consisted of untreated control animals. ATXII was administered by i.p. injection in a vehicle of 5% DMSO + 5% Kolliphor EL in DPBS. The mice were weighed and examined daily for signs of toxicity. The tumors were measured twice per week using calipers, and the tumor volume (mm^3) was calculated as length (mm) × width (mm) × height (mm). All mice were housed in an AAALAC-approved facility at UTHSCSA and given food and water ad libitum. All procedures were IACUC-approved.

3. Results

3.1. Bioassay-Guided Purification of Altertoxin II

Crude extracts and fractions prepared from fungi isolated from a combination of Great-Lakes-derived sediments [29] and the University of Oklahoma Citizen Science Soil Collection [30] were evaluated for selective cytotoxic activity in cell lines modeling five different types of pediatric solid tumors: A-673 (Ewing sarcoma), SJCRH30 (rhabdomyosarcoma), SK-N-BE(2)-C (neuroblastoma), D283 (medulloblastoma) and Hep293TT (hepatoblastoma). A fraction obtained from a soil-derived *Alternaria* sp. isolate SC5920 TV8-1 exhibited selective cytotoxic effects against A-673 ES cells as compared to the other cell lines. Bioassay-guided fractionation of the active fungal sample yielded the perylene quinone-type metabolite altertoxin II (Figure 1). Concurrent with the bioassay-guided fractionation, LC-MS analysis was performed on the biologically active fraction, alerting us to the presence of two co-eluting metabolites. Based on their estimated molecular weights and photodiode array data, the metabolites were suspected of being putative analogues of ATXII. The purification and subsequent structure analysis of those compounds led to their identification as altertoxin I (ATXI) and alteichin (Figure 1). The dereplication of ATXII, ATXI and alteichin was carried out by comparing experimentally-derived data to published accounts of their mass spectrometry data and spectroscopic (i.e., 1D and 2D NMR, ECD, and optical rotation) properties [31–35].

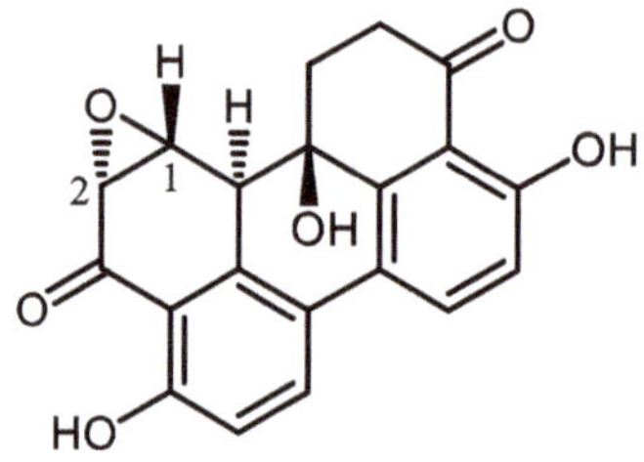

altertoxin II (ATXII)

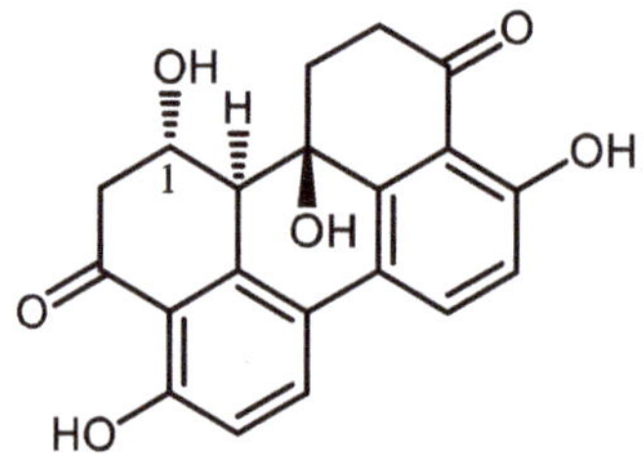

altertoxin I (ATXI)

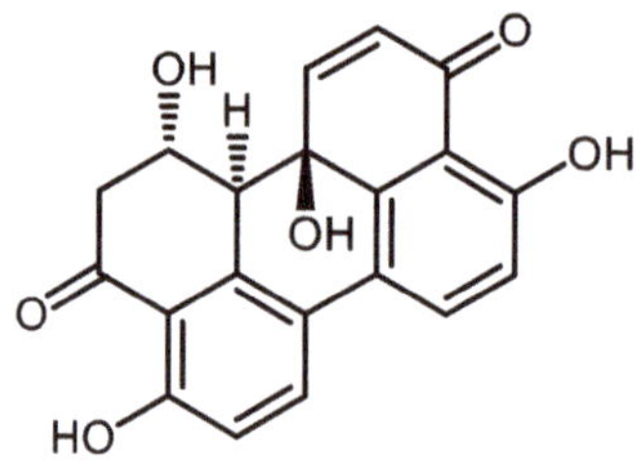

alteichin

Figure 1. Chemical structures of ATXII and its analogues ATXI and alteichin purified from *Alternaria* sp. SC5920 TV8-1.

ATXII, ATXI and alteichin were further evaluated in eight cell lines from our pediatric solid tumor panel representing ES (RD-ES, SK-ES-1, A-673) and non-ES (D283, A204, SK-N-BE(2)-C, SH-SY-5Y, and SJCRH30) cell lines. These cell lines were chosen because they represent four different types of pediatric solid tumors and demonstrated differential sensitivity to the fraction from which ATXII was isolated. The potent, ES-specific effects of ATXII were recognized to be a distinguishing feature of this metabolite, whereas ATXI and alteichin did not exhibit selective cytotoxic activity in ES cells compared to other pediatric cancer cell lines (Figure 2A–C). Considering that the only structural difference between the compounds is the presence of a 1,2-oxirane system versus a C-2 hydroxy group in ATXII and ATXI, respectively, we speculate that the epoxide group is critical for the ES-selective activity of ATXII.

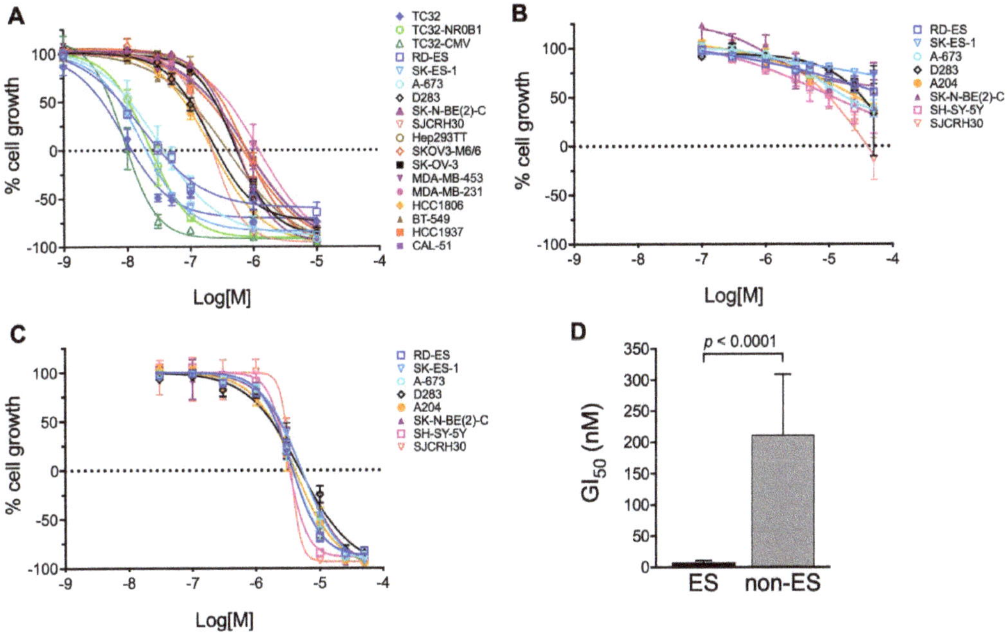

Figure 2. ATXII, but not altertoxin I or alteichin, has selective activity against Ewing sarcoma cells in vitro. (**A**) Sulforhodamine B (SRB) concentration-response curves for the inhibition of cell growth by ATXII, (**B**) altertoxin I and (**C**) alteichin. Results represent mean ± SE; $n \geq 3$ independent experiments with each concentration tested in triplicate. (**D**) Comparison of GI_{50} (concentration resulting in 50% inhibition of cell growth relative to vehicle-treated control) for ATXII in EWS and non-EWS pediatric cancer cell lines. Results represent mean ± SD. Groups compared by two-tailed *t*-test.

ATXII was next evaluated for antiproliferative and cytotoxic activities in a larger panel of six ES cell lines and 12 non-ES cell lines (four different pediatric solid tumor cell lines; two ovarian cancer cell lines, including SK-OV-3-MDR-1-6/6 cells that expresses P-glycoprotein; and six triple-negative breast cancer cell lines). ATXII showed a highly selective antiproliferative and cytotoxic effects in all of the ES cell lines with lower potency in each of the non-ES cells (Figure 2A; Table 1). The average concentration that caused a 50% inhibition of cell proliferation (GI_{50}) was significantly different ($p < 0.0001$) between ES and non-ES cells (Figure 2D). The mean GI_{50} for ATXII was 8 ± 3 nM in the six ES cell lines and 200 ± 100 nM in non-ES pediatric cancer and adult cancer cell lines, indicating an average 25-fold selectivity for ES cells compared to non-ES cancer cells. In contrast, the GI_{50} values for ATXI and alteichin were 1.9 and 1.1-fold higher in ES cells compared to non-ES cells, respectively, indicating no selectivity for ES versus non-ES cells (Figure 2B,C). Similarly, the average concentration of ATXII that caused total growth inhibition (TGI) was significantly different ($p < 0.0001$) for ES and non-ES cells (Table 1). The mean TGI for ATXII was 20 ± 10 nM in the six ES cell lines and 600 ± 300 nM in non-ES cell lines, indicating an average of 30-fold selectivity for ES cells. The concentration of ATXII resulting in 50% cell death (LC_{50}) was also significantly different ($p = 0.0001$) between ES and non-ES cells (Table 1), with a mean LC_{50} of 100 ± 100 nM in the six ES cell lines and 2000 ± 1000 nM in non-ES cell lines, demonstrating that ATXII has, on average, 20-fold cytotoxic selectivity for ES cells. ATXI and alteichin were not evaluated in this larger panel of cell lines because these compounds did not show any selectivity for ATXII when evaluated in the smaller panel of eight cell lines. The activity of ATXII was further assessed in A-673 ES cells by measuring its ability to inhibit colony formation. ATXII potently inhibited the colony formation of A-673 cells after a short, 4-h treatment followed by drug washout (Figure 3A,B). These data indicate that the effects of ATXII are highly persistent because these relatively short treatments with concentrations as low as 10 nM were sufficient to significantly inhibit the A-673 colony formation.

Table 1. Sulforhodamine B assay potency measurements for ATXII. ES, Ewing sarcoma; RMS, rhabdomyosarcoma; Med, medulloblastoma; NB, neuroblastoma; HB, hepatoblastoma; OV, ovarian, BR, breast.

Cell Line	Type	GI_{50} (nM)	TGI (nM)	LC_{50} (nM)
RD-ES	ES	7.8	32	380
SK-ES-1	ES	7.6	20	58
A-673	ES	11	37	150
TC32	ES	4.0	12	48
TC32-NR0B1	ES	10	23	57
TC32-CMV	ES	5.0	10.0	20
SJCRH30	RMS	120	220	430
D283	Med	100	260	880
SK-N-BE(2)-C	NB	270	570	1200
Hep293TT	HB	240	540	1300
SK-OV-3	OV	250	550	1300
SK-OV-3-MDR1-6/6	OV	320	770	1900
MDA-MB-453	BR	230	790	2700
MDA-MB-231	BR	400	1200	3700
HCC1806	BR	80	220	650
BT-549	BR	100	380	1500
HCC1937	BR	200	680	2400
CAL-51	BR	230	810	2800

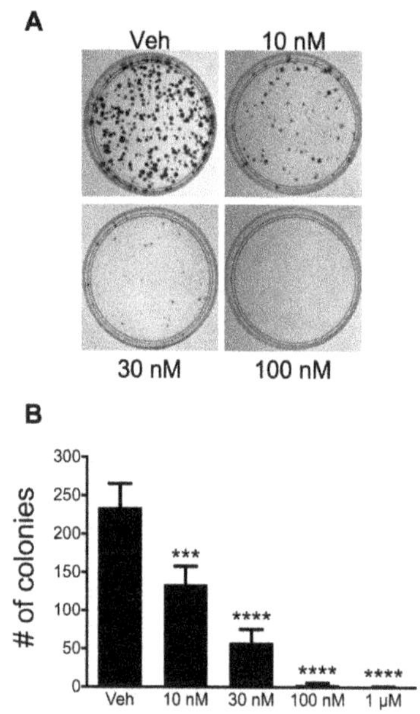

Figure 3. Altertoxin II has persistent activity in ES cells. (**A**) Inhibition of colony formation by ATXII after drug washout. A-673 cells were treated with ATXII for 4 h, the drug was washed out, and the cells were allowed to form colonies for 14 days. (**B**) Quantification of colony number after treatment with ATXII at the indicated concentration for 4 h followed by drug washout. $n = 3$; *** $p \leq 0.001$, **** $p \leq 0.0001$ compared to vehicle; one-way ANOVA with Tukey's post-hoc test.

3.2. ATXII Does Not Inhibit EWS-FLI1 Protein Expression or Transcriptional Activity

Due to the high degree of selectivity for ES cells, we assessed whether ATXII affects the abundance or transcriptional activity of EWS-FLI1, the oncogenic fusion protein that is the primary driver of ES. A relatively short time point (8 h) after treatment was evaluated to interrogate the early effects of ATXII on ES cells. The treatment of SK-ES-1 and RD-ES cells with 100 nM ATXII did not alter the EWS-FLI1 protein abundance, as assessed by immunoblotting (Figure 4A). Similarly, the treatment of A-673 cells with 10 nM–1 μM ATXII for 24 h did not affect the protein levels of EWS-FLI1 (Figure 4B). Together, these data indicate that ATXII does not alter the EWS-FLI1 protein abundance in ES cells and suggest that the depletion of EWS-FLI1 is not responsible for the selective cytotoxic effects of ATXII on ES cells.

EWS-FLI1 acts as a transcription factor to globally modulate gene expression and drive an oncogenic phenotype in ES [1,36–38]. To determine if ATXII affects the transcriptional activity of EWS-FLI1, we evaluated the effects of ATXII on the promoter activity of NR0B1, a major downstream target of EWS-FLI1, using a luciferase reporter assay. TC32 ES cells stably expressing luciferase reporters under the control of either the *NR0B1* or cytomegalovirus (CMV) promoters were treated with ATXII for 6 h at concentrations ranging from 1 nM to 1 μM before evaluating the promoter activity (Figure 4C). The 1 and 10 nM concentrations of ATXII, the latter of which is sufficient to inhibit ES growth and colony formation (Figures 2A and 3), had no effects on either the CMV or the *NR0B1* promoter activity as compared to vehicle-treated controls (Figure 4C). A small but statistically significant decrease in the *NR0B1* promoter activity was observed after treatment with higher concentrations, 100 nM or 1 μM ATXII, which was also accompanied by a decrease in the CMV promoter activity (Figure 4C). These data suggest that the decreased EWS-FLI transcriptional activity is not a major driver of ES growth inhibition and that the effects of higher concentrations of ATXII on the *NR0B1* promoter are more likely due to a generalized downregulation of transcription.

3.3. ATXII Activates DNA Damage Response Pathways and Induces Double-Strand DNA Breaks in ES Cells

ATXII has been isolated from other fungi of the *Alternaria* genus, which are plant pathogens that cause the spoilage of food products, including grains and fruit [33,39,40]. Early studies showed that ATXII causes DNA damage in mammalian cells at concentrations of 250–750 nM, significantly higher than those where we observe ES-selective cytotoxicity [40,41]. Triggering such an effect would be highly relevant because previous studies have shown that ES cells are very sensitive to DNA damaging agents and exhibit high levels of DNA replication stress [42]. To determine if the induction of DNA damage is involved in the mechanism of action of ATXII in ES cells, the phosphorylation of checkpoint kinases 1 and 2 (Chk1 and Chk2) and p53 in A-673 and RD-ES ES cells was measured. The phosphorylation of Chk1 at S345, Chk2 at T68, and p53 at S15 is indicative of a cellular DNA damage response. The phosphorylation status of Chk1, Chk2, and p53 was assessed in ES cells after treatment with concentrations of ATXII ranging from 10 to 300 nM for 6 h to capture the acute effects of ATXII rather than secondary effects due to cell death (Figure 5A). In A-673 ES cells, the phosphorylation of Chk1 and Chk2 was observed at 6 h with 10 nM ATXII with a maximum phosphorylation of Chk1 occurring with 50 nM. The phosphorylation of Chk1 and Chk2 was also observed in the RD-ES cells, with the maximal phosphorylation obtained at 6 h with 50 nM ATXII (Figure 5A). An increased Chk2 phosphorylation was observed with concentrations as low as 10 nM. Interestingly, higher concentrations of ATXII (100 and 300 nM) resulted in lower levels of total Chk2 protein in RD-ES cells. A robust phosphorylation of p53 at S15 in RD-ES cells was observed at all the concentrations tested. The total levels of p53 decreased after treatment with 100 and 300 nM, although these may be secondary effects due to cell death at these higher concentrations. We did not detect total or P-S15-p53 in A-673 cells, which is consistent with previous studies showing that this cell line is p53-null [43].

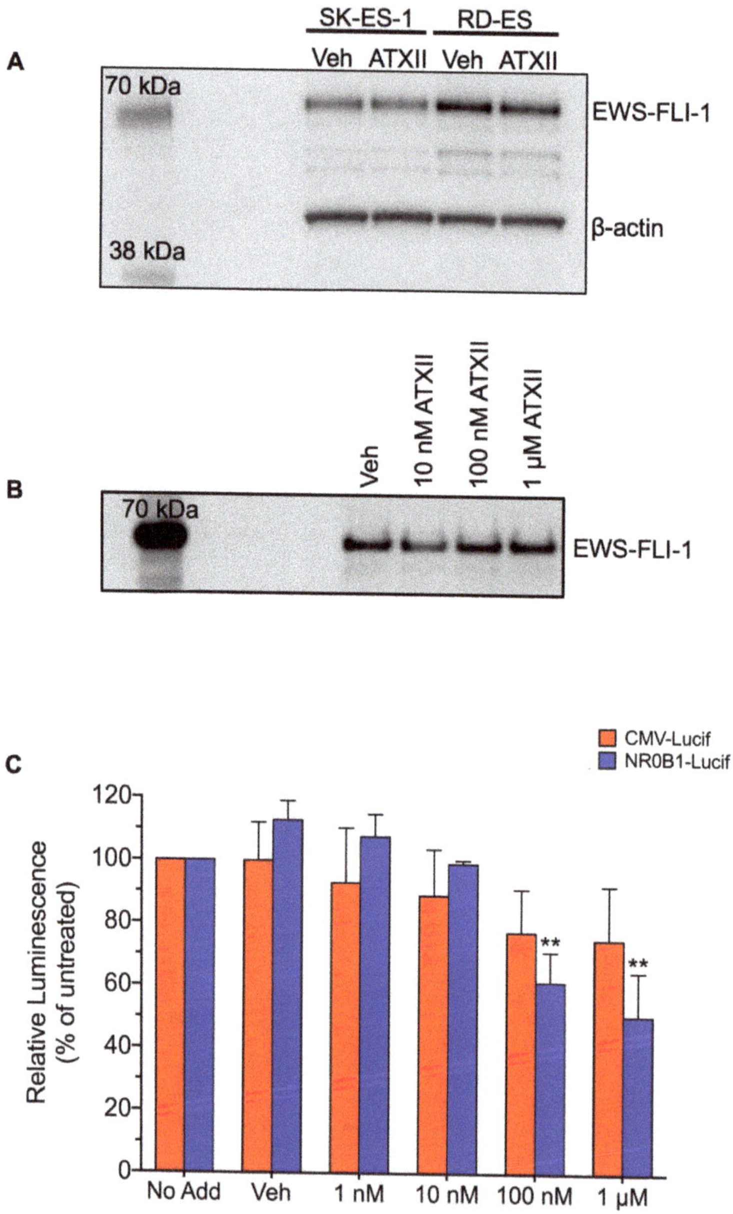

Figure 4. Effects of ATXII on EWS-FLI1 protein levels and transcriptional activity. (**A**) Immunoblotting for EWS-FLI1 (anti-FLI1) and β-actin in SK-ES-1 and RD-ES lysates following treatment with 100 nM ATXII for 8 h. (**B**) Immunoblotting for EWS-FLI1 (anti-FLI1) in A-673 lysates following treatment with a range of concentrations for 24 h. (**C**) Effects of ATXII on the promoter activity of the EWS-FLI1 target gene NR0B1. The cells were treated for 6 h with the indicated concentration of ATXII, and the activity was measured by a luciferase reporter assay. $n = 3$ independent experiments, with all concentrations tested in triplicate. ** $p \leq 0.01$ compared to vehicle; one-way ANOVA with Tukey's post-hoc test.

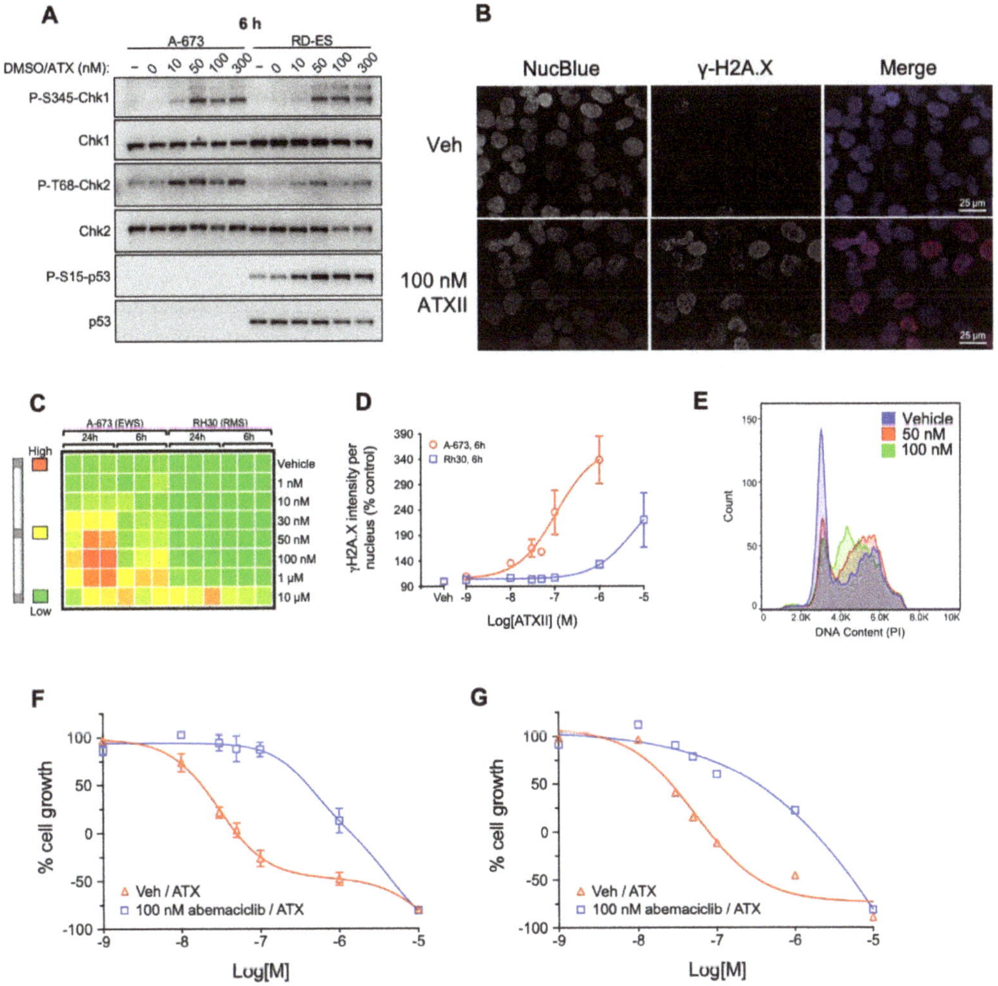

Figure 5. ATXII selectively induces DNA double-strand breaks, inhibits DNA synthesis, and is optimally potent in proliferating cells. (**A**) Immunoblotting for P-S345-Chk1, P-T68-Chk2, and P-S15-p53 in ES cell lines A-673 and RD-ES after a 6-h treatment with ATXII. (**B**) Indirect immunofluorescence microscopy of γ-H2A.X in A-673 cells after an 18-h treatment with ATXII. (**C**) Representative heatmap and (**D**) concentration-response curves for ATXII-induced γ-H2A.X accumulation in A-673 and Rh30 cells. The cells were treated in triplicate for 6 or 24 h with increasing concentrations of ATXII, and γ-H2A.X was measured by automated immunofluorescence imaging. Results represent mean ± SE; $n = 2$. (**E**) Analysis of cell cycle distribution by flow cytometry. A-673 cells were treated for 24 h with indicated concentrations of ATXII, and the DNA content was determined by PI staining of permeabilized cells. (**F**) Concentration-response curves for ATXII in RD-ES and (**G**) A-673 cells after pretreatment with the CDK4/6 inhibitor abemaciclib for 24 h. Results represent mean ± SE for $n = 3$ (RD-ES) or $n = 1$ (A-673) independent experiment(s) with each concentration tested in triplicate.

ATXII and related compounds have been evaluated using the Ames assay, where they were shown to be mutagenic in *Salmonella typhimurium* strains [39]. Our results show that ATXII causes DNA damage in ES cells at concentrations much lower than those previously shown to induce general toxicity in cancer cells [40,41]. Consistently with these activities, other DNA-damaging drugs used to treat children with cancer were also positive in the Ames assay [44]. We evaluated the ES-selectivity of a panel of other DNA damage-inducing agents. including gemcitabine, etoposide, SN38 (active metabolite of irinotecan), melphalan, and the PARP1 inhibitor olaparib, but none of these agents show the degree of selectivity for ES cells that we see with ATXII (Supplementary Figure S1). These results suggest that ATXII induces DNA damage through a unique mechanism that cannot be repaired by ES cells. ES cells are understood to be highly sensitive to genotoxic stress, and they express low levels of key DNA repair genes, including *ATM* and *BRCA1* [45]. However, our results demonstrate that ATXII has a much greater selectivity for ES cells in vitro compared to other clinically relevant DNA-damaging agents.

EWS-FLI1 initiates DNA damage and transcriptional stress as measured by high levels of H2A.X phosphorylation at S139 (γH2A.X) and a slow replication fork progression [46]. Given the effects of the fusion protein and our finding that ATXII activates DNA damage response pathways in ES cells, studies were conducted to evaluate if ATXII induces the phosphorylation of H2A.X at S-139 (γ-H2A.X), a marker of DNA double-strand breaks. The ability of 100 nM ATXII to cause H2A.X phosphorylation was evaluated in A-673 cells following an 18-h treatment. The cells were stained with a DNA marker (NucBlue) and for γ-H2A.X with a phospho-specific antibody. The results show a robust increase in γ-H2A.X in the nuclei of ATXII-treated ES cells (Figure 5B). High-content immunofluorescence microscopy was performed in A-673 ES and SJCRH30 rhabdomyosarcoma (RMS) cells treated with 1 nM–10 μM ATXII for 6 or 24 h and the γ-H2A.X intensity per nuclei quantified for each condition. A heat map presentation of the results shows that γ-H2A.X was detected in A-673 ES cells at concentrations as low as 100 nM after 6 h and 30 nM after 24 h of treatment with ATXII (Figure 5C). In contrast, γ-H2A.X phosphorylation could only be detected in SJCRH30 RMS cells after treatment with 10 μM ATXII for 6 or 24 h. After 6 h of treatment, the EC_{50} (half maximal effective concentration) for ATXII-induced γ-H2A.X accumulation was 111 nM in A-673 cells and 5.3 μM in SJCRH30 cells, while after 24 h of treatment the EC_{50} was 67 nM in A-673 cells and 8 μM in SJCRH30 cells (Figure 5D). Interestingly, the concentration-response curves for ATXII-induced DNA damage (Figure 5D) and ATXII cytotoxicity (Figure 2A) in A-673 cells show that DNA damage and cell death occur over the same concentration range, with maximal effects achieved at 1 μM, suggesting that DNA damage is the primary driver of ATXII-induced cell death in ES cells.

A common consequence of DNA damage is cell cycle accumulation, either in the late G_1- or S-phase, and thus the effects of ATXII on cell cycle distribution were evaluated in A-673 cells. The cells were treated with 50 or 100 nM ATXII for 24 h, resulting in a concentration-dependent accumulation of cells in the S-phase of the cell cycle (Figure 5E), suggesting that the ATXII-initiated DNA damage inhibits DNA synthesis, leading to a cell cycle checkpoint response and cell cycle arrest. Similar effects on cell cycle distribution were also observed in SK-ES-1 ES cells (Supplementary Figure S2).

Based on these results, we evaluated whether the potent cytotoxic effects of ATXII in ES cells require cell proliferation. RD-ES and A-673 cells were treated with the CDK4/6 inhibitor abemaciclib-mesylate, which caused the accumulation of cells in G_1 at concentrations as low as 100 nM. Abemaciclib-mesylate did not initiate cytotoxicity at concentrations as high as 1 μM (Supplementary Figure S3). RD-ES and A-673 cells were first arrested in G_1 by treatment with abemaciclib-mesylate for 24 h at 100 nM, a concentration that modestly inhibited growth but did not promote cytotoxicity on its own, followed by treatment with ATXII for 48 h. A pretreatment of RD-ES and A-673 cells with 100 nM abemaciclib-mesylate resulted in a rightward shift in the ATXII concentration-response curves (Figure 5F,G), increasing the GI_{50} 22.3-fold and 12.4-fold in RD-ES and A-673 cells, respectively, which were essentially equal to the potency in some non-ES cell lines. These data demonstrate that ES cells must be actively proliferating for ATXII to have maximal potency and further support the hypothesis that the potent cytotoxic effects of ATXII in ES cells are due to its ability to cause a selective induction of DNA damage in these cells at low concentrations.

To further probe the mechanisms of action of ATXII in ES cells, RNA sequencing (RNA-seq) was performed to assess changes in global gene expression in TC32 ES cells after 4, 8, and 12 h of treatment with 10 nM ATXII. TC32 cells were utilized for these studies in order to determine if ATXII has similar effects in multiple ES cell lines. We identified 43 genes that exhibit significant differences in expression (fold change > 2; adjusted p-value < 0.1) in ATXII-treated cells as compared to vehicle-treated cells (Supplementary Figure S4). Gene set enrichment analysis (GSEA) indicated significantly enriched gene sets related to DNA damage, including those that occur in response to ionizing radiation and cisplatin, as early as 4 h after ATXII treatment (Supplementary Figure S5). Collectively, these data strongly suggest that ATXII induces DNA damage in ES cells at concentrations as low as 10 nM. Interestingly, GSEA and gene ontology (GO) also indicate an enrichment in pathways related to type I interferon signaling and inflammatory response (Supplementary Figure S4), suggesting that these pathways might also play a role in the cytotoxic effects of ATXII. Induction of type I interferons can occur as a consequence of DNA damage due to ionizing radiation [47], and so the activation of these pathways is consistent with the induction of DNA damage by ATXII.

3.4. *ATXII Does Not Directly Bind to DNA*

Based on our observations that ATXII induces DNA damage and activates DNA damage response pathways, we sought to determine if these effects were the result of the direct binding of ATXII to DNA. We recently developed a technique for identifying DNA-binding molecules in complex mixtures called lickety-split ligand-affinity-based molecular angling system (LLAMAS) [48]. Using this assay, no binding of ATXII to purified DNA was observed, suggesting that ATXII-induced DNA damage is not due to direct DNA binding (Figure 6A,B). To confirm these results, we performed biolayer interferometry assays to assess the binding of ATXII to GC and AT-rich DNA sequences. Whereas cisplatin showed detectable binding to DNA, we did not observe any binding of ATXII to the streptavidin-bound double-stranded DNA (Figure 6C). Collectively, these results indicate that ATXII does not directly bind to DNA and suggest that the ATXII-induced DNA damage is likely a consequence of the inhibition of a protein target, potentially one involved in DNA synthesis or damage repair.

3.5. *In Vivo Antitumor Efficacy of ATXII*

Our studies show that ATXII exhibits a highly selective cytotoxic activity against ES cells in vitro. To determine if ATXII has in vivo antitumor efficacy, we evaluated the effects of ATXII in an A-673 ES murine xenograft model. Mice bearing subcutaneous A-673 xenograft tumors of ~250 mm^3 were treated with 20 mg/kg ATXII on days 1, 3, 5, 8, 10, and 12 or with 40 mg/kg ATXII on days 1, 3, and 5 for a total dose of 120 mg/kg in both cohorts. We observed a modest inhibition of tumor growth and minimal weight loss in the mice treated with the 20 mg/kg ATXII dose/schedule over the course of the 17-day trial (Figure 7A,B; Supplementary Figure S6). On day 17, the tumor volume was significantly smaller ($p < 0.05$) in this group compared to the control mice (Figure 7B). A less frequent dosing with 40 mg/kg ATXII on days 1, 3, and 5 resulted in a greater inhibition of tumor growth over the 17-day trial (Figure 7A,B). On day 17, tumors in mice in the 40 mg/kg ATXII group were significantly smaller than control tumors ($p < 0.0001$; Figure 7B). This dose and schedule caused more weight loss, although the mice recovered by the end of the trial (Supplementary Figure S6). Interestingly, the 40 mg/kg dose of ATXII produced long-lasting antitumor effects after the final dose was administered on day 5 of the trial, indicating a highly persistent antitumor activity. These results demonstrate that ATXII has antitumor efficacy against ES xenografts with an acceptable therapeutic window, and that ATXII is a possible lead molecule for the development of ES-specific therapies.

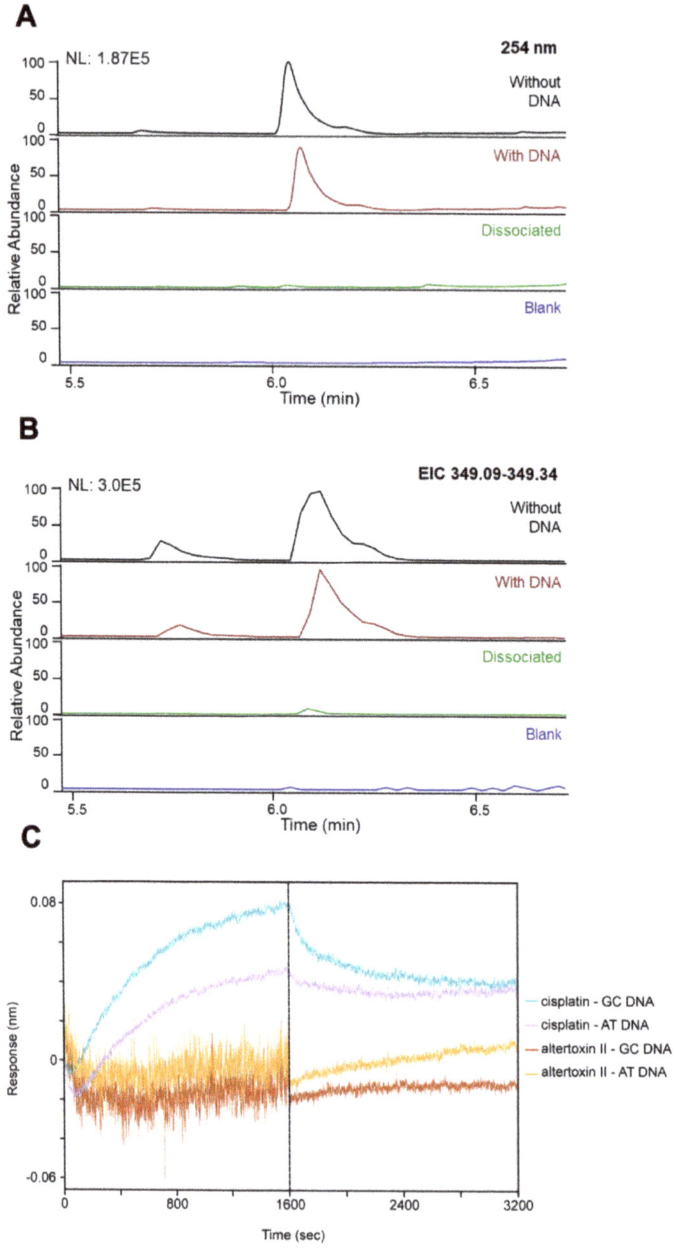

Figure 6. Altertoxin II does not directly bind to DNA. (**A**,**B**) DNA binding assay results of ATXII using LLAMAS. The UV chromatograms (λ 254 nm) and the extracted ion chromatogram (EIC) analysis revealed that ATXII does not show a recognizable DNA binding activity. (**C**) BioLayer interferometry binding sensorgrams for ATXII and cisplatin against GC- and AT-rich DNA (association step 0–1600 s and buffer dissociation step 1600–3200 s). Compounds evaluated at 125 μM against biotinylated, double-stranded DNA immobilized on streptavidin.

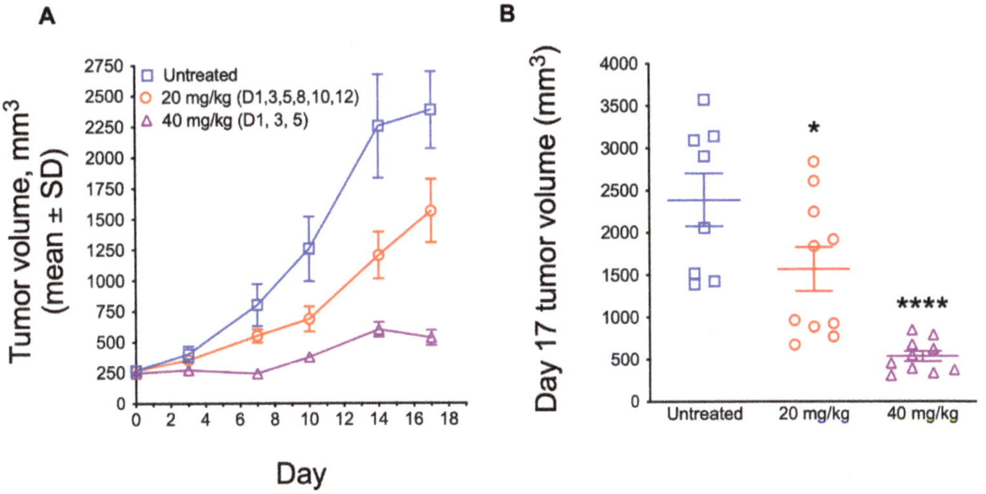

Figure 7. ATXII shows dose-dependent antitumor efficacy against an A-673 xenograft model. (**A**) Tumor growth curves for untreated control and ATXII-treated mice. The mice were injected i.p. with 20 mg/kg ATXII on days 1, 3, 5, 8, 10, and 12 or 40 mg/kg on days 1, 3, and 5. n = 8–10 tumors per group. (**B**) Comparison of final tumor volumes on day 17. * $p \leq 0.05$, **** $p \leq 0.0001$ compared to untreated control; one-way ANOVA with Tukey's post-hoc test.

4. Discussion

Unlike most adult cancers, that are caused by a lifetime of accumulated genetic changes, ES is caused by the expression of an aberrant transcription factor, EWS-FLI1, initiated most commonly by a t(11;22) (q24;q12) chromosomal translocation. Current therapies are effective for most children but are accompanied with long-term treatment-inducted morbidities [49,50]. New therapies that can improve survivors' quality of life and the efficacy for patients with metastatic or recurrent disease are needed. The potential to target the unique vulnerabilities of ES caused by fusion protein expression suggest that targeted therapies for ES can be identified. Significant challenges remain, however, because EWS-FLI1 is a highly disordered transcription factor, making direct targeting unlikely. Additionally, the chimeric fusion protein causes both transcriptional activation and repression, which will be difficult to differentially regulate using a single therapeutic strategy to inhibit EWS-FLI1 transcription [2,51]. While ES is characterized as having a "quiet" mutational landscape, recent studies on the genetic vulnerabilities of ES cell lines show that they have complex dependencies, rivaling those of adult cancers [52]. Interestingly, these genetic dependences are different from the common vulnerabilities seen in adult cancers, highlighting the need to discover ES-specific therapies.

Natural products have a long history of use as pharmaceuticals, and several natural products, including mithramycin, englarin A, and trabectedin were identified in a screen for inhibitors of EWS-FLI1 transcription [5–7]. The success of this screen suggests that the investigation of additional natural sources could be fruitful. We embarked on a discovery project using a high-throughput phenotypic screen of natural product extracts and fractions that is based on identifying mixtures containing constituents with selective cytotoxic activities against ES cells. A mechanism-blind phenotypic screen has the advantage of being unbiased towards the target and provides the opportunity to identify new unanticipated targets. The initial detection of an active fungal fraction, followed by bioassay-guided fractionation, identified ATXII as a highly selective cytotoxin in ES models. Our results confirm the continuing value of screening natural product libraries for compounds with selective activities against ES.

The advantage of a differential sensitivity screen is that compounds with high selectivity can be identified from complex mixtures. This was held to be true in that ATXII has exquisite potency against every ES cell line that we tested as compared to all other childhood and adult solid tumor cell lines. In our decades of experience, this highly consistent degree of ES selectivity, greater than 20-fold, is noteworthy. While the cytotoxic effects of ATXII in cancer cell lines have been previously described, the potencies reported are in the range of the non-ES cell lines in our study, reinforcing that ES lines are uniquely sensitive to ATXII [40,41]. Mutagenic effects of ATXII in the Ames assay have been reported, but the mutagenic effects in mammalian cells also occur at concentrations higher than those that inhibit the proliferation and promote cytotoxicity in ES cell lines [39–41]. There is good reason to be cautious about the use of a mutagen in children and young adults, but it should be noted that the drugs used for the treatment of ES, including doxorubicin, cyclophosphamide, and melphalan [2], are also mutagenic in the Ames assay [44].

The structure–activity relationships among ATXII, ATXI, and alteichin demonstrate the critical importance of the epoxide moiety in the potency and ES selectivity of ATXII. Considering the bias that epoxide groups are indiscriminately reactive, we were nevertheless surprised when our studies showed that ATXII does not bind directly to DNA. Yet while ATXII does not bind to DNA, it causes a rapid DNA damage response at concentrations as low as 10 nM, specifically in ES cells. Interestingly, previous studies have demonstrated that ES cells exhibit high levels of DNA replication stress and are particularly susceptible to DNA damaging agents [42]. Our results are consistent with these findings. However, we demonstrate that many clinically relevant agents that induce DNA damage, including topoisomerase inhibitors and the poly (ADP-ribose) polymerase (PARP) inhibitor olaparib, do not show the same degree of selectivity for ES that we observe with ATXII, suggesting that ATXII induces a unique type of damage that is particularly difficult for ES cells to repair. Our findings also indicate that the cytotoxic effects of ATXII are partially dependent on cellular proliferation, suggesting the ability of ATXII to interfere with DNA replication.

The high levels of DNA replication stress are a defining feature of ES, although the direct molecular causes of this replication stress are not clear. This phenotype of elevated replication stress induces genomic instability and is thought to confer sensitivity to DNA damaging agents. Replication stress can be caused by a variety of factors, including the deregulation of nucleotide synthesis, the inhibition of DNA polymerase/helicase, and the accumulation of DNA–RNA hybrids(R-loops), among others [53,54]. Indeed, ES shows elevated levels of R-loops compared to other cancer types, and it is thought that these R-loops induce genomic instability and a sensitivity to DNA damaging agents [42]. Our results suggest that the high sensitivity of ES cells to ATXII results from an undefined yet selective induction of DNA damage. However, our data additionally demonstrate that ATXII does not directly bind to DNA, suggesting that it induces DNA damage through an indirect mechanism. ATXII might inhibit DNA repair through the direct inhibition of a protein involved in the cellular DNA damage response or inhibition of DNA replication, leading to an accumulation of stalled replication forks. Interestingly, we observed the accumulation of ES cells in the S phase after treatment with low concentrations of ATXII, and the effects of ATXII are at least partially dependent on cellular proliferation. These data are consistent with ATXII inhibiting DNA replication, which could increase the replication stress and lead to an accumulation of DNA damage. In the future, we plan to conduct additional binding studies with radiolabeled ATXII, which will be used to identify its direct molecular target(s). Additionally, a CRISPR-Cas9-mediated gene knockout screen will be employed to further define the mechanism of ATXII-induced DNA damage in ES cells [55]. Lastly, the development of an ATXII-resistant ES cell line, followed by whole-exome and RNA sequencing, would be useful in identifying both genetic and epigenetic mechanisms of ATXII resistance, and may provide further insight into its mechanisms of action.

We demonstrate that altertoxin II has in vivo antitumor efficacy against an A-673 xenograft model of ES. It is likely that an increased antitumor efficacy will be observed by combining ATXII with other agents that act through a different mechanism of action.

Specifically, ES is known to exhibit a "BRCAness" phenotype with deficiencies in homologous recombination [42], making these tumors particularly sensitive to PARP inhibitors. It is possible that ATXII will show increased efficacy in the combination with PARP inhibitors, or that a lower dose of ATXII may be utilized to minimize the general toxicity. Our future studies will be aimed at identifying optimal drug combinations with ATXII both in vitro and in vivo to treat ES tumors.

5. Conclusions

We identified the fungal metabolite ATXII as a compound with highly selective cytotoxic activity against ES cells. Mechanism of action studies indicate that ATXII selectively induces DNA double-strand breaks and S-phase accumulation but does not directly bind to DNA. The high degree of selectivity for ES cells suggests that ATXII acts through a unique mechanism of action compared to other clinically relevant DNA-damaging agents. Overall, the efficacy of ATXII in an ES xenograft model, combined with its unique mechanistic effects, demonstrates that ATXII will be a valuable chemical probe for identifying ES-specific vulnerabilities and new drug targets for ES and as a potential therapeutic lead for this pediatric cancer.

Supplementary Materials: The following are available online at https://www.mdpi.com/article/10.3390/cancers13246176/s1, Figure S1: Clinically used DNA-damaging agents do not show selectivity for Ewing sarcoma cells, Figure S2: Analysis of cell cycle distribution in SK-ES-1 cells by flow cytometry, Figure S3: Analysis of abemaciclib-induced cytotoxicity and cell cycle arrest in RD-ES cells, Figure S4: Heatmap of differentially expressed genes in TC32 cells following ATXII treatment, Figure S5: Gene set enrichment analysis of TC32 cells treated with ATXII, Figure S6: Mean percent change in mouse weights over the course of the trial.

Author Contributions: Conceptualization, A.J.R., A.L.R., S.L., P.J.H., R.H.C. and S.L.M.; methodology, A.J.R., A.L.R., S.L., P.J.H., R.H.C. and S.L.M.; formal Analysis, A.J.R., H.M., V.M.A., A.L.R., S.L., R.H.C. and S.L.M.; investigation, A.J.R., W.D., S.H., H.M., V.M.A. and R.D.O.; resources, S.L.M., R.H.C., S.L. and P.J.H.; writing—original draft preparation, A.J.R., A.L.R., S.L., R.H.C. and S.L.M.; writing—review & editing, A.J.R., H.M., V.M.A., R.D.O., A.L.R., S.L., R.H.C., P.J.H. and S.L.M.; Visualization, A.J.R., V.M.A. and R.D.O.; supervision, S.L.M., R.H.C. and S.L.; project administration, S.L.M., R.H.C. and S.L.; funding acquisition, S.L.M. and R.H.C. All authors have read and agreed to the published version of the manuscript.

Funding: This research was funded by the National Institutes of Health, grant number R01GM107490, to R.H.C. and S.L.M., and the Greehey Endowment, to S.L.M. A.J.R. was supported by the Cancer Prevention and Research Institute of the Texas funded UTHSCSA Cancer Biology Training Program, grant number RP170345, and the National Cancer Institute T32 training grant program, grant number T32CA148724. The LC-MS instrument used for these studies was provided in part by a Challenge Grant from the Office of the Vice President for Research, University of Oklahoma, Norman Campus, and an award through the Shimadzu Equipment Grant Program to R.H.C. The data were generated in the Genome Sequencing Facility, which is supported by the National Institutes of Health—National Cancer Institute, grant number P30 CA054174 (Mays Cancer Center at UTHSCSA), National Institutes of Health Shared Instrument grant, grant number 1S10OD021805-01 (S10 grant), and the Cancer Prevention and Research Institute of Texas' Core Facility Award, grant number RP160732.

Institutional Review Board Statement: The study was conducted according to the guidelines of the Declaration of Helsinki and approved by the Institutional Animal Care and Use Committee of the University of Texas' Health Science Center at San Antonio (protocol #20120018AR, approved on 20 January 2017).

Informed Consent Statement: Not applicable.

Data Availability Statement: RNA sequencing data will be made publicly available at the Gene Expression Omnibus (GEO) of the NCBI. All other data are available upon request.

Acknowledgments: We extend our special thanks to Zhao Lai and Yidong Chen of the Computational Biology and Bioinformatics Initiative of the Greehey Children's Cancer Research Institute for the RNA-seq data and analysis.

Conflicts of Interest: A.J.R., A.L.R., P.J.H., R.H.C. and S.L.M. are inventors on a patent describing altertoxin II as a selective inhibitor of Ewing sarcoma cells.

References

1. Delattre, O.; Zucman, J.; Plougastel, B.; Desmaze, C.; Melot, T.; Peter, M.; Kovar, H.; Joubert, I.; de Jong, P.; Rouleau, G.; et al. Gene fusion with an ETS DNA-binding domain caused by chromosome translocation in human tumours. *Nature* **1992**, *359*, 162–165. [CrossRef]
2. Grünewald, T.G.P.; Cidre-Aranaz, F.; Surdez, D.; Tomazou, E.M.; de Álava, E.; Kovar, H.; Sorensen, P.H.; Delattre, O.; Dirksen, U. Ewing sarcoma. *Nat. Rev. Dis. Prim.* **2018**, *4*, 1–22. [CrossRef] [PubMed]
3. Yu, H.; Ge, Y.; Guo, L.; Huang, L. Potential approaches to the treatment of Ewing's sarcoma. *Oncotarget* **2017**, *8*, 5523–5539. [CrossRef] [PubMed]
4. Üren, A.; Tcherkasskaya, O.; Toretsky, J.A. Recombinant EWS-FLI1 oncoprotein activates transcription. *Biochemistry* **2004**, *43*, 13579–13589. [CrossRef]
5. Grohar, P.J.; Woldemichael, G.M.; Griffin, L.B.; Mendoza, A.; Chen, Q.-R.; Yeung, C.; Currier, D.G.; Davis, S.; Khanna, C.; Khan, J.; et al. Identification of an inhibitor of the EWS-FLI1 oncogenic transcription factor by high-throughput screening. *J. Natl. Cancer Inst.* **2011**, *103*, 962–978. [CrossRef]
6. Grohar, P.J.; Griffin, L.B.; Yeung, C.; Chen, Q.-R.; Pommier, Y.; Khanna, C.; Khan, J.; Helman, L.J. Ecteinascidin 743 interferes with the activity of EWS-FLI1 in Ewing sarcoma cells. *Neoplasia* **2011**, *13*, 145–153. [CrossRef] [PubMed]
7. Caropreso, V.; Darvishi, E.; Turbyville, T.J.; Ratnayake, R.; Grohar, P.J.; McMahon, J.B.; Woldemichael, G.M. Englerin A inhibits EWS-FLI1 DNA binding in Ewing sarcoma cells. *J. Biol. Chem.* **2016**, *291*, 10058–10066. [CrossRef]
8. Grohar, P.J.; Glod, J.; Peer, C.J.; Sissung, T.M.; Arnaldez, F.I.; Long, L.; Figg, W.D.; Whitcomb, P.; Helman, L.J.; Widemann, B.C. A phase I/II trial and pharmacokinetic study of mithramycin in children and adults with refractory Ewing sarcoma and EWS-FLI1 fusion transcript. *Cancer Chemother. Pharmacol.* **2017**, *80*, 645–652. [CrossRef]
9. Baruchel, S.; Pappo, A.; Krailo, M.; Baker, K.S.; Wu, B.; Villaluna, D.; Lee-Scott, M.; Adamson, P.C.; Blaney, S.M. A phase 2 trial of trabectedin in children with recurrent rhabdomyosarcoma, Ewing sarcoma and non-rhabdomyosarcoma soft tissue sarcomas: A report from the Children's Oncology Group. *Eur. J. Cancer* **2012**, *48*, 579–585. [CrossRef] [PubMed]
10. Robles, A.J.; Cai, S.; Cichewicz, R.H.; Mooberry, S.L. Selective activity of deguelin identifies therapeutic targets for androgen receptor-positive breast cancer. *Breast Cancer Res. Treat.* **2016**, *157*, 475–488. [CrossRef]
11. Robles, A.J.; Du, L.; Cichewicz, R.H.; Mooberry, S.L. Maximiscin Induces DNA Damage, Activates DNA Damage Response Pathways, and Has Selective Cytotoxic Activity against a Subtype of Triple-Negative Breast Cancer. *J. Nat. Prod.* **2016**, *79*, 1822–1827. [CrossRef]
12. Robles, A.J.; McCowen, S.; Cai, S.; Glassman, M.; Ruiz, F.; Cichewicz, R.H.; McHardy, S.F.; Mooberry, S.L. Structure-Activity Relationships of New Natural Product-Based Diaryloxazoles with Selective Activity against Androgen Receptor-Positive Breast Cancer Cells. *J. Med. Chem.* **2017**, *60*, 9275–9289. [CrossRef]
13. Grant, C.V.; Carver, C.M.; Hastings, S.D.; Ramachandran, K.; Muniswamy, M.; Risinger, A.L.; Beutler, J.A.; Mooberry, S.L. Triple-negative breast cancer cell line sensitivity to englerin A identifies a new, targetable subtype. *Breast Cancer Res. Treat.* **2019**, *177*, 345–355. [CrossRef] [PubMed]
14. Kil, Y.S.; Risinger, A.L.; Petersen, C.L.; Mooberry, S.L.; Cichewicz, R.H. Leucinostatins from *Ophiocordyceps* spp. and *Purpureocillium* spp. Demonstrate Selective Antiproliferative Effects in Cells Representing the Luminal Androgen Receptor Subtype of Triple Negative Breast Cancer. *J. Nat. Prod.* **2020**, *83*, 2010–2024. [CrossRef]
15. Pederson, P.J.; Cai, S.; Carver, C.; Powell, D.R.; Risinger, A.L.; Grkovic, T.; O'Keefe, B.R.; Mooberry, S.L.; Cichewicz, R.H. Triple-Negative Breast Cancer Cells Exhibit Differential Sensitivity to Cardenolides from Calotropis gigantea. *J. Nat. Prod.* **2020**, *83*, 2269–2280. [CrossRef]
16. Cai, S.; Risinger, A.L.; Petersen, C.L.; Grkovic, T.; O'Keefe, B.R.; Mooberry, S.L.; Cichewicz, R.H. Anacolosins A–F and Corymbulosins X and Y, Clerodane Diterpenes from Anacolosa clarkii Exhibiting Cytotoxicity toward Pediatric Cancer Cell Lines. *J. Nat. Prod.* **2019**, *82*, 928–936. [CrossRef]
17. Carter, A.C.; Petersen, C.L.; Wendt, K.L.; Helff, S.K.; Risinger, A.L.; Mooberry, S.L.; Cichewicz, R.H. In Situ Ring Contraction and Transformation of the Rhizoxin Macrocycle through an Abiotic Pathway. *J. Nat. Prod.* **2019**, *82*, 886–894. [CrossRef]
18. Tinley, T.L.; Randall-Hlubek, D.A.; Leal, R.M.; Jackson, E.M.; Cessac, J.W.; Quada, J.C.; Hemscheidt, T.K.; Mooberry, S.L. Taccalonolides E and A: Plant-derived steroids with microtubule-stabilizing activity. *Cancer Res.* **2003**, *63*, 3211–3220. [PubMed]
19. Risinger, A.L.; Li, J.; Bennett, M.J.; Rohena, C.C.; Peng, J.; Schriemer, D.C.; Mooberry, S.L. Taccalonolide binding to tubulin imparts microtubule stability and potent in vivo activity. *Cancer Res.* **2013**, *73*, 6780–6792. [CrossRef]
20. Skehan, P.; Storeng, R.; Scudiero, D.; Monks, A.; McMahon, J.; Vistica, D.; Warren, J.T.; Bokesch, H.; Kenney, S.; Boyd, M.R. New colorimetric cytotoxicity assay for anticancer-drug screening. *J. Natl. Cancer Inst.* **1990**, *82*, 1107–1112. [CrossRef] [PubMed]
21. Krishan, A. Rapid flow cytofluorometric analysis of mammalian cell cycle by propidium iodide staining. *J. Cell Biol.* **1975**, *66*, 188–193. [CrossRef] [PubMed]
22. Trapnell, C.; Roberts, A.; Goff, L.; Pertea, G.; Kim, D.; Kelley, D.R.; Pimentel, H.; Salzberg, S.L.; Rinn, J.L.; Pachter, L. Differential gene and transcript expression analysis of RNA-seq experiments with TopHat and Cufflinks. *Nat. Protoc.* **2012**, *7*, 562–578. [CrossRef] [PubMed]

23. Anders, S.; Pyl, P.T.; Huber, W. HTSeq—A Python framework to work with high-throughput sequencing data. *Bioinformatics* **2015**, *31*, 166–169. [CrossRef]
24. Anders, S.; Huber, W. Differential expression analysis for sequence count data. *Genome Biol.* **2010**, *11*, R106. [CrossRef] [PubMed]
25. Subramanian, A.; Tamayo, P.; Mootha, V.K.; Mukherjee, S.; Ebert, B.L.; Gillette, M.A.; Paulovich, A.; Pomeroy, S.L.; Golub, T.R.; Lander, E.S.; et al. Gene set enrichment analysis: A knowledge-based approach for interpreting genome-wide expression profiles. *Proc. Natl. Acad. Sci. USA* **2005**, *102*, 15545–15550. [CrossRef]
26. Overacker, R.D.; Plitzko, B.; Loesgen, S. Biolayer interferometry provides a robust method for detecting DNA binding small molecules in microbial extracts. *Anal. Bioanal. Chem.* **2021**, *413*, 1159–1171. [CrossRef]
27. Shah, N.B.; Duncan, T.M. Bio-layer interferometry for measuring kinetics of protein-protein interactions and allosteric ligand effects. *J. Vis. Exp.* **2014**, *84*, e51383. [CrossRef]
28. Nguyen, B.; Tanious, F.A.; Wilson, W.D. Biosensor-surface plasmon resonance: Quantitative analysis of small molecule-nucleic acid interactions. *Methods* **2007**, *42*, 150–161. [CrossRef]
29. Wahl, H.E.; Raudabaugh, D.B.; Bach, E.M.; Bone, T.S.; Luttenton, M.R.; Cichewicz, R.H.; Miller, A.N. What lies beneath? Fungal diversity at the bottom of Lake Michigan and Lake Superior. *J. Great Lakes Res.* **2018**, *44*, 263–270. [CrossRef]
30. Du, L.; Robles, A.J.; King, J.B.; Powell, D.R.; Miller, A.N.; Mooberry, S.L.; Cichewicz, R.H. Crowdsourcing natural products discovery to access uncharted dimensions of fungal metabolite diversity. *Angew. Chem. Int. Ed. Engl.* **2014**, *53*, 804–809. [CrossRef]
31. Hradil, C.M.; Hallock, Y.F.; Clardy, J.; Kenfield, D.S.; Strobel, G. Phytotoxins from Alternaria cassiae. *Phytochemistry* **1989**, *28*, 73–75. [CrossRef]
32. Arnone, A.; Nasini, G.; Merlini, L.; Assante, G. Secondary mould metabolites. Part 16. Stemphyltoxins, new reduced perylenequinone metabolites from *Stemphylium botryosum* var. lactucum. *J. Chem. Soc. Perkin Trans.* **1986**, *1*, 525–530. [CrossRef]
33. Stack, M.E.; Mazzola, E.P.; Page, S.W.; Pohland, A.E.; Highet, R.J.; Tempesta, M.S.; Corley, D.G. Mutagenic Perylenequinone Metabolites of Alternaria Alternata: Altertoxins I., II, and III. *J. Nat. Prod.* **1986**, *49*, 866–871. [CrossRef]
34. Okuno, T.; Natsume, I.; Sawai, K.; Sawamura, K.; Furusaki, A.; Matsumoto, T. Structure of antifungal and phytotoxic pigments produced by alternaria sps. *Tetrahedron Lett.* **1983**, *24*, 5653–5656. [CrossRef]
35. Podlech, J.; Fleck, S.C.; Metzler, M.; Bürck, J.; Ulrich, A.S. Determination of the Absolute Configuration of Perylene Quinone-Derived Mycotoxins by Measurement and Calculation of Electronic Circular Dichroism Spectra and Specific Rotations. *Chem. A Eur. J.* **2014**, *20*, 11463–11470. [CrossRef] [PubMed]
36. Delattre, O.; Zucman, J.; Melot, T.; Garau, X.S.; Zucker, J.-M.; Lenoir, G.M.; Ambros, P.F.; Sheer, D.; Turc-Carel, C.; Triche, T.J.; et al. The Ewing Family of Tumors—A Subgroup of Small-Round-Cell Tumors Defined by Specific Chimeric Transcripts. *N. Engl. J. Med.* **1994**, *331*, 294–299. [CrossRef] [PubMed]
37. Bailly, R.A.; Bosselut, R.; Zucman, J.; Cormier, F.; Delattre, O.; Roussel, M.; Thomas, G.; Ghysdael, J. DNA-binding and transcriptional activation properties of the EWS-FLI-1 fusion protein resulting from the t(11;22) translocation in Ewing sarcoma. *Mol. Cell. Biol.* **1994**, *14*, 3230–3241. [CrossRef]
38. May, W.A.; Gishizky, M.L.; Lessnick, S.L.; Lunsford, L.B.; Lewis, B.C.; Delattre, O.; Zucman, J.; Thomas, G.; Denny, C.T. Ewing sarcoma 11;22 translocation produces a chimeric transcription factor that requires the DNA-binding domain encoded by FLI1 for transformation. *Proc. Natl. Acad. Sci. USA* **1993**, *90*, 5752–5756. [CrossRef]
39. Stack, M.E.; Prival, M.J. Mutagenicity of the Alternaria metabolites altertoxins I, II, and III. *Appl. Environ. Microbiol.* **1986**, *52*, 718–722. [CrossRef]
40. Schwarz, C.; Tiessen, C.; Kreutzer, M.; Stark, T.; Hofmann, T.; Marko, D. Characterization of a genotoxic impact compound in alternaria alternata infested rice as altertoxin II. *Arch. Toxicol.* **2012**, *86*, 1911–1925. [CrossRef]
41. Fleck, S.C.; Burkhardt, B.; Pfeiffer, E.; Metzler, M. Alternaria toxins: Altertoxin II is a much stronger mutagen and DNA strand breaking mycotoxin than alternariol and its methyl ether in cultured mammalian cells. *Toxicol. Lett.* **2012**, *214*, 27–32. [CrossRef] [PubMed]
42. Gorthi, A.; Romero, J.C.; Loranc, E.; Cao, L.; Lawrence, L.A.; Goodale, E.; Iniguez, A.B.; Bernard, X.; Masamsetti, V.P.; Roston, S.; et al. EWS–FLI1 increases transcription to cause R-loops and block BRCA1 repair in Ewing sarcoma. *Nature* **2018**, *555*, 387–391. [CrossRef] [PubMed]
43. May, W.A.; Grigoryan, R.S.; Keshelava, N.; Cabral, D.J.; Christensen, L.L.; Jenabi, J.; Ji, L.; Triche, T.J.; Lawlor, E.R.; Reynolds, C.P. Characterization and Drug Resistance Patterns of Ewing's Sarcoma Family Tumor Cell Lines. *PLoS ONE* **2013**, *8*, e80060. [CrossRef] [PubMed]
44. Benedict, W.F.; Baker, M.S.; Haroun, L.; Choi, E.; Ames, B.N. Mutagenicity of cancer chemotherapeutic agents in the Salmonella/microsome test. *Cancer Res.* **1977**, *37*, 2209–2213.
45. Stewart, E.; Goshorn, R.; Bradley, C.; Griffiths, L.M.; Benavente, C.; Twarog, N.R.; Miller, G.M.; Caufield, W.; Freeman, B.B.; Bahrami, A.; et al. Targeting the DNA repair pathway in Ewing sarcoma. *Cell Rep.* **2014**, *9*, 829–841. [CrossRef]
46. Nieto-Soler, M.; Morgado-Palacin, I.; Lafarga, V.; Lecona, E.; Murga, M.; Callen, E.; Azorin, D.; Alonso, J.; Lopez-Contreras, A.J.; Nussenzweig, A.; et al. Efficacy of ATR inhibitors as single agents in Ewing sarcoma. *Oncotarget* **2016**, *7*, 58759–58767. [CrossRef]
47. Burnette, B.C.; Liang, H.; Lee, Y.; Chlewicki, L.; Khodarev, N.N.; Weichselbaum, R.R.; Fu, Y.X.; Auh, S.L. The efficacy of radiotherapy relies upon induction of type I interferon-dependent innate and adaptive immunity. *Cancer Res.* **2011**, *71*, 2488–2496. [CrossRef]

48. Ma, H.; Liang, H.; Cai, S.; O'Keefe, B.R.; Mooberry, S.L.; Cichewicz, R.H. An Integrated Strategy for the Detection, Dereplication, and Identification of DNA-Binding Biomolecules from Complex Natural Product Mixtures. *J. Nat. Prod.* **2020**, *84*, 750–761. [CrossRef]
49. Marina, N.M.; Liu, Q.; Donaldson, S.S.; Sklar, C.A.; Armstrong, G.T.; Oeffinger, K.C.; Leisenring, W.M.; Ginsberg, J.P.; Henderson, T.O.; Neglia, J.P.; et al. Longitudinal follow-up of adult survivors of Ewing sarcoma: A report from the Childhood Cancer Survivor Study. *Cancer* **2017**, *123*, 2551–2560. [CrossRef]
50. Friedman, D.N.; Chastain, K.; Chou, J.F.; Moskowitz, C.S.; Adsuar, R.; Wexler, L.H.; Chou, A.J.; DeRosa, A.; Candela, J.; Magnan, H.; et al. Morbidity and mortality after treatment of Ewing sarcoma: A single-institution experience. *Pediatric Blood Cancer* **2017**, *64*, e26562. [CrossRef]
51. Sankar, S.; Bell, R.; Stephens, B.; Zhuo, R.; Sharma, S.; Bearss, D.J.; Lessnick, S.L. Mechanism and relevance of EWS/FLI-mediated transcriptional repression in Ewing sarcoma. *Oncogene* **2013**, *32*, 5089–5100. [CrossRef] [PubMed]
52. Dharia, N.V.; Kugener, G.; Guenther, L.M.; Malone, C.F.; Durbin, A.D.; Hong, A.L.; Howard, T.P.; Bandopadhayay, P.; Wechsler, C.S.; Fung, I.; et al. A first-generation pediatric cancer dependency map. *Nat. Genet.* **2021**, *53*, 529–538. [CrossRef] [PubMed]
53. Zeman, M.K.; Cimprich, K.A. Causes and consequences of replication stress. *Nat. Cell Biol.* **2013**, *16*, 2–9. [CrossRef] [PubMed]
54. Helmrich, A.; Ballarino, M.; Nudler, E.; Tora, L. Transcription-replication encounters, consequences and genomic instability. *Nat. Struct. Mol. Biol.* **2013**, *20*, 412–418. [CrossRef]
55. Grant, C.V.; Cai, S.; Risinger, A.L.; Liang, H.; O'Keefe, B.R.; Doench, J.G.; Cichewicz, R.H.; Mooberry, S.L. CRISPR-Cas9 Genome-Wide Knockout Screen Identifies Mechanism of Selective Activity of Dehydrofalcarinol in Mesenchymal Stem-like Triple-Negative Breast Cancer Cells. *J. Nat. Prod.* **2020**, *83*, 3080–3092. [CrossRef]

cancers

MDPI

Article

The Effect of Direct and Indirect EZH2 Inhibition in Rhabdomyosarcoma Cell Lines

Andreas Schmidt [1,*], Lucas Behrendt [1], Jana Eybe [1], Steven W. Warmann [1], Sabine Schleicher [2], Joerg Fuchs [1] and Evi Schmid [1]

[1] Department of Pediatric Surgery and Pediatric Urology, University Children's Hospital, Eberhard Karls University Tuebingen, Hoppe-Seyler-Strasse 3, 72076 Tuebingen, Germany; lucasbehrendt@me.com (L.B.); jana.eybe@student.uni-tuebingen.de (J.E.); steven.warmann@med.uni-tuebingen.de (S.W.W.); joerg.fuchs@med.uni-tuebingen.de (J.F.); evi.schmid@med.uni-tuebingen.de (E.S.)

[2] Department of Pediatric Hematology and Oncology, University Children's Hospital, Eberhard Karls University Tuebingen, Hoppe-Seyler-Strasse 1, 72076 Tuebingen, Germany; sb.schleicher@outlook.de

* Correspondence: andreas.schmidt@med.uni-tuebingen.de

Simple Summary: Rhabdomyosarcoma is the most common soft tissue tumor in children. Its two major subtypes show epigenetic alterations that are associated with poor prognosis. Therefore, targeting these epigenetic alterations by pharmacological intervention could be a therapeutic approach. We investigated two different types of substances that interfere with the epigenetic process of histone methylation. We performed studies in two cell lines that carry characteristics of the major rhabdomyosarcoma subtypes. The aim of this study was to find out if the substances differ in their effect on tumor-related cellular functions and to find out if the tumor subtypes differ in their response to the substances. These findings may contribute to a better assessment of the feasibility of pharmacological intervention directed against histone methylation in subtypes of rhabdomyosarcoma.

Abstract: Enhancer of Zeste homolog 2 (EZH2) is involved in epigenetic regulation of gene transcription by catalyzing trimethylation of histone 3 at lysine 27. In rhabdomyosarcoma (RMS), increased EZH2 protein levels are associated with poor prognosis and increased metastatic potential, suggesting EZH2 as a therapeutic target. The inhibition of EZH2 can be achieved by direct inhibition which targets only the enzyme activity or by indirect inhibition which also affects activities of other methyltransferases and reduces EZH2 protein abundance. We assessed the direct inhibition of EZH2 by EPZ005687 and the indirect inhibition by 3-deazaneplanocin (DZNep) and adenosine dialdehyde (AdOx) in the embryonal RD and the alveolar RH30 RMS cell line. EPZ005687 was more effective in reducing the cell viability and colony formation, in promoting apoptosis induction, and in arresting cells in the G1 phase of the cell cycle than the indirect inhibitors. DZNep was more effective in decreasing spheroid viability and size in both cell lines than EPZ005687 and AdOx. Both types of inhibitors reduced cell migration of RH30 cells but not of RD cells. The results show that direct and indirect inhibition of EZH2 affect cellular functions differently. The alveolar cell line RH30 is more sensitive to epigenetic intervention than the embryonal cell line RD.

Keywords: rhabdomyosarcoma; RH30; RD; EZH2; epigenetic; EPZ005687; DZNep; AdOx

Citation: Schmidt, A.; Behrendt, L.; Eybe, J.; Warmann, S.W.; Schleicher, S.; Fuchs, J.; Schmid, E. The Effect of Direct and Indirect EZH2 Inhibition in Rhabdomyosarcoma Cell Lines. *Cancers* **2022**, *14*, 41. https://doi.org/10.3390/cancers14010041

Academic Editors: Saurabh Agarwal and Jianhua Yang

Received: 29 October 2021
Accepted: 21 December 2021
Published: 23 December 2021

1. Introduction

Epigenetic dysregulation may be involved in the development, growth, and metastasis of a variety of tumors [1,2]. Rhabdomyosarcoma (RMS) show increased abundance of Enhancer of Zeste Homologue 2 protein (EZH2) [3–6], a catalytic subunit of the polycomb repressive complex 2 (PCR2), and altered epigenetic markers compared to skeletal muscle [7,8].

RMS is the most common soft tissue tumor in children originating from mesenchymal precursors and expressing features of early myogenic differentiation [9]. Based on histology and genetics different subtypes can be distinguished [10–12]. The most common histological subtypes are the embryonal and alveolar subtype which account for approximately 60% and 20% of RMS [13,14]. More important for prognosis than histology is the translocations t(2;13)(q35;q14) and t(1;13)(p36;q14), which occur in 55% and 22% of alveolar RMS [15]. They result in the expression of *PAX3:FOXO1* and *PAX7:FOXO1* fusion proteins and are associated with poor prognosis [15,16]. The embryonal RMS is characterized by the loss of heterozygosity at the 11p15.5 locus, different chromosomal gains and losses, and somatic mutations [17,18]. Fusion-negative alveolar RMS share cellular features such as the loss of heterozygosity and clinical characteristics including good prognosis with the embryonal RMS [19]. The cell lines RH30 and RD are developed from a metastasized alveolar RMS and a recurrent embryonal RMS, respectively [20,21]. Although substantial achievements in the therapy of RMS have been made in the past decades, the overall survival rate of patients with high-risk and metastasized RMS amounts only to about 30% [22,23].

The abundance of EZH2 protein in RMS and its association with poor prognosis, increased metastasis and lymph node involvement make EZH2 an interesting therapeutic target [24–26]. Different approaches to affect EZH2 have been investigated [27]. Direct inhibitors of the methyltransferase activity of EZH2 such as EPZ005687 bind to the catalytic S-adenosylmethionine (SAM) pocket of the SET (Su(var)3-9, Enhancer of Zeste, Trithorax) domain of EZH2 [28,29]. Indirect inhibitors such as 3-deazaneplanocin (DZNep) and adenosine dialdehyde (AdOx) inhibit the S-adenosyl-L-homocysteine (SAH) hydrolase, which leads to increased SAH levels and a feedback inhibition of methyltransferase activity [30–32]. Indirect inhibitors reduce not only the methyltransferase activity of EZH2 but also globally of enzymes that use S-adenosyl-L-methionine (SAM) as a methyl donor [32,33]. In addition, they decrease EZH2 protein levels [31,32] and thus may affect the canonical activity of methylating H3K27 and also multiple noncanonical activities of EZH2 such as activating or inhibiting transcription factors [34,35]. Therefore, the cellular effects of indirect and direct inhibitors may differ.

It was shown that the direct inhibitor MC1945 and DZNep reduce the proliferation and xenograft growth of RD and RH30 cells but differ in inducing apoptosis [4,5]. We investigated and compared the effect of the direct inhibitor EPZ005687 and the indirect inhibitors DZNep and AdOx on a variety of cellular functions which characterize tumor behavior such as cell viability, colony formation, migration, apoptosis, cell cycle, and spheroid size and viability. The aim of this study was to characterize the two types of inhibitors in more detail to better assess the feasibility of this therapeutic approach by these types of inhibitors in the two major RMS subtypes.

2. Materials and Methods

2.1. Cell Lines and Reagents

The embryonal RMS cell line RD (ATCC, Manassas, VA, USA) and the alveolar RMS cell line RH30 (DSMZ, Braunschweig, Germany) as well as the primary human skeletal muscle cells (SkMC) (PromoCell, Heidelberg, Germany) were cultured in DMEM high glucose 4.5 g/L medium (Sigma Aldrich Chemie GmbH, Taufkirchen, Germany) supplemented with 1% L-glutamine (Biochrom, Berlin, Germany), 10% heat-inactivated fetal bovine serum (Biochrom, Berlin, Germany), and 1% penicillin/streptomycin (Biochrom, Berlin, Germany) in a humidified atmosphere containing 5% CO_2 at 37 °C. Only early passages (after purchase or authentication) which were tested to be negative for mycoplasma contamination (MycoAlert; Lonza, Cologne, Germany) were used for the current study. EZH2 inhibitors EPZ005687 (Cayman Chemical Company, Ann Arbor, MI, USA), DZNep (Cayman Chemical Company, MI, USA), and AdOx (Sigma Aldrich Chemie GmbH, Taufkirchen, Germany) were used as indicated.

2.2. Western Blotting

EZH2 protein abundance in RMS cell lines, treated with EPZ005687, DZNep, and AdOx for 24 h was detected by Western blotting as previously described [36]. The membranes were incubated with primary rabbit EZH2 antibody (1:1000, Cell Signaling Technology, Inc. (Danvers, MA, USA), New England Biolabs (Ipswich, MA, USA), 98 kDa) overnight at 4 °C. Incubation with rabbit monoclonal GAPDH antibody (1:1000, Cell Signaling Technology, Inc., New England Biolabs) served as a loading control.

2.3. Cell Viability Assay

In a humidified atmosphere containing 5% CO_2 at 37 °C, 8×10^3 (RH30) or 1.5×10^4 (RD and SkMC) cells were seeded in 96-well plates in a final volume of 100 µL culture medium per well. After overnight adherence of the cells, 72 h treatment with EPZ005687, DZNep, and AdOx was started. Every 24 h, the medium and treatment drug was renewed. The viability assays were performed as previously described [37].

2.4. Clonogenic Assay

RMS cancer cell lines were plated in 6-well plates at 750 cells per well. After incubation with or without EPZ005687, DZNep, and AdOx for 72 h, cells were washed twice with PBS and fresh medium was added. The colonies grew for 7–10 days before being fixed with 99.9% methanol for 5 min and stained with 1% (*w*/*v*) crystal violet for 30 min for RD cells, and 20 min for RH30 cells at room temperature. Images were captured using a phase-contrast microscope Zeiss Axiovert 135 microscope (original magnification, ×5; Carl Zeiss Microscopy GmbH, Jena, Germany). The number of colonies (>50 cells) was counted microscopically [38]. Dividing the number of colonies by the number of plated cells and multiplying by 100 yielded the colony formation rate according to Franken et al. [39].

2.5. Wound Healing Assay

For would healing assay, 6×10^5 cells per well were plated onto 12-well dishes. A single scratch wound was inflicted using a sterile micropipette tip in each confluent monolayer. Cells were washed with PBS to remove cell debris and incubated with or without EPZ005687, DZNep, and AdOx for 72 h (every 24 h medium and treatment substance were renewed) and monitored by photographs directly after the scratch was performed and 24 h later. Images (three per well) were captured using a Zeiss inverted microscope (Axiovert 135) with a 10× objective lens and Canon EOS 550D digital camera. The wound width was measured using Axio-Vision 3.1 Software and expressed as a percentage of the initial wound width.

2.6. Flow Cytometry

Apoptosis assays of RD and RH30 cancer cell lines were analyzed using flow cytometry with Annexin V staining with allophycocyanin (APC) conjugation (BD Biosciences, Franklin Lakes, NJ, USA) and propidium iodide (PI) staining after incubation in the presence or absence of EPZ005687, DZNep, and AdOx for 72 h with a renewal of medium and treatment substance every 24 h.

After incubation, the adherent cells were collected and stained with Annexin V (BioLegend, Koblenz, Germany)/Propidium Iodide (Sigma Aldrich Chemie GmbH, Taufkirchen, Germany) in Annexin Binding Buffer according to the manufacturer's recommendations. Acquisition and analysis of data were conducted with a BD FACS CANTO II flow cytometer and FACS Diva Software Version 8.0 (Becton Dickinson, Heidelberg, Germany).

2.7. Spheroids

For the spheroids, 2×10^5 RD and RH30 cells per well in 100 µL previously filtered (Easytrainer™ 40 µm, Greiner Bio-One GmbH, Kremsmünster, Austria) cell culture medium were pipetted onto a low-attachment round bottom 96-well plate (Thermo Fisher Scientific, Waltham, MA, USA). Centrifugation at 200× *g*, 5 min at room temperature and incubation for 72 h allow the cells to form spherical aggregates. The cancer cells were treated in the presence or absence of EPZ005687, DZNep, and AdOx for 72 h. Every 24 h, a change in treatment was performed, during which only 50 µL of the cell culture medium was aspirated. The growth behavior of the spheroids was documented photographically every 24 h. With the AxioVision 3.1 software, the change in the spheroid size was analyzed. Finally, the spheroids were incubated with 25 µL methylene blue per well for 24 h and measured on the Multiple Plate Reader (Victor X, PerkinElmer Inc., Waltham, MA, USA) with the wavelengths 486 nm and 535 nm (fluorescein). The analysis of the data was performed with GraphPad Prism.

2.8. Cell Cycle

For cell cycle, 4×10^6 cells (RD) and 3×10^6 cells (RH30) per well were plated onto 6-well dishes. After attachment, cells were incubated over night with a temperature of 37 °C, 95% humidity, and 5% CO_2. Cancer cells then were treated with EPZ005687, DZNep, and AdOx for 72 h. Every 24 h, a change in treatment was performed (during which only 50 µL of the cell culture medium was aspirated). After resuspension, cells were transferred into a Neubauer chamber and were counted. The lowest cell number was noted. For staining, for all cells, this number was used, and the corresponding volume was transferred into a FACS tube. All tubes were filled with medium to the same volume. Centrifugation at was performed at 1.500 U/min, 5 min at room temperature, with a tilt overhang. Cells were washed twice with 2 mL of cold BPS and then centrifuged. After the second wash, the supernatant was removed, and the tubes were placed on ice. Then, the tubes were carefully resuspended and fixed with 1 mL of ice-cold 80% ethanol for 30 min on ice. Again, cells were washed twice with cold PBS and separated by centrifugation. Finally, cells were incubated with 500 µL FXCycle PI/Rnase staining solution (Invitrogen, Waltham, MA, USA) per tube at room temperature for 20 min. The acquisition of data was conducted with a BD FACS CANTO II flow cytometer (Becton Dickinson, Heidelberg, Germany) and analysis was performed with FlowJo software.

2.9. Statistical Analysis

GraphPad Prism 8 (GraphPad Software, La Jolla, CA, USA) was used for statistical analyses. All data were tested for significance using ANOVA (Dunnett correction). Only results with $p \leq 0.05$ were considered statistically significant. Data are presented as means ± standard error of the mean (SEM) unless otherwise specified. All data are representative of at least three experiments.

3. Results

3.1. EZH2 Protein Abundance Is More Reduced by Indirect Inhibitors Than by the Direct Inhibitor

The Western blots revealed significantly higher EZH2 protein abundance in the RD and RH30 cell line than in the primary skeletal muscle cells (SkMC). The indirect inhibitors DZNep and AdOx significantly reduced the EZH2 protein abundance in both cell lines at all concentrations tested, more in the RH30 cell line than in the RD cell line. DZNep was more effective in both cell lines than AdOx. A significant decrease in EZH2 abundance in the range of 20–27% by EPZ005687 was detected in RD cells after 72 h of treatment. Only at a high concentration of 15 µM was a non-significant decrease in EH2 abundance observed in RH30 cells (Figure 1). The uncropped Western blots of EZH2 abundance in RMS cell lines in the presence or absence of EZH2 inhibitors can be found in the Supplementary Materials S1.

In summary, the indirect inhibitors (DZNep and AdOx) have a stronger impact on EZH2 protein abundance in RMS cell lines than the direct inhibitor EPZ005687.

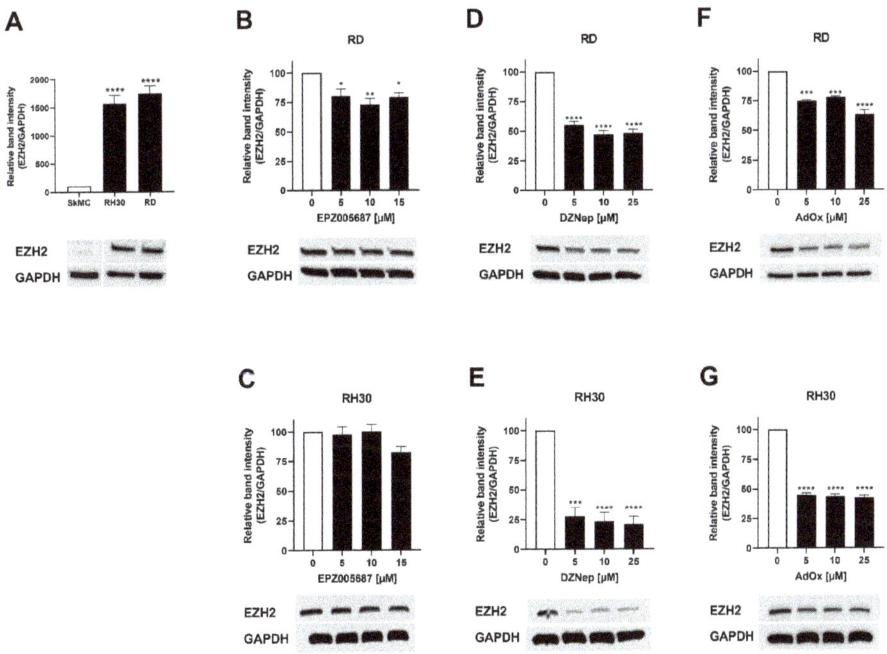

Figure 1. EZH2 abundance in RMS cell lines in the presence or absence of EZH2 inhibitors. (**A**) Representative Western blot of EZH2 and densitometric quantification of SkMC, RD, and RH30 cells. Relative ratio of EZH2/GAPDH density was normalized to ratio obtained in SkMC. Effect of EPZ005687 (**B**,**C**), DZNep (**D**,**E**), and AdOx (**F**,**G**) on RD (upper row) and RH30 cells (lower row) on EZH2 abundance. Relative ratio of EZH2/GAPDH density was normalized to ratio obtained in untreated control cultures. GAPDH was used as loading control. Error bars represent mean +/− SEM (n = 3). p * ≤ 0.05, p ** ≤ 0.01, p *** ≤ 0.001, p **** ≤ 0.0001 indicates statistical significance.

3.2. Both Types of Inhibitors Reduced RMS Cell Viability in Both Cell Lines

The reduction of cell viability by EPZ005687 reached significance at a concentration of 20.5 µM and 20 µM in the RD and RH30 cell line, respectively. It was 86% in the RD cell line and 76% in the RH30 cell line at the highest concentration of 21 µM EPZ005687. The reduction of cell viability by DZNep and AdOx was significant at 5 µM in both cell lines (range 22–43%). It increased slightly with increasing concentration but not as strongly as when treated with EPZ005687. At 25 µM, the reduction by DZNep was 37% in the RD and 56% in the RH30 cell line, and by AdOx 45% and 42%, respectively (Figure 2). From these results, it can be concluded that much lower concentrations of the two indirect inhibitors (DZNep and AdOx) can affect the viability of RD and RH30 cells than of EPZ005687.

3.3. Migration Was Inhibited by Both Types of Inhibitors in the RH30 Cells but Not in the RD Cells

The migration of cells may indicate their potential to spread and metastasize. We investigated the effect of the inhibitors on migration using the wound healing assay. In the RH30 cell line, 20 µM EPZ005687 decreased the migration significantly by 26%, and 10 µM and 25 µM AdOx significantly by 16% and 14%, respectively. The inhibition of 13% achieved by DZNep did not reach statistical significance. Neither of the substances impaired cell migration of RD cells (Figure 3). The data indicate that both types of inhibitors can inhibit migration in RH30 cells but not in RD cells.

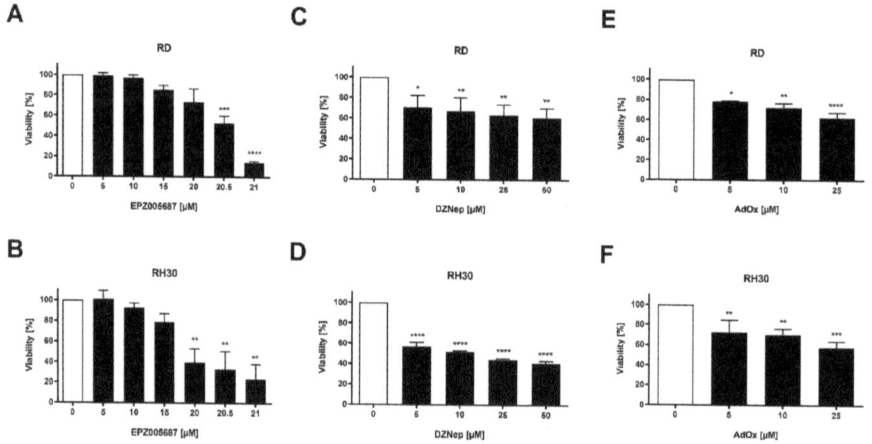

Figure 2. Effect of the EZH2 inhibitors on cell viability in RMS cell lines. Relative numbers of viable RD (upper row) and RH30 (lower row) cells following a 72 h incubation in the absence (white bars) and presence (black bars) of increasing concentrations of EPZ005687 (**A**,**B**), DZNep (**C**,**D**), and AdOx (**E**,**F**). The untreated control was set as 100%. Error bars represent mean +/− SEM ($n \geq 3$). p * ≤ 0.05, p ** ≤ 0.01, p *** ≤ 0.001, p **** ≤ 0.0001 indicates statistical significance.

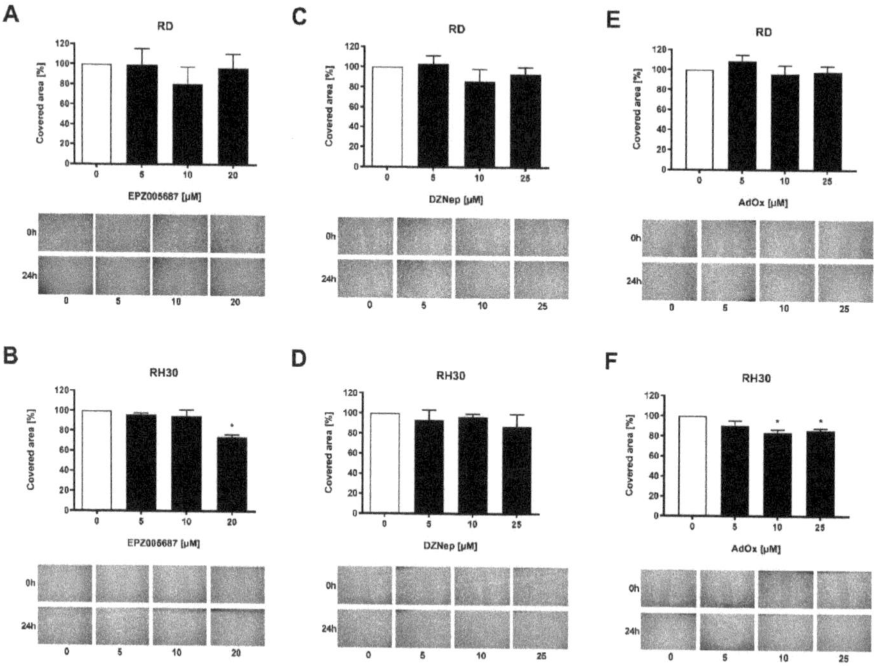

Figure 3. Migration behavior of RMS cells in response to EZH2 inhibitors. A wound healing assay was performed to assess migration. Relative numbers of migrated RD (upper row) and RH30 (lower row) cells following a 72 h incubation in the absence (white bars) and presence (black bars) of increasing concentrations of EPZ005687 (**A**,**B**), DZNep (**C**,**D**), and AdOx (**E**,**F**). The untreated control was set as 100%. Error bars represent mean +/− SEM (n = 4). p * ≤ 0.05 indicates statistical significance. A representative image of the wound healing assay is shown for each concentration of the inhibitors.

3.4. Both Types of Inhibitors Reduced Colony Formation of RD and RH30 Cells

A further characteristic of cells which is linked to the ability to metastasize is colony formation. All three inhibitors reduced the colony formation concentration-dependently. In both cell lines, the inhibition was stronger by EPZ005687 than by DZNep or AdOx and stronger in RD cells than in RH30 cells (Figure 4). In comparison to cell viability (Figure 2), colony formation was in general more impaired, e.g., for 10 µM EPZ005687, which had no effect on cell viability but showed a significant effect on colony formation at this concentration in both cell lines. DZNep and AdOx also had a more pronounced effect on colony formation than on cell viability. Obviously, colony formation is more sensitive to EZH2 inhibition than cell viability.

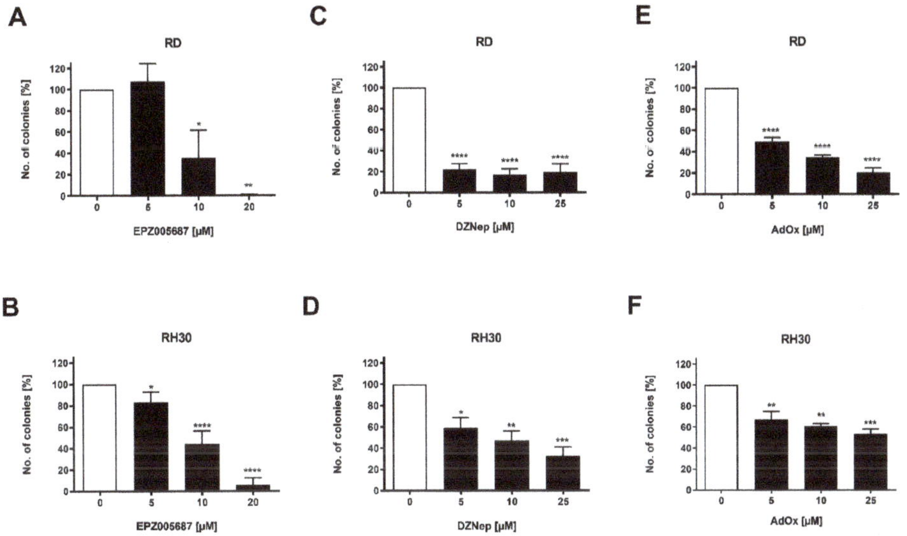

Figure 4. Effect of the EZH2 inhibitors on colony formation. Arithmetic means ± SEM (n = 3) of the percentage of evolving clones of RD (upper row) and RH30 (lower row) cells following a 72 h incubation in the presence (black bars) of EZH2 inhibitors EPZ005687 (**A**,**B**), DZNep (**C**,**D**), and AdOx (**E**,**F**) relative to the clones in the absence of the inhibitors (white bars). The untreated control was set as 100%. p * $\leq$ 0.05, p ** $\leq$ 0.01, p *** $\leq$ 0.001, p **** $\leq$ 0.0001 indicates statistical significance.

3.5. Apoptosis Was More Induced by the Inhibitors in the RH30 Cell Line Than in the RD Cell Line

In the apoptosis assay, a significant dose-dependent increase in apoptosis was detected in RH30 cells after 72 h of treatment with the inhibitors. In the RH30 cell line, 20 µM EPZ005687 increased apoptosis significantly to 85%, 25 µM DZNep and AdOx significantly to 9% and 10% of total cells, respectively. In RD cells, 20 µM EPZ005687 increased apoptosis significantly to 59% of total cells. DZNep and AdOx showed a small increase which, however, did not reach statistical significance (Figure 5).

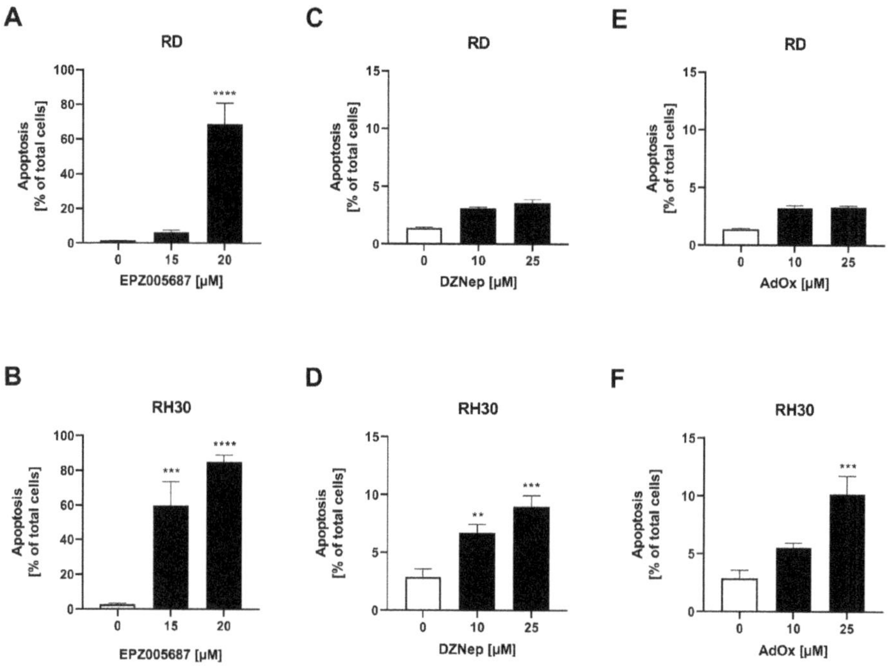

Figure 5. Flow cytometric analysis of apoptosis in RMS cells after treatment with EZH2 inhibitors. Arithmetic means ± SEM (n = 3) of the number of Annexin V-positive cells after 72 h incubation with EPZ005687 (**A**,**B**), DZNep (**C**,**D**), and AdOx (**E**,**F**) in RD cells (upper row) and in RH30 (lower row) cells. The untreated control was set as 100%. p ** ≤ 0.01, p *** ≤ 0.001, p **** ≤ 0.0001 indicates statistical significance.

3.6. The Direct Inhibitor but Not the Indirect Inhibitors Arrested RD and RH30 Cells in the G1 Phase

We further investigated the effect of 72 h of treatment with the inhibitors on the distribution of cells within the cell cycle. The proportion of cells in the G1 phase in untreated RD cells and untreated RH30 cells was 42% and 52%, respectively. This proportion did not change significantly under treatment with DZNep or AdOx in either the RD cell line or the RH30 cell line. However, EPZ005687 had a significant concentration-dependent effect in both cell lines. At 15 µM EPZ005687, a further 20% of the cell population of RD cells and 13% of the cell population of RH30 cells were arrested in the G1 phase (Figure 6).

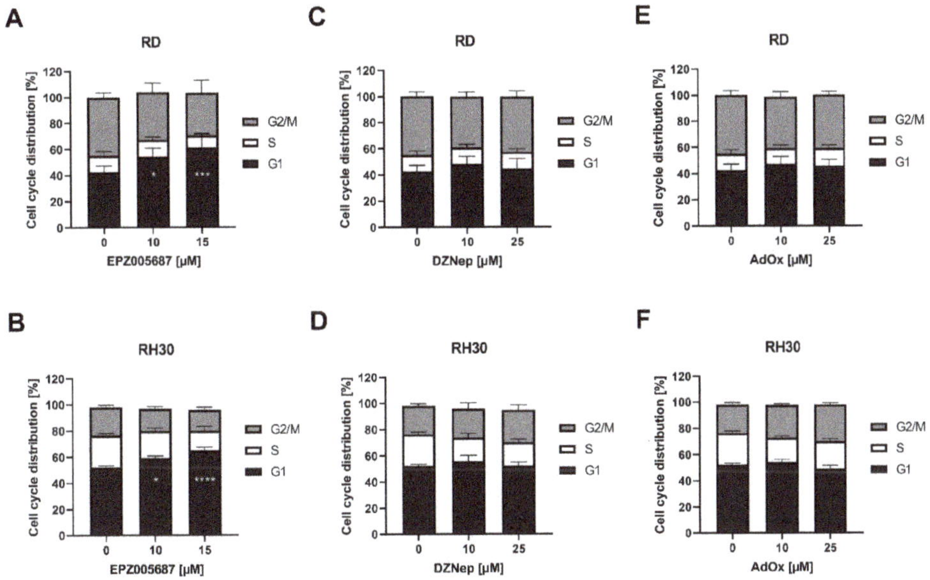

Figure 6. Flow cytometric analysis of cell cycle in RMS cell lines. Effect on cell cycle arrest following a 72 h incubation in the presence (black bars) or absence (white bars) of EPZ005687 (**A**,**B**), DZNep (**C**,**D**), and AdOx (**E**,**F**) on RD (upper row) and RH30 cells (lower row). Error bars represent mean +/− SEM (n = 4). p * ≤ 0.05, p *** ≤ 0.001, p **** ≤ 0.0001 indicates statistical significance.

3.7. DZNep Decreased Spheroid Viability and Reduced Spheroid Circumference

Spheroids more closely resemble the three-dimensional situation in tumors than two dimensional cell cultures. As cell–cell interactions and the microenvironment may modulate cellular functions, spheroids can provide additional information on the susceptibility of cells to pharmacological intervention. We therefore investigated the effect of the inhibitors of EZH2 on spheroid viability and circumference.

DZNep reduced the spheroid viability significantly in the RD cell line by 18% and in the RH30 cell line by 38%. In contrast, neither EPZ005687 nor AdOx impaired the spheroid viability in any cell line (Figure 7). In RD cells, the reduction of spheroid circumference at different DZNep concentrations was about 8% but not significant, whereas in RH30 cells, the reduction of 25% was significant. AdOx had no effect on the spheroid circumference in any cell line. EPZ005687 reduced the circumference at the concentration of 20 µM by 19% (Figure 8).

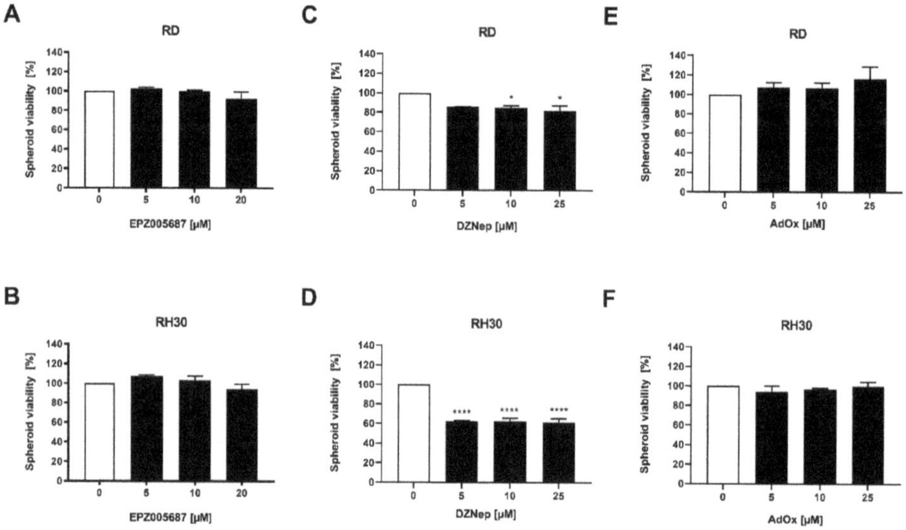

Figure 7. Modulation of EZH2 inhibitors on spheroid viability. Relative numbers of viable RD (upper row) and RH30 (lower row) cells following a 72 h incubation with increasing concentrations of EPZ005687 (**A**,**B**), DZNep (**C**,**D**), and AdOx (**E**,**F**). The untreated control was set as 100%. Error bars represent mean +/− SEM (n = 3). p * ≤ 0.05, p **** ≤ 0.0001 indicates statistical significance.

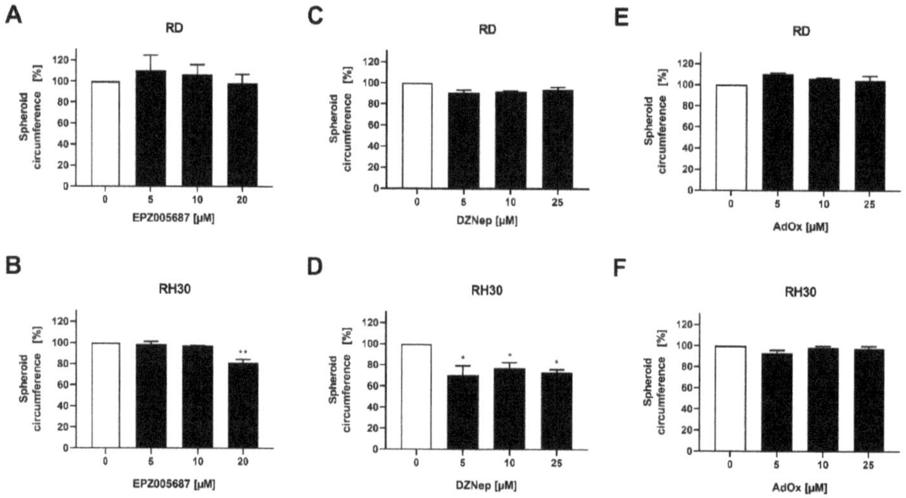

Figure 8. Effect of the EZH2 inhibitors on spheroid circumference in RMS cell lines. Inhibitory effect of EPZ005687 (**A**,**B**), DZNep (**C**,**D**), and AdOx (**E**,**F**) on RD (upper row) and RH30 (lower row) cells on spheroid circumference. The untreated control was set as 100%. Error bars represent mean +/− SEM (n = 3.) p * ≤ 0.05, p ** ≤ 0.01 indicates statistical significance.

In the studies on cell viability in two-dimensional cell cultures (Figure 2), EPZ005687 and DZNep showed a comparable inhibition pattern as in the studies on spheroid viability: EPZ005687 exerted no effect in RD cells up to 20 µM and RH30 cells up to 15 µM, and DZNep had an effect even at low concentrations in both cell lines. Contrary to the similarity in the effect of EPZ and DZNep on cell viability and spheroid viability in both cell lines, AdOx had no effect on the spheroid viability, although it impaired the cell viability at all concentrations tested.

4. Discussion

In the present study, we investigated and compared direct and indirect inhibitors of the catalytic activity of EZH2 in the embryonal RMS cell line RD and the alveolar RMS cell line RH30. The results demonstrate that the two types of inhibitors differ in the extent to which they affect cellular functions, and the two cell lines differ in their response to the inhibitors.

The effect of the two types of inhibitors may be related to their mode of action. Both types inhibit the methyltransferase of EZH2. However, indirect inhibitors additionally inhibit SAM-dependent methyltransferases which methylate DNA, lysine residues other than H3K27, and arginine residues in histones [32,33]. Additionally, they reduce the EZH2 abundance and affect other non-canonical activities of EZH2, which include the methylation of nonhistone substrates, protein and microRNA binding, and interactions with transcription factors [34,35]. The significance of methyltransferases other than H3K27 methyltransferase of EZH2 has been shown in chondrosarcoma cell lines. In these cells, DZNep effects were correlated to SAH hydrolase inhibition but not to EZH2 expression or reduced H3K27me3 methylation. The authors conclude that other methyltransferases than EZH2 are involved in the DZNep effect [40]. The significance of non-canonical activities of EZH2 is further demonstrated for castration-resistant prostate cancer cells in which the oncogenicity of EZH2 is related to its role as transcription factor [41,42].

In various cell lines, inhibition of the EZH2 methyltransferase did not copy the reduction in EZH2 abundance [43–46]. Selective EZH2 degraders have been developed to reduce EZH2 protein. In EZH2-dependent triple-negative breast cancer cells, the newly developed EZH2 specific degrader MS1943 reduced cell proliferation and induced apoptosis effectively in contrast to selective inhibitors of the enzymatic activity of EZH2 [46].

The RH30 cell line and RD cell line carry characteristics of the fusion positive and fusion negative RMS subtype, respectively. The subtypes differ in DNA and histone methylation [47], micro-RNA expression, and gene expression [48–51], including genes of signaling pathways which are involved in complex cellular functions such as motility, invasion, and metastasis [52,53]. Any difference in response of the cell lines to interference with EZH2 function may be attributed to the differing signal network established in the cells.

In our experiments, migration was inhibited in the RH30 cell line but not in the RD cell line. Regulation of invasion and metastasis involves various signaling pathways and is subject to epigenetic regulation [53]. PAX3-FOXO1 fusion protein available in RH30 cells affects different downstream targets which are involved in cell motility and invasion [54], and it interacts with epigenetic modifiers to regulate transcriptional activities which affect multiple cellular processes including motility [55,56]. The promoter of the GEFT gene is hypomethylated in RH30 cells, leading to an increased activity of the GEFT-Rho-GTPase signaling pathway, which is involved in the regulation of cell motility [57]. Genes regulating cell motility such as the gene of the transcription factor F0XF1 and LMO4 [52], cell motility gene 1 (ELMO1) and NEL-like 1 gene (NELL1) are overexpressed in the RH30 cell line compared to the RD cell line [52,58]. Silencing the genes had a greater effect on migration in the RH30 cell line than in RD cells. The MET receptor is overexpressed in RH30 cells [59–61] and may activate the ERK/MAPK pathway [61] which is involved in regulating cell migration. Activating the MET/ERK2 pathway by the hepatocyte growth factor stimulated cell motility of alveolar RMS but not embryonal RMS cells [60]. The SNAIL transcription factor is overexpressed in RH30 cells compared to RD cells [62]. SNAIL is regulating cellular processes including migration by interacting with the PI3K/AKT signaling pathway and micro-RNA [63]. In summary, the RH30 cell line showed marked changes in the network controlling motility compared to RD cells, which may explain their response to the inhibitors.

The spheroid size and viability showed a better response to DZNep, which was not to be expected considering the results of the studies on two-dimensional cell culture. This indicates the limitations of the study. Obviously, epigenetic mechanisms are particularly influenced by cellular interactions so that data of models more complex than two-dimensional models may provide more reliable results. The specific inhibitor MC1945 reduced the growth of xenografts of RD cells, and MC1945 and DZNep reduced the growth of xenografts of RH30 cells in mice [4,5]. Data on more cellular function in complex models and in vivo studies could be helpful to better assess whether a therapeutic application of EZH2 inhibitors could be reasonable.

Ciarapica et al. investigated the direct inhibitor MC1945 and the indirect inhibitor DZNep. She showed that at the concentration of 5 μM of both substances, MC1945 had no effect on EZH2 protein abundance except DZNep, that both types of inhibitors inhibit migration in RH30 cells but not in RD cells and induce apoptosis in RH30 cells but not in RD cells [4,5]. A strict comparison of these findings with our findings faces some limitations as the direct inhibitors EPZ005687 and MC1945 have differing properties such as inhibition constants, as the incubation period in the reported apoptosis experiments was 24 h longer than in our experiments, and as we used higher concentrations of the substances for the apoptosis experiment. Despite these limitations for comparison, it may be concluded that the results of the two studies are in good agreement. Our results also do not indicate an effect on migration and apoptosis at a concentration of 5 μM in RD cells but do in the RH30 cells. The EZH2 protein abundance of RD cells and RH30 cells is reduced by DZNep in both studies. However, whereas both studies show no effect of MC1945 or EPZ005687 on EZH2 protein abundance in RH30 cells, EPZ005687 but not MC1945 decreased it by a small amount in RD cells. It has been shown in cell lines of various tumors that direct inhibitors did not reduce EPZ protein abundance. So, the use of EPZ005687 in our experiments instead of MC1945 seems rather not to explain the different result we observed.

5. Conclusions

In RMS, direct and indirect inhibitors of EZH2 differ in the effects they exert on cellular functions, which may be due to multiple effects by indirect inhibitors. Differences also exist between the embryonal RD cell line and the alveolar RH30 cell line in their response to the inhibitors due to divergence in epigenetic features, gene expression and signaling pathways which are involved in the regulation of cellular functions. It may be concluded that the alveolar and embryonal RMS subtypes are different entities regarding EZH2 inhibition. Cellular interactions seem to influence epigenetic processes as different results of two- and three-dimensional culture systems indicate. Therefore, data of two-dimensional models should be supported by data from more complex models and in vivo studies.

Supplementary Materials: The following is available online at https://www.mdpi.com/article/10.3390/cancers14010041/s1, Material S1: Uncropped Western blots of EZH2 abundance in RMS cell lines in the presence or absence of EZH2 inhibitors. Addition to Figure 1.

Author Contributions: Conceptualization, A.S. and E.S.; methodology, L.B. and J.E.; data curation, L.B. and J.E.; writing—original draft preparation, A.S.; writing—review and editing, L.B., J.E., S.W.W., S.S., J.F. and E.S.; visualization, A.S., L.B., J.E. and E.S.; supervision, S.S., S.W.W. and J.F.; project administration, A.S. and E.S.; funding acquisition, E.S. All authors have read and agreed to the published version of the manuscript.

Funding: This research was funded by Else Übelmesser Stiftung and the Open Access Publishing Fund of Tuebingen University.

Institutional Review Board Statement: Not applicable.

Informed Consent Statement: Not applicable.

Data Availability Statement: The data presented in this study are available in this article.

Conflicts of Interest: The authors declare no conflict of interest.

References

1. Nebbioso, A.; Tambaro, F.P.; Dell'Aversana, C.; Altucci, L. Cancer epigenetics: Moving forward. *PLoS Genet.* **2018**, *14*, e1007362. [CrossRef]
2. Darwiche, N. Epigenetic mechanisms and the hallmarks of cancer: An intimate affair. *Am. J. Cancer Res.* **2020**, *10*, 1954–1978. [PubMed]
3. Ciarapica, R.; Russo, G.; Verginelli, F.; Raimondi, L.; Donfrancesco, A.; Rota, R.; Giordano, A. Deregulated expression of miR-26a and Ezh2 in rhabdomyosarcoma. *Cell Cycle* **2009**, *8*, 172–175. [CrossRef] [PubMed]
4. Ciarapica, R.; De Salvo, M.; Carcarino, E.; Bracaglia, G.; Adesso, L.; Leoncini, P.P.; Dall'Agnese, A.; Walters, Z.S.; Verginelli, F.; De Sio, L.; et al. The Polycomb group (PcG) protein EZH2 supports the survival of PAX3-FOXO1 alveolar rhabdomyosarcoma by repressing FBXO32 (Atrogin1/MAFbx). *Oncogene* **2014**, *33*, 4173–4184. [CrossRef] [PubMed]
5. Ciarapica, R.; Carcarino, E.; Adesso, L.; De Salvo, M.; Bracaglia, G.; Leoncini, P.P.; Dall'agnese, A.; Verginelli, F.; Milano, G.M.; Boldrini, R.; et al. Pharmacological inhibition of EZH2 as a promising differentiation therapy in embryonal RMS. *BMC Cancer* **2014**, *14*, 139. [CrossRef] [PubMed]
6. Walters, Z.S.; Villarejo-Balcells, B.; Olmos, D.; Buist, T.W.; Missiaglia, E.; Allen, R.; Al-Lazikani, B.; Garrett, M.D.; Blagg, J.; Shipley, J. JARID2 is a direct target of the PAX3-FOXO1 fusion protein and inhibits myogenic differentiation of rhabdomyosarcoma cells. *Oncogene* **2014**, *33*, 1148–1157. [CrossRef]
7. Kurmasheva, R.T.; Peterson, C.A.; Parham, D.M.; Chen, B.; McDonald, R.E.; Cooney, C.A. Upstream CpG island methylation of the PAX3 gene in human rhabdomyosarcomas. *Pediatr. Blood Cancer* **2005**, *44*, 328–337. [CrossRef]
8. Sun, W.; Chatterjee, B.; Wang, Y.; Stevenson, H.S.; Edelman, D.C.; Meltzer, P.S.; Barr, F.G. Distinct methylation profiles characterize fusion-positive and fusion-negative rhabdomyosarcoma. *Mod. Pathol.* **2015**, *28*, 1214–1224. [CrossRef]
9. Dasgupta, R.; Fuchs, J.; Rodeberg, D. Rhabdomyosarcoma. *Semin. Pediatr. Surg.* **2016**, *25*, 276–283. [CrossRef]
10. Seki, M.; Nishimura, R.; Yoshida, K.; Shimamura, T.; Shiraishi, Y.; Sato, Y.; Kato, M.; Chiba, K.; Tanaka, H.; Hoshino, N.; et al. Integrated genetic and epigenetic analysis defines novel molecular subgroups in rhabdomyosarcoma. *Nat. Commun* **2015**, *6*, 7557. [CrossRef]
11. Parham, D.M.; Barr, F.G. Classification of rhabdomyosarcoma and its molecular basis. *Adv. Anat. Pathol.* **2013**, *20*, 387–397. [CrossRef] [PubMed]
12. Rudzinski, E.R.; Kelsey, A.; Vokuhl, C.; Linardic, C.M.; Shipley, J.; Hettmer, S.; Koscielniak, E.; Hawkins, D.S.; Bisogno, G. Pathology of childhood rhabdomyosarcoma: A consensus opinion document from the Children's Oncology Group, European Paediatric Soft Tissue Sarcoma Study Group, and the Cooperative Weichteilsarkom Studiengruppe. *Pediatr. Blood Cancer* **2021**, *68*, e28798. [CrossRef]
13. Ognjanovic, S.; Linabery, A.M.; Charbonneau, B.; Ross, J.A. Trends in childhood rhabdomyosarcoma incidence and survival in the United States, 1975–2005. *Cancer* **2009**, *115*, 4218–4226. [CrossRef] [PubMed]
14. Amer, K.M.; Thomson, J.E.; Congiusta, D.; Dobitsch, A.; Chaudhry, A.; Li, M.; Chaudhry, A.; Bozzo, A.; Siracuse, B.; Aytekin, M.N.; et al. Epidemiology, Incidence, and Survival of Rhabdomyosarcoma Subtypes: SEER and ICES Database Analysis. *J. Orthop. Res.* **2019**, *37*, 2226–2230. [CrossRef] [PubMed]
15. Sorensen, P.H.; Lynch, J.C.; Qualman, S.J.; Tirabosco, R.; Lim, J.F.; Maurer, H.M.; Bridge, J.A.; Crist, W.M.; Triche, T.J.; Barr, F.G. PAX3-FKHR and PAX7-FKHR gene fusions are prognostic indicators in alveolar rhabdomyosarcoma: A report from the children's oncology group. *J. Clin. Oncol.* **2002**, *20*, 2672–2679. [CrossRef] [PubMed]
16. Shern, J.F.; Selfe, J.; Izquierdo, E.; Patidar, R.; Chou, H.C.; Song, Y.K.; Yohe, M.E.; Sindiri, S.; Wei, J.; Wen, X.; et al. Genomic Classification and Clinical Outcome in Rhabdomyosarcoma: A Report From an International Consortium. *J. Clin. Oncol.* **2021**, *39*, 2859–2871. [CrossRef]
17. Weber-Hall, S.; Anderson, J.; McManus, A.; Abe, S.; Nojima, T.; Pinkerton, R.; Pritchard-Jones, K.; Shipley, J. Gains, losses, and amplification of genomic material in rhabdomyosarcoma analyzed by comparative genomic hybridization. *Cancer Res.* **1996**, *56*, 3220–3224.
18. Shern, J.F.; Chen, L.; Chmielecki, J.; Wei, J.S.; Patidar, R.; Rosenberg, M.; Ambrogio, L.; Auclair, D.; Wang, J.; Song, Y.K.; et al. Comprehensive genomic analysis of rhabdomyosarcoma reveals a landscape of alterations affecting a common genetic axis in fusion-positive and fusion-negative tumors. *Cancer Discov.* **2014**, *4*, 216–231. [CrossRef]
19. Williamson, D.; Missiaglia, E.; de Reynies, A.; Pierron, G.; Thuille, B.; Palenzuela, G.; Thway, K.; Orbach, D.; Lae, M.; Freneaux, P.; et al. Fusion gene-negative alveolar rhabdomyosarcoma is clinically and molecularly indistinguishable from embryonal rhabdomyosarcoma. *J. Clin. Oncol.* **2010**, *28*, 2151–2158. [CrossRef]
20. McAllister, R.M.; Melnyk, J.; Finkelstein, J.Z.; Adams, E.C., Jr.; Gardner, M.B. Cultivation in vitro of cells derived from a human rhabdomyosarcoma. *Cancer* **1969**, *24*, 520–526. [CrossRef]
21. Hinson, A.R.; Jones, R.; Crose, L.E.; Belyea, B.C.; Barr, F.G.; Linardic, C.M. Human rhabdomyosarcoma cell lines for rhabdomyosarcoma research: Utility and pitfalls. *Front. Oncol.* **2013**, *3*, 183. [CrossRef] [PubMed]
22. Oberlin, O.; Rey, A.; Lyden, E.; Bisogno, G.; Stevens, M.C.; Meyer, W.H.; Carli, M.; Anderson, J.R. Prognostic factors in metastatic rhabdomyosarcomas: Results of a pooled analysis from United States and European cooperative groups. *J. Clin. Oncol.* **2008**, *26*, 2384–2389. [CrossRef]

23. Hawkins, D.S.; Spunt, S.L.; Skapek, S.X.; Committee, C.O.G.S.T.S. Children's Oncology Group's 2013 blueprint for research: Soft tissue sarcomas. *Pediatr. Blood Cancer* **2013**, *60*, 1001–1008. [CrossRef] [PubMed]
24. Ramaglia, M.; D'Angelo, V.; Iannotta, A.; Di Pinto, D.; Pota, E.; Affinita, M.C.; Donofrio, V.; Errico, M.E.; Lombardi, A.; Indolfi, C.; et al. High EZH2 expression is correlated to metastatic disease in pediatric soft tissue sarcomas. *Cancer Cell Int.* **2016**, *16*, 59. [CrossRef] [PubMed]
25. Cho, Y.J.; Kim, S.H.; Kim, E.K.; Han, J.W.; Shin, K.H.; Hu, H.; Kim, K.S.; Choi, Y.D.; Kim, S.; Lee, Y.H.; et al. Prognostic implications of polycomb proteins ezh2, suz12, and eed1 and histone modification by H3K27me3 in sarcoma. *BMC Cancer* **2018**, *18*, 158. [CrossRef]
26. Zhang, N.; Zeng, Z.; Li, S.; Wang, F.; Huang, P. High expression of EZH2 as a marker for the differential diagnosis of malignant and benign myogenic tumors. *Sci. Rep.* **2018**, *8*, 12331. [CrossRef]
27. Duan, R.; Du, W.; Guo, W. EZH2: A novel target for cancer treatment. *J. Hematol. Oncol.* **2020**, *13*, 104. [CrossRef]
28. Knutson, S.K.; Wigle, T.J.; Warholic, N.M.; Sneeringer, C.J.; Allain, C.J.; Klaus, C.R.; Sacks, J.D.; Raimondi, A.; Majer, C.R.; Song, J.; et al. A selective inhibitor of EZH2 blocks H3K27 methylation and kills mutant lymphoma cells. *Nat. Chem. Biol.* **2012**, *8*, 890–896. [CrossRef]
29. Kouznetsova, V.L.; Tchekanov, A.; Li, X.; Yan, X.; Tsigelny, I.F. Polycomb repressive 2 complex-Molecular mechanisms of function. *Protein Sci.* **2019**, *28*, 1387–1399. [CrossRef] [PubMed]
30. Chen, D.H.; Wu, K.T.; Hung, C.J.; Hsieh, M.; Li, C. Effects of adenosine dialdehyde treatment on in vitro and in vivo stable protein methylation in HeLa cells. *J. Biochem.* **2004**, *136*, 371–376. [CrossRef] [PubMed]
31. Tan, J.; Yang, X.; Zhuang, L.; Jiang, X.; Chen, W.; Lee, P.L.; Karuturi, R.K.; Tan, P.B.; Liu, E.T.; Yu, Q. Pharmacologic disruption of Polycomb-repressive complex 2-mediated gene repression selectively induces apoptosis in cancer cells. *Genes Dev.* **2007**, *21*, 1050–1063. [CrossRef]
32. Miranda, T.B.; Cortez, C.C.; Yoo, C.B.; Liang, G.; Abe, M.; Kelly, T.K.; Marquez, V.E.; Jones, P.A. DZNep is a global histone methylation inhibitor that reactivates developmental genes not silenced by DNA methylation. *Mol. Cancer Ther.* **2009**, *8*, 1579–1588. [CrossRef] [PubMed]
33. Chiang, P.K. Biological effects of inhibitors of S-adenosylhomocysteine hydrolase. *Pharmacol. Ther.* **1998**, *77*, 115–134. [CrossRef]
34. Wang, J.; Wang, G.G. No Easy Way Out for EZH2: Its Pleiotropic, Noncanonical Effects on Gene Regulation and Cellular Function. *Int. J. Mol. Sci.* **2020**, *21*, 9501. [CrossRef]
35. Kim, K.H.; Roberts, C.W. Targeting EZH2 in cancer. *Nat. Med.* **2016**, *22*, 128–134. [CrossRef] [PubMed]
36. Regenbogen, S.; Stagno, M.J.; Schleicher, S.; Schilbach, K.; Bosmuller, H.; Fuchs, J.; Schmid, E.; Seitz, G. Cytotoxic drugs in combination with the CXCR4 antagonist AMD3100 as a potential treatment option for pediatric rhabdomyosarcoma. *Int. J. Oncol.* **2020**, *57*, 289–300. [CrossRef] [PubMed]
37. Schmid, E.; Stagno, M.J.; Yan, J.; Schleicher, S.; Yu, W.; Honisch, S.; Lang, F.; Fuchs, J.; Seitz, G. Serum and Glucocorticoid Inducible Kinase 1-Sensitive Survival, Proliferation and Migration of Rhabdomyosarcoma Cells. *Cell. Physiol. Biochem.* **2017**, *43*, 1301–1308. [CrossRef]
38. Schmid, E.; Stagno, M.J.; Yan, J.; Stournaras, C.; Lang, F.; Fuchs, J.; Seitz, G. Store-operated Ca^{2+} entry in rhabdomyosarcoma cells. *Biochem. Biophys. Res. Commun.* **2016**, *477*, 129–136. [CrossRef]
39. Franken, N.A.; Rodermond, H.M.; Stap, J.; Haveman, J.; van Bree, C. Clonogenic assay of cells in vitro. *Nat. Protoc.* **2006**, *1*, 2315–2319. [CrossRef]
40. Aury-Landas, J.; Girard, N.; Lhuissier, E.; Adouane, D.; Delepee, R.; Boumediene, K.; Bauge, C. The Antitumoral Effect of the S-Adenosylhomocysteine Hydrolase Inhibitor, 3-Deazaneplanocin A, is Independent of EZH2 but is Correlated with EGFR Downregulation in Chondrosarcomas. *Cell. Physiol. Biochem.* **2019**, *53*, 731–745. [CrossRef]
41. Kim, J.; Lee, Y.; Lu, X.; Song, B.; Fong, K.W.; Cao, Q.; Licht, J.D.; Zhao, J.C.; Yu, J. Polycomb- and Methylation-Independent Roles of EZH2 as a Transcription Activator. *Cell Rep.* **2018**, *25*, 2808–2820.e2804. [CrossRef] [PubMed]
42. Xu, K.; Wu, Z.J.; Groner, A.C.; He, H.H.; Cai, C.; Lis, R.T.; Wu, X.; Stack, E.C.; Loda, M.; Liu, T.; et al. EZH2 oncogenic activity in castration-resistant prostate cancer cells is Polycomb-independent. *Science* **2012**, *338*, 1465–1469. [CrossRef] [PubMed]
43. Wee, Z.N.; Li, Z.; Lee, P.L.; Lee, S.T.; Lim, Y.P.; Yu, Q. EZH2-mediated inactivation of IFN-gamma-JAK-STAT1 signaling is an effective therapeutic target in MYC-driven prostate cancer. *Cell Rep.* **2014**, *8*, 204–216. [CrossRef] [PubMed]
44. Kim, K.H.; Kim, W.; Howard, T.P.; Vazquez, F.; Tsherniak, A.; Wu, J.N.; Wang, W.; Haswell, J.R.; Walensky, L.D.; Hahn, W.C.; et al. SWI/SNF-mutant cancers depend on catalytic and non-catalytic activity of EZH2. *Nat. Med.* **2015**, *21*, 1491–1496. [CrossRef] [PubMed]
45. Lawrence, C.L.; Baldwin, A.S. Non-Canonical EZH2 Transcriptionally Activates RelB in Triple Negative Breast Cancer. *PLoS ONE* **2016**, *11*, e0165005. [CrossRef]
46. Ma, A.; Stratikopoulos, E.; Park, K.S.; Wei, J.; Martin, T.C.; Yang, X.; Schwarz, M.; Leshchenko, V.; Rialdi, A.; Dale, B.; et al. Discovery of a first-in-class EZH2 selective degrader. *Nat. Chem. Biol.* **2020**, *16*, 214–222. [CrossRef]
47. Mahoney, S.E.; Yao, Z.; Keyes, C.C.; Tapscott, S.J.; Diede, S.J. Genome-wide DNA methylation studies suggest distinct DNA methylation patterns in pediatric embryonal and alveolar rhabdomyosarcomas. *Epigenetics* **2012**, *7*, 400–408. [CrossRef]
48. De Pitta, C.; Tombolan, L.; Albiero, G.; Sartori, F.; Romualdi, C.; Jurman, G.; Carli, M.; Furlanello, C.; Lanfranchi, G.; Rosolen, A. Gene expression profiling identifies potential relevant genes in alveolar rhabdomyosarcoma pathogenesis and discriminates PAX3-FKHR positive and negative tumors. *Int. J. Cancer* **2006**, *118*, 2772–2781. [CrossRef]

49. Lae, M.; Ahn, E.H.; Mercado, G.E.; Chuai, S.; Edgar, M.; Pawel, B.R.; Olshen, A.; Barr, F.G.; Ladanyi, M. Global gene expression profiling of PAX-FKHR fusion-positive alveolar and PAX-FKHR fusion-negative embryonal rhabdomyosarcomas. *J. Pathol.* **2007**, *212*, 143–151. [CrossRef]
50. Wachtel, M.; Dettling, M.; Koscielniak, E.; Stegmaier, S.; Treuner, J.; Simon-Klingenstein, K.; Bühlmann, P.; Niggli, F.K.; Schäfer, B.W. Gene expression signatures identify rhabdomyosarcoma subtypes and detect a novel t(2;2)(q35;p23) translocation fusing PAX3 to NCOA1. *Cancer Res.* **2004**, *64*, 5539–5545. [CrossRef]
51. Missiaglia, E.; Shepherd, C.J.; Aladowicz, E.; Olmos, D.; Selfe, J.; Pierron, G.; Delattre, O.; Walters, Z.; Shipley, J. MicroRNA and gene co-expression networks characterize biological and clinical behavior of rhabdomyosarcomas. *Cancer Lett.* **2017**, *385*, 251–260. [CrossRef] [PubMed]
52. Armeanu-Ebinger, S.; Bonin, M.; Habig, K.; Poremba, C.; Koscielniak, E.; Godzinski, J.; Warmann, S.W.; Fuchs, J.; Seitz, G. Differential expression of invasion promoting genes in childhood rhabdomyosarcoma. *Int. J. Oncol.* **2011**, *38*, 993–1000. [CrossRef] [PubMed]
53. Ramadan, F.; Fahs, A.; Ghayad, S.E.; Saab, R. Signaling pathways in Rhabdomyosarcoma invasion and metastasis. *Cancer Metastasis Rev.* **2020**, *39*, 287–301. [CrossRef]
54. Loupe, J.M.; Miller, P.J.; Bonner, B.P.; Maggi, E.C.; Vijayaraghavan, J.; Crabtree, J.S.; Taylor, C.M.; Zabaleta, J.; Hollenbach, A.D. Comparative transcriptomic analysis reveals the oncogenic fusion protein PAX3-FOXO1 globally alters mRNA and miRNA to enhance myoblast invasion. *Oncogenesis* **2016**, *5*, e246. [CrossRef]
55. Gryder, B.E.; Yohe, M.E.; Chou, H.C.; Zhang, X.; Marques, J.; Wachtel, M.; Schaefer, B.; Sen, N.; Song, Y.; Gualtieri, A.; et al. PAX3-FOXO1 Establishes Myogenic Super Enhancers and Confers BET Bromodomain Vulnerability. *Cancer Discov.* **2017**, *7*, 884–899. [CrossRef] [PubMed]
56. Bohm, M.; Wachtel, M.; Marques, J.G.; Streiff, N.; Laubscher, D.; Nanni, P.; Mamchaoui, K.; Santoro, R.; Schafer, B.W. Helicase CHD4 is an epigenetic coregulator of PAX3-FOXO1 in alveolar rhabdomyosarcoma. *J. Clin. Investig.* **2016**, *126*, 4237–4249. [CrossRef]
57. Liu, C.; Zhang, L.; Cui, W.; Du, J.; Li, Z.; Pang, Y.; Liu, Q.; Shang, H.; Meng, L.; Li, W.; et al. Epigenetically upregulated GEFT-derived invasion and metastasis of rhabdomyosarcoma via epithelial mesenchymal transition promoted by the Rac1/Cdc42-PAK signalling pathway. *EBioMedicine* **2019**, *50*, 122–134. [CrossRef]
58. Rapa, E.; Hill, S.K.; Morten, K.J.; Potter, M.; Mitchell, C. The over-expression of cell migratory genes in alveolar rhabdomyosarcoma could contribute to metastatic spread. *Clin. Exp. Metastasis* **2012**, *29*, 419–429. [CrossRef]
59. Skrzypek, K.; Kusienicka, A.; Szewczyk, B.; Adamus, T.; Lukasiewicz, E.; Miekus, K.; Majka, M. Constitutive activation of MET signaling impairs myogenic differentiation of rhabdomyosarcoma and promotes its development and progression. *Oncotarget* **2015**, *6*, 31378–31398. [CrossRef]
60. Otabe, O.; Kikuchi, K.; Tsuchiya, K.; Katsumi, Y.; Yagyu, S.; Miyachi, M.; Iehara, T.; Hosoi, H. MET/ERK2 pathway regulates the motility of human alveolar rhabdomyosarcoma cells. *Oncol. Rep.* **2017**, *37*, 98–104. [CrossRef]
61. Saini, M.; Verma, A.; Mathew, S.J. SPRY2 is a novel MET interactor that regulates metastatic potential and differentiation in rhabdomyosarcoma. *Cell Death Dis.* **2018**, *9*, 237. [CrossRef] [PubMed]
62. Skrzypek, K.; Kusienicka, A.; Trzyna, E.; Szewczyk, B.; Ulman, A.; Konieczny, P.; Adamus, T.; Badyra, B.; Kortylewski, M.; Majka, M. SNAIL is a key regulator of alveolar rhabdomyosarcoma tumor growth and differentiation through repression of MYF5 and MYOD function. *Cell Death Dis.* **2018**, *9*, 643. [CrossRef] [PubMed]
63. Skrzypek, K.; Kot, M.; Konieczny, P.; Nieszporek, A.; Kusienicka, A.; Lasota, M.; Bobela, W.; Jankowska, U.; Kedracka-Krok, S.; Majka, M. SNAIL Promotes Metastatic Behavior of Rhabdomyosarcoma by Increasing EZRIN and AKT Expression and Regulating MicroRNA Networks. *Cancers (Basel)* **2020**, *12*, 1870. [CrossRef] [PubMed]

cancers

MDPI

Systematic Review

A Systematic Review and Meta-Analysis of Malignant Rhabdoid and Small Cell Undifferentiated Liver Tumors: A Rational for a Uniform Classification

Juri Fuchs [1,2], Anastasia Murtha-Lemekhova [1,2], Markus Kessler [2,3], Fabian Ruping [3], Patrick Günther [2,3], Alexander Fichtner [2,4], Dominik Sturm [5,6] and Katrin Hoffmann [1,2,*]

1 Department of General, Visceral and Transplantation Surgery, University Hospital Heidelberg, 69120 Heidelberg, Germany; juri.fuchs@med.uni-heidelberg.de (J.F.); anastasia.lemekhova@med.uni-heidelberg.de (A.M.-L.)

2 Generating Evidence for Diagnosis and Therapy of RarE LIVEr Disease: The RELIVE Initiative for Systematic Reviews and Meta-Analyses, University Hospital Heidelberg, 69120 Heidelberg, Germany; markus.kessler@med.uni-heidelberg.de (M.K.); patrick.guenther@med.uni-heidelberg.de (P.G.); alexander.fichtner@med.uni-heidelberg.de (A.F.)

3 Department of General, Visceral and Transplantation Surgery, Division of Pediatric Surgery, University Hospital Heidelberg, 69120 Heidelberg, Germany; fabian.ruping@med.uni-heidelberg.de

4 Department of Pediatrics I, Division of Pediatric Gastroenterology, University Children's Hospital Heidelberg, 69120 Heidelberg, Germany

5 Department of Pediatric Hematology and Oncology, Heidelberg University Hospital, 69120 Heidelberg, Germany; dominik.sturm@med.uni-heidelberg.de

6 Hopp Children's Cancer Center (KiTZ), 69120 Heidelberg, Germany

* Correspondence: katrin.hoffmann@med.uni-heidelberg.de; Tel.: +49-6221-566110; Fax: +49-6221-564215

Citation: Fuchs, J.; Murtha-Lemekhova, A.; Kessler, M.; Ruping, F.; Günther, P.; Fichtner, A.; Sturm, D.; Hoffmann, K. A Systematic Review and Meta-Analysis of Malignant Rhabdoid and Small Cell Undifferentiated Liver Tumors: A Rational for a Uniform Classification. *Cancers* **2022**, *14*, 272. https://doi.org/10.3390/cancers14020272

Academic Editors: Saurabh Agarwal and Jianhua Yang

Received: 10 December 2021
Accepted: 4 January 2022
Published: 6 January 2022

Simple Summary: Malignant rhabdoid tumors of the liver are very rare pediatric liver tumors with a devastating prognosis. It is currently unclear which histological subtypes of pediatric liver tumors belong to this entity and how these tumors should be treated. In this systematic review with meta-analysis, we analyzed all reports on pediatric patients with malignant rhabdoid liver tumors, but also with so-called small cell undifferentiated liver tumors. This is another rare liver tumor subtype that has recently been regarded to belong to the entity of rhabdoid tumors by some authors. The main result of this study is that these two tumor subtypes show large overlap on several levels and even mixtures of both histological patterns have been documented. Our meta-analysis provides an evidence base for the recommendation to classify these two tumor subtypes as one entity. We showed that treatment of these tumors with hepatoblastoma directed chemotherapy is ineffective and that a therapy with chemotherapy regimens initially applied for soft tissue sarcoma is associated with a significantly better survival. This study represents the highest level of evidence available for these rare liver tumors.

Abstract: Background: Rhabdoid liver tumors in children are rare and have a devastating prognosis. Reliable diagnosis and targeted treatment approaches are urgently needed. Immunohistochemical and genetic studies suggest that tumors formerly classified as small cell undifferentiated hepatoblastoma (SCUD) belong to the entity of malignant rhabdoid tumors of the liver (MRTL), in contrast to hepatoblastomas with focal small cell histology (F-SCHB). This may have relevant implications on therapeutic approaches. However, studies with larger cohorts investigating the clinical relevance of the histological and genetic similarities for patients are lacking. **Purpose:** To analyze possible similarities and differences in patient characteristics, tumor biology, response to treatment, and clinical course of patients with MRTL, SCUD and F-SCHB. Applied therapeutic regimens and prognostic factors are investigated. **Methods:** A systematic literature search of MEDLINE, Web of Science, and CENTRAL was performed for this PRISMA-compliant systematic review. All studies of patients with MRTL, SCUD and F-SCHB that provided individual patient data were included. Demographic, histological, and clinical characteristics of the three subgroups were compared. Overall survival (OS) was estimated with the Kaplan–Meier method and prognostic factors investigated in a multivariable

Cox regression model. **Protocol registered:** PROSPERO 2021 CRD42021258760. **Results:** Fifty-six studies with a total of 118 patients were included. The two subgroups MRTL and SCUD did not differ significantly in baseline patient characteristics. However, heterogenous diagnostic and therapeutic algorithms were applied. Large histological and clinical overlap between SCUD and MRTL could be shown. Two-year OS was 22% for MRTL and 13% for SCUD, while it was significantly better in F-SCHD (86%). Chemotherapeutic regimens for hepatoblastoma proved to be ineffective for both SCUD and MRTL, but successful in F-SCHB. Soft tissue sarcoma chemotherapy was associated with significantly better survival for MRTL and SCUD, but was rarely applied in SCUD. Patients who did not undergo surgical tumor resection had a significantly higher risk of death. **Conclusions:** While F-SCHB is a subtype of HB, SCUD should be classified and treated as a type of MRTL. Surgical tumor resection in combination with intensive, multi-agent chemotherapy is the only chance for cure of these tumors. Targeted therapies are highly needed to improve prognosis. Currently, aggressive regimens including soft tissue sarcoma chemotherapy, extensive resection, radiotherapy or even liver transplantation are the only option for affected children.

Keywords: malignant rhabdoid tumor; hepatoblastoma; small cell undifferentiated (SCUD) hepatoblastoma; *SMARCB1*; INI1; pediatric liver tumors

1. Introduction

Article titles like "Small Cell Undifferentiated (SCUD) Hepatoblastomas: All Malignant Rhabdoid Tumors?" [1] or "Malignant Rhabdoid Tumor, an Aggressive Tumor Often Misclassified as Small Cell Variant of Hepatoblastoma" [2] highlight a recent and pressing question in the research field of pediatric liver tumors: should malignant rhabdoid liver tumors (MRTL) and small cell undifferentiated hepatoblastomas be classified and treated as two different entities, or do they share vital features that justify merging into one entity?

Both tumors are rare primary liver neoplasms in children and show highly aggressive behavior. MRTL account for about 3% of all primary liver malignancies of childhood [1,2]. First described as a distinct liver tumor by Gonzalez-Crussi et al. in 1982 [3], the prognosis of affected children remains poor [4,5]. Historically, the diagnosis of MRTL was based on typical histological appearance of the tumor cells, with a so-called rhabdoid morphology. Genetic analyses of MRTL cells revealed typical mutations of the *SMARCB1* gene on chromosome band 22q11.2. This causes loss of function of a chromatin remodeling complex, which acts as an important tumor suppressor. By immunohistochemistry, mutations in this gene can be detected as a loss of INI1 protein expression [2]. In contrast, hepatoblastomas (HB) are typically positive for INI1.

In contrast to the devastating survival rates of MRTL, the prognosis of pediatric patients with HB has dramatically improved within the last 40 years, and long-term OS of over 80% has been achieved [6]. However, a certain group of patients with tumors showing small cell histology is still associated with poor prognosis: Historically subsumed under the entity of "anaplastic HB", the term small cell undifferentiated HB (SCUD) was later introduced for this specific tumor. This term refers to the histological appearance of the predominant cells [7]. In historical studies, SCUD accounted for about 2% of all pediatric liver tumors classified as HB [8,9].

Recently, expert liver pathologists re-analyzed small patient cohorts or tissue biobanks and showed that SCUD share the important loss of INI1 with MRTL [1,2,10,11]. Furthermore, they found a vast predominance of small undifferentiated cells and absence of typical HB histology. Thus, some authors suggested that SCUD are in fact MRTL of the liver with a non-rhabdoid morphology. Based on these findings, they proposed to abandon the classification of SCUD as a HB subtype [1,2]. The question of whether SCUD belongs to the entity of MRTL is clinically highly relevant, as tumors classified as SCUD are still treated under regimens conceptualized for HB with inadequately low response rates. In contrast, progress in the treatment of MRTL has been made by applying aggressive treatment regimens different from those for HB, including chemotherapy regimens initially targeted against soft tissue sarcoma, radiation, and extensive surgery.

Recently, most experts came to the opinion that tumors formerly classified as SCUD should be regarded as subtype of MRTL, when INI1-staining is negative [12]. However, in the current Children's Oncology Group (COG) classification of pediatric liver tumors [13] as well as in the most recent one by the American College of Pathologists [14], SCUD are continuously classified as a HB subtype. While new case reports have shown that not only MRTL, but also most SCUD have a loss of INI1, the clinical relevance and consequences have not been investigated yet. The current unclear situation leads to unstandardized diagnoses and therapies of these tumors. For example, some authors classify tumors with non-rhabdoid morphology but negative INI1-staining as MRTL, whereas others still adhere to the classification as SCUD [5,10,13,14]. Overall, there is an evident lack of clinical data on both MRTL and SCUD.

In this context, HB with focal small cell histology must also be addressed (F-SCHB). Since the methods of histological analysis have improved over the years, it was found that a subset of predominantly typical HB contain areas or nests with small cell histology. Moreover, for those F-SCHB where testing for INI1-expression is available, INI1 positivity even in the small tumor cells was documented [15]. Depending on whether neoadjuvant chemotherapy is applied, or upfront resection is performed, the percentage of F-SCHB differs between 1% and 12% [1,9]. This difference can be explained by the generally favorable response of HB to chemotherapy, resulting in partial tumor necrosis. This in turn may hide potential nests of small tumor cells in the specimen at the time of resection. Clinically, patients with F-SCHB usually respond to standard HB chemotherapy and tend to have a substantially better prognosis than those with MRTL or SCUD [9,15]. First evidence suggests that F-SCHB are a distinct subgroup of HB and do not belong to the entity of MRTL or SCUD. However, clinical data comparing these tumors is completely lacking.

In this systematic review with meta-analysis, we summarize all reports on pediatric patients with MRTL, SCUD and F-SCHB, analyze and compare patient characteristics, immunohistological features, applied therapies, and outcomes of these subgroups. The main research question is whether MRTL and SCUD should be classified and treated as one entity. In addition, prognostic factors are investigated, including analyses of the applied chemotherapy, surgery, and radiotherapy.

2. Materials and Methods

2.1. Review Structure and Search Strategy

This systematic review (SR) is based on a structured methodology that had been conceptualized previously by the authors, with the specific aim of generating evidence for rare diseases. It is part of the *RELIVE* research initiative (Generating evidence for RarE LIVEr Disease). The methods are in accordance with the *Preferred Reporting Items for Systematic Reviews and Meta-Analyses for Individual Patient Data (PRISMA-IPD)* guidelines. Before starting the analysis, the SR was registered with the International *Prospective Register of Systematic Review* (PROSPERO 2021 CRD42021258760).

The following free text and medical subject heading (MeSH) terms were used for a systematic literature search of three different databases (MEDLINE via PubMed, Web of Science, and CENTRAL): Rhabdoid Tumor, rhabdoid, small cell undifferentiated, small cell variant, small cell tumor, SCU, SCUD, liver, hepatic, hepatoblastoma, liver neoplasms. The detailed search algorithm is provided in the Supplementary Material (File S1). Moreover, references of the relevant studies were screened for eligible articles. The last search was performed on 11th October 2021.

2.2. Study Selection Criteria and Selection Process

All study types were eligible for this systematic review. The following inclusion criteria were applied:

Patient age < 18 years;
Primary liver tumor;
Histological diagnosis of malignant rhabdoid tumor or small cell undifferentiated hepatoblastoma or focal small cell histology in hepatoblastoma;
Individual patient data on age, therapy, and outcome available.

Historic reports of patients with, at the time, so-called "anaplastic hepatoblastoma" were only included if the information of histological tumor characteristics allowed for a reliable classification as SCUD. Moreover, studies with only aggregated data accessible, and no report of subgroup outcomes of patients with MRTL, SCUD or F-SCHB, were not suitable for this analysis and thus excluded.

The list of studies retrieved by the systematic search was screened for eligible studies by two reviewers independently (JF and AML). After a first selection based on the abstracts, the two reviewers worked through the full texts of all eligible studies and decision on in- or exclusion was made. Dissent between the two reviewers was resolved after consulting with a third reviewer (KH).

2.3. Data Extraction and Investigated Variables

Data was extracted by means of a standardized form, that had been validated with data extraction of the first five studies. The collected data items were compiled after reviewing the relevant literature on pediatric liver tumors. Independently from each other, the two reviewers extracted the data based on the predefined form.

2.4. Risk of Bias Assessment

Given the rarity of the investigated tumors, no randomized trials were expected to be found. For observational studies, the validated MINORS tool was applied for the risk of bias (RoB) assessment [16]. As case reports/series were included in our analysis, a RoB tool specifically designed for this methodology was used [17].

2.5. Statistical Analyses and Certainty of Evidence

R (version 3.6.2, Vienna, Austria) [18] was used for all statistical analyses, with the the survminer [19] and survival [20] packages for survival curves and the forestmodel [21] package for forest plots. Patient data were entered individually into a database. Patient data were pooled, and patients were divided into three subgroups according to diagnosis made in the studies: MRTL, SCUD, or F-SCHB. For descriptive statistics of continuous data, means, medians, standard deviations (SD) or interquartile ranges (IQR) were calculated. For categorial data, numbers with percentages are given. Chi-squared tests (without Yate's correction) or Mann–Whitney U tests were used for univariate analyses, at a level of significance of 5%. Overall survival (OS) and progression free survival (PFS) were calculated with the Kaplan–Meier method. Univariate significance of variables for OS/PFS was tested with the log rank test (level of significance 5%). Factors that showed univariate significance were included in multivariable cox regression models for calculating the independent hazard ratios (HR) of predictive variables.

The GRADE criteria were applied for determining the certainty of evidence and strength of recommendations [22].

2.6. Applied Terms and Definitions

Staging. The widely applied and validated PRETEXT system for pediatric liver tumors was used for staging the local extent of disease [23].

Histology. Liver tumors were considered to have rhabdoid morphology when large, polygonal tumor cells with pronounced eosinophilic cytoplasm, and large eccentric nuclei with centrally visible nucleoli were found. Intracellular inclusions with positivity for vimentin or other intermediate filament proteins are typically found in a subset of rhabdoid tumor cells. The term small cell undifferentiated was used for liver tumors with round or oval cells with sparse cytoplasm, faint nuclei with unremarkable nucleoli. These cells may express vimentin or different types of cytokeratin and are negative for alpha-feto protein. F-SCHB was defined as liver tumor with focal small cell histology mixed with other hepatoblastoma-typical tumor cells, and less than 50% of areas with small undifferentiated cells in the specimen.

Surgery. Upfront surgery was defined as operation with intent of primary tumor resection before starting a chemotherapy. R0 was defined as microscopically tumor-free resection margins. R1 as macroscopic tumor resection but microscopically positive margins. R2 was defined as macroscopic tumor residual.

Chemotherapy. Chemotherapy with neoadjuvant intent was defined as treatment before a planned surgical tumor resection. In many cases, the intended surgery was not performed due to rapid disease progression during neoadjuvant chemotherapy. The term *Hepatoblastoma chemotherapy* (HB CT) was defined as regimens used in HB trials or protocols. Soft tissue sarcoma chemotherapy (STS CT) was defined as regimens that had been first conceptualized or applied in trials or protocols for soft tissue sarcoma. No response was defined as no reduction or increase in tumor size during chemotherapy. Partial response was defined as reduction of tumor size of 20–50%. Good response was defined as reduction of tumor size >50% and/or disappearance of synchronous metastatic lesions on imaging.

Outcome. Disease related death (DRD) was defined as death of a patient either caused by progressive disease and consecutive complications or by complications of the treatment (e.g., postoperative death, toxicity of chemotherapy). Complete remission was defined as no evidence of disease after the end of treatment. Partial remission was defined as reduction of tumor size that changed the staging (e.g., change of unresectable to resectable, complete resection of primary tumor with persisting distant disease). Overall survival (OS) was defined as time from diagnosis to the last follow-up or death of a patient. Progression free survival (PFS) was defined as time from diagnosis to increase of tumor size (primary or metastatic), occurrence of new metastatic lesions or serious complications directly associated with the disease, or treatment side effects.

3. Results

3.1. Literature Search and Study Selection

The study selection process is depicted in the PRISMA flow diagram (Figure 1). The 444 records found by the systematic literature search were screened for eligibility, and another 13 studies were found by going through the reference lists of relevant studies. Fifty-six studies were eventually included in our analysis, of which seven studies were retrospective observational studies and 49 case reports or series. These studies provided information for 118 patients, of which 55 were diagnosed with MRTL, 41 with SCUD and 22 with F-SCHB. The full list of all included studies is provided in the Supplementary Material (File S2).

3.2. Critical Appraisal of Included Studies and Risk of Bias

In light of the rarity of the investigated liver tumors, conduct of prospective or large studies is challenging, demands extensive organization and international coordination. Therefore, no randomized controlled trials and no prospective studies that specifically investigated MRTL, SCUD or F-SCHB have been conducted yet. Moreover, no study was found that investigated clinical similarities or differences of patients with MRTL and SCUD.

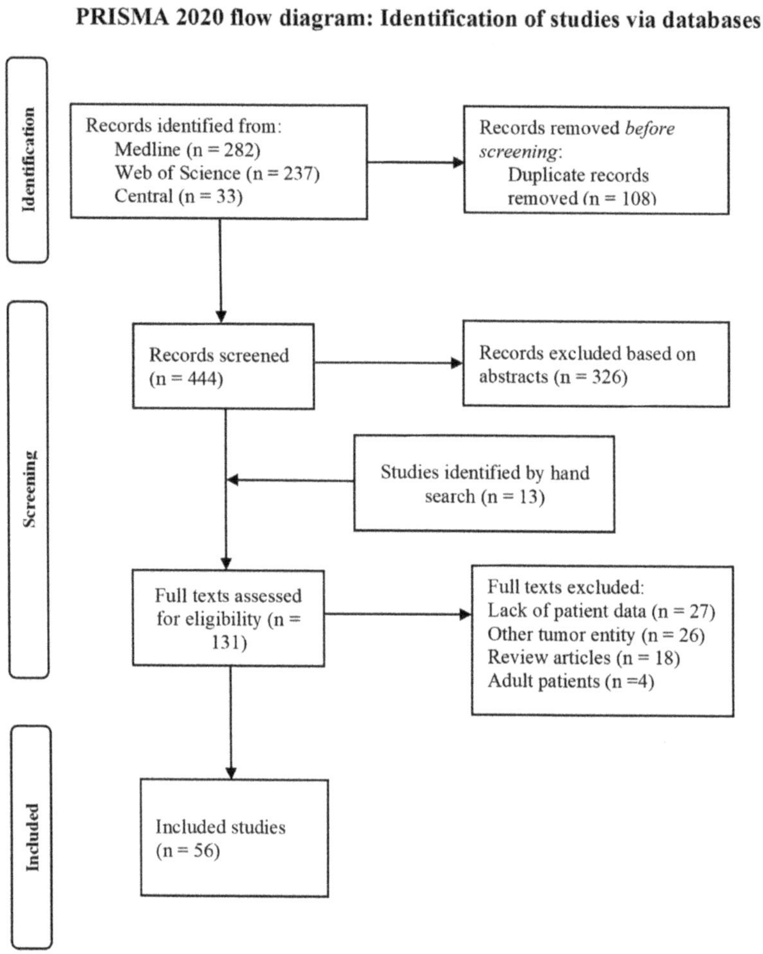

Figure 1. Prisma flow chart of study selection process.

The certainty concerning the diagnosis of either MRTL, SCUD or F-SCHB was high throughout all included studies. For the majority of the patients, inclusion in an oncologic trial registry was explicitly reported (55%). This implied that the therapeutic regimens were administered after consultation with a coordinating study center or patients received predefined therapies according to a study protocol. Nevertheless, a large variety of therapeutic approaches was found in the included studies, mirroring a lack of standardized interventions for MRTL and SCUD. In contrast, all patients with F-SCHB received standard regimens according to HB trials. Most of the included studies are case reports or small series of pediatric patients. Case reports inevitably introduced selection bias. However, the high percentage of reports with fatal outcome suggests that reporting bias for positive results was limited. The quality of the case reports differed substantially. Five case series had a low RoB. Detailed description of cases with complete data on patient characteristics, therapy and outcome was provided. Most case reports had a moderate RoB (27/49), with high selection bias, but detailed data on patient characteristics and outcome, with varying quality concerning information of the applied therapies as well as follow-up data. Seven studies were retrospective observational studies. One of them was a subgroup analysis of patients treated in the SIOPEL studies 1, 2 and 3 [24] with a low overall risk of bias (13/16 points in MINORS). Despite the low number of patients (six cases), the studies by Cornet et al. and Fazlollahi et al. provided interesting insights with detailed and comprehensive data on staging, histology, applied therapy and outcome [2,5]. The studies by Haas et al. and Zhou et al. were important concerning the differentiation of F-SCHB from SCUD and MRTL and had an acceptable RoB [9,15]. The studies by Lautz et al. and Bajpai et al. had not been specifically designed to investigate MRTL or SCUD; however, they provided adequate data for individual patients with MRTL/SCUD within their series [25,26]. Overall, most studies provided detailed descriptions of cases and an acceptable to good level of data integrity. However, a lack of studies with higher caseloads is obvious, as well as the absence of studies investigating the best suited classification and treatment of SCUD and MRTL. The tables with RoB for all included studies are provided in the Supplementary Material (Files S3 and S4). The Prisma IPD-checklist can be found in the Supplementary Material (File S5).

3.3. Composition of Subgroups and Patient Characteristics

Of the 118 patients included in this analysis, 55 were diagnosed with MRTL, 41 with SCUD and 22 with F-SCHB. The median age at diagnosis did not differ significantly between patients with MRTL and those with SCUD (7 vs. 10 months, z-score −1.86, $p = 0.061$). The median age of patients with F-SCHB was significantly higher than this of patients with MRTL and SCUD (14 months, z-score −1.97, $p = 0.049$). In all three subgroups, there was a predominance of male gender. Initial misdiagnosis, mostly as hepatoblastoma, was frequently reported for patients who were diagnosed in the final pathology report with MRTL (20 cases, 36%). Elevated alpha-feto protein (AFP) was typical for patients with F-SCHB (73%) in contrast to patients with MRTL and SCUD (9% and 5%).

The vast majority of MRTL and SCUD patients presented with large and locally advanced tumors (≥PRETEXT III in 71% and 78%, respectively). Salient is the high number of patients with tumor rupture (before start of any treatment) in the MRTL subgroup (9 cases, 16%). An overview of the patients characteristics is given in Table 1.

Table 1. Patient baseline characteristics.

	MRTL $n = 55$	**SCUD** $n = 41$	**F-SCHB** $n = 22$
Median age [months]	7 (IQR 4–15, range 0–180)	10 (IQR 7–24, range 0–138)	14 (IQR 6–20, range 4–48)
Congenital tumor	3 (5%)	1 (2%)	0
Gender	33 males (60%)	31 males (76%)	15 males (71%)
Comorbidity	0	2 (5%)	3 (14%)
Initial misdiagnosis	20 (36%)	3 (7%)	1 (5%)
Elevated AFP	4 (7%)	2 (5%)	16 (73%)
	Histology		
Rhabdoid morphology	47 (85%)	4 (10%)	0
No rhabdoid morphology	8 (15%)	37 (90%)	22 (100%)
INI1-staining			
Negative	30 (55%)	11 (27%)	0
Positive	0	1 (2%)	7 (32%)
Not available	25 (45%)	29 (71%)	15 (68%)
	Staging		
PRETEXT I	1 (2%)	0	1 (5%)
PRETEXT II	4 (7%)	5 (12%)	6 (27%)
PRETEXT III	20 (36%)	14 (34%)	0
PRETEXT IV	19 (35%)	18 (44%)	0

Table 1. Cont.

	MRTL n = 55	SCUD n = 41	F-SCHB n = 22
PREXTEXT not available	13 (24%)	4 (10%)	15 (71%)
Multifocal tumor	8 (15%)	4 (10%)	1 (5%)
Tumor rupture (before treatment)	9 (16%)	1 (2%)	0
Mean tumor size (largest diameter)	12 cm	9 cm	8 cm
Metastasis at diagnosis	22 (40%)	16 (39%)	1 (5%)
Invasion of major hepatic veins/portal vein (reported)	9 (16%)	2 (5%)	1 (5%)

3.4. *Differentiation of MRTL/SCUD from F-SCHB*

While the analysis showed large overlap concerning patient characteristics, tumor biology, and outcomes between MRTL and SCUD (see the following paragraphs), several results showed clear distinction of F-SCHB from the other two. The median age of 14 months of patients with F-SCHB in our analysis is within the range of the typical age of onset for hepatoblastoma that is described in the literature [27]. Patients with F-SCHB had significantly less often metastasis at diagnosis compared to those with MRTL and SCUD (MRTL vs. F-SCHB: OR 14.0, 95% CI 1.8–303.6, $p = 0.002$; SCUD vs. F-SCHB: OR 13.4, 95% CI 1.6–298.8, $p = 0.003$). In all F-SCHB with INI1 status available, INI1-staining was positive for all tumor cells. In contrast, 100% of MRTL and 92% of SCUD with available data on INI1 status were negative for INI1. While almost all F-SCHB showed response to standard HB chemotherapy, most patients with MRTL and SCUD had disease progression during chemotherapy (Response rate of MRTL vs. F-SCHB: 27% vs. 86%, OR 0.1, 95% CI 0.0–0.3, $p < 0.001$; SCUD vs. F-SCHB: 24% vs. 86%, OR 0.1, 95% CI 0.0–0.24, $p < 0.001$). The survival rate of patients with F-SCHB differed significantly from the outcome of patients with MRTL and SCUD. The mortality was 14% in the F-SCHB subgroup, versus 76% in the MRTL, and 93% in the SCUD subgroup (MRTL vs. F-SCHB: OR 20.5, 95% CI 4.6–104.5, $p < 0.001$; SCUD vs. F-SCHB: OR 80.2, 95% CI 12.1–709.5, $p < 0.001$, see Table 2 for details). Two-year OS was significantly higher for F-SCHB compared to MRTL and SCUD (F-SCHB: 86%, MRTL: 22%, SCUD: 13%, $p < 0.001$, Figure 2). Given these significant differences between F-SCHB on one side, and MRTL/SCUD on the other, F-SCHB was regarded as a subtype of hepatoblastoma and as a distinct entity different from MRTL/SCUD in our study. Consequently, the analysis focused on the investigation of possible similarities and differences of MRTL and SCUD, with the main question of whether these tumors should be classified as one entity.

Table 2. F-SCHB vs. MRTL/SCUD.

	Synchronous Metastasis		Response to Chemotherapy		DRD	
F-SCHB	**1 (5%)**		19 (86%)		3 (14%)	
MRTL	22 (40%)	**OR 14.0, $p = 0.002$**	15 (27%)	OR 0.1, $p < 0.001$	42 (76%)	OR 20.5, $p < 0.001$
SCUD	16 (39%)	**OR 13.4, $p = 0.003$**	10 (24%)	OR 0.03, $p < 0.001$	38 (93%)	OR 80.2, $p < 0.001$

3.5. *Comparison of MRTL and SCUD*

Baseline patient data. The median age at diagnosis did not differ significantly between the two subgroups of MRTL and SCUD (z-score = −1.87, $p = 0.061$). The rates of male gender, non-elevated AFP, PRETEXT III or IV tumors, multifocal tumors, and metastasis at diagnosis were all similar among the two subgroups, without any significant differences (See Figure 3 for details).

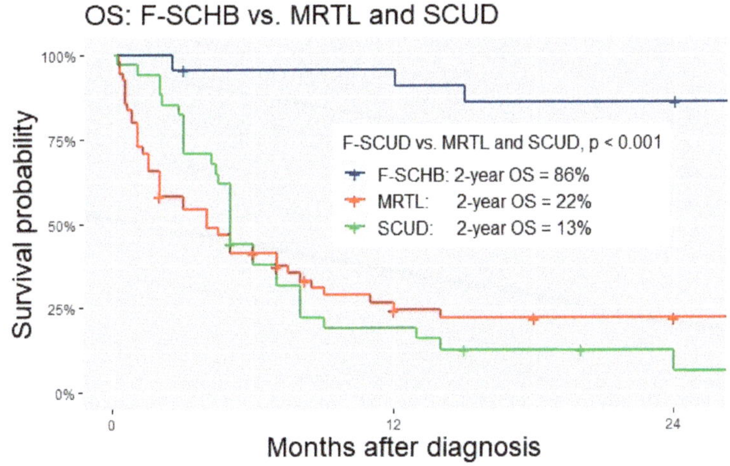

Figure 2. OS of patients with F-SCHD vs. MRTL vs. SCUD.

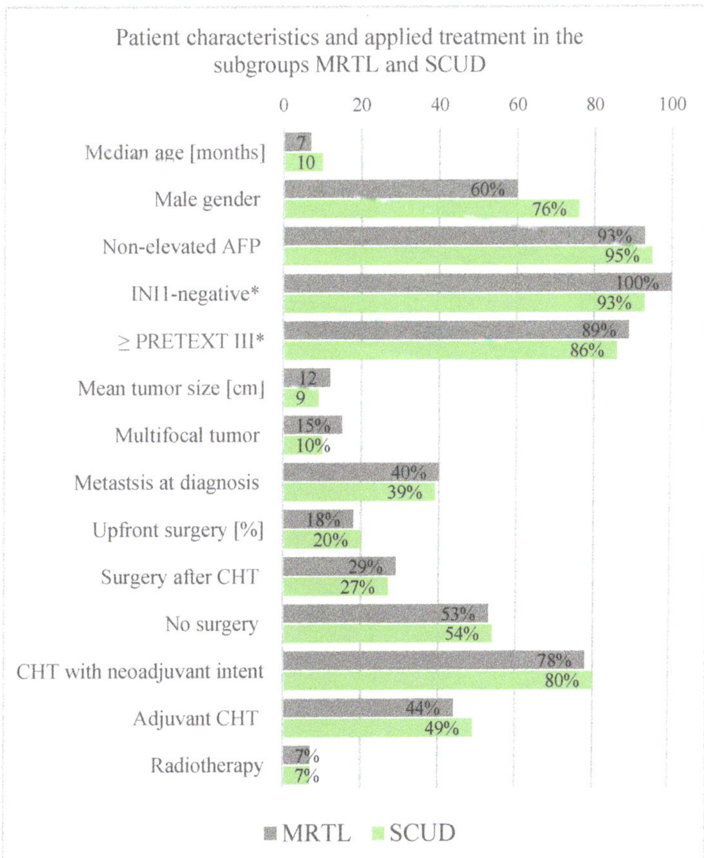

Figure 3. Overview of patient characteristics in the subgroup of MRTL and SCUD. * Percentages of those with available data on this variable.

Histology. The vast majority of tumor diagnosed as MRTL had rhabdoid morphology. However, eight patients (15%) were diagnosed with MRTL despite the absence of typical rhabdoid tumor cells. Interestingly, there were four tumors with rhabdoid cells that were classified as SCUD. All MRTL with available INI1 status were negative. Similarly, all but one SCUD with available INI1 staining were negative. Important to note is that tumors with a mixture of both, typical rhabdoid and small undifferentiated cells were reported.

Classification as MRTL or SCUD. The diagnostic criteria for MRTL and SCUD differed across the included studies and case reports. In studies conducted before the identification of *SMARCB1* mutations with consecutive loss of INI1-positivity in immunohistochemistry, the classification of liver tumors as MRTL were solely based on cytomorphology (i.e., predominance of rhabdoid tumor cells). With INI1 staining available and the discovery of loss of INI1 in liver tumors without rhabdoid morphology, some centers and authors changed their classification algorithm. In recent studies, some authors defined MRTL as liver tumors with loss of INI1 independent of the presence of rhabdoid or small undifferentiated tumor cells [2,5]. Others still adhered to the classification of small cell undifferentiated HB for tumors without rhabdoid tumor cells, but INI1-negativity of undifferentiated cells [10,28,29].

Chemotherapy regimens. Many different chemotherapy regimens were applied in patients with MRTL and SCUD. Regimens that had been conceptualized, tested, or applied mostly for soft tissue sarcoma were significantly more often applied in patients with tumors diagnosed as MRTL compared to those with SCUD (20 vs. 4, OR 5.3, 95% CI 1.3–25.3, $p = 0.007$, Figure 2). Most patients diagnosed with SCUD received hepatoblastoma-targeted chemotherapy (26 patients, 65%. PLADO in 21 patients, 53%). A variety of different regimen were applied in both subgroups and many patients were treated with unstandardized combinations of chemotherapeutic agents. See Table 3, Figures 4 and 5.

Table 3. Applied chemotherapy regimens.

	MRTL $n = 55$	SCUD $n = 41$
	Hepatoblastoma-targeted regimens	
PLADO	3	21
CDDP	2	0
C5V	0	2
C5VD	2	3
CarboDV	2	0
	Soft-tissue sarcoma-derived regimens	
CEVAIE (+Cyclo)	5	0
ICE	5	0
VAC-IE	5	0
Other sarcoma regimens	5	4
	Other regimens	
DV5Cyclo	1	1
DVCyclo	2	0
EAP	3	0
Other (non-standardized)	18	9

Surgery. Upfront surgery was performed in 18% and 20% of the patient with MRTL and SCUD, respectively. Overall, 29% (MRTL) and 27% (SCUD) underwent surgery after neoadjuvant chemotherapy. In addition, 53% (MRTL) and 54% (SCUD) did not undergo any kind of surgical tumor resection. In one patient with MRTL and in two patients with SCUD, total hepatectomy with liver transplantation was performed. All other surgeries were major liver resections. R0 resection was achieved in 46% of the operations in MRTL, and in 68% of the operations in SCUD patients. Patients with MRTL and SCUD that did not undergo surgical tumor resection had a significantly worse OS than those with upfront resection or surgery after neoadjuvant chemotherapy (Table 4, Figures 6 and 7).

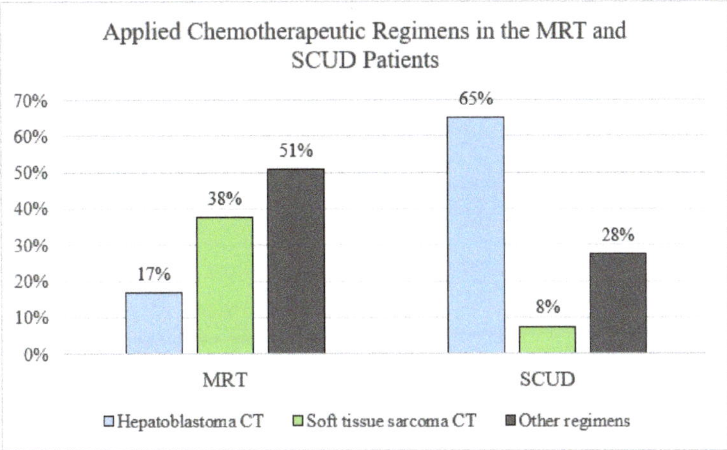

Figure 4. Overview of applied chemotherapy regiments in patients with MRTL vs. SCUD.

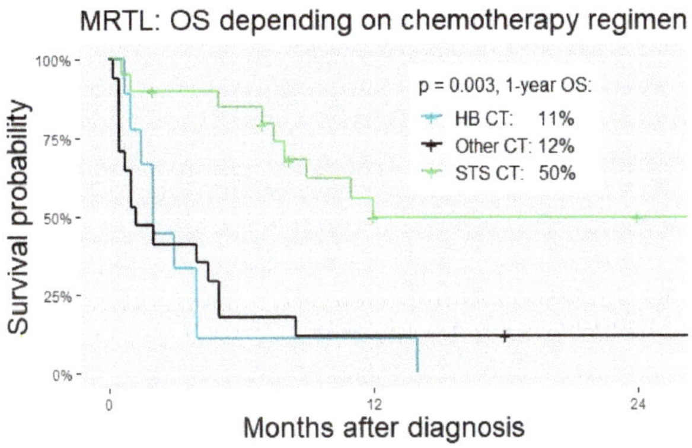

(**a**)

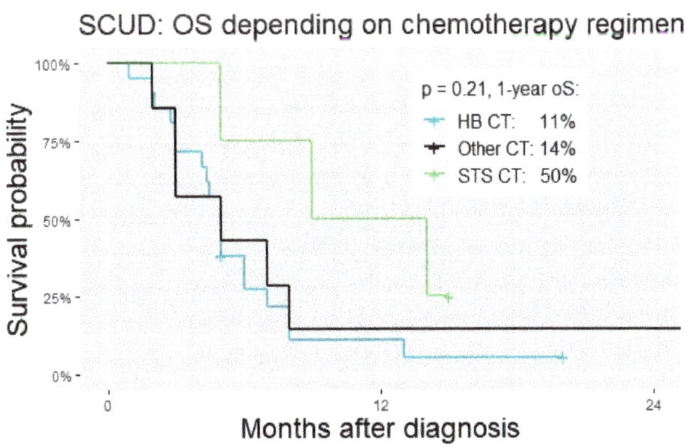

(**b**)

Figure 5. OS of patients with MRTL and SCUD according to different applied CT regimens. (**a**) OS of patients with MRTL and different CT regimens (**b**) OS of patients with SCUD and different CT regimens.

Table 4. Therapy and details of surgery.

	MRTL *n* = 55	SCUD *n* = 41	F-SCHB *n* = 22
Treated as HB	6 (11%)	40 (98%)	22 (100%)
Treated as MRTL	47 (85%)	0	0
Chemotherapy			
No oncological treatment	2 (4%)	1 (2%)	0
CHT with neoadjuvant intent	43 (78%)	33 (80%)	3 (14%)
Adjuvant chemotherapy	24 (44%)	20 (49%)	21 (95%)
PLADO regime (only)	1 (2%)	20 (49%)	15 (68%)
Soft tissue sarcoma regime			
No response	35 (64%)	29 (71%)	2 (9%)
Partial response	8 (15%)	9 (22%)	3 (14%)
Good response	7 (13%)	1 (2%)	16 (73%)
Data on response not available	5 (9%)	2 (5%)	1 (5%)
Surgery			
Upfront surgery	10 (18%)	8 (20%)	19 (86%)
Surgery after chemotherapy	16 (29%)	11 (27%)	2 (9%)
No surgical tumor resection	29 (%53)	22 (54%)	1 (5%)
Reoperations	2 (4%)	3 (7%)	2 (9%)
Liver resection	25 (45%)	18 (44%)	20 (91%)
Liver transplantation	1 (2%)	2 (5%)	1 (5%)
R-status (Biopsies excluded)			
R0	12 (22% of all, 46% of those with surgery)	13 (32% of all, 68% of those with surgery)	21 (95% of all, 100% of those with surgery)
R1	10 (18% of all, 38% of those with surgery)	5 (12% of all, 26% of those with surgery)	0
R2	4 (7% of all, 15% of those with surgery)	1 (2% of all, 5% of those with surgery)	0
Radiotherapy			
Adjuvant radiotherapy	3 (5%)	3 (7%)	0
Definitive radiotherapy	1 (2%)	0	0

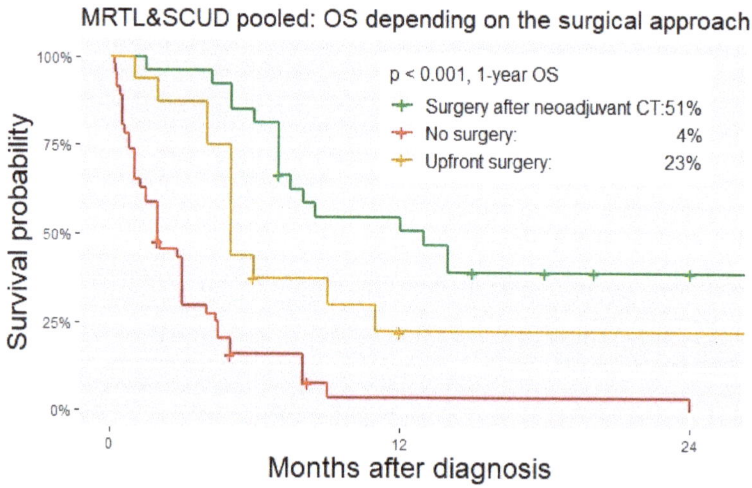

Figure 6. OS of the pooled group (MRTL and SCUD) for different surgical approaches.

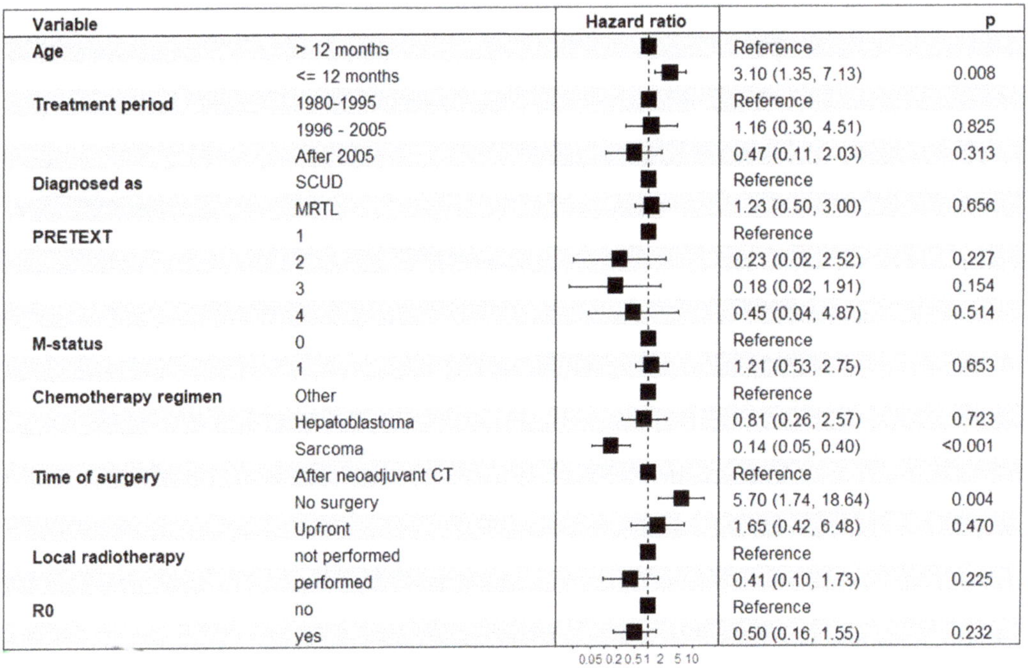

Figure 7. Multivariable Cox regression analysis with hazard ratios.

Outcome. Mean follow-up was 35 and 37 months for patients with MRTL and SCUD, respectively (median 24 and 18 months, SD 34 and 41). Overall, 76% succumbed to the disease in the MRTL subgroup (42 patient with DRD), and 93% died in the SCUD subgroup (38 patients with DRD). Kaplan–Meier estimates for 2-year OS were 22% for MRTL (95% CI 13–37) and 13% for SCUD (95% CI 5–31). This difference was not significant (p = 0.750). Two-year OS for the two groups pooled was 16% (95% CI 10–27). Disease progression during the initially therapy was observed in 56% of MRTL and 61% of SCUD patients. Relapse occurred in 11 patients in the MRTL subgroup (20%) and in 16 patients in the SCUD subgroup (39%). Mean time from (partial) remission to relapse was 5 months in the MRTL and 4 months in the SCUD subgroup. See Table 5 for details on the outcome.

Table 5. Outcome.

	MRTL n = 55	SCUD n = 41	F-SCHB n = 22
Mean/median Follow-up	35/24 months (SD 34)	37/18 months (SD 41)	85/84 months (SD 47)
Complete remission	14 (25%)	2 (5%)	21
Partial remission	10 (18%)	14 (34%)	0
Disease progression during initial therapy	31 (56%)	25 (61%)	1 (5%)
Relapse	11 (20%)	16 (39%)	10 (45%)
Local relapse	9 (16%)	7 (17%)	0
Distant relapse	5 (9%)	9 (22%)	10 (45%)
Mean time to relapse	5 months	4 months	10 months
DRD	42 (76%)	38 (93%)	3 (14%)
Alive with disease	2 (4%)	1 (2%)	0
Disease free at last follow-up	11 (20%)	2 (5%)	19 (86%)

Prognostic factors. Given the large overlap of patient characteristics, histopathologic features, and tumor behavior of MRTL and SCUD shown in the previous section of this study, the analysis of prognostic factors was performed under the hypothesis that MRTL and SCUD are subtypes of one tumor entity. The classification of tumors as either MRTL or SCUD was included in a multivariable cox regression analysis as a control variable, as well as PRETEXT staging. The following factors were significant predictors of survival in univariate analysis and thus included in the multivariable cox regression model (Figure 7): Age ($\leq$12 months/>12 months), Treatment period (1980–1995/1996–2005/After 2005), M-status (no metastasis/metastasis present), Chemotherapy regimen (Hepatoblastoma/Soft tissue sarcoma/Other), Timing of surgery (Upfront/After neoadjuvant Chemotherapy/No surgery), R0 resection status (achieved/not achieved), Local radiotherapy (Yes/No). Age $\leq$ 12 months and absence of surgical tumor resection were significantly associated with a higher risk of death. Patients who were treated with STS CT had a significantly reduced risk of death (HR 0.14, 95% CI 0.05–0.40, $p < 0.001$). The classification as either MRTL or SCUD was not an independent predictor of death.

In both subgroups, all patients that survived (disease free at last follow-up) underwent surgical tumor resection with chemotherapy (mostly neoadjuvant and adjuvant). While in both subgroups (total 96 patients) only seven patients received radiotherapy (RT) (7%), three patients among the 11 disease free survivors were treated with adjuvant RT (27%).

4. Discussion

This is the first systematic review on MRTL and SCUD and the first study with clinical comparison of those two very rare and highly aggressive liver tumors in children. The largest number of cases with MRTL and SCUD hitherto reported in the literature was analyzed. Both tumors have a devastating prognosis with mortality rates of 76% for MRTL and 93% for SCUD. Our study demonstrates large overlap in onset and course of the disease, demographic patient characteristics, response to therapy, and outcomes between both tumor subtypes. In combination with the biological and genetic similarities between MRTL and SCUD, that have been shown previously by expert pathologists [1,2,11,30], our results support the hypothesis that SCUD belong the group of rhabdoid tumors rather than being a subtype of hepatoblastoma. However, it must be acknowledged that further genetical analyses of the two histological tumor types are necessary to gain further certainty concerning the possible classification as one entity. While some improvement in the management of MRTL has been achieved by applying chemotherapeutic regimens that are specific for soft tissue sarcoma, classical hepatoblastoma chemotherapy proved to be ineffective for both MRTL and SCUD. Moreover, surgical tumor resection is essential for preserving a chance of cure for the affected patients.

4.1. *Importance of INI1-Status and Its Clinical Relevance*

The diagnosis of MRTL is difficult, which is mirrored by more than one third of patients that were initially misdiagnosed as HB. INI1-status emerged as an important diagnostic tool to corroborate a suspected diagnosis of MRTL, which otherwise relied on cytomorphology [31]. Pathology reviews and analysis of tumor biobanks have repeatedly described the histopathological overlap of MRTL and SCUD, and in particular the loss of the INI1 also in tumors classified as SCUD [1,2,32]. These findings are confirmed by our meta-analyses. Our study showed that some centers concluded that SCUD are in fact rhabdoid tumors and have already changed classification and treatment algorithm accordingly, while others stuck to the classification as *"small cell undifferentiated hepatoblastoma"* [10]. As mentioned above, current classifications of pediatric liver tumors do not yet recommend classifying and treating SCUD as MRTL, even when loss of INI1 is demonstrated [12–14]. Furthermore, our results show that HB with focal small cell histology have preserved INI1 expression within the small cell areas and are associated with a significantly better prognosis than those liver tumors with loss of INI1. F-SCHB typically respond to HB chemotherapy and must be strictly distinguished from MRTL or SCUD. F-SCHB should be classified as a subtype of HB and be treated within HB trials. The clinical relevance and prognostic value of small cell areas needs to be further investigated.

4.2. Why SCUD Should Be Classified as MRTL

Several studies on large cohorts of HB patients confirmed the significantly worse prognosis for patients with SCUD treated in HB-trials compared to all other histological subtypes [8,33]. Similarities between SCUD and MRTL haven been mentioned for already over 10 years. In 2009, Trobaugh-Lotrario et al. presented a review of historical cases of SCUD together with selected cases from personal communication [34]. The authors did not separately analyze F-SCHB and SCUD, however, they suggested that at least a subset of patients with liver tumors presenting small cell undifferentiated histology, might be more similar to rhabdoid tumors than to HB. Other authors supported these findings, albeit mainly based on histological investigations rather than analyses of clinical and therapeutical relevance [1,11,30]. For rhabdoid tumors of the central nervous system, it has long been recognized that rhabdoid morphology is only present in a subset of all rhabdoid tumors and small cell types are also encountered [35]. However, the classifications of pediatric liver tumors has not been changed concerning SCUD so far [13]. Our study demonstrated not only (immuno)histological, but also large clinical overlap between MRTL and SCUD. Several insights of our analyses substantiate the hypothesis that MRTL and SCUD belong to the same tumor entity and that SCUD may not be treated as a subtype of HB. Detailed histological examinations of tumors showed that mixed forms with typical rhabdoid morphology and small undifferentiated cells exist within one tumor [1,36]. We showed that the two patient groups do not differ significantly in relevant categories, such as patient age, elevation of AFP, INI1 status, and tumor stage at diagnosis. Significant differences were only found in the applied treatments: most patients with tumors classified as SCUD were treated with HB chemotherapy, while MRTL has increasingly been treated with STS CT in recent years. It is important to note that some tumors without rhabdoid morphology, but loss of INI1 were classified and treated as MRTL by some authors. At the same time, others classified and treated such tumors as SCUD [10]. These unstandardized classifications and treatments emphasize the need for recommendations concerning the best suited classification and therapy for these rare tumors. Our results clearly prove that classical HB chemotherapy is not only unsuited for MRTL, but also largely ineffective in SCUD. Less than 10% of patients with SCUD and therapy with HB chemotherapy showed treatment response. The rate of patients treated with HB chemotherapy was significantly higher in the SCUD than in the MRTL subgroup. Based on the hypothesis that HB chemotherapy is not a suited treatment in SCUD, the even worse outcome in the SCUD subgroup compared to patients with tumors classified as MRTL (mortality 93% vs. 76%) might be explained to some extent. In the MRTL subgroup, a significant higher number of patients were treated with STS CT, which proved to significantly reduce the risk of death. 50% of the MRTL patients who were treated with STS CT survived the disease. Notably, two of three patients with tumors missing typical rhabdoid morphology but showing INI1 negativity, survived disease free after therapy with STS CT. These findings strongly reinforce the recommendation of classifying and treating tumors with small cell undifferentiated histology and loss of INI1 as malignant rhabdoid liver tumors.

4.3. Predictive Factors and Recommendations for Therapy

Age at diagnosis. Early onset of the disease was associated with poorer outcome in our analyses. While this significant result is described for malignant rhabdoid tumor of the liver for the first time, younger age has already been identified as a negative predictor of survival in previous studies on rhabdoid tumors at various primary sites [4]. To some extent, this can be explained by a subset of cases with congenital or neonatal tumors in children with germline mutations of the *SMARCB1* gene and consecutive rhabdoid tumor syndrome. Moreover, aggressive therapies are less likely to be administered in neonatal or very young patients, which might further decrease the chances for survival of these children.

PRETEXT staging. Interestingly, PRETEXT staging was not predictive for survival of patients with MRTL and SCUD. However, given the rapid progression of these tumors, the vast majority of patients are diagnosed with large and locally advanced tumors of PRETEXT III or IV. This weakens the validity of this staging method that has been validated for HB. Only 10% of MRTL and SCUD were staged as PRETEXT I or II.

The role of surgery. A disease-free survival was only observed in 13 patients with MRTL or SCUD, all of which were treated with chemotherapy and surgical tumor resection. Three of them received additional radiotherapy. To date, a chance of cure for these aggressive liver neoplasms is only preserved if multimodal therapeutic regimens are applied, in which surgical tumor resection is an integral part. This finding was further emphasized by the fact that patients without any surgical tumor resection had a significantly increased risk of death in the multivariable cox regression analysis. The role of radiotherapy in the treatment of MRTL and SCUD has to be further investigated. Initial evidence presented by our study suggests that there is in fact an important role for radiation in the treatment of these tumors.

Soft tissue sarcoma chemotherapy. Patients with MRTL and SCUD who were treated with STS CT had a significantly reduced risk of death. The majority of the disease-free survivors were treated with STS CT. Our results show that these regimens are to date the best suited chemotherapeutic treatments available for MRTL. Our study strongly suggests that also for patients with small cell undifferentiated, INI1 negative liver tumors should be treated with these protocols. However, further studies are highly needed given the small number of these cases treated with STS CT. The most applied STS CT regimens were CEVAIE, ICE and VAC-IE. CEVAIE and VAC-IE achieved better results than ICE. However, this can only be regarded as initial evidence and further investigations of the best suited regimens are required. Within the European Paediatric Soft Tissue Sarcoma Study Group Non-Rhabdomyosarcoma Soft Tissue Sarcoma 2005 Study (EpSSG NRSTS 2005, Padua, Italy), the largest number of patients with extra-cranial rhabdoid tumors hitherto reported (100 patients) were treated according to a soft tissue sarcoma protocol, of which 43 completed the treatment protocol [4]. Three-year OS was 38% among the whole cohort, who were treated with a regimen consisting of vincristine, cyclophosphamide, doxorubicin, carboplatin, and etoposide. Surgery and radiotherapy were not performed in a standardized algorithm, which limited the possible conclusions from this study, that only provide aggregated patient data [4]. Moreover, given the very different primary tumor sites included in this study, particularities of the manifestation in certain locations, such as the liver, were not investigated. Currently, many patients with rhabdoid tumors in Europe are treated according to the protocol of the European Rhabdoid Registry (EU-RHAB). Rhabdoid liver tumors are subsumed under the category "Rhabdoid Tumors of Soft Tissue". The standard chemotherapy regimen for these tumors in EU-RHAB consists of Doxorubicin plus Ifosfamide-Carboplatinum-Etoposide (ICE) plus Vincristine-Actinomycin D-Cyclophosphamide (VAC) [37]. Subgroup outcomes of patients with MRTL with the treatments prescribed by EU-RHAB have not been investigated or published yet.

The role of radiotherapy. While the importance of radiation has been shown to atypical teratoid/rhabdoid tumors of the CNS [38,39], the potential harms or benefits of radiotherapy have not been clearly demonstrated for extracranial, and in particular rhabdoid tumors of the liver. Results of comparatively large series of patients with extracranial rhabdoid tumors of various primary sites suggest that local radiotherapy improves the prognosis [40,41], albeit the results being heterogenous and optimal timing, dosage and modalities remain unclear [42]. Local radiotherapy was only applied in 7 of 96 patients with MRTL/SCUD in our analysis (7%), which limited the possible conclusions concerning the impact of radiotherapy on the outcome of children with rhabdoid tumors of the liver. Among the 11 disease free survivors, three patients received adjuvant radiotherapy (27%). Further studies are needed to elucidate the role of radiation in the treatment of these tumors.

4.4. Status Quo and Prospect for Therapeutic Approaches

MRTL and SCUD are very rare liver tumors in children with a grim prognosis until the present. Initial misdiagnosis and unstandardized therapies slow down progress in the treatment of these challenging cases. Based on the hypothesis that MRTL and SCUD belong to the same entity, they account for about 5% of all primary liver tumors in children. At present, no treatment protocol or trial specifically designed for MRTL or SCUD, is being conducted. The classification, diagnostic algorithm, and therapy for these tumors has not been standardized internationally.

The European Rhabdoid Registry (EU-RHAB) has been founded in 2007 with the aim of establishing a central database for all cases of rhabdoid tumors in Europe in order to develop structured diagnostic and therapeutic protocols [37]. Remarkable insights have been yielded by this initiative, that formed the basis for improvement of the prognosis of patients with rhabdoid tumors. However, oncological registries that include tumors with various primary sites run the risk of missing details and important characteristics of certain primary locations. Concerning the case of rhabdoid liver tumors, this is exemplified by the fact that the important question of whether INI1-negative SCUD should be classified and treated as MRTL has not been specifically addressed and investigated in EU-RHAB so far. Similar concerns apply to the EpSSG NRSTS 2005 study. We strongly encourage clinicians and researcher that future trials investigating MRTL should include patients with tumors formerly classified as SCUD.

While improvement in the prognosis has been achieved by multimodal therapy including STS CT or high-dose chemotherapy, extensive surgery and radiotherapy, there is still a large part of patients with rhabdoid tumors that do not respond to conventional multimodal therapy [43]. Thus, new approaches are desperately needed. The molecular biology of rhabdoid tumors is comparatively well described and understood. There is a substantial number of potential molecular targets and possible inhibitors, mostly focusing on the *SMARCB1*-related pathways, that have even been investigated in in vitro studies [43,44]. This promised dramatic improvements in the therapy and prognosis of patients suffering from rhabdoid tumors. However, breakthrough therapeutical innovations have failed to materialize so far [43]. The reasons for this shortcoming are manifold and complex, among them being the rarity of the disease and the additional constraints of conducting clinical trials in pediatric populations. Clinical data with promising results is lacking or studies did not meet efficacy endpoints and were terminated early [43,45]. For the time being, expert opinion strongly recommends treating pediatric patients with rhabdoid tumor according to existing multimodal treatments regimens, given that a part of these patient will respond well to this therapy [43]. For those with primary or secondary resistance to standard regimens, well designed studies with new agents are highly needed.

4.5. Certainty of Evidence and Strength of Recommendations

The certainty of evidence produced by this systematic review must be evaluated in the context of an extremely rare disease and the low level of existing evidence. All existing and included studies were non-comparative, thus limiting the evidence to a low level and the strength of recommendations based on this study is conditional. However, our study provides a comprehensive overview of all reported cases of MRTL and SCUD and thus an evidence base for several recommendations and future trials. In particular, a uniform classification of MRTL and SCUD with consecutive standardized treatment can be based on our results. Our study represents the highest level of evidence existing for these rare pediatric liver tumors.

5. Key Insights and Messages

- ➢ Results of this meta-analysis strongly suggest that SCUD should be treated with the same therapeutic strategies as MRTL;
- ➢ The term small cell undifferentiated hepatoblastoma is misleading and should be abandoned;

- SCUD and MRTL are associated with alarming mortality rates;
- Hepatoblastoma regimens are unsuited for both SCUD and MRTL;
- Soft tissue sarcoma chemotherapy is the best suited therapy to date for MRTL;
- Surgical resection is vital to maintain a chance of cure for children with MRTL and SCUD.

Supplementary Materials: The following supporting information can be downloaded at: https://www.mdpi.com/article/10.3390/cancers14020272/s1, File S1: Complete search strategy applied for the systematic literature search; File S2: Full list of included studies; File S3: Risk of bias assessment for all included case reports; File S4: Risk of bias assessment for all included observational studies; File S5: PRISMA-IPD checklist.

Author Contributions: J.F., A.M.-L., M.K., F.R., P.G., A.F., D.S. and K.H. contributed to conception and design of the study. J.F., A.M.-L. and K.H. conducted the systematic literature search, study selection, and extracted the data. J.F. and A.M.-L. performed the statistical analysis. J.F. wrote the initial draft of the manuscript. A.M.-L., K.H. and A.F. wrote sections of the manuscript. All authors substantially contributed to manuscript revision. All authors have read and agreed to the published version of the manuscript.

Funding: This research received no external funding.

Institutional Review Board Statement: All included studies stated consent for publication of the patient data by either approval from the relevant committees or the parents. As only anonymized data were part of our analyses, the institutional review board of the Medical Faculty of the University of Heidelberg approved the data collection and conduct of the present study without additional consent being necessary (Sign July 2013 and Section 15, paragraph 1 of the code of medical ethics of the federal state of Baden-Wuerttemberg, Germany).

Informed Consent Statement: All included studies stated that informed consent was obtained from all subjects. Only patient data that had already been published was included in this meta-analysis and no additional patient consent was necessary.

Data Availability Statement: Publicly available datasets were analyzed in this study. The full search strategy used for this study is provided.

Acknowledgments: We thank all authors of the included studies and case reports for their work, that formed the basis of this analysis.

Conflicts of Interest: The authors declare that the research was conducted in the absence of any commercial or financial relationships that could be construed as a potential conflict of interest.

Abbreviations

AFP	Alfa-fetoprotein
C5V	Chemotherapy regimen consisting of Carboplatin, 5-flururacil and Vincristine
C5VD	Chemotherapy regimen consisting of Carboplatin, 5-flururacil, Vincristine and Doxorubicin
CarboDV	Chemotherapy regimen consisting of Carboplatin, Doxorubicin and Vincristine
CDDP	Chemotherapy regimen consisting of Carboplatin
CEVAIE	Chemotherapy regimen consisting of Carboplatin, Epirubicine, Vincristine, Ifosfamide (with Mesna), Actinomycin D and Etoposide.
CNS	Central nervous system
COG	Children's Oncology Group
CT	Chemotherapy
DRD	Disease related death
DV5Cyclo	Chemotherapy regimen consisting of Doxorubicin, Vincristine, 5-flururacil and Cyclophosphamid
DVCyclo	Chemotherapy regimen consisting of Doxorubicin, Vincristine and Cyclophosphamid

EAP	Chemotherapy regimen consisting of Etoposide, Doxorubicin and Cisplatin
EPSSG	European Paediatric Soft Tissue Sarcoma Study Group
EU-RHAB	European Rhabdoid Registry
F-SCHB	Hepatoblastoma with focal small cell histology
HB	Hepatoblastoma
HB CT	Hepatoblastoma chemotherapy
HR	Hazard ratio
ICE	Chemotherapy regimen consisting of Ifosfamide-Carboplatinum-Etoposide
IQR	Interquartile range
MRTL	Malignant rhabdoid liver tumor
OS	Overall survival
PFS	Progression free survival
PLADO	Chemotherapy regimen consisting of Cisplatin and Doxorubicin
PRETEXT	Pretreatment extent of disease
RoB	Risk of bias
SCUD	Small cell undifferentiated liver tumor
SD	Standard deviation
SR	Systematic review
STS CT	Soft tissue sarcoma chemotherapy
VAC	Chemotherapy regimen consisting of Vincristine, Actinomycin-D and Cyclophosphamide
VAC-IE	Chemotherapy regimen consisting of Vincristine, Doxorubicin, Cyclophosphamide, Ifosfamide and Etoposid

References

1. Vokuhl, C.; Oyen, F.; Häberle, B.; von Schweinitz, D.; Schneppenheim, R.; Leuschner, I. Small cell undifferentiated (SCUD) hepatoblastomas: All malignant rhabdoid tumors? *Genes Chromosomes Cancer* **2016**, *55*, 925–931. [CrossRef] [PubMed]
2. Fazlollahi, L.; Hsiao, S.J.; Kochhar, M.; Mansukhani, M.M.; Yamashiro, D.J.; Remotti, H.E. Malignant Rhabdoid Tumor, an Aggressive Tumor Often Misclassified as Small Cell Variant of Hepatoblastoma. *Cancers* **2019**, *11*, 1992. [CrossRef] [PubMed]
3. Gonzalez-Crussi, F.; Goldschmidt, R.A.; Hsueh, W.; Trujillo, Y.P. Infantile sarcoma with intracytoplasmic filamentous inclusions: Distinctive tumor of possible histiocytic origin. *Cancer* **1982**, *49*, 2365–2375. [CrossRef]
4. Brennan, B.; De Salvo, G.L.; Orbach, D.; De Paoli, A.; Kelsey, A.; Mudry, P.; Francotte, N.; Van Noesel, M.; Bisogno, G.; Casanova, M.; et al. Outcome of extracranial malignant rhabdoid tumours in children registered in the European Paediatric Soft Tissue Sarcoma Study Group Non-Rhabdomyosarcoma Soft Tissue Sarcoma 2005 Study—EpSSG NRSTS 2005. *Eur. J. Cancer* **2016**, *60*, 69–82. [CrossRef]
5. Cornet, M.; De Lambert, G.; Pariente, D.; Planchon, J.M.; Guettier, C.; Martelli, H.; Guérin, F.; Branchereau, S. Rhabdoid tumor of the liver: Report of 6 pediatric cases treated at a single institute. *J. Pediatr. Surg.* **2018**, *53*, 567–571. [CrossRef]
6. Aronson, D.C.; Czauderna, P.; Maibach, R.; Perilongo, G.; Morland, B. The treatment of hepatoblastoma: Its evolution and the current status as per the SIOPEL trials. *J. Indian Assoc. Pediatr. Surg.* **2014**, *19*, 201–207. [CrossRef] [PubMed]
7. Gonzalez-Crussi, F. Case 1 undifferentiated small cell ("anaplastic") hepatoblastoma. *Pediatric Pathol.* **1991**, *11*, 155–161. [CrossRef] [PubMed]
8. Maibach, R.; Roebuck, D.; Brugieres, L.; Capra, M.; Brock, P.; Dall'Igna, P.; Otte, J.-B.; de Camargo, B.; Zsiros, J.; Zimmermann, A.; et al. Prognostic stratification for children with hepatoblastoma: The SIOPEL experience. *Eur. J. Cancer* **2012**, *48*, 1543–1549. [CrossRef]
9. Haas, J.E.; Feusner, J.H.; Finegold, M.J. Small cell undifferentiated histology in hepatoblastoma may be unfavorable. *Cancer* **2001**, *92*, 3130–3134. [CrossRef]
10. Bharti, S.; Bharti, J.N.; Sinha, A.; Yadav, T. Common and Rare Histological Variants of Hepatoblastoma in Children: A Pathological Diagnosis and Review of the Literature. *Gastrointest. Tumors* **2021**, *8*, 41–46. [CrossRef]
11. Tomlinson, G.E. Cytogenetics of hepatoblastoma. *Front. Biosci.* **2012**, *4*, 1287–1292. [CrossRef]
12. Ranganathan, S.; Lopez-Terrada, D.; Alaggio, R. Hepatoblastoma and Pediatric Hepatocellular Carcinoma: An Update. *Pediatr. Dev. Pathol.* **2020**, *23*, 79–95. [CrossRef] [PubMed]
13. López-Terrada, D.; Alaggio, R.; De Dávila, M.T.; Czauderna, P.; Hiyama, E.; Katzenstein, H.M.; Leuschner, I.; Malogolowkin, M.H.; Meyers, R.L.; Ranganathan, S.; et al. Towards an international pediatric liver tumor consensus classification: Proceedings of the Los Angeles COG liver tumors symposium. *Mod. Pathol.* **2013**, *27*, 472–491. [CrossRef]
14. Rudzinski, E.; Ranganathan, S.; Hicks, J.; Kim, G.; College of American Pathologists. Protocol for the Examination of Resection Specimens from Patients with Hepatoblastoma. 2019. Available online: https://documents.cap.org/protocols/cp-pediatric-hepatoblast-resection-19-4000.pdf (accessed on 8 December 2021).
15. Zhou, S.; Gomulia, E.; Mascarenhas, L.; Wang, L. Is INI1-retained small cell undifferentiated histology in hepatoblastoma unfavorable? *Hum. Pathol.* **2015**, *46*, 620–624. [CrossRef] [PubMed]

16. Slim, K.; Nini, E.; Forestier, D.; Kwiatkowski, F.; Panis, Y.; Chipponi, J. Methodological index for non-randomized studies (MINORS): Development and validation of a new instrument. *ANZ J. Surg.* **2003**, *73*, 712–716. [CrossRef]
17. Murad, M.H.; Sultan, S.; Haffar, S.; Bazerbachi, F. Methodological quality and synthesis of case series and case reports. *BMJ Evid.-Based Med.* **2018**, *23*, 60–63. [CrossRef]
18. R Core Team. R language definition. In *R Foundation for Statistical Computing*; R Core Team: Vienna, Austria, 2000.
19. Kassambara, A.; Kosinski, M.; Biecek, P.; Fabian, S. Drawing Survival Curves using 'ggplot2'. In *Package 'Survminer', R package version 03 1*; R Core Team: Vienna, Austria, 2017.
20. Therneau, T.; Lumley, T. *R Survival Package*; R Core Team: Vienna, Austria, 2013.
21. Kennedy, N. *Forestmodel: Forest Plots from Regression Models, R Package Version 06*; R Core Team: Vienna, Austria, 2020.
22. Guyatt, G.H.; Oxman, A.D.; Vist, G.E.; Kunz, R.; Falck-Ytter, Y.; Alonso-Coello, P.; Schünemann, H.J. GRADE: An emerging consensus on rating quality of evidence and strength of recommendations. *BMJ* **2008**, *336*, 924–926. [CrossRef]
23. Towbin, A.J.; Meyers, R.L.; Woodley, H.; Miyazaki, O.; Weldon, C.B.; Morland, B.; Hiyama, E.; Czauderna, P.; Roebuck, D.J.; Tiao, G.M. 2017 PRETEXT: Radiologic staging system for primary hepatic malignancies of childhood revised for the Paediatric Hepatic International Tumour Trial (PHITT). *Pediatr. Radiol.* **2018**, *48*, 536–554. [CrossRef] [PubMed]
24. De Ioris, M.; Brugieres, L.; Zimmermann, A.; Keeling, J.; Brock, P.; Maibach, R.; Pritchard, J.; Shafford, L.; Zsiros, J.; Czaudzerna, P.; et al. Hepatoblastoma with a low serum alpha-fetoprotein level at diagnosis: The SIOPEL group experience. *Eur. J. Cancer* **2008**, *44*, 545–550. [CrossRef] [PubMed]
25. Lautz, T.; Ben-Ami, T.; Tantemsapya, N.; Gosiengfiao, Y.; Superina, R.A. Successful nontransplant resection of POST-TEXT III and IV hepatoblastoma. *Cancer* **2010**, *117*, 1976–1983. [CrossRef] [PubMed]
26. Bajpai, M.; Pal, K.; Agarwala, S.; Seth, T.; Gupta, A.K. Midterm results with hepatectomy after preoperative chemotherapy in hepatoblastoma. *Pediatr. Surg. Int.* **2005**, *21*, 364–368. [CrossRef]
27. Haeberle, B.; Rangaswami, A.; Krailo, M.; Czauderna, P.; Hiyama, E.; Maibach, R.; Lopez-Terrada, D.; Aronson, D.C.; Alaggio, R.; Ansari, M.; et al. The importance of age as prognostic factor for the outcome of patients with hepatoblastoma: Analysis from the Children's Hepatic tumors International Collaboration (CHIC) database. *Pediatr. Blood Cancer* **2020**, *67*, e28350. [CrossRef]
28. Vlajnic, T.; Brisse, H.J.; Aerts, I.; Freneaux, P.; Cellier, C.; Fabre, M.; Klijanienko, J. Fine needle aspiration in the diagnosis and classification of hepatoblastoma: Analysis of 21 New Cases. *Diagn. Cytopathol.* **2017**, *45*, 91–100. [CrossRef] [PubMed]
29. Shi, Y.; Commander, S.J.; Masand, P.M.; Heczey, A.; Goss, J.A.; Vasudevan, S.A. Vascular invasion is a prognostic indicator in hepatoblastoma. *J. Pediatr. Surg.* **2017**, *52*, 956–961. [CrossRef]
30. Tanaka, Y.; Inoue, T.; Horie, H. International pediatric liver cancer pathological classification: Current trend. *Int. J. Clin. Oncol.* **2013**, *18*, 946–954. [CrossRef] [PubMed]
31. Wu, X.; Dagar, V.; Algar, E.; Muscat, A.; Bandopadhayay, P.; Ashley, D.; Chow, C.W. Rhabdoid tumour: A malignancy of early childhood with variable primary site, histology and clinical behaviour. *Pathology* **2008**, *40*, 664–670. [CrossRef]
32. Al Nassan, A.; Sughayer, M.; Matalka, I.; Ghandour, K.; Masarweh, M.; Zimmermann, A.; Sultan, I. INI1 (BAF 47) Immunohistochemistry is an Essential Diagnostic Tool for Children with Hepatic Tumors and Low Alpha Fetoprotein. *J. Pediatric Hematol. Oncol.* **2010**, *32*, E79–E81. [CrossRef]
33. Meyers, R.L.; Rowland, J.R.; Krailo, M.; Chen, Z.; Katzenstein, H.M.; Malogolowkin, M.H. Predictive Power of Pretreatment Prognostic Factors in Children with Hepatoblastoma: A Report from the Children's Oncology Group. *Pediatr. Blood Cancer* **2009**, *53*, 1016–1022. [CrossRef]
34. Trobaugh-Lotrario, A.D.; Tomlinson, G.E.; Finegold, M.J.; Gore, L.; Feusner, J.H. Small cell undifferentiated variant of hepatoblastoma: Adverse clinical and molecular features similar to rhabdoid tumors. *Pediatr. Blood Cancer* **2009**, *52*, 328–334. [CrossRef] [PubMed]
35. Louis, D.N.; Ohgaki, H.; Wiestler, O.D.; Cavenee, W.K.; Burger, P.C.; Jouvet, A.; Scheithauer, B.W.; Kleihues, P. The 2007 WHO Classification of Tumours of the Central Nervous System. *Acta Neuropathol.* **2007**, *114*, 97–109. [CrossRef] [PubMed]
36. Russo, P.; Biegel, J.A. SMARCB1/INI1 Alterations and Hepatoblastoma: Another Extrarenal Rhabdoid Tumor Revealed? *Pediatr. Blood Cancer* **2009**, *52*, 312–313. [CrossRef] [PubMed]
37. Frühwald, M.C.; Krefeld, B.; Benesch, B.; Büchner, J.; Boos, J.; Ebetsberger, E.; Graf, N.; Kortmann, R.; Nysom, K.; Rutkowski, S.; et al. A comprehensive approach towards biology and clinical management. In *The European Rhabdoid Registry (EU-RHAB), Proceedings of the 14th International Symposium on Pediatric Neuro-Oncology 2010, Vienna, Austria, 20–23 June 2010*; Available online: https://www.skion.nl/workspace/uploads/eurhab_protocol_atrt_versie_15-11-2010.pdf (accessed on 9 December 2021).
38. Lee, J.; Kim, D.-S.; Han, J.W.; Suh, C.-O. Atypical teratoid/rhabdoid tumors in children treated with multimodal therapies: The necessity of upfront radiotherapy after surgery. *Pediatr. Blood Cancer* **2017**, *64*, e26663. [CrossRef]
39. Yang, W.-C.; Yen, H.-J.; Liang, M.-L.; Chen, H.-H.; Lee, Y.-Y.; Chang, F.-C.; Lin, S.-C.; Wong, T.-T.; Hu, Y.-W.; Chen, Y.-W. Effect of early radiotherapy initiation and high-dose chemotherapy on the prognosis of pediatric atypical teratoid rhabdoid tumors in different age groups. *J. Neuro-Oncol.* **2020**, *147*, 619–631. [CrossRef]
40. Cheng, H.; Yang, S.; Cai, S.; Ma, X.; Qin, H.; Zhang, W.; Fu, L.; Zeng, Q.; Wen, M.; Peng, X.; et al. Clinical and Prognostic Characteristics of 53 Cases of Extracranial Malignant Rhabdoid Tumor in Children. A Single-Institute Experience from 2007 to 2017. *Oncologist* **2019**, *24*, e551–e558. [CrossRef]

41. Sultan, I.; Qaddoumi, I.; Rodríguez-Galindo, C.; Al Nassan, A.; Ghandour, K.; Al-Hussaini, M. Age, stage, and radiotherapy, but not primary tumor site, affects the outcome of patients with malignant rhabdoid tumors. *Pediatr. Blood Cancer* **2010**, *54*, 35–40. [CrossRef] [PubMed]
42. Puri, D.R.; Meyers, P.A.; Kraus, D.H.; LaQuaglia, M.P.; Wexler, L.H.; Wolden, S.L. Radiotherapy in the multimodal treatment of extrarenal extracranial malignant rhabdoid tumors. *Pediatr. Blood Cancer* **2008**, *50*, 167–169. [CrossRef] [PubMed]
43. Nemes, K.; Frühwald, M.C. Emerging therapeutic targets for the treatment of malignant rhabdoid tumors. *Expert Opin. Ther. Targets* **2018**, *22*, 365–379. [CrossRef] [PubMed]
44. Kurmasheva, R.T.; Erickson, S.W.; Earley, E.; Smith, M.A.; Houghton, P.J. In vivo evaluation of the EZH2 inhibitor (EPZ011989) alone or in combination with standard of care cytotoxic agents against pediatric malignant rhabdoid tumor preclinical models—A report from the Pediatric Preclinical Testing Consortium. *Pediatr. Blood Cancer* **2021**, *68*, e28772. [CrossRef] [PubMed]
45. Upadhyaya, S.; Campagne, O.; Robinson, G.W.; Onar-Thomas, A.; Orr, B.; Billups, C.A.; Tatevossian, R.G.; Broniscer, A.; Kilburn, L.B.; Baxter, P.A.; et al. Phase II study of alisertib as a single agent in recurrent or progressive atypical teratoid rhabdoid tumors. *J. Clin. Oncol.* **2020**, *38*, 10542. [CrossRef]

cancers

Article

Reduction of LPAR1 Expression in Neuroblastoma Promotes Tumor Cell Migration

Xiangjun Liu [1,†], Mengmiao Pei [1,†], Yongbo Yu [2], Xiaolin Wang [1,*] and Jingang Gui [1,*]

1 Laboratory of Tumor Immunology, Beijing Pediatric Research Institute, Beijing Children's Hospital, Capital Medical University, National Center for Children's Health (NCCH), Beijing 100045, China; somusl@126.com (X.L.); mengmiao_pei@163.com (M.P.)
2 Beijing Key Laboratory for Pediatric Diseases of Otolaryngology, Head and Neck Surgery, Beijing Pediatric Research Institute, Beijing Children's Hospital, Capital Medical University, National Center for Children's Health (NCCH), Beijing 100045, China; yuyongbo1688@126.com
* Correspondence: wangxl19891012@163.com (X.W.); guijingang@bch.com.cn (J.G.); Tel.: +86-010-59616176 (X.W.); +86-010-59616176 (J.G.)
† These authors contributed equally to this work.

Simple Summary: The tumor metastasis in the bone marrow or other organs in high-risk neuroblastoma patients is a serious problem to tackle and strongly impairs the survival of patients. Novel and effective targets for the treatment of neuroblastoma, especially tumor metastasis, need to be explored. Using multiple databases and analysis methods, LPAR1 was screened out through our comprehensive bioinformatics analysis and found to be positively associated with survival of neuroblastoma patients. LPAR1 was proved to be reduced in neuroblastoma cells compared with non-mailgant cells. LPA-LPAR1 axis showed migration-inhibitory effects on neuroblastoma cells, suggesting that LPAR1 may be a potential target for future treatment of neuroblastoma.

Abstract: Neuroblastoma is the most common extracranial solid tumor in children. Tumor metastasis in high-risk NB patients is an essential problem that impairs the survival of patients. In this study, we aimed to use a comprehensive bioinformatics analysis to identify differentially expressed genes between NB and control cells, and to explore novel prognostic markers or treatment targets in tumors. In this way, *FN1*, *PIK3R5*, *LPAR6* and *LPAR1* were screened out via KEGG, GO and PPI network analysis, and we verified the expression and function of LPAR1 experimentally. Our research verified the decreased expression of LPAR1 in NB cells, and the tumor migration inhibitory effects of LPA on NB cells via LPAR1. Moreover, knockdown of LPAR1 promoted NB cell migration and abolished the migration-inhibitory effects mediated by LPA-LPAR1. The tumor-suppressing effects of the LPA-LPAR1 axis suggest that LPAR1 might be a potential target for future treatment of NB.

Keywords: LPAR1; neuroblastoma; LPA; tumor metastasis; bioinformatics analysis

Citation: Liu, X.; Pei, M.; Yu, Y.; Wang, X.; Gui, J. Reduction of LPAR1 Expression in Neuroblastoma Promotes Tumor Cell Migration. *Cancers* **2022**, *14*, 3346. https://doi.org/10.3390/cancers14143346

Academic Editors: Saurabh Agarwal and Jianhua Yang

Received: 8 June 2022
Accepted: 7 July 2022
Published: 9 July 2022

1. Introduction

Neuroblastoma (NB), an embryonic tumor of the sympathetic nervous system that arises in the fetus or early after birth from sympathetic cells produced by the neural crest, is a significant cause of childhood death [1]. Although some NBs automatically degenerate and have a good prognosis, tumor metastasis in the bone marrow or other organs in high-risk NB patients is still an essential problem to tackle [2,3]. It is necessary to find novel, effective targets for the treatment of NB, especially tumor metastasis. Microarray technology and bioinformatics analysis are increasingly used to explore the significant genetic or epigenetic variations in tumors and determine cancer diagnoses and prognoses, as well as determine treatment targets [4]. In this study, we aimed to identify novel diagnostic biomarkers or therapeutic targets and determine the pathogenesis in NB using bioinformatics analysis and experimental verification.

The Gene Expression Omnibus (GEO) is an international public knowledge base for archiving and distributing microarrays, next-generation sequences and other forms of high-throughput functional genomic data free of charge [5]. We accessed publicly available data on NB cells and non-malignant control cells and screened differentially expressed genes (DEGs) subjected to Kyoto Encyclopedia of Genes and Genomes (KEGG) and Gene Ontology (GO) analyses. The Protein-Protein Interaction (PPI) network and Molecular Complex Detection (MCODE) plug-in were further employed to boil down the critical DEGs including *FN1, PIK3R5, LPAR6* and *LPAR1*. Through an investigation into the expression patterns and potential regulating functions of the screened genes in NB cells, our research mainly focused on exploring the expression and function of LPAR1.

LPAR1 is a member of the G protein-coupled receptor family of lysophosphatidic acid (LPA) receptors (LPARs), including LPAR1 to LPAR6 [6]. LPA is a small phospholipid generally found in serum, ascitic effusions and inflammatory fluids [7]. LPA acts as an extracellular signaling molecule by binding to and activating its receptors LPARs, thereby exerting regulating functions in cellular proliferation/migration/survival, vascular homeostasis, stromal remodeling, lymphocyte trafficking and immune regulation [8–10]. Aberrant LPAR1 expression was evident in a variety of cancer cell lines and primary tumors [6]. LPAR1 was significantly downregulated in prostate cancer, and high LPAR1 expression was correlated with a favorable overall survival [11]. Furthermore, LPAR1 was reported to mediate migration- or invasion-inhibiting signals in prostate cancer [7], gastric cancer [12] and pancreatic cancer [13]. In a rat neuroblastoma cell line or mouse fibroblast cell line, overexpression of LPAR1 also markedly decreased intrinsic cell motility and invasion [14,15].

Our results show the decreased expression of LPAR1 in NB cells, demonstrating that LPA can exert tumor migration-inhibitory effects on NB cells via LPAR1. Knockdown of LPAR1 also promotes NB cell migration and abolishes the migration-inhibitory effects mediated by LPA-LPAR1.

2. Materials and Methods

2.1. Microarray Data Collection and Preprocessing

We searched the microarray gene expression datasets associated with neuroblastoma from GEO (https://www.ncbi.nlm.nih.gov/geo/ (accessed on 27 November 2021) for the study. GEO, NCBI's publicly available genomics database, which collects submitted high-throughput gene expression data, was thoroughly queried for all datasets involving studies on NB. Datasets were related to a neuroblastoma group and a negative control group in humans. Our inclusion criteria were as follows: (1) Expression profiling data by microarray; (3) Complete microarray normalized data. Ultimately, we chose six NB cells from GSE28019, GSE16477 and GSE90804 and three non-malignant control cells from GSE10592, GSE24733 and GSE57864, using the raw data in our study. The data were based on the [HG-U133_Plus_2] Affymetrix Human Genome U133 Plus 2.0 Array Plate. First, the dataset was quality-controlled before differential genetic screening analysis, which included use of the "affyPLM" package in R software (R version 4.0.4) to verify the conformance of parallel trials. Then, a robust multi-array averaging (RMA) algorithm was applied using the "affy" package in R to convert the raw array of data into expression values and to perform background correction, normalization and probe summarization [16,17]. Both a p-value < 0.01 and log2 fold change >2.1 were considered critical for DEG screening based on the paired t-test of the "limma" R package [18].

2.2. Functional and Pathway Enrichment Analysis of Neuroblastoma-Specific DEGs

GO is a community-based bioinformatics resource that supplies information about gene product functions, using ontologies to represent biological knowledge [19], thereby informing us what kinds of biological functions genes have. It mainly consists of three categories: cell composition (CC), molecular function (MF) and biological process (BP). KEGG is a knowledge base for the systematic analysis of gene functioning, linking genomic information with higher-order functional information. Genomic information is stored in the Gene Database, a collection of gene catalogs of all sequenced genomes and partial genomes, with updated annotations of gene functions [20]. GO and KEGG analyses can be found in the DAVID database (DAVID version 6.8; https://david.ncifcrf.gov/ (accessed on 27 November 2021)), which is a fully functional annotation tool providing a comprehensive set of functional annotation tools for investigators to use to understand the biological meaning behind a long list of genes [21]. A p-value < 0.05 was taken as the critical value when identifying DEGs using official gene symbols.

2.3. PPI Networks

An online biological database STRING (https://string-db.org (accessed on 27 November 2021), Version 11.0), from which we obtained information on protein co-expression, is well known for supporting protein co-expression prediction based on known and predicted gene PPI networks for the analysis of functional interactions between proteins [22]. In this work, PPI networks of co-expressed genes were established using the STRING database, and we considered interaction with a joint score >0.4 to be statistically significant. Then, the resulting network data were imported into local software Cytoscape (https://cytoscape.org/ (accessed on 27 November 2021), version 3.8.2) to be further visually analyzed. The functional interactions between proteins provide insights into the mechanisms of disease development, which we can access by visualizing molecular interaction networks and biological pathways and integrating these networks with annotations [23], gene expression profiles and other state data.

2.4. The PPI Networks and Module Selection

Clustering coefficients were calculated by the Molecular Complex Detection (MCODE) plugin in Cytoscape, and modularity was used to identify modules in the co-occurrence networks [24]. The Degree and Betweenness are factors of the topological algorithm and shortest path, respectively. We used degree cut-off = 2, node score cut-off = 0.2, k-core = 2 and max. depth = 100 as the MCODE plug-in default parameters.

2.5. Cell Lines

Neuroblastoma cell lines SH-SY5Y, SK-N-BE2 and IMR-32, and non-malignant cell lines RPE-1, HBE and HEK293T, were purchased from the American Type Culture Collection (Manassas, VA, USA). CHLA-255 cells were kindly provided by Prof. Shahab Asgharzadeh from the Children's Hospital Los Angeles. SH-SY5Y, SK-N-BE2, RPE-1, HBE and HEK293T were cultured in Dulbecco's modified Eagle's medium (DMEM) (Corning Incorporated, Corning, NY, USA) supplemented with 10% FBS (fetal bovine serum, Gibco, Invitrogen, Carlsbad, CA, USA) and 1% penicillin/streptomycin, IMR-32 was cultured in minimum essential medium (MEM) (Corning Incorporated, Corning, NY, USA) and CHLA-255 was cultured in Iscove's DMEM (IMEM) (Corning Incorporated, Corning, NY, USA). Cells were cultured at 37 °C in a humidified cell incubator with 5% CO_2.

2.6. Transient Transfection of siRNA

Lipofectamine RNAiMAX (Invitrogen, Carlsbad, CA, USA) was used to transfect the cells with LPAR1 siRNA. To facilitate transfection, the cells were seeded to 60% confluence on a six-well plate during transfection. The next day, siRNA was transfected using RNAiMAX and Opti-MEM according to the manufacturer's instructions. Cells were harvested or subjected to other experiments after 48 h.

2.7. PCR and Real-Time PCR

The total RNA from cells was isolated using TRIzol reagent (Invitrogen, Carlsbad, CA, USA). Reverse transcription was performed according to standard protocols using a RevertAid™ II First-Strand cDNA Synthesis Kit (Thermo Fisher Scientific Inc., Waltham, MA, USA). PCR and real-time PCR were performed as previously described [25]. The amplification conditions of real-time PCR were as follows: 10 min initial denaturation at 95 °C, then 40 cycles of 15 s at 95 °C and 1 min at 60 °C. The relative quantity (RQ) was calculated by the $2^{\Delta\Delta Ct}$ method. GAPDH was amplified as an internal standard.

The primer sequences for PCR are listed below:
LPAR1-F, 5′-AATCTATGTCAACCGCCGCT-3′
LPAR1-R, 5′-GTCAATGAGGCCCTGACGAA-3′
LPAR3-F, 5′-TTAGGGGCGTTTGTGGTATG-3′
LPAR3-R, 5′-CCTTGTAGGAGTAGATGATGGGGT-3′
LPAR6-F, 5′-CTGCGTCCTCAAAGTCCGAA-3′
LPAR6-R, 5′-CCAAATGGCCAATTCCGTGT-3′
The primer sequences for real-time PCR are listed below:
LPAR1-F, 5′-TCAACTCTGCCATGAACCCC-3′
LPAR1-R, 5′-ACTCCAGCCAAGATGGTGTG-3′

2.8. Cell Proliferation Assay

A Cell Counting Kit-8 (CCK8) detection kit (Dojindo Molecular Technologies, Japan) was used for measuring the cell proliferation; the cell number was directly proportional to the amount of formazan dye detected by the absorbance at 450 nm. Cells were seeded in 96-well plates at a concentration of 8000 cells per well (10,000 cells per well for CHLA-255 cells) and cultured in a complete culture medium with 10 μM LPA, LPA plus 10 μM Ki16425 or Ki16424 alone. At the indicated times, 10 μL of CCK-8 solution was added to each well. The plate was then incubated at 37 °C for 120 min, and the absorbance was detected.

2.9. Wound-Healing Assay

Cells were seeded in a six-well plate with 10 μM LPA, LPA plus 10 μM Ki16425 or Ki16425 alone, grown to about 80% confluence and a wound was carefully scraped with a sterilized pipette tip in the cell monolayer. After replacement with a fresh complete culture medium, photomicrographs were taken immediately, as well as 72 h after scraping. The wound widths in the pictures were measured using ImageJ software. The percentage of cell migration was calculated based on the ratio of wound width at 72 h and the initial wound width at 0 h.

2.10. Cell Migration Assay

Transwell chambers (8 μm pore size, BD Biosciences, NJ, USA) were used to measure cell migration. The cells were cultured in a serum-free culture medium for 12 h and then seeded in the upper chamber at a density of 1×10^5 cells per well in 250 μL serum-free DMEM medium with 10 μM LPA, LPA plus 10 μM Ki16425 or Ki16425 alone. The appropriate complete culture medium was added to the lower chamber. After incubation at 37 °C with 5% CO_2 for 24 h, the chambers' contents were collected. The membranes were then fixed with 4% paraformaldehyde in PBS and stained with 2% crystal violet for 10 min. Photomicrographs were taken and the absolute cell numbers were counted from images captured by a microscope (100× magnification) (IX73, Olympus, Tokyo, Japan).

2.11. Statistical Analysis

Statistical comparisons were performed using GraphPad Prism software (version 8.0) (GraphPad Software Inc., San Diego, CA, USA). Student's *t*-test was used to analyze the data. Error bars represented the SEM. Significant differences between groups were represented by * $p < 0.05$, ** $p < 0.01$, and *** $p < 0.001$.

3. Results

3.1. Identification of DEGs Using mRNA Microarray Data Analysis and GO/KEGG Enrichment Analysis

After collecting the mRNA microarray data of six NB cell lines (data for SH-SY5Y, SK-N-BE2 and IMR-32 cells from three different datasets) and three non-malignant cell lines (data of PRE-1, HBE and HEK293T cells from three different datasets) from the GEO database, we first performed relative log expression (RLE) boxplots analysis, and the results suggested the normalized raw data (Figure 1A). All DEGs were screened using R software (R version 4.0.4) based on an adjusted p-value < 0.01 and log2 fold change >2.1. Clustering analysis of these DEGs was performed using volcano plots (Figure 1B). A total of 5492 DEGs were identified from the six NB samples and the other three non-malignant control cell samples, including 38 upregulated DEGs and 5454 downregulated DEGs (Supplementary Table S1). To further our understanding of the functions of the screened DEGs, we conducted GO/KEGG enrichment analysis. All DEGs were included in the functional enrichment analysis using the DAVID database and visualized using R software. The results showed that the NB sample group had a unique GO condition. As shown in Figure 1C–E and Table 1, pathways related to extracellular matrix organization, angiogenesis, cell adhesion, positive regulation of NF-κB signaling, regulation of cell proliferation, regulation of PI3K signaling and activation of MAPK activity were enriched in GO BP analysis, and the plasma membrane, cell surface, proteinaceous extracellular matrix, cell-cell junction, focal adhesion for GO CC, calcium ion binding, receptor activity, PIK3Ca activity and cytokine receptor activity for GO MF. In terms of KEGG pathway analysis, in Figure 1F and Table 2, the NB group enriched unique pathways such as cytokine-cytokine receptor interaction, cell adhesion molecules (CAMs), the Jak-STAT signaling pathway, NF-kappa B signaling pathway, focal adhesion, the PI3K-Akt signaling pathway, pathways in cancer and the MAPK signaling pathway.

Table 1. GO analysis of DEGs.

Category	GO ID	Term	Gene ID	p-Value
BP	GO:0030198	Extracellular matrix organization	VIT, ITGB4, TNC, F11R, TNF, DAG1	1.08×10^{-8}
BP	GO:0001525	Angiogenesis	CIB1, CTGF, LEPR, SYK, EREG, TGFA	1.22×10^{-7}
BP	GO:0007155	Cell adhesion	TNC, COMP, TNR, FEZ1, CD151, LPP	2.70×10^{-6}
BP	GO:0043123	Positive regulation of NF-κB signaling	TNF, CCR7, LTF, IRF3, LPAR1, NEK6	1.89×10^{-5}
BP	GO:0042127	Regulation of cell proliferation	TES, CXCL3, JAK1, FGR, ACE2, LCK	1.06×10^{-5}
BP	GO:0014066	Regulation of PI3K signaling	KLB, EGFR, IER3, BTC, NRG4, IRS1	5.35×10^{-5}
BP	GO:0030335	Positive regulation of cell migration	ILK, PLAU, FGR, LEF1, CCL7, DAB2	3.02×10^{-4}
BP	GO:0000187	Activation of MAPK activity	LPAR1, PLCE1, GRM4, WNT5A, MOS	0.003828
CC	GO:0005886	Plasma membrane	SLA2, LIPH, AR, ACE2, FPR3, MYO6	2.53×10^{-25}
CC	GO:0009986	Cell surface	LIPG, KRT4, BST2, TF, CALR, SHH	1.58×10^{-4}
CC	GO:0005578	Proteinaceous extracellular matrix	GLDN, TNR, LOX, PI3, CILP, CALR	1.66×10^{-10}
CC	GO:0031090	Organelle membrane	FAAH, TFPI, FMO1, FA2H, CYP2S1	3.28×10^{-7}
CC	GO:0005911	Cell-cell junction	MLC1, KRT8, DSG2, TLN1, VCL	9.37×10^{-5}
CC	GO:0005925	Focal adhesion	TNC, PVR, TNS4, EZR, PXN, CALR	0.002148
CC	GO:0005789	Endoplasmic reticulum membrane	ALG1, POR, HPD, RCE1, PIGS, PIGZ	0.008747
MF	GO:0005509	Calcium ion binding	SYTL2, REG4, AIF1L, EHD1, CALR	2.32×10^{-6}
MF	GO:0004872	Receptor activity	PVR, THBD, TLR1, LRP1, CALCR	1.06×10^{-6}
MF	GO:0046934	PIK3Ca activity	KLB, PIK3R5, EGF, BTC, LCK, NRG4	7.22×10^{-5}
MF	GO:0004896	Cytokine receptor activity	FLT3, MPL, CSF2RB, OSMR, CD44	0.001221

As summarized in our results, the PI3K pathway was enriched and activated in the NB group, and the PI3K pathway is generally known to activate Akt and further mediate multiple biological effects [26], including those involved in cell proliferation, apoptosis inhibition, cell migration and cell cancerous transformation, contributing much to tumorigenesis. Therefore, we subsequently screened overlapped DEGs that were statistically significant in both the PI3K-Akt signaling pathway and pathways in cancer, and 51 further screened DEGs were shown in the heatmap (Figure 1G).

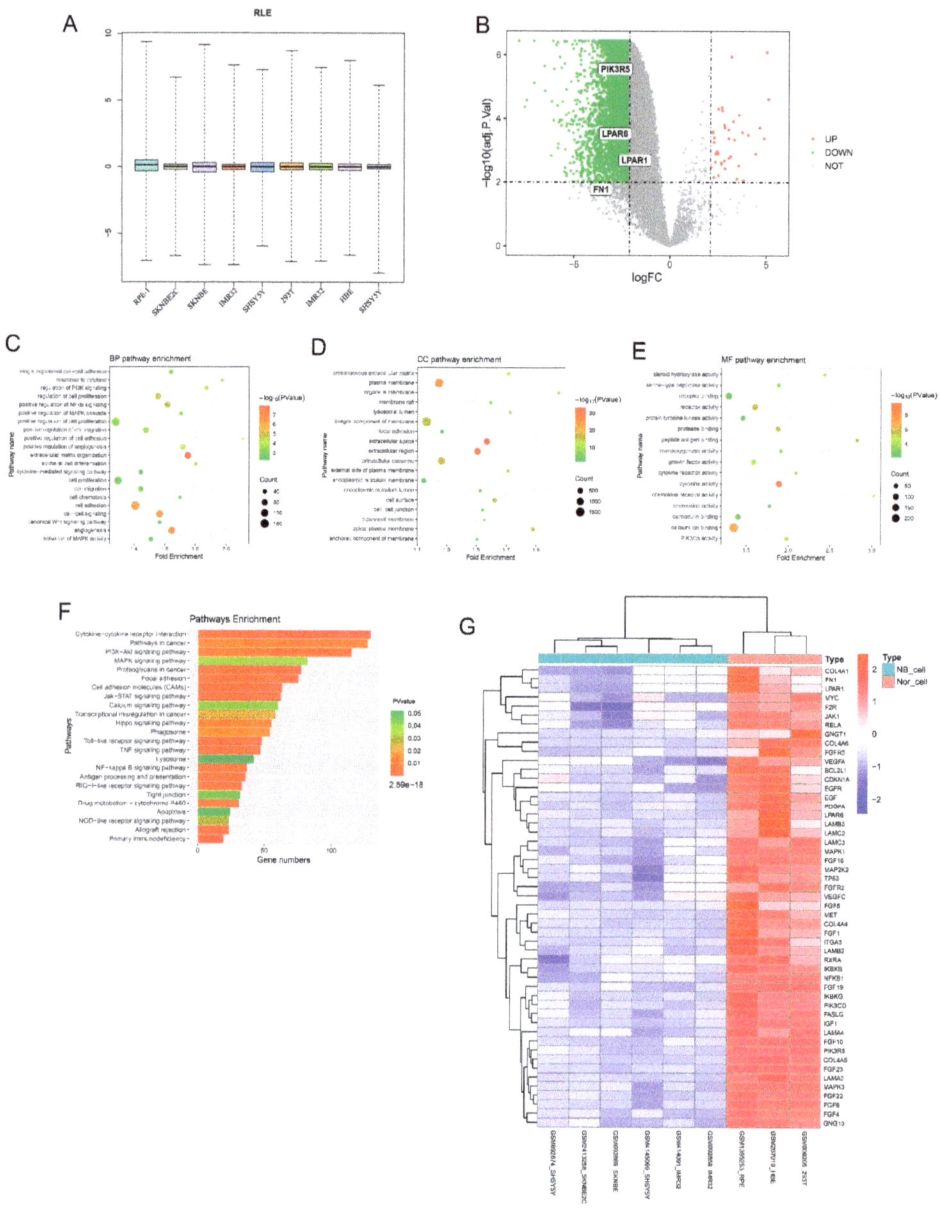

Figure 1. Identification of DEGs using mRNA microarray data analysis and GO/KEGG enrichment analysis. (**A**) Boxplots of RLE indicate the normalized raw data of microarray gene expression datasets. (**B**) Volcano plot distribution of all DEGs, with red points for the screened upregulated DEGs and green points for the screened downregulated DEGs. (**C–E**) Bubble chart visualization for GO analysis of all DGEs in NB cells and non-malignant cells. GO BP analysis (**C**), GO CC analysis (**D**) and GO MF analysis (**E**). (**F**) KEGG pathway analysis of unique DEGs in NB cells and non-malignant cells. (**G**) Hierarchical clustering analysis (heatmap) of 51 DEGs overlapping between PI3K-Akt pathways and pathways in cancer.

Table 2. KEGG pathway analysis of DEGs.

KEGG ID	Term	Gene ID	*p*-Value
hsa04060	Cytokine-cytokine receptor interaction	MPL, EDAR, NGFR, LIF, EDA, PRL	2.59×10^{-18}
hsa04514	Cell adhesion molecules (CAMs)	PVR, SPN, CTLA4, CD8A, SELP	1.01×10^{-5}
hsa04630	Jak-STAT signaling pathway	OXTR, LEPR, LPAR1, MC2R, PLG	4.98×10^{-5}
hsa04668	TNF signaling pathway	RELA, JUN, EDN1, JAG1, MLKL	2.38×10^{-6}
hsa04064	NF-kappa B signaling pathway	PTGS2, RELA, PLAU, SYK, LTBR	0.002268
hsa04510	Focal adhesion	MYLK, TNR, VWF, VCL, SRC, SPP1	0.002427
hsa04151	PI3K-Akt signaling pathway	TP53, LPAR1, CHAD, PCK1, PRL	0.005575
hsa05200	Pathways in cancer	MITF, TP53, LPAR1, LPAR6, FLT3	0.011031
hsa04010	MAPK signaling pathway	FOS, TP53, RRAS, FLNB, NTRK2	0.037591
hsa04020	Calcium signaling pathway	RYR1, OXTR, PLCE1, ORAI1, ITPR3	0.040378

3.2. PPI Network Construction, Module Analysis and Hub Gene Determination

PPI network analysis plays a major role in predicting the functionality of interacting genes or proteins and gives an insight into the functional relationships and evolutionary conservation of interactions among genes. Based on the screened DEGs, a PPI network was generated in the STRING protein interaction database and imported into the bioinformatics software platform Cytoscape (Version 3.8.2) for visualization and further analysis (Figure 2A). Then, the MCODE plug-in was used to select important functional modules of protein interaction networks for the identified DEGs (Figure 2B,C), and critical genes were defined according to the degree level. *FN1*, *PIK3R5*, *LPAR6* and *LPAR1* were determined to have a high degree of network connectivity. The expression levels of these four genes were shown to be decreased in NB cells (Figure 2D).

3.3. Hub Gene Expression and Survival Analysis

The association between hub gene expression and NB patients' survival was analyzed using the Kaplan-Meier survival curves [27]. These were generated based on the mRNA expression levels of *FN1*, *PIK3R5*, *LPAR6* and *LPAR1*, with the log-rank test *p*-value indicated using the R2: Genomics Analysis and Visualization Platform and using Tumor Neuroblastoma-SEQC-498-custom-ag44kcwolf datasets. As shown in Figure 3A–D, survival analysis revealed that a poor prognosis was significantly associated with low *LPAR1* mRNA levels in NB patients (bonf $p < 0.05$), which was the same for *FN1*, *PIK3R5* and *LPAR6*.

The results of LPAR1 expression analysis at different stages (the International Neuroblastoma Staging System (INSS)) indicated that LPAR1 showed the lowest expression level in st4 NB tumors with metastasis, rather than st4s with limited metastasis, both of which expressed lower LPAR1 levels than st1, st2 and st3 NB tumors (Figure 3E). High-risk NB tumors also showed a lower level of LPAR1 (Figure 3F). In addition, NB tumors leading to patients' death showed significantly lower LPAR1 expression (Figure 3G), consistent with the survival curve. Expression analyses of the other three genes at different stages, risk levels and death events were performed, and the results are shown in Supplementary Figure S1. PIK3R5 and LPAR6 showed similar expression patterns in NB tumors. FN1, meanwhile, demonstrated the lowest expression level in st4 NB tumors with limited metastasis, and showed no significant differences between high-risk NB tumors and NB tumors leading to patients' death.

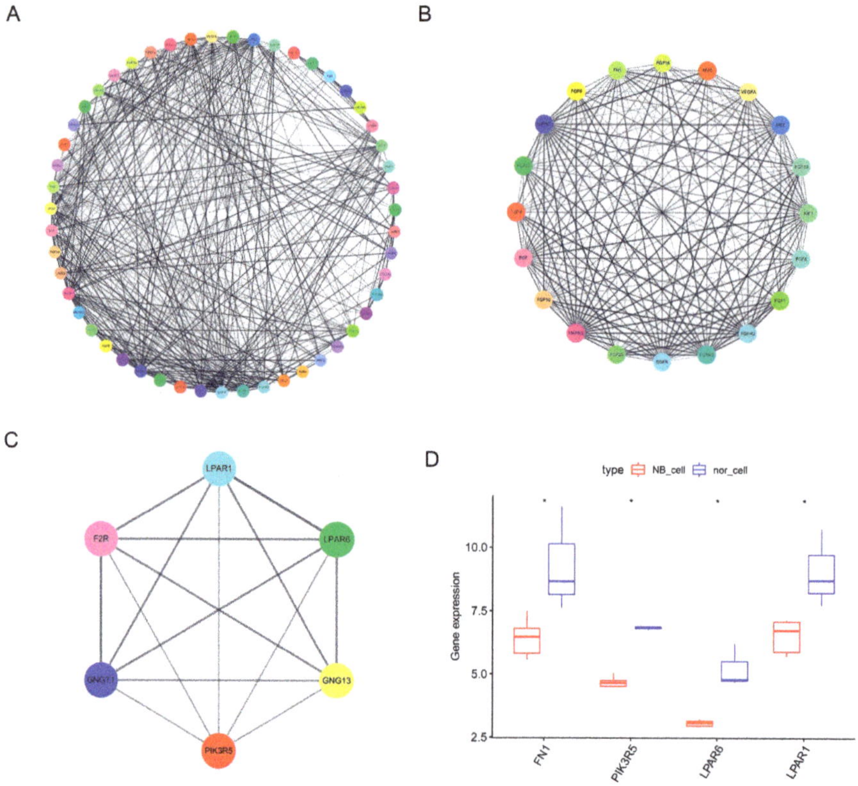

Figure 2. PPI network construction, module analysis and hub gene determination. (**A**) PPI network of screened genes was analyzed using STRING and Cytoscape for visualization. (**B**,**C**) Hub genes of protein interaction networks selected using MCODE. (**D**) Boxplot analysis was performed to identify the decreased expression of *FN1*, *PIK3R5*, *LPAR6* and *LPAR1*, with a high degree of network connectivity in the NB cells compared to the non-malignant cells. * $p < 0.05$.

LPARs are the receptors of LPA and mediate the regulating function involved in multiple tumor-related cellular processes, such as proliferation/migration/survival and vascular homeostasis [8–10]. Our analysis suggested that both LPAR1 and LPAR6 expression were beneficial to NB patients' survival, possibly involved in the regulation of tumor metastasis mediated by LPA.

3.4. NB Cells Showed Low Expression Level of LPAR1 Compared to Non-Malignant Cell Lines

According to the bioinformatics analysis results, we examined the expression of LPAR1 and LPAR6 in non-malignant cells and NB cells. The results in Supplementary Figure S2A show that both NB cells and non-malignant cells expressed extremely low levels of LPAR6, which made it difficult for the ligand LPA to exert functions via LPAR6. Yet, our real-time PCR and PCR results in Figure 4A,B indicate that all detected cell lines expressed LPAR1 to some extent, and NB cells, including SH-SY5Y, SK-N-BE2 and IMR-32 cells, expressed lower levels of LPAR1 compared to non-malignant cells (PRE-1, HBE and HEK293T cells). Only NB cell line CHLA-255 showed a relatively high expression of LPAR1. Therefore, we focused mainly on the expression and function of LPAR1 in NB.

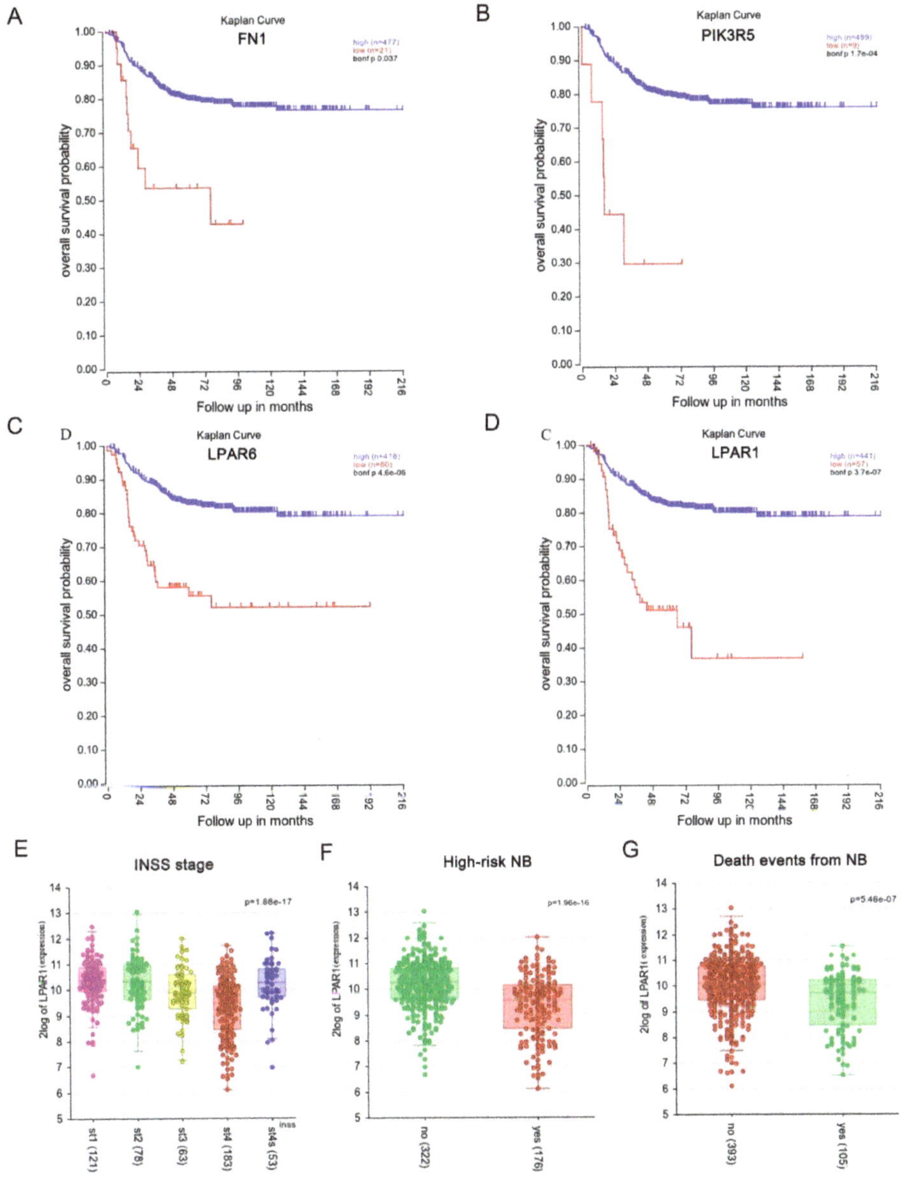

Figure 3. Hub gene expression and survival analysis. (**A–D**) Kaplan–Meier survival analysis for the SEQC datasets of 498 NB patients based on the average mRNA expression. Survival curves of FN1 (**A**), PIK3R5 (**B**), LPAR6 (**C**) and LPAR1 (**D**) in NB are shown, where $p < 0.05$ is regarded as the critical point with statistical significance. (**E–G**) R2 database view-a-gene was used to analyze the association between the LPAR1 expression and the NB INSS stage (**E**), likelihood of being high-risk (**F**) and likelihood of a death event (**G**) based on the average mRNA expression of the 498 NB SEQC datasets, with $p < 0.05$ regarded as the critical point with statistical significance.

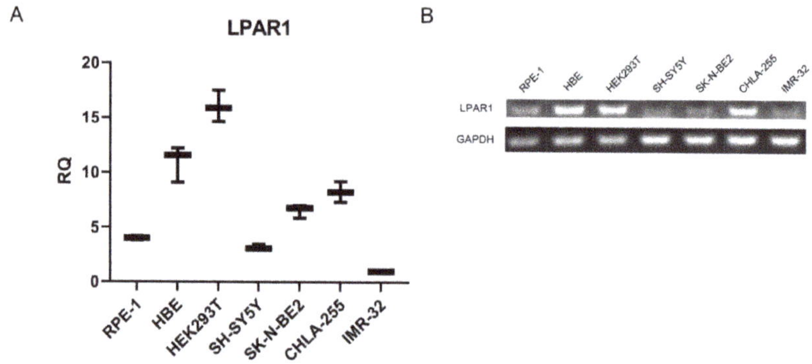

Figure 4. NB cells showed low expression level of LPAR1 compared to non-malignant cell lines. (**A**,**B**) The expression of LPAR1 at the mRNA level was analyzed by real-time PCR (**A**) and PCR (**B**) in NB cells and non-malignant cells. Original blots see File S1.

3.5. LPA Suppressed the Migration of NB Cells via LPAR1

We had identified a relatively low LPAR1 level in NB cells and a positive correlation between LPAR1 expression and NB patient survival. To investigate the function of LPAR1 in NB, then we examined the effect of LPA, mediating intracellular actions mainly via LPARs, on NB cell proliferation and migration with or without LPAR1/LPAR3 inhibitor Ki16425. The expression levels of LPAR1 and LPAR3 were detected, and the results showed that SH-SY5Y, SK-N-BE2 and CHLA-255 cells expressed LPAR1 but barely expressed LPAR3, indicating the main inhibitory effect of Ki16425 was against LPAR1 (Supplementary Figure S2B). Using three NB cell lines SH-SY5Y, SK-N-BE2 and CHLA-255, the proliferation was assessed by CCK-8 assays, while the migration was assessed either using Transwell or wound-healing assays. As shown in Figure 5A, LPA treatment with or without Ki16425 showed no effect on the proliferation of SH-SY5Y cells. In contrast, the decreased migrated cells of the LPA treatment group in Transwell assays and retarded wound closures of the LPA treatment groups in wound-healing assays (especially under the condition of unaffected proliferation) (Figure 5B,C) suggested that LPA could significantly inhibit the migration of SH-SY5Y cells. While Ki16425 exhibited no effects alone, LPAR1 inhibitor Ki16425 treatment abolished the inhibitory effects of LPA, suggesting the indispensable role of LPAR1 in the migration-inhibitory function of LPA. We also performed the same assays in other NB cells, SK-N-BE2 and CHLA-255, and obtained consistent results with those for SH-SY5Y (Figure 5D–I). The above results indicated that LPA suppressed the migration of NB cells via LPAR1.

3.6. Knockdown of LPAR1 Promoted the Migration of NB Cells

Subsequently, we knocked down LPAR1 in NB cells to identify its function. The efficiency of siRNA was first examined in HEK293T cells, and LPAR1-siRNA1 with a high knockdown efficiency was screened out (Supplementary Figure S2C). The expression of LPAR1 decreased significantly in siRNA-transfected SH-SY5Y, SK-N-BE2 and CHLA-255 cells (Figure 6A,D,G). Using these cells, CCK-8 and Transwell assays were performed. Our results in Figure 6B,E,H show that knockdown of LPAR1 had no effect on NB cell proliferation. Of note, knockdown of LPAR1 could promote the migration of NB cells, and LPA treatment hardly reversed the migration-promoting effect (Figure 6C,F,I), which verifies the significant role of LPAR1 in LPA for mediating the migration-suppressing function.

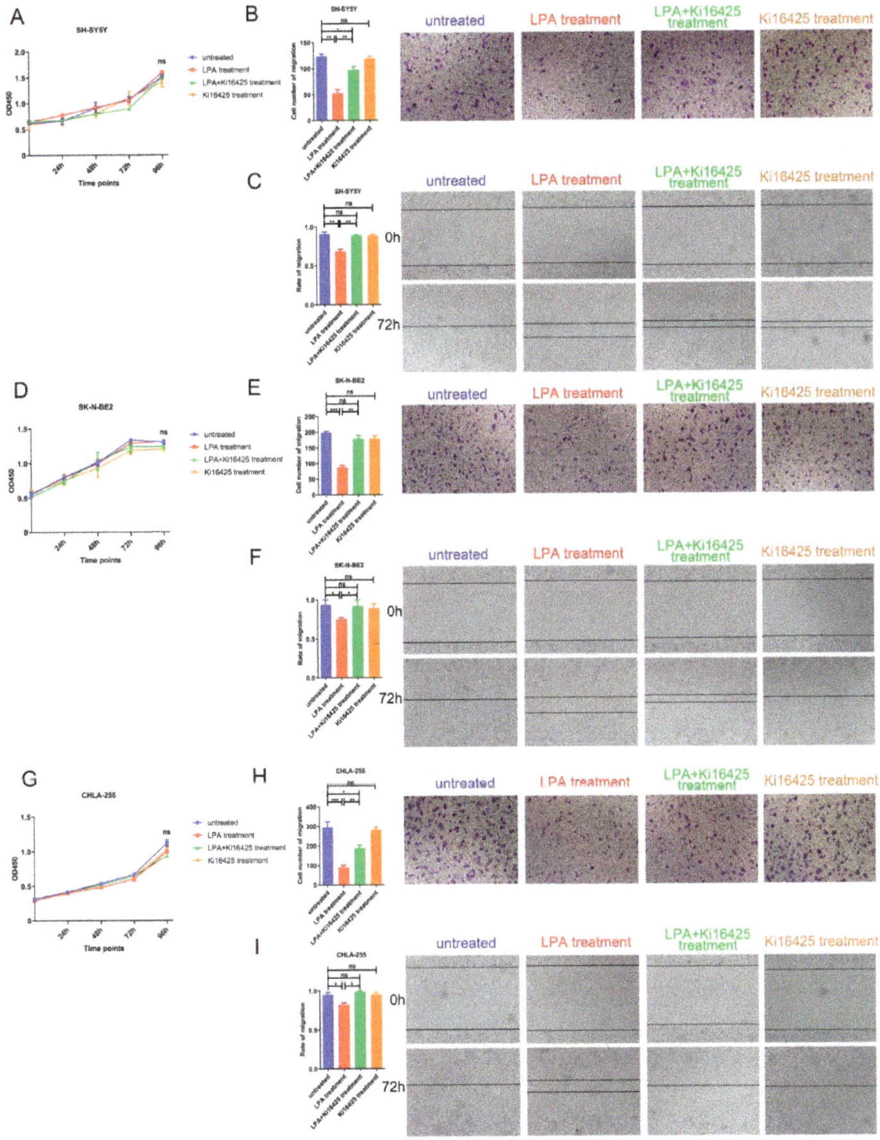

Figure 5. LPA suppressed the migration of NB cells via LPAR1. (**A**,**D**,**G**) CCK-8 assays were performed using SH-SY5Y, SK-N-BE2 and CHLA-255 cells treated with 10 μM LPA, LPA plus 10 μM Ki16425 or Ki16425 alone. (**B**,**E**,**H**) Transwell assays were performed using SH-SY5Y, SK-N-BE2 and CHLA-255 cells treated with 10 μM LPA, LPA plus 10 μM Ki16425 in the upper chamber or Ki16425 alone. Representative images of migrated cells obtained from the Transwell (magnification ×200) are shown (**right**). The cell numbers obtained from the Transwell assays were counted (**left**). (**C**,**F**,**I**) Wound-healing assays were performed and representative images (magnification ×100) are shown (**right**). The relative migration rate obtained from the wound-healing assays was calculated by dividing the change in the distance between the scratch edges by the initial distance (**left**). The results are expressed as the means ± SEMs from three independent experiments conducted in triplicate. * $p < 0.05$, ** $p < 0.01$ and *** $p < 0.001$ compared to the controls.

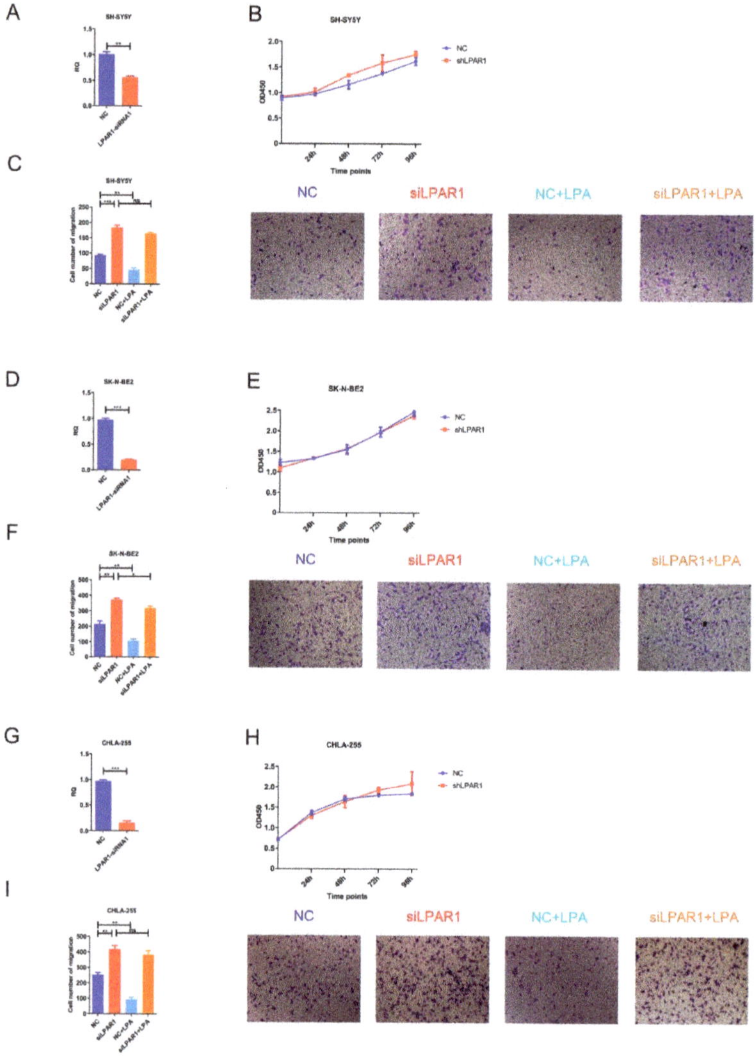

Figure 6. Knockdown of LPAR1 promoted the migration of NB cells. (**A**,**D**,**G**) The LPAR1 knockdown efficiency was analyzed by real-time PCR. (**B**,**E**,**H**) CCK-8 assays were performed using SH-SY5Y, SK-N-BE2 and CHLA-255 control cells and LPAR1 knockdown cells treated with 10 μM LPA. (**C**,**F**,**I**) Transwell assays were performed using SH-SY5Y, SK-N-BE2 and CHLA-255 control cells and LPAR1 knockdown cells treated with 10 μM LPA in the upper chamber. Representative images of migrated cells obtained from the Transwell (magnification ×200) are shown (**right**). The cell numbers obtained from the Transwell assays were counted (**left**). The results are expressed as the means ± SEMs from three independent experiments conducted in triplicate. * $p < 0.05$, ** $p < 0.01$ and *** $p < 0.001$ compared to the controls.

4. Discussion

Using bioinformatics analysis, mRNA microarray data analysis and experimental verification, our study aimed to identify DEGs between NB and control cells to further our understanding of the pathogenesis of NB and potentially provide diagnostic biomarkers and therapeutic targets. *FN1*, *PIK3R5*, *LPAR6* and *LPAR1* were screened out via KEGG, GO and PPI network analysis, and the research mainly focused on exploring the expression and function of LPAR1. We verified the lower expression level of LPAR1 in NB cells and further demonstrated that the LPA-LPAR1 axis suppressed the migration of NB cells.

LPAR1 was reported to be closely associated with the PI3K-Akt signaling pathway and tumor development [28], supporting the DEG screening procedures in our study. Accumulated research has revealed the decreased expression of LPAR1 and its migration-inhibiting effects in tumors including prostate cancer, gastric cancer and pancreatic cancer, which is consistent with our results [7,12,13]. However, LPAR1 expression has also been reported to be significantly increased in other tumors, such as human hepatic cancer [29], osteosarcoma [30] and ovarian cancer [31], and to exert tumor-promoting effects directly or mediated by chemotherapy resistance [30,32]. The controversial research results about LPAR1 suggest its different signaling transduction pathways and functions in different types of tumor cells. Some clinical trials of LPAR1 antagonists in cancer therapy were conducted, though there were no therapeutic trials or positive results reported [6]. Besides the different expression levels and signaling transduction patterns of LPAR1 in different tumors, another significant reason for the controversy around LPAR1's function or failed clinical trials of LPAR1 antagonists is the mutations of *LPAR1* in cancer tissues. A study on metastatic neuroblastoma revealed an accumulation of *de novo* mutations, including a mutation of *LAPR1*, and identified that cells expressing the *LPAR1* R163W mutant showed significantly increased motility [33]. Several missense mutations of *LPAR1* were also found in rat cancer tissues [34]. When inducing MMP-2 expression and cell migration [35,36], or failing to show LPA-induced cellular responses [34], these *LPAR1* mutations resulted in changes to LPAR1's function.

FN1, *PIK3R5* and *LPAR6* were also screened out by our bioinformatics analysis. It was reported that downregulation of FN1 (fibronectin 1) had no significant effects on NB cell proliferation, but it partially blocked ATRA-induced inhibition of cell migration and invasion in NB cells [37]. However, FN1 expression, when analyzed in our study, was not closely related to NB tumor stages and did not show significantly lower levels in high-risk NB tumors or NB tumors leading to patients' death. LPAR6, another member of the LPAR family, showed extremely low expression levels in both NB cells and non-malignant cells, suggesting its minor function in NB. PIK3R5 is the regulatory subunit of PI3Kγ responsible for phosphorylating membrane lipids to activate the Akt pathway, and it is involved in tumorigenesis and progression. Suppressing the expression of PIK3R5 by miRNAs resulted in the promotion of epithelial-mesenchymal transition and oncogenic autophagy by regulating the Akt-mTOR signaling pathway in tumor cells [38]. Since there is a close association of LPAR1 with the PI3K-Akt signaling pathway, LPAR1 may cooperate with PIK3R5 to exert tumor-suppressing effects, which needs further exploration. Given that it is easier to apply the LPA-LPARs axis to clinical therapeutics for NB than it is to apply PIK3R5, we mainly focused on the expression and function of LPAR1 in NB in our study. Manipulating the ligand LPA could be a potential approach to NB therapy according to the function of the LPA-LPAR1 axis in our study. However, it is very difficult to apply just a ligand-protein LPA clinically, especially if the mutation of *LPAR1* and the function of other LPARs in NB tumor cells remain unclear.

Recently, the heterogeneity of neuroblastoma cells was defined by super-enhancer-associated transcription factors, such as *MYCN* and *PHOX2B*, and different tumor-cell subpopulations showed different characteristics of tumor development and metastasis [39–41]. Exploring the expression patterns and functions of LPAR1 in different subpopulations will be necessary in further studies. Beyond this, the heterogeneity of neuroblastoma, especially of metastasis-related changes in the bone marrow environment, was identified by RNA-sequencing analysis and single-cell analysis. A study revealed great diversity among disseminated NB tumor cells, and suggested that FAIM2 (Fas apoptotic inhibitory molecule 2) might be a complementary marker to capture metastatic tumor cells [42]. Beyond that study and our analysis based on data from the bone marrow of NB patients, the expression and function of LPAR1 in metastatic NB tumor cells remain to be further explored. Whether its expression is heterogeneous in particular subpopulations, and whether a particular subpopulation with extremely low LPAR1 expression plays the determining role in chemotherapy/radiotherapy resistance, are worthwhile investigating.

5. Conclusions

Taken together, our findings demonstrate the downregulation of LPAR1 in NB cells and the tumor-suppressing effects of the LPA-LPAR1 axis. We suggest that LPAR1 may represent a potential target for future treatment of NB.

Supplementary Materials: The following supporting information can be downloaded at: https://www.mdpi.com/article/10.3390/cancers14143346/s1, Figure S1: The expression pattern of FN1, PIK3R5 and LPAR6 in NB. The R2 database view-a-gene was used to analyze the expression pattern of screened genes based on the average mRNA expression of the 498 NB SEQC dataset, with $p < 0.05$ regarded as the critical point with statistical significance. (A–C) The association between the FN1 expression and the NB INSS stage (A), likelihood of being high-risk (B) and likelihood of a death event (C). (D–F). The association between the PIK3R5 expression and the NB INSS stage (D), likelihood of being high-risk (E) and likelihood of a death event (F). (G–I) The association between the LPAR6 expression and the NB INSS stage (G), likelihood of being high-risk (H) and likelihood of a death event (I); Figure S2: The expression of LPAR1, LPAR3 and LPAR6 in NB cell lines and non-malignant cell lines. (A,B) The LAPR6 (A), LPAR1 and LPAR3 (B) expression were detected by PCR. (C) The knockdown efficiency of LPAR1 siRNAs was detected in HEK293T cells by real-time PCR; Table S1: Total DEGs identified between NB samples and non-malignant control cell samples. File S1: Original blots.

Author Contributions: Conceptualization, X.W. and J.G.; methodology, X.W.; software, X.L.; validation, X.W., M.P. and J.G.; formal analysis, X.L.; investigation, M.P.; resources, Y.Y.; data curation, X.L.; writing—original draft preparation, X.L.; writing—review and editing, X.W.; visualization, M.P.; supervision, J.G.; project administration, J.G.; funding acquisition, J.G. All authors have read and agreed to the published version of the manuscript.

Funding: This research was funded by National Natural Science Foundation of China [No. 82173084 and No. 82002637]. And The APC was funded by [No. 82002637].

Institutional Review Board Statement: Not applicable.

Informed Consent Statement: Not applicable.

Data Availability Statement: The data presented in this study is available on request from the corresponding author.

Acknowledgments: This work was supported by grants from the National Natural Science Foundation of China (nos. 82173084 and 82002637).

Conflicts of Interest: The authors declare no conflict of interest.

References

1. Newman, E.A.; Abdessalam, S.; Aldrink, J.H.; Austin, M.; Heaton, T.E.; Bruny, J.; Ehrlich, P.; Dasgupta, R.; Baertschiger, R.M.; Lautz, T.B.; et al. Update on neuroblastoma. *J. Pediatr. Surg.* **2019**, *54*, 383–389. [CrossRef]
2. Siegel, R.L.; Miller, K.D.; Fuchs, H.E.; Jemal, A. Cancer statistics. *CA Cancer J. Clin.* **2022**, *72*, 7–33. [CrossRef] [PubMed]
3. Lundberg, K.I.; Treis, D.; Johnsen, J.I. Neuroblastoma Heterogeneity, Plasticity, and Emerging Therapies. *Curr. Oncol. Rep.* **2022**. [CrossRef]
4. Toro-Dominguez, D.; Martorell-Marugan, J.; Lopez-Dominguez, R.; Garcia-Moreno, A.; Gonzalez-Rumayor, V.; Alarcon-Riquelme, M.E.; Carmona-Saez, P. ImaGEO: Integrative gene expression meta-analysis from GEO database. *Bioinformatics* **2019**, *35*, 880–882. [CrossRef] [PubMed]
5. Barrett, T.; Troup, D.B.; Wilhite, S.E.; Ledoux, P.; Rudnev, D.; Evangelista, C.; Kim, I.F.; Soboleva, A.; Tomashevsky, M.; Marshall, K.A.; et al. NCBI GEO: Archive for high-throughput functional genomic data. *Nucleic Acids Res.* **2009**, *37*, D885–D890. [CrossRef]
6. Lin, Y.H.; Lin, Y.C.; Chen, C.C. Lysophosphatidic Acid Receptor Antagonists and Cancer: The Current Trends, Clinical Implications, and Trials. *Cells* **2021**, *10*, 1629. [CrossRef] [PubMed]
7. Harma, V.; Knuuttila, M.; Virtanen, J.; Mirtti, T.; Kohonen, P.; Kovanen, P.; Happonen, A.; Kaewphan, S.; Ahonen, I.; Kallioniemi, O.; et al. Lysophosphatidic acid and sphingosine-1-phosphate promote morphogenesis and block invasion of prostate cancer cells in three-dimensional organotypic models. *Oncogene* **2012**, *31*, 2075–2089. [CrossRef]

8. Geraldo, L.; Spohr, T.; Amaral, R.; Fonseca, A.; Garcia, C.; Mendes, F.A.; Freitas, C.; Dossantos, M.F.; Lima, F. Role of lysophosphatidic acid and its receptors in health and disease: Novel therapeutic strategies. *Signal Transduct. Target Ther.* **2021**, *6*, 45. [CrossRef]
9. Taniguchi, R.; Inoue, A.; Sayama, M.; Uwamizu, A.; Yamashita, K.; Hirata, K.; Yoshida, M.; Tanaka, Y.; Kato, H.E.; Nakada-Nakura, Y.; et al. Structural insights into ligand recognition by the lysophosphatidic acid receptor LPA6. *Nature* **2017**, *548*, 356–360. [CrossRef]
10. Choi, J.W.; Herr, D.R.; Noguchi, K.; Yung, Y.C.; Lee, C.W.; Mutoh, T.; Lin, M.E.; Teo, S.T.; Park, K.E.; Mosley, A.N.; et al. LPA receptors: Subtypes and biological actions. *Annu. Rev. Pharmacol. Toxicol.* **2010**, *50*, 157–186. [CrossRef]
11. Shi, J.; Jiang, D.; Yang, S.; Zhang, X.; Wang, J.; Liu, Y.; Sun, Y.; Lu, Y.; Yang, K. LPAR1, Correlated With Immune Infiltrates, Is a Potential Prognostic Biomarker in Prostate Cancer. *Front. Oncol.* **2020**, *10*, 846. [CrossRef]
12. Ma, X.; Feng, J.; Lu, M.; Tang, W.; Han, J.; Luo, X.; Zhao, Q.; Yang, L. Microrna-501-5p promotes cell proliferation and migration in gastric cancer by downregulating LPAR1. *J. Cell. Biochem.* **2020**, *121*, 1911–1922. [CrossRef] [PubMed]
13. Kato, K.; Yoshikawa, K.; Tanabe, E.; Kitayoshi, M.; Fukui, R.; Fukushima, N.; Tsujiuchi, T. Opposite roles of LPA1 and LPA3 on cell motile and invasive activities of pancreatic cancer cells. *Tumour. Biol.* **2012**, *33*, 1739–1744. [CrossRef] [PubMed]
14. Hayashi, M.; Okabe, K.; Kato, K.; Okumura, M.; Fukui, R.; Fukushima, N.; Tsujiuchi, T. Differential function of lysophosphatidic acid receptors in cell proliferation and migration of neuroblastoma cells. *Cancer Lett.* **2012**, *316*, 91–96. [CrossRef]
15. Hirane, M.; Araki, M.; Dong, Y.; Honoki, K.; Fukushima, N.; Tsujiuchi, T. Inhibitory effects of LPA1 on cell motile activities stimulated by hydrogen peroxide and 2,3-dimethoxy-1,4-naphthoquinone in fibroblast 3T3 cells. *Biochem. Biophys. Res. Commun.* **2013**, *441*, 47–52. [CrossRef]
16. Irizarry, R.A.; Hobbs, B.; Collin, F.; Beazer-Barclay, Y.D.; Antonellis, K.J.; Scherf, U.; Speed, T.P. Exploration, normalization, and summaries of high density oligonucleotide array probe level data. *Biostatistics* **2003**, *4*, 249–264. [CrossRef] [PubMed]
17. Gautier, L.; Cope, L.; Bolstad, B.M.; Irizarry, R.A. Affy—analysis of Affymetrix GeneChip data at the probe level. *Bioinformatics* **2004**, *20*, 307–315. [CrossRef] [PubMed]
18. Ritchie, M.E.; Phipson, B.; Wu, D.; Hu, Y.; Law, C.W.; Shi, W.; Smyth, G.K. Limma powers differential expression analyses for RNA-sequencing and microarray studies. *Nucleic Acids Res.* **2015**, *43*, e47. [CrossRef] [PubMed]
19. Ding, R.; Qu, Y.; Wu, C.H.; Vijay-Shanker, K. Automatic gene annotation using GO terms from cellular component domain. *BMC Med. Inform. Decis. Mak.* **2018**, *18* (Suppl. S5), 119. [CrossRef]
20. Kanehisa, M.; Goto, S. KEGG: Kyoto encyclopedia of genes and genomes. *Nucleic Acids Res.* **2000**, *28*, 27–30. [CrossRef]
21. Dennis, G.J.; Sherman, B.T.; Hosack, D.A.; Yang, J.; Gao, W.; Lane, H.C.; Lempicki, R.A. DAVID: Database for Annotation, Visualization, and Integrated Discovery. *Genome Biol.* **2003**, *4*, 3. [CrossRef]
22. Feng, H.; Gu, Z.Y.; Li, Q.; Liu, Q.H.; Yang, X.Y.; Zhang, J.J. Identification of significant genes with poor prognosis in ovarian cancer via bioinformatical analysis. *J. Ovarian Res.* **2019**, *12*, 35. [CrossRef] [PubMed]
23. Shannon, P.; Markiel, A.; Ozier, O.; Baliga, N.S.; Wang, J.T.; Ramage, D.; Amin, N.; Schwikowski, B.; Ideker, T. Cytoscape: A software environment for integrated models of biomolecular interaction networks. *Genome Res.* **2003**, *13*, 2498–2504. [CrossRef] [PubMed]
24. Islam, S.; Kitagawa, T.; Baron, B.; Abiko, Y.; Chiba, I.; Kuramitsu, Y. ITGA2, LAMB3, and LAMC2 may be the potential therapeutic targets in pancreatic ductal adenocarcinoma: An integrated bioinformatics analysis. *Sci. Rep.* **2021**, *11*, 10563. [CrossRef]
25. Zhang, H.; Chai, W.; Yang, W.; Han, W.; Mou, W.; Xi, Y.; Chen, X.; Wang, H.; Wang, W.; Qin, H.; et al. The increased IL-17-producing gammadeltaT cells promote tumor cell proliferation and migration in neuroblastoma. *Clin. Immunol.* **2020**, *211*, 108343. [CrossRef] [PubMed]
26. Noorolyai, S.; Shajari, N.; Baghbani, E.; Sadreddini, S.; Baradaran, B. The relation between PI3K/AKT signalling pathway and cancer. *Gene* **2019**, *698*, 120–128. [CrossRef] [PubMed]
27. Rich, J.T.; Neely, J.G.; Paniello, R.C.; Voelker, C.C.; Nussenbaum, B.; Wang, E.W. A practical guide to understanding Kaplan-Meier curves. *Otolaryngol. Head Neck. Surg.* **2010**, *143*, 331–336. [CrossRef] [PubMed]
28. Zhang, X.; Li, M.; Yin, N.; Zhang, J. The Expression Regulation and Biological Function of Autotaxin. *Cells* **2021**, *10*, 939. [CrossRef]
29. Zuckerman, V.; Sokolov, E.; Swet, J.H.; Ahrens, W.A.; Showlater, V.; Iannitti, D.A.; Mckillop, I.H. Expression and function of lysophosphatidic acid receptors (LPARs) 1 and 3 in human hepatic cancer progenitor cells. *Oncotarget* **2016**, *7*, 2951–2967. [CrossRef]
30. Takagi, S.; Sasaki, Y.; Koike, S.; Takemoto, A.; Seto, Y.; Haraguchi, M.; Ukaji, T.; Kawaguchi, T.; Sugawara, M.; Saito, M.; et al. Platelet-derived lysophosphatidic acid mediated LPAR1 activation as a therapeutic target for osteosarcoma metastasis. *Oncogene* **2021**, *40*, 5548–5558. [CrossRef]
31. Ha, J.H.; Ward, J.D.; Radhakrishnan, R.; Jayaraman, M.; Song, Y.S.; Dhanasekaran, D.N. Lysophosphatidic acid stimulates epithelial to mesenchymal transition marker Slug/Snail2 in ovarian cancer cells via Galphai2, Src, and HIF1alpha signaling nexus. *Oncotarget* **2016**, *7*, 37664–37679. [CrossRef]
32. Liu, J.; Rebecca, V.W.; Kossenkov, A.V.; Connelly, T.; Liu, Q.; Gutierrez, A.; Xiao, M.; Li, L.; Zhang, G.; Samarkina, A.; et al. Neural Crest-Like Stem Cell Transcriptome Analysis Identifies LPAR1 in Melanoma Progression and Therapy Resistance. *Cancer Res.* **2021**, *81*, 5230–5241. [CrossRef] [PubMed]

33. Wei, J.S.; Johansson, P.; Chen, L.; Song, Y.K.; Tolman, C.; Li, S.; Hurd, L.; Patidar, R.; Wen, X.; Badgett, T.C.; et al. Massively parallel sequencing reveals an accumulation of de novo mutations and an activating mutation of LPAR1 in a patient with metastatic neuroblastoma. *PLoS ONE* **2013**, *8*, e77731. [CrossRef] [PubMed]
34. Ishii, S.; Tsujiuchi, T.; Fukushima, N. Functional characterization of lysophosphatidic acid receptor 1 mutants identified in rat cancer tissues. *Biochem. Biophys. Res. Commun.* **2017**, *486*, 767–773. [CrossRef] [PubMed]
35. Kato, K.; Fukui, R.; Okabe, K.; Tanabe, E.; Kitayoshi, M.; Fukushima, N.; Tsujiuchi, T. Constitutively active lysophosphatidic acid receptor-1 enhances the induction of matrix metalloproteinase-2. *Biochem. Biophys. Res. Commun.* **2012**, *417*, 790–793. [CrossRef] [PubMed]
36. Kitayoshi, M.; Kato, K.; Tanabe, E.; Yoshikawa, K.; Fukui, R.; Fukushima, N.; Tsujiuchi, T. Enhancement of endothelial cell migration by constitutively active LPA(1)-expressing tumor cells. *Biochem. Biophys. Res. Commun.* **2012**, *422*, 339–343. [CrossRef]
37. Tan, X.; Gong, W.; Chen, B.; Gong, B.; Hua, Z.; Zhang, S.; Chen, Y.; Li, Q.; Li, Z. Downregulation of fibronectin 1 attenuates ATRA-induced inhibition of cell migration and invasion in neuroblastoma cells. *Mol. Cell. Biochem.* **2021**, *476*, 3601–3612. [CrossRef]
38. Liu, W.; Jiang, D.; Gong, F.; Huang, Y.; Luo, Y.; Rong, Y.; Wang, J.; Ge, X.; Ji, C.; Fan, J.; et al. MiR-210-5p promotes epithelial-mesenchymal transition by inhibiting PIK3R5 thereby activating oncogenic autophagy in osteosarcoma cells. *Cell Death Dis.* **2020**, *11*, 93. [CrossRef]
39. Boeva, V.; Louis-Brennetot, C.; Peltier, A.; Durand, S.; Pierre-Eugene, C.; Raynal, V.; Etchevers, H.C.; Thomas, S.; Lermine, A.; Daudigeos-Dubus, E.; et al. Heterogeneity of neuroblastoma cell identity defined by transcriptional circuitries. *Nat. Genet.* **2017**, *49*, 1408–1413. [CrossRef]
40. van Groningen, T.; Koster, J.; Valentijn, L.J.; Zwijnenburg, D.A.; Akogul, N.; Hasselt, N.E.; Broekmans, M.; Haneveld, F.; Nowakowska, N.E.; Bras, J.; et al. Neuroblastoma is composed of two super-enhancer-associated differentiation states. *Nat. Genet.* **2017**, *49*, 1261–1266. [CrossRef]
41. Gartlgruber, M.; Sharma, A.K.; Quintero, A.; Dreidax, D.; Jansky, S.; Park, Y.G.; Kreth, S.; Meder, J.; Doncevic, D.; Saary, P.; et al. Super enhancers define regulatory subtypes and cell identity in neuroblastoma. *Nat. Cancer* **2021**, *2*, 114–128. [CrossRef]
42. Lazic, D.; Kromp, F.; Rifatbegovic, F.; Repiscak, P.; Kirr, M.; Mivalt, F.; Halbritter, F.; Bernkopf, M.; Bileck, A.; Ussowicz, M.; et al. Landscape of Bone Marrow Metastasis in Human Neuroblastoma Unraveled by Transcriptomics and Deep Multiplex Imaging. *Cancers* **2021**, *13*, 4311. [CrossRef] [PubMed]

cancers

MDPI

Article

SOX4 Mediates ATRA-Induced Differentiation in Neuroblastoma Cells

Dongyang Zhang [1,2], Baocheng Gong [3], Qiang Zhao [3], Zhijie Li [1,2], Xiaolin Tan [4,*,†] and Zhongyan Hua [1,2,*,†]

1 Department of Pediatrics, Shengjing Hospital of China Medical University, Shenyang 110004, China
2 Liaoning Key Laboratory of Research and Application of Animal Models for Environmental and Metabolic Diseases, Medical Research Center, Shengjing Hospital of China Medical University, Shenyang 110004, China
3 Department of Pediatric Oncology, Tianjin Medical University Cancer Institute and Hospital, Tianjin 300060, China
4 Department of Rehabilitation, Shengjing Hospital of China Medical University, Shenyang 110004, China
* Correspondence: changchuntengtan@163.com or xitan@ukaachen.de (X.T.); huazhongyan_123@sina.com or zyhua@cmu.edu.cn (Z.H.); Tel.: +86-18940255251 (Z.H.)
† These authors contributed equally to this work.

Simple Summary: Neuroblastoma (NB) is considered to be caused by the differentiation failure of neural crest cells. Researchers are working on exploring the mechanisms of NB cell differentiation to improve the cure rate. Here, our results show that SOX4 has a significant effect on NB cell proliferation, cells' neurites, and the cell cycle and that SOX4 mediates the effect of RA in NB cells. All indicate that SOX4 may be a target to induce NB cell differentiation.

Abstract: Neuroblastoma (NB), which is considered to be caused by the differentiation failure of neural crest cells, is the most common extracranial malignant solid tumor in children. The degree of tumor differentiation in patients with NB is closely correlated with the survival rate. To explore the potential targets that mediate NB cell differentiation, we analyzed four microarray datasets from GEO, and the overlapping down- or upregulated DEGs were displayed using Venn diagrams. SOX4 was one of the overlapping upregulated DEGs and was confirmed by RT-qPCR and Western blot in ATRA-treated NGP, SY5Y, and BE2 cells. To clarify whether SOX4 was the target gene regulating NB cell differentiation, the correlation between the expression of SOX4 and the survival of clinical patients was analyzed via the R2 database, SOX4 overexpression plasmids and siRNAs were generated to change the expression of SOX4, RT-qPCR and Western blot were performed to detect SOX4 expression, cell confluence or cell survival was detected by IncuCyte Zoom or CCK8 assay, immunocytochemistry staining was performed to detect cells' neurites, and a cell cycle analysis was implemented using Flow cytometry after PI staining. The results showed that the survival probabilities were positively correlated with SOX4 expression, in which overexpressing SOX4 inhibited NB cell proliferation, elongated the cells' neurite, and blocked the cell cycle in G1 phase, and that knockdown of the expression of SOX4 partially reversed the ATRA-induced inhibition of NB cell proliferation, the elongation of the cells' neurites, and the blocking of the cell cycle in the G1 phase. These indicate that SOX4 may be a target to induce NB cell differentiation.

Keywords: neuroblastoma; cell differentiation; ATRA; SOX4

Citation: Zhang, D.; Gong, B.; Zhao, Q.; Li, Z.; Tan, X.; Hua, Z. SOX4 Mediates ATRA-Induced Differentiation in Neuroblastoma Cells. *Cancers* **2022**, *14*, 5642. https://doi.org/10.3390/cancers14225642

Academic Editors: Saurabh Agarwal and Jianhua Yang

Received: 19 September 2022
Accepted: 12 November 2022
Published: 17 November 2022

1. Introduction

Neuroblastoma (NB), which derives from neural crest cells, is the most common extracranial malignant solid tumor in children and is considered to be caused by the differentiation failure of neural crest cells [1–3]. Patients with low-risk NB have a very good prognosis; however, high-risk patients have consistently poor prognosis. Despite the availability of multiple types of chemotherapy and stem cell transplantation, the long-term survival ratio of high-risk patients is less than 50% [2]. NB is a highly heterogenous tumor, and some tumors can spontaneously heal or differentiation, while others are highly invasive or exhibit a therapeutic resistance phenotype. According to the INSS (International Neuroblastoma Phased System), NB is divided into Phases 1, 2a, 2b, 3, 4, and 4S. Phase 4S NB cells have the potential for differentiation. When the patient is diagnosed, the symptoms are serious, manifesting as multiple tumor metastases; later, the tumor can spontaneously differentiate, and prognosis is good [4]. Another study also found that the degree of tumor differentiation in patients with NB is closely related to its overall survival rate [5]. Therefore, exploring the molecular mechanisms that regulate NB cell differentiation is essential to improve the cure ratio and life quality of patients.

Retinoids are a family of signaling molecules that are related to vitamin A (retinol) in terms of their chemical structures. There are six biologically active isoforms of retinoids: they are all-trans retinoic acid (ATRA); 11-cis retinoic acid; 13-cis retinoic acid; 9, 13-di-cis retinoic acid; 9-cis retinoic acid; and 11, 13-di-cis retinoic acid. Retinoids have been reported to have an effect on cell differentiation, cell proliferation, and cell apoptosis. All-trans retinoic acid (ATRA) is one of the most common retinoids, and it is widely used in the study and treatment of leukemia, lymphoma, neuroblastoma, and glioblastoma [6–8]. However, high doses of ATRA are associated with side effects, including teratogenicity and chemical hepatitis [9,10]. Therefore, researchers and clinicians are working on exploring more selective targets and less toxic compounds, which could function as ATRA [7,11,12].

SOX4 is a member of the SOX (SRY-related HMG-box) transcription factor family, which contains a high mobility group (HMG) DNA binding domain (DBD), a glycine-rich domain, and a serine-rich domain [13,14]. Members of the SOX transcription factor family contribute to the development of many organs and tissues, including central nervous system (CNS), retina, bone, hematopoietic system, lymphatic system, and so on. As a member of this family, SOX4 is closely associated with both normal development and various cancers, such as lung cancer, breast cancer, leukemias, glioblastoma, and medulloblastoma [14]. SOX4 also contributes to the differentiation of skeletal myoblast differentiation [15], and divergent sarcomatous differentiation in uterine carcinosarcoma [16]. Studies show that SOX4 cooperates with neurogenin 3 to determine the cell fate during the development of pancreatic beta cells [17], and SOX4 partners with neurogenin 2 to activate Tbrain2 to determine the cell fate of intermediate progenitors during the neuronal differentiation [18]. Studies also show that SOX4 regulates cell survival and metastasis of cancer cells, such as breast cancer [13,19,20].

In this study, we analyzed four microarray datasets from GEO and identified the overlapping differentially expressed genes (DEGs), and the results showed that SOX4 was one of these overlapping upregulated genes, a result that is in accordance with our previous study [21]. Furthermore, our results showed that the survival probabilities of clinical patients have a positive correlation with SOX4 expression, in which overexpressing SOX4 inhibited NB cell proliferation; elongated cells' neurites; and blocked the cell cycle in the G1 phase and that knockdown of the expression of SOX4 partially reversed the function of ATRA in NB cell proliferation, cells' neurites, and the cell cycle. All these indicate that SOX4 plays an important role in the differentiation of NB cells, and prompt that SOX4 may be a potential new therapy target for the clinical treatment of patients with NB.

2. Materials and Methods

2.1. Reagents

All-trans retinoic acid (ATRA) was from Sigma (R2625, Ronkonkoma, NY, USA). Fetal bovine serum was from Gibco (Brisbane, Australia). RPMI-1640 medium, glutamine, antibiotics (penicillin 100 units/mL, streptomycin 100 μg/mL), and PBS were from Bioind (Beit-Haemek, Israel). Trizol was from Solarbio (R1100, China), the GoScript^{TM} Reverse Transcription system was from Promega (A5001, Madison, WI, USA), and TB Green was from TAKARA (RR820A, Beijing, China). Antibody against SOX4 (ab70598) and Goat polyclonal Secondary Antibody to Rabbit IgG-H&L (Alexa Fluor 488, ab150077) were from Abcam, and the antibody against GAPDH and α-Tubulin was from Proteintech (Wuhan,

China). Cell counting kit-8 (CCK8) was from Bimake (B34304, Shanghai, China). The cell cycle analysis kit was from Beyotime (C1052, Shanghai, China).

2.2. Cell Culture and Treatment

The NB cell lines (NGP, SY5Y, and BE2) used in this study were obtained from Dr. Carol J. Thiele (Cellular and Molecular Biology Section, Pediatric Oncology Branch, National Cancer Institute, National Institutes of Health, Bethesda, MD, USA). All three cell lines (NGP, SY5Y, and BE2) are neuroblastic-type cell lines and can be induced to differentiate. The cell lines were cultured in RPMI-1640 medium with 10% fetal bovine serum, 2 mM glutamine, and antibiotics at 37 °C in 5% CO2 incubator. NGP, SY5Y, and BE2 cells were treated with or without ATRA (5 μM) for 48 h and were collected for RT-qPCR and Western Blotting.

2.3. Microarray Data and Data Analysis

Four microarray datasets (GSE45587: NB cell line BE(2)-C cells treated with RA for 24 h and 72 h [22], GSE16451: BE(2)-C cells treated with RA for 3 weeks [23], and GSE87784: Sphere cells from TH-MYCN mice treated with RA for 3 weeks [24]) were from the NCBI Gene Expression Synthesis (GEO) database (http://www.ncbi.nlm.nih.gov/geo/, accessed on 13 January 2021). The volcano plots were generated to display the differentially expressed genes (DEGs) via GEO2R according to the instruction (https://www.ncbi.nlm.nih.gov/geo/info/geo2r.html, accessed on 13 January 2021), and adjusted p value (Padj) < 0.05 and $|\log2(\text{fold change})| > 1$ were considered to be statistically significant. The overlapping down- or up-regulated DEGs in the four microarray datasets were displayed using Venn diagrams, which were generated in R language.

The relationship between the expression of SOX4 and clinical patients with NB was analyzed via R2 database (https://hgserver1.amc.nl/cgi-bin/r2/main.cgi?species=hs, accessed on 13 January 2021), and four datasets were used (88, 283, 498, and 649 samples of patients with NB). The relationship between the expression of SOX4 and the differentiation degree of tumors originating from sympathetic nervous system (ganglioneuroblastoma, ganglioneuroma, and neuroblastoma) was analyzed through the Oncomine database (https://www.oncomine.org/resource/login.html, accessed on 13 January 2021).

2.4. Cell Transfection

SOX4 expression plasmids were isolated using the HiSpeed Plasmid Maxi Kit (Qiagen, Germany) according to the manufacturer's instructions. NGP cells were seeded into a 6-well plate (4×10^5/well) and cultured overnight. The SOX4 expression plasmids were transfected into cells using jetPRIME (Polyplus Transfection, Illkirsch, France). After 16 h, cells were collected and seeded into 96-well plates or 6-well plates. After 24 h, the cells were collected for Western Blotting, and after 48 h, they were collected for CCK8 assay, cell confluence analysis using IncuCyte Zoom, immunocytochemistry staining (ICC) assay, and cell cycle analysis.

Small interfering RNAs (siRNAs) (Tongyong, Shanghai, China) were used to knockdown the expression of SOX4. The sequences of siRNAs were SOX4 siRNA #1: 5′-GCAAGCACCUGGCGGAGAATT-3′, 5′-UUCUCCGCCAGGUGCUUGCTT-3′; SOX4 siRNA #2: 5′-GCUGGAAGCUGCUCAAAGATT-3′, 5′-UCUUUGAGCAGCUUCCAGCTT-3′; SOX4 siRNA #3: 5′-CCAACAAUGCCGAGAACACTT-3′, and 5′-GUGUUCUCGGCA UUGUUGGTT-3′. NGP cells were seeded into 6-well plates (4×10^5/well) and cultured overnight. The siRNAs were transfected into cells using jetPRIME. After 16 h, the cells were collected and seeded into 96-well plates or 6-well plates and treated with RA (5 μM), 24 h for RT-qPCR and Western Blotting and 48 h for CCK8 assay, cell confluence analysis using IncuCyte Zoom, ICC assay, and cell cycle analysis.

2.5. *Quantitative RT-PCR*

Total RNA was isolated from NGP cells treated with RA or transfected with siRNAs by using Trizol reagent according to the manufacturer's instructions and our previous study. The GoScript^{TM} Reverse Transcription System kit was used to generate cDNA. Quantitative PCR was performed with the cDNA by using TB Green according to the manufacturer's instructions. Beta-actin expression served as an internal control. The relative quantification of gene expression was performed with the $2(-\Delta\Delta Ct)$ method. Details of the PCR primers sequences were as follows: SOX4 sense 5′-GTGGTCCTCAAAGCCAGACACT-3′, SOX4 anti-sense 5′-GCAATGCGCTTTCTGCCGTAGT-3′; β-actin sense 5′- AACT GGGACGA-CATGGAGAAA -3′, β-actin anti-sense 5′- AGGGATAGCACAGCCTGGATA -3′.

2.6. *Western Blotting*

Total protein was extracted with RIPA Buffer (Beyotime, Shanghai, China) according to the manufacturer's protocol. The concentration was quantified by using the Bradford reagent (Beyotime, China). Under each condition, 30 μg of total protein was loaded and separated by 10% gel (Epizyme Biomedical Technology Co., Shanghai, China) and then transferred to a PVDF membrane (Millipore, Burlington, MA, USA). The membranes were blocked with 5% skim milk in TBST buffer for 1–2 h and then incubated with primary antibody (SOX4) at 4 °C overnight. The membranes were washed with TBST 3 times and then incubated with the peroxidase-conjugated goat anti-rabbit (1:5000) or anti-mouse (1:5000) antibodies for 2 h at room temperature. The binds were detected using enhanced chemiluminescence (ECL) reagents (Thermo Scientific, Rockford, IL, USA).

2.7. *Cell Survival Analysis*

CCK8 (Cell Counter Kit 8) was used to detect the cell survival, according to the manufacturer's specification and our previous study. NGP cells were seeded in a 96-well plate 16 h after transfection with SOX4 expression plasmids or siRNAs. CCK8 was added to each well and incubated for 1 h after treated with or without RA 48 h. Optical density was measured at 450 nm using ELISA reader. Cell confluence (%) was calculated and analyzed by using Incucyte Zoom software (Essen BioScince, MI, USA) according to the phase-contrast images, as in our previous study.

2.8. *Immunocytochemistry Staining*

The cell slices were fixed using 4% paraformaldehyde in PBS pH 7.4 for 30 min at room temperature and then washed in PBS 3 times for 3 min. The cell slices were incubated with 0.5% Triton X-100 for 20 min at room temperature and then washed in PBS 3 times. The cell slices were blocked with goat serum for 30 min at room temperature. The goat serum was sucked up with absorbent paper, and the cell slices were incubated with primary antibody (β-Tubulin) overnight at 4 °C. The cell slices were washed in PBST (PBS + 0.1% Tween 20) 3 times for 3 min and incubated with secondary antibody for 1 h at room temperature in the dark. The cell slices were washed in PBST 3 times, and the slices were incubated with DAPI for 5 min at room temperature in the dark. The cell slices were washed in PBST 4 times, the liquid was sucked up with absorbent paper, and the cell slices were observed under fluorescence microscope.

2.9. *Cell Cycle Analysis*

After treated with different conditions, all the cells were harvested, washed with PBS, and then resuspended and fixed with cold 70% ethyl alcohol at 4 °C overnight. Then, the cells were washed again with PBS and incubated with RNase A (100 μg/mL) and PI (50 μg/mL) at room temperature in the dark for 30 minutes. Then, the stained cells were analyzed by Flow cytometry (Becton, Dickinson and Company, Franklin Lakes, NJ, USA). The percentage of cells in each phase of cell cycle was analyzed using the Software of the Flow cytometry system.

2.10. Statistical Analyses

Means ± SD of independent experiments were analyzed by Student's *t*-test. *p* values less than 0.05 were considered as statistically significant. Data were analyzed by using GraphPad Prism software.

3. Results

3.1. The Differentially Expressed Genes Obtained from the Microarray Data Analysis

All-trans retinoic acid (ATRA or RA) has been widely used to induce cell differentiation of NB, while the mechanisms still need further study. To explore the potential target gene of RA, we analyzed four microarray datasets from GEO (GSE45587: NB cell line BE(2)-C cells treated with RA for 24 h and 72 h, GSE16451: BE(2)-C cells treated with RA for 3 weeks, and GSE87784: Sphere cells from TH-MYCN mice treated with RA for 3 weeks). The differentially expressed genes (DEGs) were shown in volcano plot form in Figure 1A–C. Additionally, the overlapping DEGs were identified, and there were 13 overlapping upregulated DEGs (Figure 1D) and 2 overlapping downregulated DEGs (Figure 1E). The 13 overlapping upregulated DEGs were SOX4, SOX9, ADD3, ATP7A, CAMK2N1, CTSB, RET, CYP26A1, CYP26B1, DDAH2, LTBP3, MEIS1, and NCOA3 (Table 1). The two overlapping downregulated DEGs were FHL2 and BRCA2 (Table 2).

Table 1. Overlapping upregulated DEGs of the four datasets from GEO.

Overlapping Upregulated-DEGs	
SOX4	CYP26A1
SOX9	CYP26B1
ADD3	DDAH2
ATP7A	LTBP3
CAMK2N1	MEIS1
CTSB	NCOA3
RET	

Table 2. Overlapping downregulated DEGs of the four datasets from GEO.

Overlapping Downregulated-DEGs
FHL2
BRCA2

3.2. SOX4 Has a Positive Correlation with the Survival Rate of Patients with NB

SOX4 is a member of the SRY-related HMG-box (SOX) family of transcription factors and has been reported to involved in the regulation of embryonic development and in the determination of cell fate. To identify if SOX4 plays a role in NB, we treated the cells (NGP, SY5Y, and BE2) with RA (5 μM) for 48 h first and then detected the expression of SOX4 at the mRNA level and protein level using RT-qPCR and Western blotting. The results showed that the expression of SOX4 was significantly increased at both the mRNA level and the protein level (Figure 2B,C), which was in consonance with the microarray data from our previous data (Figure 2A) [21] and the other four microarray datasets (Table 1). Then, we analyzed the data with 88, 283, 498, and 649 samples of patients with NB from the R2 database, and the results showed that patients with higher expressions of SOX4 have better overall survival probability and relapse-free/progression-free/event-free survival probability compared to those with lower expressions of SOX4 (Figure 3A–D). To identify if SOX4 was correlated with the differentiation degree of tumors originating from the sympathetic nervous system, we analyzed the expression of SOX4 in ganglioneuroblastoma, ganglioneuroma, and neuroblastoma through the Oncomine database, and the results showed that there have no significant difference between the three tumors (Figure 3E). These results indicated that SOX4 is positively correlated with the survival rate of patients with

NB but did not demonstrate the exact relationship between SOX4 and the differentiation degree of sympathetic nervous system-originating tumors.

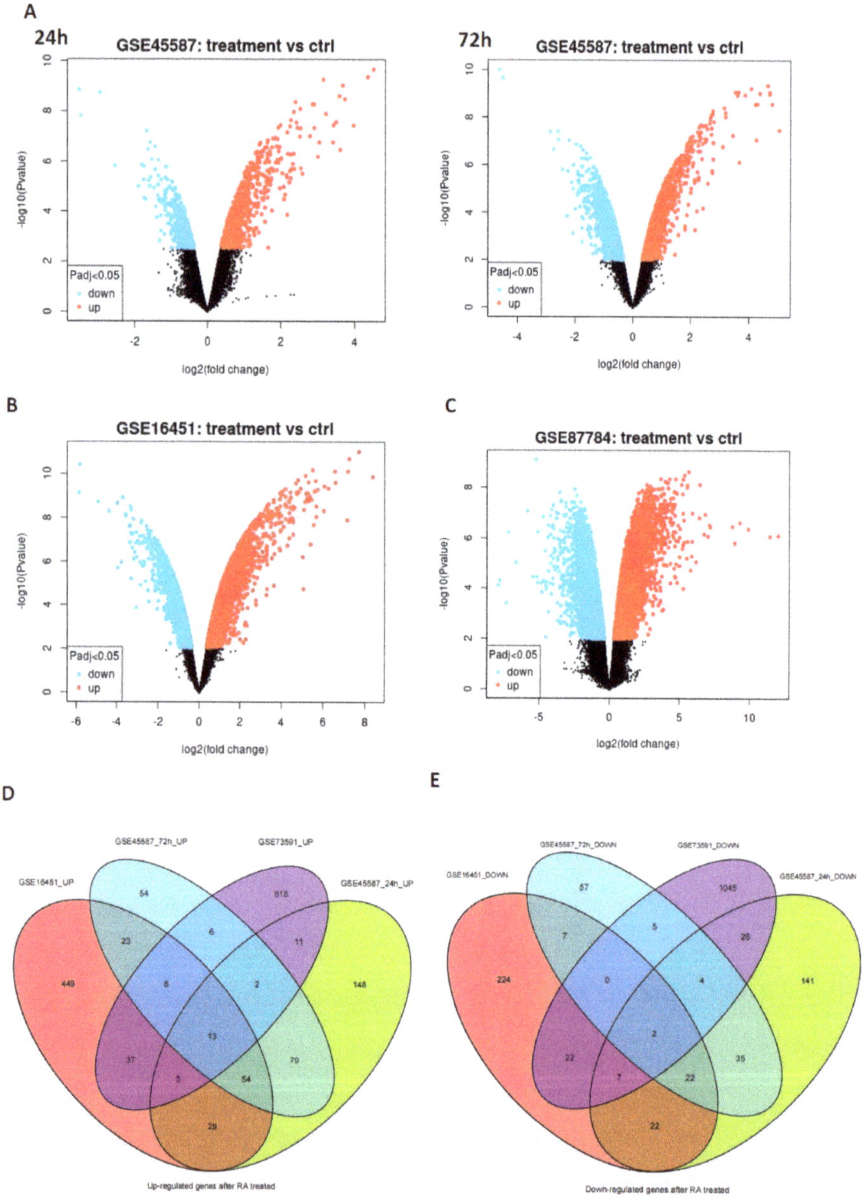

Figure 1. Volcano plot distribution of DEGs and the overlapping DEGs of the four GEO datasets. The volcano plot of (**A**) GSE45587: NB cell line BE(2)-C cells treated with RA for 24 h and 72 h, (**B**) GSE16451: BE(2)-C cells treated with RA for 3 weeks, and (**C**) GSE87784: Sphere cells from TH-MYCN mice treated with RA for 3 weeks. The blue points indicate the downregulated DEGs, red points indicate the upregulated DEGs, and the gray points indicate the genes without significant changes. All DEGs were screened based on an adjusted p value < 0.05 and $|\log2(\text{fold change})| > 1$. (**D**) The overlapping upregulated DEGs and (**E**) the overlapping downregulated DEGs.

A

Gene	log2(FC)	*P* value	FDR
SOX4	1.627004272	1.93×10^-8^	4.27×10^-6^

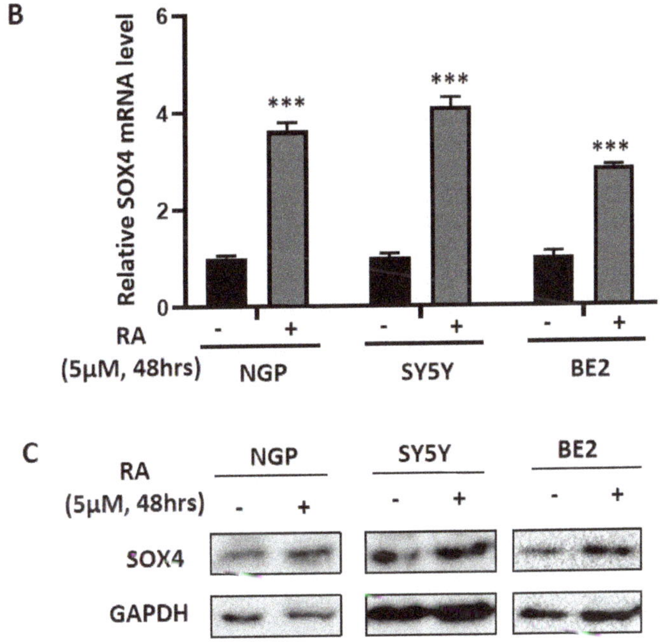

Figure 2. RA treatment increased the expression of SOX4 in NB cells. (**A**) The expression of SOX4 was increased from our previous RNA-Seq data (NGP cells treated with RA for 48 h). RT-qPCR (**B**) and Western blot (**C**) were performed to detect the expression of SOX4 in RA-treated (5 μM, 48 h) NGP, SY5Y, and BE2 cells. Control vs. RA treatment, *** $p < 0.001$. The uncropped western blots have been shown in Figure S3.

3.3. Overexpression of SOX4 Has the Potential to Induce the Differentiation of NB Cells

RA has been widely used clinically to induce the differentiation of NB tumor cells, and our results showed that RA treatment induced the increased expression of SOX4 and that the expression of SOX4 has a positive correlation with the survival rate of patients with NB. To detect if SOX4 mediated the differentiation of NB cells, we transfected SOX4 overexpression plasmids into NGP cells, and the results showed that the expression of SOX4 was significantly overexpressed (Figure 4A). Additionally, the cell confluence (% of the surface area of cells) or cell survival of SOX4-overexpressed NGP cells was detected by IncuCyte Zoom or CCK8 assay, and the results showed that both the cell confluence and the cell survival were inhibited by the overexpression of SOX4 (Figure 4B–D). The cell survival of SOX4-overexpressed NGP cells was 82.11% compared to the empty-vector transfected cells (100%) ($p < 0.01$) (Figure 4D). These indicated that overexpressing SOX4 decreased the NGP cell survival. One of the important characteristics of differentiated NB cells was elongated neuritic projections. Next, ICC staining was performed to show the morphological changes in NB cells after overexpressing SOX4, and elongated neurites can be observed in SOX4-overexpressed NGP cells (Figure 4E). It has been reported that G1 phase blockage correlates with cell differentiation [25,26], so we detected the cell cycle and compared the percentage of cells in the G1 phase between SOX4-overexpressed and empty-vector transfected NGP cells, and the results showed that the percentage of cells in G1 phase was 63.47% in SOX4-overexpressed NGP cells and 57.91% in empty-vector transfected cells, which indicated that overexpressing SOX4 blocked the cells in the G1 phase ($p < 0.05$) (Figure 4F). Similar effects on cell proliferation, cells' neurites, and the cell cycle were observed in BE2 cells after overexpressing SOX4 (Figure S1A–F). To confirm the change in the cell cycle in the G1 phase, we performed a Western blot to detect the expression changes in Cyclin D1 and CDK4, both of them are key regulators of the G1 phase. The results showed that the expressions of Cyclin D1 and CDK4 decreased after overexpressing SOX4 (Figure S1G) in both the NGP and BE2 cells.

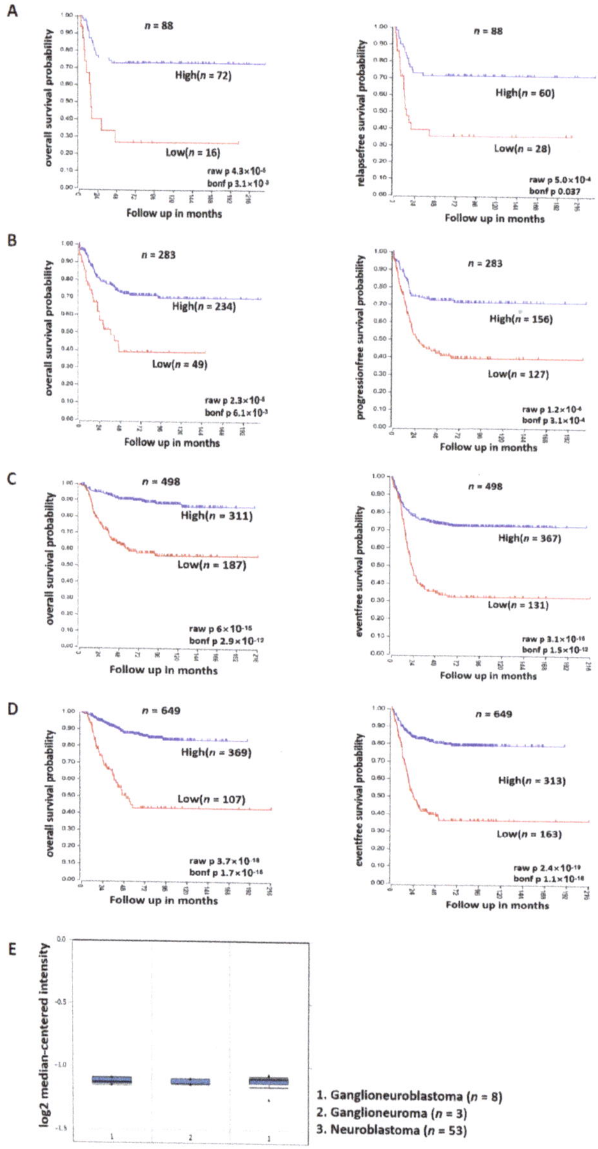

Figure 3. SOX4 has a positive correlation with the survival rate of patients with NB. Four datasets with different clinical samples ((**A**): 88 samples, (**B**): 283 samples, (**C**): 498 samples, and (**D**): 649 samples)

are from the R2 database, and the relationship between the expression of SOX4 and overall survival probability and relapse-free/progression-free/event-free survival probability was analyzed. Bonf $p < 0.05$ was considered as statistically significant. (**E**). The relationship between the expression of SOX4 and the differentiation degree of tumors originating from the sympathetic nervous system (ganglioneuroblastoma (n = 8), ganglioneuroma (n = 3), and neuroblastoma (n = 53)) was analyzed through the Oncomine database. Box-plots represent the expression levels of SOX4 (Log2 median-centered intensity).

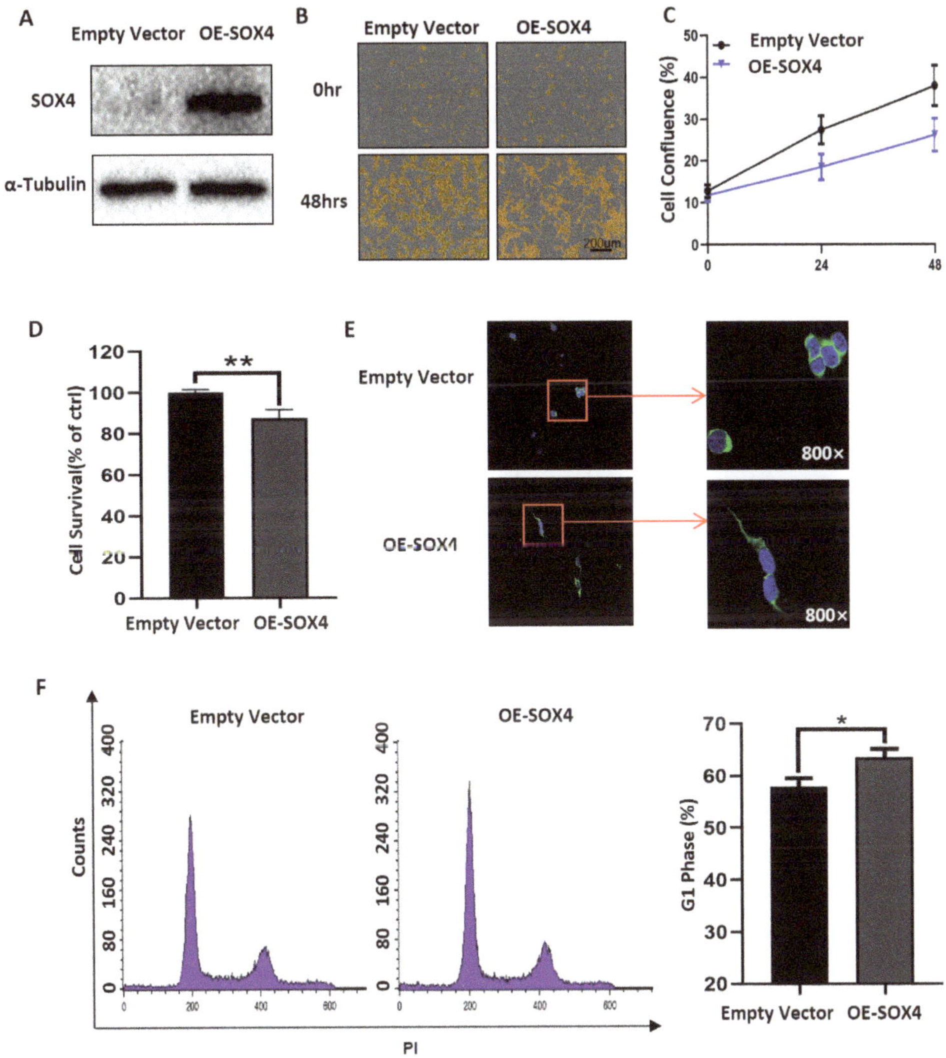

Figure 4. Overexpression of SOX4 has the potential to induce the differentiation of NB cells. (**A**) The expression of SOX4 in NGP cells was detected by Western blot 24 h after transfection with SOX4 overexpression plasmids. The cell confluence (% of the surface area of cells) or cell survival of SOX4-overexpressed NGP cells was detected by IncuCyte Zoom (**B**,**C**) or CCK8 assay (**D**). Empty vector vs. OE-SOX4, ** $p < 0.01$. (**E**) ICC staining was performed to show the morphological changes in NB cells

48 h after overexpressing SOX4. (**F**) Cell cycle analysis was performed 48 h after overexpressing SOX4, and the percentage of cells in G1 phase was analyzed. Empty vector vs. OE-SOX4, * $p < 0.05$. The uncropped Western blots have been shown in Figure S4.

3.4. *Downregulation of SOX4 Partially Blocks the Function of RA in NB Cells*

To further evaluate the role of SOX4 in RA-induced NB cell differentiation, we developed three siRNAs to knockdown the expression of SOX4 and then detected whether downregulated SOX4 could block the function of RA. The results showed that all three siRNAs decreased the expression of SOX4 at the protein level in NGP cells (Figure 5A); furthermore, RA-induced increased expression of SOX4 in NGP cells was downregulated after transfection with SOX4 siRNAs, both at the protein and mRNA levels (Figure 5A,B). Additionally, the cell confluence or cell survival of NGP cells transfected with SOX4 siRNAs or/and treated with RA was detected by IncuCyte Zoom or CCK8 assay, and the results showed that both the decreased cell confluence and the cell survival induced by RA were blocked by SOX4 siRNAs (Figure 5C–E). The cell survival of RA-treated NGP cells was 80.77%, and that of RA-treated and SOX4 siRNA-transfected cells was 100.14% (RA + SOX4 siRNA#1), 105.21% (RA + SOX4 siRNA#2), and 109.82% (RA + SOX4 siRNA#3) compared to the empty-vector transfected cells (100%) ($p < 0.01$) (Figure 5E). Then, ICC staining was performed to show the morphological changes of NGP cells, and the results showed that RA-induced elongated neurites could be blocked by SOX4 siRNAs (Figure 5F). We also detected the cell cycle and compared the percentage of cells in the G1 phase between RA-treated and SOX4 siRNAs-transfected NGP cells, and the results showed that the percentage of cells in the G1 phase was 65.67% in RA-treated NGP cells, 60.30% in RA + SOX4 siRNA#1-treated cells, 61.80% in RA + SOX4 siRNA#2-treated cells, and 61.68% in RA + SOX4 siRNA#3-treated cells ($p < 0.01$) (Figure 5G). These results indicated that knocking down the expression of SOX4 could reverse the function of RA in NB cells. Similar effects on cell proliferation, cells' neurites, and the cell cycle were observed in RA-treated BE2 cells after the knockdown of SOX4 (Figure S2A–F). To confirm the change in the cell cycle in the G1 phase, we performed a Western blot to detect the expression changes of Cyclin D1 and CDK4. The results showed that the knockdown of SOX4 reversed the decreased expressions of Cyclin D1 and CDK4 induced by RA (Figure S2G) in both NGP and BE2 cells.

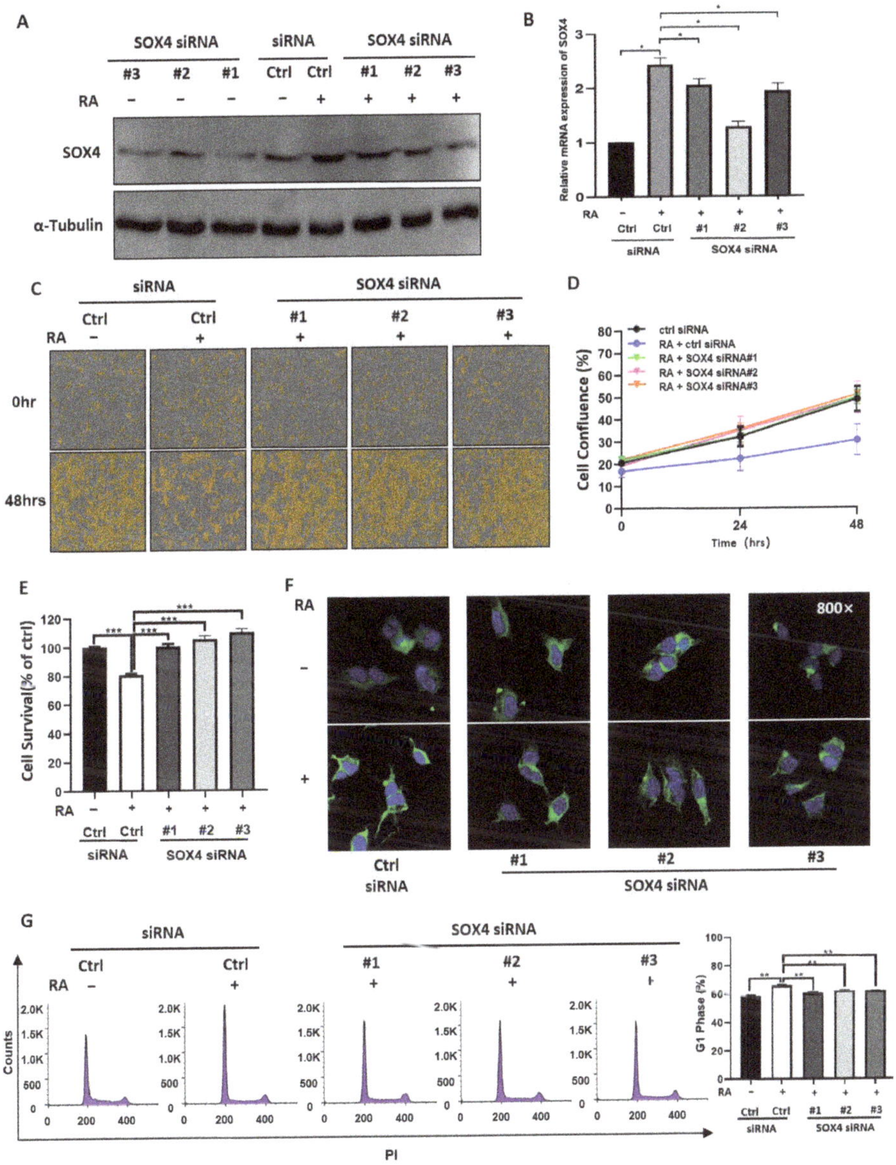

Figure 5. Downregulation of SOX4 partially blocks the function of RA in NB cells. (**A**). The expression of SOX4 in NGP cells transfected with SOX4 siRNAs or/and treated with RA was detected by Western blot (**A**) and RT-qPCR (**B**). Ctrl siRNA vs. Ctrl siRNA + RA, SOX4 siRNA + RA vs. Ctrl siRNA + RA, * $p < 0.05$. The cell confluence or cell survival of NGP cells transfected with SOX4 siRNAs or/and treated with RA was detected by IncuCyte Zoom (**C**,**D**) or CCK8 assay (**E**,**F**) ICC staining was performed to show the morphological changes in NGP cells transfected with SOX4 siRNAs or/and treated with RA. Ctrl siRNA vs. Ctrl siRNA + RA, SOX4 siRNA + RA vs. Ctrl siRNA + RA, *** $p < 0.001$. (**G**) Cell cycle analysis was performed and the percentage of cells in G1 phase was analyzed in RA-treated or/and SOX4 siRNAs-transfected NGP cells. Ctrl siRNA vs. Ctrl siRNA + RA, SOX4 siRNA + RA vs. Ctrl siRNA + RA, ** $p < 0.01$. The uncropped Western blots have been shown in Figure S5.

4. Discussion

To explore the potential targets that mediate RA-induced NB cell differentiation, we analyzed four microarray datasets from GEO, which were performed after RA treatment, and then, the overlapping DEGs were identified; there were 13 overlapping upregulated DEGs and 2 overlapping downregulated DEGs. SOX4 was one of the overlapping upregulated DEGs, which is in accordance with our previous study [21], and was confirmed in RA-treated NB cell lines (NGP, SY5Y, and BE2). Our results also showed that the clinical patients' survival probabilities was positively correlated with SOX4 expression, in which overexpressing SOX4 inhibited NB cell proliferation, elongated the cells' neurites, and blocked the cell cycle in the G1 phase. Furthermore, knockdown of the expression of SOX4 partially reversed the RA-induced inhibition of NB cell proliferation, the elongation of the cells' neurites, and the blocking of the cell cycle in the G1 phase. These indicate that SOX4 may be a target to induce NB cell differentiation.

NB is considered to be caused by the differentiation failure of neural crest cells, and it has been reported that the degree of tumor differentiation in patients with NB is closely related to its overall survival rate [1,2]. However, the mechanisms that regulate the differentiation of NB are still not very clear. In this study, we analyzed four microarray datasets from GEO (GSE45587: NB cell line BE(2)-C cells treated with RA for 24 h and 72 h, GSE16451: BE(2)-C cells treated with RA for 3 weeks, and GSE87784: Sphere cells from TH-MYCN mice treated with RA for 3 weeks), and 13 overlapping upregulated DEGs (SOX4, SOX9, ADD3, ATP7A, CAMK2N1, CTSB, RET, CYP26A1, CYP26B1, DDAH2, LTBP3, and MEIS1) and 2 overlapping downregulated DEGs (FHL2 and BRCA2) were identified via Venn diagrams. Both SOX4 and SOX9 are from the SOX family of transcription factors, and SOX transcription factor family members have been reported play important roles in the development of many organs and tissues [13,14]. SOX9 has been reported to be one of the important regulators in neural crest cell development [27], ATP7A plays important roles in neuronal differentiation [28] and glial differentiation [29], CTSB promotes the differentiation of preadipocytes [30], RET has been reported regulate the differentiation of NB cells [31], CYP26A1 and CYP26B1 are retinoic acid catabolic enzymes [32,33], DDAH2 is a biomarker for neural stem cell differentiation [34], LTBP3 regulates the differentiation of mesenchymal stem cells [35,36], and MEIS1 regulates cell proliferation and differentiation during cell fate commitment in different neoplasms [37]. FHL2 has been reported to play important roles in different cells' differentiation, including neuronal cells, gastric and colon cancer cells, and limb mesodermal progenitors [38–40], and BRCA2 regulates the differentiation of both normal tissues and different cancers [41–44]. These indicate that the overlapping up- and downregulated DEGs from the four GEO datasets may have essential roles in RA-induced cell differentiation.

In this study, we focused on the role of SOX4 in NB cells. As a member of the SOX transcription factor family, SOX4 has been shown to have an essential relationship with not only normal development but also cancers, such as lung cancer, breast cancer, leukemias, glioblastoma, and medulloblastoma [14–16]. Firstly, we found that patients with higher expressions of SOX4 exhibited higher overall survival probability and relapse-free/progression-free/event-free survival probability compared to patients with lower expressions of SOX4 by analyzing the data from the R2 database, while the results from the Oncomine database could not demonstrate the exact relationship between SOX4 and the differentiation degree of sympathetic nervous system-originating tumors. Then, our results showed that overexpressing SOX4 inhibited NGP cell proliferation, elongated the cells' neurites, and blocked the cell cycle in the G1 phase and that the downregulation of SOX4 partially reversed the RA-induced inhibition of NGP cell proliferation, the elongation of the cells' neurites, and the blocking of the cell cycle in the G1 phase. Guanhua Song et al.'s study evaluated the function of SOX4 in ATRA-induced differentiation of acute promyelocytic leukemia (APL) and, similar to the role of SOX4 in NB cells from our study, showed that SOX4 is essential for the differentiation and regulated by PAD4 [45]. Yuyin Yi et al.'s study found that SOX4 promotes the BMP2regulated differentiation of invasive trophoblast [46], while Wei Han et al.'s study found that FHL3 blocked glioma cell proliferation via the downregulation of SOX4 [47], and Dong Chen et al.'s study shows that the knockdown of SOX4 decreased cell proliferation, migration, and invasion and induced apoptosis in osteosarcoma cell lines [48]. These indicate that SOX4 play different roles in different types of cells. The limitation of this study is that the mechanisms underlying how SOX4 regulates the differentiation of NB cells need to be well-studied, which are the focus of our future studies.

In summary, our present study clearly provided evidence that SOX4 plays an important role in the differentiation of NB cells.

5. Conclusions

In conclusion, our study demonstrated that NB patients with higher expressions of SOX4 had good prognosis. Overexpressing SOX4 inhibited NB cell proliferation, elongated the cells' neurites, and blocked the cell cycle in the G1 phase, and the knockdown of the expression of SOX4 partially reversed the function of RA in NB cells, which provided evidence that SOX4 mediates the differentiation of NB cells.

Supplementary Materials: The following supporting information can be downloaded at: https://www.mdpi.com/article/10.3390/cancers14225642/s1, Figure S1: Overexpression of SOX4 inhibited BE2 cell proliferation, elongated the cell's neuritic, and blocked the cell cycle in G1 phase; Figure S2: Knockdown SOX4 partially reserved RA-induced inhibition of BE2 cell proliferation, elongated the cell's neuritic, and blocking the cell cycle in G1 phase; Figure S3: Uncropped western blot of Figure 2C; Figure S4: Uncropped western blot of Figure 4A; Figure S5: Uncropped western blot of Figure 5A.

Author Contributions: Conceptualization, Z.H. and X.T.; methodology, D.Z.; software, B.G.; validation, D.Z., Z.H. and X.T.; formal analysis, D.Z.; investigation, D.Z.; resources, Z.L.; data curation, D.Z.; writing—original draft preparation, Z.H. and X.T.; writing—review and editing, B.G., Q.Z., Z.H. and X.T.; visualization, D.Z.; supervision, Z.L., Z.H. and X.T.; project administration, Z.L.; funding acquisition, Z.L. and Z.H. All authors have read and agreed to the published version of the manuscript.

Funding: This work was supported by the National Science Foundation of China (82002641 and 81972515), the China Postdoctoral Science Foundation (2020M681015), the Key Research and Development foundation of Liaoning Province (2019JH8/10300024), the "Xingliao Talents Program" of Liaoning Province (XLYC2008010), the 2013 Liaoning Climbing Scholar Foundation, and the 345 Talent Project of Shengjing Hospital of China Medical University.

Institutional Review Board Statement: Not applicable.

Informed Consent Statement: Not applicable.

Data Availability Statement: The data presented in this study are available from the corresponding author upon request.

Conflicts of Interest: The authors declare no conflict of interest.

References

1. Brodeur, G.M. Neuroblastoma: Biological insights into a clinical enigma. *Nat. Rev. Cancer* **2003**, *3*, 203–216. [CrossRef] [PubMed]
2. Shohet, J.; Foster, J. Neuroblastoma. *BMJ* **2017**, *357*, j1863. [CrossRef] [PubMed]
3. Tomolonis, J.A.; Agarwal, S.; Shohet, J.M. Neuroblastoma pathogenesis: Deregulation of embryonic neural crest development. *Cell Tissue Res.* **2018**, *372*, 245–262. [CrossRef] [PubMed]
4. Edsjo, A.; Holmquist, L.; Pahlman, S. Neuroblastoma as an experimental model for neuronal differentiation and hypoxia-induced tumor cell dedifferentiation. *Semin. Cancer Biol.* **2007**, *17*, 248–256. [CrossRef] [PubMed]
5. Gheisari, S.; Catchpoole, D.R.; Charlton, A.; Kennedy, P.J. Convolutional Deep Belief Network with Feature Encoding for Classification of Neuroblastoma Histological Images. *J. Pathol. Inform.* **2018**, *9*, 17. [CrossRef] [PubMed]
6. Bushue, N.; Wan, Y.J. Retinoid pathway and cancer therapeutics. *Adv. Drug Deliv. Rev.* **2010**, *62*, 1285–1298. [CrossRef]
7. di Masi, A.; Leboffe, L.; De Marinis, E.; Pagano, F.; Cicconi, L.; Rochette-Egly, C.; Lo-Coco, F.; Ascenzi, P.; Nervi, C. Retinoic acid receptors: From molecular mechanisms to cancer therapy. *Mol. Asp. Med.* **2015**, *41*, 1–115. [CrossRef]
8. Tang, X.H.; Gudas, L.J. Retinoids, retinoic acid receptors, and Cancer. *Annu. Rev. Pathol.* **2011**, *6*, 345–364. [CrossRef]

9. Lammer, E.J.; Chen, D.T.; Hoar, R.M.; Agnish, N.D.; Benke, P.J.; Braun, J.T.; Curry, C.J.; Fernhoff, P.M.; Grix, A.W., Jr.; Lott, I.T.; et al. Retinoic acid embryopathy. *N. Engl. J. Med.* **1985**, *313*, 837–841. [CrossRef]
10. Roenigk, H.H., Jr. Liver toxicity of retinoid therapy. *Pharmacol. Ther.* **1989**, *40*, 145–155. [CrossRef]
11. Altucci, L.; Leibowitz, M.D.; Ogilvie, K.M.; de Lera, A.R.; Gronemeyer, H. RAR and RXR modulation in cancer and metabolic disease. *Nat. Rev. Drug Discov.* **2007**, *6*, 793–810. [CrossRef] [PubMed]
12. Kagechika, H.; Shudo, K. Synthetic retinoids: Recent developments concerning structure and clinical utility. *J. Med. Chem.* **2005**, *48*, 5875–5883. [CrossRef] [PubMed]
13. Grimm, D.; Bauer, J.; Wise, P.; Kruger, M.; Simonsen, U.; Wehland, M.; Infanger, M.; Corydon, T.J. The role of SOX family members in solid tumours and metastasis. *Semin. Cancer Biol.* **2020**, *67*, 122–153. [CrossRef] [PubMed]
14. Moreno, C.S. SOX4: The unappreciated oncogene. *Semin. Cancer Biol.* **2020**, *67 Pt 1*, 57–64. [CrossRef]
15. Jang, S.M.; Kim, J.W.; Kim, C.H.; An, J.H.; Johnson, A.; Song, P.I.; Rhee, S.; Choi, K.H. KAT5-mediated SOX4 acetylation orchestrates chromatin remodeling during myoblast differentiation. *Cell Death Dis.* **2015**, *6*, e1857. [CrossRef]
16. Inoue, H.; Takahashi, H.; Hashimura, M.; Eshima, K.; Akiya, M.; Matsumoto, T.; Saegusa, M. Cooperation of Sox4 with beta-catenin/p300 complex in transcriptional regulation of the Slug gene during divergent sarcomatous differentiation in uterine carcinosarcoma. *BMC Cancer* **2016**, *16*, 53. [CrossRef]
17. Xu, E.E.; Krentz, N.A.; Tan, S.; Chow, S.Z.; Tang, M.; Nian, C.; Lynn, F.C. SOX4 cooperates with neurogenin 3 to regulate endocrine pancreas formation in mouse models. *Diabetologia* **2015**, *58*, 1013–1023. [CrossRef]
18. Chen, C.; Lee, G.A.; Pourmorady, A.; Sock, E.; Donoghue, M.J. Orchestration of Neuronal Differentiation and Progenitor Pool Expansion in the Developing Cortex by SoxC Genes. *J. Neurosci.* **2015**, *35*, 10629–10642. [CrossRef]
19. Liu, P.; Ramachandran, S.; Ali Seyed, M.; Scharer, C.D.; Laycock, N.; Dalton, W.B.; Williams, H.; Karanam, S.; Datta, M.W.; Jaye, D.L.; et al. Sex-determining region Y box 4 is a transforming oncogene in human prostate cancer cells. *Cancer Res.* **2006**, *66*, 4011–4019. [CrossRef]
20. Lee, H.; Goodarzi, H.; Tavazoie, S.F.; Alarcon, C.R. TMEM2 Is a SOX4-Regulated Gene That Mediates Metastatic Migration and Invasion in Breast Cancer. *Cancer Res.* **2016**, *76*, 4994–5005. [CrossRef]
21. Tan, X.; Gong, W.; Chen, B.; Gong, B.; Hua, Z.; Zhang, S.; Chen, Y.; Li, Q.; Li, Z. Downregulation of fibronectin 1 attenuates ATRA-induced inhibition of cell migration and invasion in neuroblastoma cells. *Mol. Cell. Biochem.* **2021**, *476*, 3601–3612. [CrossRef] [PubMed]
22. Frumm, S.M.; Fan, Z.P.; Ross, K.N.; Duvall, J.R.; Gupta, S.; VerPlank, L.; Suh, B.C.; Holson, E.; Wagner, F.F.; Smith, W.B.; et al. Selective HDAC1/HDAC2 inhibitors induce neuroblastoma differentiation. *Chem. Biol.* **2013**, *20*, 713–725. [CrossRef] [PubMed]
23. Peltier, D.C.; Simms, A.; Farmer, J.R.; Miller, D.J. Human neuronal cells possess functional cytoplasmic and TLR-mediated innate immune pathways influenced by phosphatidylinositol-3 kinase signaling. *J. Immunol.* **2010**, *184*, 7010–7021. [CrossRef]
24. Tsubota, S.; Kishida, S.; Shimamura, T.; Ohira, M.; Yamashita, S.; Cao, D.; Kiyonari, S.; Ushijima, T.; Kadomatsu, K. PRC2-Mediated Transcriptomic Alterations at the Embryonic Stage Govern Tumorigenesis and Clinical Outcome in MYCN-Driven Neuroblastoma. *Cancer Res.* **2017**, *77*, 5259–5271. [CrossRef] [PubMed]
25. Wang, J.G.; Barsky, L.W.; Davicioni, E.; Weinberg, K.I.; Triche, T.J.; Zhang, X.K.; Wu, L. Retinoic acid induces leukemia cell G1 arrest and transition into differentiation by inhibiting cyclin-dependent kinase-activating kinase binding and phosphorylation of PML/RARalpha. *FASEB J.* **2006**, *20*, 2142–2144. [CrossRef] [PubMed]
26. Zarrilli, R.; Pignata, S.; Apicella, A.; Di Popolo, A.; Memoli, A.; Ricchi, P.; Salzano, S.; Acquaviva, A.M. Cell cycle block at G1-S or G2-M phase correlates with differentiation of Caco-2 cells: Effect of constitutive insulin-like growth factor II expression. *Gastroenterology* **1999**, *116*, 1358–1366. [CrossRef]
27. Ponzoni, M.; Bachetti, T.; Corrias, M.V.; Brignole, C.; Pastorino, F.; Calarco, E.; Bensa, V.; Giusto, E.; Ceccherini, I.; Perri, P. Recent advances in the developmental origin of neuroblastoma: An overview. *J. Exp. Clin. Cancer Res.* **2022**, *41*, 92. [CrossRef]
28. Hatori, Y.; Yan, Y.; Schmidt, K.; Furukawa, E.; Hasan, N.M.; Yang, N.; Liu, C.N.; Sockanathan, S.; Lutsenko, S. Neuronal differentiation is associated with a redox-regulated increase of copper flow to the secretory pathway. *Nat. Commun.* **2016**, *7*, 10640. [CrossRef]
29. Chakraborty, K.; Kar, S.; Rai, B.; Bhagat, R.; Naskar, N.; Seth, P.; Gupta, A.; Bhattacharjee, A. Copper dependent ERK1/2 phosphorylation is essential for the viability of neurons and not glia. *Metallomics* **2022**, *14*, mfac005. [CrossRef]
30. Zhang, Z.Y.; Mai, Y.; Yang, H.; Dong, P.Y.; Zheng, X.L.; Yang, G.S. CTSB promotes porcine preadipocytes differentiation by degrading fibronectin and attenuating the Wnt/beta-catenin signaling pathway. *Mol. Cell. Biochem.* **2014**, *395*, 53–64. [CrossRef]
31. Peterson, S.; Bogenmann, E. The RET and TRKA pathways collaborate to regulate neuroblastoma differentiation. *Oncogene* **2004**, *23*, 213–225. [CrossRef] [PubMed]
32. Liu, T.; Bohlken, A.; Kuljaca, S.; Lee, M.; Nguyen, T.; Smith, S.; Cheung, B.; Norris, M.D.; Haber, M.; Holloway, A.J.; et al. The retinoid anticancer signal: Mechanisms of target gene regulation. *Br. J. Cancer* **2005**, *93*, 310–318. [CrossRef]
33. Stoney, P.N.; Fragoso, Y.D.; Saeed, R.B.; Ashton, A.; Goodman, T.; Simons, C.; Gomaa, M.S.; Sementilli, A.; Sementilli, L.; Ross, A.W.; et al. Expression of the retinoic acid catabolic enzyme CYP26B1 in the human brain to maintain signaling homeostasis. *Brain Struct. Funct.* **2016**, *221*, 3315–3326. [CrossRef]
34. Backdahl, L.; Herberth, M.; Wilson, G.; Tate, P.; Campos, L.S.; Cortese, R.; Eckhardt, F.; Beck, S. Gene body methylation of the dimethylarginine dimethylamino-hydrolase 2 (Ddah2) gene is an epigenetic biomarker for neural stem cell differentiation. *Epigenetics* **2009**, *4*, 248–254. [CrossRef] [PubMed]

35. Goessler, U.R.; Bugert, P.; Bieback, K.; Deml, M.; Sadick, H.; Hormann, K.; Riedel, F. In-vitro analysis of the expression of TGFbeta -superfamily-members during chondrogenic differentiation of mesenchymal stem cells and chondrocytes during dedifferentiation in cell culture. *Cell. Mol. Biol. Lett.* **2005**, *10*, 345–362. [PubMed]
36. Koli, K.; Ryynanen, M.J.; Keski-Oja, J. Latent TGF-beta binding proteins (LTBPs)-1 and -3 coordinate proliferation and osteogenic differentiation of human mesenchymal stem cells. *Bone* **2008**, *43*, 679–688. [CrossRef]
37. Jiang, M.; Xu, S.; Bai, M.; Zhang, A. The emerging role of MEIS1 in cell proliferation and differentiation. *Am. J. Physiol. Cell Physiol.* **2021**, *320*, C264–C269. [CrossRef]
38. Kim, S.Y.; Volkl, S.; Ludwig, S.; Schneider, H.; Wixler, V.; Park, J. Deficiency of Fhl2 leads to delayed neuronal cell migration and premature astrocyte differentiation. *J. Cell Sci.* **2019**, *132*, jcs228940. [CrossRef]
39. Lorda-Diez, C.I.; Montero, J.A.; Sanchez-Fernandez, C.; Garcia-Porrero, J.A.; Chimal-Monroy, J.; Hurle, J.M. Four and a half domain 2 (FHL2) scaffolding protein is a marker of connective tissues of developing digits and regulates fibrogenic differentiation of limb mesodermal progenitors. *J. Tissue Eng. Regen. Med.* **2018**, *12*, e2062–e2072. [CrossRef]
40. Wang, J.; Yang, Y.; Xia, H.H.; Gu, Q.; Lin, M.C.; Jiang, B.; Peng, Y.; Li, G.; An, X.; Zhang, Y.; et al. Suppression of FHL2 expression induces cell differentiation and inhibits gastric and colon carcinogenesis. *Gastroenterology* **2007**, *132*, 1066–1076. [CrossRef]
41. Yoshikawa, Y.; Kimura, S.; Soga, A.; Sugiyama, M.; Ueno, A.; Kondo, H.; Zhu, Z.; Ochiai, K.; Nakayama, K.; Hakozaki, J.; et al. Plasmodium berghei Brca2 is required for normal development and differentiation in mice and mosquitoes. *Parasites Vectors* **2022**, *15*, 244. [CrossRef] [PubMed]
42. Wang, C.Y.; Deneen, B.; Tzeng, S.F. BRCA1/BRCA2-containing complex subunit 3 controls oligodendrocyte differentiation by dynamically regulating lysine 63-linked ubiquitination. *Glia* **2019**, *67*, 1775–1792. [CrossRef] [PubMed]
43. Rajan, J.V.; Wang, M.; Marquis, S.T.; Chodosh, L.A. Brca2 is coordinately regulated with Brca1 during proliferation and differentiation in mammary epithelial cells. *Proc. Natl. Acad. Sci. USA* **1996**, *93*, 13078–13083. [CrossRef] [PubMed]
44. Ding, L.; Su, Y.; Fassl, A.; Hinohara, K.; Qiu, X.; Harper, N.W.; Huh, S.J.; Bloushtain-Qimron, N.; Jovanovic, B.; Ekram, M.; et al. Perturbed myoepithelial cell differentiation in BRCA mutation carriers and in ductal carcinoma in situ. *Nat. Commun.* **2019**, *10*, 4182. [CrossRef] [PubMed]
45. Song, G.; Shi, L.; Guo, Y.; Yu, L.; Wang, L.; Zhang, X.; Li, L.; Han, Y.; Ren, X.; Guo, Q.; et al. A novel PAD4/SOX4/PU.1 signaling pathway is involved in the committed differentiation of acute promyelocytic leukemia cells into granulocytic cells. *Oncotarget* **2016**, *7*, 3144–3157. [CrossRef]
46. Yi, Y.; Zhu, H.; Klausen, C.; Leung, P.C.K. Transcription factor SOX4 facilitates BMP2-regulated gene expression during invasive trophoblast differentiation. *FASEB J.* **2021**, *35*, e22028. [CrossRef]
47. Han, W.; Hu, P.; Wu, F.; Wang, S.; Hu, Y.; Li, S.; Jiang, T.; Qiang, B.; Peng, X. FHL3 links cell growth and self-renewal by modulating SOX4 in glioma. *Cell Death Differ.* **2019**, *26*, 796–811. [CrossRef]
48. Chen, D.; Hu, C.; Wen, G.; Yang, Q.; Zhang, C.; Yang, H. DownRegulated SOX4 Expression Suppresses Cell Proliferation, Migration, and Induces Apoptosis in Osteosarcoma In Vitro and In Vivo. *Calcif. Tissue Int.* **2018**, *102*, 117–127. [CrossRef]

cancers

Review

Immune Microenvironment and Immunotherapies for Diffuse Intrinsic Pontine Glioma

Yujia Chen, Chao Zhao, Shenglun Li, Jun Wang and Hongwei Zhang *

Department of Neurosurgery, Sanbo Brain Hospital, Capital Medical University, Beijing 100093, China
* Correspondence: zhanghongwei@ccmu.edu.cn; Tel.: +86-010-62856705

Simple Summary: Diffuse intrinsic pontine glioma (DIPG) is a malignant primary glial tumor that occurs in all age groups but predominates in children and is estimated to account for approximately 10–15% of pediatric brain tumors. The median age at diagnosis was 6–7 years, and the median survival of children is less than 12 months. At present, there is no effective treatment method for DIPG in the clinic. The continuous progress in immunotherapy has brought new prospects for the treatment of DIPG. In this review, we summarize the knowledge about the immune profile in DIPG and existing clinical trial results of DIPG, hoping to clarify the development of novel immunotherapies for DIPG treatment.

Abstract: Diffuse intrinsic pontine glioma (DIPG) is a primary glial glioma that occurs in all age groups but predominates in children and is the main cause of solid tumor-related childhood mortality. Due to its rapid progression, the inability to operate and insensitivity to most chemotherapies, there is a lack of effective treatment methods in clinical practice for DIPG patients. The prognosis of DIPG patients is extremely poor, with a median survival time of no more than 12 months. In recent years, there have been continuous breakthroughs for immunotherapies in various hematological tumors and malignant solid tumors with extremely poor prognoses, which provides new insights into tumors without effective treatment strategies. Meanwhile, with the gradual development of stereotactic biopsy techniques, it is gradually becoming easier and safer to obtain live DIPG tissue, and the understanding of the immune properties of DIPG has also increased. On this basis, a series of immunotherapy studies of DIPG are under way, some of which have shown encouraging results. Herein, we review the current understanding of the immune characteristics of DIPG and critically reveal the limitations of current immune research, as well as the opportunities and challenges for immunological therapies in DIPG, hoping to clarify the development of novel immunotherapies for DIPG treatment.

Keywords: diffuse intrinsic pontine glioma; immunotherapy; immune microenvironment; immune checkpoint inhibitor; vaccine; oncolytic virus

Citation: Chen, Y.; Zhao, C.; Li, S.; Wang, J.; Zhang, H. Immune Microenvironment and Immunotherapies for Diffuse Intrinsic Pontine Glioma. *Cancers* **2023**, *15*, 602. https://doi.org/10.3390/cancers15030602

Academic Editors: Saurabh Agarwal and Jianhua Yang

Received: 6 January 2023
Accepted: 12 January 2023
Published: 18 January 2023

1. Introduction

Brain tumors are the most common cause of solid cancer mortality in children [1]. Diffuse intrinsic pontine glioma (DIPG) is a primary glial tumor that occurs in all age groups but predominates in children and is estimated to account for approximately 10–15% of pediatric brain tumors [2,3]. The median age at diagnosis is 6–7 years and the median survival of patients in children is less than 12 months [3]. The standard treatment for DIPG is performed fractionated radiotherapy with or without chemotherapeutic drugs. Although this treatment can achieve transient clinical improvement, it is not curative [4,5]. Due to the complexity of the brainstem, DIPGs are not amenable to surgical resection. Other treatment options, including numerous chemotherapy and targeted therapy drugs, have not been shown to prolong the survival of DIPG patients alone in clinical trials.

With the continuous improvement of biopsy technology, the understanding of the genetic and epigenetic characteristics of DIPG is increasing. In recent years, the greatest breakthrough of DIPG was the discovery of changes in H3K27M based on epigenetic studies [6–8]. The hypomethylation of H3K27 is common in DIPG, and it is also found in the other midline gliomas located in the thalamus, spine, or other parts of the brainstem [9]. This finding led the World Health Organization (WHO) to reclassify this type of glioma with the alterations in H3K27 as "diffuse midline glioma, H3K27-altered (DMG)" in the 5th edition of the central nervous system (CNS) tumor classification [10]. However, it should be noted that DMG is not the histological counterpart of DIPG and the two are not inclusive. DIPG is a clinical diagnosis based on clinical symptoms and typical imaging features. About 70–85% of DIPG patients have H3K27 alternation and would be diagnosed as "DMG, H3 K27-altered" [11,12]. Patients with pontine high-grade glioma but no IDH mutation and H3K27 alternation would be diagnosed as "pediatric diffuse high-grade glioma, *H3* wild-type and *IDH* wild-type" [10].

The further understanding of genetic and epigenetic features of DIPG inspired us to identify innovative and more effective therapeutic approaches. Immunotherapy is based on different techniques aimed at redirecting the patient's own immune system to specifically fight cancer cells and has become a powerful clinical strategy for treating cancer [13]. However, immunotherapy research for DIPG is relatively lacking compared with other tumors partly because of the limited understanding of the immune microenvironment of DIPG. Therefore, this review clarifies the current understanding and limitations of the immune characteristics of DIPG, outlines the limitations and possible opportunities for immunotherapy treatment methods, and introduces current clinical trials of immune therapies in DIPG. Since most patients with DIPG have H3K27 alterations, some DMG studies, including DIPG, were included in this review. We believe such a review could strengthen our understanding of the progress in DIPG immunotherapy and thus guide the development of novel immunotherapies for DIPG treatment.

2. Immune Characteristics of DIPG

Since DIPG cannot be operated and biopsy is not warranted in most cases because of its clear characteristics in imaging, there has long been a lack of research on tumor tissue, and immunological research is even rarer. With the continuous progress in immunotherapy, researchers have become increasingly interested in the immune characteristics of DIPG, especially immune cell infiltration, immune checkpoint expression and other characteristics that may be applied for immunotherapy, and the study of immunological characteristics in DIPG is ongoing (Figure 1).

Immune cells are important parts of the tumor immune microenvironment and the cellular underpinnings of immunotherapy. Bailey et al. analyzed raw RNA-seq data about 220 high-grade glioma (HGG) patients published by Mackay et al. [14] using the CIBERSORT algorithm with the standard LM22 matrix. Individual patients were classified according to tumor location (brainstem, midline and hemispheric) and histone mutation status (wildtype and K27M). Compared with pediatric hemisphere HGG, there were higher proportions of CD4+ Treg cells; eosinophils; activated dendritic cell (DC) cells; more DCs and neutrophils; and fewer CD8+ T cells, NK cells, M1 cells and activated mast cells in DIPG [15]. Unlike CD4+ Treg cells, activated DC cells, eosinophils and other cell types that correlate positively with prognosis in hemisphere HGG, all these cells do not correlate significantly with prognosis in DIPG, except neutrophils, which are negatively associated with prognosis in both hemispheric HGG and DIPG [15]. Another study using immunohistochemistry (IHC) to compare immune cell infiltration in pediatric tumors showed that although the median number of total CD68+ cells accounted for 10% of total cells in pediatric low-grade glioma (pLGG), pediatric high-grade glioma (pHGG) and DIPG, the number of CD163+ macrophages were 10.4 times and 5.9 times higher than normal tissues in pLGG and pHGG samples, but there was no significant increase in DIPG. The ratio of CD163+ to CD68+ macrophages was 8.0-fold larger than the control in pHGG but was not different in DIPG [16]. Nevertheless, there was no significant increase in any subset of the T cells in DIPG tumors compared to the control, whereas T cell infiltration was significantly increased in pLGG and pHGG, especially CD8+ T cells (6.5- and 5.1-fold) [16]. In another study, Lin et al. used flow cytometry to detect immune cell infiltration in DIPG and adult GBM, and the results showed that compared with adult GBM, there were more CD11b+ myeloid cells (DIPG: 94.80% ± 0.92% vs. adult GBM: 70.34% ± 7.20%) and fewer CD3+ lymphocytes (DIPG: 1.72–2.65% vs. adult GBM: 7.09–50.2%) in DIPG samples [17].

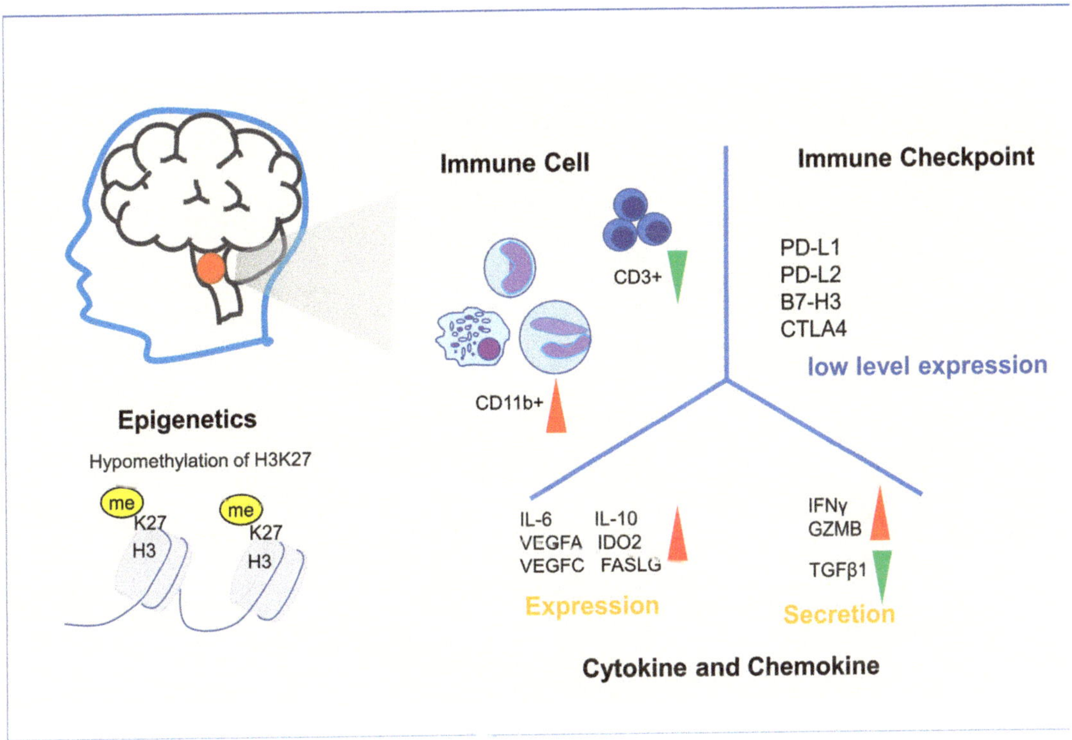

Figure 1. The epigenetic and immune microenvironment characteristics of DIPG. Most DIPGs have hypomethylation of H3K27. Compared with other high-grade gliomas, DIPG has lower $CD3^+$T cell infiltration and a higher proportion of $CD11b^+$ cell infiltration in its immune microenvironment. There is little expression of inhibitory immune checkpoints in DIPG, the expression of immunosuppressive cytokines is relatively low and the expression of proinflammatory factors is relatively high compared with high grade glioma.

In addition to the immune cell infiltration, we also focused on the expression of immune checkpoints, cytokines and chemokines in DIPG. There are few studies on the immune checkpoints in DIPG. A study that focused on the immune checkpoint expressed in the DIPG immune microenvironment showed that PD-L1 was only present at low levels in the DIPG tumor microenvironment detected by IHC staining [16]. Although B7-H3 was 2.4-fold higher than that in the control, both PD-L1and B7-H3 were significantly lower than those in glioblastoma cells [16]. However, in other studies with whole-brain HGG, the results showed that there was no increased expression of PD-L1 in DIPG [16,18]. In a study of 28 DIPG patients with radiation therapy, RNA-Seq of DIPG tissue found that the expression levels of PD-L1 and CTLA-4 in tumors were close to those in normal brain tissue, although PD-L2 levels tended to increase. If "upregulated expression" was defined as being a more than two-fold increase in RNA level, then 54%, 11%, 21% and 39% of patients showed upregulated expression of PD-1, PD-L1, PD-L2 and CTLA-4, respectively [19]. Although the studies of immunosuppressive checkpoints in DIPG were not sufficient, the expression of PD-L1was not significantly upregulated in DIPG tissues, which is consistent with the result of existing studies.

In a study on immunosuppressive gene expression and cytokine secretion in DIPG, RNA-Seq data analysis showed that there were higher expression levels of IDO2, IL10, FASLG, IL6, VEGFA and VEGFC; higher secretion levels of IFNG and GZMB; and lower secretion levels of TGFβ1 in K27M DIPG samples ($n = 23$) than in wildtype hemispheric HGG samples ($n = 85$) [15]. In addition, it has been shown that macrophages in DIPG secrete fewer chemokines and cytokines than macrophages in adult HGG [17]. Notably, studies also found that DIPG cells do not express inflammatory cytokines and chemokines that may recruit other immune cells to the tumor microenvironment, such as IL-2 [16,17].

According to the results of existing studies, DIPG has the characteristics of no increase in T cell infiltration, no increase in the number and proportion of CD163+ macrophages, no increase of inhibitory immune checkpoints and low expression of cytokines, which suggests that its internal immune microenvironment presents a "cold" or "inert" status [15–17,19]. However, the "cold" immune state does not mean that DIPG is completely devoid of specific immune responses. H3.3K27M-specific CD8+ T cells have been found in the peripheral blood of DIPG patients, suggesting that the immune system can generate specific cellular immunity against DIPG cells [20]. Additionally, the "cold" immune state does not mean that there is an immunosuppressive state in DIPG. Studies showed DIPG cells and macrophages in DIPG did not express a large number of immunosuppressive factors, and most DIPG cell cultures do not repolarize macrophages but can be lysed by NK cells [16].

Of course, the immune characteristics of DIPG still need further study. The current understanding of the immune characteristics of DIPG is often based on RNA-seq and IHC staining, which is far from enough. Some novel research technologies, such as single-cell sequencing technology and spatial transcriptome technology, may help us better understand the infiltration of immune cells, immune checkpoints and the true expression of immune-related genes in DIPG. However, if the assumption of the inert immune microenvironment in DIPG is correct, then immunotherapies that could supplement cytotoxic immune cells into DIPG directly or stimulate immune cells to infiltrate into DIPG, such as adaptive cell transfer therapy, oncolytic viruses and vaccines, are all reasonable strategies. On this basis, further use of the immune checkpoint inhibitor (ICI) therapy may yield better results.

3. Immunotherapy Research in DIPG

Immunotherapy for DIPG is currently under investigation. According to the registration data on the ClinicalTrials.gov website (https://www.clinicaltrials.gov/), there were 108 ongoing DIPG registration clinical trials as of 14 June 2022, including 25 immunotherapy clinical trials, mainly involving adoptive cell transfer therapy, vaccines, oncolytic virus therapy, immune checkpoint inhibitor (ICI) therapy and combination therapy (Figure 2, Table 1).

Table 1. The clinical trials of immunotherapies for DIPG.

Classification	NCT Number	Title	Status	Interventions	Age (Years)	Phases	Enrollment	Combination with Other Therapy
Adoptive Transfer Cell Therapy	NCT04196413	GD2 CAR T Cells in Diffuse Intrinsic Pontine Gliomas (DIPG) and Spinal Diffuse Midline Glioma (DMG)	Recruiting	GD2 CAR T cells + Fludarabine +Cyclophosphamide	2–30	Phase 1	54	Combination with Chemotherapy
	NCT04099797	C7R-GD2.CAR T Cells for Patients with GD2-expressing Brain Tumors (GAIL-B)	Recruiting	(C7R)-GD2.CART cells + Cyclophosphamide+ Fludarabine	1–21	Phase 1	34	Combination with Chemotherapy
	NCT05298995	GD2-CAR T Cells for Pediatric Brain Tumors	Not yet recruiting	GD2-CART01 (iC9-GD2-CAR T cells)	0.5–30	Phase 1	54	No
	NCT04185038	Study of B7-H3-Specific CAR T-Cell Locoregional Immunotherapy for Diffuse Intrinsic Pontine Glioma/Diffuse Midline Glioma and Recurrent or Refractory Pediatric Central Nervous System Tumors	Recruiting	SCRI-CARB7H3(s); B7H3-specific chimeric antigen receptor (CAR) T cells	1–26	Phase 1	90	No
Vaccine	NCT02960230	H3.3K27M Peptide Vaccine With Nivolumab for Children With Newly Diagnosed DIPG and Other Gliomas	Active, not recruiting	K27M peptide+ Nivolumab	3–21	Phase 1 | Phase 2	50	Combination with an Anti-PD-1 mAb
	NCT04749641	Neoantigen Vaccine Therapy Against H3.3-K27M Diffuse Intrinsic Pontine Glioma	Recruiting	Histone H3.3-K27M Neoantigen Vaccine	5-	Phase 1	30	No
	NCT04808245	A MultIceNTER Phase I Peptide VaCcine Trial for the Treatment of H3-Mutated Gliomas	Not yet recruiting	Tecentriq +H3K27M peptide vaccine+ Imiquimod (5%)	18-	Phase 1	15	Combination with an Anti-PD-L1 mAb
	NCT01058850	Phase I Rindopepimut After Conventional Radiation in Children w/Diffuse Intrinsic Pontine Gliomas	Terminated	Rindopepimut	3–18	Phase 1	3	No
	NCT04978727	A Pilot Study of SurVaxM in Children Progressive or Relapsed Medulloblastoma, High Grade Glioma, Ependymoma and Newly Diagnosed Diffuse Intrinsic Pontine Glioma	Recruiting	SurVaxM	1–21	Phase 1	35	Part of patients combination with radiotherapy
	NCT04943848	rHSC-DIPGVax Plus Checkpoint Blockade for the Treatment of Newly Diagnosed DIPG and DMG	Recruiting	rHSC-DIPGVax + Balstilimab + Zalifrelimab	1–18	Phase 1	36	Combination an Anti-PD-1 mAb and an Anti-CTLA4 mAb
	NCT02750891	A Study of DSP-7888 in Pediatric Patients with Relapsed or Refractory High-grade Gliomas	Completed	DSP-7888	0–19	Phase 1 | Phase 2	18	No

Table 1. *Cont.*

Classification	NCT Number	Title	Status	Interventions	Age (Years)	Phases	Enrollment	Combination with Other Therapy
	NCT02840123	Safety Study of DIPG Treatment with Autologous Dendritic Cells Pulsed with Lysates from Allogenic Tumor Lines	Unknown status	Autologous dendritic cells	3–18	Phase 1	10	No
	NCT03914768	Immune Modulatory DC Vaccine Against Brain Tumor	Enrolling by invitation	Immunomodulatory DC vaccine to target DIPG and GBM	1–75	Phase 1	10	No
	NCT04911621	Adjuvant Dendritic Cell Immunotherapy for Pediatric Patients With High-grade Glioma or Diffuse Intrinsic Pontine Glioma	Recruiting	Dendritic cell vaccination + temozolomide-based chemoradiation+- conventional next-line treatment	1–17	Phase 1| Phase 2	10	Combination with Temozolomide-based Chemoradiation
	NCT03396575	Brain Stem Gliomas Treated with Adoptive Cellular Therapy During Focal Radiotherapy Recovery Alone or with Dose-intensified Temozolomide (Phase I)	Recruiting	TTRNA-DC vaccines with GM-CSF/TTRNA-xALT/cyclophosphamide + fludarabine lymphodepletive conditioning/dose-intensified TMZ/Td vaccine/autologous hematopoietic stem cells (HSCs)	3–30	Phase 1	21	Combination with GM-CSF or Chemotherapy or Autologous Hematopoietic Stem Cells
	NCT02750891	A Study of DSP-7888 in Pediatric Patients with Relapsed or Refractory High-grade Gliomas	Completed	DSP-7888	0–19	Phase 1| Phase 2	18	No
Oncolytic Virus Therapy	NCT03330197	A Study of Ad-RTS-hIL-12 + Veledimex in Pediatric Subjects with Brain Tumors Including DIPG	Terminated	Ad-RTS-hIL-12+ Oral Veledimex	0–21	Phase 1| Phase 2	6	No
	NCT03178032	Oncolytic Adenovirus, DNX-2401, for Naive Diffuse Intrinsic Pontine Gliomas	Active, not recruiting	DNX-2401	1–18	Phase 1	12	No
	NCT02444546	Wild-Type Reovirus in Combination with Sargramostim in Treating Younger Patients With High-Grade Relapsed or Refractory Brain Tumors	Active, not recruiting	Sargramostim+ Wild-type Reovirus	10–21	Phase 1	6	Combination with Sargramostim
	NCT04758533	Clinical Trial to Assess the Safety and Efficacy of AloCELYVIR with Newly Diagnosed Diffuse Intrinsic Pontine Glioma (DIPG) in Combination with Radiotherapy or Medulloblastoma in Monotherapy	Recruiting	AloCELYVIR	1–21	Phase 1| Phase 2	12	No
	NCT05096481	PEP-CMV Vaccine Targeting CMV Antigen to Treat Newly Diagnosed Pediatric HGG and DIPG and Recurrent Medulloblastoma	Not yet recruiting	PEP-CMV+ Temozolomide+ Tetanus Diphtheria Vaccine	3–25	Phase 2	120	Combination with Temozolomide

Table 1. *Cont.*

Classification	NCT Number	Title	Status	Interventions	Age (Years)	Phases	Enrollment	Combination with Other Therapy
Immune Checkpoint Inhibitor Therapy	NCT01952769	Anti PD1 Antibody in Diffuse Intrinsic Pontine Glioma	Unknown status	MDV9300	3–21	Phase 1 | Phase 2	50	No
	NCT02359565	Pembrolizumab in Treating Younger Patients With Recurrent, Progressive, or Refractory High-grade Gliomas, Diffuse Intrinsic Pontine Gliomas, Hypermutated Brain Tumors, Ependymoma or Medulloblastoma	Recruiting	Pembrolizumab	1–29	Phase 1	110	No
	NCT03690869	REGN2810 in Pediatric Patients with Relapsed, Refractory Solid, or Central Nervous System (CNS) Tumors and Safety and Efficacy of REGN2810 in Combination With Radiotherapy in Pediatric Patients With Newly Diagnosed or Recurrent Glioma	Recruiting	Cemiplimab	0–25	Phase 1 | Phase 2	130	No
Other	NCT05063357	^{131}I-omburtamab Delivered by Convection-Enhanced Delivery in Patients with Diffuse Intrinsic Pontine Glioma	Not yet recruiting	^{131}I-Omburtamab	3–21	Phase 1	36	No
	NCT01502917	Convection-Enhanced Delivery of 124I-Omburtamab for Patients with Non-Progressive Diffuse Pontine Gliomas Previously Treated with External Beam Radiation Therapy	Completed	Radioactive iodine-labeled monoclonal antibody omburtamab+ external beam radiotherapy	2–21	Phase 1	50	Radiotherapy

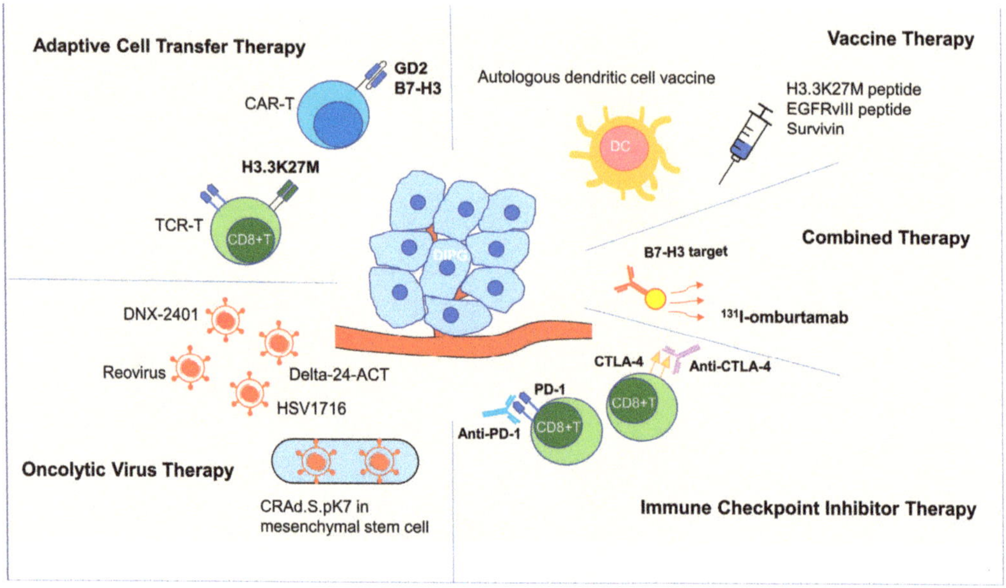

Figure 2. Immunotherapy strategies for DIPG. CAR T cells and TCR T cells are designed to target DIPG cells. Oncolytic viruses are designed to infect DIPG cells, kill them directly or attenuate their progression. A vaccine is injected to activate the adaptive immune system, especially CD8+ cytotoxic T cells. Anti-PD-1 and anti-CTLA-4 antibodies bind to PD-1 and CTLA-4, impeding the inactivation of CD8+ T cells, respectively. ^{131}I-omburtamab can kill DIPG cells by targeting B7-H3 and internal irradiation. CAR T, chimeric antigen receptor T cells; TCR T, T-cell receptor-gene engineered T cells; CTLA-4, cytotoxic T-lymphocyte-associated protein 4; PD-1, programmed cell death protein 1; DC, dendritic cell.

3.1. *Adoptive Cell Transfer Therapy*

Current research shows that there is an immunodeficiency microenvironment in DIPGs [15–19]. Therefore, many current studies attempt to directly supplement immune cells into DIPG to change its immune status and kill tumor cells. Chimeric antigen receptor-modified T (CAR-T) therapy and T cell receptor-gene engineered T (TCR-T) therapy were the two main therapies for this purpose.

3.1.1. CAR-T Therapy

A chimeric antigen receptor (CAR) is a fusion protein comprised of an antigen recognition moiety and T-cell signaling domains. The clinical trials of CAR T cells have shown clear efficacy in multiple hematologic malignancies [21].

Disialoganglioside (GD2), a proper target for CAR-T therapy, is widely and specifically expressed in DIPG [22–25]. An animal experiment with CAR-T cells targeting GD2 showed that the systemic administration of GD2-targeted CAR-T cells could kill almost all transplanted tumor cells in vivo, except for a small number of tumor cells without GD2 expression [22]. Although GD2 CAR-T cells have shown good therapeutic effects in vivo, some data have proven that GD2 CAR-T-cell treatment may result in severe neuroinflammation in the acute phase and lead to fatal brainstem edema and hydrocephalus in experimental animals [22]. Fortunately, some drugs and operations can be used to treat these severe complications. In a clinical study of GD2 CAR-T therapy for H3K27 M-mutated diffuse midline gliomas (DMGs) (clinicaltrials.gov: NCT04196413), four patients (three DIPG patients and one spine DMG patient) received 1 × 106 GD2-CAR T cells per kg administered intravenously and three exhibited significant radiographic and clinical benefits. Neither on-target nor off-tumor toxicity was observed. Although proinflammatory cytokine levels were increased in the plasma and cerebrospinal fluid, brainstem inflammation induced by CAR-T-cells was reversible with intensive supportive care. The systemic inflammatory response was suppressed by clinical drugs (IL-6R inhibitor, IL-1R inhibitor, corticosteroids), and hydrocephalus was controlled by releasing cerebrospinal fluid (CSF) (hypertonic saline, corticosteroids or an Ommaya reservoir) [24,26]. They also identified that multiple doses could provide a greater radiographic and clinical benefit than a single dose and different modes of administration could influence the neuroinflammatory response, whereby intracerebroventricular injection resulted in higher levels of proinflammatory cytokines, lower levels of immunosuppressive cells in the CSF, and fewer neurotoxic effects than intravenous injection [24]. Another study confirmed that IGF1R/IR in combination with GD2 CAR-T cells could further enhance the anti-tumor efficacy and increase the T-cell central memory profile in DMG/DIPG patients [23]. These results underscored the promise of GD2 CAR-T therapy for patients with H3K27M-mutated DIPG.

In addition to GD2 CAR-T cells, B7-H3 CAR-T cells are being explored in an ongoing clinical trial related to DIPG (clinicaltrials.gov: NCT04185038). B7-H3, also known as CD276, has been found to be highly expressed in most neuroectodermal tumors. B7-H3-targeting CAR-T therapy has shown favorable safety and efficacy in children with stage IV CNS metastasis and high expression of B7-H3 neuroblastoma [27]. A previous study showed that the mRNA level of B7-H3 in DIPG tissues ($n = 9$) was significantly higher than that in non-DIPG tumors and normal brain tissues. IHC staining results were consistent with mRNA results, demonstrating abnormally high expression of B7-H3 in DIPG [2]. However, there were no results of B7-H3-targeting CAR-T therapy on DIPG published before, and the effect needs to be further explored.

3.1.2. TCR-T Therapy

The main target of current TCR-T research is the H3.3K27M mutation. Marie-Anne reported that the substitution from lysine (K) to methionine (M) at position 27 of H3.3 (H3.3K27M mutation) was present in more than 70% of DIPG cases [12]. In animal experiments, TCR-transduced HLA-A2+ CD8+ T cells were able to efficiently kill HLA-A2+ H3.3K27M+ glioma cells in an antigen-dependent manner and significantly inhibit tumor progression [20]. In addition, Hideho et al. illustrated that the key amino acid residues required for recognition by the H3.3K27M-targeted TCR were absent in all known human proteins, suggesting that this H3.3K27M-targeted TCR-T therapy could be safely administered to patients. However, to date, no clinical data on H3.3K27M-targeted TCR-T therapy have been reported, and further clinical studies are needed.

3.2. *Vaccine Therapy*

Vaccine therapy is an important aspect of immunotherapy. Using tumor-specific biological macromolecules as antigens, enhancing the specific recognition and killing the ability of tumors to attack the immune system are the main ways in which vaccine immunity exerts an effect. In cancer vaccine immunotherapy experiments, life-threatening and lethal events are mainly caused by cross-reactivity of on- or off-target T cells against normal cells. This requires that the antigens used in tumor immunotherapy be tumor-specific as much as possible, such as mutation-derived antigens, to achieve safety and efficiency. However, there are few molecules that can act as specific antigens for DIPG. Specific antigens discovered in the field of glioma in recent years mainly include epidermal growth factor receptor vIII (EGFRvIII) and mutant isocitrate dehydrogenase 1 (IDH1) [28–30]. The K27M mutation (H3.1 and H3.3) is the predominant mutation in DMG and DIPG and is also the most concerning specific vaccine antigen of DIPG at present [9,12,14,31,32].

3.2.1. H3K27M Peptide Vaccine

A prominent feature of DIPG is the presence of the K27M mutation of H3.3 or H3.1. This specific molecular epigenetic change is uniformly found in DIPG but not in normal cells, constituting an excellent target for vaccine therapy [33]. In MHC-humanized mice (HLA-A*0201, HLA-DRA*0101 and HLA-DRB1*0101), vaccination with the 27-amino acid H3K27M peptide fragment was able to effectively induce an immune response and drive IFNγ immune responses in cytotoxic T cells and Th1 cells, whereas no immune responses to H3K27 WT peptides were observed. The induced immune response effectively suppressed the growth of subcutaneous H3K27M tumors [34]. Although there was a lack of therapeutic studies on orthotopic H3K27M tumors, previous studies have shown that H3.3K27M-specific cytotoxic T cells could be isolated from the peripheral blood of DIPG patients [20], which indicates that H3.3K27M may induce an immune response in DIPG patients. A clinical trial (NCT02960230) on H3.3K27M-specific vaccine responses in diffuse midline glioma patients that included 19 DIPG patients showed that the H3.3K27M peptide vaccine is safe for DIPG patients with HLA-A*0201 H3.3K27M characteristics, with no grade 4 treatment-related adverse reactions observed [35]. In terms of the treatment effect, 39% (7/18) of all patients developed an immune response to the H3.3K27M peptide vaccine, resulting in detectable H3.3K27M-specific CD8+ T cells in peripheral blood, and the overall survival (OS) in responders was prolonged by 6.1 months compared with nonresponders (16.1 months vs. 10 months, $n = 6$) [35]. This result is encouraging. At present, three clinical studies related to the H3K27M vaccine (clinicaltrials.gov: NCT02960230, NCT04749641, NCT04808245) are still in progress, and further results are expected to be released.

3.2.2. EGFRvIII and Survivin Vaccines

In addition to the H3K27M vaccine, some vaccine studies targeting other antigens also explore whether they have therapeutic effects on DIPG, such as the EGFRvIII vaccine and Survivin vaccine. EGFR variant III is the most common EGFR gene rearrangement and the most common gene mutation in GBM, which can be detected in approximately 25–33% of GBM patients [36]. A study exploring the expression of EGFRvIII in eleven DIPG samples showed that it could be detected in 54.5% (6/11) of DIPG tissue [37]. As this mutation is only present in tumors, it is considered a good tumor-specific antigen for vaccine immunotherapy. Accordingly, a peptide-specific vaccine targeting EGFRvIII, Rindopepimut, a synthetic mutant EGFRvIII neoantigen-specific peptide conjugated to KLH and administered granulocyte-macrophage colony-stimulating factor (GM-CSF) was designed and tested [38]. Both phase I and phase II clinical trials of Rindopepimut in GBM have been successful, with good safety and efficacy [39,40]. The PFS rate was 66% at 5.5 months, and the median overall survival (OS) was 21.8 months in 65 patients [40]. Based on this success, the U.S. Food and Drug Administration (FDA) granted Rindopepimut the breakthrough therapy designation in February 2015 [38]. Encouraged by these results, a phase I clinical trial of the EGFRvIII vaccine in DIPG patients (clinicaltrials.gov: NCT01058850) is ongoing.

Survivin is a protein expressed during fetal development. There is low or even undetectable expression of survivin in normal human terminal tissues and high expression in multiple malignant tumors, including approximately 85% of GBM patients [41–43]. Additionally, the high expression of survivin was confirmed in most DIPG tissues, as detected by RNA-seq, Western blotting and RT-PCR [44]. This provides a rationale for clinical trials using the survivin vaccine for DIPG. A clinical trial of SurVaxM (a kind of survivin vaccine, clinicaltrials.gov: NCT04978727) for the treatment of various childhood malignant intracranial tumors, including DIPG, is currently underway, although the results have not yet been published.

3.2.3. DC Vaccines

DC cells, as the main antigen-presenting cells in the body, have been studied in the vaccine therapy of tumors for many years and have shown effects in the treatment of glioma in vitro [45]. Stimulating DCs with tumor-specific peptides, tumor lysates or vectors encoding specific tumor antigens could enhance the antitumor effects in GBM and other brain tumors [46–48], and some studies have shown that the autologous dendritic cell vaccine (ADCV) could satisfactorily induce a sustained antitumor immune response in high-grade glioma and DIPG [49,50]. For DIPG, a clinical trial (clinicaltrials.gov: NCT02840123) enrolled nine patients with newly diagnosed DIPG and treated them with ADCV prepared from monocytes obtained by leukapheresis. Five ADCV doses were administered intradermally during the induction phase. In the absence of tumor progression, patients received three boosts of tumor lysate every 3 months during the maintenance phase. The results showed that none of the nine patients experienced obvious toxicity or neuroinflammation with this schema. The subcutaneous inoculation of autologous DC cells activated by the DIPG cell lysate resulted in nonspecific antitumor responses in all nine patients and specific antitumor responses in eight patients. Immunological responses were also confirmed in T lymphocytes isolated from the cerebrospinal fluid (CSF) of two patients [50]. This experimental result shows that DC vaccine therapy has a promising role in the immune treatment of DIPG.

3.3. Oncolytic Virus Therapy

With the development of immunotherapy, oncolytic virus therapy has attracted increasing attention. As oncolytic viruses can preferentially replicate in tumor cells, they can be designed as a multifaceted therapeutic platform to directly kill tumors, release tumor-specific antigens, express transgenes to enhance direct cytotoxicity and alter the tumor microenvironment to optimize immune-mediated tumor clearance [51]. In addition, many oncolytic viruses have a favorable safety profile, and some of them, such as Talimogene laherparepvec (T-Vec), have been approved by the FDA for the treatment of certain tumors due to their excellent levels of safety and efficacy [52].

The preclinical and clinical studies of oncolytic viruses have shown promising results in DIPG [44,53,54]. DNX-2401 is a modified replicative oncolytic adenovirus that specifically kills tumors and has good antitumor effects in a variety of tumors [55]. Moreover, DNX-2401 shows good antitumor effects in both immunodeficient and immunocompetent animal models of HGG and DIPG [54]. After the in vivo administration of DNX-2401, the infiltration degree of CD3+ lymphocytes and CD8+ lymphocytes in DIPG increased significantly and the DIPG murine cells were killed, leading to a significant increase in the survival of tumor-bearing animals [54]. Encouraged by this result, the same institution expedited a phase 1 clinical trial of DNX-2401 in DIPG (clinicaltrials.gov: NCT03178032) [56]. A total of 12 newly diagnosed DIPG patients were enrolled in this clinical trial and received 1*1010 or 5*1010 DNX-2401 viral particles by intratumoral injection, 11 of whom continued to receive subsequent radiation therapy. On safety assessment, there were some severe adverse complications. Two patients developed serious adverse reactions, one developed hemiplegia and one developed quadriplegia during DNX-2401 treatment. For the efficiency of the 12 patients who were followed for a median of 17.8 months, MRI-assessed tumor shrinkage was observed in 9 patients. Three patients achieved partial response (PR), and eight patients achieved stable disease (SD). The median OS was 17.8 months, and the survival rate was 75% at 12 months, 50% at 18 months and 25% at 24 months. The tumor microenvironment and T-cell activity were explored in one patient, and the results showed that there were significant alterations in the immune cell infiltration. Tumor samples were collected before DNX-2401 treatment, after treatment and during autopsy. Comparing the immune cell infiltration among the three DIPG samples, the infiltration of immune cells inside the tumor was low before the start of treatment, increased to a certain extent after oncolytic virus treatment and decreased again at autopsy. It was also found that the patient's T-cell receptor clonally increased in peripheral blood at 2 months after oncolytic virus treatment, indicating that T cells were involved in the tumor-killing process [53,57]. These results showed that intratumorally injected DNX-2401 combined with radiotherapy in DIPG patients resulted in changes in T-cell activity and a reduction in or stabilization of tumor size in some patients but was associated with adverse events. How to balance efficacy and safety needs further research and exploration in DNX-2401.

In addition to DNX-2401, reovirus had previously shown good safety in a phase I clinical trial for recurrent GBM and in an animal orthotopic model study of DIPG [58,59].

In the clinical trial of recurrent GBM, no treatment-related grade 4 adverse events occurred, and anti-glioma activity was observed in some patients after 1*1010 reovirus treatment [58]. In an animal orthotopic model study of DIPG, the use of reovirus did not cause severe neurological symptoms but prevented tumor development without toxicity, yet this effect only occurred when tumor cells had been inoculated with the virus before implantation. If reovirus was injected into the DIPG in situ xenograft implantation, it did not prolong survival, even though CD3+ T cells in the tumor increased. However, if the DIPG in situ xenograft implantation was treated with reovirus combined with ICI therapy, the survival of tumor-bearing animals could be significantly prolonged [59].

The synergistic effect of ICIs and oncolytic viruses can be achieved not only through combination therapy but also through viral transgene expression [60]. A recently published article on the anti-DIPG effect of the engineered oncolytic adenovirus Delta-24-ACT showed that it could elevate the expression of functional 4-1BBL on the membrane of infected DIPG cells to enhance the stimulation of CD8+ T lymphocytes in the tumor microenvironment. In mouse model studies, Delta-24-ACT not only improved the survival rate but also led to long-term immune memory in surviving mice and increased the number of infiltrating immune cells in tumors, with a good safety profile and no local or systemic toxicity [60]. In addition to directly killing DIPG cells, some oncolytic viruses can also inhibit tumor spread [61]. Oncolytic herpes simplex virus 1716 (HSV1716) shows good capacity to inhibit the migration and infiltration of DIPG cells in vitro and in vivo [61].

Although oncolytic virus therapy has shown some effect in the treatment of DIPG in preclinical and clinical trials, there are some limitations of current oncolytic virus therapy in DIPG, and an important limitation is the difficulty of its administration. Currently, oncolytic viruses must be administered stereotactically to tumors through invasive surgery. Maciej et al. utilized mesenchymal stem cells as carriers to carry the modified oncolytic virus CRAd.S.pK7 and administered it through subcutaneous injection or intraperitoneal injection. The homing capacity of mesenchymal stem cells made it possible to cross the blood–brain barrier and enter the tumor. In addition, the stem cell-mediated delivery of CRAd.S.pK7 ensured virus dissemination throughout the tumor, delayed virus clearance by the host immune system and reduced neuroinflammatory responses, thereby providing neuroprotection to normal brain tissue [44,62,63]. In animal orthotopic model experiments, the results showed that the injection of CRAd.S.pK7-loaded mesenchymal stem cells after radiotherapy could lead to better survival in animals than either treatment alone [44].

3.4. *ICI Therapy*

Immune checkpoints are essentially costimulatory molecules that affect T-cell responses. They regulate or inhibit T-cell function by binding to corresponding ligands or receptors on the surface of T cells [64]. Inhibitory immune checkpoints, such as PD-L1, PD-L2 and CTLA-4, are highly expressed in many tumors, and they impair the function of cytotoxic T cells in the tumor [65–67]. Currently, the FDA has approved the ICIs CTLA-4 and PD-1, as well as their ligands PD-L1 and PD-L2, for specific tumor treatment [68,69]. Inhibitors targeting more immune checkpoints, such as LAG3, TIGIT, TIM3, B7H3, CD39, CD73, adenosine A2A receptor and CD47, are still under investigation.

Great progress has been made in ICI therapy in recent years. Previous studies have shown that ICIs can exert therapeutic effects on various tumors, including malignant melanoma, non-small cell lung cancer, renal cell carcinoma, gastric cancer and esophageal cancer [70–76]. However, the studies of ICI therapy in brain tumors are few, and the efficiency is not significant [77–81].

For clinical research on the role of ICIs in DIPG, the current research data are limited and the results are not consistent. In a retrospective study involving two DIPG patients who received both PD-1 inhibitor and CTLA-4 inhibitor therapy for a short duration before the discontinuation of ICI used to curtail disease progression, the results showed that there was no obvious drug toxicity in patients, but the survival time was also not significantly prolonged [82]. In a retrospective study of 31 recurrent DIPG patients, 8 patients received both reradiation and PD-1 inhibitor combination therapy, and their survival time was not significantly different but slightly prolonged compared with those in the control group, and none developed acute or chronic toxicity [83]. However, in another clinical trial that enrolled five DIPG children with a median age of 3.5 years, the PD-1 inhibitor Pembrolizumab was intravenously administered at 2 mg/kg every three weeks, and the patients experienced clinical or radiographic deterioration after 1 or 2 doses of treatment. Their PFS was only 1.02 months (range, 0.5–1.7), and OS from initial treatment was 1.7 months (range, 0.5–6.2), which was much shorter than in previous clinical studies. All patients clinically and/or radiographically worsened after one (n = 1) or two (n = 4) doses of pembrolizumab. In addition, 40% of enrolled patients (n = 2) had grade 3 or 4 treatment-related adverse events, including fatigue (n = 2) and new or increased grade neurologic symptoms [84].

3.5. Combination Therapy

Overall, the intersection and combination of immunotherapy with other treatments have increased in clinical trials of DIPG. Since the standard treatment for DIPG is radiotherapy, there is often a combination of radiation therapy and immunotherapy. The use of ^{131}I-omburtamab to treat DIPG is one kind of radioimmunotherapy that is currently involved in a clinical trial for DIPG (clinicaltrials.gov: NCT05063357). Omburtamab is a monoclonal antibody that targets B7-H3 on the surface of DIPG, and ^{131}I labeling can kill tumor cells through the effect of internal irradiation. In addition, there are many other combination therapies, including combinations of multiple immunotherapies or combinations of immunotherapy and chemotherapy, which are also involved in clinical trials (clinicaltrials.gov: NCT04943848, NCT02960230).

3.6. Summary of Immunotherapy Research on DIPG

Above all, the current immunotherapy research on DIPG mainly includes five parts: adaptive cell transfer therapy, oncolytic virus therapy, vaccine therapy, ICI therapy and combined therapy. From the results of clinical trials, the first three are the most promising new treatment options for DIPG.

Among them, GD2 CAR-T showed a good effect on prolonging the OS and obvious neuroinflammatory response [24]. How to weaken its neuroinflammatory response and intensify its anti-tumor effect may be the focus of future research and development.

For oncolytic viruses, DNX-2401 experiments have shown that it has good safety and efficacy [53,56,57]. However, studies have also shown that its changes in immunity are very short-lived and that it will increase adverse effects when combined with radiotherapy. In addition, Oncolytic viruses still require in situ inoculation, which limits their clinical use.

For vaccines, the H3.3K27M peptide vaccine significantly prolongs the OS of patients with immune response, but its response rate is only 39% [35]. How to improve the response rate of vaccines may be an important issue in future research.

In terms of ICI therapy, it is not a reasonable strategy for DIPG as immune characterization studies have shown that the expression of immune checkpoints is not significantly elevated in DIPG. ICI treatment should be considered in combination with other therapies to improve the therapeutic effect.

4. Conclusions

The development of immunotherapy has brought new potential prospects for DIPG, which is unresectable and lacks effective treatment strategies. The existing research shows that the immune microenvironment of DIPG presents a state of cold or immunodeficiency status, specifically manifested as a lack of T lymphocyte infiltration, noninflammatory performance of macrophages, low levels of regulating cytokine secretion and low levels of immune checkpoint expression [17]. Some clinical trials have shown significant effects on prolonging the OS of DIPG patients, which could supplement T cells directly or stimulate the adaptive immune system to infiltrate specific cytotoxic cells into tumors, involving adaptive T-cell transfer therapy, vaccine therapy and oncolytic virus therapy [24,35,56].

Although studies on the immune microenvironment are still insufficient and most clinical trials involving DIPG are still in phase 1 or phase 2, the results presented in this research provide a key background for future DIPG research.

Author Contributions: Methodology: C.Z. and J.W.; Investigation and Writing—original draft: Y.C.; Visualization: S.L.; Conceptualization, Funding acquisition, Project administration, Supervision and Writing—review and editing: H.Z. All authors have read and agreed to the published version of the manuscript.

Funding: This research was supported by the National Key R&D Program of China (grant number 2019YFC1316104, FNL).

Conflicts of Interest: The authors declare no conflict of interest.

References

1. Ostrom, Q.T.; Fahmideh, M.A.; Cote, D.J.; Muskens, I.S.; Schraw, J.M.; Scheurer, M.E.; Bondy, M.L. Risk factors for childhood and adult primary brain tumors. *Neuro Oncol.* **2019**, *21*, 1357–1375. [CrossRef] [PubMed]
2. Zhou, Z.; Luther, N.; Ibrahim, G.M.; Hawkins, C.; Vibhakar, R.; Handler, M.H.; Souweidane, M.M. B7-H3, a potential therapeutic target, is expressed in diffuse intrinsic pontine glioma. *J. Neurooncol.* **2013**, *111*, 257–264. [CrossRef] [PubMed]
3. Grimm, S.A.; Chamberlain, M.C. Brainstem glioma: A review. *Curr. Neurol. Neurosci. Rep.* **2013**, *13*, 346. [CrossRef] [PubMed]
4. Warren, K.E. Diffuse intrinsic pontine glioma: Poised for progress. *Front. Oncol.* **2012**, *2*, 205. [CrossRef]
5. Perrone, M.G.; Ruggiero, A.; Centonze, A.; Carrieri, A.; Ferorelli, S.; Scilimati, A. Diffuse Intrinsic Pontine Glioma (DIPG): Breakthrough and Clinical Perspective. *Curr. Med. Chem.* **2021**, *28*, 3287–3317. [CrossRef]
6. Chan, K.-M.; Fang, D.; Gan, H.; Hashizume, R.; Yu, C.; Schroeder, M.; Gupta, N.; Mueller, S.; James, C.D.; Jenkins, R.; et al. The histone H3.3K27M mutation in pediatric glioma reprograms H3K27 methylation and gene expression. *Genes Dev.* **2013**, *27*, 985–990. [CrossRef] [PubMed]
7. Persson, M.L.; Douglas, A.M.; Alvaro, F.; Faridi, P.; Larsen, M.R.; Alonso, M.M.; Vitanza, N.A.; Dun, M.D. The intrinsic and microenvironmental features of diffuse midline glioma; implications for the development of effective immunotherapeutic treatment strategies. *Neuro Oncol.* **2022**, *24*, 1408–1422. [CrossRef]
8. Yu, J.-R.; LeRoy, G.; Bready, D.; Frenster, J.D.; Saldaña-Meyer, R.; Jin, Y.; Descostes, N.; Stafford, J.M.; Placantonakis, D.G.; Reinberg, D. The H3K36me2 writer-reader dependency in H3K27M-DIPG. *Sci. Adv.* **2021**, *7*, eabg7444. [CrossRef]
9. Schwartzentruber, J.; Korshunov, A.; Liu, X.-Y.; Jones, D.T.W.; Pfaff, E.; Jacob, K.; Sturm, D.; Fontebasso, A.M.; Quang, D.-A.K.; Tönjes, M.; et al. Driver mutations in histone H3.3 and chromatin remodelling genes in paediatric glioblastoma. *Nature* **2012**, *482*, 226–231. [CrossRef]
10. WHO Classification of Tumours Editorial Board (Ed.) *Central Nervous System Tumours WHO Classification of Tumours*; IARC Press: Lyon, France, 2021.
11. Wiese, M.; Hamdan, F.H.; Kubiak, K.; Diederichs, C.; Gielen, G.H.; Nussbaumer, G.; Carcaboso, A.M.; Hulleman, E.; Johnsen, S.A.; Kramm, C.M. Combined treatment with CBP and BET inhibitors reverses inadvertent activation of detrimental super enhancer programs in DIPG cells. *Cell Death Dis.* **2020**, *11*, 673. [CrossRef]
12. Khuong-Quang, D.-A.; Buczkowicz, P.; Rakopoulos, P.; Liu, X.-Y.; Fontebasso, A.M.; Bouffet, E.; Bartels, U.; Albrecht, S.; Schwartzentruber, J.; Letourneau, L.; et al. K27M mutation in histone H3.3 defines clinically and biologically distinct subgroups of pediatric diffuse intrinsic pontine gliomas. *Acta Neuropathol.* **2012**, *124*, 439–447. [CrossRef] [PubMed]
13. Zhang, Y.; Zhang, Z. The history and advances in cancer immunotherapy: Understanding the characteristics of tumor-infiltrating immune cells and their therapeutic implications. *Cell. Mol. Immunol.* **2020**, *17*, 807–821. [CrossRef]
14. Mackay, A.; Burford, A.; Carvalho, D.; Izquierdo, E.; Fazal-Salom, J.; Taylor, K.R.; Bjerke, L.; Clarke, M.; Vinci, M.; Nandhabalan, M.; et al. Integrated Molecular Meta-Analysis of 1000 Pediatric High-Grade and Diffuse Intrinsic Pontine Glioma. *Cancer Cell* **2017**, *32*, 520–537.e525. [CrossRef] [PubMed]
15. Bailey, C.P.; Wang, R.; Figueroa, M.; Zhang, S.; Wang, L.; Chandra, J. Computational immune infiltration analysis of pediatric high-grade gliomas (pHGGs) reveals differences in immunosuppression and prognosis by tumor location Computational and Systems. *Oncology* **2021**, *1*, e1016. [CrossRef]
16. Lieberman, N.A.P.; DeGolier, K.; Kovar, H.M.; Davis, A.; Hoglund, V.; Stevens, J.; Winter, C.; Deutsch, G.; Furlan, S.N.; Vitanza, N.A.; et al. Characterization of the immune microenvironment of diffuse intrinsic pontine glioma: Implications for development of immunotherapy. *Neuro Oncol.* **2019**, *21*, 83–94. [CrossRef] [PubMed]
17. Lin, G.L.; Nagaraja, S.; Filbin, M.G.; Suvà, M.L.; Vogel, H.; Monje, M. Non-inflammatory tumor microenvironment of diffuse intrinsic pontine glioma. *Acta Neuropathol. Commun.* **2018**, *6*, 51. [CrossRef] [PubMed]
18. Bockmayr, M.; Klauschen, F.; Maire, C.L.; Rutkowski, S.; Westphal, M.; Lamszus, K.; Schüller, U.; Mohme, M. Immunologic Profiling of Mutational and Transcriptional Subgroups in Pediatric and Adult High-Grade Gliomas. *Cancer Immunol. Res.* **2019**, *7*, 1401–1411. [CrossRef]

19. Zhu, X.; Lazow, M.A.; Schafer, A.; Bartlett, A.; Kumar, S.S.; Mishra, D.K.; Dexheimer, P.; DeWire, M.; Fuller, C.; Leach, J.L.; et al. A pilot radiogenomic study of DIPG reveals distinct subgroups with unique clinical trajectories and therapeutic targets. *Acta Neuropathol. Commun.* **2021**, *9*, 14. [CrossRef]
20. Chheda, Z.S.; Kohanbash, G.; Okada, K.; Jahan, N.; Sidney, J.; Pecoraro, M.; Yang, X.; Carrera, D.A.; Downey, K.M.; Shrivastav, S.; et al. Novel and shared neoantigen derived from histone 3 variant H3.3K27M mutation for glioma T cell therapy. *J. Exp. Med.* **2018**, *215*, 141–157. [CrossRef]
21. Brudno, J.N.; Kochenderfer, J.N. Recent advances in CAR T-cell toxicity: Mechanisms, manifestations and management. *Blood Rev* **2019**, *34*, 45–55. [CrossRef]
22. Mount, C.W.; Majzner, R.G.; Sundaresh, S.; Arnold, E.P.; Kadapakkam, M.; Haile, S.; Labanieh, L.; Hulleman, E.; Woo, P.J.; Rietberg, S.P.; et al. Potent antitumor efficacy of anti-GD2 CAR T cells in H3-K27M diffuse midline gliomas. *Nat. Med.* **2018**, *24*, 572–579. [CrossRef] [PubMed]
23. de Billy, E.; Pellegrino, M.; Orlando, D.; Pericoli, G.; Ferretti, R.; Businaro, P.; Ajmone-Cat, M.A.; Rossi, S.; Petrilli, L.L.; Maestro, N.; et al. Dual IGF1R/IR inhibitors in combination with GD2-CAR T-cells display a potent anti-tumor activity in diffuse midline glioma H3K27M-mutant. *Neuro Oncol.* **2021**, *24*, 1150–1163. [CrossRef] [PubMed]
24. Majzner, R.G.; Ramakrishna, S.; Yeom, K.W.; Patel, S.; Chinnasamy, H.; Schultz, L.M.; Richards, R.M.; Jiang, L.; Barsan, V.; Mancusi, R.; et al. GD2-CAR T cell therapy for H3K27M-mutated diffuse midline gliomas. *Nature* **2022**, *603*, 934–941. [CrossRef] [PubMed]
25. Franson, A.; Koschmann, C. Enhancing GD2 CAR-T Therapy with IGF1R Blockade: Are DIPG CAR-Ts ready for combinatorial therapy? *Neuro Oncol.* **2022**, 24. [CrossRef]
26. Majzner, R.G.; Ramakrishna, S.; Mochizuki, A.; Patel, S.; Chinnasamy, H.; Yeom, K.; Schultz, L.; Richards, R.; Campen, C.; Reschke, A.; et al. Abstract CT031: GD2 CAR T cells mediate clinical activity and manageable toxicity in children and young adults with DIPG and H3K27M-mutated diffuse midline gliomas. *Cancer Res.* **2021**, *81*, CT031. [CrossRef]
27. Kramer, K.; Kushner, B.H.; Modak, S.; Pandit-Taskar, N.; Smith-Jones, P.; Zanzonico, P.; Humm, J.L.; Xu, H.; Wolden, S.L.; Souweidane, M.M.; et al. Compartmental intrathecal radioimmunotherapy: Results for treatment for metastatic CNS neuroblastoma. *J. Neurooncol.* **2010**, *97*, 409–418. [CrossRef] [PubMed]
28. Schumacher, T.; Bunse, L.; Pusch, S.; Sahm, F.; Wiestler, B.; Quandt, J.; Menn, O.; Osswald, M.; Oezen, I.; Ott, M.; et al. A vaccine targeting mutant IDH1 induces antitumour immunity. *Nature* **2014**, *512*, 324–327. [CrossRef]
29. Brennan, C.W.; Verhaak, R.G.W.; McKenna, A.; Campos, B.; Noushmehr, H.; Salama, S.R.; Zheng, S.; Chakravarty, D.; Sanborn, J.Z.; Berman, S.H.; et al. The somatic genomic landscape of glioblastoma. *Cell* **2013**, *155*, 462–477. [CrossRef]
30. Thorne, A.H.; Zanca, C.; Furnari, F. Epidermal growth factor receptor targeting and challenges in glioblastoma. *Neuro Oncol.* **2016**, *18*, 914–918. [CrossRef]
31. Wu, G.; Broniscer, A.; McEachron, T.A.; Lu, C.; Paugh, B.S.; Becksfort, J.; Qu, C.; Ding, L.; Huether, R.; Parker, M.; et al. Somatic histone H3 alterations in pediatric diffuse intrinsic pontine gliomas and non-brainstem glioblastomas. *Nat. Genet.* **2012**, *44*, 251–253.
32. Buczkowicz, P.; Hoeman, C.; Rakopoulos, P.; Pajovic, S.; Letourneau, L.; Dzamba, M.; Morrison, A.; Lewis, P.; Bouffet, E.; Bartels, U.; et al. Genomic analysis of diffuse intrinsic pontine gliomas identifies three molecular subgroups and recurrent activating ACVR1 mutations. *Nat. Genet.* **2014**, *16*, 451–456. [CrossRef] [PubMed]
33. Solomon, D.A.; Wood, M.D.; Tihan, T.; Bollen, A.W.; Gupta, N.; Phillips, J.J.J.; Perry, A. Diffuse Midline Gliomas with Histone H3-K27M Mutation: A Series of 47 Cases Assessing the Spectrum of Morphologic Variation and Associated Genetic Alterations. *Brain Pathol.* **2016**, *26*, 569–580. [CrossRef] [PubMed]
34. Ochs, K.; Ott, M.; Bunse, T.; Sahm, F.; Bunse, L.; Deumelandt, K.; Sonner, J.K.; Keil, M.; von Deimling, A.; Wick, W.; et al. K27M-mutant histone-3 as a novel target for glioma immunotherapy. *Oncoimmunology* **2017**, *6*, e1328340. [CrossRef] [PubMed]
35. Mueller, S.; Taitt, J.M.; Villanueva-Meyer, J.E.; Bonner, E.R.; Nejo, T.; Lulla, R.R.; Goldman, S.; Banerjee, A.; Chi, S.N.; Whipple, N.S.; et al. Mass cytometry detects H3.3K27M-specific vaccine responses in diffuse midline glioma. *J. Clin. Invest.* **2020**, *130*, 6325–6337. [CrossRef] [PubMed]
36. Aldape, K.D.; Ballman, K.; Furth, A.; Buckner, J.C.; Giannini, C.; Burger, P.C.; Scheithauer, B.W.; Jenkins, R.B.; James, C.D. Immunohistochemical detection of EGFRvIII in high malignancy grade astrocytomas and evaluation of prognostic significance. *J. Neuropathol. Exp. Neurol.* **2004**, *63*, 700–707. [CrossRef] [PubMed]
37. Li, G.; Mitra, S.S.; Monje, M.; Henrich, K.N.; Bangs, C.D.; Nitta, R.T.; Wong, A.J. Expression of epidermal growth factor variant III (EGFRvIII) in pediatric diffuse intrinsic pontine gliomas. *J. Neurooncol.* **2012**, *108*, 395–402. [CrossRef]
38. Dunn-Pirio, A.M.; Vlahovic, G. Immunotherapy approaches in the treatment of malignant brain tumors. *Cancer* **2017**, *123*, 734–750. [CrossRef]
39. Sampson, J.H.; Archer, G.E.; Mitchell, D.A.; Heimberger, A.B.; Herndon, J.E.; Lally-Goss, D.; McGehee-Norman, S.; Paolino, A.; Reardon, D.A.; Friedman, A.H.; et al. An epidermal growth factor receptor variant III-targeted vaccine is safe and immunogenic in patients with glioblastoma multiforme. *Mol. Cancer Ther.* **2009**, *8*, 2773–2779. [CrossRef]
40. Schuster, J.; Lai, R.K.; Recht, L.D.; Reardon, D.A.; Paleologos, N.A.; Groves, M.D.; Mrugala, M.M.; Jensen, R.; Baehring, J.M.; Sloan, A.; et al. A phase II, multicenter trial of rindopepimut (CDX-110) in newly diagnosed glioblastoma: The ACT III study. *Neuro Oncol.* **2015**, *17*, 854–861. [CrossRef]
41. Altieri, D.C. Validating survivin as a cancer therapeutic target. *Nat. Rev. Cancer* **2003**, *3*, 46–54. [CrossRef]

42. Fenstermaker, R.A.; Ciesielski, M.J. Challenges in the development of a survivin vaccine (SurVaxM) for malignant glioma. *Exp. Rev. Vaccines* **2014**, *13*, 377–385. [CrossRef] [PubMed]
43. Adida, C.; Crotty, P.L.; McGrath, J.; Berrebi, D.; Diebold, J.; Altieri, D.C. Developmentally regulated expression of the novel cancer anti-apoptosis gene survivin in human and mouse differentiation. *Am. J. Pathol.* **1998**, *152*, 43–49.
44. Chastkofsky, M.I.; Pituch, K.C.; Katagi, H.; Zannikou, M.; Ilut, L.; Xiao, T.; Han, Y.; Sonabend, A.M.; Curiel, D.T.; Bonner, E.R.; et al. Mesenchymal Stem Cells Successfully Deliver Oncolytic Virotherapy to Diffuse Intrinsic Pontine Glioma. *Clin. Cancer Res.* **2021**, *27*, 1766–1777. [CrossRef] [PubMed]
45. Heimberger, A.B.; Crotty, L.E.; Archer, G.E.; McLendon, R.E.; Friedman, A.; Dranoff, G.; Bigner, D.D.; Sampson, J.H. Bone marrow-derived dendritic cells pulsed with tumor homogenate induce immunity against syngeneic intracerebral glioma. *J. Neuroimmunol.* **2000**, *103*, 16–25. [CrossRef]
46. Weller, M.; Roth, P.; Preusser, M.; Wick, W.; Reardon, D.A.; Platten, M.; Sampson, J.H. Vaccine-based immunotherapeutic approaches to gliomas and beyond. *Nat. Rev. Neurol.* **2017**, *13*, 363–374. [CrossRef] [PubMed]
47. Grauer, O.M.; Sutmuller, R.P.M.; van Maren, W.; Jacobs, J.F.M.; Bennink, E.; Toonen, L.W.J.; Nierkens, S.; Adema, G.J. Elimination of regulatory T cells is essential for an effective vaccination with tumor lysate-pulsed dendritic cells in a murine glioma model. *Int. J. Cancer* **2008**, *122*, 1794–1802. [CrossRef]
48. Fedorova, L.; Mudry, P.; Pilatova, K.; Selingerova, I.; Merhautova, J.; Rehak, Z.; Valik, D.; Hlavackova, E.; Cerna, D.; Faberova, L.; et al. Assessment of Immune Response Following Dendritic Cell-Based Immunotherapy in Pediatric Patients with Relapsing Sarcoma. *Front. Oncol.* **2019**, *9*, 1169. [CrossRef]
49. Garg, A.D.; Vandenberk, L.; Koks, C.; Verschuere, T.; Boon, L.; Van Gool, S.W.; Agostinis, P. Dendritic cell vaccines based on immunogenic cell death elicit danger signals and T cell-driven rejection of high-grade glioma. *Sci. Transl. Med.* **2016**, *8*, 328ra327. [CrossRef]
50. Benitez-Ribas, D.; Cabezón, R.; Flórez-Grau, G.; Molero, M.C.; Puerta, P.; Guillen, A.; Paco, S.; Carcaboso, A.M.; Santa-Maria Lopez, V.; Cruz, O.; et al. Immune Response Generated with the Administration of Autologous Dendritic Cells Pulsed with an Allogenic Tumoral Cell-Lines Lysate in Patients with Newly Diagnosed Diffuse Intrinsic Pontine Glioma. *Front. Oncol.* **2018**, *8*, 127. [CrossRef]
51. Harrington, K.; Freeman, D.J.; Kelly, B.; Harper, J.; Soria, J.-C. Optimizing oncolytic virotherapy in cancer treatment. *Nat. Rev. Drug Discov.* **2019**, *18*, 689–706. [CrossRef]
52. Grigg, C.; Blake, Z.; Gartrell, R.; Sacher, A.; Taback, B.; Saenger, Y. Talimogene laherparepvec (T-Vec) for the treatment of melanoma and other cancers. *Semin. Oncol.* **2016**, *43*, 638–646. [CrossRef]
53. Gállego Pérez-Larraya, J.; Garcia-Moure, M.; Labiano, S.; Patiño-García, A.; Dobbs, J.; Gonzalez-Huarriz, M.; Zalacain, M.; Marrodan, L.; Martinez-Velez, N.; Puigdelloses, M.; et al. Oncolytic DNX-2401 Virus for Pediatric Diffuse Intrinsic Pontine Glioma. *New Engl. J. Med.* **2022**, *386*, 2471–2481. [CrossRef] [PubMed]
54. Martínez-Vélez, N.; Garcia-Moure, M.; Marigil, M.; González-Huarriz, M.; Puigdelloses, M.; Gallego Pérez-Larraya, J.; Zalacaín, M.; Marrodán, L.; Varela-Guruceaga, M.; Laspidea, V.; et al. The oncolytic virus Delta-24-RGD elicits an antitumor effect in pediatric glioma and DIPG mouse models. *Nat. Commun.* **2019**, *10*, 2235. [CrossRef] [PubMed]
55. Lang, F.F.; Conrad, C.; Gomez-Manzano, C.; Yung, W.K.A.; Sawaya, R.; Weinberg, J.S.; Prabhu, S.S.; Rao, G.; Fuller, G.N.; Aldape, K.D.; et al. Phase I Study of DNX-2401 (Delta-24-RGD) Oncolytic Adenovirus: Replication and Immunotherapeutic Effects in Recurrent Malignant Glioma. *J. Clin Oncol.* **2018**, *36*, 1419–1427. [CrossRef] [PubMed]
56. Tejada, S.; Alonso, M.; Patiño, A.; Fueyo, J.; Gomez-Manzano, C.; Diez-Valle, R. Phase I Trial of DNX-2401 for Diffuse Intrinsic Pontine Glioma Newly Diagnosed in Pediatric Patients. *Neurosurgery* **2018**, *83*, 1050–1056. [CrossRef] [PubMed]
57. Tejada, S.; Díez-Valle, R.; Domínguez, P.D.; Patiño-García, A.; González-Huarriz, M.; Fueyo, J.; Gomez-Manzano, C.; Idoate, M.A.; Peterkin, J.; Alonso, M.M. DNX-2401, an Oncolytic Virus, for the Treatment of Newly Diagnosed Diffuse Intrinsic Pontine Gliomas: A Case Report. *Front. Oncol.* **2018**, *8*, 61. [CrossRef]
58. Kicielinski, K.P.; Chiocca, E.A.; Yu, J.S.; Gill, G.M.; Coffey, M.; Markert, J.M. Phase 1 clinical trial of intratumoral reovirus infusion for the treatment of recurrent malignant gliomas in adults. *Mol. Ther. J. Am. Soc. Gene Ther.* **2014**, *22*, 1056–1062. [CrossRef]
59. Schuelke, M.R.; Wongthida, P.; Thompson, J.; Kottke, T.; Driscoll, C.B.; Huff, A.L.; Shim, K.G.; Coffey, M.; Pulido, J.; Evgin, L.; et al. Diverse immunotherapies can effectively treat syngeneic brainstem tumors in the absence of overt toxicity. *J. Immunother Cancer* **2019**, *7*, 188. [CrossRef]
60. Laspidea, V.; Puigdelloses, M.; Labiano, S.; Marrodán, L.; Garcia-Moure, M.; Zalacain, M.; Gonzalez-Huarriz, M.; Martínez-Vélez, N.; Ausejo-Mauleon, I.; de la Nava, D.; et al. Exploiting 4-1BB immune checkpoint to enhance the efficacy of oncolytic virotherapy for diffuse intrinsic pontine gliomas. *JCI Insight* **2022**, *7*, e154812. [CrossRef]
61. Cockle, J.V.; Brüning-Richardson, A.; Scott, K.J.; Thompson, J.; Kottke, T.; Morrison, E.; Ismail, A.; Carcaboso, A.M.; Rose, A.; Selby, P.; et al. Oncolytic Herpes Simplex Virus Inhibits Pediatric Brain Tumor Migration and Invasion. *Mol. Ther. Oncolytics* **2017**, *5*, 75–86. [CrossRef]
62. Ahmed, A.U.; Rolle, C.E.; Tyler, M.A.; Han, Y.; Sengupta, S.; Wainwright, D.A.; Balyasnikova, I.V.; Ulasov, I.V.; Lesniak, M.S. Bone marrow mesenchymal stem cells loaded with an oncolytic adenovirus suppress the anti-adenoviral immune response in the cotton rat model. *Mol. Ther. J. Am. Soc. Gene Ther.* **2010**, *18*, 1846–1856. [CrossRef] [PubMed]

63. Morshed, R.A.; Gutova, M.; Juliano, J.; Barish, M.E.; Hawkins-Daarud, A.; Oganesyan, D.; Vazgen, K.; Yang, T.; Annala, A.; Ahmed, A.U.; et al. Analysis of glioblastoma tumor coverage by oncolytic virus-loaded neural stem cells using MRI-based tracking and histological reconstruction. *Cancer Gene. Ther.* **2015**, *22*, 55–61. [CrossRef] [PubMed]
64. Leach, D.R.; Krummel, M.F.; Allison, J.P. Enhancement of antitumor immunity by CTLA-4 blockade. *Science* **1996**, *271*, 1734–1736. [CrossRef]
65. Vranic, S.; Cyprian, F.S.; Gatalica, Z.; Palazzo, J. PD-L1 status in breast cancer: Current view and perspectives. *Semin. Cancer Biol.* **2021**, *72*, 146–154. [CrossRef]
66. Yaghoubi, N.; Soltani, A.; Ghazvini, K.; Hassanian, S.M.; Hashemy, S.I. PD-1/PD-L1 blockade as a novel treatment for colorectal cancer. *Biomed Pharm.* **2019**, *110*, 312–318. [CrossRef] [PubMed]
67. Chen, G.; Huang, A.C.; Zhang, W.; Zhang, G.; Wu, M.; Xu, W.; Yu, Z.; Yang, J.; Wang, B.; Sun, H.; et al. Exosomal PD-L1 contributes to immunosuppression and is associated with anti-PD-1 response. *Nature* **2018**, *560*, 382–386. [CrossRef]
68. Hodi, F.S.; O'Day, S.J.; McDermott, D.F.; Weber, R.W.; Sosman, J.A.; Haanen, J.B.; Gonzalez, R.; Robert, C.; Schadendorf, D.; Hassel, J.C.; et al. Improved survival with ipilimumab in patients with metastatic melanoma. *N. Engl. J. Med.* **2010**, *363*, 711–723. [CrossRef] [PubMed]
69. Bagcchi, S. Pembrolizumab for treatment of refractory melanoma. *Lancet. Oncol.* **2014**, *15*, e419. [CrossRef] [PubMed]
70. Paz-Ares, L.; Ciuleanu, T.-E.; Cobo, M.; Schenker, M.; Zurawski, B.; Menezes, J.; Richardet, E.; Bennouna, J.; Felip, E.; Juan-Vidal, O.; et al. First-line nivolumab plus ipilimumab combined with two cycles of chemotherapy in patients with non-small-cell lung cancer (CheckMate 9LA): An international, randomised, open-label, phase 3 trial. *Lancet. Oncol.* **2021**, *22*, 198–211. [CrossRef]
71. Motzer, R.J.; Tannir, N.M.; McDermott, D.F.; Arén Frontera, O.; Melichar, B.; Choueiri, T.K.; Plimack, E.R.; Barthélémy, P.; Porta, C.; George, S.; et al. Nivolumab plus Ipilimumab versus Sunitinib in Advanced Renal-Cell Carcinoma. *New Engl. J. Med.* **2018**, *378*, 1277–1290. [CrossRef]
72. Motzer, R.J.; Rini, B.I.; McDermott, D.F.; Arén Frontera, O.; Hammers, H.J.; Carducci, M.A.; Salman, P.; Escudier, B.; Beuselinck, B.; Amin, A.; et al. Nivolumab plus ipilimumab versus sunitinib in first-line treatment for advanced renal cell carcinoma: Extended follow-up of efficacy and safety results from a randomised, controlled, phase 3 trial. *Lancet. Oncol.* **2019**, *20*, 1370–1385. [CrossRef] [PubMed]
73. Larkin, J.; Chiarion-Sileni, V.; Gonzalez, R.; Grob, J.-J.; Rutkowski, P.; Lao, C.D.; Cowey, C.L.; Schadendorf, D.; Wagstaff, J.; Dummer, R.; et al. Five-Year Survival with Combined Nivolumab and Ipilimumab in Advanced Melanoma. *New Engl. J. Med.* **2019**, *381*, 1535–1546. [CrossRef] [PubMed]
74. Janjigian, Y.Y.; Shitara, K.; Moehler, M.; Garrido, M.; Salman, P.; Shen, L.; Wyrwicz, L.; Yamaguchi, K.; Skoczylas, T.; Campos Bragagnoli, A.; et al. First-line nivolumab plus chemotherapy versus chemotherapy alone for advanced gastric, gastro-oesophageal junction, and oesophageal adenocarcinoma (CheckMate 649): A randomised, open-label, phase 3 trial. *Lancet* **2021**, *398*, 27–40. [CrossRef] [PubMed]
75. Tawbi, H.A.; Forsyth, P.A.; Algazi, A.; Hamid, O.; Hodi, F.S.; Moschos, S.J.; Khushalani, N.I.; Lewis, K.; Lao, C.D.; Postow, M.A.; et al. Combined Nivolumab and Ipilimumab in Melanoma Metastatic to the Brain. *New Engl. J. Med.* **2018**, *379*, 722–730. [CrossRef]
76. Wolchok, J.D.; Chiarion-Sileni, V.; Gonzalez, R.; Grob, J.-J.; Rutkowski, P.; Lao, C.D.; Cowey, C.L.; Schadendorf, D.; Wagstaff, J.; Dummer, R.; et al. Long-Term Outcomes with Nivolumab Plus Ipilimumab or Nivolumab Alone Versus Ipilimumab in Patients with Advanced Melanoma. *J. Clin. Oncol.* **2022**, *40*, 127–137. [CrossRef]
77. Reardon, D.A.; Kaley, T.J.; Dietrich, J.; Clarke, J.L.; Dunn, G.; Lim, M.; Cloughesy, T.F.; Gan, H.K.; Park, A.J.; Schwarzenberger, P.; et al. Phase II study to evaluate safety and efficacy of MEDI4736 (durvalumab) + radiotherapy in patients with newly diagnosed unmethylated MGMT glioblastoma (new unmeth GBM). *J. Clin. Oncol.* **2019**, *37*, 2032. [CrossRef]
78. Wen, P.Y.; Weller, M.; Lee, E.Q.; Alexander, B.M.; Barnholtz-Sloan, J.S.; Barthel, F.P.; Batchelor, T.T.; Bindra, R.S.; Chang, S.M.; Chiocca, E.A.; et al. Glioblastoma in adults: A Society for Neuro-Oncology (SNO) and European Society of Neuro-Oncology (EANO) consensus review on current management and future directions. *Neuro Oncol.* **2020**, *22*, 1073–1113. [CrossRef]
79. Nayak, L.; Standifer, N.; Dietrich, J.; Clarke, J.L.; Dunn, G.P.; Lim, M.; Cloughesy, T.; Gan, H.K.; Flagg, E.; George, E.; et al. Circulating Immune Cell and Outcome Analysis from the Phase II Study of PD-L1 Blockade with Durvalumab for Newly Diagnosed and Recurrent Glioblastoma. *Clin. Cancer Res.* **2022**, *28*, 2567–2578. [CrossRef]
80. Lee, E.Q. Immune checkpoint inhibitors in GBM. *J. Neuro Oncol.* **2021**, *155*, 1–11. [CrossRef]
81. Reardon, D.A.; Brandes, A.A.; Omuro, A.; Mulholland, P.; Lim, M.; Wick, A.; Baehring, J.; Ahluwalia, M.S.; Roth, P.; Bähr, O.; et al. Effect of Nivolumab vs Bevacizumab in Patients with Recurrent Glioblastoma: The CheckMate 143 Phase 3 Randomized Clinical Trial. *JAMA Oncol.* **2020**, *6*, 1003–1010. [CrossRef]
82. Cacciotti, C.; Choi, J.; Alexandrescu, S.; Zimmerman, M.A.; Cooney, T.M.; Chordas, C.; Clymer, J.; Chi, S.; Yeo, K.K. Immune checkpoint inhibition for pediatric patients with recurrent/refractory CNS tumors: A single institution experience. *J. Neuro Oncol.* **2020**, *149*, 113–122. [CrossRef] [PubMed]

83. Kline, C.; Liu, S.J.; Duriseti, S.; Banerjee, A.; Nicolaides, T.; Raber, S.; Gupta, N.; Haas-Kogan, D.; Braunstein, S.; Mueller, S. Reirradiation and PD-1 inhibition with nivolumab for the treatment of recurrent diffuse intrinsic pontine glioma: A single-institution experience. *J. Neuro Oncol.* **2018**, *140*, 629–638. [CrossRef] [PubMed]
84. Hwang, E.; Onar, A.; Young-Poussaint, T.; Mitchell, D.; Kilburn, L.; Margol, A.; Gilheeny, S.; Lin, T.; Dunkel, I.; Fouladi, M. Immu-09. outcome of patients with recurrent diffuse intrinsic pontine glioma (DIPG) treated with pembrolizumab (ANTI-PD-1): A pediatric brain tumor consortium study (PBTC045). *Neuro Oncol.* **2018**, *20*, i100. [CrossRef]

cancers

MDPI

Review

Pediatric Myeloid Sarcoma, More than Just a Chloroma: A Review of Clinical Presentations, Significance, and Biology

Kristin E. Zorn [1,2], Ashley M. Cunningham [3], Alison E. Meyer [2], Karen Sue Carlson [2,4] and Sridhar Rao [1,2,5,*]

1 Department of Pediatrics, Division of Hematology/Oncology/Transplantation, Medical College of Wisconsin, Milwaukee, WI 53226, USA
2 Versiti Blood Research Institute, Milwaukee, WI 53226, USA
3 Department of Pathology, Medical College of Wisconsin, Milwaukee, WI 53226, USA
4 Department of Medicine, Division of Hematology/Oncology, Medical College of Wisconsin, Milwaukee, WI 53226, USA
5 Department of Cell Biology, Neurobiology, and Anatomy, Medical College of Wisconsin, Milwaukee, WI 53226, USA
* Correspondence: sridhar.rao@versiti.org; Tel.: +1-414-937-3841

Simple Summary: Childhood acute myeloid leukemia (AML) remains a cancer with poor overall outcomes. Myeloid sarcomas (MS) are extramedullary masses of leukemia cells that can develop in patients with AML. In children, MS occurs more frequently than described in adults. Their clinical significance in both pediatric and adult patients with AML is unclear. In this review, we aim to summarize the current knowledge of MS in children and its underlying biology in the hopes of sparking future studies and ultimately improving treatment options for children with AML.

Abstract: Myeloid sarcomas (MS), commonly referred to as chloromas, are extramedullary tumors of acute myeloid leukemia (AML) with varying incidence and influence on outcomes. Pediatric MS has both a higher incidence and unique clinical presentation, cytogenetic profile, and set of risk factors compared to adult patients. Optimal treatment remains undefined, yet allogeneic hematopoietic stem cell transplantation (allo-HSCT) and epigenetic reprogramming in children are potential therapies. Importantly, the biology of MS development is poorly understood; however, cell-cell interactions, epigenetic dysregulation, cytokine signaling, and angiogenesis all appear to play key roles. This review describes pediatric-specific MS literature and the current state of knowledge about the biological determinants that drive MS development. While the significance of MS remains controversial, the pediatric experience provides an opportunity to investigate mechanisms of disease development to improve patient outcomes. This brings the hope of better understanding MS as a distinct disease entity deserving directed therapeutic approaches.

Keywords: myeloid sarcoma; chloroma; acute myeloid leukemia; pediatric

Citation: Zorn, K.E.; Cunningham, A.M.; Meyer, A.E.; Carlson, K.S.; Rao, S. Pediatric Myeloid Sarcoma, More than Just a Chloroma: A Review of Clinical Presentations, Significance, and Biology. *Cancers* **2023**, *15*, 1443. https://doi.org/10.3390/cancers15051443

Academic Editors: Saurabh Agarwal and Jianhua Yang

Received: 22 December 2022
Revised: 14 February 2023
Accepted: 21 February 2023
Published: 24 February 2023

1. Introduction

Myeloid sarcomas (MS) are extramedullary tumors of myeloid blasts forming masses disrupting normal tissue architecture in patients with acute myeloid leukemia (AML) [1–3]. They are also known as myeloblastomas, granulocytic sarcomas, chloroleukemia, and chloromas given the historically green appearance of the tumors resulting from myeloperoxidase exposure to air. Importantly, there is no clearly accepted definition of what qualifies as MS. Most agree that discrete tumor masses of myeloid blasts are MS; however, whether gingival infiltration and masses within lymph nodes, the liver, or spleen should also be considered MS is debated given their propensity for generalized infiltration. Central nervous system (CNS) leukemia is also a challenge with categorization including both cerebral spinal fluid (CSF)-positive disease and CNS infiltrates/masses on imaging. This has made clear, consistent reporting of clinical presentation and outcome data difficult given the lack of consensus within the literature. Given this limitation, the following terms will be used for this review: extramedullary disease and MS. Extramedullary disease will more broadly refer to leukemic disease outside the bone marrow/peripheral blood, while MS will be specific to myeloid blast tumors. These terms are not used interchangeably and are used as defined to more accurately portray the referenced literature, with extramedullary disease as an umbrella term that also includes MS.

MS most frequently presents with a mass in subcutaneous/soft tissue, bone, and skin (also known as leukemia cutis). Case reports include masses and infiltrative involvement in nearly every conceivable tissue including the GI tract, reproductive organs, CNS, heart, lungs, kidneys, and breast [4]. Interestingly, MS, while most often seen concurrently with intramedullary AML, can occur in isolation in the absence of bone marrow disease. MS can also occur in the setting of a preceding hematologic disease such as myelodysplastic syndrome (MDS) or myeloproliferative neoplasms (MPN). Finally, MS can develop as a relapse following a hematopoietic malignancy, including after allogeneic hematopoietic stem cell transplantation (allo-HSCT) [1,2].

Although AML is seen primarily in older adults with a median age at diagnosis of 68 years, AML accounts for 10 to 15% of acute leukemias in children [5,6]. Pediatric AML differs from AML in adults in terms of clinical course, outcomes, and genomic landscape [7–9]. MS in pediatrics represent an inadequately understood aspect of AML. MS presentations offer another distinction between pediatric and adult AML with opportunities for improvement in diagnosis, management, and further investigation into the biological mechanisms of development and treatment resistance.

This review will discuss the clinical presentations and reported outcomes of pediatric patients with MS including post-allo-HSCT, imaging approaches to diagnosis, and finally, the biology of MS will be addressed. While this review focuses on pediatric MS, important comparisons with adults will also be discussed.

2. Pediatric Clinical Presentation, Incidence, and Outcomes

Although generally considered a rare presentation, MS and extramedullary AML are common in children with AML. Numerous cooperative groups and large institution studies have reported both characteristics and outcomes associated with extramedullary disease (Table 1). The inconsistent terminology surrounding MS and extramedullary disease prevents direct comparison across these studies. Despite this limitation, these studies provide helpful data about the clinical features and associations seen as well as insight into outcomes.

Table 1. Summary of survival outcomes of pediatric patients with AML and extramedullary disease. Described terminology as per original report. Abbreviations: YO, years old; EMD, extramedullary disease; CSF, cerebral spinal fluid; EML, extramedullary leukemia; EMI, extramedullary infiltration; MS, myeloid sarcoma; EFS, event free survival; OS, overall survival; RFS, relapse free survival; SE, standard error; AML, acute myeloid leukemia; CI, confidence interval [10–23].

Study/Publication	Age	Extramedullary Disease Involvement Study Definitions	Population	Incidence	5-Year Estimated EFS (±SE) or (95% CI)		5-Year Estimated OS (±SE) or (95% CI)	
POG8821 (Chang, et. al., 2000) [10]	<21 yo	EMD: including CSF disease, not defined	n = 492	Any EMD 10.4% CSF only 4.7% Non-CSF EMD 5.7%	**4-year EFS** CSF only: 34.8 ± 9.9% Non-CSF EMD: 21.6 ± 8.6% No EMD: 34.4 ± 2.5%	$p = 0.91$ $p = 0.043$ $p = 0.18$	Not available	
DCLSG (Bisschop et. al., 2001) [11]	0–16 yo	EML: Clinically obvious infiltrate in soft tissues, skin, muscles or bone, gingiva, CSF or brain	n = 477	EML in 25.1%	No EML 38% ± 3% Myeloblastoma (MS) 43 ± 13% Skin infiltrates 45 ± 21%	$p = 0.85$	Not available	
Children's Cancer Group, CCG AML 213 and 213P, 2861 and 2891 (Dusenbery et. al., 2003) [12]	0–21 yo	"Chloroma" on data entry form yes or no, gum only not included "Skin involvement" yes or no	n = 1832	Skin EML ± other 5.9% Non skin EML 4.9% EML 10.9%	Skin ± other: 26% (17–35%) Non skin EML: 46% (34–58%) Non EML: 29% (27–32%)	$p = 0.005$	Not available	
Single Center—Turkey (AML-90 and AML-94 protocols) (Hiçsönmez et. al., 2004) [13]	<17 yo	EMI: involvement of gingiva, CNS, orbit, soft tissue, bone, pleura	n = 127	EMI total in 40% Gingiva only in 11% Orbital in 10% MS in 21%	**4-year EFS:** AML-90 therapy: MS = 0% Without EMI = 37 ± 11% AML-94 therapy: MS = 56 ± 17% Without EMI = 31 ± 1%	$p < 0.05$ $p > 0.05$	Not available	
Japanese childhood AML cooperative study group (Kobayashi et. al., 2007) [14]	<16 yo	CNS disease (>5 WBC/µL with blasts) EMI: leukemic infiltration in organs other than liver, spleen, lymph nodes (including CNS disease)	n = 240	EMI in 23.3% (Excluding CSF only: 20.4%)	**3-year estimate EFS** EMI: 53.3 ± 6.7% No EMI: 62.5 ± 3.6% EMI + WBC > 100 × 10^9/L: 23.8 ± 12.9% No EMI or EMI + WBC < 100 × 10^9/L: 60 ± 3.5%	$p = 0.11$ $p = 0.0052$	**3-year OS** EMI: 77.3% No EMI: 77.6%	
Children's Oncology Group (CCG 2861, 2891, 2941, 2961) (Johnston et. al., 2012) [15]	0–21 yo	CNS3 (≥5 WBC/µL with blasts) CNS MS (brain or spinal cord tumor)	n = 1459	CNS3 11% CNS MS 1% Orbital MS 2% Non CNS MS 4%	No MS 40 ± 3% CNS MS 52 ± 21% Orbital MS 76 ± 17% Non CNS MS 34 ± 13%	$p = 0.005$	No MS 50 ± 3% CNS MS 73 ± 19% Orbital MS 92 ± 11% Non CNS MS 38 ± 13%	$p < 0.001$
European AML Study Groups (Creutzig et. al., 2017) [16]	0–17 yo	CNS involvement (CSF with >5 WBC/µL with blasts or intracranial infiltrates on imaging or neurologic symptoms)	n = 2365	CNS 11.0%	CNS + 48 ± 3% CNS—52 ± 2%	$p = 0.11$	CNS + 64 ± 3% CNS—67 ± 1%	$p = 0.23$
NOPHO AML 2004 (Støve et. al., 2017) [17]	0–17 yo	MS: myeloblast tumor CNS disease (≥5 WBC/µL with blasts or new neurologic symptoms) EML: MS or CNS disease	n = 322	MS (±CNS disease) 15.8% CNS only an additional 7%	EML: 54% (42–65%) No EML: 45% (37–51%)	$p = 0.57$	EML: 64% (51–74%) No EML: 73% (66–78%)	$p = 0.008$

Table 1. *Cont.*

Study/Publication	Age	Extramedullary Disease Involvement Study Definitions	Population	Incidence	5-Year Estimated EFS (±SE) or (95% CI)		5-Year Estimated OS (±SE) or (95% CI)	
Single Center—India (Pramanik et. al. 2018) [18]	0–18 yo	MS (did not include CSF only disease)	n = 570	MS in 21.2%	**Median EFS:**	$p = 0.002$	**Median OS:**	$p = 0.002$
					AML with MS: 21.6 months		With MS: 26.3 months	
					AML without MS: 11.1 months		Without MS: 12.7 months	
TARGET dataset (COG-NCI) (COG AAML03P1, AAML0531, CCG-2961) (Xu et. al., 2020) [19]	<18 yo	MS on biopsy diagnosis, excluding CSF disease	n = 884	MS in 12.3%	MS: 35.4 ± 4.6%	$p = 0.001$	MS: 53.4 ± 4.8%	$p = 0.008$
					Non-MS: 48.5 ± 1.8%		Non-MS: 64.0 ± 1.8%	
Single Center—Korea (Lee et. al., 2020) [20] **Only *RUNX1-RUNX1T1* AML**	<18 yo	EMI: excluded CSF only	n = 40	EMI in 30%	EMI: 50.0 ± 14.4%	$p = 0.022$	Not available	
					No EMI: 78.6 ± 7.8%			
Single Center—China (Hu et. al., 2020) [21] **Only Low Risk AML (includes Hu et. al., 2021 study)**	≤18 yo	MS: including lymph nodes >2cm, excluded CNSL	n = 214	MS in 20.6%	**3-year RFS**	$p = 0.000$	**3-year OS**	$p = 0.01$
					With MS: 62.6 ± 7.5%		With MS 73.5 ± 7.1%	
					Without MS: 87.0 ± 2.8%		Without MS 88.8 ± 2.6%	
Single Center—China (Hu et. al., 2021) [22]	1–18 y	MS: clinical, biopsy, radiology findings	n = 127	MS in 23.6%	**3-year RFS**	$p = 0.004$	**3-year OS**	$p = 0.249$
Only t(8;21) AML		CNS MS: dura deposits or paraspinal tumor			With MS: 68.8 ± 8.8%		With MS: 78.1 ± 8.1%	
	o				Without MS: 88.0 ± 3.4%		Without MS: 86.4 ± 3.7%	
Polish Pediatric Leukemia and Lymphoma Study Group (Samborska et. al., 2022) [23]	0–18 yo	MS: pathology diagnosis or extramedullary tumor and concurrent bone marrow disease (AML, MDS)	n = 43	MS in 100%	De novo: 0.56 ± 0.12	$p = 0.0247$	**pOS**	$p = 0.0251$
				De novo/isolated in 37.2% Concurrent in 55.8%	Concurrent: 0.82 ± 0.08		De novo: 0.56 ± 0.12 Concurrent: 0.84 ± 0.09	

2.1. Incidence

The incidence of MS varies widely, particularly when comparing adults and children. This is predominantly related to the lack of consistency in MS evaluation and reporting. There is no standard recommendation for patients with AML to undergo screening evaluation for MS and the true incidence is likely higher than that reported given the potential for asymptomatic occult tumors. Pediatric studies describe an incidence of MS ranging from 5.7% to as high as 40% with the expansive definition of extramedullary disease, although most commonly it is between 10 and 25% [10,13]. The distribution of common anatomic sites in children is illustrated in Figure 1. A lower incidence of MS is generally reported in adults (4–9%) with newly diagnosed AML, although this is likely an underestimate considering a recent prospective study [24]. It is unclear why such differences exist between children and adults, but this may be related to differences in diagnostic evaluations performed or inherent differences in the leukemias that children develop compared to adults in terms of mutational spectrum [7–9].

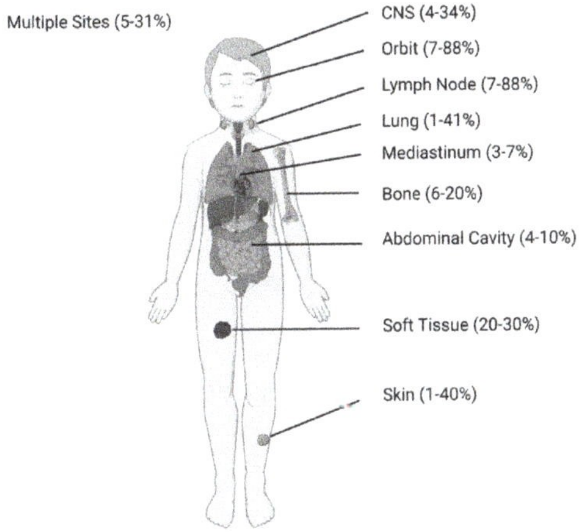

Figure 1. Common locations of initial presentation of MS in children with percentage of MS site involvement. Abbreviations: CNS, central nervous system; MS, myeloid sarcoma [11–18,21–23,25]. Figure created with BioRender.com.

2.2. Clinical Associations

In children, MS is typically associated with a younger age at diagnosis, particularly in infants, and more frequently in males, with 55–75% of patients with MS being male [19,21,22]. Higher WBC counts and hepatosplenomegaly are often seen in the presence of MS, although less consistently. Additionally, the FAB M4 (acute myelomonocytic) and M5 (acute monocytic) subtypes are most commonly associated with MS. Cytogenetically, the most frequently described associations are inv(16), t(8;21), and chromosome 11 abnormalities, namely 11q23, with both deletions and rearrangements [12,15,17–19,26]. By contrast, in adults, a recent study of 1,583 patients identified increased odds of extramedullary disease in patients with *PTPN11*, *NPM1*, and *FLT3-ITD* mutations with no association with inv(16) or t(8;21) and decreased odds of extramedullary disease with *IDH2* and *CEBPA* mutations [27]. Pediatric AML has distinct mutational profiles in comparison to adult AML, and similarly, the AML genetic profile associated with pediatric MS appears distinct from that of adults. This suggests different biologic drivers of pediatric AML and MS or extramedullary disease development, and potential significance for treatment and outcomes.

2.3. Outcomes

There is no consensus on the influence of MS on prognosis in children with AML. Table 1 summarizes many of the larger studies, with notably conflicting results. Although typically considered a presentation of advanced disease, event free survival (EFS) and relapse free survival (RFS) do not consistently demonstrate worse outcomes for MS.

Reports from Children's Oncology Group (COG) demonstrate an improved EFS in subsets of MS patients with non-skin MS, later defined as orbital MS and CNS MS, with otherwise similar outcomes to non-MS patients in those with other sites of MS disease including skin [12,15]. A single center report from India similarly describes improved EFS and overall survival (OS) in patients with MS excluding CSF-only disease [18].

By contrast, other studies describe no significant association between MS or more general extramedullary disease and EFS [11,16,17]. A Turkish single center study found significant effects on outcomes for patients with MS only when less intensive treatment was given [13]. The Japanese childhood AML cooperative study group also only found inferior EFS with extramedullary disease in the setting of WBC count > $100 \times 10^9/L$ [14]. By contrast, other reports describe significantly worse EFS and OS for children with AML and MS compared to those without MS [10,19,25]. Collectively, this indicates that whether MS is a critical driver of outcomes remains an important unanswered question within the field.

Favorable cytogenetics, including core binding factor mutations, are common in patients with MS. In evaluation of children with low risk AML (e.g., inv(16), t(8;21), *NPM1* mutated without FLT3-ITD mutation, and *CEBPA* mutation), reports demonstrated worse RFS and EFS in patients with MS present compared to those without MS present [20–22]. This suggests that even in otherwise favorable AML, the presence of MS could be relevant for both prognosis and potentially risk stratification.

Although controversy remains in broadly assigning prognostic impact to the presence or absence of MS in pediatric leukemia, the significance should not be simply ignored. Particularly in patients with t(8;21), inv(16), or chromosome 11/*KMT2A* abnormalities, evaluating for the presence of MS may be significant in considering therapeutic approaches. Additional studies are needed to prospectively identify patients with MS, as defined by clear criteria, to determine the impact on prognosis, the potential need for altering risk stratification, and defining remission status. This will be important to identify which patients may benefit from specific or intensified therapy regimens to improve outcomes.

3. Significance of Extramedullary Disease and Myeloid Sarcomas Post-Allogeneic Hematopoietic Stem Cell Transplant

Relapse of AML remains the predominant cause of treatment failure and death with allo-HSCT as the only curative option for many patients. The role of allo-HSCT in the setting of MS is a moving target in children. However, new data are emerging to address this important point because anecdotal evidence suggests isolated MS relapse, in the absence of bone marrow relapse, is a common occurrence post-allo-HSCT.

The Japan Society for Hematopoietic Cell Transplantation (JSHCT) used their national database to identify pediatric AML patients that underwent allo-HSCT and found that the presence of extramedullary disease (both CNS disease and MS) had no impact on OS or leukemia-free survival (LFS) after transplant. However, the patients with extramedullary disease prior to transplant were more likely to have extramedullary relapses after transplant, with 41% of relapses being extramedullary. In comparison, those without prior extramedullary disease had only 6% extramedullary relapse, although the overall rates of recurrence were the same between the two groups [28]. Relapse with isolated MS was not separated out as a group and was therefore difficult to directly assess.

The Turkish Pediatric Bone Marrow Transplantation Registry recently reported on their experience with isolated extramedullary relapse (iEMR) in children following allo-HSCT, although they included both acute lymphoblastic leukemia (ALL) and AML. They found different risk factors for medullary relapse post-allo-HSCT versus iEMR. Transplant in CR2 or later or active disease at time of transplant and matched sibling donor transplants were independently associated with increased risk of medullary relapse as well as iEMR. The presence of chronic graft versus host disease (cGVHD) was conversely associated with decreased risk of medullary relapse with no impact on the risk of iEMR. iEMR rates were, however, independently higher in those with prior extramedullary disease [29]. A higher rate of second iEMR was also seen following a first iEMR at 58.8% versus after a first medullary relapse at 13% [29]. Local radiotherapy of extramedullary disease sites prior to transplantation and the presence of cGVHD had no impact on post-allo-HSCT iEMR, while cGVHD was protective in preventing medullary relapse [29]. This suggests that although a graft versus leukemia effect is helpful in preventing medullary relapse, this immune-mediated mechanism is not effective against MS masses and extramedullary sites of leukemia. A single site report from the University of Michigan found that children were also three times more likely than adults to experience an extramedullary relapse with an associated higher pretransplant extramedullary disease incidence [30]. The significance of extramedullary relapse, particularly in these settings, provides insight into mechanisms of disease resistance specific to the MS phenotype following allo-HSCT. Despite these concerns, allo-HSCT remains the best disease management for patients with high-risk AML.

Extramedullary disease prior to transplant is consistently associated with increased risk of extramedullary relapse after HSCT in both children and adults [30,31]. The presence of prior extramedullary disease (including CNS disease and MS) in adults with AML was not found to be an independent risk factor for post-allo-HSCT relapse, DFS, or OS in both a large CIBMTR analysis and a Canadian report [32,33]. This confirms the anecdotal clinical concern that allo-HSCT is more effective for medullary versus extramedullary disease.

Adults with iEMR post-allo-HSCT are more likely to have had prior extramedullary disease and GVHD present compared to those with medullary relapse [30,31,34,35]. Additionally, extramedullary relapse has a higher incidence following allo-HSCT than intensive chemotherapy alone [36]. These iEMRs may represent sanctuary sites in which immune-based therapies may be less effective and may result from a different mechanism of pathogenesis compared to medullary relapse. There is currently no treatment consensus for iEMR post-allo-HSCT, with a range of treatment approaches taken including local radiotherapy and systemic chemotherapy [37–40]. The significance of iEMR on outcomes compared to medullary relapse is more controversial, with conflicting studies limited by inclusion of both ALL and AML patients with known differences in the efficacy of graft versus leukemia effect between the two diseases [31,34]. One retrospective adult study, however, did report that allo-HSCT was an effective treatment for patients with MS compared to chemotherapy alone, although lack of complete MS remission prior to transplant had independently worse OS and PFS [41].

In summary, the presence of known or occult MS prior to allo-HSCT may be clinically important for a subset of patients, although additional studies are needed to define which groups may benefit. Furthermore, identifying the increased risk for iEMR following allo-HSCT can inform evaluations and management of patients during their post-transplant course. This emphasizes the importance of identifying and following MS in patients with AML prior to allo-HSCT and remaining vigilant to the possibility of iEMR.

4. Imaging Evaluation of Myeloid Sarcomas

The use of imaging to identify occult MS as well as re-evaluation of disease presence has remained inconsistent, both in frequency and modality, particularly in children. Ultrasounds of MS lesions typically show homogenously hypoechoic lesions with hypervascularity [42]. Computed tomography (CT) scans identify MS as isodense lesions compared to muscle with moderate enhancement with IV contrast media. Enhancement is more commonly homogenous (65%) versus inhomogenous (35%) [43]. MRI scans demonstrate predominantly T2 hyperintense (82%) or isointense (18%) lesions compared to muscle and T1 isointense (61%) or hypointense (39%) lesions with homogenous contrast enhancement and a mean apparent diffusion coefficient (ADC) on diffusion weighted imaging (DWI) of 0.57×10^{-3} mm^2/s [43].

Fluorodeoxyglucose (FDG)-positron emission tomography (PET) scans are increasingly being used for diagnosing extramedullary disease, with MS lesions displaying moderate uptake of FDG [44]. A retrospective study including pediatric patients showed a sensitivity of 93% and a specificity of 71.4% limited by difficulty differentiating extramedullary leukemia disease from infectious/inflammatory entities [45]. The recent PETAML trial however prospectively evaluated adult patients with AML prior to therapy initiation with total body ^{18}FDG PET/CT scans to determine prevalence of extramedullary disease. This showed a prevalence of 22% with a sensitivity of 77% and specificity of 97% [24]. Interestingly, leukemia cutis and CNS meningeal involvement were not necessarily ^{18}FDG-PET-avid [24,45]. In addition, there were four patients who remained with residual ^{18}FDG-PET-positive lesions despite complete marrow remission. Three of those four subsequently relapsed, suggesting there may be a specific role for ^{18}FDG-PET imaging for remission evaluation of patients with AML [24]. Consideration should be given to prospectively evaluating patients with AML to identify MS lesions requiring focused follow-up and possible treatment modifications and may be of value in designing de novo AML clinical trials, particularly in pediatrics with a high incidence of MS.

5. Pathology of Myeloid Sarcomas

MS are infiltrative tumor masses of myeloid blasts that efface or disrupt the normal architecture of the involved organ. The leukemic blasts found in MS have heterogeneous morphology; however, monocytic differentiation is common where the blasts will show either myelomonocytic or monoblastic morphology [46]. An example of the histology of MS is shown in Figure 2. Immunophenotypic profiling by flow cytometry or immunohistochemical stains is often necessary for a definitive diagnosis, as many of these tumor masses may resemble carcinoma. Evaluation of markers of immaturity, including CD34 and CD117, are helpful in addition to other markers which are variably expressed on myeloid blasts including CD13, CD33, CD68 (KP1), CD45, and myeloperoxidase (MPO). In the setting of monocytic differentiation, expression of monocytic markers such as CD68, CD163, CD14, and/or non-specific esterase (NSE) can be seen.

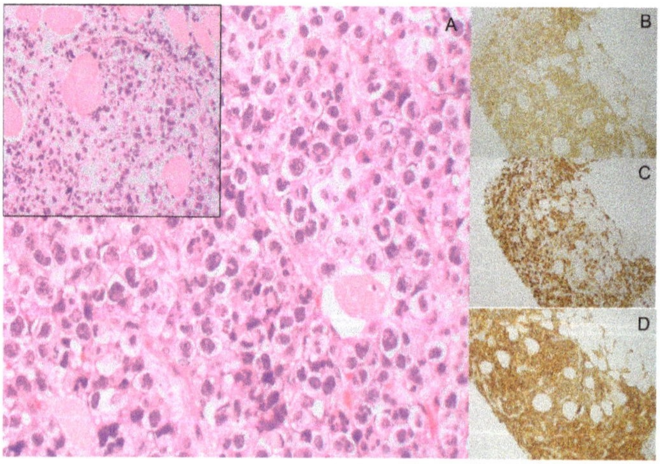

Figure 2. Histology of myeloid sarcoma involving the psoas muscle. (**A**, 500×). The tumor cells are blasts with irregular/convoluted nuclei, finely dispersed chromatin, and small distinct nucleoli. The inset (400×) shows the tumor cells infiltrating between skeletal muscle fibers. The tumor cells are immunoreactive for CD33 (**B**, 100×), CD163 (**C**, 100×), and CD45 (**D**, 100×).

Several translational studies have begun to investigate the genetic landscape of MS lesions in isolation as well as in comparison to their paired intramedullary leukemia counterparts. One recent study evaluated 7 adult trios of AML, MS, and normal tissue using capture-based next generation sequencing (NGS) of 479 cancer genes. Genes recurrently altered in these patients included *KMT2A*, *FLT3*, *NRAS*, *CEBPA*, *TP53*, *WT1*, and *NPM1*, with 84% of variants found in the AML also present in the MS [47]. Three of the seven patients had additional variants detected in the MS compared to the AML including additional *FLT3*, *SETD2*, and *NF1* mutations in the MS, while two had additional variants of *U2AF1* and *RAD21* in the AML but not the MS [47]. In the relapsed MS samples, there were increased single nucleotide variants (SNV) in the MS [47]. Another study evaluated 6 isolated MS tumors (without concurrent AML) and performed a 21 gene targeted panel of AML and MDS associated genes. They found recurrent variants in the genes for *FLT3* (50%), *NPM1* (33%), and *KIT* (67%) and additional variants in *WT1*, *SF3B1*, *EZH2*, *ASXL1*, and *TET2* in one MS each [48]. The genomic reports of patient-derived MS are all limited by targeted NGS sequencing without exploration of novel gene variants that may be specific to MS pathogenesis. Additionally, RNA transcriptome analysis of MS is lacking in the literature and provides an opportunity for investigation of transcriptome-based changes that may contribute to MS development outside of genetic mutations. Such studies may also facilitate the identification of cryptic translocations, which are common in pediatric AML [7]. Although, typically, the genomic profile of MS is in concordance with the AML and marrow, this is not always true. Particularly in cases of isolated MS, NGS and molecular evaluation may inform targeted treatment options and should be included in diagnostic evaluation of patients.

6. Biological Understandings of Pathogenesis

The biology underlying development of MS remains poorly defined with no clear molecular determinants. Biological features such as cytogenetic changes, molecular abnormalities, and cell surface marker expression are not consistent across studies. Much of the work on MS development surrounds the invasiveness of AML as studied using in vitro transwell assays and infiltration in the spleen and liver. The simple infiltration of hematopoietic organs, however, appears distinct from MS development in non-hematopoietic sites with no clear biological explanation. The development of MS appears to require leukemia mobilization/release from the marrow environment, tissue invasion, and further changes leading to a tumor/mass phenotype, as illustrated in Figure 3. These steps will be further discussed below.

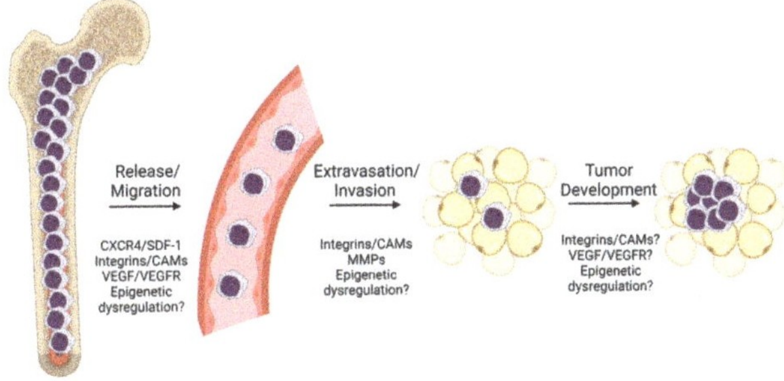

Figure 3. Depiction of MS development in AML with proposed mediators. Initial release/migration of leukemic blasts from the bone marrow into the peripheral blood circulation, followed by extravasation and invasion into distant tissue spaces (e.g., subcutaneous tissue). This results in organization of those extravasated cells into a mass with subsequent tissue architectural distortion and gross observation and clinical symptoms. Abbreviations: CAMs, cell adhesion molecules; MMPs, matrix metalloproteinases; VEGF, vascular endothelial growth factor; VEGFR, vascular endothelial growth factor receptor. Figure created with BioRender.com.

6.1. CXCR4

CXCR4 (CXC chemokine receptor 4, CD184) is the receptor for the chemokine matrix cell derivative-1 (SDF-1/CXCL12) and is expressed by most tissues as well as hematopoietic stem cells and leukemic blasts wherein it facilitates the retention of hematopoietic stem cells in the bone marrow niche [49,50]. The CXCR4/SDF-1 axis may contribute to chemoresistance through downstream signaling cascade dysregulation within leukemia cells [49,51]. Higher *CXCR4* expression has been seen in AML patients with extramedullary infiltration at diagnosis and extramedullary infiltration in childhood ALL [49,52]. The proposed mechanism of extramedullary involvement in acute leukemias is altered bone marrow homing and increased peripheral blood dissemination via a chemotactic gradient of SDF-1 with increased CXCR4 expression on the leukemia cells [53]. CXCR4/SDF-1 can promote the retention of AML cells within the skin of children with AML; however, CXCR4 expression by peripheral blood blasts was no different in patients with or without skin involvement [54]. Furthermore, a lack of association between SDF-1 polymorphisms and MS implies that small variants do not contribute to extramedullary disease development, although these have been previously of interest and described [55]. More common in adults, *NPM1*-mutated AML is associated with extramedullary disease and is associated with downregulation of *CXCL12* and *CXCR4* gene pathways [50].

While CXCR4 expression and signaling may be a contributing factor to extramedullary disease, its impact appears limited to initial release and migration of leukemia cells from the marrow and is not specific to MS development. Further work is needed to better characterize this mechanism and whether or how CXCR4 is contributing to discrete MS formation.

6.2. CD56

CD56 (also known as neural cell adhesion molecule-1 or NCAM1) is normally expressed by natural killer (NK) cells and other immune cell subtypes and is housed on chromosome 11q23.1. It is frequently described as part of the immunophenotype of AML with MS [1,56,57]. Expression patterns of CD56 are not consistently described, however, and in a population of adult t(8;21) AML patients, there was no association between CD56 expression and presence of extramedullary disease [58]. AML in adults with CD56 positivity is more commonly associated with worse 5-year EFS and OS; however, a report in low-risk patients shows no association with outcome [21,25,59]. Additionally, post-allo-HSCT CD56 positivity is not associated with extramedullary relapse [30]. Despite the frequent CD56+ immunophenotype, there is no described mechanism or in vitro data to suggest the significance of this finding. Additional experimental studies are required to determine if CD56 is simply a biomarker of MS or is required for MS development.

6.3. Integrins and Cell Adhesion Molecules (CAMs)

An AML-extracellular matrix interaction is likely critical to the development of MS. This is illustrated in transcriptome analysis of adult patient-derived AMLs demonstrating enrichment of cell surface gene sets in those AMLs with concomitant MS [60]. This includes integrin-α7 (*ITGA7*), which showed a higher expression in AML with associated MS in addition to high expression in MS samples [60]. Laminin 211 is a specific ligand of integrin-α7 that signals through the ERK signaling cascade [60]. While there is much described about the role of integrins and selectins in migration and homing of hematopoietic stem cells, there remains no clear mechanism by which these molecules facilitate MS formation [61,62]. Further study is needed to better evaluate the role of cell adhesion molecules in MS development and whether targeting these cell interactions may provide therapeutic benefit for patients with MS.

6.4. Vascular Endothelial Growth Factor (VEGF) and Receptor (VEGFR)

Angiogenesis plays a notable role in acute leukemia with increased microvascular density in AML and adult MS [63]. VEGFR2, the major mediator of the mitogenic, angiogenic, and permeability effects of VEGF, may contribute to the development of MS [64]. VEGF signaling via the PI3K/Akt pathway in the setting of *hERG1* expression was necessary for an in vitro migratory phenotype in AML cells [65]. In adults, the small molecule VEGFR2 tyrosine kinase inhibitor apatinib (also known as TN968D1) demonstrated enhanced antileukemic effects in ex vivo cytotoxicity studies from patient-derived AML samples with associated extramedullary disease [66]. Angiogenesis is well-described in the pathogenesis of other malignancies and it is reasonable to think that a unique perturbation may play a role in the migration or tumor formation of MS.

6.5. Matrix Metalloproteinases (MMP)

In vitro studies have described the role of MMP secretion (MMP-2 and MMP-9) by leukemia cells contributing to invasion capacity, most notably of the blood-brain barrier, with upstream regulation by mitogen-activated protein kinases (MAPKs) and phosphoinositide 3-kinase (PI3-K)/AKT pathways [67,68]. Additionally, TIMP-2 (tissue inhibitor of metalloproteinase 2) upregulation has been seen with increased leukemia cell line (i.e., SHI-1) invasion both in vitro and in vivo with more extensive and severe extramedullary infiltration through both MMP-2-dependent and independent activities [69,70]. Other in vitro studies propose a role for the β2 integrin-proMMP-9 complex in the extramedullary phenotype of AML [71]. Type IV collagenase secretion enhanced by TNFα and TGFβ from a patient-derived MS cell line increased in vitro cell invasion with collagenase secretion demonstrated in the MS AML cell line but not other leukemia cell lines [72]. While tissue invasion by leukemia cells is likely required for MS development, not all of the critical players have been identified.

6.6. Epigenetic Dysregulation

Epigenetic dysregulation has been reported in the context of extramedullary disease and infiltration in AML. Enhancer of zeste homolog 2 (EZH2) is a histone methyltransferase and is the catalytic subunit of the Polycomb Repressive Complex 2 (PRC2), which deposits Histone 3 Lysine 27 trimethylation (H3K27me3). High *EZH2* expression is correlated with higher peripheral blood blast percentages as well as extramedullary infiltration in patients with AML with numerous well-established biological roles. In vitro studies suggest that migration of AML cells appears to be regulated by EZH2/p-ERK/p-cmyc/MMP-2 and E-cadherin signaling pathways [73]. *EZH2* is a frequently mutated gene in AML; however, *EZH2* has a variety of biologic influences and a unique role in MS formation remains undefined [7,9].

Altered DNA methylation is another described mechanism in the development of extramedullary disease with key enzymes frequently mutated in AML. DNA methyltransferase 3A (*DNMT3A*) mutations contribute to altered DNA methylation, subsequently resulting in increased expression of a subset of genes with specific roles in myeloproliferation and extramedullary hematopoiesis [74]. *DNMT3A* mutation appears to contribute to extramedullary CNS infiltration mediated by overexpression of *TWIST1*, a key epithelial mesenchymal transition transcription factor, which is not otherwise well described in AML [75]. Furthermore, *TET2* is a member of the ten-eleven translocation (TET) gene family and is a key enzyme for DNA demethylation and a critical regulator for hematopoietic stem cell homeostasis. Models using *TET2*-deficient mice demonstrated not only high incidence of MS development but also transplant ability of the MS cells as well as an in vivo response to azacitidine treatment [76]. Decreased *TET2* expression was also seen in patient-derived MS samples with further suggestion of methylation changes impacting MS development [77].

AML has many examples of mutations in epigenetic pathways that are enriched in AML more than many other disease entities and may not be directly related to their involvement in MS [9]. The role of epigenetic dysregulation in leukemia migration and invasion with described MS phenotypes is intriguing yet requires further study. Additional research may uncover future targetable pathways for MS treatment, and as noted below, case reports have demonstrated the safety and efficacy of hypomethylating agent use for patients with MS.

6.7. Other Biological Associations

Mesothelin (MSLN) is a cell surface protein hypothesized to be involved in cell adhesion and is overexpressed in a subset of AML patients. MSLN overexpression was strongly associated with KMT2A-R, t(8;21), and inv(16) as well as the presence of extramedullary disease in children and young adults with AML. Methylation profiling further demonstrated an inverse association between MSLN promoter methylation and MSLN expression, suggesting another impact of epigenetic dysregulation [78].

Versican (VCAN) overexpression in the setting of *NPM1*-mutated AML is associated with an invasive phenotype and higher expression levels in patients with skin infiltration [79]. Lysyl oxidase (LOX), which has roles in pediatric acute megakaryoblastic leukemia and in the creation of a growth permissive fibrotic microenvironment, was associated with increased extramedullary disease in adults with AML and high plasma LOX activity [80]. *WT1* overexpression has also been described in MS cases as well as in extramedullary relapsed disease [81,82]. *ERG* transcription factor overexpression, similar to that of Ewing sarcoma, has been seen in patient-derived MS samples [83]. Multiple studies describe other associations observed in AML and extramedullary disease, including increased expression of amyloid precursor protein (APP) in *AML1/ETO* leukemia cells perhaps mediating the p-ERK/c-Myc/MMP-2 pathway, expression of miR-29c&b2, circular RNA expression patterns, and expression of CD25 and CD117 [84–90]. Polo-like kinase 1 (PLK1), which is involved in cell cycle control, was effectively inhibited in vivo using a patient-derived leukemia in mice with improvement in extramedullary disease [91]. PD-1 and PD-L1 have been investigated given the described efficacy of checkpoint inhibitors; however, there was no difference in expression of *PD-1*/*PD-L1* in MS tested, and they may instead have more impact in the surrounding tumor microenvironment [92,93]. Using mouse models, others observe a maturation plasticity of leukemia cells, with potential implications for chemotherapy resistance as a mechanism for extramedullary relapse [94]. A *PIM2/MYC* co-expressed mouse model demonstrated consistent and lethal in vivo MS development with *MYC* expression likely contributing to the phenotype [95]. Mouse models have also demonstrated cooperation between *MLL/AF10* and activating *KRAS* mutations, with increased cell adhesion properties contributing to in vivo MS formation via *Adgra3* and *Hoxa11* [96,97].

While many different mechanisms have been suggested in the development of MS, there remains no clear understanding of the pathogenesis. As such, it is hard to definitively identify the potential molecular determinants causing MS formation is some AML patients but not others. While the pathogenesis remains to be fully elucidated, prior studies suggest that there are likely multiple steps leading to MS development, including release from the bone marrow (which may be represented by higher WBC counts associated with MS), tissue invasion, and discrete mass formation, with the latter being the most consequential with regards to leukemia and the least described. Investigating how these different steps may cooperate and ultimately how the leukemia cells aggregate and sustain an aggregated phenotype requires dedicated study. Furthermore, the immune evasive or immunosuppressive microenvironment of MS illustrated in the post-allo-HSCT setting highlights that there is much more to learn about the pathogenesis of MS and its uniqueness with respect to its systemic/intramedullary AML counterpart.

7. Treatment Considerations

There is no consensus on the best treatment approach for management of MS, particularly isolated extramedullary disease. In pediatrics, systemic chemotherapy has been favored with consideration of allo-HSCT independent of the presence of MS.

The general approach to management in pediatric patients has evolved over the last three decades. In prior large cooperative group study treatment protocols (e.g., CCG

2961), children with MS would receive radiation therapy to the affected sites following initial induction chemotherapy, given MS responsiveness to irradiation. Although part of protocol therapy, many patients were not irradiated and outcomes demonstrated no difference in 5-year EFS, similar to smaller cohorts [12,98]. In adults, radiation therapy is more commonly used to treat isolated relapses, but the effects are not typically sustained and both localized and medullary relapse following radiotherapy are common [30,99]. Although prior COG studies included radiotherapy for treatment of MS sarcomas, this is no longer standard of care in the US; however, it continues to be recommended for MS in Berlin–Frankfurt–Munster (BFM) studies [17,100].

Given the epigenetic basis of AML development, inclusion of novel epigenetic approaches in treatment are increasing in utility for AML [9,101,102]. The hypomethylating agents azacitidine and decitabine have demonstrated efficacy in management of extramedullary disease in AML, including in pediatric patients [103]. Case reports in pediatrics have demonstrated complete response to monotherapy with azacitidine in MDS patients who received allo-HSCT as consolidation [103]. In the setting of post-allo-HSCT relapses of MS, multiple case reports in adults demonstrate efficacy of azacitidine or decitabine including complete response [104,105]. Hypomethylating agents in adults with AML and extramedullary disease showed improvement after one–two cycles and complete or near complete resolution of MS following four–five cycles [106–109]. Venetoclax has also shown activity against MS [109–111]. The utility of venetoclax and hypomethylating agents suggests a role for epigenetic reprogramming as a means for MS treatment, although DNA methylation-based mutations including *DNMT3A*, *IDH1*, *IDH2*, and *TET2* are far less common in children than in adults and translation of utility is more challenging [9,112]. While no consensus exists regarding optimal treatment of MS, particularly in pediatrics, radiation therapy is unlikely to contribute to durable remission of disease and expanding chemotherapeutic options to include hypomethylating agents and venetoclax in both initial chemotherapy regimens or as maintenance therapy following allo-HSCT should be considered and deserves further investigation.

8. Conclusions: Knowledge Gaps and Areas for Improvement

Children with AML and MS are distinct from adults. Given these differences, it is necessary to further study MS in the context of the driver lesions specific to children. MS remains a known clinical presentation with unclear impact on prognosis, risk stratification, and potential consequences in the setting of allo-HSCT. The advancement of imaging techniques and data for MS provides the opportunity for more directed and prospective evaluation in children. Collectively, this highlights the need for further large-scale cooperative group studies with clear criteria for the identification of MS in children.

While there is no specific treatment approach for children with MS, the use of intensive systemic chemotherapy remains at the forefront. However, additional studies are required to determine if epigenetic or immuno-oncology therapies may be beneficial. The role of allo-HSCT continues to be important as a curative option for many patients with high-risk AML; although, considering its potential lack of efficacy in the setting of extramedullary disease, the role of an immunosuppressive microenvironment in MS requires additional study. Further investigation into the potentially immunosuppressive MS microenvironment will be crucial to improving efficacy of allo-HSCT and managing isolated extramedullary relapses post-HSCT.

Although many different biological associations exist, there is an ongoing lack of clarity as to how leukemic blasts can not just invade tissues but form discrete tumors. Experiments delineating the potential epigenetic and transcriptomic differences between medullary AML disease and MS are required to identify the underlying molecular mechanisms that promote MS development. Understanding how leukemic blasts transform into and sustain an MS phenotype is critical to identifying specific targetable mechanisms. Understanding and combatting chemotherapy resistance and immune escape will ultimately improve survival in patients with AML and MS.

Author Contributions: Conceptualization, K.E.Z. and S.R.; writing—original draft preparation, K.E.Z., A.M.C. and S.R.; writing—review and editing, K.E.Z., A.M.C., A.E.M., K.S.C. and S.R.; visualization, K.E.Z. and A.M.C.; supervision, S.R.; project administration, S.R.; funding acquisition, S.R. All authors have read and agreed to the published version of the manuscript.

Funding: S.R. is funded by CA204231 and HL149620.

Conflicts of Interest: The authors declare no conflict of interest.

References

1. Wilson, C.S.; Medeiros, L.J. Extramedullary Manifestations of Myeloid Neoplasms. *Am. J. Clin. Pathol.* **2015**, *144*, 219–239. [CrossRef] [PubMed]
2. Campidelli, C.; Agostinelli, C.; Stitson, R.; Pileri, S.A. Myeloid sarcoma: Extramedullary manifestation of myeloid disorders. *Am. J. Clin. Pathol.* **2009**, *132*, 426–437. [CrossRef] [PubMed]
3. Arber, D.A.; Orazi, A.; Hasserjian, R.; Thiele, J.; Borowitz, M.J.; Le Beau, M.M.; Bloomfield, C.D.; Cazzola, M.; Vardiman, J.W. The 2016 revision to the World Health Organization classification of myeloid neoplasms and acute leukemia. *Blood* **2016**, *127*, 2391–2405. [CrossRef]
4. Pileri, S.A.; Ascani, S.; Cox, M.C.; Campidelli, C.; Bacci, F.; Piccioli, M.; Piccaluga, P.P.; Agostinelli, C.; Asioli, S.; Novero, D.; et al. Myeloid sarcoma: Clinico-pathologic, phenotypic and cytogenetic analysis of 92 adult patients. *Leukemia* **2007**, *21*, 340–350. [CrossRef] [PubMed]
5. De Rooij, J.D.; Zwaan, C.M.; van den Heuvel-Eibrink, M. Pediatric AML: From Biology to Clinical Management. *J. Clin. Med.* **2015**, *4*, 127–149. [CrossRef]
6. Shallis, R.M.; Wang, R.; Davidoff, A.; Ma, X.; Zeidan, A.M. Epidemiology of acute myeloid leukemia: Recent progress and enduring challenges. *Blood Rev.* **2019**, *36*, 70–87. [CrossRef]
7. Bolouri, H.; Farrar, J.E.; Triche, T., Jr.; Ries, R.E.; Lim, E.L.; Alonzo, T.A.; Ma, Y.; Moore, R.; Mungall, A.J.; Marra, M.A.; et al. The molecular landscape of pediatric acute myeloid leukemia reveals recurrent structural alterations and age-specific mutational interactions. *Nat. Med.* **2018**, *24*, 103–112. [CrossRef]
8. Papaemmanuil, E.; Gerstung, M.; Bullinger, L.; Gaidzik, V.I.; Paschka, P.; Roberts, N.D.; Potter, N.E.; Heuser, M.; Thol, F.; Bolli, N.; et al. Genomic Classification and Prognosis in Acute Myeloid Leukemia. *N. Engl. J. Med.* **2016**, *374*, 2209–2221. [CrossRef]
9. Ley, T.J.; Miller, C.; Ding, L.; Raphael, B.J.; Mungall, A.J.; Robertson, A.; Hoadley, K.; Triche, T.J., Jr.; Laird, P.W.; Baty, J.D.; et al. Genomic and epigenomic landscapes of adult de novo acute myeloid leukemia. *N. Engl. J. Med.* **2013**, *368*, 2059–2074. [CrossRef]
10. Chang, M.; Raimondi, S.C.; Ravindranath, Y.; Carroll, A.J.; Camitta, B.; Gresik, M.V.; Steuber, C.P.; Weinstein, H. Prognostic factors in children and adolescents with acute myeloid leukemia (excluding children with Down syndrome and acute promyelocytic leukemia): Univariate and recursive partitioning analysis of patients treated on Pediatric Oncology Group (POG) Study 8821. *Leukemia* **2000**, *14*, 1201–1207. [CrossRef]
11. Bisschop, M.M.; Révész, T.; Bierings, M.; van Weerden, J.F.; van Wering, E.R.; Hählen, K.; van der Does-van den Berg, A. Extramedullary infiltrates at diagnosis have no prognostic significance in children with acute myeloid leukaemia. *Leukemia* **2001**, *15*, 46–49. [CrossRef] [PubMed]
12. Dusenbery, K.E.; Howells, W.B.; Arthur, D.C.; Alonzo, T.; Lee, J.W.; Kobrinsky, N.; Barnard, D.R.; Wells, R.J.; Buckley, J.D.; Lange, B.J.; et al. Extramedullary leukemia in children with newly diagnosed acute myeloid leukemia: A report from the Children's Cancer Group. *J. Pediatr. Hematol. Oncol.* **2003**, *25*, 760–768. [CrossRef] [PubMed]
13. Hiçsönmez, G.; Cetin, M.; Tuncer, A.M.; Yenicesu, I.; Aslan, D.; Ozyürek, E.; Unal, S. Children with acute myeloblastic leukemia presenting with extramedullary infiltration: The effects of high-dose steroid treatment. *Leuk. Res.* **2004**, *28*, 25–34. [CrossRef] [PubMed]
14. Kobayashi, R.; Tawa, A.; Hanada, R.; Horibe, K.; Tsuchida, M.; Tsukimoto, I. Extramedullary infiltration at diagnosis and prognosis in children with acute myelogenous leukemia. *Pediatr. Blood Cancer* **2007**, *48*, 393–398. [CrossRef]
15. Johnston, D.L.; Alonzo, T.A.; Gerbing, R.B.; Lange, B.J.; Woods, W.G. Superior outcome of pediatric acute myeloid leukemia patients with orbital and CNS myeloid sarcoma: A report from the Children's Oncology Group. *Pediatr. Blood Cancer* **2012**, *58*, 519–524. [CrossRef]
16. Creutzig, U.; Dworzak, M.N.; Zimmermann, M.; Reinhardt, D.; Sramkova, L.; Bourquin, J.P.; Hasle, H.; Abrahamsson, J.; Kaspers, G.; van den Heuvel, M.M.; et al. Characteristics and outcome in patients with central nervous system involvement treated in European pediatric acute myeloid leukemia study groups. *Pediatr. Blood Cancer* **2017**, *64*, e26664. [CrossRef]
17. Støve, H.K.; Sandahl, J.D.; Abrahamsson, J.; Asdahl, P.H.; Forestier, E.; Ha, S.Y.; Jahnukainen, K.; Jónsson, Ó.G.; Lausen, B.; Palle, J.; et al. Extramedullary leukemia in children with acute myeloid leukemia: A population-based cohort study from the Nordic Society of Pediatric Hematology and Oncology (NOPHO). *Pediatr. Blood Cancer* **2017**, *64*, e26520. [CrossRef]
18. Pramanik, R.; Tyagi, A.; Chopra, A.; Kumar, A.; Vishnubhatla, S.; Bakhshi, S. Myeloid Sarcoma Predicts Superior Outcome in Pediatric AML; Can Cytogenetics Solve the Puzzle? *Clin. Lymphoma Myeloma Leuk.* **2018**, *18*, e249–e254. [CrossRef]
19. Xu, L.H.; Wang, Y.; Chen, Z.Y.; Fang, J.P. Myeloid sarcoma is associated with poor clinical outcome in pediatric patients with acute myeloid leukemia. *J. Cancer Res. Clin. Oncol.* **2020**, *146*, 1011–1020. [CrossRef]

20. Lee, J.W.; Kim, S.; Jang, P.S.; Chung, N.G.; Cho, B.; Im, S.A.; Kim, M. Prognostic Role of Postinduction Minimal Residual Disease and Myeloid Sarcoma Type Extramedullary Involvement in Pediatric RUNX1-RUNX1T1 (+) Acute Myeloid Leukemia. *J. Pediatr. Hematol. Oncol.* **2020**, *42*, e132–e139. [CrossRef]
21. Hu, G.H.; Lu, A.D.; Jia, Y.P.; Zuo, Y.X.; Wu, J.; Zhang, L.P. Prognostic Impact of Extramedullary Infiltration in Pediatric Low-risk Acute Myeloid Leukemia: A Retrospective Single-center Study Over 10 Years. *Clin. Lymphoma Myeloma Leuk.* **2020**, *20*, e813–e820. [CrossRef] [PubMed]
22. Hu, G.; Lu, A.; Wu, J.; Jia, Y.; Zuo, Y.; Ding, M.; Zhang, L. Characteristics and prognosis of pediatric myeloid sarcoma in the cytogenetic context of t(8;21). *Pediatr. Hematol. Oncol.* **2021**, *38*, 14–24. [CrossRef] [PubMed]
23. Samborska, M.; Barańska, M.; Wachowiak, J.; Skalska-Sadowska, J.; Thambyrajah, S.; Czogała, M.; Balwierz, W.; Kołtan, S.; Peszyńska-Żelazny, K.; Wysocki, M.; et al. Clinical Characteristics and Treatment Outcomes of Myeloid Sarcoma in Children: The Experience of the Polish Pediatric Leukemia and Lymphoma Study Group. *Front. Oncol.* **2022**, *12*, 935373. [CrossRef] [PubMed]
24. Stölzel, F.; Lüer, T.; Löck, S.; Parmentier, S.; Kuithan, F.; Kramer, M.; Alakel, N.S.; Sockel, K.; Taube, F.; Middeke, J.M.; et al. The prevalence of extramedullary acute myeloid leukemia detected by ^{18}FDG-PET/CT: Final results from the prospective PETAML trial. *Haematologica* **2020**, *105*, 1552–1558. [CrossRef]
25. Xin, X.; Zhu, H.; Chang, Z.; Feng, M.; Gao, S.; Hou, L.; Su, X. Risk factors and prognosis analysis of acute myeloid leukemia in children. *J. Balk. Union Oncol.* **2021**, *26*, 166–172.
26. Zhou, T.; Bloomquist, M.S.; Ferguson, L.S.; Reuther, J.; Marcogliese, A.N.; Elghetany, M.T.; Roy, A.; Rao, P.H.; Lopez-Terrada, D.H.; Redell, M.S.; et al. Pediatric myeloid sarcoma: A single institution clinicopathologic and molecular analysis. *Pediatr. Hematol. Oncol.* **2020**, *37*, 76–89. [CrossRef]
27. Eckardt, J.N.; Stölzel, F.; Kunadt, D.; Röllig, C.; Stasik, S.; Wagenführ, L.; Jöhrens, K.; Kuithan, F.; Krämer, A.; Scholl, S.; et al. Molecular profiling and clinical implications of patients with acute myeloid leukemia and extramedullary manifestations. *J. Hematol. Oncol.* **2022**, *15*, 60. [CrossRef]
28. Sakaguchi, H.; Miyamura, T.; Tomizawa, D.; Taga, T.; Ishida, H.; Okamoto, Y.; Koh, K.; Yokosuka, T.; Yoshida, N.; Sato, M.; et al. Effect of extramedullary disease on allogeneic hematopoietic cell transplantation for pediatric acute myeloid leukemia: A nationwide retrospective study. *Bone Marrow Transplant.* **2021**, *56*, 1859–1865. [CrossRef]
29. Hazar, V.; Öztürk, G.; Yalçın, K.; Uygun, V.; Aksoylar, S.; Küpesiz, A.; Ok Bozkaya, İ.; Karagün, B.; Bozkurt, C.; İleri, T.; et al. Different Kinetics and Risk Factors for Isolated Extramedullary Relapse after Allogeneic Hematopoietic Stem Cell Transplantation in Children with Acute Leukemia. *Transplant Cell Ther* **2021**, *27*, 859.e1–859.e10. [CrossRef]
30. Harris, A.C.; Kitko, C.L.; Couriel, D.R.; Braun, T.M.; Choi, S.W.; Magenau, J.; Mineishi, S.; Pawarode, A.; Yanik, G.; Levine, J.E. Extramedullary relapse of acute myeloid leukemia following allogeneic hematopoietic stem cell transplantation: Incidence, risk factors and outcomes. *Haematologica* **2013**, *98*, 179–184. [CrossRef]
31. Shem-Tov, N.; Saraceni, F.; Danylesko, I.; Shouval, R.; Yerushalmi, R.; Nagler, A.; Shimoni, A. Isolated Extramedullary Relapse of Acute Leukemia after Allogeneic Stem Cell Transplantation: Different Kinetics and Better Prognosis than Systemic Relapse. *Biol. Blood Marrow Transplant.* **2017**, *23*, 1087–1094. [CrossRef] [PubMed]
32. Goyal, S.D.; Zhang, M.J.; Wang, H.L.; Akpek, G.; Copelan, E.A.; Freytes, C.; Gale, R.P.; Hamadani, M.; Inamoto, Y.; Kamble, R.T.; et al. Allogeneic hematopoietic cell transplant for AML: No impact of pre-transplant extramedullary disease on outcome. *Bone Marrow Transplant.* **2015**, *50*, 1057–1062. [CrossRef] [PubMed]
33. Bourlon, C.; Lipton, J.H.; Deotare, U.; Gupta, V.; Kim, D.D.; Kuruvilla, J.; Viswabandya, A.; Thyagu, S.; Messner, H.A.; Michelis, F.V. Extramedullary disease at diagnosis of AML does not influence outcome of patients undergoing allogeneic hematopoietic cell transplant in CR1. *Eur. J. Haematol.* **2017**, *99*, 234–239. [CrossRef] [PubMed]
34. Frietsch, J.J.; Hunstig, F.; Wittke, C.; Junghanss, C.; Franiel, T.; Scholl, S.; Hochhaus, A.; Hilgendorf, I. Extra-medullary recurrence of myeloid leukemia as myeloid sarcoma after allogeneic stem cell transplantation: Impact of conditioning intensity. *Bone Marrow Transplant.* **2021**, *56*, 101–109. [CrossRef]
35. Huang, Q.; Reddi, D.; Chu, P.; Snyder, D.S.; Weisenburger, D.D. Clinical and pathologic analysis of extramedullary tumors after hematopoietic stem cell transplantation. *Hum. Pathol.* **2014**, *45*, 2404–2410. [CrossRef]
36. Shimizu, H.; Saitoh, T.; Hatsumi, N.; Takada, S.; Handa, H.; Jimbo, T.; Sakura, T.; Miyawaki, S.; Nojima, Y. Prevalence of extramedullary relapses is higher after allogeneic stem cell transplantation than after chemotherapy in adult patients with acute myeloid leukemia. *Leuk. Res.* **2013**, *37*, 1477–1481. [CrossRef]
37. Solh, M.; DeFor, T.E.; Weisdorf, D.J.; Kaufman, D.S. Extramedullary relapse of acute myelogenous leukemia after allogeneic hematopoietic stem cell transplantation: Better prognosis than systemic relapse. *Biol. Blood Marrow Transplant.* **2012**, *18*, 106–112. [CrossRef]
38. Kikushige, Y.; Takase, K.; Sata, K.; Aoki, K.; Numata, A.; Miyamoto, T.; Fukuda, T.; Gondo, H.; Harada, M.; Nagafuji, K. Repeated relapses of acute myelogenous leukemia in the isolated extramedullary sites following allogeneic bone marrow transplantations. *Intern. Med.* **2007**, *46*, 1011–1014. [CrossRef]
39. Ando, T.; Mitani, N.; Matsui, K.; Yamashita, K.; Nomiyama, J.; Tsuru, M.; Yujiri, T.; Tanizawa, Y. Recurrent extramedullary relapse of acute myelogenous leukemia after allogeneic hematopoietic stem cell transplantation in a patient with the chromosomal abnormality t(8;21) and CD56-positivity. *Int. J. Hematol.* **2009**, *90*, 374–377. [CrossRef]

40. Ando, T.; Mitani, N.; Matsunaga, K.; Nakazora, T.; Gondo, T.; Yujiri, T.; Tanizawa, Y. Gemtuzumab ozogamicin therapy for isolated extramedullary AML relapse after allogeneic hematopoietic stem-cell transplantation. *Tohoku J. Exp. Med.* **2010**, *220*, 121–126. [CrossRef]
41. Shan, M.; Lu, Y.; Yang, M.; Wang, P.; Lu, S.; Zhang, L.; Qiu, H.; Chen, S.; Xu, Y.; Zhang, X.; et al. Characteristics and transplant outcome of myeloid sarcoma: A single-institute study. *Int. J. Hematol.* **2021**, *113*, 682–692. [CrossRef] [PubMed]
42. Kim, J.Y.; Im, S.A.; Lee, J.H.; Lee, J.W.; Chung, N.G.; Cho, B. Extramedullary Relapse of Acute Myeloid and Lymphoid Leukemia in Children: A Retrospective Analysis. *Iran. J. Pediatr.* **2016**, *26*, e1711. [CrossRef] [PubMed]
43. Meyer, H.J.; Beimler, M.; Borte, G.; Pönisch, W.; Surov, A. Radiological and clinical patterns of myeloid sarcoma. *Radiol. Oncol.* **2019**, *53*, 213–218. [CrossRef] [PubMed]
44. Lee, E.Y.; Anthony, M.P.; Leung, A.Y.; Loong, F.; Khong, P.L. Utility of FDG PET/CT in the assessment of myeloid sarcoma. *Am. J. Roentgenol.* **2012**, *198*, 1175–1179. [CrossRef] [PubMed]
45. Zhou, W.L.; Wu, H.B.; Wang, L.J.; Tian, Y.; Dong, Y.; Wang, Q.S. Usefulness and pitfalls of F-18-FDG PET/CT for diagnosing extramedullary acute leukemia. *Eur. J. Radiol.* **2016**, *85*, 205–210. [CrossRef]
46. Swerdlow, S.H.; Campo, E.; Harris, N.L.; Jaffe, E.S.; Pileri, S.A.; Stein, H.; Thiele, J. *WHO Classification of Tumours of Haematopoietic and Lymphoid Tissues*; IARC: Lyon, France, 2017; Volume 2.
47. Greenland, N.Y.; Van Ziffle, J.A.; Liu, Y.C.; Qi, Z.; Prakash, S.; Wang, L. Genomic analysis in myeloid sarcoma and comparison with paired acute myeloid leukemia. *Hum. Pathol.* **2021**, *108*, 76–83. [CrossRef]
48. Li, Z.; Stölzel, F.; Onel, K.; Sukhanova, M.; Mirza, M.K.; Yap, K.L.; Borinets, O.; Larson, R.A.; Stock, W.; Sasaki, M.M.; et al. Next-generation sequencing reveals clinically actionable molecular markers in myeloid sarcoma. *Leukemia* **2015**, *29*, 2113–2116. [CrossRef]
49. Du, W.; Lu, C.; Zhu, X.; Hu, D.; Chen, X.; Li, J.; Liu, W.; Zhu, J.; He, Y.; Yao, J. Prognostic significance of CXCR4 expression in acute myeloid leukemia. *Cancer Med.* **2019**, *8*, 6595–6603. [CrossRef]
50. Chou, S.H.; Ko, B.S.; Chiou, J.S.; Hsu, Y.C.; Tsai, M.H.; Chiu, Y.C.; Yu, I.S.; Lin, S.W.; Hou, H.A.; Kuo, Y.Y.; et al. A knock-in Npm1 mutation in mice results in myeloproliferation and implies a perturbation in hematopoietic microenvironment. *PLoS ONE* **2012**, *7*, e49769. [CrossRef]
51. Peled, A.; Klein, S.; Beider, K.; Burger, J.A.; Abraham, M. Role of CXCL12 and CXCR4 in the pathogenesis of hematological malignancies. *Cytokine* **2018**, *109*, 11–16. [CrossRef]
52. Crazzolara, R.; Kreczy, A.; Mann, G.; Heitger, A.; Eibl, G.; Fink, F.M.; Möhle, R.; Meister, B. High expression of the chemokine receptor CXCR4 predicts extramedullary organ infiltration in childhood acute lymphoblastic leukaemia. *Br. J. Haematol.* **2001**, *115*, 545–553. [CrossRef] [PubMed]
53. Zheng, Q.; Shuai, X.; Ye, Y.; Jin, Y.; Jiang, N.; Chen, X.; Su, J. The role of polymorphisms of stromal-derived factor-1 and CXC receptor 4 in acute myeloid leukemia and leukemia cell dissemination. *Gene* **2016**, *588*, 103–108. [CrossRef]
54. Faaij, C.M.; Willemze, A.J.; Révész, T.; Balzarolo, M.; Tensen, C.P.; Hoogeboom, M.; Vermeer, M.H.; van Wering, E.; Zwaan, C.M.; Kaspers, G.J.; et al. Chemokine/chemokine receptor interactions in extramedullary leukaemia of the skin in childhood AML: Differential roles for CCR2, CCR5, CXCR4 and CXCR7. *Pediatr. Blood Cancer* **2010**, *55*, 344–348. [CrossRef] [PubMed]
55. Ponziani, V.; Mannelli, F.; Bartalucci, N.; Gianfaldoni, G.; Leoni, F.; Antonioli, E.; Guglielmelli, P.; Ciolli, S.; Bosi, A.; Vannucchi, A.M. No role for CXCL12-G801A polymorphism in the development of extramedullary disease in acute myeloid leukemia. *Leukemia* **2008**, *22*, 669–671. [CrossRef]
56. Chang, H.; Brandwein, J.; Yi, Q.L.; Chun, K.; Patterson, B.; Brien, B. Extramedullary infiltrates of AML are associated with CD56 expression, 11q23 abnormalities and inferior clinical outcome. *Leuk. Res.* **2004**, *28*, 1007–1011. [CrossRef] [PubMed]
57. Deeb, G.; Baer, M.R.; Gaile, D.P.; Sait, S.N.; Barcos, M.; Wetzler, M.; Conroy, J.M.; Nowak, N.J.; Cowell, J.K.; Cheney, R.T. Genomic profiling of myeloid sarcoma by array comparative genomic hybridization. *Genes Chromosomes Cancer* **2005**, *44*, 373–383. [CrossRef] [PubMed]
58. Park, S.S.; Yoon, J.H.; Kim, H.J.; Jeon, Y.W.; Lee, S.E.; Cho, B.S.; Eom, K.S.; Kim, Y.J.; Lee, S.; Min, C.K.; et al. Characteristics and Survival Outcome Analysis of Extramedullary Involvement in Adult Patients with t(8;21) Acute Myeloid Leukemia. *Clin. Lymphoma Myeloma Leuk.* **2017**, *17*, 38–45.e32. [CrossRef]
59. Alegretti, A.P.; Bittar, C.M.; Bittencourt, R.; Piccoli, A.K.; Schneider, L.; Silla, L.M.; Bó, S.D.; Xavier, R.M. The expression of CD56 antigen is associated with poor prognosis in patients with acute myeloid leukemia. *Rev. Bras. Hematol. Hemoter.* **2011**, *33*, 202–206. [CrossRef]
60. Kobayashi, N.; Oda, T.; Takizawa, M.; Ishizaki, T.; Tsukamoto, N.; Yokohama, A.; Takei, H.; Saitoh, T.; Shimizu, H.; Honma, K.; et al. Integrin α7 and Extracellular Matrix Laminin 211 Interaction Promotes Proliferation of Acute Myeloid Leukemia Cells and Is Associated with Granulocytic Sarcoma. *Cancers* **2020**, *12*, 363. [CrossRef]
61. Kim, H.N.; Ruan, Y.; Ogana, H.; Kim, Y.M. Cadherins, Selectins, and Integrins in CAM-DR in Leukemia. *Front. Oncol.* **2020**, *10*, 592733. [CrossRef]
62. Valent, P.; Sadovnik, I.; Eisenwort, G.; Herrmann, H.; Bauer, K.; Mueller, N.; Sperr, W.R.; Wicklein, D.; Schumacher, U. Redistribution, homing and organ-invasion of neoplastic stem cells in myeloid neoplasms. *Semin. Cancer Biol.* **2020**, *60*, 191–201. [CrossRef] [PubMed]
63. Piccaluga, P.P.; Paolini, S.; Navari, M.; Etebari, M.; Visani, G.; Ascani, S. Increased angiogenesis seems to correlate with inferior overall survival in myeloid sarcoma patients. *Pol. J. Pathol.* **2018**, *69*, 254–265. [CrossRef] [PubMed]

64. Schuch, G.; Machluf, M.; Bartsch, G., Jr.; Nomi, M.; Richard, H.; Atala, A.; Soker, S. In vivo administration of vascular endothelial growth factor (VEGF) and its antagonist, soluble neuropilin-1, predicts a role of VEGF in the progression of acute myeloid leukemia in vivo. *Blood* **2002**, *100*, 4622–4628. [CrossRef]
65. Pillozzi, S.; Brizzi, M.F.; Bernabei, P.A.; Bartolozzi, B.; Caporale, R.; Basile, V.; Boddi, V.; Pegoraro, L.; Becchetti, A.; Arcangeli, A. VEGFR-1 (FLT-1), beta1 integrin, and hERG K+ channel for a macromolecular signaling complex in acute myeloid leukemia: Role in cell migration and clinical outcome. *Blood* **2007**, *110*, 1238–1250. [CrossRef] [PubMed]
66. Deng, M.; Zha, J.; Zhao, H.; Jia, X.; Shi, Y.; Li, Z.; Fu, G.; Yu, L.; Fang, Z.; Xu, B. Apatinib exhibits cytotoxicity toward leukemia cells by targeting VEGFR2-mediated prosurvival signaling and angiogenesis. *Exp. Cell Res.* **2020**, *390*, 111934. [CrossRef]
67. Xian, J.; Shao, H.; Chen, X.; Zhang, S.; Quan, J.; Zou, Q.; Jin, H.; Zhang, L. Nucleophosmin Mutants Promote Adhesion, Migration and Invasion of Human Leukemia THP-1 Cells through MMPs Up-regulation via Ras/ERK MAPK Signaling. *Int. J. Biol. Sci.* **2016**, *12*, 144–155. [CrossRef]
68. Song, J.H.; Kim, S.H.; Cho, D.; Lee, I.K.; Kim, H.J.; Kim, T.S. Enhanced invasiveness of drug-resistant acute myeloid leukemia cells through increased expression of matrix metalloproteinase-2. *Int. J. Cancer* **2009**, *125*, 1074–1081. [CrossRef]
69. Wang, C.; Cai, X.; Chen, B.; He, Z.; Chen, Z.; Cen, J.; Li, Z. Up-regulation of tissue inhibitor of metalloproteinase-2 promotes SHI-1 cell invasion in nude mice. *Leuk. Lymphoma* **2013**, *54*, 2707–2711. [CrossRef]
70. Wang, C.; Chen, Z.; Li, Z.; Cen, J. The essential roles of matrix metalloproteinase-2, membrane type 1 metalloproteinase and tissue inhibitor of metalloproteinase-2 in the invasive capacity of acute monocytic leukemia SHI-1 cells. *Leuk. Res.* **2010**, *34*, 1083–1090. [CrossRef]
71. Stefanidakis, M.; Karjalainen, K.; Jaalouk, D.E.; Gahmberg, C.G.; O'Brien, S.; Pasqualini, R.; Arap, W.; Koivunen, E. Role of leukemia cell invadosome in extramedullary infiltration. *Blood* **2009**, *114*, 3008–3017. [CrossRef]
72. Kobayashi, M.; Hamada, J.; Li, Y.Q.; Shinobu, N.; Imamura, M.; Okada, F.; Takeichi, N.; Hosokawa, M. A possible role of 92 kDa type IV collagenase in the extramedullary tumor formation in leukemia. *Jpn. J. Cancer Res.* **1995**, *86*, 298–303. [CrossRef]
73. Zhu, Q.; Zhang, L.; Li, X.; Chen, F.; Jiang, L.; Yu, G.; Wang, Z.; Yin, C.; Jiang, X.; Zhong, Q.; et al. Higher EZH2 expression is associated with extramedullary infiltration in acute myeloid leukemia. *Tumour Biol.* **2016**, *37*, 11409–11420. [CrossRef]
74. Guryanova, O.A.; Lieu, Y.K.; Garrett-Bakelman, F.E.; Spitzer, B.; Glass, J.L.; Shank, K.; Martinez, A.B.; Rivera, S.A.; Durham, B.H.; Rapaport, F.; et al. Dnmt3a regulates myeloproliferation and liver-specific expansion of hematopoietic stem and progenitor cells. *Leukemia* **2016**, *30*, 1133–1142. [CrossRef]
75. Xu, J.; Zhang, W.; Yan, X.J.; Lin, X.Q.; Li, W.; Mi, J.Q.; Li, J.M.; Zhu, J.; Chen, Z.; Chen, S.J. DNMT3A mutation leads to leukemic extramedullary infiltration mediated by TWIST1. *J. Hematol. Oncol.* **2016**, *9*, 106. [CrossRef] [PubMed]
76. Wang, J.; Miao, Z.; Jiang, Y.; Zou, P.; Li, W.; Tang, X.; Lv, Y.; Xing, D.; Chen, S.; Yang, F.; et al. Characteristics of myeloid sarcoma in mice and patients with TET2 deficiency. *Oncol. Lett.* **2020**, *19*, 3789–3798. [CrossRef] [PubMed]
77. Xiao, D.; Shi, Y.; Fu, C.; Jia, J.; Pan, Y.; Jiang, Y.; Chen, L.; Liu, S.; Zhou, W.; Zhou, J.; et al. Decrease of TET2 expression and increase of 5-hmC levels in myeloid sarcomas. *Leuk. Res.* **2016**, *42*, 75–79. [CrossRef]
78. Kaeding, A.J.; Barwe, S.P.; Gopalakrishnapillai, A.; Ries, R.E.; Alonzo, T.A.; Gerbing, R.B.; Correnti, C.; Loken, M.R.; Broderson, L.E.; Pardo, L.; et al. Mesothelin is a novel cell surface disease marker and potential therapeutic target in acute myeloid leukemia. *Blood Adv.* **2021**, *5*, 2350–2361. [CrossRef] [PubMed]
79. Yang, L.; Wang, L.; Yang, Z.; Jin, H.; Zou, Q.; Zhan, Q.; Tang, Y.; Tao, Y.; Lei, L.; Jing, Y.; et al. Up-regulation of EMT-related gene VCAN by NPM1 mutant-driven TGF-β/cPML signalling promotes leukemia cell invasion. *J. Cancer* **2019**, *10*, 6570–6583. [CrossRef]
80. Kunadt, D.; Kramer, M.; Dill, C.; Altmann, H.; Wagenführ, L.; Mohr, B.; Thiede, C.; Röllig, C.; Schetelig, J.; Bornhäuser, M.; et al. Lysyl oxidase expression is associated with inferior outcome and Extramedullary disease of acute myeloid leukemia. *Biomark. Res.* **2020**, *8*, 20. [CrossRef]
81. Kwon, M.; Martínez-Laperche, C.; Infante, M.; Carretero, F.; Balsalobre, P.; Serrano, D.; Gayoso, J.; Pérez-Corral, A.; Anguita, J.; Díez-Martín, J.L.; et al. Evaluation of minimal residual disease by real-time quantitative PCR of Wilms' tumor 1 expression in patients with acute myelogenous leukemia after allogeneic stem cell transplantation: Correlation with flow cytometry and chimerism. *Biol. Blood Marrow Transplant.* **2012**, *18*, 1235–1242. [CrossRef]
82. Al-Adnani, M.; Williams, S.; Anderson, J.; Ashworth, M.; Malone, M.; Sebire, N.J. Immunohistochemical nuclear positivity for WT1 in childhood acute myeloid leukemia. *Fetal Pediatr. Pathol.* **2007**, *26*, 193–197. [CrossRef] [PubMed]
83. Vargas, A.C.; Turner, J.; Burchett, I.; Ho, L.L.; Zumbo, R.; Gill, A.J.; Maclean, F.M. Myeloid sarcoma and extramedullary hematopoiesis expand the spectrum of ERG-positive proliferations: An ancillary tool in the diagnosis. *Hum. Pathol.* **2022**, *124*, 1–13. [CrossRef] [PubMed]
84. Wang, C.L.; Ding, B.J.; Jiang, L.; Yin, C.X.; Zhong, Q.X.; Yu, G.P.; Li, X.D.; Meng, F.Y. Increased expression of amyloid precursor protein promotes proliferation and migration of AML1/ETO-positive leukemia cells and be inhibited by panobinostat. *Neoplasma* **2015**, *62*, 864–871. [CrossRef] [PubMed]
85. Wei, Y.; Lu, W.; Yu, Y.; Zhai, Y.; Guo, H.; Yang, S.; Zhao, C.; Zhang, Y.; Liu, J.; Liu, Y.; et al. miR-29c&b2 encourage extramedullary infiltration resulting in the poor prognosis of acute myeloid leukemia. *Oncogene* **2021**, *40*, 3434–3448. [CrossRef]
86. Lv, C.; Sun, L.; Guo, Z.; Li, H.; Kong, D.; Xu, B.; Lin, L.; Liu, T.; Guo, D.; Zhou, J.; et al. Circular RNA regulatory network reveals cell-cell crosstalk in acute myeloid leukemia extramedullary infiltration. *J. Transl. Med.* **2018**, *16*, 361. [CrossRef]

87. Lin, L.; Wang, Y.; Bian, S.; Sun, L.; Guo, Z.; Kong, D.; Zhao, L.; Guo, D.; Li, Q.; Wu, M.; et al. A circular RNA derived from PLXNB2 as a valuable predictor of the prognosis of patients with acute myeloid leukaemia. *J. Transl. Med.* **2021**, *19*, 123. [CrossRef]
88. Chen, J.; Yanuck, R.R., 3rd; Abbondanzo, S.L.; Chu, W.S.; Aguilera, N.S. c-Kit (CD117) reactivity in extramedullary myeloid tumor/granulocytic sarcoma. *Arch. Pathol. Lab. Med.* **2001**, *125*, 1448–1452. [CrossRef]
89. Jiang, L.; Meng, W.; Yu, G.; Yin, C.; Wang, Z.; Liao, L.; Meng, F. MicroRNA-144 targets APP to regulate AML1/ETO$^+$ leukemia cell migration via the p-ERK/c-Myc/MMP-2 pathway. *Oncol. Lett.* **2019**, *18*, 2034–2042. [CrossRef]
90. Isshiki, Y.; Ohwada, C.; Togasaki, E.; Shimizu, R.; Hasegawa, N.; Yamazaki, A.; Sugita, Y.; Kawaguchi, T.; Tsukamoto, S.; Sakai, S.; et al. Acute myeloid leukemia concurrent with spinal epidural extramedullary myeloid sarcoma accompanied by a high CD25 expression and the FLT3-ITD mutation. *Intern. Med.* **2014**, *53*, 1159–1164. [CrossRef]
91. Casolaro, A.; Golay, J.; Albanese, C.; Ceruti, R.; Patton, V.; Cribioli, S.; Pezzoni, A.; Losa, M.; Texido, G.; Giussani, U.; et al. The Polo-Like Kinase 1 (PLK1) inhibitor NMS-P937 is effective in a new model of disseminated primary CD56+ acute monoblastic leukaemia. *PLoS ONE* **2013**, *8*, e58424. [CrossRef]
92. Meleveedu, K.S.; Chen, D.; Nadiminti, K.; Sidiqi, H.; Khan, S.; Alkhateeb, H.; Shah, M.V.; Patnaik, M.; Hogan, W.J.; Begna, K.; et al. PD-1/PD-L1 expression in extramedullary lesions of acute myeloid leukemia. *Leuk. Lymphoma* **2021**, *62*, 764–767. [CrossRef] [PubMed]
93. Kawamoto, K.; Miyoshi, H.; Suzuki, T.; Kiyasu, J.; Yokoyama, S.; Sasaki, Y.; Sone, H.; Seto, M.; Takizawa, J.; Ohshima, K. Expression of programmed death ligand 1 is associated with poor prognosis in myeloid sarcoma patients. *Hematol. Oncol.* **2018**, *36*, 591–599. [CrossRef] [PubMed]
94. Ngo, S.; Oxley, E.P.; Ghisi, M.; Garwood, M.M.; McKenzie, M.D.; Mitchell, H.L.; Kanellakis, P.; Susanto, O.; Hickey, M.J.; Perkins, A.C.; et al. Acute myeloid leukemia maturation lineage influences residual disease and relapse following differentiation therapy. *Nat. Commun.* **2021**, *12*, 6546. [CrossRef] [PubMed]
95. Jang, S.H.; Chung, H.Y. MYC and PIM2 co-expression in mouse bone marrow cells readily establishes permanent myeloid cell lines that can induce lethal myeloid sarcoma in vivo. *Mol. Cells* **2012**, *34*, 201–208. [CrossRef] [PubMed]
96. Fu, J.F.; Yen, T.H.; Chen, Y.; Huang, Y.J.; Hsu, C.L.; Liang, D.C.; Shih, L.Y. Involvement of Gpr125 in the myeloid sarcoma formation induced by cooperating MLL/AF10(OM-LZ) and oncogenic KRAS in a mouse bone marrow transplantation model. *Int. J. Cancer* **2013**, *133*, 1792–1802. [CrossRef]
97. Fu, J.F.; Wen, C.J.; Yen, T.H.; Shih, L.Y. Hoxa11-mediated reduction of cell migration contributes to myeloid sarcoma formation induced by cooperation of MLL/AF10 with activating KRAS mutation in a mouse transplantation model: Hoxa11 in myeloid sarcoma formation. *Neoplasia* **2022**, *29*, 100802. [CrossRef] [PubMed]
98. Halahleh, K.; Alhalaseh, Y.; Al-Rimawi, D.; Da'na, W.; Alrabi, K.; Kamal, N.; Muradi, I.; Abdel-Razeq, H. Extramedullary acute myeloid leukemia (eAML): Retrospective single center cohort study on clinico-pathological, molecular analysis and survival outcomes. *Ann. Med. Surg.* **2021**, *72*, 102894. [CrossRef]
99. Choi, K.H.; Song, J.H.; Kwak, Y.K.; Lee, J.H.; Jang, H.S. Analysis of PET parameters predicting response to radiotherapy for myeloid sarcoma. *PLoS ONE* **2021**, *16*, e0261550. [CrossRef]
100. Creutzig, U.; van den Heuvel-Eibrink, M.M.; Gibson, B.; Dworzak, M.N.; Adachi, S.; de Bont, E.; Harbott, J.; Hasle, H.; Johnston, D.; Kinoshita, A.; et al. Diagnosis and management of acute myeloid leukemia in children and adolescents: Recommendations from an international expert panel. *Blood* **2012**, *120*, 3187–3205. [CrossRef]
101. Takahashi, S. Current Understandings of Myeloid Differentiation Inducers in Leukemia Therapy. *Acta Haematol.* **2021**, *144*, 380–388. [CrossRef]
102. Melnick, A.M. Epigenetics in AML. *Best Pract. Res. Clin. Haematol.* **2010**, *23*, 463–468. [CrossRef] [PubMed]
103. Shabanova, I.; Cada, M.; Johnston, D.L.; Abbott, L.S.; Leung, E.W.; Schechter, T.; Dror, Y.; Klaassen, R.J. Reduction of Extramedullary Complications in Patients with Acute Myeloid Leukemia/Myelodysplastic Syndrome Treated with Azacitidine. *J. Pediatr. Hematol. Oncol.* **2020**, *42*, 170–174. [CrossRef] [PubMed]
104. Antar, A.; Otrock, Z.K.; Kharfan-Dabaja, M.; Salem, Z.; Aractingi, S.; Mohty, M.; Bazarbachi, A. Azacitidine in the treatment of extramedullary relapse of AML after allogeneic hematopoietic cell transplantation. *Bone Marrow Transplant.* **2013**, *48*, 994–995. [CrossRef] [PubMed]
105. Okamoto, H.; Kamitsuji, Y.; Komori, Y.; Sasaki, N.; Tsutsumi, Y.; Miyashita, A.; Tsukamoto, T.; Mizutani, S.; Shimura, Y.; Kobayashi, T.; et al. Durable Remission of Chemotherapy-Refractory Myeloid Sarcoma by Azacitidine. *Tohoku J. Exp. Med.* **2021**, *254*, 101–105. [CrossRef]
106. Gornicec, M.; Wölfler, A.; Stanzel, S.; Sill, H.; Zebisch, A. Evidence for a role of decitabine in the treatment of myeloid sarcoma. *Ann. Hematol.* **2017**, *96*, 505–506. [CrossRef]
107. Singh, S.N.; Cao, Q.; Gojo, I.; Rapoport, A.P.; Akpek, G. Durable complete remission after single agent decitabine in AML relapsing in extramedullary sites after allo-SCT. *Bone Marrow Transplant.* **2012**, *47*, 1008–1009. [CrossRef] [PubMed]
108. Niscola, P.; Abruzzese, E.; Trawinska, M.M.; Palombi, M.; Tendas, A.; Giovannini, M.; Scaramucci, L.; Cupelli, L.; Fratoni, S.; Noguera, N.I.; et al. Decitabine treatment of multiple extramedullary acute myeloid leukemia involvements after essential thrombocytemia transformation. *Acta Oncol.* **2017**, *56*, 1331–1333. [CrossRef]
109. Shatilova, A.; Girshova, L.; Zaytsev, D.; Budaeva, I.; Mirolyubova, Y.; Ryzhkova, D.; Grozov, R.; Bogdanov, K.; Nikulina, T.; Motorin, D.; et al. The myeloid sarcoma treated by Venetoclax with hypomethylating agent followed by stem cell transplantation: Rare case report. *BMC Womens Health* **2021**, *21*, 184. [CrossRef]

110. Otoukesh, S.; Zhang, J.; Nakamura, R.; Stein, A.S.; Forman, S.J.; Marcucci, G.; Pullarkat, V.; Aldoss, I. The efficacy of venetoclax and hypomethylating agents in acute myeloid leukemia with extramedullary involvement. *Leuk. Lymphoma* **2020**, *61*, 2020–2023. [CrossRef]
111. Pan, W.; Zhao, X.; Shi, W.; Jiang, Z.; Xiao, H. Venetoclax induced complete remission in extramedullary relapse of AML co-harboring NPM1, TET2, and NRAS mutations after haploidentical hematopoietic stem cell transplantation. *Leuk. Lymphoma* **2020**, *61*, 2756–2759. [CrossRef]
112. Farrar, J.E.; Bolouri, H.; Ries, R.E.; Triche, T.J.; Lim, E.L.; Alonzo, T.A.; Ma, Y.; Moore, R.; Mungall, A.J.; Marra, M.A.; et al. Marked Differences in the Genomic Landscape of Pediatric Compared to Adult Acute Myeloid Leukemia: A Report from the Children's Oncology Group and NCI/COG Therapeutically Applicable Research to Generate Effective Treatments (TARGET) Initiative. *Blood* **2016**, *128*, 595. [CrossRef]

cancers

MDPI

Review

The Battlefield of Chemotherapy in Pediatric Cancers

Letao Bo [1], Youyou Wang [1], Yidong Li [1], John N. D. Wurpel [1], Zoufang Huang [2,*] and Zhe-Sheng Chen [1,3,*]

1 Department of Pharmaceutical Sciences, College of Pharmacy and Health Sciences, St. John's University, Queens, NY 11439, USA
2 Ganzhou Key Laboratory of Hematology, Department of Hematology, The First Affiliated Hospital of Gannan Medical University, Ganzhou 341000, China
3 Institute for Biotechnology, St. John's University, Queens, NY 11439, USA
* Correspondence: nfyyjsjj@126.com (Z.H.); chenz@stjohns.edu (Z.-S.C.); Tel.: +86-138-797-27439 (Z.H.); +1-718-990-1432 (Z.-S.C.); Fax: +1-718-990-1877 (Z.-S.C.)

Simple Summary: The survival rate for pediatric cancers has improved significantly over the last decades. Conventional chemotherapies play a vital role in pediatric cancer treatment, especially in low- and middle-income countries, and the roster of chemo drugs for use in children has expanded. However, patients suffer from chemotherapy as a result of its countless side effects. Furthermore, multidrug resistance (MDR) continues to be an insurmountable obstacle that limits survival for a considerable number of patients. In this review, we discuss severe side effects in pediatric chemotherapies such as doxorubicin-induced cardiotoxicity (DIC) and vincristine-induced peripheral neuropathy (VIPN). Here, MDR mechanisms in chemotherapy are elucidated with the aim of improving survival, while also reducing the intensity and toxicity of chemotherapy. Furthermore, we focus on various drug transporters in common types of pediatric tumors, which could provide different therapeutic strategies.

Abstract: The survival rate for pediatric cancers has remarkably improved in recent years. Conventional chemotherapy plays a crucial role in treating pediatric cancers, especially in low- and middle-income countries where access to advanced treatments may be limited. The Food and Drug Administration (FDA) approved chemotherapy drugs that can be used in children have expanded, but patients still face numerous side effects from the treatment. In addition, multidrug resistance (MDR) continues to pose a major challenge in improving the survival rates for a significant number of patients. This review focuses on the severe side effects of pediatric chemotherapy, including doxorubicin-induced cardiotoxicity (DIC) and vincristine-induced peripheral neuropathy (VIPN). We also delve into the mechanisms of MDR in chemotherapy to the improve survival and reduce the toxicity of treatment. Additionally, the review focuses on various drug transporters found in common types of pediatric tumors, which could offer different therapeutic options.

Keywords: chemotherapy; pediatric cancers; VIPN; DIC; MDR; drug transporters

Citation: Bo, L.; Wang, Y.; Li, Y.; Wurpel, J.N.D.; Huang, Z.; Chen, Z.-S. The Battlefield of Chemotherapy in Pediatric Cancers. *Cancers* **2023**, *15*, 1963. https://doi.org/10.3390/cancers15071963

Academic Editor: David Wong

Received: 10 February 2023
Revised: 12 March 2023
Accepted: 21 March 2023
Published: 24 March 2023

1. Introduction

Pediatric cancer is relatively rare and has a high survival rate compared with adult cancer. Pediatric cancer includes 12 major types and over 100 subtypes [1]. In the United States, pediatric cancer therapies have made remarkable advances over the last 70 years, and the overall survival (OS) rate for pediatric cancer patients has increased to 80% [2,3]. However, cancer is still the top cause of death by disease in children. In 2023, diagnosed pediatric cancer patients in the USA will include ~9910 children (birth to age 14) and ~5280 adolescents (aged 15–19 years). In addition, approximately 1040 children and 550 adolescents in the USA will die from cancer in 2023 [4]. An improvement in these amounts can be achieved by an increased understanding of the molecular basis of cancer through international collaborative efforts.

Despite these successes, the incidences and mortalities between countries are greatly different. In 2017, the 5-year survival rate of childhood cancer was 80% in high-income countries. However, data suggest that far fewer children survive in most low- and middle-income countries—the 5-year survival rate of children is ~40%, while more than 90% of children at risk of potential pediatric cancer live in these countries [5,6]. Because of major advances in modern science over the past several decades, the improvement in pediatric cancer in high-income countries has not translated to most low- and middle-income countries. There are still gaps among these countries regarding recognition, diagnosis, and treatment [7].

Despite financial hardship or treatment results, conventional chemotherapy is still a standard option for childhood cancer treatments in low- and middle-income countries, where molecular-targeted drugs are broadly accessible [8–10]. Severe side effects and resistance mechanisms against cancer therapies remain the primary reasons limiting chemotherapy outcomes [11]. Here, we consider the advancement of chemotherapy in pediatric cancer, including drugs and specific considerations. Beyond the previous focus on adult cancer alone, in the current review, we discuss severe side effects such as neuropathy and myopathy, the effects of the MDR phenomenon, and the role of various drug transporters in common types of pediatric tumors that could lead to more efficient pediatric chemotherapy.

2. Current Chemotherapy in Pediatric Cancer

There are currently various cancer treatment approaches for cancer. The main four lines of treatment are surgical removal, immunotherapy, radiotherapy, and chemotherapy [12]. Surgical removal has the longest history in cancer treatment. It is an essential treatment option and often is often performed together with other cancer treatments, depending on the type of cancer [13]. Immunotherapy is generally considered to have been applied in 1890 by Dr. Coley, who injected a mixture of live and inactivated *Streptococcus* and *Serratia* to achieve responses [14]. It modulates the immune system to act better against cancer, and it includes immune checkpoint inhibitors, T-cell transfer therapy, monoclonal antibodies, and treatment vaccines. Radiotherapy precisely delivers radiation to kill cancer cells and shrink tumors, and was started in 1896 by Emile Grubbé, who treated a breast cancer patient using X-rays [15]. Chemotherapy was first introduced after the Second World War, and works by interfering with cell proliferation. Other cancer treatment approaches include stem cell transplantation, targeted therapy, photodynamic therapy, and hormone therapy.

The efficient treatment of pediatric cancer started after 1945. Sidney Farber treated a 3-year-old patient with acute lymphoblastic leukemia (ALL) with aminopterin [16]. As a conventional cancer therapy, chemotherapy has played an important role in pediatric cancer over the past several decades. In pediatric diagnosis, ALL is the most common type of cancer. The OS of ALL increased from 57% in the 1970s to 96% in recent years [17]. This enormous progress is partly attributed to advances in chemotherapy [3]. Currently, chemotherapy has reached its limit, as severe toxicities have been observed in pediatric patients during treatment. Modifications to chemotherapy are made in order to reduce multiple toxicities. For instance, studies have provided evidence that chronomodulated chemotherapy, based on the body's intrinsic circadian clock, might minimize toxicity while maintaining an anticancer activity [18]. High-dose chemotherapy has been explored for treating childhood malignant glioma so as to pass the blood–brain barrier (BBB), reduce cell chemoresistance, and achieve a more comprehensive response [19].

Chemo drugs are grouped by function, chemical structure, and interaction with other medications. Types of chemo drugs include alkylating agents, antimetabolites, antibiotics, topoisomerase inhibitors, mitotic inhibitors, and corticosteroids [20,21]. In 1997, under the Food and Drug Administration Modernization Act2 (FDAMA), a pediatric exclusivity provision was enacted by the U.S. Congress. Few chemo drugs were explicitly approved for treating pediatric cancer. The FDAMA later reauthorized as the Best Pharmaceuticals Act for Children (BPCA), which encourages pediatric drug studies in companies by providing a financial incentive. A total of 16 chemo drugs have been approved by the FDA with pediatric indications in the U.S. Before the FDAMA, only nine chemo drugs were approved for pediatric indications, see Table 1. After the FDAMA (1997–2022), seven chemo drugs were approved to treat childhood cancer (see Table 2).

There are less drugs with pediatric indications worldwide compared with adult indications [22]. For some chemo drugs, the initial approval was only for adult indications. On 20 May 2022, Azacitidine was newly approved for childhood cancers, while it first approved on 19 May 2004 for treating all subtypes of myelodysplastic syndrome. Moreover, FDA approved these chemo drugs (Clofarabine, Nelarabine, Erwinia, and Mercaptopurine) for both adult and pediatric indications at the same time (Table 2). Although efforts have been made in pediatric chemotherapy, only a few chemo drugs have gained pediatric indications—there is a gap between pediatric and adult approval.

The therapy to treat pediatric cancers requires specific prerequisites and considerations, as patients are still growing and developing. Some adverse effects of cancer treatments are more severe for children than adults, as developing organs are more susceptible. Infection-related serious adverse events are more common in ALL patients receiving chemotherapy, which enhances the risk of death [23]. Therefore, it is necessary to identify drug resistance mechanisms to reduce chemotherapy intensity and toxicity. The counterstrategies against drug resistance are discussed in the following section.

Table 1. Chemo drugs approved prior to FDAMA, for which labeling includes pediatric indications [24].

Drug	Original Approval	Indications
Doxorubicin Hydrochloride	7 August 1974	Wilm's Tumor and Other Childhood Kidney Cancers
Vincristine Sulfate	10 July 1963	ALL, Neuroblastoma, Non-Hodgkin Lyphoma, Rhabdomyosarcoma, Wilm's tumor and other childhood kidney cancers
Cytarabine	17 June 1969	Acute Nonlymphocytic Leukemia
Cyclophosphamide	16 November 1959	ALL
Methotrexate Sodium (Trexall)	10 August 1959	ALL
Mercaptopurine (Purinethol, Purixan)	11 September 1953	ALL
Daunorubicin Hydrochloride (Rubidomycin)	19 December 1979	ALL
Procarbazine Hydrochloride (Matulane)	22 July 1969	Hodgkin Lymphoma
Dactinomycin (Cosmegen)	10 December 1964	Ewing sarcoma, gestational trophoblastic disease

Data provided by National Cancer Institute (https://www.cancer.gov/), accessed on 30 January 2023, updated: 20 December 2022, and U.S. FDA (https://www.fda.gov/). The drugs and drug combinations are not listed here.

Table 2. Chemo drugs approved post FDAMA with pediatric specific indications (1997–2022) [24].

Drugs	Original Approval	Pediatric Approval	Indications for Pediatric Cancer
(Drugs approved post FDAMA with pediatric specific indications (1997–2022))			
Azacitidine (Vidaza)	19 May 2004	20 May 2022	JMML
Calaspargase Pegol-mknl (Asparlas)	same	20 December 2018	ALL
Everolimus	1 November 2010	29 August 2012	Giant Cell Astrocytoma
Asparaginase Erwinia Chrysanthemi (Erwinaze)	same	18 November 2011	ALL
Clofarabine (Clolar)	same	28 December 2004	ALL
Pegaspargase (Oncaspar)	same	24 July 2006	ALL
Nelarabine (Arranon)	same	28 October 2005	Non-Hodgkin Lymphoma

Data provided by National Cancer Institute (https://www.cancer.gov/), updated: 20 December 2022, and U.S. FDA (https://www.fda.gov/). JMML: juvenile myelomonocytic leukemia. The drugs and drug combinations are not listed here.

3. Severe Neuropathy and Myopathy Side Effects in Chemotherapy

In cancer treatment, chemotherapeutic agents are considered a double edge sword, as it kills cancer cells and other healthy, fast-growing cells, without differentiation. As a matter of fact, patients suffer from various non-negligible side effects such as hair loss, fatigue, nausea, vomiting, and diarrhea. Currently, doxorubicin (DOX) and vincristine (VCR) are two commonly used chemotherapeutic agents in pediatric cancer treatment [25,26]. However, the potential harm of neuropathy and myopathy has become a considerate risk factor that limits the effect of DOX and VCR [27,28]. The pathway mechanism of neuropathy and myopathy induced by DOX and VCR in illustrated in Figure 1. Thus, it is necessary to discuss the opposing side and corresponding solutions for these two chemo drugs, mainly when applied to pediatric cancer patients.

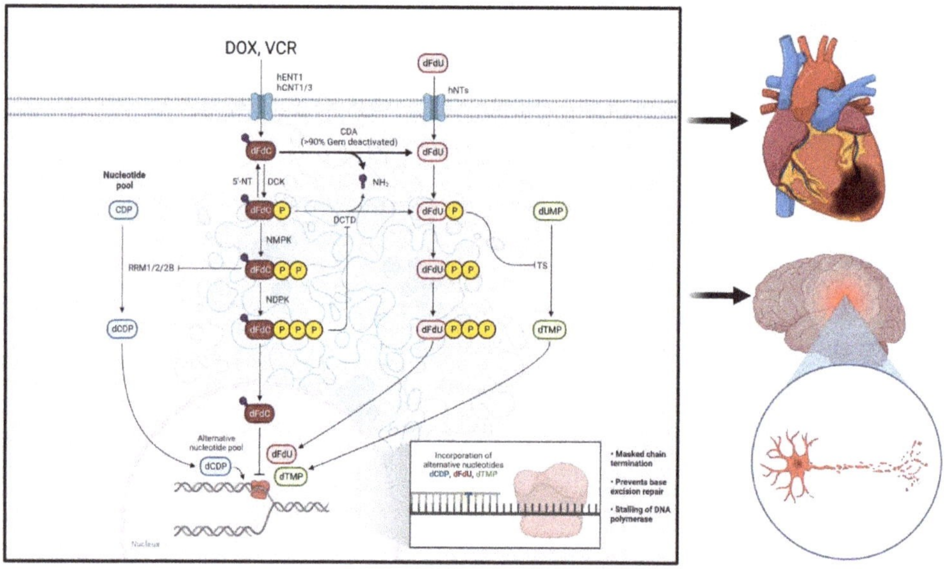

Figure 1. Neuropathy and myopathy induced by two commonly used chemotherapeutic agents (DOX and VCR) in pediatric cancer treatment.

DOX, an anthracycline drug to treat solid tumors in children, such as Wilm's tumor, was approved by the FDA in 1974 to become the first clinical liposomal encapsulated anticancer drug [29]. DOX acts as an intercalation that inhibits topoisomerase II and further obstructs DNA replication [30]. The recommended dosage (2 mg/mL) is the same in both pediatric and adult patients. Symptoms of the side effects may include abnormality in electrocardiography, rhythm disturbances, and even left ventricular hypertrophy that leads to a reduction in ejection fraction [31].

DIC is a severe dose-dependent risk factor in cancer treatment [32]. It has been reported that SNP (rs2229774) in retinoic acid receptor-γ (RARG) has a significant impact on the increased occurrence of DIC. Further investigation found that a RARG agonist CD1530 showed a cardioprotective effect in an in vivo mouse model of DIC [33]. Other drug transporter genomic variants in the *SLC28A3* locus were also identified. Based on that, a single dosage of 3 µmol/L desipramine per day before the administration of DOX is recommended to prevent DIC [34]. Moreover, circular RNA (circRNA)-based therapy is a novel approach for treating DIC [35]. The overexpression of insulin receptor encoded circRNA (Circ-INSR) successfully prevented and reversed DIC in an in vivo mouse model [36]. Han. et al. reported another promising therapeutic target for DIC, namely tumor-suppressive human circular RNA CircITCH [37]. Other recent potential strategies that possibly reverse DIC include *atg7*-based autophagy activation [38], meteorin-like (METRNL) protein [39], sirtuin 1 (SIRT1) [40], berberine [41], ADAR2 [42], elabela (ELA) [43], phenylalanine-butyramide (FBA) [44], gasdermin D [45], melatonin [46], levosimendan [47], paeonol [48], SNX17 [49], irisin [50], isorhapontigenin [51], liensinine [52], etc.

VCR, known initially as Leurocristine, is the first-line chemotherapeutic medication often administered in the combination chemotherapeutic treatment of pediatric hematologic malignancies and solid tumors [53]. VCR acts as mitotic inhibitors by binding to the β-tubulin subunit of αβ-tubulin heterodimers, thus functionally destabilizing microtubule fibers, which ultimately leads to the termination of cancer cell division [54]. Previous studies have shown a cumulative effect for its neurotoxicity and overdosage may cause very serious or fatal outcomes [55,56]. Until 6 June 2022, it has been approved by FDA in the treatment of pediatric ALL, neuroblastoma, non-Hodgkin lymphoma, rhabdomyosarcoma, Wilm's tumor, and other childhood kidney cancers. USP recommended dose of VCR injection for pediatric patients is 1.5–2 mg/m^2 compared to adults is 1.4 mg/m^2. For neonate and infant patients weighing 10 kg or less, the first dose should be 0.05 mg/kg, administered once a week. The latest research showed in the treatment of ALL and Wilm's tumors, compared with older children, neonates or infants have similar clearance in vincristine. Thus, doses less than 0.05 mg/kg should not be applied in neonate and infant patients due to inappropriate suboptimal VCR exposures [57].

The common side effect encountered with VCR use is VIPN [58]. Clinical patterns of VIPN can be classified as sensory neuropathy (e.g., numbness and paresthesia), motor neuropathy (e.g., extremity weakness and walking difficulties), and autonomic neuropathy (e.g., constipation and urinary retention) [59]. VIPN can be assessed by the National Cancer Institute Common Terminology Criteria for Adverse Events (CTCAE), the pediatric-modified Total Neuropathy Scale (ped-m TNS), and the Total Neuropathy Score-Pediatric version (TNS-PV) [28,60]. There are currently no effective strategies for reducing vincristine-induced neurotoxicity. Also, whether VIPN exists chronically in the survivors remains ambiguous [61]. Some studies showed that a significant proportion of patients receiving VCR would undergo a certain extent VIPN [62,63]. In 2017, Tay et al. reported that ~16% of pediatric ALL survivors suffer from VIPN [64]. Nevertheless, in 2020, an investigation among 150 pediatric patients with ALL and Wilm's tumors showed that significant side effects of the vincristine regimen are mostly neurotoxic, which is at a mild to moderate level [65]. In recent years, more and more studies have revealed the deep connection between genetic polymorphisms and VIPN [66–68]. The *CEP72* genetic variant is an optimistic VIPN signature marker since the *CEP72* gene encodes centrosome proteins that participate in the development of microtubules. Patients that carry *CEP72* TT genotype take potentially higher risk and severity of VIPN than CC or CT genotype patients [69]. Besides, the enzyme CYP3A5 contributes to hepatic clearance of VCR, which means CYP3A5 genetic polymorphisms, and its allelic variants are assumed to be associated with VCR neurotoxicity in different human populations. CYP3A5 genotyping analysis results showed over 70% of African Americans had been found to have one or more CYP3A5*1 alleles (such as CYP3A5 expresser), which is about five times higher compared to Caucasians [70]. Moreover, VCR is transported by some members of ATP-binding cassette (ABC) transporter superfamily such as ABCB1, ABCC1, ABCC2, and ABCB4 [71]. Lopez-Lopez et al. reported that ABCC1 is the critical mediator acts transporting VCR into the blood, while ABCB1 and ABCC2 are indispensable in the biliary excretion of VCR. The genotypes rs3740066 GG and rs12826 GG of ABCC2 were identified with significant associations with increased VIPN, which suggested that ABCC2 polymorphism could be used as a potential biomarker for VIPN in screening and diagnosis of pediatric ALL [72].

Recently, numerous updates have been related to the strategy to overcome VIPN. Zhou et al. reported levo-corydalmine ameliorates VIPN in mice by inhibiting Cx43 expression and NFκB-dependent CXCL1/CXCR2 signalling pathway [73,74]. Also, both VIPN and tumor growth were alleviated by the inhibition of histone deacetylase 6 (HDAC6) in mice. Other prospective targets that possibly prevent VIPN include mitoquinone [75], puerarin [76], nerve growth factor (NGF) monoclonal antibody DS002 [77], minocycline [78], bergapten [79], etc.

4. MDR: The Challenge in Pediatric Cancer Chemotherapy

Cancer cells can develop resistance to one chemotherapeutic drug, as well as other chemotherapeutic drugs that may have different chemical properties and mechanisms of action, which is called multi-drug resistance (MDR) [80]. MDR can occur in both adult and pediatric cancer chemotherapy. Although the mechanisms of MDR have been studied for a few decades, MDR is a very limiting factor to the success of cancer chemotherapies. MDR exists not only in chemotherapy and radiation therapy, but also in newly developed therapies such as targeted therapy and immunotherapy [81,82]. Drug resistance can be divided into intrinsic resistance or extrinsic resistance based on the cause of its occurrence. Cancer cells may have inherent drug resistance before receiving chemotherapy. However, various adaptive responses of cancer cells during the treatment cause extrinsic or acquired drug resistance of cancer cells. Several MDR mechanisms have been discovered, yet they are not yet fully understood. As shown in Figure 2, the factors of MDR include increased drug inactivation, reduction of influx and increased efflux of drugs, decreased activation of prodrugs, epigenetic dysregulation, changes in cell surface markers, tumor microenvironment (TME), epithelial–mesenchymal transition (EMT), altered miRNA, disruption of responses to DNA damage, inhibition of apoptotic pathways, tumor heterogeneity, and cancer stem cells [80,83,84].

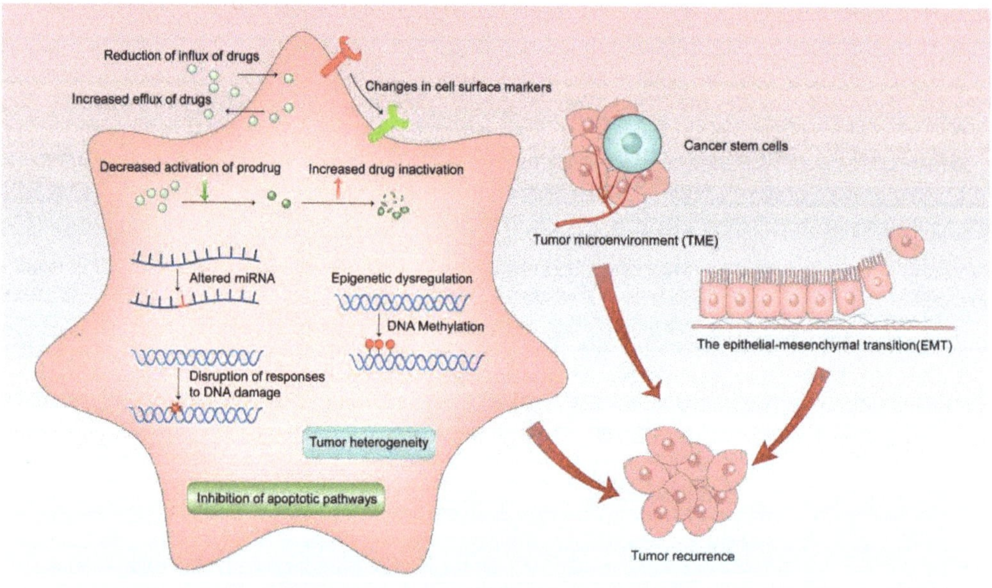

Figure 2. Mechanisms of MDR in cancer treatment. MDR occurs as a result of various causes, which ultimately lead to tumor recurrence.

5. MDR-Related Drug Transporters and Their Roles in Pediatric Cancers

Drug transporters are membrane proteins involved in the absorption, distribution, and excretion of drugs. Two transporter superfamilies have been identified in humans: the solute carrier (SLC) superfamily and the ABC superfamily [85]. ABC transporters are generally involved with the efflux of drugs, and SLC transporters have been chiefly described as influx transporters [86]. In addition to transporting therapeutic drugs across membranes, these transporters also mediate the transport of endogenous compounds.

There is considerable interest in transporters from both families, as they are known to confer MDR to cancer cells.

5.1. SLC Family Transporters

SLC family transporters are a family that includes more than 300 membrane-bound proteins involved in the influx and efflux of a wide array of substrates, such as ions, metabolites, and drugs [87]. SLC transports substrates based on the electrochemical potential difference between the biological membrane or the ion gradient originally generated by the primary active transporters. Genetic variants of SLC transporters and clinical outcomes of methotrexate (MTX) have been studied in pediatric patients with ALL [88]. Although SLC transporter families are essential in human health, related studies focusing on pediatric cancer therapy are rare. We limited our discussion to ABC transporters.

5.2. ABC Transporters

The ABC transporter superfamily, one of the most prominent transporter families, is responsible for MDR by mediating drug efflux, which subsequently leads to a low intracellular concentration of antineoplastic agents in cancer cells and deteriorates therapeutic outcome [89]. In addition, more than efflux pumps, other critical roles in cancer development of this transporter superfamily have been revealed step by step [90–94]. So far, 7 subfamilies (ABC-A to ABC-G) and at least 48 additional subfamily members have been found and characterized depending on their structural differences and similarities [95,96]. Among these subfamily members, three members have been demonstrated that are closely related to MDR in chemotherapy, including P-glycoprotein (P-gp/ABCB1/MDR1), multidrug resistance protein 1 (MRP1/ABCC1), and breast cancer resistance protein (BCRP/ABCG2) [97,98].

5.2.1. MDR1

The ABCB subfamily involves four full transporters (ABCB1/4/5/11) and 7 half transporters (ABCB2/3/6/7/8/9/10) that can transport a vast variety of molecules, including peptides, drugs, and ions [99]. Half transporters include two polypeptides, each having a transmembrane binding domain (TMD) and (nucleotide-binding domains) NBD to form a homo- or hetero-dimer. Full transporters are characterized as all four domains reside on a single polypeptide [97].

P-gp (also named ABCB1 or MDR1) is the first discovered and well-studied ABCB subfamily transporter that mediates MDR in cancer cells [100]. It is most expressed in the blood–brain barrier, liver, placenta, gallbladder, and endocrine tissues. In 2020, Nosol et al. revealed the protein structure of P-gp and found that inhibitors are bound in pairs and interact with structural features to block the function of P-gp [101]. As P-gp is the substrate of a broad range of antineoplastic agents, overexpressed P-gp has been demonstrated to lead to the development of drug resistance, including chemotherapeutic agents Vinca alkaloids (vinblastine and vincristine), Taxanes (paclitaxel and docetaxel), Anthracyclines (doxorubicin, daunorubicin, and epirubicin), and imatinib mesylate [102,103]. Meanwhile, a high-level expression of P-gp that causes cancer drug resistance has also been found in some tyrosine kinase inhibitors (Imatinib [104], GSK1070916 [105], and WYE354 [106]).

Altered expression levels of MDR1 have been detected in various tumors, such as neuroblastoma, rhabdomyosarcoma, and Wilm's tumor. Neuroblastoma is the most common pediatric solid tumor, which accounts for 7–8% of childhood malignancies and 15% of all pediatric cancer deaths [107]. A study showed that the mRNA expression of MDR1 was increased in neuroblastoma patients with previous chemotherapy [108]. In addition, Qiu et al. [109] found that MDR1 hypermethylation expression can be associated with the pathogenesis and progression of neuroblastoma. MDR1 expression was observed to increase after chemotherapy in rhabdomyosarcoma and Wilm's tumor [110]. These findings may be helpful to understand the role of MDR1 in pediatric malignancies regarding drug resistance and allow researchers to come up with strategies for therapeutic intervention.

5.2.2. MRPs

Multidrug resistance proteins (MRPs) include 9 transporters from 13 members in the ABCC subfamily due to their ability to mediate cancer MDR [111].

Although each of the MRPs have slight differences in structures and amino acid compositions, the mechanism of transport driven by ATP hydrolysis is much the same. Unlike P-gp, which extrudes mostly xenobiotics, MRPs account for the extruding of both endo- and xenobiotics, thus showing its crucial role in regulating MDR processes in cancer development [112]. The structure of MRP1 has shown a novel substrate recruitment mechanism in that substrates are recruited straight from the cytoplasm, whereas P-gp attaches substrates from the inner leaflet of the lipid bilayer [113]. MRPs are distributed in the human body in various tissues, including the blood−brain barrier, brain, lung, kidney, liver, etc. [111].

The expression of MRPs has been investigated in several pediatric malignant tumors as they are a vital factor causing cytotoxic drug resistance and chemotherapy failure. Abnormal MDR expressions have been observed in pediatric malignancies, such as ALL, neuroblastoma, rhabdomyosarcoma, Wilm's tumor, and retinoblastoma [114]. After chemotherapy, MRP1 expression has been observed to be upregulated in neuroblastoma, hepatoblastoma, and rhabdomyosarcoma patients [110]. Increased expression of MRP2-6 and decreased expression of MRP1 and MRP10 have been observed in ALL patients, which are associated with high doses of three chemotherapies [115]. Henderson et al. found that the inhibition of MPR1 is associated with reduced neuroblastoma development in transgenic mice [116]. These studies indicate their role in the chemotherapeutic drug efflux of MPRs and cancer prognosis.

5.2.3. BCRP

The human BCRP has 665 amino acid residues with a molecular weight of 72 kDa. It is a half-transporter that is prominently expressed in various tissues, including, but not limited to, the brain, placenta, testis, liver, breast, and BBB [117]. A high overexpression of BCRP can be observed in different drug resistance cancer types, including in solid tumors and hematopoietic malignancies [118,119]. Although the clinical significance of BCRP-mediated drug resistance remains unclear, many studies have shown strong evidence to support that modulating the expression of BCRP could enhance drug sensitivity in chemotherapy [120,121].

BCRP was first identified in 1998, followed by expression studies to explore its potential role in chemoresistance [122]. The expression of the *BCRP* gene in childhood ALL has been observed at a low expression level [123]. However, compared with a diagnosis when the co-expression of BCRP and MDR1 was observed, a higher RNA level of BCRP was expressed at the relapsed/refractory state in acute myeloid leukemia (AML) [124]. Correlations between BCRP and MRPs have also been reported. The combined high expression of BCRP and MRP4 is correlated with reduced antileukemia drug methotrexate accumulation. Similarly, evaluated expression of the *BCRP* gene was found in primary neuroblastoma mitoxantrone-resistant cells [125]. These results underscore the potential value of BCRP as a predictor of chemoresistance drug efflux.

Essentially, MDR can arise because of altered targeted proteins or cell signaling pathways. Changes in cellular or non-cellular processes is also a significant factor of MDR. However, most encountered MDRs are related to drug efflux, which is mainly caused by ABC transporters. Once MDR transporters are overexpressed, the efflux of a chemo drug can increase. For example, the overexpression of P-gp, MRP1, and BCRP decreases the chemosensitivity of cancer cells by limiting exposure to anticancer drugs [126]. The overexpression of P-gp in cancer cells is associated with increased drug resistance to DOX and paclitaxel [127,128]. After anticancer agent treatment, the overexpression of P-gp has been found in acute myeloid leukemia [129]. Decreased chemosensitivity and a high expression of P-gp and BCRP were noticed in MDR patients with chronic lymphocytic leukemia, metastatic breast cancer, and multiple myeloma [130]. BCRP is regulated by proteins such as TGF-β1 and VEGFR-2 [131,132].

6. Tackling Strategies Regarding Drug Transporters

MDR is one of the barrier mechanisms against chemotherapy and thus is considered a key factor leading to the failure of chemotherapeutics. The overexpression of MDR1 that can enhance the efflux of cytotoxic agents is one of the targets to improve the effect of chemotherapeutics. One of the strategies is to reverse resistance mechanisms [133]. Several P MDR1 inhibitors have been investigated and shown successful MDR reversal on a drug-resistant prostate cancer cell line, without exhibiting a toxic potential [134]. Lei et al. [135] observed intracellular accumulation of paclitaxel and decreased drug efflux activity in the MDR1-knockout colorectal cancer cell line.

Furthermore, applying a nano-drug delivery system (NDDS) is a practical approach to enhance chemotherapy validity due to the targeted co-delivery, reduced sides affected, and long-time blood circulation achieved [136]. Curcumin has an antitumor activity and reverses the tumor MDR effect by regulating the MDR1 protein [137]. Degradable poly (lactic-co-glycolic acid) (PLGA) nanoparticles coloaded with curcumin and DOX could directly target cells or xenografted tumors and inhibit the growth of DOX-resistant esophageal carcinoma with a high biosafety [138]. NDDS-containing compounds have been developed that can inhibit MDR1, including tariquidar (XR9576), tetrandrine, verapamil, and cyclosporin A [139].

Therapy regimens can be different for treating children and adults as they have different drug-resistance profiles. The triple combination of fludarabine, ara-C, and G-CSF has been used in the treatment of childhood AML and caused an additive cell kill [140]. Tipifarnib can target the malignancies, such as leukemia, by activating RAS proteins (HRAS, KRAS, and NRAS) [141].

7. Special Considerations of Resistance in Pediatric Cancer Treatment

Pediatric malignancies have significant differences in their treatment compared with adult tumors, and thus require special considerations. Pan-cancer analyses have shown that pediatric cancers have comparatively lower mutation frequencies compared with adult cancers [142]. Epigenetic dysregulation, however, seems to be a particular factor in many types of pediatric cancers [143–145]. These genetic and non-genetic changes suggest therapeutic implications regarding chemotherapy.

Children with ALL have better prognoses and outcomes than adult patients with ALL. The 5-year OS rate is 87% for children aged 0–15 years, as opposed to 44% for adults aged 20–29 years [146]. A plethora of factors are responsible for the different outcomes, including socio-economic factors, resistance, disease heterogeneity, host responses, therapeutic treatment, etc. [147]. Resistance is one of the main factors leading to variables among ALL patients of different ages. For example, the activation of P-gp has a higher expression in adults [148]. In addition, the accumulated mutations in the *p53* gene and lower methotrexate polyglutamate may also contribute to the differences in the responses of drug resistance mechanisms [148,149]. Genetic lesions of polycomb repressor complex 2 (PRC2) have been reported in pediatric T-cell ALL (T-ALL), which promotes mutations of the IL7R/JAK/STAT pathway. Poor prednisone response and persistent MRD have been connected to adult T-ALL patients with loss-of-function alterations of PRC2 [150]. Genescà et al. reported that a complex karyotype (≥3 cytogenetic alterations) in adult T-ALL was associated with a minimal residual disease (MRD) level, but no correlation regarding the prednisone response [151]. Protein tyrosine phosphatase nonreceptor type 2 (PTPN2) is a phosphatase suppressing a gene in T-cell ALL. Deletions of PTPN2 in pediatric patients were associated with a higher glucocorticoid response and improved survival in children, yet these trends were not found in adults [149]. Other studies have suggested that a single subclone with additional mutations confers resistance to therapy, although half of the leukemia patients had multiple subclonal mutations [152]. Because children are not simply small adults, the difference in outcomes across age categories should be considered in diagnosis and therapy.

8. Clinical Trials and Recommendations for Risk Surveillance

With typically more aggressive protocols being used in children than in adults, chemotherapy is the standard treatment in cancers such as ALL, AML, and Hodgkin lymphoma [153]. The number of clinical trials related to chemotherapy-induced side effects is on the rise.

Notably, VIPN is being studied in two clinical trials. A Phase 4 clinical trial (NCT02923388) is testing Vitamin B12 and vitamin B6 in ALL patients treated with VCR.

The results from a study of 102 patients showed that vitamin B6 and B12 significantly reduced the incidence, relative risk, and severity of VIPN. The amount needed to treat was encouragingly low, and vitamin B6 and B12 were recommended as promising neuroprotective agents against VIPN [154]. A clinical trial (NCT02796365) is currently underway to evaluate the effectiveness of exercise rehabilitation as a preventive measure to DIC. The studies are summarized in Table 3.

Table 3. Clinical trials of DIC and VIPN.

Toxicity	Study Title	NCT Identifier	Phase	Patient Number	Disease	Status	Treatment/Method
DIC *	Protective Role of Vitamin D in Breast Cancer Patients Treated with Doxorubicin	NCT04166253	Phase 2	100	Breast cancer	Completed	Vitamin D
DIC	99mTc-rhAnnexin V-128 Imaging and Cardiotoxicity in Patients with Early Breast Cancer	NCT02677714	Phase 2	14	Breast cancer	Terminated	Radiation: 99mTc-rhAnnexin V-128
DIC	Prevention Using Exercise Rehabilitation to Offset Cardiac Toxicities Induced Via Chemotherapy (HF-PROACTIVE)	NCT02796365	Not Applicable	29	Breast cancer, Gastric cancer, Leukemia	Completed	Exercise
DIC	Evaluation of Myocardial Injury After Anthracycline Chemotherapy in Osteosarcoma Patients Using CMR	NCT04461223	Not Applicable	55	Osteosarcoma Myocardial Injury	Unknown	Contrast-enhanced cardiac magnetic resonance imaging, observational Study
VIPN #	Neuroprotective Effect of Vitamin B12 and Vitamin B6 Against Vincristine Induced Peripheral Neuropathy	NCT02923388	Phase 4	88	Acute Lymphoblastic Leukemia (ALL)	Completed	Vitamin B12 and vitamin B6
VIPN	Physiologic Measure of VIPN	NCT04786977	Not Applicable	40	Chemotherapy-induced Peripheral Neuropathy	Recruiting	No Intervention, observational Study

Data provided by Clinical Trials (https://clinicaltrials.gov). * DIC-doxorubicin-induced cardiotoxicity; # Vincristine Induced Peripheral Neuropathy.

Significantly, four clinical trials are investigating or studying DIC. A Phase 2 clinical trial (NCT04166253) is testing vitamin D in breast cancer patients treated with DOX. A clinical trial (NCT02796365) is currently underway to evaluate the effectiveness of exercise rehabilitation as a preventive measure for DIC. The studies are summarized in Table 3.

PanCare is Pan-European Network for care of pediatric cancer survivors. The risk for significant and potentially life-threatening late effects can be identified by certain long-term follow-up projects such as PanCareSurPass and PanCareFollowUp. Based on the reports from 10 countries, ototoxicity following platinum-based chemotherapy has been evaluated regarding the quality of evidence. The ototoxicity surveillance recommendations for pediatric cancer survivors' future studies should focus on the evaluation of otoprotectants and the identification of optimal threshold doses to prevent ototoxicity [155]. In addition, candidate genetic markers are useful for identifying childhood cancer patients at risk of severe late effects, such as *SLC22A2*, which detects those at risk of platinum-induced hearing loss [156]. Moreover, adequate knowledge of cancer history, subsequent treatment exposure, and potential risks of late effects are needed to enhance survivors' health and self-management skills. Accessible and reliable information is essential to increase awareness about late effects, which is necessary for providing personal recommendations for surveillance and prevention [157].

9. Conclusions

Together with the significant advancements in target therapy and immunotherapy, chemotherapy remains the primary treatment option for childhood cancer patients in most low- and middle-income countries. Recognizing the absolute validity of chemo drugs is crucial as they have shown to be highly effective for obtaining long-term survival in childhood cancers. Until now, in total, 16 chemo drugs have been approved by the FDA for pediatric chemotherapy. Unlike adult cancer patients, children diagnosed with cancer are less tolerant of chemotherapy. The numerous adverse effects and occurrence of MDR have become a significant obstacle in pediatric chemotherapy. It is crucial to alleviate severe side effects such as VIPN and DIC. Meanwhile, a comprehensive understanding of MDR and its reversal mechanism is essential. To improve follow-up care quality, the electronic document summarizes the clinical history of childhood/adolescent cancer survivors, including treatments received, and provides personalized follow-up and screening recommendations. By overcoming the challenges of traditional chemotherapy, we will hopefully be able to improve both the survival rate and overall quality of life of pediatric cancer patients.

Author Contributions: L.B. and Y.W. wrote the original draft of the manuscript; Y.L. was responsible for preparing figures and tables; J.N.D.W. and Z.H. reviewed and edited the manuscript; Z.-S.C. supervised the paper. All authors have read and agreed to the published version of the manuscript.

Funding: Science and Technology Plan of Jiangxi Provincial Health Commission (no: 202210918); the Guiding Science and Technology Programme of Ganzhou city (no: GZ2022ZSF231).

Conflicts of Interest: The authors declare no conflict of interest.

References

1. Ward, E.; DeSantis, C.; Robbins, A.; Kohler, B.; Jemal, A. Childhood and adolescent cancer statistics, 2014. *CA Cancer J. Clin.* **2014**, *64*, 83–103. [CrossRef] [PubMed]
2. McEachron, T.A.; Helman, L.J. Recent Advances in Pediatric Cancer Research. *Cancer Res.* **2021**, *81*, 5783–5799. [CrossRef]
3. Shah, N. Dodging the bullet: Therapeutic resistance mechanisms in pediatric cancers. *Cancer Drug Resist.* **2019**, *2*, 428–446. [CrossRef] [PubMed]
4. Siegel, R.L.; Miller, K.D.; Wagle, N.S.; Jemal, A. Cancer statistics, 2023. *CA Cancer J. Clin.* **2023**, *73*, 17–48. [CrossRef] [PubMed]
5. GBD 2017 Childhood Cancer Collaborators. The global burden of childhood and adolescent cancer in 2017: An analysis of the Global Burden of Disease Study 2017. *Lancet Oncol.* **2019**, *20*, 1211–1225. [CrossRef]
6. Rodriguez-Galindo, C.; Friedrich, P.; Alcasabas, P.; Antillon, F.; Banavali, S.; Castillo, L.; Israels, T.; Jeha, S.; Harif, M.; Sullivan, M.J.; et al. Toward the Cure of All Children with Cancer Through Collaborative Efforts: Pediatric Oncology as a Global Challenge. *J. Clin. Oncol.* **2015**, *33*, 3065–3073. [CrossRef]
7. Pramesh, C.S.; Badwe, R.A.; Bhoo-Pathy, N.; Booth, C.M.; Chinnaswamy, G.; Dare, A.J.; de Andrade, V.P.; Hunter, D.J.; Gopal, S.; Gospodarowicz, M.; et al. Priorities for cancer research in low- and middle-income countries: A global perspective. *Nat. Med.* **2022**, *28*, 649–657. [CrossRef]
8. Selmouni, F.; Zidouh, A.; Belakhel, L.; Sauvaget, C.; Bennani, M.; Khazraji, Y.C.; Benider, A.; Wild, C.P.; Bekkali, R.; Fadhil, I.; et al. Tackling cancer burden in low-income and middle-income countries: Morocco as an exemplar. *Lancet Oncol.* **2018**, *19*, e93–e101. [CrossRef]
9. Patel, A.; Goldstein, D.A.; Tannock, I.F. Improving access to immunotherapy in low- and middle-income countries. *Ann. Oncol.* **2022**, *33*, 360–361. [CrossRef]

10. Chantada, G.; Lam, C.G.; Howard, S.C. Optimizing outcomes for children with non-Hodgkin lymphoma in low- and middle-income countries by early correct diagnosis, reducing toxic death and preventing abandonment. *Br. J. Haematol.* **2019**, *185*, 1125–1135. [CrossRef]
11. Yalcin-Ozkat, G. Molecular Modeling Strategies of Cancer Multidrug Resistance. *Drug Resist. Updat.* **2021**, *59*, 100789. [CrossRef] [PubMed]
12. Kattner, P.; Strobel, H.; Khoshnevis, N.; Grunert, M.; Bartholomae, S.; Pruss, M.; Fitzel, R.; Halatsch, M.E.; Schilberg, K.; Siegelin, M.D.; et al. Compare and contrast: Pediatric cancer versus adult malignancies. *Cancer Metastasis Rev.* **2019**, *38*, 673–682. [CrossRef] [PubMed]
13. Nonnenmacher, L.; Hasslacher, S.; Zimmermann, J.; Karpel-Massler, G.; La Ferla-Bruhl, K.; Barry, S.E.; Burster, T.; Siegelin, M.D.; Bruhl, O.; Halatsch, M.E.; et al. Cell Death Induction in Cancer Therapy—Past, Present, and Future. *Crit. Rev. Oncog.* **2016**, *21*, 253–267. [CrossRef] [PubMed]
14. Abbott, M.; Ustoyev, Y. Cancer and the Immune System: The History and Background of Immunotherapy. *Semin. Oncol. Nurs.* **2019**, *35*, 150923. [CrossRef]
15. Citrin, D.E. Recent Developments in Radiotherapy. *N. Engl. J. Med.* **2017**, *377*, 1065–1075. [CrossRef]
16. Fletcher, J.I.; Ziegler, D.S.; Trahair, T.N.; Marshall, G.M.; Haber, M.; Norris, M.D. Too many targets, not enough patients: Rethinking neuroblastoma clinical trials. *Nat. Rev. Cancer* **2018**, *18*, 389–400. [CrossRef]
17. Malczewska, M.; Kosmider, K.; Bednarz, K.; Ostapinska, K.; Lejman, M.; Zawitkowska, J. Recent Advances in Treatment Options for Childhood Acute Lymphoblastic Leukemia. *Cancers* **2022**, *14*, 2021. [CrossRef]
18. Printezi, M.I.; Kilgallen, A.B.; Bond, M.J.G.; Stibler, U.; Putker, M.; Teske, A.J.; Cramer, M.J.; Punt, C.J.A.; Sluijter, J.P.G.; Huitema, A.D.R.; et al. Toxicity and efficacy of chronomodulated chemotherapy: A systematic review. *Lancet Oncol.* **2022**, *23*, e129–e143. [CrossRef]
19. Massimino, M.; Biassoni, V. Use of high-dose chemotherapy in front-line therapy of childhood malignant glioma. *Expert Rev. Anticancer. Ther.* **2006**, *6*, 709–717. [CrossRef]
20. Mir, M.A.; Sofi, S.; Qayoom, H. Chapter 4—Conventional adjuvant chemotherapy in combination with surgery, radiotherapy, and other specific targets. In *Combinational Therapy in Triple Negative Breast Cancer*; Mir, M., Ed.; Academic Press: Cambridge, MA, USA, 2022; pp. 95–120.
21. Principe, D.R.; Kamath, S.D.; Korc, M.; Munshi, H.G. The immune modifying effects of chemotherapy and advances in chemo-immunotherapy. *Pharmacol. Ther.* **2022**, *236*, 108111. [CrossRef]
22. Nishiwaki, S.; Ando, Y. Gap between pediatric and adult approvals of molecular targeted drugs. *Sci. Rep.* **2020**, *10*, 17145. [CrossRef]
23. Xu, F.L.; Guan, X.M.; Wen, X.H.; Shen, Y.L.; Xiao, J.W.; Guo, Y.X.; Deng, M.Y.; Yu, J. Serious adverse events associated with chemotherapy in children with acute lymphoblastic leukemia. *Zhongguo Dang Dai Er Ke Za Zhi* **2020**, *22*, 828–833. [PubMed]
24. Barone, A.; Casey, D.; McKee, A.E.; Reaman, G. Cancer drugs approved for use in children: Impact of legislative initiatives and future opportunities. *Pediatr. Blood Cancer* **2019**, *66*, e27809. [CrossRef] [PubMed]
25. Wu, C.Y.; Li, G.T.; Chu, C.C.; Guo, H.L.; Fang, W.R.; Li, T.; Wang, Y.R.; Xu, J.; Hu, Y.H.; Zhou, L.; et al. Proactive therapeutic drug monitoring of vincristine in pediatric and adult cancer patients: Current supporting evidence and future efforts. *Arch. Toxicol.* **2022**, *97*, 377–392. [CrossRef] [PubMed]
26. Bhagat, A.; Kleinerman, E.S. Anthracycline-Induced Cardiotoxicity: Causes, Mechanisms, and Prevention. *Adv. Exp. Med. Biol.* **2020**, *1257*, 181–192. [PubMed]
27. Mancilla, T.R.; Iskra, B.; Aune, G.J. Doxorubicin-Induced Cardiomyopathy in Children. *Compr. Physiol.* **2019**, *9*, 905–931. [PubMed]
28. van de Velde, M.E.; Kaspers, G.L.; Abbink, F.C.H.; Wilhelm, A.J.; Ket, J.C.F.; van den Berg, M.H. Vincristine-induced peripheral neuropathy in children with cancer: A systematic review. *Crit. Rev. Oncol. Hematol.* **2017**, *114*, 114–130. [CrossRef]
29. Rivankar, S. An overview of doxorubicin formulations in cancer therapy. *J. Cancer Res. Ther.* **2014**, *10*, 853–858. [CrossRef]
30. Garbayo, E.; Pascual-Gil, S.; Rodriguez-Nogales, C.; Saludas, L.; Estella-Hermoso de Mendoza, A.; Blanco-Prieto, M.J. Nanomedicine and drug delivery systems in cancer and regenerative medicine. *Wiley Interdiscip. Rev. Nanomed. Nanobiotechnol.* **2020**, *12*, e1637. [CrossRef]
31. Liu, C.; Ma, X.; Zhuang, J.; Liu, L.; Sun, C. Cardiotoxicity of doxorubicin-based cancer treatment: What is the protective cognition that phytochemicals provide us? *Pharmacol. Res.* **2020**, *160*, 105062. [CrossRef]
32. Kalyanaraman, B. Teaching the basics of the mechanism of doxorubicin-induced cardiotoxicity: Have we been barking up the wrong tree? *Redox Biol.* **2020**, *29*, 101394. [CrossRef] [PubMed]
33. Magdy, T.; Jiang, Z.; Jouni, M.; Fonoudi, H.; Lyra-Leite, D.; Jung, G.; Romero-Tejeda, M.; Kuo, H.H.; Fetterman, K.A.; Gharib, M.; et al. RARG variant predictive of doxorubicin-induced cardiotoxicity identifies a cardioprotective therapy. *Cell Stem Cell* **2021**, *28*, 2076–2089.e7. [CrossRef] [PubMed]
34. Magdy, T.; Jouni, M.; Kuo, H.H.; Weddle, C.J.; Lyra-Leite, D.; Fonoudi, H.; Romero-Tejeda, M.; Gharib, M.; Javed, H.; Fajardo, G.; et al. Identification of Drug Transporter Genomic Variants and Inhibitors That Protect Against Doxorubicin-Induced Cardiotoxicity. *Circulation* **2022**, *145*, 279–294. [CrossRef] [PubMed]
35. Lim, G.B. Circular RNA prevents doxorubicin-induced cardiotoxicity. *Nat. Rev. Cardiol.* **2022**, *19*, 574. [CrossRef]

36. Lu, D.; Chatterjee, S.; Xiao, K.; Riedel, I.; Huang, C.K.; Costa, A.; Cushman, S.; Neufeldt, D.; Rode, L.; Schmidt, A.; et al. A circular RNA derived from the insulin receptor locus protects against doxorubicin-induced cardiotoxicity. *Eur. Heart J.* **2022**, *43*, 4496–4511. [CrossRef]
37. Han, D.; Wang, Y.; Wang, Y.; Dai, X.; Zhou, T.; Chen, J.; Tao, B.; Zhang, J.; Cao, F. The Tumor-Suppressive Human Circular RNA CircITCH Sponges miR-330-5p to Ameliorate Doxorubicin-Induced Cardiotoxicity Through Upregulating SIRT6, Survivin, and SERCA2a. *Circ. Res.* **2020**, *127*, e108–e125. [CrossRef]
38. Wang, Y.; Lu, X.; Wang, X.; Qiu, Q.; Zhu, P.; Ma, L.; Ma, X.; Herrmann, J.; Lin, X.; Wang, W.; et al. atg7-Based Autophagy Activation Reverses Doxorubicin-Induced Cardiotoxicity. *Circ. Res.* **2021**, *129*, e166–e182. [CrossRef]
39. Hu, C.; Zhang, X.; Song, P.; Yuan, Y.P.; Kong, C.Y.; Wu, H.M.; Xu, S.C.; Ma, Z.G.; Tang, Q.Z. Meteorin-like protein attenuates doxorubicin-induced cardiotoxicity via activating cAMP/PKA/SIRT1 pathway. *Redox Biol.* **2020**, *37*, 101747. [CrossRef]
40. Wang, A.J.; Tang, Y.; Zhang, J.; Wang, B.J.; Xiao, M.; Lu, G.; Li, J.; Liu, Q.; Guo, Y.; Gu, J. Cardiac SIRT1 ameliorates doxorubicin-induced cardiotoxicity by targeting sestrin 2. *Redox Biol.* **2022**, *52*, 102310. [CrossRef]
41. Wu, Y.Z.; Zhang, L.; Wu, Z.X.; Shan, T.T.; Xiong, C. Berberine Ameliorates Doxorubicin-Induced Cardiotoxicity via a SIRT1/p66Shc-Mediated Pathway. *Oxid. Med. Cell. Longev.* **2019**, *2019*, 2150394. [CrossRef]
42. Wu, X.; Wang, L.; Wang, K.; Li, J.; Chen, R.; Wu, X.; Ni, G.; Liu, C.; Das, S.; Sluijter, J.P.G.; et al. ADAR2 increases in exercised heart and protects against myocardial infarction and doxorubicin-induced cardiotoxicity. *Mol. Ther.* **2022**, *30*, 400–414. [CrossRef]
43. Chen, D.; Yu, W.; Zhong, C.; Hong, Q.; Huang, G.; Que, D.; Wang, Y.; Yang, Y.; Rui, B.; Zhuang, Z.; et al. Elabela ameliorates doxorubicin-induced cardiotoxicity by promoting autophagic flux through TFEB pathway. *Pharmacol. Res.* **2022**, *178*, 106186. [CrossRef] [PubMed]
44. Russo, M.; Guida, F.; Paparo, L.; Trinchese, G.; Aitoro, R.; Avagliano, C.; Fiordelisi, A.; Napolitano, F.; Mercurio, V.; Sala, V.; et al. The novel butyrate derivative phenylalanine-butyramide protects from doxorubicin-induced cardiotoxicity. *Eur. J. Heart Fail.* **2019**, *21*, 519–528. [CrossRef] [PubMed]
45. Ye, B.; Shi, X.; Xu, J.; Dai, S.; Xu, J.; Fan, X.; Han, B.; Han, J. Gasdermin D mediates doxorubicin-induced cardiomyocyte pyroptosis and cardiotoxicity via directly binding to doxorubicin and changes in mitochondrial damage. *Transl. Res.* **2022**, *248*, 36–50. [CrossRef] [PubMed]
46. Yang, G.; Song, M.; Hoang, D.H.; Tran, Q.H.; Choe, W.; Kang, I.; Kim, S.S.; Ha, J. Melatonin prevents doxorubicin-induced cardiotoxicity through suppression of AMPKalpha2-dependent mitochondrial damage. *Exp. Mol. Med.* **2020**, *52*, 2055–2068. [CrossRef]
47. Efentakis, P.; Varela, A.; Chavdoula, E.; Sigala, F.; Sanoudou, D.; Tenta, R.; Gioti, K.; Kostomitsopoulos, N.; Papapetropoulos, A.; Tasouli, A.; et al. Levosimendan prevents doxorubicin-induced cardiotoxicity in time- and dose-dependent manner: Implications for inotropy. *Cardiovasc. Res.* **2020**, *116*, 576–591. [CrossRef]
48. Ding, M.; Shi, R.; Fu, F.; Li, M.; De, D.; Du, Y.; Li, Z. Paeonol protects against doxorubicin-induced cardiotoxicity by promoting Mfn2-mediated mitochondrial fusion through activating the PKCepsilon-Stat3 pathway. *J. Adv. Res.* 2022; *in press*.
49. Zhang, Y.; Ni, L.; Lin, B.; Hu, L.; Lin, Z.; Yang, J.; Wang, J.; Ma, H.; Liu, Y.; Yang, J.; et al. SNX17 protects the heart from doxorubicin-induced cardiotoxicity by modulating LMOD2 degradation. *Pharmacol. Res.* **2021**, *169*, 105642. [CrossRef]
50. Pan, J.A.; Zhang, H.; Lin, H.; Gao, L.; Zhang, H.L.; Zhang, J.F.; Wang, C.Q.; Gu, J. Irisin ameliorates doxorubicin-induced cardiac perivascular fibrosis through inhibiting endothelial-to-mesenchymal transition by regulating ROS accumulation and autophagy disorder in endothelial cells. *Redox Biol.* **2021**, *46*, 102120. [CrossRef]
51. Wang, P.; Wang, M.; Hu, Y.; Chen, J.; Cao, Y.; Liu, C.; Wu, Z.; Shen, J.; Lu, J.; Liu, P. Isorhapontigenin protects against doxorubicin-induced cardiotoxicity via increasing YAP1 expression. *Acta Pharm. Sin. B* **2021**, *11*, 680–693. [CrossRef]
52. Liang, X.; Wang, S.; Wang, L.; Ceylan, A.F.; Ren, J.; Zhang, Y. Mitophagy inhibitor liensinine suppresses doxorubicin-induced cardiotoxicity through inhibition of Drp1-mediated maladaptive mitochondrial fission. *Pharmacol. Res.* **2020**, *157*, 104846. [CrossRef]
53. Martino, E.; Casamassima, G.; Castiglione, S.; Cellupica, E.; Pantalone, S.; Papagni, F.; Rui, M.; Siciliano, A.M.; Collina, S. Vinca alkaloids and analogues as anti-cancer agents: Looking back, peering ahead. *Bioorg. Med. Chem. Lett.* **2018**, *28*, 2816–2826. [CrossRef] [PubMed]
54. Keenan, B.; Finol-Urdaneta, R.K.; Hope, A.; Bremner, J.B.; Kavallaris, M.; Lucena-Agell, D.; Oliva, M.A.; Diaz, J.F.; Vine, K.L. N-alkylisatin-based microtubule destabilizers bind to the colchicine site on tubulin and retain efficacy in drug resistant acute lymphoblastic leukemia cell lines with less in vitro neurotoxicity. *Cancer Cell Int.* **2020**, *20*, 170. [CrossRef] [PubMed]
55. Harris, C.M.; Blanchaert, R.H. Bilateral recurrent laryngeal nerve palsy resulting from treatment with vincristine. *J. Oral Maxillofac. Surg.* **2006**, *64*, 738–739. [CrossRef] [PubMed]
56. Nama, N.; Barker, M.K.; Kwan, C.; Sabarre, C.; Solimano, V.; Rankin, A.; Raabe, J.; Ross, C.J.; Carleton, B.; Zwicker, J.G.; et al. Vincristine-induced peripheral neurotoxicity: A prospective cohort. *Pediatr. Hematol. Oncol.* **2020**, *37*, 15–28. [CrossRef] [PubMed]
57. Barnett, S.; Hellmann, F.; Parke, E.; Makin, G.; Tweddle, D.A.; Osborne, C.; Hempel, G.; Veal, G.J. Vincristine dosing, drug exposure and therapeutic drug monitoring in neonate and infant cancer patients. *Eur. J. Cancer* **2022**, *164*, 127–136. [CrossRef]
58. Uittenboogaard, A.; Neutel, C.L.G.; Ket, J.C.F.; Njuguna, F.; Huitema, A.D.R.; Kaspers, G.J.L.; van de Velde, M.E. Pharmacogenomics of Vincristine-Induced Peripheral Neuropathy in Children with Cancer: A Systematic Review and Meta-Analysis. *Cancers* **2022**, *14*, 612. [CrossRef]

59. Triarico, S.; Romano, A.; Attina, G.; Capozza, M.A.; Maurizi, P.; Mastrangelo, S.; Ruggiero, A. Vincristine-Induced Peripheral Neuropathy (VIPN) in Pediatric Tumors: Mechanisms, Risk Factors, Strategies of Prevention and Treatment. *Int. J. Mol. Sci.* **2021**, *22*, 4112. [CrossRef]
60. Smolik, S.; Arland, L.; Hensley, M.A.; Schissel, D.; Shepperd, B.; Thomas, K.; Rodgers, C. Assessment Tools for Peripheral Neuropathy in Pediatric Oncology: A Systematic Review from the Children's Oncology Group. *J. Pediatr. Oncol. Nurs.* **2018**, *35*, 267–275. [CrossRef]
61. Al-Mahayri, Z.N.; AlAhmad, M.M.; Ali, B.R. Long-Term Effects of Pediatric Acute Lymphoblastic Leukemia Chemotherapy: Can Recent Findings Inform Old Strategies? *Front. Oncol.* **2021**, *11*, 710163. [CrossRef]
62. Smith, E.M.L.; Kuisell, C.; Cho, Y.; Kanzawa-Lee, G.A.; Gilchrist, L.S.; Park, S.B.; Scott, M.R.; Alberti, P.; Toxic Neuropathy Consortium of the Peripheral Nerve Society. Characteristics and patterns of pediatric chemotherapy-induced peripheral neuropathy: A systematic review. *Cancer Treat. Res. Commun.* **2021**, *28*, 100420. [CrossRef]
63. Madsen, M.L.; Due, H.; Ejskjaer, N.; Jensen, P.; Madsen, J.; Dybkaer, K. Aspects of vincristine-induced neuropathy in hematologic malignancies: A systematic review. *Cancer Chemother. Pharmacol.* **2019**, *84*, 471–485. [CrossRef] [PubMed]
64. Tay, C.G.; Lee, V.W.M.; Ong, L.C.; Goh, K.J.; Ariffin, H.; Fong, C.Y. Vincristine-induced peripheral neuropathy in survivors of childhood acute lymphoblastic leukaemia. *Pediatr. Blood Cancer* **2017**, *64*, e26471. [CrossRef] [PubMed]
65. Adil, M.K.; Ali, Z.; Arshad, U.; Fawad, U. Vincristine induced neurotoxicity in children who underwent chemotherapy for acute lymphoblastic leukemia and Wilm's tumor. *Pak. J. Med. Sci.* **2021**, *37*, 1331–1334. [CrossRef] [PubMed]
66. Li, G.Z.; Hu, Y.H.; Li, D.Y.; Zhang, Y.; Guo, H.L.; Li, Y.M.; Chen, F.; Xu, J. Vincristine-induced peripheral neuropathy: A mini-review. *Neurotoxicology* **2020**, *81*, 161–171. [CrossRef] [PubMed]
67. Yang, Q.Y.; Hu, Y.H.; Guo, H.L.; Xia, Y.; Zhang, Y.; Fang, W.R.; Li, Y.M.; Xu, J.; Chen, F.; Wang, Y.R.; et al. Vincristine-Induced Peripheral Neuropathy in Childhood Acute Lymphoblastic Leukemia: Genetic Variation as a Potential Risk Factor. *Front. Pharmacol.* **2021**, *12*, 771487. [CrossRef]
68. Pozzi, E.; Fumagalli, G.; Chiorazzi, A.; Canta, A.; Cavaletti, G. Genetic factors influencing the development of vincristine-induced neurotoxicity. *Expert Opin. Drug Metab. Toxicol.* **2021**, *17*, 215–226. [CrossRef]
69. Stock, W.; Diouf, B.; Crews, K.R.; Pei, D.; Cheng, C.; Laumann, K.; Mandrekar, S.J.; Luger, S.; Advani, A.; Stone, R.M.; et al. An Inherited Genetic Variant in CEP72 Promoter Predisposes to Vincristine-Induced Peripheral Neuropathy in Adults with Acute Lymphoblastic Leukemia. *Clin. Pharmacol. Ther.* **2017**, *101*, 391–395. [CrossRef]
70. Egbelakin, A.; Ferguson, M.J.; MacGill, E.A.; Lehmann, A.S.; Topletz, A.R.; Quinney, S.K.; Li, L.; McCammack, K.C.; Hall, S.D.; Renbarger, J.L. Increased risk of vincristine neurotoxicity associated with low CYP3A5 expression genotype in children with acute lymphoblastic leukemia. *Pediatr. Blood Cancer* **2011**, *56*, 361–367. [CrossRef]
71. Zgheib, N.K.; Ghanem, K.M.; Tamim, H.; Aridi, C.; Shahine, R.; Tarek, N.; Saab, R.; Abboud, M.R.; El-Solh, H.; Muwakkit, S.A. Genetic polymorphisms in candidate genes are not associated with increased vincristine-related peripheral neuropathy in Arab children treated for acute childhood leukemia: A single institution study. *Pharmacogenet. Genom.* **2018**, *28*, 189–195. [CrossRef] [PubMed]
72. Lopez-Lopez, E.; Gutierrez-Camino, A.; Astigarraga, I.; Navajas, A.; Echebarria-Barona, A.; Garcia-Miguel, P.; Garcia de Andoin, N.; Lobo, C.; Guerra-Merino, I.; Martin-Guerrero, I.; et al. Vincristine pharmacokinetics pathway and neurotoxicity during early phases of treatment in pediatric acute lymphoblastic leukemia. *Pharmacogenomics* **2016**, *17*, 731–741. [CrossRef]
73. Zhou, L.; Ao, L.; Yan, Y.; Li, C.; Li, W.; Ye, A.; Liu, J.; Hu, Y.; Fang, W.; Li, Y. Levo-corydalmine Attenuates Vincristine-Induced Neuropathic Pain in Mice by Upregulating the Nrf2/HO-1/CO Pathway to Inhibit Connexin 43 Expression. *Neurotherapeutics* **2020**, *17*, 340–355. [CrossRef]
74. Zhou, L.; Hu, Y.; Li, C.; Yan, Y.; Ao, L.; Yu, B.; Fang, W.; Liu, J.; Li, Y. Levo-corydalmine alleviates vincristine-induced neuropathic pain in mice by inhibiting an NF-kappa B-dependent CXCL1/CXCR2 signaling pathway. *Neuropharmacology* **2018**, *135*, 34–47. [CrossRef] [PubMed]
75. Chen, X.J.; Wang, L.; Song, X.Y. Mitoquinone alleviates vincristine-induced neuropathic pain through inhibiting oxidative stress and apoptosis via the improvement of mitochondrial dysfunction. *Biomed. Pharmacother.* **2020**, *125*, 110003. [CrossRef]
76. Xie, H.; Chen, Y.; Du, K.; Wu, W.; Feng, X. Puerarin alleviates vincristine-induced neuropathic pain and neuroinflammation via inhibition of nuclear factor-kappaB and activation of the TGF-beta/Smad pathway in rats. *Int. Immunopharmacol.* **2020**, *89*, 107060. [CrossRef] [PubMed]
77. Liang, Z.J.; Tan, J.; Tang, L.; Xie, Z.B.; Chen, G.J.; Liu, G.J.; Yuan, L.; Wang, K.X.; Ding, H.P.; Qiu, H.; et al. NGF monoclonal antibody DS002 alleviates chemotherapy-induced peripheral neuropathy in rats. *Acta Pharmacol. Sin.* **2022**, *43*, 2841–2847. [CrossRef] [PubMed]
78. Starobova, H.; Mueller, A.; Allavena, R.; Lohman, R.J.; Sweet, M.J.; Vetter, I. Minocycline Prevents the Development of Mechanical Allodynia in Mouse Models of Vincristine-Induced Peripheral Neuropathy. *Front. Neurosci.* **2019**, *13*, 653. [CrossRef] [PubMed]
79. Singh, G.; Singh, A.; Singh, P.; Bhatti, R. Bergapten Ameliorates Vincristine-Induced Peripheral Neuropathy by Inhibition of Inflammatory Cytokines and NFkappaB Signaling. *ACS Chem. Neurosci.* **2019**, *10*, 3008–3017. [CrossRef]
80. Vasan, N.; Baselga, J.; Hyman, D.M. A view on drug resistance in cancer. *Nature* **2019**, *575*, 299–309. [CrossRef] [PubMed]
81. O'Donnell, J.S.; Teng, M.W.L.; Smyth, M.J. Cancer immunoediting and resistance to T cell-based immunotherapy. *Nat. Rev. Clin. Oncol.* **2019**, *16*, 151–167. [CrossRef]

82. Zhang, C.; Liu, X.; Jin, S.; Chen, Y.; Guo, R. Ferroptosis in cancer therapy: A novel approach to reversing drug resistance. *Mol. Cancer* **2022**, *21*, 47. [CrossRef]
83. Singh, M.S.; Tammam, S.N.; Shetab Boushehri, M.A.; Lamprecht, A. MDR in cancer: Addressing the underlying cellular alterations with the use of nanocarriers. *Pharmacol. Res.* **2017**, *126*, 2–30. [CrossRef] [PubMed]
84. Wu, Q.; Yang, Z.; Nie, Y.; Shi, Y.; Fan, D. Multi-drug resistance in cancer chemotherapeutics: Mechanisms and lab approaches. *Cancer Lett.* **2014**, *347*, 159–166. [CrossRef]
85. Lin, L.; Yee, S.W.; Kim, R.B.; Giacomini, K.M. SLC transporters as therapeutic targets: Emerging opportunities. *Nat. Rev. Drug Discov.* **2015**, *14*, 543–560. [CrossRef] [PubMed]
86. Girardi, E.; Cesar-Razquin, A.; Lindinger, S.; Papakostas, K.; Konecka, J.; Hemmerich, J.; Kickinger, S.; Kartnig, F.; Gurtl, B.; Klavins, K.; et al. A widespread role for SLC transmembrane transporters in resistance to cytotoxic drugs. *Nat. Chem. Biol.* **2020**, *16*, 469–478. [CrossRef] [PubMed]
87. Fang, X.; Liu, Y.; Xiao, W.; Zhao, N.; Zhu, C.; Yu, D.; Zhao, Y. Prognostic SLC family genes promote cell proliferation, migration, and invasion in hepatocellular carcinoma. [illegible] **2021**, *50*, 1065–1075. [CrossRef] [PubMed]
88. Hu, Y.H.; Zhou, L.; Wang, S.S.; Jing, X.; Guo, H.L.; Sun, F.; Zhang, Y.; Chen, F.; Xu, J.; Ji, X. Methotrexate Disposition in Pediatric Patients with Acute Lymphoblastic Leukemia: What Have We Learnt from the Genetic Variants of Drug Transporters. *Curr. Pharm. Des.* **2019**, *25*, 627–634. [CrossRef] [PubMed]
89. Robey, R.W.; Pluchino, K.M.; Hall, M.D.; Fojo, A.T.; Bates, S.E.; Gottesman, M.M. Revisiting the role of ABC transporters in multidrug-resistant cancer. *Nat. Rev. Cancer* **2018**, *18*, 452–464. [CrossRef] [PubMed]
90. Fletcher, J.I.; Haber, M.; Henderson, M.J.; Norris, M.D. ABC transporters in cancer: More than just drug efflux pumps. *Nat. Rev. Cancer* **2010**, *10*, 147–156. [CrossRef]
91. Muriithi, W.; Macharia, L.W.; Heming, C.P.; Echevarria, J.L.; Nyachieo, A.; Filho, P.N.; Neto, V.M. ABC transporters and the hallmarks of cancer: Roles in cancer aggressiveness beyond multidrug resistance. *Cancer Biol. Med.* **2020**, *17*, 253–269. [CrossRef]
92. Domenichini, A.; Adamska, A.; Falasca, M. ABC transporters as cancer drivers: Potential functions in cancer development. *Biochim. Biophys. Acta Gen. Subj.* **2019**, *1863*, 52–60. [CrossRef]
93. Pasello, M.; Giudice, A.M.; Scotlandi, K. The ABC subfamily A transporters: Multifaceted players with incipient potentialities in cancer. *Semin. Cancer Biol.* **2020**, *60*, 57–71. [PubMed]
94. Cui, Q.; Yang, Y.; Ji, N.; Wang, J.Q.; Ren, L.; Yang, D.H.; Chen, Z.S. Gaseous signaling molecules and their application in resistant cancer treatment: From invisible to visible. *Future Med. Chem.* **2019**, *11*, 323–336. [CrossRef] [PubMed]
95. Nobili, S.; Lapucci, A.; Landini, I.; Coronnello, M.; Roviello, G.; Mini, E. Role of ATP-binding cassette transporters in cancer initiation and progression. *Semin. Cancer Biol.* **2020**, *60*, 72–95. [PubMed]
96. Juan-Carlos, P.M.; Perla-Lidia, P.P.; Stephanie-Talia, M.M.; Monica-Griselda, A.M.; Luz-Maria, T.E. ABC transporter superfamily. An updated overview, relevance in cancer multidrug resistance and perspectives with personalized medicine. *Mol. Biol. Rep.* **2021**, *48*, 1883–1901.
97. Wang, J.Q.; Wu, Z.X.; Yang, Y.; Teng, Q.X.; Li, Y.D.; Lei, Z.N.; Jani, K.A.; Kaushal, N.; Chen, Z.S. ATP-binding cassette (ABC) transporters in cancer: A review of recent updates. *J. Evid. Based Med.* **2021**, *14*, 232–256.
98. Kadioglu, O.; Saeed, M.E.M.; Munder, M.; Spuller, A.; Greten, H.J.; Efferth, T. Effect of ABC transporter expression and mutational status on survival rates of cancer patients. *Biomed. Pharmacother.* **2020**, *131*, 110718.
99. Szollosi, D.; Rose-Sperling, D.; Hellmich, U.A.; Stockner, T. Comparison of mechanistic transport cycle models of ABC exporters. *Biochim. Biophys. Acta Biomembr.* **2018**, *1860*, 818–832.
100. Sarkadi, B.; Homolya, L.; Szakacs, G.; Varadi, A. Human multidrug resistance ABCB and ABCG transporters: Participation in a chemoimmunity defense system. *Physiol. Rev.* **2006**, *86*, 1179–1236. [CrossRef]
101. Nosol, K.; Romane, K.; Irobalieva, R.N.; Alam, A.; Kowal, J.; Fujita, N.; Locher, K.P. Cryo-EM structures reveal distinct mechanisms of inhibition of the human multidrug transporter ABCB1. *Proc. Natl. Acad. Sci. USA* **2020**, *117*, 26245–26253. [CrossRef]
102. Szakacs, G.; Paterson, J.K.; Ludwig, J.A.; Booth-Genthe, C.; Gottesman, M.M. Targeting multidrug resistance in cancer. *Nat. Rev. Drug Discov.* **2006**, *5*, 219–234.
103. Choudhuri, S.; Klaassen, C.D. Structure, function, expression, genomic organization, and single nucleotide polymorphisms of human ABCB1 (MDR1), ABCC (MRP), and ABCG2 (BCRP) efflux transporters. *Int. J. Toxicol.* **2006**, *25*, 231–259. [PubMed]
104. Rajamani, B.M.; Benjamin, E.S.B.; Abraham, A.; Ganesan, S.; Lakshmi, K.M.; Anandan, S.; Karathedath, S.; Varatharajan, S.; Mohanan, E.; Janet, N.B.; et al. Plasma imatinib levels and ABCB1 polymorphism influences early molecular response and failure-free survival in newly diagnosed chronic phase CML patients. *Sci. Rep.* **2020**, *10*, 20640. [CrossRef] [PubMed]
105. Wu, Z.X.; Yang, Y.; Wang, J.Q.; Zhou, W.M.; Chen, J.; Fu, Y.G.; Patel, K.; Chen, Z.S.; Zhang, J.Y. Elevated ABCB1 Expression Confers Acquired Resistance to Aurora Kinase Inhibitor GSK-1070916 in Cancer Cells. *Front. Pharmacol.* **2020**, *11*, 615824. [CrossRef] [PubMed]
106. Wang, J.; Yang, D.H.; Yang, Y.; Wang, J.Q.; Cai, C.Y.; Lei, Z.N.; Teng, Q.X.; Wu, Z.X.; Zhao, L.; Chen, Z.S. Overexpression of ABCB1 Transporter Confers Resistance to mTOR Inhibitor WYE-354 in Cancer Cells. *Int. J. Mol. Sci.* **2020**, *21*, 1387. [CrossRef] [PubMed]
107. Zafar, A.; Wang, W.; Liu, G.; Wang, X.; Xian, W.; McKeon, F.; Foster, J.; Zhou, J.; Zhang, R. Molecular targeting therapies for neuroblastoma: Progress and challenges. *Med. Res. Rev.* **2021**, *41*, 961–1021. [PubMed]
108. Bourhis, J.; Benard, J.; Hartmann, O.; Boccon-Gibod, L.; Lemerle, J.; Riou, G. Correlation of MDR1 gene expression with chemotherapy in neuroblastoma. *J. Natl. Cancer Inst.* **1989**, *81*, 1401–1405. [CrossRef]

109. Qiu, Y.Y.; Mirkin, B.L.; Dwivedi, R.S. MDR1 hypermethylation contributes to the progression of neuroblastoma. *Mol. Cell. Biochem.* **2007**, *301*, 131–135. [CrossRef]
110. Oue, T.; Yoneda, A.; Uehara, S.; Yamanaka, H.; Fukuzawa, M. Increased expression of multidrug resistance-associated genes after chemotherapy in pediatric solid malignancies. *J. Pediatr. Surg.* **2009**, *44*, 377–380. [CrossRef]
111. Wang, J.Q.; Yang, Y.; Cai, C.Y.; Teng, Q.X.; Cui, Q.; Lin, J.; Assaraf, Y.G.; Chen, Z.S. Multidrug resistance proteins (MRPs): Structure, function and the overcoming of cancer multidrug resistance. *Drug Resist. Updat.* **2021**, *54*, 100743.
112. Cole, S.P. Targeting multidrug resistance protein 1 (MRP1, ABCC1): Past, present, and future. *Annu. Rev. Pharmacol. Toxicol.* **2014**, *54*, 95–117. [CrossRef]
113. Arana, M.R.; Altenberg, G.A. ATP-binding Cassette Exporters: Structure and Mechanism with a Focus on P-glycoprotein and MRP1. *Curr. Med. Chem.* **2019**, *26*, 1062–1078. [PubMed]
114. Fruci, D.; Cho, W.C.; Nobili, V.; Locatelli, F.; Alisi, A. Drug Transporters and Multiple Drug Resistance in Pediatric Solid Tumors. *Curr. Drug Metab.* **2016**, *17*, 308–316. [CrossRef] [PubMed]
115. Mehrvar, N.; Abolghasemi, H.; Rezvany, M.R.; Esmaeil Akbari, M.; Saberynejad, J.; Mehrvar, A.; Ehsani, M.A.; Nourian, M.; Qaddoumi, I.; Movafagh, A. Pattern of ABCC Transporter Gene Expression in Pediatric Patients with Relapsed Acute Lymphoblastic Leukemia. *Rep. Biochem. Mol. Biol.* **2019**, *8*, 184–193. [PubMed]
116. Henderson, M.J.; Haber, M.; Porro, A.; Munoz, M.A.; Iraci, N.; Xue, C.; Murray, J.; Flemming, C.L.; Smith, J.; Fletcher, J.I.; et al. ABCC multidrug transporters in childhood neuroblastoma: Clinical and biological effects independent of cytotoxic drug efflux. *J. Natl. Cancer Inst.* **2011**, *103*, 1236–1251. [PubMed]
117. Pena-Solorzano, D.; Stark, S.A.; Konig, B.; Sierra, C.A.; Ochoa-Puentes, C. ABCG2/BCRP: Specific and Nonspecific Modulators. *Med. Res. Rev.* **2017**, *37*, 987–1050.
118. Khot, M.I.; Downey, C.L.; Armstrong, G.; Svavarsdottir, H.S.; Jarral, F.; Andrew, H.; Jayne, D.G. The role of ABCG2 in modulating responses to anti-cancer photodynamic therapy. *Photodiagnosis Photodyn. Ther.* **2020**, *29*, 101579.
119. Toyoda, Y.; Takada, T.; Suzuki, H. Inhibitors of Human ABCG2: From Technical Background to Recent Updates with Clinical Implications. *Front. Pharmacol.* **2019**, *10*, 208. [CrossRef]
120. Jain, H.D.; Zhang, C.; Zhou, S.; Zhou, H.; Ma, J.; Liu, X.; Liao, X.; Deveau, A.M.; Dieckhaus, C.M.; Johnson, M.A.; et al. Synthesis and structure-activity relationship studies on tryprostatin A, an inhibitor of breast cancer resistance protein. *Bioorg. Med. Chem.* **2008**, *16*, 4626–4651.
121. Zattoni, I.F.; Delabio, L.C.; Dutra, J.P.; Kita, D.H.; Scheiffer, G.; Hembecker, M.; Pereira, G.D.S.; Moure, V.R.; Valdameri, G. Targeting breast cancer resistance protein (BCRP/ABCG2): Functional inhibitors and expression modulators. *Eur. J. Med. Chem.* **2022**, *237*, 114346.
122. Doyle, L.A.; Yang, W.; Abruzzo, L.V.; Krogmann, T.; Gao, Y.; Rishi, A.K.; Ross, D.D. A multidrug resistance transporter from human MCF-7 breast cancer cells. *Proc. Natl. Acad. Sci. USA* **1998**, *95*, 15665–15670. [CrossRef]
123. Sauerbrey, A.; Sell, W.; Steinbach, D.; Voigt, A.; Zintl, F. Expression of the BCRP gene (ABCG2/MXR/ABCP) in childhood acute lymphoblastic leukaemia. *Br. J. Haematol.* **2002**, *118*, 147–150. [CrossRef] [PubMed]
124. van den Heuvel-Eibrink, M.M.; Wiemer, E.A.; Prins, A.; Meijerink, J.P.; Vossebeld, P.J.; van der Holt, B.; Pieters, R.; Sonneveld, P. Increased expression of the breast cancer resistance protein (BCRP) in relapsed or refractory acute myeloid leukemia (AML). *Leukemia* **2002**, *16*, 833–839. [PubMed]
125. Hirschmann-Jax, C.; Foster, A.E.; Wulf, G.G.; Nuchtern, J.G.; Jax, T.W.; Gobel, U.; Goodell, M.A.; Brenner, M.K. A distinct "side population" of cells with high drug efflux capacity in human tumor cells. *Proc. Natl. Acad. Sci. USA* **2004**, *101*, 14228–14233. [PubMed]
126. Choi, Y.H.; Yu, A.M. ABC transporters in multidrug resistance and pharmacokinetics, and strategies for drug development. *Curr. Pharm. Des.* **2014**, *20*, 793–807. [PubMed]
127. Vaidyanathan, A.; Sawers, L.; Gannon, A.L.; Chakravarty, P.; Scott, A.L.; Bray, S.E.; Ferguson, M.J.; Smith, G. ABCB1 (MDR1) induction defines a common resistance mechanism in paclitaxel- and olaparib-resistant ovarian cancer cells. *Br. J. Cancer* **2016**, *115*, 431–441. [CrossRef]
128. Lal, S.; Wong, Z.W.; Sandanaraj, E.; Xiang, X.; Ang, P.C.; Lee, E.J.; Chowbay, B. Influence of ABCB1 and ABCG2 polymorphisms on doxorubicin disposition in Asian breast cancer patients. *Cancer Sci.* **2008**, *99*, 816–823.
129. Wang, X.; Zhang, H.; Chen, X. Drug resistance and combating drug resistance in cancer. *Cancer Drug Resist.* **2019**, *2*, 141–160. [CrossRef]
130. Wu, C.P.; Hsiao, S.H.; Huang, Y.H.; Hung, L.C.; Yu, Y.J.; Chang, Y.T.; Hung, T.H.; Wu, Y.S. Sitravatinib Sensitizes ABCB1- and ABCG2-Overexpressing Multidrug-Resistant Cancer Cells to Chemotherapeutic Drugs. *Cancers* **2020**, *12*, 195.
131. Ahmed, F.; Haass, N.K. Microenvironment-Driven Dynamic Heterogeneity and Phenotypic Plasticity as a Mechanism of Melanoma Therapy Resistance. *Front. Oncol.* **2018**, *8*, 173. [CrossRef]
132. Allen, J.D.; Brinkhuis, R.F.; Wijnholds, J.; Schinkel, A.H. The mouse Bcrp1/Mxr/Abcp gene: Amplification and overexpression in cell lines selected for resistance to topotecan, mitoxantrone, or doxorubicin. *Cancer Res.* **1999**, *59*, 4237–4241.
133. Wei, G.; Wang, Y.; Yang, G.; Wang, Y.; Ju, R. Recent progress in nanomedicine for enhanced cancer chemotherapy. *Theranostics* **2021**, *11*, 6370–6392. [CrossRef] [PubMed]
134. Nanayakkara, A.K.; Follit, C.A.; Chen, G.; Williams, N.S.; Vogel, P.D.; Wise, J.G. Targeted inhibitors of P-glycoprotein increase chemotherapeutic-induced mortality of multidrug resistant tumor cells. *Sci. Rep.* **2018**, *8*, 967.

135. Lei, Z.N.; Teng, Q.X.; Wu, Z.X.; Ping, F.F.; Song, P.; Wurpel, J.N.D.; Chen, Z.S. Overcoming multidrug resistance by knockout of ABCB1 gene using CRISPR/Cas9 system in SW620/Ad300 colorectal cancer cells. *MedComm* **2021**, *2*, 765–777. [CrossRef] [PubMed]
136. Yee Kuen, C.; Masarudin, M.J. Chitosan Nanoparticle-Based System: A New Insight into the Promising Controlled Release System for Lung Cancer Treatment. *Molecules* **2022**, *27*, 473. [CrossRef] [PubMed]
137. Xu, T.; Guo, P.; He, Y.; Pi, C.; Wang, Y.; Feng, X.; Hou, Y.; Jiang, Q.; Zhao, L.; Wei, Y. Application of curcumin and its derivatives in tumor multidrug resistance. *Phytother. Res.* **2020**, *34*, 2438–2458. [PubMed]
138. Gao, Y.; Zhu, Y.; Xu, X.; Wang, F.; Shen, W.; Leng, X.; Zhao, J.; Liu, B.; Wang, Y.; Liu, P. Surface PEGylated Cancer Cell Membrane-Coated Nanoparticles for Codelivery of Curcumin and Doxorubicin for the Treatment of Multidrug Resistant Esophageal Carcinoma. *Front. Cell Dev. Biol.* **2021**, *9*, 688070.
139. Das, T.; Anand, U.; Pandey, S.K.; Ashby, C.R., Jr.; Assaraf, Y.G.; Chen, Z.S.; Dey, A. Therapeutic strategies to overcome taxane resistance in cancer. *Drug Resist. Updat.* **2021**, *55*, 100754.
140. Styczynski, J. Drug resistance in childhood acute myeloid leukemia. *Curr. Pharm. Biotechnol.* **2007**, *8*, 59–75. [CrossRef]
141. Roskoski, R., Jr. Blockade of mutant RAS oncogenic signaling with a special emphasis on KRAS. *Pharmacol. Res.* **2021**, *172*, 105806.
142. Grobner, S.N.; Worst, B.C.; Weischenfeldt, J.; Buchhalter, I.; Kleinheinz, K.; Rudneva, V.A.; Johann, P.D.; Balasubramanian, G.P.; Segura-Wang, M.; Brabetz, S.; et al. The landscape of genomic alterations across childhood cancers. *Nature* **2018**, *555*, 321–327. [CrossRef]
143. Drozak, P.; Brylinski, L.; Zawitkowska, J. A Comprehensive Overview of Recent Advances in Epigenetics in Pediatric Acute Lymphoblastic Leukemia. *Cancers* **2022**, *14*, 5384. [CrossRef] [PubMed]
144. Filbin, M.; Monje, M. Developmental origins and emerging therapeutic opportunities for childhood cancer. *Nat. Med.* **2019**, *25*, 367–376. [CrossRef] [PubMed]
145. Nordlund, J.; Syvanen, A.C. Epigenetics in pediatric acute lymphoblastic leukemia. *Semin. Cancer Biol.* **2018**, *51*, 129–138. [PubMed]
146. Kenderian, S.S.; Adults, Y. Acute lymphoblastic leukemia in adolescents and young adults—From genomics to the clinics. *Clin. Oncol. Adolesc. Young Adults* **2013**, *3*, 49–62.
147. Neaga, A.; Jimbu, L.; Mesaros, O.; Bota, M.; Lazar, D.; Cainap, S.; Blag, C.; Zdrenghea, M. Why Do Children with Acute Lymphoblastic Leukemia Fare Better Than Adults? *Cancers* **2021**, *13*, 3886. [CrossRef]
148. Styczynski, J.; Wysocki, M. In vitro drug resistance profiles of adult acute lymphoblastic leukemia: Possible explanation for difference in outcome to similar therapeutic regimens. *Leuk. Lymphoma* **2002**, *43*, 301–307. [CrossRef]
149. Alcantara, M.; Simonin, M.; Lhermitte, L.; Touzart, A.; Dourthe, M.E.; Latiri, M.; Grardel, N.; Cayuela, J.M.; Chalandon, Y.; Graux, C.; et al. Clinical and biological features of PTPN2-deleted adult and pediatric T-cell acute lymphoblastic leukemia. *Blood Adv.* **2019**, *3*, 1981–1988. [CrossRef]
150. Andrieu, G.P.; Kohn, M.; Simonin, M.; Smith, C.L.; Cieslak, A.; Dourthe, M.E.; Charbonnier, G.; Graux, C.; Huguet, F.; Lheritier, V.; et al. PRC2 loss of function confers a targetable vulnerability to BET proteins in T-ALL. *Blood* **2021**, *138*, 1855–1869.
151. Genesca, E.; Morgades, M.; Gonzalez-Gil, C.; Fuster-Tormo, F.; Haferlach, C.; Meggendorfer, M.; Montesinos, P.; Barba, P.; Gil, C.; Coll, R.; et al. Adverse prognostic impact of complex karyotype (>/=3 cytogenetic alterations) in adult T-cell acute lymphoblastic leukemia (T-ALL). *Leuk. Res.* **2021**, *109*, 106612. [CrossRef]
152. Ma, X.; Edmonson, M.; Yergeau, D.; Muzny, D.M.; Hampton, O.A.; Rusch, M.; Song, G.; Easton, J.; Harvey, R.C.; Wheeler, D.A.; et al. Rise and fall of subclones from diagnosis to relapse in pediatric B-acute lymphoblastic leukaemia. *Nat. Commun.* **2015**, *6*, 6604. [CrossRef]
153. Miller, K.D.; Nogueira, L.; Devasia, T.; Mariotto, A.B.; Yabroff, K.R.; Jemal, A.; Kramer, J.; Siegel, R.L. Cancer treatment and survivorship statistics, 2022. *CA Cancer J. Clin* **2022**, *72*, 409–436. [CrossRef] [PubMed]
154. Jeenia, F.T.; Sojib, F.A.; Rahman, M.S.; Ara, T.; Khan, R.; Tanin, M.J.U.J.M. Neuroprotective effect of vitamin B_6 and vitamin B_{12} against vincristine-induced peripheral neuropathy: A randomized, double-blind, placebo controlled, multicenter trial. *medRxiv* 2021. [CrossRef]
155. Clemens, E.; van den Heuvel-Eibrink, M.M.; Mulder, R.L.; Kremer, L.C.M.; Hudson, M.M.; Skinner, R.; Constine, L.S.; Bass, J.K.; Kuehni, C.E.; Langer, T.; et al. Recommendations for ototoxicity surveillance for childhood, adolescent, and young adult cancer survivors: A report from the International Late Effects of Childhood Cancer Guideline Harmonization Group in collaboration with the PanCare Consortium. *Lancet Oncol.* **2019**, *20*, e29–e41. [CrossRef] [PubMed]
156. Langer, T.; Clemens, E.; Broer, L.; Maier, L.; Uitterlinden, A.G.; de Vries, A.C.H.; van Grotel, M.; Pluijm, S.F.M.; Binder, H.; Mayer, B.; et al. Usefulness of current candidate genetic markers to identify childhood cancer patients at risk for platinum-induced ototoxicity: Results of the European PanCareLIFE cohort study. *Eur. J. Cancer* **2020**, *138*, 212–224. [CrossRef] [PubMed]
157. van Kalsbeek, R.J.; Mulder, R.L.; Haupt, R.; Muraca, M.; Hjorth, L.; Follin, C.; Kepak, T.; Kepakova, K.; Uyttebroeck, A.; Mangelschots, M.; et al. The PanCareFollowUp Care Intervention: A European harmonised approach to person-centred guideline-based survivorship care after childhood, adolescent and young adult cancer. *Eur. J. Cancer* **2022**, *162*, 34–44. [CrossRef]

cancers

MDPI

Article

Hyperleukocytosis in Childhood Acute Leukemia: Early Complications and Survival Outcomes

Sirinthip Kittivisuit [1,†], Nichanan Jongthitinon [1,†], Pornpun Sripornsawan [1], Natsaruth Songthawee [1], Shevachut Chavananon [1], Chompoonut Limratchapong [1], Edward B. McNeil [2] and Thirachit Chotsampancharoen [1,*]

1 Department of Pediatrics, Faculty of Medicine, Prince of Songkla University, Hat Yai 90110, Thailand; chompoonut.llm@gmail.com (C.L.)
2 Epidemiology Unit, Faculty of Medicine, Prince of Songkla University, Hat Yai 90110, Thailand
* Correspondence: cthirachit@yahoo.com; Tel.: +66-74451250
† These authors contributed equally to this work.

Simple Summary: Hyperleukocytosis (WBC > 100 × 10^9/L) has been associated with unfavorable outcomes. Few studies have focused on childhood acute leukemia with hyperleukocytosis in developing countries, where the treatment outcomes remain poorer than in developed countries. Our study reviewed the medical records of 690 children who were diagnosed with acute leukemia between January 1998 and December 2017. The incidence of hyperleukocytosis was 16.6% in acute lymphoblastic leukemia (ALL) patients and 20.3% in acute myeloid leukemia (AML) patients. Hyperleukocytosis, extreme hyperleukocytosis (WBC > 200 × 10^9/L), age less than 1 year, age greater than 10 years, and male sex were independently associated with overall survival in the ALL group, while extreme hyperleukocytosis and age less than 1 year were independently associated with overall survival in the AML group. We believe that to improve survival outcomes for children with hyperleukocytosis, the recommended treatment regimen must be modified, while early treatment-related complications, which are more likely to develop, should raise concerns.

Abstract: Hyperleukocytosis and extreme hyperleukocytosis, defined as initial white blood cell counts greater than 100 × 10^9/L and 200 × 10^9/L, respectively, have been associated with unfavorable outcomes. This study aimed to determine the early complications and survival outcomes of childhood leukemia patients with hyperleukocytosis. The medical records of 690 children newly diagnosed with acute leukemia between January 1998 and December 2017 were retrospectively reviewed. The Kaplan–Meier method and log-rank test were used to assess and compare the survival outcomes. Multivariate Cox proportional hazards regression was used to determine associated risk factors for overall survival. We found that 16.6% of 483 childhood acute lymphoblastic leukemia (ALL) patients and 20.3% of 207 childhood acute myeloid leukemia (AML) patients had hyperleukocytosis at diagnosis. ALL patients with hyperleukocytosis had more early complications than those without hyperleukocytosis ($p < 0.05$). Among the ALL group, the 5-year overall survival rate for those with hyperleukocytosis was significantly lower than for those without hyperleukocytosis (37.2% vs. 67.8%, $p < 0.0001$), while the difference was not statistically significant in the AML group (19.0% vs. 30.2%, respectively, $p = 0.26$). Hyperleukocytosis (hazard ratio [HR]: 2.04), extreme hyperleukocytosis (HR: 2.71), age less than 1 year (HR: 3.05), age greater than 10 years (HR: 1.64), and male sex (HR: 1.37) were independently associated with poorer overall survival in childhood ALL patients. Extreme hyperleukocytosis (HR: 2.63) and age less than 1 year (HR: 1.82) were independently associated with poorer overall survival in AML patients. Hyperleukocytosis was associated with adverse survival outcomes in childhood leukemia.

Keywords: childhood acute leukemia; early complication; hyperleukocytosis; risk factor; survival outcome

Citation: Kittivisuit, S.; Jongthitinon, N.; Sripornsawan, P.; Songthawee, N.; Chavananon, S.; Limratchapong, C.; McNeil, E.B.; Chotsampancharoen, T. Hyperleukocytosis in Childhood Acute Leukemia: Early Complications and Survival Outcomes. *Cancers* **2023**, *15*, 3072. https://doi.org/10.3390/cancers15123072

Academic Editors: Saurabh Agarwal and Jianhua Yang

Received: 15 May 2023
Revised: 2 June 2023
Accepted: 3 June 2023
Published: 6 June 2023

1. Introduction

Acute leukemia is the most common malignancy in childhood, accounting for one-third of all malignancies [1–3]. Hyperleukocytosis, defined as an initial white blood cell (WBC) count greater than 100 $\times$ 10^9/L, is a serious presenting feature, with reported incidences ranging from 10.2 to 19.2% in childhood acute lymphoblastic leukemia (ALL) patients [4–12], and 12.6 to 21.7% in childhood acute myeloid leukemia (AML) patients [4,6,13–16]. Childhood ALL with hyperleukocytosis has been associated with a number of characteristics that are known to increase the likelihood of adverse outcomes, including age at diagnosis, male gender, T-cell immunophenotype, massive hepatosplenomegaly, and high blood lactate dehydrogenase (LDH) levels [5,7,8,17]. Patients presenting with hyperleukocytosis are at risk for developing early treatment-related complications secondary to leukostasis, including seizures, intracranial bleeding, respiratory problems, coagulopathy, renal failure, and metabolic abnormalities related to tumor lysis syndrome [5–8,18]. Previous studies have reported that childhood acute leukemia with hyperleukocytosis have poor survival outcomes [5,7,8,15]. However, most of these studies were conducted in developed countries; few have been conducted in developing countries, where treatment outcomes remain worse than in developed countries [19]. This study aimed to determine the clinical course and survival outcomes of childhood acute leukemia with hyperleukocytosis in Thailand, a developing country in South-East Asia.

2. Materials and Methods

We retrospectively reviewed the medical records of children aged less than 15 years who had been diagnosed with acute leukemia and had received chemotherapy between January 1998 and December 2017 at the Oncology Clinic, Department of Pediatrics, Faculty of Medicine, Prince of Songkla University, the principal tertiary referral center in southern Thailand. The demographic characteristics, clinical symptoms at diagnosis, initial laboratory investigations, subtype of acute leukemia (ALL or AML), and treatment-related complications were recorded. Diagnosis of subtype was made according to the French–American–British (FAB) classification by morphological examination of bone marrow staining. Three cytochemical stains were used in the diagnostic process: periodic acid-Schiff, peroxidase, and α-naphthyl acetate esterase. After 2000, immunophenotyping of cell surface markers was additionally used to differentiate subtypes (i.e., AML from ALL, and T-cell from B-cell ALL). Prior to 2006, our chemotherapy treatment protocols were based on the Children's Cancer Study Group's modified ALL treatment protocols [20–22] and the AML treatment protocols of the Berlin–Frankfurt–Munster group [23,24]. In 2006, nationwide standardized protocols for treating childhood leukemia were developed by the Thai Pediatric Oncology Group [21,25–27]. Our treatment protocols were based on the national protocols according to the subtype of leukemia and risk stratification of the patient at presentation. In 2014, an updated version of the Thai national protocols according to the subtypes of leukemia and risk stratification of patients at presentation was implemented [28,29].

Hyperleukocytosis was defined as a WBC count at initial presentation greater than 100 $\times$ 10^9/L, and extreme hyperleukocytosis was defined as a WBC count at initial presentation greater than 200 $\times$ 10^9/L. Treatment-related complications which occurred during the induction phase of chemotherapy, namely tumor lysis syndrome, seizure, intracranial hemorrhage, acute kidney injury (AKI), septic shock, disseminated intravascular coagulation (DIC), endotracheal tube (ETT) intubation, and intensive care unit (ICU) admission, were recorded. Tumor lysis syndrome was diagnosed according to the Cairo and Bishop criteria, namely, the presence of two or more of the following abnormal conditions: hyperkalemia (serum potassium level $\geq$ 6 mmol/L), hyperuricemia (serum uric acid level $\geq$ 8 mmol/L), hyperphosphatemia (serum phosphate level $\geq$ 6.5 mmol/L), and/or hypocalcemia (serum calcium level < 7 mmol/L) within 3 days before or 7 days after the start of chemotherapy [30]. Acute kidney injury was defined according to the Kidney Disease: Improving Global Outcomes (KDIGO) criteria [31]. Septic shock was defined according to international pediatric sepsis consensus guidelines [32]. Disseminated intravascular coagulation was defined according to the International Society on Thrombosis and Hemostasis (ISTH) scoring system [33].

Statistical Analysis

Descriptive statistics are presented using mean and standard deviation (SD) or median and interquartile range (IQR) for continuous variables as appropriate. Categorical variables are presented using frequency and percentage. We compared continuous variables between children with and without hyperleukocytosis using Student's *t*-test for normally distributed variables and the rank-sum test otherwise. Categorical variables were compared using the chi-square or Fisher's exact tests as appropriate. Overall and event-free survival were compared using log-rank tests and survival distributions were depicted visually with Kaplan–Meier curves. Risk factors for overall survival were analyzed using Cox regression analysis. Variables having a p value less than 0.2 from the univariate analysis were included in the initial multivariate Cox regression model for assessment of independent risk factors. The risk factors for overall survival are presented as adjusted hazard ratios (HR) with 95% confidence intervals (CI). R version 4.2.1 was used for all analyses. A p-value less than 0.05 was considered significant.

3. Results

3.1. Study Population

During the 20-year study period, a total of 779 children were diagnosed with acute leukemia in our center. We excluded 39 who had incomplete data and 50 whose parents refused to allow their child to receive standard chemotherapy, leaving a total of 690 children for analysis. Of these, 483 were diagnosed with ALL and 207 with AML. The median (IQR) age at diagnosis of all patients was 55.5 (34.0–105.0) months and the majority (57.5%) were male. The median (IQR) survival time was 4.36 (1.11–12.90) years. The 5-year and 10-year overall survival rates were 52.3% and 47.7%, respectively. Most (77.9%) ALL patients had B-cell subtype. Approximately 60% of ALL patients had cytogenetic studies, and one-fourth of these displayed aberrant cytogenetic findings such as hyperdiploidy, hypodiploidy, and complex karyotype. Cytogenetic studies were performed on almost 70% of AML patients, and around 50% of these showed abnormal cytogenetic findings such as trisomy 21, trisomy 8, monosomy 7, t(8; 21), and complex karyotype.

3.2. Hyperleukocytosis in Acute Lymphoblastic Leukemia

Among the 483 ALL patients, 80 (16.6%) had hyperleukocytosis at initial presentation and 40 (8.3%) had extreme hyperleukocytosis. Comparisons of demographic characteristics and outcomes between children with and without hyperleukocytosis are presented in Table 1. Those with hyperleukocytosis were significantly older than those without hyperleukocytosis (91 months vs. 52 months, $p = 0.013$). A higher proportion with hyperleukocytosis had T-cell subtype compared to those without hyperleukocytosis (31.2% vs. 9.4%, $p = 0.04$). Mediastinal mass, hepatomegaly and splenomegaly were significantly more common in the hyperleukocytosis group ($p < 0.05$). Those with hyperleukocytosis had a significantly lower average platelet count and a higher percentage of peripheral blast cells ($p < 0.05$). Average serum calcium, phosphorus and LDH levels were significantly lower, whereas the average uric acid level was significantly higher, in the hyperleukocytosis group ($p < 0.05$). Regarding treatment-related complications during the induction phase of chemotherapy, a significantly higher proportion of those with hyperleukocytosis developed tumor lysis syndrome, seizure, acute kidney injury, septic shock, and disseminated intravascular coagulation required endotracheal tube intubation and intensive care unit admission. Those with hyperleukocytosis also had significantly lower remission rates of induction and higher mortality rates compared with those without hyperleukocytosis ($p = 0.011$ and $p < 0.001$, respectively).

Table 1. Comparison of demographic characteristics and outcomes between childhood acute lymphoblastic leukemia patients with and without hyperleukocytosis.

Variables	Total (N = 483)	Hyperleukocytosis (N = 80)	No Hyperleukocytosis (N = 403)	*p* Value
Demographic and clinical characteristics, n (%)				
Age (months), median (IQR)	54.0 (36.0–97.0)	91.0 (34.0–137.0)	52.0 (36.0–86.0)	0.013
Sex				0.171
Male	272 (56.3)	39 (48.8)	233 (57.8)	
Female	211 (43.7)	41 (51.2)	170 (42.2)	
Immunophenotype				<0.001
T-cell	63 (13.0)	25 (31.2)	38 (9.4)	
B-cell	376 (77.9)	51 (63.8)	325 (80.7)	
FAB classification	44 (9.1)	4 (5.0)	40 (9.9)	
Fever	360 (74.5)	62 (77.5)	298 (73.9)	0.599
Mediastinal mass	23 (4.8)	8 (10.0)	15 (3.7)	0.037
Hepatomegaly	448 (92.8)	80 (100.0)	368 (91.3)	0.012
Splenomegaly	332 (68.7)	72 (90.0)	260 (64.5)	<0.001
Laboratory parameters, mean (SD)				
Hemoglobin (g/dL)	7.5 (3.5)	7.0 (2.7)	7.6 (3.7)	0.223
Platelet count ($\times 10^9$/L)	75.5 (92.3)	38.0 (29.9)	82.9 (98.5)	<0.001
Blast cells (%)	50.2 (34.8)	90.4 (11.5)	42.2 (32.2)	<0.001
Calcium (mmol/L)	9.6 (1.1)	9.3 (0.9)	9.6 (1.1)	0.018
Phosphorus (mmol/L)	4.8 (1.3)	4.1 (1.5)	5 (1.3)	<0.001
Uric acid (mmol/L)	6.1 (3.5)	7.5 (4.1)	5.9 (3.3)	<0.001
Lactate dehydrogenase (U/L)	2776.3 (5339.4)	4971.5 (7105.8)	2350.3 (4820.7)	<0.001
Treatment-related complications, n (%)				
Tumor lysis syndrome	67 (13.9)	24 (30.0)	43 (10.7)	<0.001
Seizure	21 (4.3)	9 (11.3)	12 (3.0)	0.003
Intracranial hemorrhage	6 (1.2)	3 (3.8)	3 (0.7)	0.056
Acute kidney injury	27 (5.6)	9 (11.3)	18 (4.5)	0.025
Septic shock	106 (21.9)	26 (32.5)	80 (19.9)	0.009
DIC	65 (13.5)	21 (26.3)	44 (10.9)	<0.001
ETT intubation	38 (7.9)	14 (17.5)	24 (6.0)	<0.001
ICU admission	50 (10.4)	22 (27.5)	28 (6.9)	<0.001
Treatment outcomes, n (%)				
Induction of remission	428 (96.8)	63 (91.3)	365 (97.9)	0.011
Relapse	162 (33.5)	31 (38.8)	131 (32.5)	0.342
Mortality	207 (42.9)	53 (66.3)	154 (38.2)	<0.001

Values are expressed as n (%), mean (SD), or median (IQR). DIC, disseminated intravascular coagulopathy; ETT, endotracheal tube; FAB, French–American–British classification system; ICU, intensive care unit.

3.3. *Hyperleukocytosis in Acute Myeloid Leukemia*

Among the 207 AML patients, 42 (20.3%) had hyperleukocytosis and 22 (10.6%) had extreme hyperleukocytosis at initial presentation. Comparisons of the demographic characteristics and outcomes between those with and without hyperleukocytosis are presented in Table 2. Those with hyperleukocytosis were significantly older than those without hyperleukocytosis (137 months vs. 58 months, $p = 0.021$). Those with hyperleukocytosis had significantly more fever, splenomegaly and lymphadenopathy than those without hyperleukocytosis ($p < 0.05$) and a significantly higher percentage of peripheral blast cells ($p < 0.05$). The average serum calcium and phosphorus levels were significantly lower, whereas the average uric acid level was significantly higher, in the hyperleukocytosis group ($p < 0.05$). The rates of treatment-related complications during the induction phase of chemotherapy were not significantly different between those with and without hyperleukocytosis, except for intracranial hemorrhage, which was only observed in a small number of patients with hyperleukocytosis. Treatment outcomes (remission, relapse and mortality) were not significantly different between the two groups.

Table 2. Comparison of demographic characteristics and outcomes between childhood acute myeloid leukemia patients with and without hyperleukocytosis.

Variables	Total (N = 207)	Hyperleukocytosis (N = 42)	No Hyperleukocytosis (N = 165)	*p* Value
Demographic and clinical characteristics, n (%)				
Age (months), median (IQR)	65.0 (25.5–133.5)	137.0 (34.0–150.5)	58.0 (25.0–116.0)	0.021
Sex				0.268
Male	125 (60.4)	29 (69.0)	96 (58.2)	
Female	82 (39.6)	13 (31.0)	69 (41.8)	
Fever	157 (75.8)	39 (92.9)	118 (71.5)	0.007
Hepatomegaly	165 (79.7)	37 (88.1)	128 (77.6)	0.194
Splenomegaly	119 (57.5)	31 (73.8)	88 (53.3)	0.026
Lymphadenopathy	142 (68.6)	35 (83.3)	107 (64.8)	0.034
Laboratory parameters, mean (SD)				
Hemoglobin (g/dL)	7.3 (2.2)	7.1 (1.6)	7.3 (2.3)	0.585
Platelet count ($\times 10^9$/L)	64.1 (93.1)	50.0 (35.3)	67.7 (102.5)	0.272
Blast cells (%)	43.4 (34.7)	80.5 (24.7)	33.9 (30.3)	<0.001
Calcium (mmol/L)	9.3 (0.7)	9.0 (0.9)	9.3 (0.7)	0.021
Phosphorus (mmol/L)	4.6 (1.2)	4.1 (1.4)	4.8 (1.0)	<0.001
Uric acid (mmol/L)	5.0 (2.1)	6.0 (2.6)	4.8 (1.9)	<0.001
Lactate dehydrogenase (U/L)	2035.6 (2210.0)	2619.7 (2425.5)	1888.6 (2135.3)	0.058
Treatment-related complications, n (%)				
Tumor lysis syndrome	11 (5.3)	4 (9.5)	7 (4.2)	0.276
Seizure	8 (3.9)	1 (2.4)	7 (4.2)	1
Intracranial hemorrhage	2 (1.0)	2 (4.8)	0 (0)	0.037
Acute kidney injury	14 (6.8)	2 (4.8)	12 (7.3)	1
Septic shock	59 (28.5)	11 (26.2)	48 (29.1)	1
DIC	49 (23.7)	13 (31.0)	36 (21.8)	0.219
ETT intubation	41 (19.8)	8 (19.0)	33 (20.0)	1
ICU admission	46 (22.2)	11 (26.2)	35 (21.2)	0.504
Treatment outcomes, n (%)				
Induction of remission	109 (67.3)	19 (59.3)	90 (69.2)	0.289
Relapse	63 (30.4)	12 (28.6)	51 (30.9)	0.915
Mortality	160 (77.3)	34 (81.0)	126 (76.4)	0.669

Values are expressed as n (%), mean (SD), or median (IQR). DIC, disseminated intravascular coagulopathy; ETT, endotracheal tube; ICU, intensive care unit.

3.4. Survival Outcomes in Acute Lymphoblastic Leukemia

The median survival time for the 483 ALL patients was 4.32 (3.31–9.27) years, and the 5-year and 10-year overall survival rates were 48.9% and 44.9%, respectively. The median survival times for the patients with hyperleukocytosis and those without hyperleukocytosis were 1.84 (1.51–4.28) years and 23.2 (22.8–∞) years, respectively. The Kaplan–Meier curves comparing overall survival (OS) between ALL patients with and without hyperleukocytosis are presented in Figure 1. The 5-year OS and EFS rates for those with hyperleukocytosis were both significantly lower than for those without (OS, 37.2% vs. 67.8%, respectively, $p < 0.0001$, and EFS, 33.7% vs. 59.1%, respectively, $p < 0.0001$).

In the multivariate Cox regression analysis, risk factors for overall survival among the ALL patients are shown in Table 3. The independent risk factors for poorer overall survival were age, sex and WBC counts at initial presentation. Age less than 1 year and age greater than 10 years had hazard ratios of 3.05 (95% CI: 1.57–5.96) and 1.64 (95% CI: 1.51–2.32), respectively, when compared to those aged between 1 and 9 years ($p = 0.001$ and $p = 0.006$, respectively). Males had a hazard ratio of 1.37 (95% CI: 1.05–1.79) when compared to females (p = 0.021). Hyperleukocytosis and extreme hyperleukocytosis had hazard ratios of 2.04 (95% CI: 1.33–3.14) and 2.71 (95% CI: 1.74–4.21) compared with those who had an initial WBC count less than 50 × 10^9/L (p = 0.001 and p <0.001, respectively). Patients who had an initial WBC count of 50 to 100 × 10^9/L had a hazard ratio of 1.59 (95% CI: 1.07–2.36) compared with those who had an initial WBC count less than 50 × 10^9/L (p = 0.022).

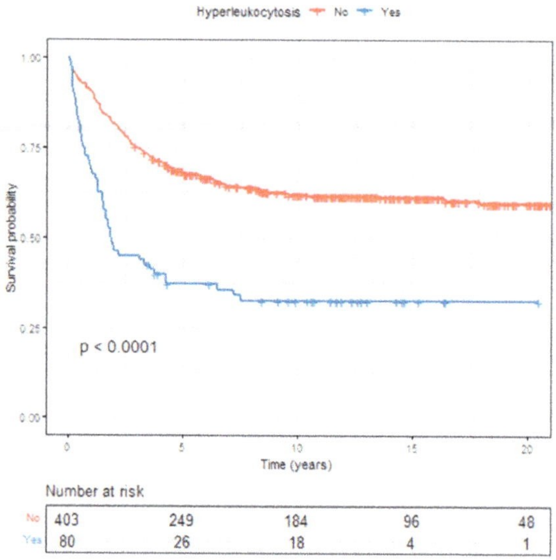

Figure 1. Comparison of overall survival between childhood acute lymphoblastic leukemia patients with and without hyperleukocytosis.

Table 3. Multivariate analysis results showing independent risk factors for overall survival in childhood acute lymphoblastic leukemia patients.

Risk Factor	Crude HR (95% CI)	Adjusted HR (95% CI)	*p* Value
Age (years)			
1–9	Reference	Reference	
<1	3.51 (1.79–6.89)	3.05 (1.57–5.96)	0.001
≥10	1.80 (1.29–2.52)	1.64 (1.15–2.32)	0.006
Male sex	1.11 (0.84–1.47)	1.37 (1.05–1.79)	0.021
Immunophenotype			
T-cell	Reference	Reference	
B-cell	0.79 (0.55–1.15)	1.37 (0.90–2.08)	0.14
FAB classification	0.56 (0.31–1.01)	0.85 (0.46–1.56)	0.6
Initial WBC count (×10^9/L)			
<50	Reference	Reference	
50–<100	1.47 (0.97–2.23)	1.59 (1.07–2.36)	0.022
≥100–<200	2.45 (1.60–4.61)	2.04 (1.33–3.14)	<0.001
≥200	3.04 (2.00–4.61)	2.71 (1.74–4.21)	<0.001
Early complication	1.16 (0.79–1.71)	0.88 (0.62–1.26)	0.5

HR, hazard ratio; CI, confidence interval; FAB, French–American–British classification system; WBC, white blood cell.

3.5. Survival Outcomes in Acute Myeloid Leukemia

The median survival times for the 42 AML patients with hyperleukocytosis and 165 AML patients without hyperleukocytosis were 1.11 (0.87–1.76) years and 0.83 (0.33–2.03) years, respectively. The Kaplan–Meier curves showing overall survival between children with and without hyperleukocytosis are presented in Figure 2. The OS and EFS rates were lower for those with hyperleukocytosis compared with those without hyperleukocytosis, but the differences were not statistically significant (p = 0.26 and 0.29, respectively). The 5-year OS rate for those with hyperleukocytosis was 19.0% compared with 30.2% for those without. The 5-year EFS rate for those with hyperleukocytosis was 19.0% compared with 27.7% for those without.

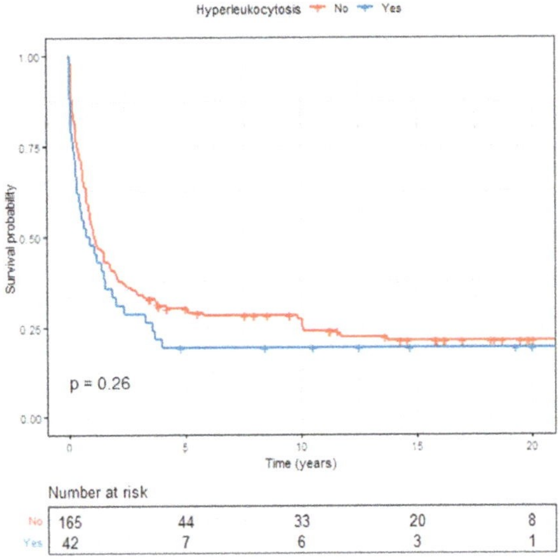

Figure 2. Comparison of overall survival between childhood acute myeloid leukemia patients with and without hyperleukocytosis.

In the multivariate Cox regression analysis, risk factors for overall survival in childhood AML patients are shown in Table 4. The independent risk factors for poorer overall survival were extreme hyperleukocytosis at initial presentation and age. Extreme hyperleukocytosis had a hazard ratio of 2.63 (95% CI: 1.56–4.43) compared with those who had initial WBC count less than 50 $\times$ 10^9/L (p < 0.001), while those aged less than 1 year had a hazard ratio of 1.82 (95% CI: 1.00–3.32) when compared to those aged between 1 and 9 years (p = 0.049).

Table 4. Multivariate analysis results showing independent risk factors for overall survival in childhood acute myeloid leukemia patients.

Risk Factor	Crude HR (95% CI)	Adjusted HR (95% CI)	*p* Value
Age (years)			
1–9	Reference	Reference	
<1	1.43 (0.80–2.55)	1.82 (1.00–3.32)	0.049
≥10	0.84 (0.59–1.20)	0.70 (0.48–1.02)	0.064
Male sex	1.06 (0.77–1.47)	1.03 (0.74–1.42)	0.9
Initial WBC count ($\times 10^9$/L)			
<50	Reference	Reference	
50–<100	1.13 (0.70–1.82)	1.25 (0.76–2.03)	0.4
≥100–<200	0.83 (0.48–1.43)	0.81 (0.46–1.44)	0.5
≥200	2.16 (1.32–3.53)	2.63 (1.56–4.43)	<0.001
Early complication	0.83 (0.50–1.40)	0.89 (0.53–1.50)	0.7

HR, hazard ratio; CI, confidence interval; WBC, white blood cell.

4. Discussion

Our long-term study covered a 20-year period and included a total of 690 childhood acute leukemia patients, consisting of 483 ALL and 207 AML patients. We found that 16.6% of ALL patients and 20.3% of AML patients had hyperleukocytosis at initial presentation. Previous studies have reported incidences of hyperleukocytosis at initial presentation of childhood ALL patients ranging from 10.2 to 19.2% [4–10] and in childhood AML patients ranging from 12.6 to 21.7% [4,6,13–16]. However, the majority of these studies were conducted in developed countries and there are limited data from developing countries. Two studies from India reported that 14.6% and 16.7% of childhood ALL patients had hyperleukocytosis at presentation [11,12]. However, these studies had limited data on the clinical characteristics and outcomes of their hyperleukocytosis patients. Our study found that the incidence of hyperleukocytosis in childhood leukemia patients was comparable to that of previous studies.

Hyperleukocytosis in childhood ALL has been associated with a variety of clinical characteristics [5,7,8]. Eguiguren et al. reported that hyperleukocytosis was significantly associated with age less than 1 year, T-cell immunophenotype, mediastinal mass, massive hepatosplenomegaly, and elevated LDH [5]. Kong et al. retrospectively reviewed 104 childhood ALL patients and found that age 1–10 years, male gender, T-cell immunophenotype, and massive splenomegaly were associated with extreme hyperleukocytosis (WBC > 200 $\times$ 10^9/L) [7]. A later multicenter study by Park et al. reported that children aged $\geq$ 10 years were more likely to develop extreme hyperleukocytosis [8]. Nguyen et al. reported that childhood ALL with extreme hyperleukocytosis was associated with age younger than 1 or older than 9 years of age, male, T-cell immunophenotype, central nervous system involvement, elevated LDH, and elevated uric acid [18]. Similarly, our study found that hyperleukocytosis in childhood ALL was associated with older age, T-cell immunophenotype, mediastinal mass, hepatomegaly, and splenomegaly. Regarding childhood AML patients, data on the associated factors that may enhance the risk of hyperleukocytosis are limited. Sung et al. reported that hyperleukocytosis in childhood AML patients was associated with an age less than 1 year, FAB M1, M4, and M5, and certain chromosomal abnormalities [14]. In contrast, our study found that older age, fever, lymphadenopathy, and splenomegaly were clinical features significantly associated with hyperleukocytosis in childhood AML. However, factors that increase the risk of hyperleukocytosis in childhood ALL and AML remain inconclusive. Further studies, including cytogenetic ones, with larger sample sizes, multicenter designs, and diverse populations are needed.

Earlier studies found that early treatment-related complications were more common in childhood ALL patients who had hyperleukocytosis at initial presentation. Bunin et al. reported that metabolic complications occurred in 13.7% of 161 childhood ALL patients with hyperleukocytosis [4]. Alba et al. found that among 84 childhood ALL cases with hyperleukocytosis, 19.0% had respiratory complications followed by neurological complications and renal dysfunction in 7.1% and 6.0% of their cases, respectively [6]. Kong et al. reported that metabolic complications occurred in 40% of 20 childhood ALL patients with hyperleukocytosis, followed by neurological and respiratory complications in 20% and 10%, respectively [7]. Park et al. found that among 72 childhood ALL cases with hyperleukocytosis, metabolic complications developed in 59.7%, followed by respiratory and neurologic complications in 11.1% and 9.7%, respectively [8]. Similarly, our study found that the rates of early treatment-related complications among childhood ALL patients with hyperleukocytosis were comparable to previous studies, namely septic shock (32.5%), metabolic complications related to tumor lysis syndrome (30.0%), disseminated intravascular coagulation (26.3%), required endotracheal tube intubation (17.5%), acute kidney injury (11.3%), seizure (11.3%), and intracranial hemorrhage (3.8%). The varying incidences of treatment-related complications reported in the literature might be due, at least partly, to differences in the definitions among the studies. Nevertheless, our study also demonstrated that early treatment-related complications were significantly higher in childhood ALL patients with hyperleukocytosis compared with those without.

In a comparable pattern to the childhood ALL population, hyperleukocytosis in children with AML has been associated with early complications. Bunin et al. reported that metabolic complications, acute renal failure, and intracerebral hemorrhage occurred in 5.5%, 5.5%, and 11% of 73 childhood AML patients with hyperleukocytosis, respectively. In another study serious hemorrhagic complications were reported in significantly more childhood AML patients with hyperleukocytosis than in childhood ALL patients with hyperleukocytosis (19% vs. 2.5%) [4]. Inaba et al. found that among 106 childhood AML patients with hyperleukocytosis, early complications developed in 42.5%, including respiratory, neurological and renal complications at rates of 18.9%, 16%, and 15%, respectively [15]. Sung et al. reported that a high WBC count in childhood AML cases was significantly associated with cerebral nervous system ischemia or hemorrhage, hypoxia, and pulmonary hemorrhage [14]. Alba et al. found that among 18 childhood ALL cases with hyperleukocytosis, respiratory, neurological, and renal complications developed in 22.2%, 22.2%, and 11.1% of their cases, respectively [6]. Our study observed that the rates of early treatment-related complications among childhood AML patients with hyperleukocytosis were less than in previous studies. However, the incidence of early complications may vary depending on the definitions applied and the limited sample sizes of childhood AML cases with hyperleukocytosis in the literature. Further multicenter prospective studies are required.

Our study found that the survival outcomes for childhood leukemia patients in the entire cohort were comparable to previous studies from developing countries but were lower than studies conducted in developed countries [34,35]. The subgroup of childhood ALL cases with hyperleukocytosis has been associated with poor survival outcomes. Eguiguren et al. reported that the 4-year EFS rate of 64 childhood ALL patients with hyperleukocytosis was significantly lower than in those without (52% vs. 75%, respectively) [5]. Kong et al. reported that the 3-year OS and EFS rates of 104 childhood ALL cases with hyperleukocytosis were 81.2% and 75%, respectively [7]. Park et al. found that the 10-year OS and EFS rates among 72 childhood ALL cases with hyperleukocytosis were 82.6% and 78.3%, respectively [8]. Maurer et al. reported that the EFS for 106 childhood ALL patients with an initial WBC count > 200 $\times$ 10^9/L was 55% at 3 years [36]. However, most of these studies were conducted in developed countries, and there have been few studies from developing countries. Our study found lower survival rates for childhood ALL with hyperleukocytosis (5-year EFS rate of 33.7%) than all studies from developed countries, and that children with hyperleukocytosis had poorer survival rates than children without hyperleukocytosis. When focusing on the subgroup of childhood AML cases with hyperleukocytosis, a study by Inaba et al. reported that those with hyperleukocytosis had a lower 10-year EFS rate compared with those without, but the difference in the 10-year OS rates was not significant. Similarly, our study found that the OS and EFS rates were lower for the AML patients with hyperleukocytosis compared with those without, however the difference was not statistically significant.

Several factors, including age, male sex, WBC count at diagnosis, central nervous system involvement, T-cell immunophenotype, race, and cytogenetics have been considered as adverse prognostic factors in childhood ALL [37–40]. Age and WBC count at initial diagnosis were factors used by The National Cancer Institute (NCI) risk classification to separate high and standard risk groups [41]. Eguiguren et al. reported that only leukocyte count was identified as a significant prognostic factor for EFS [5]. A study by Kong et al. found associations between EFS and WBC count, age, splenomegaly, mediastinal mass, immunophenotype, and chromosomal abnormalities. However, multivariate analysis was not performed [7]. Maurer et al. reported that a WBC count at diagnosis greater than 600 $\times$ 10^9/L and massive splenomegaly were the only two significant adverse prognostic factors for survival on multivariate analysis with hazard ratios of 2.60 and 2.42, respectively [36]. Our study found that on multivariate analysis, age, male sex, and WBC count at initial presentation were independently associated with overall survival in childhood ALL patients, consistent with results from other studies.

Several factors associated with early mortality in the AML population have been reported in the literature. Bunin et al. found that early mortality in childhood AML cases with hyperleukocytosis (23.3%) was significantly higher than in childhood ALL cases with hyperleukocytosis (5%) [4]. Abla et al. reported the rate of early death was 16.7% in 18 childhood AML patients with hyperleukocytosis compared with 1.2% in 634 childhood ALL patients. They also found that early death due to hyperleukocytosis in childhood leukemia was associated with neurologic complications, AML diagnosis, and initial coagulopathy [6]. Creutzig et al. found that early mortality from hemorrhage or leukostasis occurred in 30.8% of 65 childhood AML patients with hyperleukocytosis and the risk factors for early death were FAB M5, hyperleukocytosis and the presence of extramedullary organ involvement [42]. Nevertheless, the literature on the factors that affect overall survival in the population of children with AML is still sparse. Our study found that extreme hyperleukocytosis and age less than 1 year were independent associated factors for overall survival in childhood AML patients. Our opinion is that treatment recommendations for hyperleukocytosis in both AML and ALL patients should be modified to be more intensive in those who have identified risk factors to reduce complications secondary to leukostasis while being aware of the risk of developing consequences resulting from treatment by close monitoring and providing intensive care.

There is no standard treatment guideline for hyperleukocytosis in Thailand. The current practice focuses on preventing and reducing complications secondary to leukostasis by providing intensive supportive care and prompt cytoreduction by means of hydration, urine alkalinization, allopurinol administration, early chemotherapy induction, and occasionally leukapheresis. In our center, patients with hyperleukocytosis were managed by hydration, urine alkalinization, allopurinol administration, and early chemotherapy induction. Less than 10% of the ALL patients with hyperleukocytosis received pretreatment with corticosteroids. None of the AML patients with hyperleukocytosis received pretreatment with cytarabine. Leukapheresis was only performed on a small number of patients (3.5%) since the efficacy is still controversial and the procedure is constrained by age, availability, the requirement for central venous access, and the potential for adverse effects. Due to the small proportion of patients who underwent leukapheresis, our study was restricted to evaluating the efficacy of leukapheresis in patients with hyperleukocytosis. Further multicenter studies involving a larger number of patients are needed.

The notable strength of our study was that it included a large number of patients with a long follow-up period. However, the study also had some limitations. First, this study covered a 20-year period during which there were significant advancements in both treatment and supportive care for cancer patients, and these advances would undoubtedly have had some impact on the outcomes which we were unable to assess in this study. Second, the statistical analysis was also impacted by factors such as the multiple periods of diagnosis where there may have been differences in classification of ALL (i.e., from FAB to T-cell or B-cell ALL) and the method of diagnosis, which in later years additionally used immunophenotyping of cell markers. Third, our study had a limited number of cytogenetic studies and lack of molecular and minimal residual disease (MRD) studies.

5. Conclusions

Hyperleukocytosis in childhood leukemia was associated with higher rates of morbidity and worse survival outcomes. Hyperleukocytosis, extreme hyperleukocytosis, age less than 1 year, age greater than 10 years, and male sex were independently associated with a poorer overall survival in childhood ALL patients. Extreme hyperleukocytosis and age less than 1 year were independently associated with poorer overall survival in childhood AML patients.

Author Contributions: Conceptualization and methodology, S.K., N.J., P.S., N.S., S.C., C.L. and T.C.; validation, S.K., N.J. and T.C.; formal analysis and data curation, E.B.M., S.K., N.J. and T.C.; writing—original draft preparation, review and editing, S.K., N.J. and T.C. All authors have read and agreed to the published version of the manuscript.

Funding: This research received no external funding.

Institutional Review Board Statement: The study was conducted in accordance with the Declaration of Helsinki and approved by the Ethics Committee of the Faculty of Medicine, Prince of Songkla University, Songkhla, Thailand (protocol code 60-421-01-1).

Informed Consent Statement: Patient consent was waived due to retrospective nature of the study.

Data Availability Statement: The datasets generated and analyzed during the current study are available from the corresponding author on reasonable request.

Acknowledgments: We would like to thank Dave Patterson of the Office of International Affairs, Faculty of Medicine, Prince of Songkla University, for his help with English editing.

Conflicts of Interest: The authors declare no conflict of interest.

References

1. Bidwell, S.S.; Peterson, C.C.; Demanelis, K.; Zarins, K.R.; Meza, R.; Sriplung, H.; Wiangnon, S.; Chotsampancharoen, T.; Chitapanarux, I.; Pongnikorn, D.; et al. Childhood Cancer Incidence and Survival in Thailand: A Comprehensive Population-Based Registry Analysis, 1990–2011. *Pediatr. Blood Cancer* **2019**, *66*, e27428. [CrossRef] [PubMed]
2. Ward, E.; DeSantis, C.; Robbins, A.; Kohler, B.; Jemal, A. Childhood and Adolescent Cancer Statistics, 2014. *Cancer J. Clin.* **2014**, *64*, 83–103. [CrossRef]
3. Steliarova-Foucher, E.; Colombet, M.; Ries, L.A.G.; Moreno, F.; Dolya, A.; Bray, F.; Hesseling, P.; Shin, H.Y.; Stiller, C.A. IICC-3 contributors International Incidence of Childhood Cancer, 2001–2010: A Population-Based Registry Study. *Lancet Oncol.* **2017**, *18*, 719–731. [CrossRef]
4. Bunin, N.J.; Pui, C.H. Differing Complications of Hyperleukocytosis in Children with Acute Lymphoblastic or Acute Nonlymphoblastic Leukemia. *J. Clin. Oncol.* **1985**, *3*, 1590–1595. [CrossRef] [PubMed]
5. Eguiguren, J.M.; Schell, M.J.; Crist, W.M.; Kunkel, K.; Rivera, G.K. Complications and Outcome in Childhood Acute Lymphoblastic Leukemia with Hyperleukocytosis. *Blood* **1992**, *79*, 871–875. [CrossRef] [PubMed]
6. Abla, O.; Angelini, P.; Di Giuseppe, G.; Kanani, M.F.; Lau, W.; Hitzler, J.; Sung, L.; Naqvi, A. Early Complications of Hyperleukocytosis and Leukapheresis in Childhood Acute Leukemias. *J. Pediatr. Hematol. Oncol.* **2016**, *38*, 111–117. [CrossRef]
7. Kong, S.G.; Seo, J.H.; Jun, S.E.; Lee, B.K.; Lim, Y.T. Childhood Acute Lymphoblastic Leukemia with Hyperleukocytosis at Presentation. *Blood Res.* **2014**, *49*, 29–35. [CrossRef]
8. Park, K.M.; Yang, E.J.; Lee, J.M.; Hah, J.O.; Park, S.K.; Park, E.S.; Lim, J.Y.; Kim, J.Y.; Park, J.; Shim, Y.J.; et al. Treatment Outcome in Pediatric Acute Lymphoblastic Leukemia with Hyperleukocytosis in the Yeungnam Region of Korea: A Multicenter Retrospective Study. *J. Pediatr. Hematol. Oncol.* **2020**, *42*, 275–280. [CrossRef]
9. Haase, R.; Merkel, N.; Diwan, O.; Elsner, K.; Kramm, C.M. Leukapheresis and Exchange Transfusion in Children with Acute Leukemia and Hyperleukocytosis. A Single Center Experience. *Klin. Padiatr.* **2009**, *221*, 374–378. [CrossRef]
10. İrken, G.; Ören, H.; Öniz, H.; Çetingül, N.; Vergin, C.; Atabay, B.; Gülen, H.; Türker, M.; Kantar, M.; Yılmaz, Ş. Hyperleukocytosis in Childhood Acute Lymphoblastic Leukemia: Complications and Treatment Outcome. *Turk. J. Haematol.* **2006**, *23*, 142–146.
11. Kulkarni, K.P.; Marwaha, R.K.; Trehan, A.; Bansal, D. Survival Outcome in Childhood ALL: Experience from a Tertiary Care Centre in North India. *Pediatr. Blood Cancer* **2009**, *53*, 168–173. [CrossRef] [PubMed]
12. Magrath, I.; Shanta, V.; Advani, S.; Adde, M.; Arya, L.S.; Banavali, S.; Bhargava, M.; Bhatia, K.; Gutiérrez, M.; Liewehr, D.; et al. Treatment of Acute Lymphoblastic Leukaemia in Countries with Limited Resources; Lessons from Use of a Single Protocol in India over a Twenty Year Period. *Eur. J. Cancer Oxf. Engl.* **2005**, *41*, 1570–1583. [CrossRef]
13. Creutzig, U.; Ritter, J.; Riehm, H.; Langermann, H.J.; Henze, G.; Kabisch, H.; Niethammer, D.; Jürgens, H.; Stollmann, B.; Lasson, U. Improved Treatment Results in Childhood Acute Myelogenous Leukemia: A Report of the German Cooperative Study AML-BFM-78. *Blood* **1985**, *65*, 298–304. [CrossRef] [PubMed]
14. Sung, L.; Aplenc, R.; Alonzo, T.A.; Gerbing, R.B.; Gamis, A.S. AAML0531/PHIS Group Predictors and Short-Term Outcomes of Hyperleukocytosis in Children with Acute Myeloid Leukemia: A Report from the Children's Oncology Group. *Haematologica* **2012**, *97*, 1770–1773. [CrossRef] [PubMed]
15. Inaba, H.; Fan, Y.; Pounds, S.; Geiger, T.L.; Rubnitz, J.E.; Ribeiro, R.C.; Pui, C.-H.; Razzouk, B.I. Clinical and Biologic Features and Treatment Outcome of Children with Newly Diagnosed Acute Myeloid Leukemia and Hyperleukocytosis. *Cancer* **2008**, *113*, 522–529. [CrossRef]
16. Creutzig, U.; Zimmermann, M.; Reinhardt, D.; Dworzak, M.; Stary, J.; Lehrnbecher, T. Early Deaths and Treatment-Related Mortality in Children Undergoing Therapy for Acute Myeloid Leukemia: Analysis of the Multicenter Clinical Trials AML-BFM 93 and AML-BFM 98. *J. Clin. Oncol.* **2004**, *22*, 4384–4393. [CrossRef]

17. Porcu, P.; Cripe, L.D.; Ng, E.W.; Bhatia, S.; Danielson, C.M.; Orazi, A.; McCarthy, L.J. Hyperleukocytic Leukemias and Leukostasis: A Review of Pathophysiology, Clinical Presentation and Management. *Leuk. Lymphoma* **2000**, *39*, 1–18. [CrossRef] [PubMed]
18. Nguyen, R.; Jeha, S.; Zhou, Y.; Cao, X.; Cheng, C.; Bhojwani, D.; Campbell, P.; Howard, S.C.; Rubnitz, J.; Ribeiro, R.C.; et al. The Role of Leukapheresis in the Current Management of Hyperleukocytosis in Newly Diagnosed Childhood Acute Lymphoblastic Leukemia. *Pediatr. Blood Cancer* **2016**, *63*, 1546–1551. [CrossRef] [PubMed]
19. Demanelis, K.; Sriplung, H.; Meza, R.; Wiangnon, S.; Rozek, L.S.; Scheurer, M.E.; Lupo, P.J. Differences in Childhood Leukemia Incidence and Survival between Southern Thailand and the United States: A Population-Based Analysis. *Pediatr. Blood Cancer* **2015**, *62*, 1790–1798. [CrossRef]
20. Laosombat, V.; Wongchanchailert, M.; Sattayasevana, B.; Wiriyasateinkul, A.; Watana-Arepornchai, S. The Treatment of Children with Acute Lymphoblastic Leukemia in Thailand. *Med. Pediatr. Oncol.* **2002**, *38*, 266–268. [CrossRef]
21. Tubergen, D.G.; Gilchrist, G.S.; O'Brien, R.T.; Coccia, P.F.; Sather, H.N.; Waskerwitz, M.J.; Hammond, G.D. Improved Outcome with Delayed Intensification for Children with Acute Lymphoblastic Leukemia and Intermediate Presenting Features: A Childrens Cancer Group Phase III Trial. *J. Clin. Oncol.* **1993**, *11*, 527–537. [CrossRef] [PubMed]
22. Gaynon, P.S.; Steinherz, P.G.; Bleyer, W.A.; Ablin, A.R.; Albo, V.C.; Finklestein, J.Z.; Grossman, N.J.; Novak, L.J.; Pyesmany, A.F.; Reaman, G.H. Improved Therapy for Children with Acute Lymphoblastic Leukemia and Unfavorable Presenting Features: A Follow-up Report of the Childrens Cancer Group Study CCG-106. *J. Clin. Oncol.* **1993**, *11*, 2234–2242. [CrossRef]
23. Creutzig, U.; Ritter, J.; Schellong, G. Identification of Two Risk Groups in Childhood Acute Myelogenous Leukemia after Therapy Intensification in Study AML-BFM-83 as Compared with Study AML-BFM-78. AML-BFM Study Group. *Blood* **1990**, *75*, 1932–1940. [CrossRef] [PubMed]
24. Creutzig, U.; Zimmermann, M.; Ritter, J.; Reinhardt, D.; Hermann, J.; Henze, G.; Jürgens, H.; Kabisch, H.; Reiter, A.; Riehm, H.; et al. Treatment Strategies and Long-Term Results in Paediatric Patients Treated in Four Consecutive AML-BFM Trials. *Leukemia* **2005**, *19*, 2030–2042. [CrossRef] [PubMed]
25. Seksarn, P.; Wiangnon, S.; Veerakul, G.; Chotsampancharoen, T.; Kanjanapongkul, S.; Chainansamit, S.O. Outcome of Childhood Acute Lymphoblastic Leukemia Treated Using the Thai National Protocols. *Asian Pac. J. Cancer Prev.* **2015**, *16*, 4609–4614. [CrossRef]
26. Lauer, S.J.; Shuster, J.J.; Mahoney, D.H.; Winick, N.; Toledano, S.; Munoz, L.; Kiefer, G.; Pullen, J.D.; Steuber, C.P.; Camitta, B.M. A Comparison of Early Intensive Methotrexate/Mercaptopurine with Early Intensive Alternating Combination Chemotherapy for High-Risk B-Precursor Acute Lymphoblastic Leukemia: A Pediatric Oncology Group Phase III Randomized Trial. *Leukemia* **2001**, *15*, 1038–1045. [CrossRef] [PubMed]
27. Cooper, T.M.; Franklin, J.; Gerbing, R.B.; Alonzo, T.A.; Hurwitz, C.; Raimondi, S.C.; Hirsch, B.; Smith, F.O.; Mathew, P.; Arceci, R.J.; et al. AAML03P1, a Pilot Study of the Safety of Gemtuzumab Ozogamicin in Combination with Chemotherapy for Newly Diagnosed Childhood Acute Myeloid Leukemia: A Report from the Children's Oncology Group. *Cancer* **2012**, *118*, 761–769. [CrossRef] [PubMed]
28. Chotsampancharoen, T.; Songthawee, N.; Chavananon, S.; Sripornsawan, P.; McNeil, E.B. Relapsed Childhood Acute Lymphoblastic Leukemia: Experience from a Single Tertiary Center in Thailand. *Asian Pac. J. Cancer Prev.* **2022**, *23*, 3517–3522. [CrossRef]
29. Songthawee, N.; Sripornsawan, P.; Chavananon, S.; McNeil, E.B.; Chotsampancharoen, T. Relapsed Childhood Acute Myeloid Leukemia: Experience from a Single Tertiary Center in Thailand. *Asian Pac. J. Cancer Prev.* **2022**, *23*, 4079–4084. [CrossRef]
30. Cairo, M.S.; Bishop, M. Tumour Lysis Syndrome: New Therapeutic Strategies and Classification. *Br. J. Haematol.* **2004**, *127*, 3–11. [CrossRef]
31. Kellum, J.A.; Lameire, N.; Aspelin, P.; Barsoum, R.S.; Burdmann, E.A.; Goldstein, S.L.; Herzog, C.A.; Joannidis, M.; Kribben, A.; Levey, A.S.; et al. Kidney Disease: Improving Global Outcomes (KDIGO) Acute Kidney Injury Work Group. KDIGO Clinical Practice Guideline for Acute Kidney Injury. *Kidney Int. Suppl.* **2012**, *2*, 1–138.
32. Goldstein, B.; Giroir, B.; Randolph, A. Definitions for Sepsis and Organ Dysfunction in Pediatrics. *Pediatr. Crit. Care Med.* **2005**, *6*, 2–8. [CrossRef] [PubMed]
33. Taylor, F.B.; Toh, C.H.; Hoots, W.K.; Wada, H.; Levi, M. Scientific Subcommittee on Disseminated Intravascular Coagulation (DIC) of the International Society on Thrombosis and Haemostasis (ISTH). Towards Definition, Clinical and Laboratory Criteria, and a Scoring System for Disseminated Intravascular Coagulation. *Thromb. Haemost.* **2001**, *86*, 1327–1330. [CrossRef] [PubMed]
34. Bonaventure, A.; Harewood, R.; Stiller, C.A.; Gatta, G.; Clavel, J.; Stefan, D.C.; Carreira, H.; Spika, D.; Marcos-Gragera, R.; Peris-Bonet, R.; et al. Worldwide Comparison of Survival from Childhood Leukaemia for 1995–2009, by Subtype, Age, and Sex (CONCORD-2): A Population-Based Study of Individual Data for 89,828 Children from 198 Registries in 53 Countries. *Lancet Haematol.* **2017**, *4*, e202–e217. [CrossRef]
35. Ssenyonga, N.; Stiller, C.; Nakata, K.; Shalkow, J.; Redmond, S.; Bulliard, J.-L.; Girardi, F.; Fowler, C.; Marcos-Gragera, R.; Bonaventure, A.; et al. Worldwide Trends in Population-Based Survival for Children, Adolescents, and Young Adults Diagnosed with Leukaemia, by Subtype, during 2000-14 (CONCORD-3): Analysis of Individual Data from 258 Cancer Registries in 61 Countries. *Lancet Child Adolesc. Health* **2022**, *6*, 409–431. [CrossRef]
36. Maurer, H.S.; Steinherz, P.G.; Gaynon, P.S.; Finklestein, J.Z.; Sather, H.N.; Reaman, G.H.; Bleyer, W.A.; Hammond, G.D. The Effect of Initial Management of Hyperleukocytosis on Early Complications and Outcome of Children with Acute Lymphoblastic Leukemia. *J. Clin. Oncol.* **1988**, *6*, 1425–1432. [CrossRef]

37. Inaba, H.; Mullighan, C.G. Pediatric Acute Lymphoblastic Leukemia. *Haematologica* **2020**, *105*, 2524–2539. [CrossRef]
38. Friedmann, A.M.; Weinstein, H.J. The Role of Prognostic Features in the Treatment of Childhood Acute Lymphoblastic Leukemia. *Oncologist* **2000**, *5*, 321–328. [CrossRef]
39. Möricke, A.; Zimmermann, M.; Reiter, A.; Gadner, H.; Odenwald, E.; Harbott, J.; Ludwig, W.-D.; Riehm, H.; Schrappe, M. Prognostic Impact of Age in Children and Adolescents with Acute Lymphoblastic Leukemia: Data from the Trials ALL-BFM 86, 90, and 95. *Klin. Padiatr.* **2005**, *217*, 310–320. [CrossRef]
40. Vrooman, L.M.; Silverman, L.B. Treatment of Childhood Acute Lymphoblastic Leukemia: Prognostic Factors and Clinical Advances. *Curr. Hematol. Malig. Rep.* **2016**, *11*, 385–394. [CrossRef]
41. Smith, M.; Arthur, D.; Camitta, B.; Carroll, A.J.; Crist, W.; Gaynon, P.; Gelber, R.; Heerema, N.; Korn, E.L.; Link, M.; et al. Uniform Approach to Risk Classification and Treatment Assignment for Children with Acute Lymphoblastic Leukemia. *J. Clin. Oncol.* **1996**, *14*, 18–24. [CrossRef] [PubMed]
42. Creutzig, U.; Ritter, J.; Budde, M.; Sutor, A.; Schellong, G. Early Deaths Due to Hemorrhage and Leukostasis in Childhood Acute Myelogenous Leukemia. Associations with Hyperleukocytosis and Acute Monocytic Leukemia. *Cancer* **1987**, *60*, 3071–3079. [CrossRef] [PubMed]

MDPI
St. Alban-Anlage 66
4052 Basel
Switzerland
www.mdpi.com

Cancers Editorial Office
E-mail: cancers@mdpi.com
www.mdpi.com/journal/cancers

www.ingramcontent.com/pod-product-compliance
Lightning Source LLC
Chambersburg PA
CBHW061508220926
51850CB00004B/207